HANDBOOK OF STATISTICS
VOLUME 6

Handbook of Statistics

VOLUME 6

General Editors

P. R. Krishnaiah[†]

C. R. Rao

NORTH-HOLLAND
AMSTERDAM · NEW YORK · OXFORD

Sampling

Edited by

P. R. Krishnaiah[†]

C. R. Rao
Department of Mathematics and Statistics
University of Pittsburgh, Pittsburgh, PA, U.S.A.

1988

NORTH-HOLLAND
AMSTERDAM · NEW YORK · OXFORD

© ELSEVIER SCIENCE PUBLISHERS B.V., 1988

All rights reserved. No part of this publication may be reproduced, stored in a retrieval system or transmitted, in any form or by any means, electronic, mechanical, photocopying, recording or otherwise, without the prior permission of the copyright owner.

ISBN: 0 444 70289 X

Published by:
ELSEVIER SCIENCE PUBLISHERS B.V.
P.O. Box 1991
1000 BZ Amsterdam
The Netherlands

Sole distributors for the U.S.A. and Canada:
ELSEVIER SCIENCE PUBLISHING COMPANY, INC.
52 Vanderbilt Avenue
New York, N.Y. 10017
U.S.A.

LIBRARY OF CONGRESS
Library of Congress Cataloging-in-Publication Data

Sampling / edited by P.R. Krishnaiah, C.R. Rao.
  p.   cm. -- (Handbook of statistics : v. 6)
Includes bibliographies and index.
ISBN. J-444-70289-X (U.S.)
1. Sampling (Statistics) I. Krishnaiah. Paruchuri R. II. Rao.
C. Radhakrishna (Calyampudi Radhakrishna), 1920- . III. Series.
QA276.6.S334 1988   88-374
519.5'2--dc 19   CIP

Transferred to digital printing 2006

Printed and bound by Antony Rowe Ltd, Eastbourne

Preface

The series *Handbook of Statistics* was started by Professor P. R. Krishnaiah, who unfortunately passed away on August 1, 1987 at the prime age of 55 years. The object of bringing out these volumes, as mentioned by the founding editor in the preface to the first volume, is to provide 'comprehensive and self contained reference books to disseminate information on various aspects of statistical methodology and applications'. This is not an easy task and only an erudite scholar like Professor Krishnaiah with a deep knowledge of different fields of statistics and a missionary zeal could achieve it. This is the sixth volume which he edited and shortly before his death he made plans for producing six more volumes devoted to different areas of applications of statistics. These volumes have been well received by the entire statistical community, and scientists in various disciplines who use statistical methodology in their work, which is a great tribute to the imaginative efforts of Professor Krishnaiah.

The present volume, number six in the series, is devoted to the theory and practice of Sample Surveys, which is the most widely used method in statistical practice. The basic ideas of survey sampling methodology were formulated in the twenties, but its firm foundations were laid only in the thirties and forties. A brief historical account of sample surveys is given by D. R. Bellhouse. An overview of the subject and the contents of a course on survey sampling are outlined by T. Dalenius. A. Chaudhuri discusses various sampling strategies and the optimality problems associated with them.

The current developments in sample survey methodology are surveyed by the rest of the authors. P. K. Pathak throws new light on the cost-efficiency of simple random sampling, while V. P. Godambe and M. E. Thompson discuss the role of randomization in inference with special reference to single stage unequal probability sampling. Systematic sampling is the theme of contributions by D. R. Bellhouse, M. N. Murty and T. J. Rao. Repeated sampling over time is considered by D. A. Binder and M. A. Hidiroglou. Some theoretical aspects of inference in finite populations are covered in the contributions by W. A. Ericson, Gad Nathan, J. Sedransk and P. J. Smith, P. K. Sen, Ib Thomsen and Dinke Tesfu, and R. M. Royall.

J. C. Koop discusses the concept of interpenetrating subsamples introduced by P. C. Mahalanobis. D. H. Freeman discusses the analysis of contingency tables compiled from survey data. J. N. K. Rao reviews the various methods of variance estimation in sample surveys and P. S. R. S. Rao, the methodology of ratio and regression estimation.

G. P. Patil, G. J. Babu, R. C. Hennemuth, W. L. Meyers, M. B. Rajarshi, C. Tallie, M. T. Boswell, F. L. Ramsey, C. E. Gates and K. P. Burnham review special survey techniques in environmental and ecological studies, while R. Velu and G. M. Naidu review the current sampling methods in marketing research. P. V. Sukhatme discusses methods for controlling and estimating observational errors in sample surveys.

A Hedayat, C. R. Rao and H. Stufken present some new problems in the design of sample surveys. They provide sampling designs to avoid contiguous units occuring in samples in order to increase the efficiency of estimates.

This volume provides an unusual and useful collection of articles covering many theoretical and practical aspects of sample surveys in social and biological investigations. Written by experts and actual practioners of sample surveys, it would be a valuable guide to those involved in designing sample surveys for collection of data and estimation of unknown population parameters.

I would like to thank North-Holland Publishing Company for their patience and excellent cooperation in bringing out this Volume.

C. R. Rao

Table of Contents

Preface v

Contributors xv

Ch. 1. A Brief History of Random Sampling Methods 1
    D. R. Bellhouse

1. Introduction 1
2. Kiaer 2
3. Bowley 4
4. Neyman 7
5. Hansen and Hurwitz 9
6. Other developments 10
7. The paradigm challenged and defended 11
   References 13

Ch. 2. A First Course in Survey Sampling 15
    T. Dalenius

   Introduction 15
  I. The notion of survey sampling 16
 II. A review of the statistics needed 18
III. Getting observational access to the population 21
 IV. Element sampling 24
  V. Cluster sampling 32
 VI. Multi-stage sampling 39
VII. The problems of non-sampling errors 41
VIII. Survey sampling design 44

Ch. 3. Optimality of Sampling Strategies 47
    A. Chaudhuri

1. Introduction 47
2. Repeated sampling approach: Fixed population case 49

3. Optimality under super-population modelling 75
4. Likelihood approach 79
5. Prediction approach 83
6. A summing up: Efficacy of an asymptotic theory 84
7. Roles of labels and randomization: Controversies 88
   References 90

Ch. 4. Simple Random Sampling 97
P. K. Pathak

0. Summary 97
1. Introduction 97
2. Simple random sampling without replacement (SRSWOR) 99
3. Simple random sampling with replacement (SRSWR) 102
4. Fixed cost simple random sampling (SRSFC) 104
   Acknowledgment 108
   References 108

Ch. 5. On Single Stage Unequal Probability Sampling 111
V. P. Godambe and M. E. Thompson

1. Introduction 111
2. Use of randomization 111
3. Estimating finite population means and totals 114
4. Stratified random sampling 115
5. Sampling designs for regression models 117
6. Monetary unit sampling 119
7. A historical note 121
   References 122

Ch. 6. Systematic Sampling 125
D. R. Bellhouse

1. Introduction 125
2. Sampling theory based on the randomization alone 125
3. Trends in the population 128
4. Autocorrelated populations 134
5. Variance estimation 138
6. Spatial sampling 141
   References 143

Ch. 7. Systematic Sampling with Illustrative Examples 147
M. N. Murthy and T. J. Rao

1. The basic procedure 147
2. Estimation and sampling variance 149

3. Efficiency of systematic sampling   152
4. Variance estimation   158
5. Illustrative examples   161
6. Bias in systematic sampling   164
7. Variations in systematic sampling   165
8. Systematic sampling with probability proportional to size   173
9. Superpopulation models and asymptotic results   175
10. Applications and illustrations   180
11. Some of the available computer programs   181
12. Two-dimensional systematic sampling   182
    Bibliographical note   183
    Reference   183

Ch. 8. Sampling in Time   187
D. A. Binder and M. A. Hidiroglou

1. Introduction   187
2. General approaches to designs with overlapping units   188
3. The classical approach   190
4. Rotation bias   197
5. Time series approaches   200
   Acknowledgements   210
   References   210

Ch. 9. Bayesian Inference in Finite Populations   213
W. A. Ericson

1. Introduction   213
2. The basic model   213
3. Some basic results   215
4. Exchangeability and simple random sampling   217
5. Stratified sampling   220
6. Two stage sampling   223
7. Ratio and regression estimation   227
8. Response error and bias   223
   Discussion   241
   References   243

Ch. 10. Inference Based on Data from Complex Sample Designs   247
G. Nathan

1. Introduction   247
2. General methods   251
3. Regression and linear models   255
4. Categorical data analysis   258
5. Other methods of analysis   262
   References   263

Ch. 11. Inference for Finite Population Quantiles 267
J. Sedransk and P. J. Smith

1. Introduction 267
2. Confidence interval methods for complex sample designs 268
3. Point estimation for finite population quantiles 280
4. Bayesian methods 283
   Acknowledgement 288
   References 288

Ch. 12. Asymptotics in Finite Population Sampling 291
P. K. Sen

1. Introduction 291
2. Asymptotics in SRS 292
3. Some probability and moment inequalities for SRS 297
4. Jackknifing in finite population sampling 301
5. Estimation of population size: Asymptotics 304
6. Sampling with varying probabilities: Asymptotics 313
7. Successive sub-sampling with varying probabilities: Asymptotics 324
   References 328

Ch. 13. The Technique of Replicated or Interpenetrating Samples 333
J. C. Koop

1. Introductory review 333
2. Theoretical basis of the technique 337
3. Applications 353
4. Analysis of variance to study methods of investigation 360
5. Summary and comments 364
   Acknowledgements 365
   References 365

Ch. 14. On the Use of Models in Sampling from Finite Populations 369
I. Thomsen and D. Tesfu

1. Introduction 369
2. On the prediction approach 370
3. Application of the prediction approach to two-stage sampling 374
4. Estimation in election surveys 378
5. Other applications of models 381
6. Models for response errors 387
7. A probabilistic model for nonresponse 391
   References 396

Ch. 15. The Prediction Approach to Sampling Theory 399
R. M. Royall

1. Introduction 399
2. The linear prediction approach 401
3. The approach based on probability sampling distributions 410
   References 412

Ch. 16. Sample Survey Analysis: Analysis of Variance and Contingency Tables 415
D. H. Freeman, Jr.

1. Introduction 415
2. The KFF methodology for analysis of variance models 417
3. An example: Connecticut Blood Pressure Survey 421
4. Summary 424
   Acknowledgement 425
   References 425

Ch. 17. Variance Estimation in Sample Surveys 427
J. N. K. Rao

Introduction 427
1. Unified approach for linear statistics 428
2. Nonlinear statistics 436
3. Modelling mean square errors 444
   Acknowledgements 446
   References 446

Ch. 18. Ratio and Regression Estimators 449
P. S. R. S. Rao

1. Introduction 449
2. Bias and mean square error of the ratio estimator 450
3. Bias reduction 451
4. Relative merits of the estimators 453
5. Variance estimation and confidence limits 454
6. Regression through the origin and the ratio estimator 455
7. Stratification and the ratio estimators 457
8. Regression estimator 459
9. Multivariate ratio, product and regression estimators 461
10. Two phase sampling for ratio and regression estimators 464
11. Further developments 465
    Acknowledgements 465
    References 465

Ch. 19. Role and Use of Composite Sampling and Capture–Recapture Sampling in Ecological Studies 469
M. T. Boswell, K. P. Burnham and G. P. Patil

Introduction 469
1. Composite sampling 470
2. Capture–recapture sampling 478
  Acknowledgements 486
  References 486

Ch. 20. Data-based Sampling and Model-based Estimation for Environmental Resources 489
G. P. Patil, G. J. Babu, R. C. Hennemuth, W. L. Meyers, M. B. Rajarshi and C. Taillie

1. Introduction 489
2. Data-based definitions of populations 489
3. Sampling design as design of encounters—The case of living marine resources 493
4. Sampling design as design of encounters—The bias in fisheries harvest data 497
5. Sample-based modeling of populations—An approach with weighted distributions 499
6. Combining recruitment data and kernel approach 502
7. Encountered ecotoxicological data of chronic effects thresholds 507
8. Synthesis 511
  Acknowledgements 511
  References 512

Ch. 21. On Transect Sampling to Assess Wildlife Populations and Marine Resources 515
F. L. Ramsey, C. E. Gates, G. P. Patil and C. Taillie

1. Introduction 515
2. Transect surveys 517
3. Estimation procedures 518
4. Variable detectability conditions 521
5. Bob-white flushing 524
6. Deep-sea red crab 524
7. Summary 530
  Acknowledgements 530
  References 530

Ch. 22. A Review of Current Survey Sampling Methods in Marketing Research (Telephone, Mall Intercept and Panel Surveys) 533
R. Velu and G. M. Naidu

1. Introduction 533
2. Telephone Surveys 534

3. Shopping center sampling and interviewing   544
4. Consumers panels   546
   Acknowledgements   552
   References   552

Ch. 23. Observational Errors in Behavioural Traits of Man and their Implications for Genetics   555
P. V. Sukhatme

1. Introduction   555
2. Current theory and its limitations: The sample mean and its variance   556
3. Estimation of the different components   559
4. Application to longitudinal studies   561
5. Genetic implications   569
   References   572

Ch. 24. Designs in Survey Sampling Avoiding Contiguous Units   575
A. S. Hedayat, C. R. Rao and J. Stufken

1. Introduction   575
2. Sample designs exclusing contiguous   575
3. Results on the existence and construction   579
4. Implementation   582
   References   583

Subject Index   585

Contents of Previous Volumes   589

Contributors

G. J. Babu, *Center for Statistical Ecology and Environmental Statistics, The Pennsylvania State University, Dept. of Statistics, 303 Pond Laboratory, University Park, PA 16802, USA* (Ch. 20)

D. R. Bellhouse, *Dept. of Statistical & Actuarial Sciences, The University of Western Ontario, Faculty of Science, Room 3005 EMSc, London, Canada N6A 6B9* (Ch. 1, 6)

D. A. Binder, *Social Survey Methods Division, Statistics Canada, Floor 4, Jean-Talon, Tunney's Pasture, Ottawa, Ontario, K1A 0T6 Canada* (Ch. 8)

M. T. Boswell, *Center for Statistical Ecology and Environmental Statistics, The Pennsylvania State University, Dept. of Statistics, 303 Pond Laboratory, University Park, PA 16802, USA* (Ch. 19)

K. P. Burnham, *Dept. of Statistics, North Carolina State University, Raleigh, NC, USA* (Ch. 19)

A. Chaudhuri, *Computer Science Unit, India Statistical Institute, 203 Barrackpore Trunk Road, Calcutta 700 035, India* (Ch. 3)

T. Dalenius, *Dept. of Statistics, Brown University, Providence, RI 02912, USA* (Ch. 2)

W. A. Ericson, *Dept. of Statistics, University of Michigan, Ann Arbor, MI 48104, USA* (Ch. 9)

D. H. Freeman, Jr., *Associate Professor of Public Health (Biostatistics), Yale University, P.O. Box 3333, 60 College Street, New Haven, CT 06510, USA* (Ch. 16)

C. E. Gates, *Dept. of Statistics, Texas A&M University, College Station, TX 77843, USA* (Ch. 21)

V. P. Godambe, *Dept. of Statistics, University of Waterloo, Waterloo, Ontario, Canada* (Ch. 5)

S. Hedayat, *Dept. of Math. Stat. and Comp. Science, 322 Science and Engineering Offices, Box 4348, Chicago, IL 60680, USA* (Ch. 24)

R. C. Hennemuth, *Northeast Fisheries Center, National Marine Fisheries Service, Woods Hole, MA, USA* (Ch. 20)

M. A. Hidiroglou, *Business Survey Methods Division, Statistics Canada, Floor 11, RHC Bldg., Tunney's Pasture, Ottawa, Ontario, Canada K1A 0T6* (Ch. 8)

J. C. Koop, *3201 Clark Avenue, Raleigh, NC 27606, USA* (Ch. 13)

M. N. Murthy, *129 Luz Church Road, Mylapore, Madras 600 004, India* (Ch. 7)

W. L. Meyers, *Center for Statistical Ecology and Environmental Statistics, The Pennsylvania State University, Dept. of Statistics, 303 Pond Laboratory, University Park, PA 16802, USA* (Ch. 20)

G. M. Naidu, *University of Wisconsin—White Water, 800 West Main Street, Whitewater, WI 53190, USA* (Ch. 22)

G. M. Nathan, *Dept. of Statistics, The Hebrew University of Jerusalem, Jerusalem, Israel* (Ch. 10)

P. K. Pathak, *Dept. of Mathematics, University of New Mexico, Albuquerque, NM 87131, USA* (Ch. 4)

G. P. Patil, *Center for Statistical Ecology and Environmental Statistics, The Pennsylvania State University, Department of Statistics, 303 Pond Laboratory, University Park, PA 16802, USA* (Ch. 19, 20, 21)

M. B. Rajarshi, *Center for Statistical Ecology and Environmental Statistics, The Pennsylvania State University, Dept. of Statistics, 303 Pond Laboratory, University Park, PA 16802, USA* (Ch. 20)

F. L. Ramsey, *Dept. of Statistics, Oregon State University, Corvallis, OR 97331, USA* (Ch. 21)

C. R. Rao, *Dept. of Mathematics & Statistics, University of Pittsburgh, Pittsburgh, PA 15260, USA* (Ch. 24)

J. N. K. Rao, *Dept. of Mathematics, Carleton University, Colonel by Drive, Ottawa, Ontario, Canada K1S 5B6* (Ch. 17)

P. S. R. S. Rao, *Dept. of Statistics, University of Rochester, Dewey Hall 232, Rochester, NY 14627, USA* (Ch. 18)

T. J. Rao, *Math.-Stat. Division, Indian Statistical Institute, 203 B.T. Road, Calcutta 700035, India* (Ch. 7)

R. M. Royall, *The John Hopkins University, School of Hygiene & Public Health, 615 North Wolfe Street, Baltimore, MD 21205, USA* (Ch. 15)

J. Sedransk, *Dept. of Mathematics, SUNY at Albany, Albany, NY 12222* (Ch. 11)

P. K. Sen, *Dept. of Biostatistics, University of North Carolina, Chapel Hill, NC 27514, USA* (Ch. 12)

P. J. Smith, *International Pacific Halibut Commission, P.O. Box 95009, Seattle, WA 98145-2009, USA* (Ch. 11)

J. Stufken, *University of Georgia, Athens, GA, USA* (Ch. 24).

P. V. Sukhatme, *Biometry Dept., Maharashtra Assoc. for the Cultivation of Science, Law College Road, Poona 411004, India* (Ch. 23)

C. Taillie, *Center for Statistical Ecology and Environmental Statistics, The Pennsylvania State University, Dept. of Statistics, 303 Pond Laboratory, University Park, PA 16802, USA* (Ch. 20, 21)

D. Tesfu, *Central Statistical Office, P.O. Box 1143, Addis Ababa, Ethiopia* (Ch. 14)

M. E. Thompson, *Dept. of Statistics, University of Waterloo, Waterloo, Ontario, Canada* (Ch. 5)

I. Thomsen, *Central Statistical Office of Norway P.B. 8131 Dep., Oslo 1, Norway* (Ch. 14)

R. Velu, *University of Wisconsin—White Water, 800 West Main Street, Whitewater, WI 53190, USA* (Ch. 22).

A Brief History of Random Sampling Methods

D. R. Bellhouse

1. Introduction

The field of survey sampling can claim many roots. These roots include a variety of activities over the nineteenth and twentieth centuries in areas such as agriculture, forestry, government administration, and social research. It is the latter two areas, the collection of large sets of data by governments and individual social investigators, that have given the greatest stimulus to the development of the random sampling techniques discussed here. In this chapter the early work (to about 1945) in sampling is reviewed by looking at the work of those who obtained the principal results which form the basis of many of the later survey sampling textbooks. The issues which confronted these early workers are also related to some of the current discussion over model and design-based inference in survey sampling.

Several histories of sampling have been written. Among them are Chang (1976), Duncan and Shelton (1978), Hansen, Dalenius and Tepping (1985), Kruskal and Mosteller (1980), Seng (1951), Stephan (1948), Sukhatme (1966), and Yates (1948). In many ways this article is a review of the previous reviews. No new historical material is examined. What I hope is new is that in reviewing the reviews some of the motivations behind the early work in sampling are discussed along with the tension between the users of models and randomization that has been present over the whole history of the subject.

An examination of this tension is carried out with Kuhn's (1970) theory of paradigms in the history of science in mind. A paradigm or exemplar is followed by later workers because it provides a framework in which adequate answers are given to the questions being asked. New paradigms are put forward when the old framework no longer provides adequate answers. Kuhn (1970, p. 10) has described two general characteristics of paradigms that, in particular, appear in the history of random sampling. The first characteristic is that the paradigm attracts around it a loyal group of followers away from other competing modes of activity; the other is that the paradigm is open-ended enough to provide a number of unresolved problems on which adherents to the paradigm may work. With regard to the first characteristic, the attraction of a loyal group of followers does not occur in a vacuum. We shall see that in every 'sampling paradigm', either the proponent or a disciple has actively promoted his ideas.

The initial paradigm in survey sampling is that of the desire for a representative sample as first propounded by A. N. Kiaer in the 1890's. Prior to Kiaer there are several examples of sampling procedures; see, for example, Stephan (1948), Kent (1981, Ch. 1, 3), Chang (1976), and Godambe (1976). These examples illustrate the randomness exhibited in research which, according to Kuhn, is typical of preparadigmatic times. Kiaer's contribution was to provide a framework under which sampling became a reasonable activity; in particular, sampling is useful when it provides a miniature of the population, i.e., it is a representative sample. After Kiaer there emerged two competing methods to attempt to achieve representativeness; randomization and purposive selection. For the most part, in large-scale surveys the paradigm of randomization has become dominant. The reasons for this will be examined as the history of sampling is traced.

2. Kiaer

During the nineteenth century in government statistical agencies and among the social reformers of the so-called statistical movement (see for example, Cullen, 1975 and Kent, 1981, for a discussion of this movement), the generally accepted method of coverage was a complete enumeration. Some sampling was done, but it was the exception rather than the rule. The desire for a complete enumeration in social surveys may be traced back to at least Quetelet; see Stigler (1986, pp. 164–165). The position of complete enumeration was challenged by the Norwegian statistician A. N. Kiaer. Kiaer was the first director of the Norwegian Central Bureau of Statistics, a position he held from the Bureau's inception in 1876 until 1913. His responsibilities included the decennial censuses of population and agriculture and many large-scale statistical investigations. Detailed discussions of Kiaer's work and its impact on sampling methodology may be found in Seng (1951) and Kruskal and Mosteller (1980). What follows is a synopsis of the material found in these papers.

At the Berne meeting of the International Statistical Institute (ISI) in 1895 Kiaer (1895/1896) put forward the idea that a partial investigation (i.e. a sample) based on what he called the 'representative method' could provide useful information. The aim of his representative method, the new paradigm in statistical investigations, was that the sample should be an approximate miniature of the population. There had been earlier anticipations of this idea, e.g., Laplace's estimate of the population of France in 1802 (see Cochran, 1978, for a description of Laplace's methodology). However, it was Kiaer's suggestion of the idea and his subsequent campaign for its acceptance that brought about the revolution in data collection. What Kiaer meant by a 'representative sample' is best described by Kruskal and Mosteller (1980, p. 175):

'First, he thought of social and economic surveys in which one could begin by choosing districts, towns, parts of cities, streets, etc., to be followed by systematic, rather than probabilistic, choice of units (houses, families, individuals). Second, he insisted on substantial sample sizes at all levels of such a selection process. Third, he emphasized the need for spreading out the sample in a variety of ways, primarily geographically, but in other ways as well. For example, if a sample had a deficiency of cattle farmers, he would add more of them.'

Kiaer's reasoning for going the route of 'representative partial investigations' is expressed in his 1897 paper (1976, English translation, p. 37): 'The characteristic feature of this method is that in connection with the general and complete information provided by the established statistics for the field of study as a whole, *more penetrating, more detailed and more specialized surveys are instituted* (italics mine), based on certain points or limited areas, distributed over the domain of study in such a way and selected in such a manner that they will yield a sample that might be assumed to constitute a correct representation of the wole.' On the next page, when describing a survey he had carried out by the representative method, he goes on to say, 'It is easy to see that the original census material, for a survey of this kind, would be unmanageable if complete coverage had been attempted, ...'. From these quotes it may be seen that Kiaer's motivation in applying a sampling method was to increase the scope of the large-scale statistical investigation he was carrying out.

The initial reaction to Kiaer's suggestion was mainly negative although some statisticians were moderately receptive. The main opposition, led by George von Mayr of Munich University, was based on the belief that a partial investigation could never replace a complete census. In the face of strong opposition by relatively eminent statisticians Kiaer did not give up. He returned to the theme at the 1897 ISI meetings in St. Petersburg and at a conference of Scandinavian statisticians held in Stockholm in the same year. At the latter meeting he made progress in getting his ideas accepted. A conference resolution (see Dalenius, 1957, p. 30, for details) gave guarded support for the representative method.

At the 1901 and 1903 ISI meetings in Budapest and Berlin respectively, Kiaer continued to make the case for acceptance of the representative method, while at the same time clarifying his position. Kiaer was supported in a letter at the 1901 meetings by Caroll D. Wright, founder of what is now the U.S. Bureau of Labor Statistics and later by A. L. Bowley, who became a member of the ISI in 1903. Wright's impact on survey sampling was minimal. Wright had carried out a number of sample surveys in the United States; however, Kruskal and Mosteller (1980) contend that he had a 'superficial view of representative sampling' in comparison with Kiaer's. Moreover, Wright died in 1909 during the period of gradual acceptance of the idea. Any influence he may have had on sampling in the United States died with him.

Before turning to Bowley's work and its impact on survey sampling, it should be noted that at this point there was little or no discussion of the role of randomization in surveys. Kiaer (1897, English translation, p. 39) mentions the possibility of randomization, in his words a sample 'selected through the drawing of lots', but does not develop the idea any further in his writings. It was Lucien March, a French statistician, who, in the discussion to Kiaer's paper at the 1903

Berlin meeting (Kiaer, 1905, pp. 129–131), was the first to introduce concepts such as (although not in these words) simple random sampling without replacement and simple random cluster sampling. At the same time March was not a firm advocate of probability sampling. In their analysis of March's work, Kruskal and Mosteller (1980, p. 179) say, 'It strikes us as ironic, although perfectly understandable, that March should be a leader in the introduction of probability models for sampling and at the same time an advocate of caution, even skepticism, about the use of such models'.

From the 1903 meeting in Berlin to the 1925 ISI meeting in Rome there was no further discussion about the representative method of sampling. By the 1925 meeting, however, the paradigm of the representative method had been generally accepted. Moreover, there had emerged two competing paradigms concerned with the method of sample selection, randomization and purposive selection by balancing on a set of covariates. Sampling, either by randomization or purposive selection, had the characteristics by which Kuhn defines a paradigm. Several individuals were drawn toward the use of samples drawn by the 'representative method' and several individuals were inspired to make a deeper investigation into the method.

3. Bowley

The individual who brought randomization in survey sampling to the fore was Arthur Lyon Bowley. At the same time it was Bowley who developed a theory of purposive selection. Like Kiaer, Bowley actively promoted his own ideas on sampling and randomization in particular. His obituary in *The Times* (January 23, 1957) alludes to this in at least two places: 'He was at his best in the role of a wise and experienced counsellor and his reputation in this respect was very high, both in Britain and abroad.' And later, 'He travelled extensively, influencing statisticians in many parts of the world.' In terms of statistical genealogy, Bowley is a descendent of the British statistical movement of the nineteenth century. The social reformers of the 1820's, 30's and 40's had formed statistical societies in various cities in England for examining issues such as the state of the poor. They had run surveys during these years, often complete enumerations of towns or parts of cities, and had soon run into problems financing their activities (see Ashton (1977) and Cullen (1975) for discussions of the statistical societies). Later in the nineteenth century many statisticians tended to resort to secondary analyses of data collected by government bureaus. Some notable exceptions and notable English social surveyors of the late nineteenth and early twentieth centuries were Charles Booth and Seebohm Rowntree, both of whom carried out self-financed studies of poverty in London and York respectively. Bowley comes directly out of this tradition. He carried out a study of poverty in Reading in 1912 and compared his results to Rowntree's. A major difference between his and Rowntree's investigations (other than the fact that his study was not self-financed but depended on some donations of others) was that Bowley's study was based on a sample survey. In addition the survey respondents were randomly chosen.

Before discussing this survey it is of interest to examine some of Bowley's previous work, in particular his 1906 paper containing a sampling empirical study (Bowley, 1906). In this paper Bowley sought to give an empirical verification to a type of central limit theorem for simple random sampling. The motivation for the work was from a number of theoretical results of Edgeworth (Bowley, 1906, p. 550). In appealing to Edgeworth, Bowley was taking the first step in making inferences independent of any model that may apply to the population. Bowley mentions that Edgeworth's results apply to 'almost any curve of frequency satisfying simple and common conditions'. Using a crude random number table (the last digits in a table from the *Nautical Almanac*) Bowley chose a random sample (it is not stated whether it was with or without replacement) of size 400 from a population of 3878 yield rates for dividends on companies listed in *Investor's Record*. The 400 sampled items were recorded in the order in which they were chosen and then successively put into 40 groups of 10. The empirical distribution of the 40 sample means was compared to a normal curve and was found to be adequately described by it. Bowley also noted that the results he obtained were independent of the size of the population from which he was sampling. This observation is not completely accurate since he did not include the finite population correction factor in the variance of the sample mean. The correct result was obtained a few years later by Isserlis (1918). A similar empirical sampling study was repeated the following year in Bowley's textbook of statistics (Bowley, 1907, pp. 308–315). In promoting the technique of random sampling in his 1906 paper Bowly already appears to be looking ahead to his later study of poverty in the mould of Booth and Rowntree. He says:

'The method of sampling is, of course, only one of many instances of the application of the theory of probability to statistics. I have taken it at length because the method is so persistently neglected, and even when it is used the test of precision is ignored. We are thus throwing aside a very powerful weapon of research. It is frequently impossible to cover a whole area, as the census does, or as Mr. Rowntree here (the paper was from an address given at York) and Mr. Booth successfully accomplished, but it is not necessary. We can obtain as good results as we please by sampling, and very often quite small samples are enough; the only difficulty is to ensure that every person or thing has the same chance of inclusion in the investigation.'

At this point Bowley has probably equated random sampling to any sampling scheme in which the inclusion probabilities are the same for every sampling units. This idea recurs in his 1912 sample of Reading (Bowley, 1913). Here a 1 in 10 systematic sample of buildings of Reading was chosen from street listings in the local directory. Nonresidential buildings were eliminated from the sample as well as those residences of the 'Principal Residents'. What remained was a sample of 1350 working-class houses. A sample of this size was found too large to cover adequately and so every second house was sampled to obtain an initial sample of 677 houses. After eliminating houses in which the occupier was considered to be not of the working class and after some reasonable substitution for nonexistent houses which were listed in the directory, a final sample of 622 houses was obtained. In his report of the results of the sampling investigation Bowley applied the results of his 1906 paper thus equating random and systematic sampling. A year or two later Bowley came to recognize the difference in efficiencies of random and systematic samples. The Reading sample was studied along with similarly chosen systematic samples taken from three other English towns, Northampton, Warrington and Stanley. The results were published with A. R. Burnett-Hurst in 1915 in a book entitled *Livelihood and Poverty*, a book which has remained a classic sociological study to this day. In an analysis of the sampling results (Chapter VI of the book, written by Bowley), Bowley again applies his 1906 results to the sampled data. However, he goes on to make the following comment on the analyses by the use of the simple random formulae:

'Actually we have a greater security and a smaller margin of error than this formula shows, because our samples were deliberately taken from houses distributed nearly uniformly through the towns, whereas the formula applies to samples taken quite at random, as if the houses had been numbered and numbers drawn from a bag. Our distribution cannot then contain a greater number of extreme cases; it must have a relation to the distribution of the rich, the comfortably well off, the poor and the destitute, so far as these are aggregated, similar families in similar streets. We have no data for using the mathematical expression appropriate to this consideration, but it most likely reduces the margin of error to allow us to neglect all roughness and unbiased errors of record, and regard ± 2 per cent as a very safe limit.'

Bowley also checked the representativeness of his samples by comparing his sample results to known population counts for variables on which these counts were available. For two cases in which he found a discrepancy between his sample and the official statistics, on further checking he discovered that the official statistics were in error.

Bowley's best remembered contribution to sampling theory was through the commission appointed in 1924 by the International Statistical Institute to study the application of the 'representative method'. The commission reported its findings to the ISI meeting in Rome in 1925 at which a resolution was adopted which gave acceptance to certain sampling methods both by random *and* purposive selection. The work of the commission and its report are discussed in detail by Yates (1946), Seng (1951) and Kruskal and Mosteller (1980). Bowley and Adolph Jensen were the major discussants for the commission's report; Jensen provided a lengthy description of the representative method in practice while Bowley provided a theoretical monograph summarizing the known results in random and purposive selection (Bowley, 1926; Jensen, 1926a). In addition to several other ideas, the monograph contains a development of stratified sampling with proportional allocation, sometimes referred to as Bowley allocation, and a theoretical development of purposive selection through correlations between control variables and the variable of interest. This latter development included formulae for the measurement of the precision of the estimate under a purposive sampling design. The work on proportional allocation reflects Bowley's continued desire to maintain equal inclusion probabilities for all units in a randomized sampling design.

The two methods of sampling, randomization and purposive selection, both under certain rules of operation, remained the standard acceptable methods of sampling for the next decade. Jensen (1926b, 1928), for example, has described a purposively selected balanced sample which gave reasonable results. The 20 percent sample of parishes in Denmark was chosen iteratively until the sample means of ten variables including items such as numbers of farms, areas under various crops and numbers of various kinds of livestock, matched approximately the means as calculated from the Danish Agricultural Census. The estimates of the means or the frequency distributions for a large number of variables related to the 'design variables' agreed closely with the census counts. Purposive selection did not, however, always work. To obtain a sample from the 1921 Italian Census, the Italian statisticians Corrado Gini and Luigi Galvani balanced on seven important variables and made a purposive selection of 29 out of 214 administrative units in Italy. The resulting sample showed wide discrepancies with the census counts on other variables (see Gini, 1928; Gini and Galvani, 1929). There were other groups taking random samples. Margaret Hogg, for example, after working with Bowley in Britain, emigrated to the United States where she carried out some random sampling studies in the early thirties (see Stephan, 1948). She was also very critical of some purposively selected samples carried out in the United States. It is of interest to note that in the same paper she reiterated Bowley's desire to have random samples in which each unit had the same inclusion probability and expressed the opinion that tests of representativeness of the sample are possible only through randomization.

4. Neyman

Neyman's 1934 paper on sampling was immediately recognized as an important contribution to the field of statistics. More recently, Kruskal and Mosteller (1980) have described the work as 'the Neyman watershed' and Hansen et al. (1985) have commented that the 'paper played a paramount role in promoting theoretical research, methodological developments, and applications of what is now known as probability sampling'. The work is obviously recognized by many statisticians as a paradigm, according to Kuhn's terminology, in statistical work. It is of interest to examine this since others, especially Bowley, had been critical of purposive selection and since the result in the paper on optimal allocation in stratified sampling had been previously, although independently, obtained by Tschuprow (1923).

Several reasons can be cited which make this paper a classic. The major reasons are twofold. First, Neyman was able to provide cogent reasons, both theoretically and with practical examples, why randomization gave a much more reasonable solution than purposive selection to the problems that then confronted sampling statisticians. Bowley (1926) had provided a theory for purposive selection from which the shortcomings of the method *may* have been deduced. However, the material is difficult to follow and even Neyman had trouble with it (Neyman, 1934, p. 573). Kruskal and Mosteller (1980) give the following description of Bowley's work on purposive selection: "Bowley attempts to make precise the purposive sampling idea in terms of correlations between control variables and the variate of primary interest, but he fails, in our reading, through lack of clarity in his description of sampling." What Neyman provided in his discussion of purposive selection was an easily understood theory of the selection method *and* the model assumptions under which purposive selection would work well, namely a linear regression relationship between the control variables and the variate of interest. He illustrated why Gini and Galvani's sample had been a failure by providing scatterplots of the data which showed a departure from the assumption of linearity. He went on to say that the necessary assumption of linearity is probably the exception rather than the rule in practice. Neyman also provided methods of adjusting the estimates when the linearity assumption was violated and at the same time showed that these adjustments could lead to large sampling biases and sampling variances. In short, purposive selection, although it could be useful, was severely limited in its applicability to the sampling problems of the day. The second major reason why the paper is a classic and provides a paradigm in the history of sampling is that a theory of point and interval estimation is provided under randomization that breaks out of an old train of thought and opens up new areas of research. Bowley and his followers had repeatedly stuck to sampling designs in which the inclusion probabilities were the same for all population units. Their reasoning was that this method of sampling would provide a representative sample of the universe. By providing a theory of stratified sampling with optimal allocation and a theory of cluster sampling with ratio estimation, Neyman showed that 'valid' estimates of the mean were possible without requiring a strictly 'representative sample'. His result in stratified sampling showed that designs with unequal inclusion probabilities were not only reasonable but they could also be optimal. Viewed in this way, his development of confidence intervals is almost icing on a well-baked cake. On combining the two reasons one sees that in a single publication the applicability of one method is shown to be severely curtailed while the applicability of its rival is greatly expanded.

This leaves the problem of Tschuprow and his priority for optimal allocation. Why did his 1923 work not become the lauching pad for the burst of activity in random sampling theory instead of Neyman's 1934 paper? One answer is that Tschuprow gave only part of the answer that Neyman provided; he broke out of the mould of constant inclusion probabilities but did not show that his methods were superior to the main competitor, purposive selection. This answer is incomplete. Kruskal and Mosteller (1980) have touched on the completion to the rest of the answer by noting that Tschuprow's 'influence appears to have been limited because of the formalities and abstractness of his style, and because of his geographical movements.' Tschuprow's 'formalities and abstractness' may be contrasted with the easy readability of Neyman's 1934 work; and his geographical movements were many following the Russian Revolution—after May 1917 he was never again in Russia. Isserlis (1926), in his obituary of Tschuprow, points to an additional compelling reason. Following his exile from Russia, Isserlis describes Tschuprow as living the life of a hermit: '... in Berlin he found too many friends and lacked the peace and quietness he needed for his work. So he moved to Dresden, where he lived the life of a hermit, seeing only occasionally the few who, passing through Germany, visited his modest retreat.' This may be contrasted with Neyman's work which was read before a meeting of the Royal Statistical Society and followed by several discussants. Moreover, the proposer of the vote of thanks was Bowley, at that time the most influential proponent of random sampling.

Neyman's landmark paper provoked an almost immediate reaction within the statistical research community. P. V. Sukhatme (1935), who later became the Statistics Division Director in the Food and Agriculture Organization of the United Nations, examined the consequences of the major drawback to Neyman allocation, unknown strata variances. He showed that when a preliminary sample is taken to estimate the variances, Neyman allocation based on the estimated variances is almost always more efficient than proportional allocation. The increase in efficiency is more pronounced as the variances differ more from stratum to stratum.

The diffusion of Neyman's ideas on random sampling received a further stimulus when W. Edwards Deming invited Neyman to come to Washington, DC to give a series of lectures on sampling in 1937. According to Duncan and Shelton (1978) the 1934 paper had little impact on U.S. government statisticians until the visit. They attribute this and the fact that Neyman had greater influence than other visiting statisticians to 'Neyman's ability to relate theory to practice from personal experience in the economic and social fields ...'. One other advance in sampling came out of these lectures. During the lectures a question was put to Neyman about two-phase sampling. Neyman's solution to the problem (Neyman, 1938) introduced the use of cost functions into survey sampling theory.

5. Hansen and Hurwitz

The sampling seeds sown by Neyman in the United States soon bore fruit in the U.S. Bureau of Census through the work of Morris Hansen and William Hurwitz and their colleagues. Hansen joined the Bureau in 1935 and Hurwitz in 1940. Both were members of the Statistical Research Division, Hansen beginning in 1936 and Hurwitz from the time he joined the Bureau. In 1942 Hansen became chief of the Division. The Division's major mission was to develop sampling methods for use by the Bureau. It was a time when random sampling was gaining acceptance by U.S. government agencies. Duncan and Shelton (1978) note that few randomized surveys were carried out before 1933 and describe several surveys carried out in the thirties. They also provide an interesting description of the politics surrounding the introduction of sampling in the 1940 U.S. Census and the diplomacy used by Hansen to have random sampling methods accepted throughout the Bureau in the thirties and forties.

Morris Hansen's first large-scale randomized survey was a 1937 unemployment survey, a 2% sample of all nonbusiness postal routes in the United States (see Duncan and Shelton, 1978, pp. 44-45 for details of the survey). He worked under his immediate superior, Calvert Dedrick, the then Statistical Research Division Chief. Hansen was responsible for the sampling computations which appeared in the final report, published jointly by Dedrick and Hansen (1938). Duncan and Shelton (1978) point to this survey as one that reflected the immediate influence of Neyman's Washington lectures earlier in 1937. In the early 1940's Hansen and Hurwitz (especially in their 1943 paper) made some fundamental contributions to sampling theory. In their 1943 *Annals* paper they took an important step forward by extending the idea of sampling with unequal inclusion probabilities for units in different strata as put forward by Neyman to differing inclusion probabilities for all units within a stratum. This allowed the development of very complex multi-stage sampling designs that are the backbone of large-scale social and economic survey research. With these surveys large samples with acceptable (not necessarily minimal) levels of variance could be conducted at a reasonable cost.

The influence of Hansen and Hurwitz through the U.S. Bureau of Census on the future development of sampling theory and methods is not to be under-estimated. Duncan and Shelton (1978) describe the degree of their influence as '... in the history of sampling theory the only other institution which can claim more distinguished alumni and more profound influence on thought is probably University College of the University of London under the leadership of Karl Pearson, his son Egon S. Pearson, R. A. Fisher, and for a short time, Jerzy Neyman.'

6. Other developments

I have, of course, left out many of the contributions of others to survey sampling in the period up to 1945. One big gap is the influence of R. A. Fisher and the work on sampling at Rothamsted by Yates and the researchers Rothamsted produced such as W. G. Cochran. Their contributions are described in Yates (1946). I have also confined the geographical locations of this history to Scandinavia in the early years, followed in time by Britain and the United States. Admittedly, there were almost world-wide developments in sampling theory and methods which have been omitted here. Finally, the subject matter which lead to the development of some sampling ideas has been restricted to social and economic surveys. There were other subject matter areas which stimulated developments, most notably agriculture and forestry (see Yates, 1946; Dalenius, 1957; Prodan, 1958). For example, the problem of estimation of crop yields was a primary stimulus to the use of sampling techniques in India.

The tremendous growth in India of the use of sampling techniques, and of research into sampling theory and methods, may be credited to P. C. Mahalanobis, founder of the Indian Statistical Institute. Details of his contributions to survey sampling may be found in Murthy (1964). Mahalanobis was interested in survey sampling as early as 1932. By the end of that decade he had been deeply involved in the successful completion of several gradually expanding exploratory surveys to estimate the acreage under jute in Bengal (Mahalanobis, 1940, 1944). A full large-scale survey was carried ut in 1941. It was in connection with these surveys that Mahalanobis brought to full development the method of interpenetrating subsamples for variance estimation. It is of interest to note here that Fisher's work at Rothamsted was influenced by some early random sampling studies carried out in India by Hubback (1927). Fisher, in return (Box, 1978, pp. 328–329; Mahalanobis, 1944, p. 332), helped to convince the Indial Central Jute Committee, the agency responsible for collecting the appropriate agricultural statistics in the Bengal surveys, of the usefulness of a randomized survey after the first exploratory survey of 1937 had been completed.

I have two major reasons for the superficial treatment of the other contributions to sampling. The lesser of these reasons is that this is a *brief* history of the subject. The more important reason is that in this brief history I believe that I have covered the major intellectual breakthroughs in the subject to about 1945. Moreover, in narrowing the focus of the history some insights into the development of sampling techniques become apparent. A major motivating force behind many so-called classical sampling methods is the execution of large-scale social and economic surveys. There is a gradual move away from reliance on models for design and estimation toward a pure randomization approach so that the minimum of assumptions may be made about the population. As shown by Neyman (1934) sampling designs and estimation techniques based on incorrect models can lead to disastrous results. Beginning with Bowley, an appeal is made to a central limit theorem, since the sample sizes are large, in order to obtain interval estimates for the quantities of interest. By the 1940's the focus of the sampling design is on reducing or minimizing the cost to sample while at the same time retaining an acceptable, though not necessarily minimal, level of variance. One final insight is that in every new breakthrough a major 'marketing campaign' was carried out to get the new ideas accepted.

7. The paradigm challenged and defended

The first challenge to the design-based approach to survey sampling was by V. P. Godambe in 1955. Godambe may be credited with creating a new paradigm in the field, that of establishing the logical foundations of estimation theory in survey sampling. The characteristics displayed by the new paradigm are exactly the same as the 'classical' paradigm of randomization, only the motivation is different. Indeed, Godambe's (1955) result may be viewed as a paradigm in Kuhn's framework. The result was sufficiently novel and open-ended that a number of statisticians were attracted to work on theoretical problems in survey sampling. The book by Cassel, Särndal, and Wretman (1977) is an eloquent testimony to the scope and volume of the work generated by Godambe's original questioning of the foundations of the subject. There is one further parallel to the paradigm of randomization; Godambe has actively promoted his point of view both in and out of print.

One of the ways out of Godambe's theorem on the nonexistence of a unique minimum variance unbiased estimator of the finite population mean is to assume a superpopulation model on the character of interest. The formulation of superpopulation models dates back at least to Cochran's (1946) classic paper on systematic sampling. The current use of models turns the clock back to the early days of sampling when models were used in both the design and estimation stage of the sampling process. This route has been taken by Richard Royall and his school. Royall (1970) found that by purposively selecting the units associated with the largest covariate values, the model variance of the ratio estimator was minimized under a regression through the origin model. Following criticism of this approach that the estimator could be severely biased under other model assumptions, Royall and Herson (1973) put forward a method of balanced sampling to retain the model unbiased property of the ratio estimator under polynomial regression models. This is distinctly reminiscent of the balanced sampling techniques of the twenties, the difference being that Royall could show some optimality properties for his methods. Although Royall's (1970) result had been obtained earlier by Brewer (1963), Royall's school of model-based inference, following on his 1970 paper, has all the elements of a paradigm in sampling. There are a number of researchers who have been drawn away from other possible areas of activity to follow the model-based approach. This did not occur after Brewer's (1963) paper. Moreover, like Godambe and the proponents of the randomization paradigm, Royall has actively promoted his ideas. This new paradigm has not, however, replaced the older paradigm of randomization. Debate over the use of models in sampling has continued in the literature and at conferences to the present day. The most recent published criticism of Royall's model-based approach with a rebuttal from Richard Royall may be found in Hansen et al. (1983).

Is there a resolution to model and randomization approaches to sampling? I believe there is, if one looks at the factors which motivated the earlier, or classical, researchers to use randomization techniques. The key to the resolution is that the earlier workers in sampling were drawn to random sampling because they were taking *large* samples from *large* populations. The same problems with modelling that confronted them then still hold true today. Where inferences based on randomization break down, i.e. when randomization no longer adequately answers the questions being asked, is the situation in which one moves from large to small samples. This is evident in small area estimation problems within large-scale surveys. Variance estimates based on randomization are unacceptably high when estimates are required for small areas but are acceptable when looking at larger aggregates. Modelling is reasonable and desirable for inferences about small areas. The same argument can be carried over to small surveys. Provided that enough prior information has been collected on a population through previous surveys, a reasonably well-fitted model may be formulated about the population prior to the selection of the sample. A purposive sample, selected using the model, may provide substantian gains in efficiency over a randomly selected sample and therefore purposive selection would, in this case, be a reasonable strategy.

References

Ashton, T. S. (1977). *Economic and Social Investigations in Manchester 1833–1933*. Harvester, Hassocks, Sussex.
Bowley, A. L. (1906). Address to the Economic and Statistics Section of the British Association for the Advancement of Science, York, 1906. *J. Roy. Statist. Soc.* **69**, 540–558.
Bowley, A. L. (1907). *Elements of Statistics, 3rd Ed.* King and Son, London.
Bowley, A. L. (1913). Working-class households in Reading. *J. Roy. Statist. Soc.* **76**, 672–701.
Bowley, A. L. (1926). Measurement of the precision attained in sampling. *Bull. Int. Statist. Inst.* **22**, Supplement to Liv. 1, 6–62.
Bowley, A. L. and Burnett-Hurst, A. R. (1915). *Livelihood and Poverty.* G. Bell, London.
Box, J. F. (1978). *R. A. Fisher: The Life of a Scientist.* Wiley, New York.
Brewer, K. R. W. (1963). Ratio estimation and finite populations: Some results deducible from the assumption of an underlying stochastic process. *Aust. J. Statist.* **5**, 93–105.
Cassel, C.-M., Särndal, C.-E. and Wretman, J. H. (1977). *Foundations of Inference in Survey Sampling.* Wiley, New York.
Chang, W.-C. (1976). Sampling theories and practice. In: D. B. Owen, ed., *On the History of Statistics and Probability*. Dekker, New York, 299–315.
Cochran, W. G. (1946). Relative accuracy of systematic and stratified samples for a certain class of populations. *Ann. Math. Statist.* **17**, 164–177.
Cochran, W. G. (1978). Laplace's ratio estimator. In: H. A. David, ed., *Contributions to Survey Sampling and Applied Statistics: Papers in Honor of H. O. Hartley*. Academic Press, New York, 3–10.
Cullen, M. J. (1975). *The Statistical Movement in Early Victorian Britain.* Harvester, Hassocks, Sussex.
Dalenius, T. (1957). *Sampling in Sweden.* Almqvist and Wiksell, Stockholm.
Dedrick, C. L. and Hansen, M. H. (1938). *Census of Partial Employment, Unemployment and Occupations: Vol. IV. The Enumerative Check Census.* U.S. Government Printing Office, Washington.
Duncan, J. W. and Shelton, W. C. (1978). *Revolution in U.S. Government Statistics, 1926–1976.* U.S. Dept. of Commerce, Washington.
Gini, C. (1928). Une application de la méthode representative aux matériaux du dernier recensement de la population italienne (1er décembre 1921). *Bull. Int. Statist. Inst.* **23**, 198–215.
Gini, C. and Galvani, L. (1929). Di una applicazione del metodo rappresentativo all'ultimo censimento italiano della popolazione (1 decembre, 1921). *Annali di Statistica*, Series 6, **4**, 1–107.
Godambre, V. P. (1952). A unified theory of sampling from finite population. *J. Roy. Statist. Soc., Ser. B* **17**, 269–278.
Godambe, V. P. (1976). A historical perspective of the recent developments in the theory of sampling from actual populations. *Dr. Panse Memorial Lecture.* Ind. Soc. Agric. Statist., New Delhi.
Hansen, M. H., Dalenius, T. and Tepping, B. J. (1985). The development of sample surveys of finite populations. In: A. C. Atkinson and S. E. Fienberg, eds., *A Celebration of Statistics*. Springer, New York, 327–354.
Hansen, M. H. and Hurwitz, W. N. (1943). On the theory of sampling from finite populations. *Ann. Math. Statist.* **14**, 333–362.
Hansen, M. H., Madow, W. G. and Tepping, B. J. (1983). An evaluation of model-dependent and probability-sampling inferences in sample surveys. *J. Amer. Statist. Assoc.* **78**, 776–807.
Hubback, J. A. (1927). Sampling for rice yields in Bihar and Orissa. *Indian Agricultural Institute, Pysa*, Bulletin No. 166, reprinted in *Sankhyā* (1946) **7**, 281–294.

Isserlis, L. (1918). On the value of a mean as calculated from a sample. *J. Roy. Statist. Soc.* **81**, 75–81.

Isserlis, L. (1926). Alexander Alexandrovitch Tschuprow. *J. Roy. Statist. Soc.* **89**, 619–622.

Jensen, A. (1926a). Report on the representative method in statistics. *Bull. Int. Statist. Inst.* **22**, 359–380.

Jensen, A. (1926b). The representative method in practice. *Bull. Int. Statist. Inst.* **22**, 381–439.

Jensen, A. (1928). Purposive selection. *J. Roy. Statist. Soc.* **91**, 541–547.

Hogg, M. H. (1930). Sources of incomparability and error in employment-unemployment surveys. *J. Amer. Statist. Assoc.* **25**, 284–294.

Kent, R. A. (1981). *A History of British Empirical Sociology*. Gower, London.

Kiaer, A. N. (1895/1896). Observations et expériences concernant des dénombrements représentatifs. *Bull. Int. Statist. Inst.* **9**, 176–183.

Kiaer, A. N. (1897). *The Representative Method of Statistical Surveys* (1976, English translation of the original Norwegion). Central Bureau of Statistics of Norway, Oslo.

Kiaer, A. N. (1905). Untitled speech with discussion. *Bull. Int. Statist. Inst.* **14**, 119–134.

Kruskal, W. and Mosteller, F. (1980). Representative sampling, IV: The history of the concept in statistics, 1895–1939. *Int. Statist. Rev.* **48**, 169–195.

Kuhn, T. S. (1970). *The Structure of Scientific Revolutions*. 2nd Ed. University of Chicago Press, Chicago.

Mahalanobis, P. C. (1940). A sample survey of the acreage under jute in Bengal. *Sankhyā* **4**, 511–530.

Mahalanobis, P. C. (1944). On large-scale sample surveys. *Phil. Trans. Roy. Soc. (B)* **231**, 329–451.

Murthy, M. N. (1964). On Mahalanobis' contributions to the development of sample survey theory and methods. In: C. R. Rao ed., *Contributions to Statistics, Presented to Professor P. C. Mahalanobis on the occasion of his 70th Birthday*. Statistical Publishing Society, Calcutta, 283–316.

Neyman, J. (1934). On the two different aspects of the representative method: the method of stratified sampling and the method of purposive selection. *J. Roy. Statist. Soc.* **97**, 558–625.

Neyman, J. (1938). Contribution to the theory of sampling human populations. *J. Amer. Statist. Assoc.* **33**, 101–116.

Prodan, M. (1958). Untersuchungen über die durchführung von repräsentativaufnahmen. *Allgemeine Forst- und Jagdzeitung* **1**.

Royall, R. M. (1970). On finite population sampling theory under certain linear regression models. *Biometrika* **57**, 377–387.

Royall, R. M. and Herson (1973). Robust estimation in finite populations I. *J. Amer. Statist. Assoc.* **68**, 880–889.

Seng, Y. P. (1951). Historical survey of the development of sampling theories and practice. *J. Roy. Statist. Soc. Ser. A* **114**, 214–231.

Stephan, F. F. (1948). History of the uses of modern sampling procedures. *J. Amer. Statist. Assoc.* **43**, 12–39.

Stigler, S. M. (1986). *The History of Statistics: The Measurement of Uncertainty Before 1900*. Belknap Press, Cambridge, MA.

Sukhatme, P. V. (1935). Contribution to the theory of the representative ethod. *J. Roy. Statist. Soc. Supp.* **2**, 263–68.

Sukhatme, P. V. (1966). Major developments in sampling theory and practice. In: F. N. David, ed., *Research Papers in Statistics: Festschrift for J. Neyman*. Wiley, New York, 367–409.

Tschuprow, A. A. (1923). On the mathematical expectation of the moments of frequency distributions in the case of correlated observations. *Metron* **2**, 461–493, 646–680.

Yates, F. (1946). A review of recent statistical developments in sampling and sampling surveys. *J. Roy. Statist. Soc. Ser. A* **109**, 12–42.

A First Course in Survey Sampling

Tore Dalenius

Introduction

An introductory course in survey sampling may, for sure, be designed in several ways, depending upon a wide variety of relevant circumstances. The design of the course I am about to present reflects the following circumstances (a)–(d):

(a) The course is thought of as part of a statistics curriculum at a university.

(b) The teacher is assumed to be an expert in survey sampling. More specifically, the teacher is assumed not only to be knowledgeable in the methods and theory of survey sampling, but also to have experience in real-life sample surveys.

(c) The students are assumed to be 'social science' students, with some, albeit limited, background in statistical theory. In particular, they are assumed to be familiar with the elements of expected values of random variables and the principles of estimation.

(d) Moreover, the class of students is assumed to comprise two different categories: students who want to prepare themselves for vocational training in survey sampling; and students who want to become acquainted with the basic methodological ideas of survey sampling.

Given these conditions, the design of the course reflects the following considerations (e)–(i):

(e) The specification of the statistics needed is a decisive part of the survey design.

(f) How to get observational access to the population to be studied plays a key role in terms of the efficiency of the survey design.

(g) It is preferable, on educational grounds, to organize the presentation of methods and theory as follows: first, *general* methods and theory are presented, to be followed by *special* methods and theory.

(h) It is important to use terms and symbols which have a mnemonic property, hence reducing the need for memorizing 'details'.

(i) Special survey sampling techniques should be presented in the simplest possible context, where the basic ideas are perspicuous, but not necessarily in the context where they are most powerful.

In what follows, a set of eight course modules will be presented, viz.:

    I. The notion of survey sampling,
   II. A review of the statistics needed,
  III. Getting observational access to the population,
  IV. Element sampling,
   V. Cluster sampling,
  VI. Multi-stage sampling,
 VII. The problems of non-sampling errors,
VIII. Survey sampling design.

For each module, a brief account will be given of its main topics. The language I am using is explicitly narrative: I present the *topics* I think should be taught, *how* they should be taught, and in what *order* they should be taught. The presentation may be viewed as an annotated list of topics in survey sampling which should be discussed in a First Course.

I. The notion of survey sampling

1. The wide scope of statistical sampling

'Statistical sampling' is a *general* term: it denotes a statistical investigation of a part (sample) of some whole (population), for the purpose of inferring something about that whole.

Statistical sampling is being used in a great variety of fields. Hence, a number of specialities have emerged, such as:
   (i) experiments;
  (ii) Monte Carlo sampling;
 (iii) acceptance sampling;
 (iv) sampling for auditing; and
  (v) survey sampling

to give but five examples.

The nature of these specialities should be illuminated by presenting brief accounts of simple applications. Such accounts will naturally highlight the methodological differences between the specialities. The discussion may preferably be supplemented by a discussion of the distinction between 'experimental data', as generated by an experiment, and 'observational data' as generated by Nature.

In addition, it should be stated that the specialities have a common theoretical basis.

2. Surveys and survey sampling

There is no universally accepted definition of 'survey' and hence no such definition of 'sample survey'. There is, however, no shortage of suggested definitions in the literature. Two examples are:

'Many research problems require the systematic collection of data from populations or samples of populations through the use of personal interviews or other data-gathering devices. These studies are usually called surveys, especially when they are concerned with large or widely dispersed groups of people.',

Campbell and Katona (1953), and:

'An examination of an aggregate unit, usually human beings or economic or social institutions. Strictly speaking perhaps, 'survey' should relate to the whole population under consideration but it is often used to denote a sample survey, i.e., an examination of a sample in order to draw conclusions about the whole.',

Kendall and Buckland (1960). It is suggested to ask the students to provide additional definitions and then discuss their common characteristics: they are indeed rather vague, hence giving only a broad idea of what a survey is.

By the same token, there is no universally accepted definition of 'survey sampling'. The meaning of this term is, however, clear: the use of statistical sampling to collect data for a survey.

Sample surveys may be characterized with respect to the populations studied as follows:

(i) surveys of some finite population of individually identifiable objects, typically called elements; an example is a population of households in a country as of a certain date; in what follows 'objects' and 'elements' will be used interchangeable;

(ii) surveys of a finite population of objects, which are not individually identifiable; an example is the population of fishes in a lake;

(iii) surveys of a process in time; an example is a sequence of events such as car accidents in a year; and

(iv) surveys of a process in the plane; an example is the forest area in a country.

Additional examples of populations studied by sample surveys should be provided, in order to elucidate the wide scope of survey sampling.

3. The main features of the course

The *subject* of the course will be methods and theory for survey sampling from a finite population of objects, based on 'probability sampling'.

The *focus* will be on populations composed of individually identifiable elements. But parts of the content of the course should be of interest also for other types of populations.

The *emphasis* of the course, finally, will be on a variety of sample designs, i.e., means of controlling and measuring the *sampling* errors of a survey. In addition, means for coping with *non-sampling* errors will be considered.

4. The choice of symbols

It is desirable, especially in a first course, to use symbols which are informative and still simple to understand and remember. To some extent, this may be achieved by using symbols having natural mnemonic properties. The point just made will be illustrated by an example; additional examples are given subsequently.

Thus, the N objects of a *population* will be denoted as follows:

$$\{O_1, \ldots, O_J, \ldots, O_N\}$$

i.e., by the letter 'O' with the upper case letter 'J' to denote the individual object; for short, $\{O\}$ may be used to denote this population. The n objects of a *sample* from this population will be denoted by:

$$\{O_1, \ldots, O_j, \ldots, O_n\}.$$

While this convention is not perfect (it fails to distinguish between the object no. 1 in the population and object no. 1 in the sample), it is simple to understand and remember.

References

Campbell, A. A. and Katona, G. (1953). The sample survey: A technique for social science research. Ch. 1 in: L. Festinger and D. Katz, eds., *Research Methods in the Behavioral Sciences*. The Dryden Press, New York.

Kendall, M. G. and Buckland, W. R. (1960). *A Dictionary of Statistical Terms*. Oliver and Boyd, Edinburgh and London.

II. A review of the statistics needed

'Until the purpose is stated, there is no right or wrong way of going about the survey.'
W. Edwards Deming

1. The reason for this module

Notwithstanding the objective of a first course in survey sampling, as stated in the Introduction, it is desirable to impress upon the students the *decisive* importance for good survey practice of specifying—at an early stage of the survey design—the statistics needed, in order to be able to tailor the design to the contemplated uses of the statistics to be produced.

This specification may proceed in two steps. First, there should be a careful formulation of the *subject-matter problem*, to the solution of which the survey is expected to contribute; this problem may be called the user's problem. As the saying goes: it is often better to have a reasonably good solution of the proper problem than an exact solution of the wrong problem. This first step should be followed by a translation of the user's problem into a *statistical problem*, viz. the survey design which will provide the statistics to be used as a basis for tackling the user's problem.

No two users' problems may be expected to be identical. This does not mean, however, that it is not possible to systematize the discussion of the statistics needed. Thus, such a discussion may be organized as follows:
  (i) the population of interest;
  (ii) the data to be collected;
  (iii) the population characteristics ('parameters') to estimate; and
  (iv) the accuracy needed.
In Sections 2–5, these aspects will be briefly discussed.

2. The population of interest

The population of interest is some finite set of identifiable objects (elements), existing in a specified area at a specified point in time (during a specified period of time).

3. The data to be collected

There is—for each one of a set of measurable (observable) properties of interest to the user and to the statistician—a specific measurement procedure which—if applied in a uniform manner to all elements in the population—would generate data to be represented as follows:

$$Y_1, \ldots, Y_J, \ldots, Y_N,$$
$$X_1, \ldots, X_J, \ldots, X_N,$$
$$\vdots \qquad \vdots \qquad \vdots$$
$$A_1, \ldots, A_J, \ldots, A_N.$$

It may be useful to point out to the students that the choice of symbols shown above differs from conventions in mathematics, where a distinction is being made between a variable and the value it may assume.

The Y-data etc. reflect the properties of the subject-matter interest. The A-data are 'auxiliary data', which may be exploited by the statistician in the survey design.

4. The population characteristics to estimate

The purpose of the sample survey is assumed to be *descriptive*: to provide a statistical description of (parts of) the population of interest, as that population exists as of a certain date (period). This purpose may be made clear by stating that the survey is to provide answers to questions such as 'How many...?,' 'How much...?'. With today's public concerns about 'invasion of privacy' in the context of surveys and censuses, it may be added that the survey is not to provide a possibility to take action on an individual element in the population. This specification should be supplemented by stating what the purpose is *not*; it is not to make inferences about the mechanism that generated the population.

The descriptive purpose of the survey calls for estimating certain numerical characteristics. If they concern the population in its entirety, they are usually referred to as 'population characteristics' or 'population parameters'. If they concern parts of the population, they are referred to as characteristics (parameters) of 'domains of study', sub-populations.

Typically, the characteristics in which the user is interested are:
 (i) the means per element;
 (ii) proportions (percentages);
 (iii) totals; and
 (iv) ratios.

In the case of population characteristics, the following symbols may be used:
 (i) for means: $\bar{Y}, \bar{X}, \ldots$;
 (ii) for proportions: $P_Y, P_X, \ldots$;
 (iii) for totals: $T_Y, T_X, \ldots$; and
 (iv) for ratios: $\bar{Y}/\bar{X}$ (rather than $\bar{X}/\bar{Y}$).

As part of this module, the students' attention should be drawn to the fact that the user is typically interested in *several* characteristics, while most of the methods and theory of survey sampling are restricted to very *few* characteristics.

Making a list of the characteristics of interest to the user is important, but it is not necessarily sufficient as a list of the statistics *needed* by the user. To specify these statistics, the list must be subjected to an assessment with respect to the *relevance* of the statistics with respect to the uses to be made of the statistics.

The statistician will, in addition to the characteristics listed above, be interested in such characteristics as:
 (v) variances; and
 (vi) correlation coefficients.

As to the variances, symbols like σ_Y^2, σ_X^2, etc. may prove awkward for typing; an alternative would be to use Var Y, Var X,..., but not use V to denote the variance, as V is commonly used to denote the coefficient of variation.

5. The accuracy needed

This topic will be discussed here with respect to estimation of the mean $\overline{Y}$.

The fact that the user is prepared to rely on a sample survey implies that his needs would be satisfied with an approximation $\overline{Y}^*$ of $\overline{Y}$, granted that $\overline{Y}^*$ is a 'good' approximation.

When discussing the accuracy needed, the following points deserve to be considered:

   (i) it must be possible to *measure* the degree of approximation;

   (ii) a distinction must be made between two cases; the case where the *direction*, in which $\overline{Y}^*$ differs from $\overline{Y}$ does not matter and the case where the direction does matter (for example a case, where an overestimate is more serious than an underestimate, given that the absolute size of the error is the same); and

   (iii) it is important to find out how *close* $\overline{Y}^*$ has to be to $\overline{Y}$.

III. Getting observational access to the population

1. Efficiency considerations

Carrying out a sample survey calls for having some means of getting observational access to the population of elements for the purpose of collecting the data. Such a means is referred to as a *sampling frame*, or frame for short.

If it were possible to disregard the cost of collecting the data, it would be natural to prepare a list of the elements in the population and collect the data from a sample of elements selected from that list.

In most applications—and especially when dealing with large populations distributed over a large area—this approach would be prohibitively expensive on one or both of the following grounds:

   (i) the cost of preparing the list; and
   (ii) the cost of collecting the data.

Instead of preparing a list of the individual elements in the population $\{O\}$, a list (in a *broad* sense of this word) is prepared of M identifiable units of some kind, having a known association with the elements in $\{O\}$. This list will be denoted by:

$$\{U_I | I = 1, \ldots, M\} = \{U_1, \ldots, U_I, \ldots, U_M\}$$

and for short by $\{U\}$. The units in $\{U\}$ will serve as *sampling units*, and $\{U\}$ will be referred to as the sampling frame. Observational access to a sample of elements in $\{O\}$ will then be provided by selecting a sample of units from $\{U\}$ and observe all or a subsample of the elements associated with the selected units.

2. Four basic rules of association

As stated above, there must be a known association between the units in $\{U\}$ and the elements in $\{O\}$. We will consider four such rules.

2.1. Rule 1: One-to-one

Each unit in $\{U\}$ is associated with one and only one element in $\{O\}$; and every element in $\{O\}$ is associated with one and only one unit in $\{U\}$. Symbolically, this rule may be represented as follows:

$$U_J \to O_J, \quad J = 1, \ldots, N.$$

In this case, $\{U\}$ is a list of the individual elements in $\{O\}$. Hence, $M = N$.

2.2. Rule 2: Many-to-one

Each unit in $\{U\}$ is associated with one and only one element in $\{O\}$, but an element in $\{O\}$ may be associated with more than one unit in $\{U\}$, as exemplified in the following diagram:

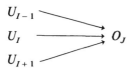

Hence, $M > N$.

2.3. Rule 3: One-to-many

Each unit in $\{U\}$ may be associated with more than one element in $\{O\}$, but no element in $\{O\}$ is associated with more than one unit in $\{U\}$, as exemplified in the following diagram:

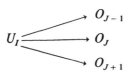

Hence, $M < N$.

2.4. Rule 4: Many-to-many

This rule may be exemplified as follows:

$U_G \begin{array}{c} \to O_{J-1} \\ \to O_J \end{array} \qquad U_I \begin{array}{c} \to O_K \\ \to O_{K+1} \end{array}$

2.5. A practical consideration

It may be informative to mention that sampling frames used in actual sample surveys may contain units which are not associated with any element in $\{O\}$. This is sometimes referred to as 'some-to-none'.

3. Important and desirable properties of a frame

The construction of a sampling frame should be guided by considerations of which properties are important and/or desirable. A list of some such properties is given below. Properties 1–3 are indeed important, while the remaining are desirable.

Property 1. The frame must be virtually complete, i.e., provide observational access to virtually all elements in $\{O\}$. This is sometimes expressed as follows: the 'sampled population' (= the elements accessible by the frame) must be sufficiently 'close' to the 'target population' (= the population of interest).

Property 2. It must be possible to identify unambiguously every element included in any selected sampling unit.

Property 3. The rule of association must be known (determinable).

Property 4. It is desirable that the frame is simple to use.

Property 5. It is desirable that the frame is organized in a systematic fashion. As an example, it is often advantageous if the frame is organized by 'geography' and/or by 'size' of the units.

Property 6. It is desirable that the frame contains 'auxiliary' data which may be used in the survey design.

These six properties—and others—should be elaborated on in some detail as part of presenting this module.

4. Techniques for frame construction

In Section 1 above, the frame $\{U\}$ was viewed as a list of a set of sampling units, where 'list' was to be understood in a broad sense. Two major types of lists in this broad sense are:
  (i) the *directory type list*: this is a list of sampling units in the every day sense of the word 'list'; and
  (ii) the *map type list*: this is a map of the area containing the elements in $\{O\}$; this area has been divided into smaller areas with which the elements are associated; these areas serve as the sampling units.

In some applications, the one or the other of these two types of frames is used. In other applications, simultaneous use may be made of the two types.

5. Three sampling systems

It is recommended to use 'sample design' to denote both the sample selection plan and the estimation procedure; these components of the sample design are like the two sides of a coin.

The 'sample selection plan' in turn may be viewed as having two components, viz. the 'sampling system' and the 'sampling scheme'. There are three basic sampling systems:
  (i) element sampling;
  (ii) cluster sampling; and
  (iii) multi-stage sampling.

Modules IV–VI will present sampling schemes for each of these systems. Throughout the assumption will be made that the sample selection is carried out by 'probability sampling' (also known as a 'measurable design'), that is with known probabilities of selection for each element in the population.

IV. Element sampling

1. The notion of element sampling

This term is not current in the literature on survey sampling. It is used here as an analogue of 'cluster sampling' to be discussed in Module V. The term is justified by the fact that selecting one unit from a frame $\{U\}$ for which the rule of association is 'one-to-one' or 'many-to-one' will select one element from $\{O\}$.

As part of this module, three classes of sampling schemes will be discussed, viz.:
  (i) simple random sampling;
  (ii) stratified sampling; and
  (iii) sampling with unequal probabilities of selection.
Within each of these three classes, two or more sample designs will be presented.

2. Simple random sampling

2.1. The starting point

For this class of sampling schemes, the frame $\{U\}$ is identical with the population $\{O\}$. Hence, the rule of association is one-to-one.

It should be emphasized from the beginning that 'simple random sampling' is not likely to be used as the sampling scheme for a real-life sample survey. Its prime role is to provide a 'tool' for use in more complicated schemes.

In some (but not all) textbooks, 'simple random sampling' is used to denote two different ways of selecting a sample of elements: 'with replacement' and 'without replacement'. This terminology reflects, of course, the tradition to present the theory by means of 'urn schemes'.

2.2. Selection with replacement

For the presentation of this scheme, it is instructive to introduce the term 'draw': the scheme calls for making n draws.

The conventional representation of the sample of data is:

$$Y_1, \ldots, Y_j, \ldots, Y_n$$

(using the lower 'j' to denote a *sample* datum). This representation has the well-known disadvantage that it does not reflect the size of the sample in terms of the number of distinct elements included; this number is usually denoted by n_d ('d' for 'distinct'); for n_d, we have $n_d \leqslant n$.

This suggests that an alternative representation of the sample data be presented, viz.:

$$Y_1 t_1, \ldots, Y_J t_J, \ldots, Y_N t_N$$

where t_J is an indicator (a random variable), which shows how many times O_J was selected for the sample of elements. Thus, O_J assumes one of the integer values $0, 1, \ldots, n$.

Simple random sampling with replacement offers an opportunity to introduce the idea of 'replicated sampling'. Thus, each single draw is viewed as a sample of size $n = 1$; in all, r draws are made. The data collected in the r draws are r estimates of the mean $\bar{Y}$, with variances equal to σ^2. These estimates are then 'pooled' to get a better estimate of $\bar{Y}$:

$$\hat{y} = \frac{1}{r} \sum_j Y_j$$

The variance of this pooled estimate is estimated by:

$$\operatorname{var} \hat{y} = \frac{1}{r} \frac{1}{r-1} \sum_j (Y_j - \hat{y})^2 .$$

It should be mentioned that these two formulas (for $\hat{y}$ and var $\hat{y}$, respectively) are perfectly general.

2.3. Selection without replacement

In this case, the representation of the sample data by:

$$Y_1, \ldots, Y_j, \ldots, Y_n$$

causes no difficulty: the data are data for $n_d = n$ distinct elements.

It may nonetheless be helpful to the students to present also the following representation:

$$Y_1 \varepsilon_1, \ldots, Y_J \varepsilon_J, \ldots, Y_N \varepsilon_N$$

where the indicator ε assumes the value $\varepsilon = 1$, if O_J is selected and else $\varepsilon_J = 0$.

Given the Y-data, the mean $\overline{Y}$ would be estimated by the sample mean $\overline{y}$. By considering a miniature population of say $N = 10$ elements with specified Y-data, and a miniature sample of $n = 2$ elements, it is easy to demonstrate the properties (unbiased and variance) of this estimate by enumerating all possible samples, and computing the estimates for each sample and the variance between these estimates.

2.4. Estimation using auxiliary data

Simple random sampling offers an opportunity to discuss the possibility to exploit auxiliary data, A-data, as a means of improving an estimate.

Applying the principle of first discussing general results and then special results (cf. the Introduction), the point of departure would be the following estimator of $\overline{Y}$:

$$\overline{y}_q = \overline{y} + q(\overline{A} - \overline{a})$$

where $\overline{y}$ and $\overline{a}$ are the unbiased estimators of $\overline{Y}$ and $\overline{A}$, respectively, and q is some quantity yet to be specified. Various choices of q will generate various estimators:

(i) $q = 0$ yields $\overline{y}$;

(ii) $q = K$, a *constant*, not depending on the sample at hand, yields the 'difference estimator';

(iii) $q = B$, the population regression coefficient, a *constant*, not depending on the sample at hand, yields the 'regression estimator';

(iv) $q = b$, an *estimator* of B, yields what also is known as a 'regression estimator'; and

(v) $q = \overline{y}/\overline{a} = r$, an *estimator* of $R = \overline{Y}/\overline{A}$, yields the 'ratio estimator' $\overline{y}_r = r\overline{A}$.

Whether or not this general approach is being used, the properties of the ratio estimator may be discussed in an intuitive manner by means of two diagrams: one for the case, where the correlation between $\overline{y}$ and $\overline{a}$ is positive, as in Fig. 1 and the other for the case with negative correlation. In this second case, the diagram will demonstrate that the ratio estimator $\overline{y}_r$ is inferior to $\overline{y}$.

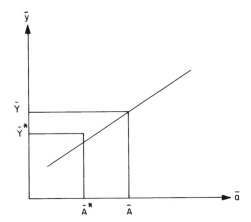

Fig. 1. Regression between $\bar{y}$ and $\bar{a}$.

As part of the discussion of the use of auxiliary data as a means of improving an estimate, it should be mentioned that using the auxiliary data in the estimator is but one way (and not necessarily the best way) of taking advantage of having access to such data,

2.5. Using a table of random numbers

The presentation above may be supplemented by a discussion of random numbers and their use as a selection mechanism.

3. Stratified sampling

3.1. The starting point

There is a frame $\{U\}$ with the 'one-to-one' rule of association with the population $\{O\}$. Hence, the discussion may be carried out in terms of sampling from $\{O\}$ without introducing any ambiguity.

The N elements in $\{O\}$ are partitioned into L groups called strata, with N_H elements in the Hth stratum. The Y-data of the elements in the Hth stratum are:

$$Y_{H1}, \ldots, H_{HJ}, \ldots, Y_{HN_H}$$

with mean $\overline{Y}_H$ and variance σ_{HY}^2. With the relative size of the Hth stratum denoted by W_H, the population mean is written as:

$$\overline{Y} = \sum_H W_H \overline{Y}_H,$$

It is instructive to derive the expression for the population variance 'from scratch':

$$\sigma^2 = \frac{1}{N} \sum_H \sum_J (Y_{HJ} - \overline{Y})^2$$

and then show how this expression may be written as:

$$\sigma_Y^2 = \sum_H W_H \sigma_{HY}^2 + \sum_H W_H (\overline{Y}_H - \overline{Y})^2$$

$$= \text{'Within stratum variance'} + \text{'Between stratum variance'}$$

or

$$\sigma_Y^2 = \sigma_W^2 + \sigma_B^2$$

for short.

3.2. General theory

By means of simple random sampling without replacement, n_H elements are selected from the Hth stratum, $H = 1, \ldots, L$. The Y-data collected is:

$$Y_{H1}, \ldots, Y_{Hj}, \ldots, H_{Hn_H}$$

where the use of the upper case 'H' and the lower case 'j' is in accordance with the aim to make a clear distinction between population values (the strata indices) and sample values (the indices for the sample elements). The means of the Y-data are $\overline{y}_H$, $H = 1, \ldots, L$.

It is now straightforward to estimate $\overline{Y}$ by:

$$\overline{y} = \sum_H W_H \overline{y}_H$$

which has the variance

$$\operatorname{Var} \overline{y} = \sum_H W_H^2 \frac{N_H - n_H}{N_H - 1} \frac{1}{n_H} \sigma_{HY}^2.$$

At his point it may be natural to remind the students of the specific role of simple random sampling as discussed in Section 2.

The presentation above of the general theory may be supplemented by considering the use of auxiliary data, A-data, to improve the estimate. Especially, two types of ratio estimators are available: Thus, if $\overline{Y}$ is written as:

$$\overline{Y} = \frac{\sum_H W_H \overline{Y}_H}{\sum_H W_H \overline{A}_H} \overline{A}$$

the corresponding combined ratio estimator is:

$$\bar{y}_r = \frac{\Sigma_H W_H \bar{y}_H}{\Sigma_H W_H \bar{a}_H} \bar{A} \; .$$

Alternatively, $\bar{Y}$ may be written as:

$$\bar{Y} = \sum_H W_H \frac{\bar{Y}_H}{\bar{A}_H} \bar{A}_H$$

with the stratum-by-stratum estimator:

$$\hat{y} = \sum_H W_H \frac{\bar{y}_H}{\bar{a}_H} \bar{A}_H = \sum_H W_H r_H \bar{A}_H \; .$$

3.3. Four design problems

The use of stratified sampling poses four *specific* design problems, viz. the choices of:

(i) the stratification variable;
(ii) the number of strata;
(iii) the mode of stratification; and
(iv) the sample sizes in each stratum, i.e. the sample allocation.

In a First Course, the emphasis should be on the sample allocation, assuming that the choices (i)–(iii) have somehow been done.

In Sections 3.4 and 3.5, the sample allocation will be discussed.

3.4. Proportional allocation

This allocation calls for making n_H proportional to N_H, which may be expressed as:

$$\frac{n_H}{n} = \frac{N_H}{N} \; .$$

The expressions for $\bar{y}$ and Var $\bar{y}$ may be arrived at by inserting $n_H = W_H \cdot n$ in the formulas given for the general theory.

The emphasis of the discussion of proportional allocation should, however, not be on the expressions for $\bar{y}$ and Var $\bar{y}$ but on a discussion of the notion of 'representativity'. Special mention should be made of the fact that stratified sampling, using proportional allocation, may resemble 'quota sampling' (often associated with the Gallup polls), but in principle it is not equivalent with that procedure, a decisive difference being that 'quota sampling' does not make use of probability sampling.

2. Optimum allocation

One reason for discussing this type of sample allocation is to draw the students' attention to the fact that 'optimum allocation' is used in the literature in two different ways, viz. to denote:

(i) minimum variance allocation with a fixed sample size n; and

(ii) minimum variance allocation with a fixed total cost C, optimum allocation proper.

It should be pointed out that in practice, the use of optimum allocation amounts to using probabilities of selection which differ between strata (but are the same within the respective strata). In general, the probabilities of selection will be larger in strata with large variances and/or low cost of data collection; the probabilities of selection will be smaller in strata with small variances and/or high cost of data collection.

3.6. The three remaining design problems

It is suggested that they are only mentioned but not elaborated on.

3.7. Estimation for domains of study

Some surveys are carried out to provide estimates of the means $\overline{Y}_H$ of the L strata. If in addition also the overall mean $\overline{Y}$ is estimated, the term 'stratified sampling' may be appropriate.

If, however, the purpose is solely to estimate the strata means, the purpose should be referred to as one of estimating characteristics of L domains of study; to be sure, it is a special, simple case, viz. the case where the domains happen to be identifiable in the frame, thus making it possible to select a sample from each domain.

4. Sampling with unequal probabilities of selection

4.1. The starting point

A natural way of introducing this sampling scheme is as follows. The students are first reminded that using optimum allocation (in the context of stratified sampling) amounts to using L classes of probabilities of selection: the probabilities differ between the strata but are equal within the respective strata. It is then pointed out that sampling with unequal probabilities of selection amounts to using N individual probabilities of selection, say:

$$\pi_1, \ldots, \pi_J, \ldots, \pi_N$$

for the N elements; in its extreme, the sampling scheme calls for N *different* probabilities of selection.

If the rule of association is 'one-to-one', the probabilities are numbers which the statistician assigns to the N elements. If the rule of association is 'many-to-one', the number of sampling units in the frame associated with O_J may be used as the basis of determining the probabilities of selection. These numbers are typically denoted by S_J, and referred to as 'measures of size'. The probabilities of selection may be proportional to these measures, i.e.:

$$\pi_J = S_J / \sum_J S_J$$

in which case the sampling scheme is called 'PPS-sampling'; but more general, the probabilities π_J are defined as some simple function of these measures, such as the square root of the measures.

Given the Y-data and the π-data, the mean $\overline{Y}$ may be defined in the usual way. It is instructive to introduce a special kind of X-data, viz.:

$$X_J = \frac{1}{N} \frac{Y_J}{\Pi_J}, \quad J = 1, \ldots, N,$$

with mean:

$$\overline{X} = \frac{1}{N} \sum_J X_J$$

In a First Course, considerations of simplicity suggest that the sampling be 'with replacement'; hence the sampling should be referred to in terms of 'draws'.

4.2. General theory

The sample will comprise n_d elements with n pairs of Y-data and π-data:

$$Y_1, \ldots, Y_j, \ldots, Y_n,$$
$$\Pi_1, \ldots, \Pi_j, \ldots, \Pi_n.$$

The mean Y is estimated by:

$$\bar{y} = \frac{1}{n} \sum_j \frac{1}{N} \frac{Y_j}{\pi_j}$$

which may be written in the following equivalent form:

$$\hat{y} = \frac{1}{n} \sum_j X_j$$

suggesting that the sampling scheme may be *viewed* as simple random sampling with replacement from a population of X-data.

Finally, the student should be told that sampling with unequal probabilities of selection is an especially powerful scheme in the context of cluster sampling and multi-stage sampling, to be discussed in the next two modules.

Clearly, if A-data are collected in each draw, $\overline{Y}$ may be estimated by means of e.g. a ratio estimator, in complete agreement with the discussion in Sections 2 and 3.

4.3. Using stratified sampling

Finally, mention should be made of the fact that the population $\{O\}$ may be stratified into L strata. Thereafter, n_H draws are made from each stratum, using unequal probabilities of selection.

V. Cluster sampling

1. The notion of cluster sampling

'Cluster sampling'—in some textbooks referred to as 'one-stage cluster sampling'—denotes sampling from a frame $\{U\}$ of M 'clusters', i.e. groups of elements; the rule of association is 'one-to-many'. *All* elements in the m selected clusters are selected for the survey.

Clearly, for the selection of the clusters, a variety of sampling schemes may be used. The emphasis of the course should be on a simple random sampling of clusters.

2. Simple random sampling—preliminary considerations

2.1. The starting point

The M clusters in the frame may be represented as follows:

$$Y_1, \ldots, Y_I, \ldots, Y_M,$$
$$N_1, \ldots, N_I, \ldots, N_M,$$

when N_I is the number of elements in the Ith cluster, the 'size' of that cluster and

$$Y_I = \sum_J Y_{IJ}$$

is the sum taken over the N_I elements in the Ith cluster.

2.2. Two cases

We will distinguish two cases:
(i) the clusters are of unequal sizes: $N_1, \ldots, N_I, \ldots, N_M$;
(ii) the clusters are of equal sizes: $N_1 = \cdots = N_I = \cdots = N_M = N_0$.

Many textbooks begin the presentation of the theory by considering clusters of equal size, and then proceed to the case of clusters of unequal sizes.

My suggestion is to reverse this order: having presented theory for the case of clusters of unequal sizes, the theory for the case of clusters of equal size follows directly/simply by substituting N_0 for the N_I in the formulas for the general theory.

In Section 3, some general theory will be presented for simple random sampling, when the clusters are of unequal sizes. In Section 4, the special theory will be presented for the use of clusters of equal size.

3. The general theory: Clusters of unequal sizes

The data associated with the $N = \Sigma_N N_I$ elements in the population may be represented by a data matrix as shown in Table 1.

Table 1

Element index	Cluster index				
	1	...	I	...	M
1	Y_{11}		Y_{I1}		Y_{M1}
$\vdots$	$\vdots$		$\vdots$		$\vdots$
J	Y_{1J}		Y_{IJ}		Y_{MJ}
$\vdots$	$\vdots$		$\vdots$		$\vdots$
	Y_{1N_1}				
					Y_{MN_M}
			Y_{IN_I}		

The different lengths of the M columns reflect the fact that, in general, the clusters are of unequal sizes.

The population characteristics may now be defined with reference to this matrix. Expecially, the mean per element in the population is:

$$\bar{Y} = \frac{1}{N} \sum_I \sum_J Y_{IJ} = T/N$$

where

$$T = \sum_I T_I .$$

The expression for the population variance follows simply from the expression for the variance of a stratified population (Module IV, Section 3.1) by viewing the clusters as strata:

$$\sigma_Y^2 = \frac{1}{N} \sum_I N_I(\bar{Y}_{IJ} - \bar{Y}_I)^2 + \frac{1}{N} \sum_I N_I(\bar{Y}_I - \bar{Y})^2$$

which may be written as

$$\sigma_Y^2 = \sigma_W^2 + \sigma_B^2$$

in analogy of the case of the stratified population.

A sample of m clusters is selected by simple random sampling without replacement. The data collected for these m clusters is:

$$T_1, \ldots, T_i, \ldots, T_m,$$
$$N_1, \ldots, N_i, \ldots, N_m,$$

where

$$T_i = \sum_J Y_{iJ}$$

using the chosen indexing convention.

For the estimation of $\bar{Y} = T/N$, two different estimators are available:

$$\bar{y} = \frac{1}{N} M \frac{1}{m} \sum_i T_i$$

for which the variance is:

$$\text{Var } \bar{y} = \frac{M-m}{M-1} \frac{1}{m} \frac{1}{M} \sum_I \left[\frac{N_I}{\bar{N}} \bar{Y}_I - \bar{Y} \right]^2$$

and

$$\hat{y} = \frac{(M/m) \sum_i T_i}{(M/m) \sum_i N_i} = \sum_i T_i / \sum_i N_i .$$

The estimator $\hat{y}$ is clearly a radio-type estimator: both the numerator and the denominator are random variables. Hence, in the general case, the expression for Var $\bar{y}$ will differ from that for Var $\hat{y}$.

The discussion of the general theory may be supplemented by some considerations of the variance properties of cluster sampling. This may be carried out by comparing for example the estimator $\bar{y}$ for simple random sampling of $n = m\bar{N}$

elements with replacement, for which the variance is:

$$\text{Var } \hat{y} = \frac{1}{m\bar{N}} \sigma_Y^2$$

with the expression for the variance of cluster sampling using the estimator $\bar{y}$ as given above. This latter variance may be expressed in terms of the intracluster correlation coefficient as follows:

$$\text{Var } \bar{y} = \left[\frac{M-m}{M-1}\right] \times \left[\frac{1}{m\bar{N}} \sigma_Y^2\right] \times [1 + \delta(\bar{N}-1)] \times \left[\frac{\alpha^2}{\sigma_Y^2}\right]$$

$$= F_1 \quad \times \quad F_2 \quad \times \quad F_3 \quad \times \quad F_4$$

where F_1 is the fpc; F_2 is the variance of $\bar{y}$ for simple random sampling with replacement of $n = m\bar{N}$ elements; F_3 reflects the degree, to which the elements in a cluster are alike, the 'homogeneity' of the clusters. In general, the more like the elements within a cluster are, the larger the variance Var $\bar{y}$ becomes. As an example, if $\delta = 0.10$ and $\bar{N} = 25$, F_3 will become 3.4. F_4, finally, reflects the impact of the unequal cluster sizes.

The role that F_3 plays may be illuminated by the following simple example. The data of interest is people's breakfast eating habits. Data may be collected from all members of say $m = 250$ families. Or alternatively, data may be collected from a sample of individual members (the same number as in the 250 families). It is intuitively clear that the alternative approach will make the variance less than the first-mentioned approach (but not necessarily make the cost less).

A somewhat artificial way of conveying this point of view may be as follows. We want to estimate the probability that a coin will show 'head' if tossed. We toss the coin once and record the outcome, 'head' or 'tail'. Looking upon the outcome of this single toss 10 times does not improve our estimate!

Finally, as part of the presentation of the general theory, the possibility of using such sampling schemes as stratified sampling and PPS-sampling should be mentioned, but not dwelled upon in any detail.

4. The special theory: Clusters of equal size

The data associated with the $N = MN_0$ elements in the population may be represented by a data matrix as shown in Table 2 with all columns having the *same* length.

The mean per element in the population is $\bar{Y} = T/N$ and the population variance is:

$$\sigma_Y^2 = \sigma_{W_0}^2 + \sigma_{B_0}^2$$

Table 2

Element index	Cluster index		
	1 ...	I ...	M
1	Y_{11}	Y_{I1}	Y_{M1}
$\vdots$	$\vdots$	$\vdots$	$\vdots$
J	Y_{1J}	Y_{IJ}	Y_{MJ}
$\vdots$	$\vdots$	$\vdots$	$\vdots$
N_0	Y_{1N_0}	Y_{IN_0}	Y_{MN_0}

A sample of m clusters is selected by simple random sampling without replacement, yielding the sample data:

$$T_1, \ldots, T_i, \ldots, T_m,$$
$$N_1, \ldots, N_i, \ldots, N_m.$$

For the estimators, we get:

$$\bar{y} = \frac{1}{N} M \frac{1}{m} \sum_i T_i = \frac{1}{N_0} \frac{1}{m} \sum_i T_i, \tag{1}$$

$$\hat{y} = \frac{(M/m) \Sigma_i T_i}{(M/m) \Sigma_i N_i} = \frac{1}{N_0} \frac{1}{m} \sum_i T_i = \bar{y}. \tag{2}$$

As $\bar{y} = \hat{y}$, Var $\bar{y}$ = Var $\hat{y}$. Replacing N_I in the formula for Var $\bar{y}$ given in Section 3 by N_0 yields:

$$\text{Var } \bar{y} = \text{Var } \hat{y} = \frac{M-m}{M-1} \frac{1}{m} \frac{1}{M} \sum_I (\bar{Y}_I - \bar{Y})^2$$

as $N_I/N = 1$ in this case. This variance may be expressed as:

$$\text{Var } \bar{y} = \text{Var } \bar{y} = \frac{M-m}{M-1} \frac{1}{m} \sigma_{B_0}^2.$$

5. Systematic sampling

Systematic sampling is discussed in most textbooks as a scheme for 'selecting every kth element for a list of N elements'. This implies that systematic sampling is a case of element sampling. I think, however, that it is more instructive to view systematic sampling as a case of cluster sampling. I will elaborate on this viewpoint here.

Consider a data matrix such as the one presented in Section 4: there are M columns, all comprising N_0 elements. The sampling scheme is as follows:

(i) one of the cluster indices $1, \ldots, I, \ldots, M$ is selected at random; let the selected index be 's';

(ii) this index identifies the 'starting element' in the first of the N_0 rows;

(iii) the sample is composed of the starting element and very kth following element, when they are enumerated row by row.

Table 3 is a pictorial presentation of the sampling scheme, with $s = 3$ as the starting element; the scheme selects the elements in the sth = 3rd row, marked with crosses ($\times$):

Table 3

Element index	Cluster index					
	1	2	3	... I ...	$M = k$
1			×		
2			×		
3			×		
4			×		
⋮			⋮		
N_0			×		

The sample column data is the data in the selected column:

$$Y_{s1}$$
$$\vdots$$
$$Y_{sJ}$$
$$\vdots$$
$$Y_{sN_0}$$

the sum of which is T_s.

The estimator:

$$\bar{y} = \frac{1}{N_0} \frac{1}{m} \sum_i T_i$$

now becomes:

$$\bar{y} = \frac{1}{N_0} T_s$$

with the variance:

$$\text{Var}\,\bar{y} = \frac{M-m}{M-1}\frac{1}{m}\sigma_{B_0}^2 = \sigma_{B_0}^2$$

or equivalently:

$$\text{Var}\,\bar{y} = \frac{\sigma_Y^2}{N_0}\{1 + \delta_0(N_0 - 1)\}.$$

It is clearly not possible to estimate the variance on the basis of the sample data alone, as there is only one sampling unit selected; the design is not self-contained. The problem that this represents may be coped with by using systematic sampling within the framework of what is known as 'the Tukey plan': instead of using the skipping interval of $1/k$, (or $1/M$ as discussed above) in combination with a single starting point, r starting points are selected in combination with the correspondingly longer skipping interval $1/rk$. This amounts to using replicated sampling, as discussed in Module IV, Section 2.2. For these starting points, we get the r estimates:

$$\bar{y}_1, \ldots, \bar{y}_i, \ldots, \bar{y}_r$$

and use the pooled estimate:

$$\hat{y} = \frac{1}{r}\sum_i \bar{y}_i$$

to estimate $\bar{Y}$. This estimate $\hat{y}$ has a variance which can be estimated by:

$$\text{Var}\,\hat{y} = \frac{1}{r}\frac{1}{r-1}\sum_i (\bar{y}_i - \hat{y})^2.$$

The discussion of systematic sampling should be supplemented by a brief account of the properties of this sampling scheme, emphasizing the fact that its performance will depend on the way in which the elements in the population (the columns in the data matrix) are ordered.

6. Additional sampling schemes

It should be mentioned that in most applications, the sample of clusters would be selected by means of stratified or sampling using unequal probabilities of selection.

VI. Multi-stage sampling

1. The notion of multi-stage sampling

This sampling system—also referred to as 'K-stage sampling', where $K \geqslant 2$—denotes the following sample selection plan.

There is a frame of M clusters, typically referred to as 'first-stage sampling units' or 'primary sampling units'. A sample of m of these clusters is selected according to some sampling scheme.

For each first-stage sampling unit selected, a frame of 'second-stage sampling units' ('secondary sampling units') is created. A sample of such units is then selected from each selected first-stage sampling unit; this is referred to as 'sub-sampling'. In principle, this selection plan is carried out for K stages.

In describing this system, it should be pointed out that the selection proceeds 'from top down' in the hierarchy of sampling units.

Multi-stage sampling may be carried out using a large number of different sampling schemes, in fact one sampling scheme for each stage. In a First Course, the presentation should be restricted to the simplest case, viz. simple random sampling at each of $K = 2$ stages.

2. Simple random sampling at each of $K = 2$ stages

2.1. The starting point

The M first-stage sampling units in the frame may be represented as follows:

$$Y_1, \ldots, Y_I, \ldots, Y_M,$$
$$N_1, \ldots, N_I, \ldots, N_M,$$

as in the case of cluster sampling (Module V, Section 2.1).

By the same token, the data associated with the N elements in the population may be represented by a data matrix such as that presented in Module V, Section 3. The mean $\overline{Y}$ and the variance σ_Y^2 are also as given in that section.

2.2. The sampling schemes

A simple random sample of m first-stage units is selected; and from each such unit a sample of n_i elements is selected, again with simple random sampling without replacement. This yields the sample data:

$$y_{i1}, \ldots, y_{ij}, \ldots, y_{in_i},$$
$$N_{i1}, \ldots, N_{ij}, \ldots, N_{in_i}, \quad (i = 1, \ldots, m).$$

While the sample selection proceeded 'from top down', the estimation will proceed in the reverse direction: 'from bottom up'. This may be illustrated by means of the following estimator.

The first step calls for estimating the totals T_i for the m selected clusters by

$$t_i = N_i \frac{1}{n_i} \sum_j y_{ij} = N_i \bar{y}_i.$$

The second step calls for computing:

$$\bar{y} = \frac{1}{N} \frac{M}{m_i} \sum t_i = \frac{1}{N} t$$

where t is the estimator of the total T.

The students should be asked to 'check' the correctness of these formulas as follows:

(i) For $m = M$, the sampling scheme becomes equal to stratified sampling as discussed in Module IV.

(ii) For $n_i = N_i$ (i.e. no sub-sampling) the sampling scheme becomes equal to cluster sampling as discussed in Module V.

An alternative estimator, which deserves to be mentioned is:

$$\hat{y} = \frac{\Sigma_i N_i \bar{y}_i}{\Sigma_i N_i}.$$

3. Additional sampling schemes

As with respect to Module V, it should be briefly mentioned that multi-stage sampling may exploit other sampling schemes than simple random sampling; especially, the sample of first-stage sampling units may be selected with PPS-sampling.

4. Expected values

It may be of interest to some students to consider expected values, more specificly $E\bar{y}$ and Var $\bar{y}$. In the context of two-stage sampling, two mnemonic expressions are:

$$E\bar{y} = \underset{1}{E}(\underset{2}{E}\bar{y}) = \underset{1}{E}\underset{2}{E}\bar{y} \tag{1}$$

which reads:

(i) first take the expected value $\underset{2}{E}\bar{y}$ over sampling of second-stage sampling units from the *selected* sample of first-stage sampling units; clearly E_2 is a short for the conditional expectation, which is usually written as $E_{2|1}$.

(ii) next, take the expected value of $E_2 \bar{y}$ over sampling of first-stage sampling units,
and

$$\text{Var } \bar{y} = E(\bar{y} - E\bar{y})^2 = E_1 \text{Var}_2 \bar{y} + \text{Var}_1 E_2 \bar{y} \tag{2}$$

with an analogous interpretation.

It should be pointed out that $E\bar{y}$ and $\text{Var } \bar{y}$ for any number of stages may be written in a similar way!

VII. The problems of non-sampling errors

1. The notion of non-sampling errors

In the proceeding Modules III–VI, the discussion has concerned the use of *sampling* as a source of errors in the estimates, and on the methods and theory to control and measure the resulting 'sampling errors'.

The use of sampling is, however, only one of several sources of errors. These additional sources of errors are usually called sources of '*non-sampling* errors'. They operate in *all* survey operations. It is convenient to discuss them in the context of the following classification of the survey operations:
(i) the data collection;
(ii) the measurement process; and
(iii) the data processing.

For three reasons, the sources of non-sampling errors are viewed as sources of 'problems'. First, the resulting errors may be much larger than the sampling errors. Second, contrary to what holds true for sampling errors, they are typically not reduced by increasing the sample size (at least not to the same extent as applies to the sampling errors); in fact, they may be larger for a large survey than for a small survey. And third, it is more difficult to *measure* (and hence control) the non-sampling errors than the sampling errors—the needed body of methods and theory is less developed than is the body of sampling methods and theory. In what follows, some illustrations will be given.

2. The problem of non-response

The relative simplicity of this problem makes it reasonable to deal with it in some detail even in a First Course.

2.1. The notion of non-response

A sample of n elements have been selected for a survey. Data is to be collected for several characteristics, which in this context often are referred to as 'items'. There may be two kinds of non-response:

(i) element non-response: no data is collected for some elements; and

(ii) item non-response: for some elements, no data is collected for one or more items.

2.2. The consequences of non-response

Clearly, non-response is at variance with the design calling for collecting data for n elements: the 'effective' sample size is reduced from n to n'.

More important, however, is the fact that non-response typically tends to *distort* the estimates, reflecting the circumstance that non-respondents tend to differ from respondents with respect to the items studied. The extent to which an estimate may be distorted is illustrated by the following example.

A sample of $n = 1000$ elements have been selected to estimate the percentage P in a population favoring a certain proposal. These 1000 elements receive a questionnaire by mail and are asked to answer a question about this proposal and return the questionnaire. Out of the 1000 elements, 700 return a completed questionnaire. $P^* = 50\%$ of these 700 respondents 'favor' the proposal. The question now comes up: is P^* a good estimate of the unknown percentage P? This question may be analysed as follows. Assume first that all 300 non-respondents do in fact favor the proposal. That means that P would be estimated by

$$P^*_{max} = (350 + 300)/1000\% = 65\%,$$

well above $P^* = 50\%$. If on the other hand none of the 300 non-respondents favor the proposal, P would be estimated by

$$P^*_{min} = (350 + 0)/1000\% = 35\%,$$

well below $P^* = 50\%$. It may be mentioned that it is no coincidence that

$$P^*_{max} - P^*_{min} = 30\%,$$

the percentage non-respondents!

2.3. Coping with the non-response problem

The discussion in Section 2.2 explains why non-response is referred to as a *problem*, which has somehow to be addressed.

It should first be pointed out that increasing the sample size with the idea of 'compensating' for the reduction in the sample size caused by the non-response is *not* a satisfactory approach. It does not address the fact that non-respondents tend to differ from respondents.

A summary account may then be given of various statistically satisfactory ways of coping with the non-response problem. It is suggested that this account comprises the following approaches:

(i) taking preventive measures, such as making the general public 'survey-minded', giving the interviewers (in an interview survey) special training etc.;

(ii) making call-backs to elements not contacted at the first visit in an interview survey, and sending reminders to aimed-at respondents in a mail survey; and

(iii) making special efforts to get cooperation from a subsample of non-respondents. In a mail survey with 700 respondents out of $n = 1000$ as discussed in Section 2.2, a sub-sample of $m = 100$ of the non-respondents may be contacted for personal interview (which is likely to yield a considerably higher rate of cooperation than a mail survey); the estimate of the population characteristic would then be computed as a weighted average of the estimate based on the 700 respondents and the estimate based on the 100 respondents.

3. The problem of measurement errors

Measurement errors are—in my view—difficult to deal with in a First Course; the relevant methods and theory are likely to be beyond the student's reach. This calls for being satisfied with a broad presentation aiming at a clarification of the meaning of measurement errors.

For some characteristics, it may be possible and meaningful to associate, with each element in a population, a 'true value' and hence view a measurement error as a deviation from that value. Some examples of characteristics amenable to this approach are:

(i) age; the phenomenon of stating one's age in round numbers (30 or 35 rather than 33), called 'age heaping', should be discussed;

(ii) education; the phenomenon of exaggerating one's education should be mentioned;

(iii) having smoked marijuana; the respondent may not want to give a true answer; other examples of 'sensitive' topics may be discussed as well; and

(iv) size of farm; again, rounding may occur.

For many other characteristics, the approach involving a 'true value' is less suitable, or not suitable at all, as a basis for defining measurement error. For these characteristics, it may be helpful to relate the discussion to the fact that respondents have been observed to change their initial responses when re-interviewed, that is to exhibit 'response variability', which implies a possible measurement error.

Finally, it should be mentioned that the occurrence of measurement errors is not restricted to the data collection operation. Coding is a special case of a measurement procedure and as such subject to measurement errors.

4. The problem of processing errors

The processing operation calls for:
(i) editing the data collected;
(ii) transfering the data to punched cards and/or tape; and
(iii) tabulating the data

to give but three examples.

In older days, these operations were largely carried out manually; this is still true in many parts of the world. As a means of reducing cost and keeping the frequency of errors low, sophisticated technical devices for automatic operations have come into use. An example is FOSDIC, which stands for Film Optical Sensing Device for Input to Computers, i.e. a device for transferring data directly from the questionnaires to tape, hence significantly reducing the need for punching cards.

The problem of processing errors may be illustrated with errors in the card punching operation. Assume that '1' is to be punched, if an element has a certain property, and '0', if it does not have its property. If the number of elements having the property is very large, i.e. '1' is to be punched very often, the punching clerk may occasionally punch '1' by routine instead of '0'; the opposite kind of error is more rare. Hence, the result—if corrective action is not taken—is that the number of elements having the property will be overestimated.

VIII. Survey sampling design

1. The purpose of this module

When discussing survey sampling design in a First Course, it is important to emphasize that the purpose is not to teach how to go about the design of a survey but only to give some non-technical illustrations.

These illustrations may be chosen to illustrate various degrees of technical difficulties. In Sections 2–3, some possible examples will be suggested.

2. Surveys of finite populations of elements

I suggest that the following kinds of surveys are illustrated in broad outline.

2.1. Preparing advance estimates of census statistics

If the results of a population census will not be available a brief time after the census date, it may be desirable to select a sample of census forms and prepare advance estimates, especially for characteristics which are subject to fast fast change.

To this end, a sample of census forms may be selected, using for example systematic sampling. In the simplest case, a skipping interval of $1/k$ is used;

alternatively, use may be made of the Tukey plan, i.e. selecting r samples, each with the skipping interval of $1/rk$.

2.2. A housing survey

The purpose of this survey may be to determine the composition of the stock of dwelling units in the capital of the nation: the average size of the dwelling units, and the number of dwelling units with adequate kitchen facilities.

Assuming that there is no up to date register of the dwelling units in the capital, a possible sampling scheme would be to select—from a map of the capital—a sample of blocks and similar areas and collect the data from all dwelling units in these blocks (areas); this would be a case of cluster sampling.

2.3. A nation-wide socio-economic survey

The purpose of this survey may be to describe the nation's households with respect to certain socio-economic characteristics, using data collected by personal interviews.

Cost considerations would suggest the use of multi-stage sampling:
 (i) in the first stage, a sample of counties would be selected;
 (ii) in the second stage, segments within the selected counties would be selected; and
 (iii) in the third stage, dwelling units in the selected segments would be selected; *all* households in the selected dwelling units would be included in the survey.

Clearly, a design such as the one just outlined will help to control two important costs: the cost of creating a frame, and the cost of collecting the data.

2.4. A survey of retail stores

The purpose of this survey may be to estimate the total turnover, 'total sales', in the nation's stores in the previous year.

Sales may be expected to have a markedly skewed distribution: a small number of large stores contribute heavily to the total; at the same time, there are many small stores which together contribute much to the total.

Experience from previous surveys may show that data from the large stores may be collected by mail, while data from the small stores must be collected by interviews.

Consequently, it is reasonable to use a combination of 'element sampling' and 'multi-stage sampling'. Thus, element sampling would be applied to a directory of large stores, which, while not comprising all large stores, comprises a substantial fraction of these stores. The data would be collected by mail. Multi-stage sampling would be applied to pick up a sample of the large stores not in the directory, and in addition, pick up a sample of the small stores. The data would be collected by interviews.

3. Surveys of processes in time and the plane

In addition to the illustrations discussed in Section 2, the use of sampling to study processes in time (such as an accident survey) and in the plane (such as a forest survey or crop estimation) may be briefly discussed.

Optimality of Sampling Strategies

Arijit Chaudhuri

1. Introduction

The problem treated here is one of estimating finite population parameters. For this, one needs a 'sampling strategy' which prescribes the selection-probability of a sample and a method of estimation using a sample to be drawn and surveyed. The choice should be optimal. To choose criteria for optimality and to identify optimal strategies are matters that call for discussions which follow in brief.

In general statistical studies other than through survey sampling the starting point is a 'sample' which is a number of observations. The sample is supposed to come from a certain population. A class of populations whence it possibly comes is described by a family of probability laws. One is then inclined to ascribe varying levels of plausibility to the memebers of the family which may be inferred to have yielded the sample at hand. Conclusions are reached through probability theory. Subsequent studies relate to the population, the sample recedes to the background. For example, postulating normality, one may use a sample of, say, twenty measurements to estimate the probability that a next measurement may be within a specified interval.

In sample surveys, on the other hand, we start with a 'population' of, say, N individuals distinguishable by 'tagged-in' labels, say, $1, \ldots, i, \ldots, N$—more complicated possibilities will not be treated here. Out of them, one chooses a sample of labels—$i_1, \ldots, i_n$, say, all not necessarily distinct. The problem is to estimate the value of some function (called an 'estimand', e.g. say, the total, mean, variance, coefficient of variation etc.) of some real-variate (y, say)-values $Y_1, \ldots, Y_i, \ldots, Y_N$ defined on the population $I_N = (1, \ldots, i, \ldots, N)$. This is done on ascertaining and using the values $Y_{i_1}, \ldots, Y_{i_n}$ (when actually these are determined for a sample $s = (i_1, \ldots, i_n)$ drawn and surveyed these will be denoted as $y_{i_1}, \ldots, y_{i_n}$) for the sampled labels combining them suitably with other relevant materials available. The values for the unsampled labels are unknown but 'fixed' for the populations under survey. Along with the y-values, certain other (say, x, z, etc.) variate-values may be defined on I_N with values $X_1, \ldots, X_N$; $Z_1, \ldots, Z_N$, which may be 'unknown' or 'known fully or partly'; these may be utilized in drawing samples and in estimation depending on availability.

In order to proceed statistically, applying probability theory, the classical way is to choose the sample $s = (i_1, \ldots, i_n)$ with any suitable probability $p(s)$, say. This is possible, unlike in general statistical studies, because the population members are concrete, rather than hypothetical entities, identified and labelled. Given a sample s chosen with probability $p(s)$ along with ascertained values y_{i_j} ($j = 1, \ldots, n$), one may speculate about what other samples might have been chosen and thereby establish a link among the y-values associated with the 'sampled' and 'unsampled' labels and thus estimate $Y = \Sigma_1^N Y_i$ using an 'estimator' t (say). This $t = t(s, Y)$ is a function of the observable survey data $d = ((i_1, \ldots, i_n), (y_{i_1}, \ldots, y_{i_n})) = ((i_1, y_{i_1}), \ldots, (i_n, y_{i_n}))$, involving Y_i's only for $i \in s$. Here $Y = (Y_1, \ldots, Y_N)$. Since s is a random variable so is $t = t(s, Y) = t(d)$. Using the probability distribution of t one may study its behaviour as an estimator for Y not from the sample data at hand but in terms of the probability p (called the 'design') with which a sample is selected and the probability statements about the performance of a strategy (p, t) used in estimating Y or any other parametric function $\tau(Y)$, say, of $Y = (Y_1, \ldots, Y_N)$, the parameter here. The relevant theory will be discussed briefly but in some details in Section 2.

With this 'fixed population approach' Y being a vector of fixed but unknown real numbers (not a random vector), the theory developed to date is not rich enough to delineate fruitful and universally acceptable courses of action in practice which may be claimed to meet theoretical and practical requirements together. It is possible, however, to regard the fixed population $I_N = (1, \ldots, i, \ldots, N)$ of labels to have an associated vector $Y = (Y_1, \ldots, Y_N)$ which is possibly one of a family of such vectors constituting a super-population whence Y is a particular sample realized through an unknown but postulated random mechanism. In terms of the probability distribution of $t = t(s, Y)$, first conditional with respect to p for a given Y and then with respect to the postulated probability distribution of Y, useful inferences are possible. Choice of specific forms of distribution of Y involves formulation of stochastic models chosen on exploiting anticipated structural features of it. The relevant theory is detailed in Section 3.

Another approach is to proceed as in the general statistical theory speculating about the family of probability laws characterizing the nature of the unobserved coordinates of Y posterior to ascertainment through a survey of the Y_i's for $i \in s$ when s is chosen with a probability $p(s)$ as above. Here the speculation is about the parametric space $\Omega = \{Y\}$ viz. the totality of all Y's but not on the sample space $\mathcal{D} = \{d = (s, y) = ((i_1, y_{i_1}), \ldots, (i_n, y_{i_n})\}$, where $s = (i_1, \ldots, i_n)$, $y = (y_{i_1}, \ldots, y_{i_n})$. Inference is to be drawn from the realized data d with the help of model-formulation to take note of the structural features of Y but not in terms of the probability distribution generated through p i.e. randomization in the choice of the sample. One course is then to study the likelihood of Y given d and proceed in a suitable Bayesian or a non-Bayesian way; another is to follow a prediction approach. These are briefly reported in Sections 4 and 5 respectively.

Our discussion will not touch on such topics as optimal construction of strata, optimal allocation of sample sizes to strata with one or more variate under study, optimal choice of sample-sizes for units in different stages in multi-stage sampling etc.

Section 6 gives a 'summing up' for the procedures narrated in Sections 2–5. Certain problems of 'robustness' are referred to and asymptotic theories are presented to indicate a currently developing area of activities.

In Section 7 we discuss certain controversies among theoreticians on the roles of labels and randomization in inference on survey populations.

In this exposition we quote selectively from published materials, offer explanatory notes and make critical comments. Completion in coverage is of course not aimed at. Inadequacy of personal knowledge and interest coupled with consideration for pressure on publication space induces limitations. Arrangement of topics in the presentation is guided by our own taste and judgement. Important omissions, if any, are regretted. No originality is claimed. For a better appreciation of this overview, supplementary reading, especially of the source material mentioned in the references is essential.

2. Repeated sampling approach: Fixed population case

2.1. Neyman's traditional approach: Strategies with controlled design-based biases and variances

Mainly our problem is to estimate the total Y or mean $\bar{Y} = Y/N$ and occasionally the variance $V = (1/N) \sum_1^N (Y_i - \bar{Y})^2$. A sample is a sequence $s = (i_1, \ldots, i_n)$ with labels from I_N, not necessarily all distinct. Its selection-probability is $p(s)$—a method of selection is a 'scheme' and the probability measure p involved here is the 'design'. Using $d = (s, y)$, and other relevant available material we are to use an estimator $t = t(s, Y)$—a random variable (through s) though Y is fixed. Both p and t are free of unobserved Y_i's. Generally, we will consider p to be totally free of Y_i's observed or not—calling it a non-informative design. If it depends on observed y_i's it is an 'informative' design. The couplet (p, t) is a strategy. The value of t for a realized 'datapoint' d for a chosen sample is an estimate for Y. We do not intend to claim how close it is to Y. Following Neyman (1934) we talk about how good is the procedure applied. In fact the strategy (p, t) is subject to appraisal. Its performance characteristics are 'design-based' quantities viz. the (design-) mean (i) $E_p(t) = \sum_s p(s) t(s)$, the sum is over all s's in $\mathscr{S}$ = the totality of all possible samples and (ii) $M_p(t) = E_p(t - Y)^2$, the mean square error which reduces to the variance $V_p(t) = \sigma_p^2(t)$ (the p-variance, rather), if the (design-) bias (p-bias, say) of t namely $B_p(t) = E_p(t - Y)$ is nil. Thus the reference is here to long term average performance of (p, t) through a hypothetically repeated process of sampling (and estimation). Between two strategies (p, t) and (p', t') we would prefer the former if $|B_p(t)| \leq |B_{p'}(t')|$ uniformly in Y and simultaneously, $M_p(t) \leq M_{p'}(t')$ uniformly in Y and both strictly so at least for one Y.

The Markov inequality viz. (calculating the probability in terms of p),

$$\text{Prob}\,[|t - Y| \geq k] \leq \frac{1}{k^2} \,(\sigma_p^2(t) + B_p^2(t),$$

for any real $k > 0$, provides an obvious justification for these requirements, whatever the structure of Y. Using this, Neyman's confidence statement is immediately feasible, e.g.,

$$\text{Prob}[t - 3\sigma_p(t) \leq Y \leq t + 3\sigma_p(t)] \geq \tfrac{8}{9} - \tfrac{1}{9}\left(\frac{|B_p(t)|}{\sigma_p(t)}\right)^2,$$

irrespective of any structure of Y confirming the virtues of controlling (design-) bias and (design-) mean square error. In practice, of course, $\sigma_p(t)$ will be unknown and an estimate for it, say, $\hat{\sigma}_p(t)$ will be needed to substitute for it to give an (i) appraisal of the long term average measure of error in the estimate for Y and (ii) a confidence interval $t \pm 3\hat{\sigma}_p(t)$ around Y say. Snags in the latter are of course possible e.g. in (ii) if, say, $Y_i = 0$ or -1 for $i = 1, \ldots, N - 1 = 99$ and $Y_{100} = 5000$ (say, when y is the gain in buying lottery tickets costing $\$1$ each) whence the conficence coefficient associated with $t \pm 3\hat{\sigma}_p(t)$ will be quite poor using t as $N\bar{y}$ (the expansion estimator; $\bar{y}$ is the sample mean) and $\hat{\sigma}_p(t)$ as $(N/\sqrt{n})s_y$ (where s_y is the sample standard deviation from a simple random sample, SRS).

Yet, following Neyman, we will aim at employing unbiased strategies (p, t), with the uniformly minimum variances (UMV), if available. Hanurav (1962, 1966) has shown that given any design p, one can find a sampling scheme giving rise to it and vice versa. So, we can develop a unified theory in terms of any arbitrary p and arbitrary t based on s chosen with probablity p—a possibility for its implementation being guaranteed in theory. We will talk about choices of (1) strategies (p, t), of (2) estimators t for a given p and of (3) designs p whatever may be the estimators.

2.2. Non-existence results

We will first consider only unbiased estimators (UE) for a fixed design p. Of two p-unbiased estimators t, t', we say, that t is better than t' $(t \succ t')$ if $V_p(t) \leq V_p(t')$, $\forall Y$ and strictly so for at least one Y. If the inequality is nowhere strict, then t is at least as good as t'. If unbiasedness is relaxed, the criterion is $M_p(t)$. An estimator $t = t_b$ (say) is a homogeneous linear unbiased estimator (HLUE) for Y if $t_b = \Sigma_{i \in s} b_{si} y_i$ with $E_p(t_b) = Y$, $\forall Y$, requiring $\Sigma_{s \ni i} b_{si} p(s) = 1$, $\forall i$). A necessary and sufficient condition for unbiasedness of t is $0 < \pi_i = \Sigma_{s \ni i} p(s)$, $\forall i$. (This π_i is called the (first order) inclusion-probability of i in a sample chosen according to a design p.) A design p is called a uni-cluster design (UCD) if for two samples s_1, s_2 with $p(s_1) > 0$, $p(s_2) > 0$, either they are disjoint or equivalent (both containing an identical set of distinct units). Other designs are non-unicluster designs (NUCD). By $\pi_{ij} = \Sigma_{s \ni ij} p(s)$ we denote inclusion-probability (second order) of the pair of units i, j. A design with $\pi_i > 0$, $\forall i$, is denoted as p^+ and one with $\pi_{ij} > 0$, $\forall i, j$ is denoted as p^{++}, when we intend to emphasize these aspects for a design p.

The development of a flourishing theory of inference in finite populations was greatly stimulated by Godambe (1955) who opened a major problem in this field by showing the 'non-existence' of a UMV estimator (UMVE) among HLUE's for Y based on most designs barring some exceptional ones. Godambe and Joshi (1965) extended the result covering all UE's. Basu (1971) gave a simple alternative proof (wrongly alleged to be wrong by Seth (1980)) of the latter (neither Basu's (1971) result implies Godambe's (1955) nor vice versa), valid even for every non-census design $\bar{p}_c$; a census design p_c is one for which $p_c(s) = 1$ (0) as $s = I_N (\neq I_N)$. Koop (1957) and Ajgaonkar (1962) demonstrated non-existence of UMVE's among some sub-classes of UE's with or without restrictions on designs. Hege (1965), Hanurav (1966) and Lanke (1975) observed (as also confirmed by Godambe (1965) himself) that Godambe's result applied only to NUCD's and also showed that a necessary and sufficient condition (NSC) for existence of a UMVE in the HLUE class is that the design is a UCD and that the resulting estimator is the Horvitz–Thompson (1952) estimator (HTE, say), viz.

$$\bar{t} = \sum_{i \in s} y_i / \pi_i,$$

assuming a p^+ design.

Examples may be checked with Roy and Chakravarti (1960), J. N. K. Rao (1975) etc. to verify that $\bar{y}$ based on SRSWOR may have a larger variance than another estimator for a certain Y. However, Kempthorne (1969) has a discernible observation that it is seldom possible to guess a correct Y for which this happens to be the case. Godambe's (1955) result has an alternative proof given by Ericson (1974) using C. R. Rao's (1952) theorem concerning UE's for 'zero'. Excepting systematic sampling design usual designs are NUCD's. These non-existence results concerning UMVUE's based on fixed designs motivate a search (i) for alternative serviceable criteria for optimality of (1) estimators for fixed designs and of (2) strategies meeting these criteria.

2.3. Admissibility and inadmissibility of estimators for a fixed design and uniform admissibility of strategies for classes of designs

If within a certain class (say, of all estimators AE, of unbiased estimators, UE, etc.) there exists, *for a fixed design p*, an estimator $t_0 = t_0 (s, Y)$, say, such that no other estimator within the class is better than it, then we say that t_0 is an 'admissible' estimator within the class. If an estimator t is such that a better one exists, then it is 'inadmissible' within the class, for the given p. Godambe (1960) and, independently, Roy and Chakravarti (1960) proved that the HTE is 'admissible' for Y among all HLUE's whatever may be a given p. Godambe and Joshi (1965) proved the stronger result that the HTE is admissible among all UE's too. Ramakrishnan (1973) provided a simpler proof.

For a sample s, by v_s we will denote its effective size viz. the number of distinct units it contains. If $p(s) > 0 \Rightarrow v_s = n$ (a fixed integer) we will denote the design p as p_n and call it a 'fixed (effective-) size' design; if instead, p is such that $E_p(v_s) = \Sigma_s p(s) v_s = n$ (a given integer again) we call it a fixed average (effective-)

size design and denote it by p'_n. By n_s we will sometimes denote the size of s namely the number of units in the sequence s when each unit is not distinct. Of course n_s may exceed n and may tend to infinity. Godambe and Joshi (1965) prove the following results. For any p_n, among all HLE's (homogeneous linear estimators, without unbiasedness restrictions), every estimator $t = \Sigma_{i \in s} b_i y_i$, with $b_i \geq 1$, $\Sigma_1^N 1/b_i = n$, is 'admissible'; as a corollary, the HTE based on p_n is also admissible in this class. For non-fixed (effective-) size designs, in the HLE class (and hence in AE class obviously too) the HTE is 'inadmissible'. Joshi (1965) proved the admissibility of (i) HTE for fixed-size designs and of (ii) $N\bar{y}$ (equivalently, of sample mean $\bar{y}$ as an estimator for $\bar{Y}$), based on any arbitrary design, among all estimators (i.e. in the AE class) for Y. Joshi (1968) considered a general loss function $\mathscr{L} = \mathscr{L}(t, Y) = \mathscr{L}(t)$, (a special case is $(t - Y)^2$, the square error) in estimating Y by t and the resulting risk $R_p(t) = E_p \mathscr{L}(t)$ (a special case is $M_p(t)$) and called t admissible for Y, given p, if no competitor t' exists with $R_p(t') \leq R_p(t)$, $\forall Y$, and strictly so for at least one Y; in the contrary case t is called 'inadmissible'. With some mild restriction on $\mathscr{L}$, Joshi (1968) proved admissibility of $\bar{y}$ (for $\bar{Y}$) in the AE class, whatever the design p may be. Joshi (1977, 1979) extended the definition of admissibility to a concept of 'joint admissibility' of several estimators used to simultaneously estimate (i) population totals (or means) of several characters each based on a single sample from the population I_N and (ii) totals (or means) of several populations 'independently' sampled using arbitrary designs, using a general loss function in (i) but one involving quantities proportional to squared errors of respective estimators in (ii). With the case (i), he (1977) proved joint admissibility of HTE's (for totals) for a fixed-size design and of sample means (for population means), using any arbitrary design—a negation of the well-known Stein's (1956) inadmissibility result for a sample mean vector (with $k \geq 3$ components) as an estimator for the mean-vector of multivariate normal populations. Later, Joshi (1980) extended the latter case considering more general loss functions with mild restrictions and a class of estimators slightly restricted with some mild regularity (measurability) conditions.

It is well known (see Horvitz and Thompson, 1952) that the variance of HTE is

$$V_p(\bar{t}) = \sum_i Y_i^2/\pi_i + \sum\sum_{i \neq j} Y_i Y_j \pi_{ij}/\pi_i \pi_j - Y^2.$$

For fixed-size designs an alternative formula (due to Yates and Grundy, 1953) is

$$V_p(\bar{t}) = \sum\sum_{i<j} (\pi_i \pi_j - \pi_{ij})(R_i - R_j)^2,$$

where $R_k = Y_k/\pi_k$, $k = 1, \ldots, N$. An unbiased estimator for $V_{p++}(\bar{t})$ is (due to Horvitz and Thompson, 1952)

$$v_1 = \sum_{i \in s} y_i^2 \left(\frac{1 - \pi_i}{\pi_i^2} \right) + \sum\sum_{i \neq j \in s} y_i y_j \left(\frac{\pi_{ij} - \pi_i \pi_j}{\pi_{ij} \pi_i \pi_j} \right).$$

For any fixed-size design p^{++}, an alternative unbiased estimator (due to Yates and Grundy, 1953 is

$$v_2 = \sum\sum_{i<j\in s} (\pi_i \pi_j - \pi_{ij}) \frac{(R_i - R_j)^2}{\pi_{ij}}.$$

Of two unbiased estimators (for a given design p^{++}) v, v' for $V_p(\bar{t})$, v is better than v' if, as usual, $V_p(v) \leq V_p(v')$, $\forall Y$, and strictly so for one Y at least; if given v, no better v' exists, then it is admissible. Godambe and Joshi (1965) proved admissibility of v_1, whatever may be the design p^{++}. Joshi (1970) proved v_2 admissibility only for a p_n^+ design with $n = 2$.

Let for a p_n design,

$$s_y^2 = \frac{1}{n-1} \sum_{i\in s}(y_i - \bar{y})^2$$

be the sample variance, $\bar{x}$, the sample mean of x,

$$b = \frac{\sum_{i\in s}(y_i - \bar{y})(x_i - \bar{x})}{\sum_{i\in s}(x_i - \bar{x})^2}.$$

the sample regression coefficient of y on x, $\hat{R} = (\bar{y}/\bar{x})\bar{X}$, the ratio estimator for $\bar{Y}$,

$$s_{\hat{R}}^2 = \frac{N-n}{Nn} \cdot \frac{1}{n-1} \sum_{i\in s}(y_i - \hat{R}x_i)^2,$$

the usual estimator for MSE($\hat{R}$), $t_R = \bar{y} + b(\bar{X} - \bar{x})$, the regression estimator for $\bar{Y}$. Joshi (1967) defined 'admissibility' of confidence intervals for $\bar{Y}$ based on estimators for p_n designs and proved this property for usual confidence intervals in terms of (1) ($\bar{y}$, s_y) and (2) ($\hat{R}$, $s_{\hat{R}}$) with very general loss functions when $n = 2$ but with slightly restricted ones when $n > 2$. With very general loss functions and slight Lebesgue measurability restrictions on estimators Joshi (1967) also proved admissibility of both $\bar{y}$ and $\hat{R}$ as estimators for $\bar{Y}$. Joshi (1965) also proved admissibility of $\hat{R}$ and t_R among all measurable estimators for $\bar{Y}$. Lanke (1973) deduces from Godambe's (1960) results that (1) for any p^+ design, the HTE is UMV among HLUE's for Y if Y is restricted to $\mathbb{R}_N^{(1)}$ viz. the sub-space of N-dimensional Euclidean space $\mathbb{R}_N$ with only one non-zero component and (2) for any NUCD, the HTE is not the UMV among HLUE's for y if $y \in \mathbb{R}_N^{(2)}$ i.e. for Y's with at most two non zero co-ordinates. He also points out that an estimator $t' = t'(s, Y)$ which is free of Y_i even if $i \in s$, (so that he calls it an estimator 'not depending on Y fully') may be 'admissable' even though intuitively it does not seem to have high efficiency (i.e. small variance)—here unbiasedness is however demanded.

Let (p, t) and (p', t') be two strategies with designs p, p' in a certain wide class c, say, (e.g. p^+, p_n, p'_n etc.) of designs. If for a loss function $\mathscr{L}(\cdot, Y)$ the risks viz. $R_p(t) = E_p \mathscr{L}(t, Y)$ and $R_{p'}(t') = E_{p'} \mathscr{L}(t', Y)$ be such that $R_p(t) \leq R_{p'}(t')$, $\forall Y$ and strictly so for some Y, then we regard (p, t) as better than (p', t'). If given a (p, t) with $p \in c$ there does not exist any other strategy (p', t') with $p' \in c$, then we say that (p, t) is 'uniformly admissible' in the class c with $p \in c$. In particular, $\mathscr{L}(t, Y) = (t - Y)^2, (t - V)^2$ etc. $E_p(t) = Y = E_{p'}(t')$, $\forall Y$; unless mentioned otherwise loss will be taken as the square error for various parametric functions $\tau(Y) = Y, \bar{Y}, V, \sqrt{V/Y}$ etc. This concept of 'uniform admissibility' is due to Joshi (1966) modifying an earlier one of 'global admissibility' introduced but not adequately studied by Godambe (1966). Joshi (1966) proved 'uniform admissibility' of $(p_n, \bar{y})$ among all strategies for estimating $\bar{Y}$ with designs in the class p'_n. Godambe (1969) proved uniform admissibility of $(p'_n, t^* = \Sigma_s y_i + \Sigma_{\bar{s}} \lambda_i)$ among all strategies with estimators for Y and p'_n designs with n fixed; here the λ_i's are arbitrary (fixed) real numbers. Ericson (1970) proved the same for the strategies $(p_n, \alpha_n \bar{y} + \beta)$ among strategies with estimators for Y with p_n designs with n fixed, α_n being any fixed number depending on n and β any real number.

While defining 'admissibility', 'uniform admissibility' etc. the vector Y is supposed to be any point in the entire Euclidean space $\mathbb{R}_N$. If there is a subset Λ of $\mathbb{R}_N$ with a Lebesgue measure zero such that on its complement $\bar{\Lambda} = \mathbb{R}_N - \Lambda$, the property holds but may be violated for Y's in Λ, then we call the corresponding property as 'weak admissibility' and 'weak uniform admissibility'. Sekkappan and Thompson (1975) proved uniform admissibility of

$$(p_n(\pi), t_2 = \sum_{i \in s} b_i y_i, \text{ with } b_i > 1, \sum_1^N 1/b_i = n)$$

among all measurable estimators for Y based on $p_n(\pi)$ designs; a $p_n(\pi)$ is a p_n design such that for every one of them the π_i's are same with $\Sigma_1^N \pi_i = n$. As a corollary, $((p_n(\pi), \bar{t} = \Sigma_{i \in s} y_i/\pi_i)$ is also uniformly admissible in this context. Sekkappan (1973) proved 'weak uniform admissibility' of

$$(p_n^+, t_3 = \sum_{i \in s} b_i y_i + \beta n, \, b_i > 1, \, \forall i, \, \sum_1^N 1/b_i = n, \, \beta_n \text{ fixed arbitrarily})$$

among all (p_n, t)'s with measurable t. Chaudhuri (1978) proved (ii) admissibility of

$$\hat{\sigma}^2 = \frac{1}{N^2} \left[\sum\sum_{i<j \in s} (y_i - y_j)^2 + (N - n) \sum_{i \in s} y_i^2 + (N - n)(N - 1)\beta \right]$$

(with β a pre-assigned positive number) as an estimator for V, based on any p_n design with $n \geq 2$ and (ii) 'uniform admissibility' of $(p_n, \hat{\sigma}^2)$ among all strategies estimating V with p_n designs with $n \geq 2$ throughout. Chaudhuri (1978) used Bayesian arguments for his derivations. Ghosh and Meeden (1983) obtained further optimality results on variance estimation. Certain results on estimation for V were reported earlier by Liu (1974).

Incidentally let us refer to Basu's (1970) comments that admissibility of the HTE depends essentially on the unrealistic width of the Ω-space which is taken as $\mathbb{R}_N$ or a subset of it wide enough to contain the origin. If, however, we can hit upon a fixed point, say, $A = (A_1, \ldots, A_i, \ldots, A_N)$ with $A = \Sigma A_i$, then for all Y's in a close neighbourhood of A, the estimator $t_A = \Sigma_{i \in s}(Y_i - A_i)/\pi_i + A$, will have a smaller variance than $V_p(\bar{t} = \Sigma_{i \in s} y_i/\pi_i)$. Thus, if Y is restriced to such a neighbourhood only, then $\bar{t}$ will be inadmissible because t_A will dominate ('beat' i.e. will be better than) $\bar{t}$ uniformly over that neighbourhood. The question whether t_A itself is admissible in UE class for any p^+ if $\Omega = \mathbb{R}_N$ is answered in the affirmative by Cassel, Särndal and Wretman (CSW, 1977). Many optimality results applicable to HTE are shown by them to extend to cover t_A, called generalized difference estimator (GDE) by Basu and popularised by CSW.

In order to describe a few more admissibility results we need to discuss the concept of 'linear invariance' and 'connected designs'. If $t = t(s, Y)$ be an estimator for Y and $t = t(s, Z)$ that for $Z = \Sigma^N Z_i$ where $Z_i = \alpha + \beta Y_i$ (with α, β's any real constants), then t is 'linear invariant' if it satisfies the condition $t(s, Z) = N\alpha + \beta t(s, Y)$. For example, the HTE is not 'linear invariant' (LI, say, in brief), it is not even origin-invariant (i.e. the special case of LI when $\beta = 1$), unless $\Sigma_{i \in s} 1/\pi_i = N$, $\forall s$, a condition satisfied for a p_n design with $\pi_i = n/N$, $\forall i$ (e.g. SRSWOR designs). However, the GDE, viz. t_A above, is linearly invariant, noting that A is a point as Y and when Y_i changes to $\alpha + \beta_i$, A_i changes to $\alpha + \beta A_i$ giving the result. Roy and Chakravarti (1960) proved the non-existence of UMVE'S among LI estimators (LIE) within the HLUE's which are of the form $t = \Sigma_{i \in s} b_{si} y_i$ with $\Sigma_{i \in s} b_{si} = N$, $\forall s$. So, they restricted within a sub-class of HLUE class such that $V_p(t) \propto V$, with constant of (unknown) proportionality k, which they call a 'regular' sub-class of HLUE class. Within this sub-class their optimal estimator (best regular estimator, BRE, say, within the HLUE with LI restrictions) is one for which k assumes the least possible value, K_0, say. A design yielding the BRE is called a 'balanced design'. They show that a BRE is of the form $t_0 = (1/v_s) \Sigma_i^N v_{si} y_i$ viz. the sample mean of the distinct units), where $v_{si} = 1(0)$ if $i \in s$ ($i \notin s$), and $v_s = \Sigma_{i=1}^N v_{si}$. They claimed $E(v_{si}/v_s) = 1/N$, $\forall i$, as sufficient to yield a 'balanced design' and obstained the formula

$$K_0 = \frac{1}{N-1}[NE(1/v_s) - 1].$$

Sinha (1976), however, rightly pointed out that two additional conditions viz.,

$$E[v_{si}/v_s^2] = \frac{1}{N} E\left(\frac{1}{v_s}\right), \quad \forall i, \text{ and}$$

$$E[v_{si} v_{sj}/v_s^2] = \frac{1}{N(N-1)}[1 - E(1/v_s)], \quad \forall i \neq j,$$

are needed to make a design 'balanced'. Unfortunately, BRE's and 'balanced designs' together constitute a very restrictive class and admissible estimators like the HTE, GDE etc. are not of the form t_0 above; also it is not known whether t_0 based on a balanced design is admissible or not among all LIE's (without the 'regularity' restriction) among all HLUE's (not to speak of all UE's). Patel and Dharmadhikary (1977) obtained NSC's on designs to yield LIE's among HLUE's and in the process they introduced what are called 'connected designs'. Two labels i, j in I_N are 'connected' if $\exists$ samples $s_1, \ldots, s_m$ in $\mathscr{S}$ with $p(s_i) > 0$, $\forall i = 1, \ldots, m$, and labels $i_1, \ldots, i_{m-1}$ in $I_N . \ni . s_1 \ni i, i_1, s_2 \ni i_1, i_2, \ldots, s_m \ni i_{m-1}, j$. This 'connectedness' is an 'equivalence relation' which splits I_N into 'equivalence classes' $U_1, \ldots, U_k$, say, called the 'components' of I_N under a design p (say). If $k = 1$, than p is a 'connected design' implying that every pair of labels in I_N is 'connected'. They define an $N \times N$ matrix $C = (C_{ij})$, called C-matrix with

$$C_{ii} = \pi_i - \sum_{i \in s} \frac{p(s)}{v_s}, \qquad C_{ij} = - \sum_{i, j \in s} p(s)/v_s,$$

such that (i) C is symmetric, (ii) $C_{ij} \leq 0$, $\forall i \neq j$. (iii) $\Sigma_j C_{ij} = 0$, $\forall i$. Suppose for a design p, $\mathscr{S}$ is partitioned into $\mathscr{S}_r$ $(r = 1, \ldots, k)$. $\ni$. $\mathscr{S}_r$ contains units exclusively from the components U_r $(r = 1, \ldots, k)$ of I_N. Let $\alpha_r = \Sigma_{s \in \mathscr{S}_r} p(s)$, $N_r = \#$ labels in U_r and $p_r = p(s)/\alpha_r$ for $s \in \mathscr{S}_r$, $r = 1, \ldots, k$. Then, p_r is a 'connected' design with respect to the population U_r and the totality $\mathscr{S}_r$ of samples from U_r. They show, inter alia that (i) A design p is connected iff Rank(C) $= N - 1$; (ii) A necessary condition (NC) for the existence of an LIE among HLUE's is that $\alpha_r = N_r/N$, $r = 1, \ldots, k$, (iii) for a design subject to $\alpha_r = N_r/N$, $r = 1, \ldots, k$, the system of linear equations $C\lambda = d$ is consistent where $\lambda = (\lambda_1, \ldots, \lambda_N)$, λ_i's are determinable numbers, $d_i = 1 - N \Sigma_{i \in s} p(s)/v_s$ and most importantly, the estimator $t_b = \Sigma_{i \in s} b_{si} y_i$ (with $b_{si} = (\lambda_i - \bar{\lambda}_s) + N/v_s$, $\bar{\lambda}_s = \Sigma_{i \in s} \lambda_i/v_s$, λ_i's are solved from $C\lambda = d$) is (iv) the unique, LIE among HLUE's for Y and (v) is 'admissible' for Y among all HLUE's. They further show, as a corollary, that a 'connected design' necessarily admits an 'admissible' LIE among HLUE's for Y. They also note that for a UCD the UMVHLUE viz. the HTE is LIE iff $p(s) = v_s/N$, $\forall s$. In a later paper, Patel and Dharmadhikary (1978) note that designs in practice are all connected, simplify the proofs and presentations and also prove admissibility of specific HLUE's applying their general theory outlined above. In order to describe the specific admissible estimators we need the following preliminaries.

So far we considered mainly the unified theory of sampling where a sampling strategy (p, t) is defined with respect to a sampling design without specifying a scheme of sampling corresponding to a design. This simplicity is available through Hanurav's (1962, 1966) results and his unit drawing mechanism. But is is possible to refer to certain specific schemes and associated strategies so as to investigate their admissibility and other optimality properties or negation of the properties.

Let us mention a few of the well-known schemes and strategies: (1) Simple random sampling with replacement (SRSWR) in n (say) draws choosing each unit with a probability $1/N$ on each draw, (2) simple random sampling without replacement (SRSWOR), choosing a sample of n units with a probability $1/\binom{N}{n}$ in a single draw or in successive draws without replacement giving a constant selection-probability to each unit left in the population after the preceding draws, (3) probability proportional to size with replacement (PPSWR) sampling, independently choosing in n draws giving on each draw a selection probability P_i ($0 < P_i < 1$, $\sum_1^N P_i = 1$) to the ith unit (P_i's are called normed size-measures given) (4) probability proportional to size without replacement (PPSWOR) sampling—as in (3) with selection without replacement, (5) πps or IPPS ('inclusion probability proportioned to size') scheme of sampling (with $\pi_i \propto P_i$), (6) Midzuno's (Lahiri, 1950; Ikeda, 1951; Midzuno, 1953; Sen, 1953; Hájek, 1949) scheme of sampling, choosing the ith unit with probability P_i on the first draw and then by SRSWOR method from the rest on $(n-1)$ subsequent draws, also (7) its modification to ensure πps property (i.e. $\pi_i \propto P_i$), (8) Rao, Hartley and Cochran (RHC, 1962) scheme of sampling which consists in random splitting of I_N into n mutually exclusive groups and selecting independently from each group one unit with a probability proportional to P_i. Usual estimators associated with scheme (3) are given by Hansen and Hurwitz (1943), with (4) are by Raj (1956) Murthy (1957), with (5) by Horvitz and Thompson (1952), with (6) by Lahiri (1951) the ratio estimator, with (8) by RHC themselves—the combinations give us the well-known strategies.

For PPSWOR sampling with $n = 2$, Murthy (1957, 1963) considered two unbiased estimators (for Y) viz.

$$t_1 = \frac{1}{(2 - P_i - P_j)} \left[(1 - P_j) \frac{Y_i}{P_i} + (1 - P_i) \frac{Y_j}{P_j} \right]$$

and

$$t_2 = \frac{1}{(2 - P_i - P_j)} \left[(1 - P_i) \frac{Y_i}{(\pi_i - P_i)} + (1 - P_j) \frac{Y_j}{\pi_j - P_j} \right]$$

when the unordered sample consists of distinct units, i, j and π_i for this scheme equals $P_i[1 + \sum_{j \neq i} P_j/1 - P_j]$.

Patel–Dharmadhikari (1978) proved that t_1, t_2 are both 'admissible' for Y among HLUE's. But Joshi's (1970) proof (i) of 'admissibility' of t_1, t_2 among UE's for Y was found wrong by Patel and Dharmadhikari (1978) who applied their 'concepts of connected designs' to prove admissibility of many other estimators among HLUE's for Y including in particular, $N\hat{R}$ and

$$t_3 = \frac{N-1}{P_i + P_j} \left[Y_i \frac{P_j}{1 - P_i} + Y_j \frac{P_i}{1 - P_i} \right]$$

due to Murthy (1963), both based on Midzuno's scheme when $n = 2$. Utilizing Patel and Dharmadhikari's (1977, 1978) and Lanke's (1975) results concerning non-negativity of unbiased estimates for $V_p(\bar{t})$, Sen Gupta (1980, 1982, 1983) proved a number of 'admissibility' results for a number of estimators which inter alia include (i) t_1, t_2 (among UE's) based on PPSWOR scheme with $n = 2$, (ii) t_1 among HLE's for PPSWOR scheme with $n = 2$, (iii) a broad class of HLE's based on arbitrary designs with $n = 2$ including $N\bar{y}$, $\hat{R}$, t_1 (P_i's not related to designs), (iv) t_1 among AE's for Y based on any designs p_2 (P_i's in t_1 not related to any designs)

(v) $\quad e = e(s, Y) = X\bar{r} + \dfrac{n(N-1)}{n-1}(\bar{y} - \bar{r}\bar{x}),$

which is Hartley–Ross's (1954) estimator (HRE) for Y based on SRSWOR (with $\bar{r} = (1/n) \Sigma_{i \in s} y_i/x_i$) among HLUE's, provided the design corresponds to the SRSWOR scheme, (vi) a class of UE's (based on arbitrary p_n designs with $n \leq 2$) including v_2 (for $V_p(\bar{t})$) based on a p_2 design,

$$v_3 = \dfrac{P_i P_j}{P_i + P_j}\left[(N-1) - \dfrac{1}{P_i + P_j}\right]\left(\dfrac{y_i}{P_i} - \dfrac{y_j}{P_j}\right)^2$$

for $V_p(R)$ based on Midzuno's scheme with $n = 2$. Though v_2 is admissible for $V_p(\bar{t})$ among UE's for a p_2 design (see Joshi, 1970), Biyani (1980) demonstrated that $\exists$ designs p_n with $n > 2$ for which v_2 is inadmissible among non-negative quadratic unbiased estimators (NNQUE) vor $V_p(\bar{t})$. Sankaranarayanan's (1980) claim that v_2 is admissible for $V_p(\bar{t})$ for any p_n among the same NNQUE class is unfortunately wrong because his arguments involve essentially a faulty assertion that positive definiteness of, say, $Q = \Sigma\Sigma_{i<j}^k a_{ij}(x_i - x_i)^2$ demands $a_{ij} > 0$, $\forall i < j$ (a violation may be checked on taking $k = 3$, $a_{12} = -1$, $a_{13} = a_{23} = 4$ though $Q > 0$, $\forall x_i$'s).

After this rather detailed survey of admissibility (and also incidentally of inadmissibility) results we will now consider construction of 'complete' classes of estimators such that given any estimator outside the 'complete' class there is one inside it which is better. A method for this is Rao–Blackwellization. We will discuss it through consideration of 'sufficiency'.

2.4. Sufficiency, minimal sufficiency of statistics and complete class of estimators

Basu (1958) showed how 'the concept sufficiency and its utilization through Rao–Blackwellization' is feasible in finite population inference. From what precedes we may write the probability of observing the data $d = ((i_1, y_{i_1}), \ldots, (i_n, y_{i_n}))$, when a sample $s = (i_1, \ldots, i_n)$ is chosen with probability $p(s)$ from $I_N = (1, \ldots, i, \ldots, N)$ and $Y = (Y_1, \ldots, Y_N)$ is a vector of reals defined on I_N as

$$P_Y(d) = p(s)$$

for every Y with $Y_{i_j} = y_{i_j}$ for $j = 1, \ldots, n$ i.e. Y is consistant with $y = (y_{i_1}, \ldots, y_{i_n})$, as observed in the survey data d; for other Y's (inconsistent with d) we take $P_Y(d) = 0$ since when d is observed such Y's are impossible to yield the part of the data y in d. Let $\hat{s} = \{j_1, \ldots, j_k\}$ denote the 'set' of (unordered) distinct labels out of $i_1, \ldots, i_n$ with $1 \leq k \leq n$; let $\hat{d} = \{(j_1, y_{j_1}), \ldots, (j_k, y_{j_k})\}$ be the 'set' of distinct labels along with the associated observed variate-values. The probability of selection of the unordered sample (set) $\hat{s}$ is $p(\hat{s}) = \Sigma_{s \to \hat{s}} p(s)$; $\Sigma_{s \to \hat{s}}$ denotes summing over all s's containing the distinct labels as in $\hat{s}$ ignoring 'order' and 'multiplicity'. The probability of observing the reduced data $\hat{d}$ corresponding to d is $P_Y(\hat{d}) = p(\hat{s})$ for every Y with $Y_{j_t} = y_{j_t}$, $t = 1, \ldots, k$, i.e. those Y's consistent with $\hat{y} = (y_{j_1}, \ldots, y_{j_k})$, a sequence arranged in a fixed but arbitrary order with coordinates as the y-values of the distinct labels $j_1, \ldots, j_k$ in s. For other Y's 'impossible' to arise when $\hat{y}$, as a part of $\hat{d}$ is observed, $P_Y(\hat{d})$ is taken as zero. We will write $I_Y(d) = 1(0)$ if Y is consistent with y (not consistent with y). Similarly, $I_Y(\hat{d}) = 1(0)$ with $\hat{y}$ in lieu of y under consideration. We may note that given s, one can determine $\hat{s}$ but not necessarily vice versa. Thus $p(\hat{s})$ is obtained by summing $p(s)$'s; but if $p(\hat{s})$ is given, a corresponding $p(s)$ may not be determinate and/or defined and many or all of them may be zero as well because one may not attach any positive selection-probability to an 'ordered' sample or to a sample with units occuring with frequencies exceeding one. So, we have $P_Y(\hat{d}) = 0 \Rightarrow P_Y(d) = 0$ but $P_Y(d) = 0 \not\Rightarrow P_Y(\hat{d}) = 0$. Also $P_Y(\hat{d}) > 0 \not\Rightarrow P_Y(d) > 0$. We assume that given a d there exists at least one $Y . \ni . P_Y(\hat{d}) = 0$. In fact we ignore the Y's for which $P_Y(\hat{d}) = 0$. We assume that only the situation '$P_Y(\hat{d}) > 0$' is relevant to our further analysis. We may write

$$P_Y(d) = P_Y(d \cap \hat{d}) = P_Y(\hat{d}) P_Y(d | \hat{d})$$

giving us the conditional probability of observing d given $\hat{d}$ when $\underline{y}$ is the underlying parameter vector as

$$P_Y(d | \hat{d}) = \frac{P_Y(d)}{P_Y(\hat{d})} = \frac{p(s)}{p(\hat{s})},$$

since Y's consistent with y and $\hat{y}$ coincide.

Assuming (as we will, throughout, except, in a specific case later) non-informative designs (i.e. p is free of $\underline{y}$) it follows that $\hat{d}$ is a sufficient statistic, given d; by a statistic we mean any function of d free of unobserved (viz. those with $i \notin s) Y_i$'s. Let $t = t(d)$ be any sufficient statistic given d. For any two data points d_1 and d_2 corresponding to two samples respectively, say, s_1, s_2 chosen with probabilities $p(s_1)$, $p(s_2)$, let $\hat{d}_1$, $\hat{d}_2$ be defined (just as $\hat{d}$ corresponds to d) corresponding to d_1, d_2.

Let $t(d_1) = t(d_2)$. It follows that $P_Y(d_1) = P_Y(t(d_1))$:

$$P_Y(d_1 | t(d_1)) = P_Y(t(d_2)) P_Y(d_1 | t(d_1)) = P_Y(t(d_2)) \frac{P_Y(d_1 | t(d_1))}{P_Y(d_2 | t(d_2))}$$

or $p(s_1)I_Y(d_1) = p(s_2)I_Y(d_2)\Psi(d_1, d_2)$, say, where $\Psi(d_1, d_2)$ is free of Y (by sufficiency of t). Assuming $P_Y(d_i) > 0$, $i = 1, 2$ (and hence $p(s_i) > 0$, $i = 1, 2$), it follows that $I_Y(d_1) = I_Y(d_2)$ which implies that $\hat{d}_1 = \hat{d}_2$. Thus, the partition induced on the sample space $\{d\}$ by t is a sub-partition of that induced by $\hat{d}$. This being the for any t, it follows that $\hat{d}$ is the minimal sufficient statistic inducing the 'thickest' partition. The above proof of (1) sufficiency and (2) minimal sufficiency is essentially due to CSW (1977). Other useful references are Basu (1969) and Basu/Ghosh (1967). So, following Rao and Blackwell (1945, 1947), given any design-unbiased estimator $t = t(d)$ for Y (or $\bar{Y}$ or any other parametric function based on Y) one may construct another unbiased estimator

$$t^* = t^*(d) = E_p[t(d)|\hat{d}] = \sum_{s \to \hat{s}} p(s)t(s, \underline{y}) / \sum_{s \to \hat{s}} p(s)$$

such that, for any convex (downwards) loss function $\mathcal{L}(., Y)$, one has $R_p(t^*) \leq R_p(t)$, $\forall \underline{y}$ and strictly so for at least one $\underline{y}$, unless t^* coincides with t with probability one. Thus the class of unbiased estimators based on $\hat{d}$ is a 'complete' sub-class C of the class of all unbiased estimators based on d, whatever parametric function $\tau(\underline{y})$ we may estimate. Basu (1958), Hájek (1959), Pathak (1961, 1962) and following them many others have made extensive use of (1) this 'minimal sufficiency' and (2) this method of construction (called Rao/Blackwellization) to improve upon many well-known estimators in popular use the latter thus turning out 'inadmissible'. However, this 'complete' class is not generally a 'minimal complete' class in the sense that any estimator in this class is not necessarily admissible and a sub-class of it may be 'complete' too. Also, t^*, though a function of the 'minimal sufficient' statistic $\hat{d}$, it is not 'complete' i.e. it is not a 'unique' unbiased estimator of its p-expectation. Hege (1972) showed that for any UCD, the class of unbiased estimators for Y based on $\hat{d}$ is a 'minimal complete class' i.e. consists only of admissible estimators among UE's for Y. She showed that for a UCD, say, p^+, every UE for Y is of the form $t = t(s, Y) = \sum_{i \in s} y_i/\pi_i + h(\hat{s})$ with $h(\hat{s})$ free of $Y. \ni . E_{p+}(h(\hat{s})) = 0$ and that each estimator is 'admissible' among UE's for Y. Hanurav (1966) earlier proved the weaker result that for any UCD, among all unbiased polynomial estimators for Y, the sub-class of estimators depending on d only through $\hat{d}$ is a 'minimal complete' class.

Roy and Chakravarti (1960) showed that given any HLUE for Y that depends on the order and/or multiplicity of the units in a sample, there exists another independent of order and/or multiplicity with a smaller variance thus giving a construction of a 'complete' class, say C_0. Godambe and Joshi (1965) and Dharmadhikari (1969) gave examples to show that Roy and Chakravarti's (1960) class though 'complete' is not 'minimal complete'. Patel and Dharmadhikari (1974) gave the condition '$\Sigma_{s \in \mathcal{S}} v_s > N(N + 1)/2$', ($\mathcal{S}$ is the support of a given design, i.e. the collection of samples with positive selection-probability), as 'sufficient' but not 'necessary' to make 'C_0 a complete sub-class' of the HLUE class such that it is 'not minimal complete'. However, they gave examples to show that C_0 may be 'minimal complete' even for a NUCD. They gave an artificial construction for a 'minimal complete' sub-class of HLUE when $N = 3$ and the design is p_2 orresponding to SRSWOR scheme. That Roy and Chakravarti's (1960) linearity restriction may be relaxed as follows to construct a more a general complete class C by Rao–Blackwellization may be seen as follows with reference to the problem of unbiased estimation of Y using the square error loss function.

Let s be an ordered sample (sequence) of not necessarily distinct units chosen with probability $p(s)$ and $t = t(s, Y)$ be an unbiased estimator for Y depending on the order and/or multiplicity of units in s. Let $\hat{s}$ be the set of distinct labels in s and $p(\hat{s}) = \Sigma_{s \to \hat{s}} p(s)$. Then,

$$Y = E_p(t) = \sum_s t(s, Y) p(s) = \sum_{\hat{s}} \left[\sum_{s \to \hat{s}} t(s, Y) \frac{p(s)}{p(\hat{s})} \right] p(\hat{s}).$$

Then, writing

$$\hat{t} = \hat{t}(s, Y) = \hat{t}(\hat{s}, Y) = \sum_{s \to \hat{s}} t(s, Y) p(s)/p(\hat{s}) \quad \forall s \to \hat{s},$$

we have

$$E_p(\hat{t}) = Y, \; E_p(t\hat{t}) = \sum_{\hat{s}} \{\hat{t}(\hat{s}, Y)\}^2 p(\hat{s}) = E_p(\hat{t})^2.$$

So,

$$V_p(t) - V_p(\hat{t}) = E_p(t^2) - E_p(\hat{t}^2) = E_p(t^2 + \hat{t}^2 - 2t\hat{t})$$
$$= E_p(t - \hat{t})^2 \geq 0.$$

Thus, $\hat{t}$ which is a function of $\hat{d}$ is better than t unless both coincide with probability one.

We may mention that Godambe and Joshi (1965) pointed out that classes of (1) linear (homogeneous or non-homogeneous) and (2) unbiased estimators are neither 'complete' in the class AE in the sense of yielding uniformly less mean square errors than for those outside the LUE or UE classes respectively.

2.5. Admissibility and other optimality properties of sampling designs

In the Horvitz–Thompson estimation of Y let p and p' be two designs with expectyed effective sizes respectively v and v'. Assuming cost of observation and analysis of every unit to be the same so that the total cost of survey is proportional to v, v' for p, p' designs it is reasonable to regard 'p' as a better design than p' if (i) $v' \leq v$ and (ii) $V_{p'}(\text{HTE}) \leq V_p(\text{HTE})$, $\forall Y$, and either (i) is strict or (ii) is strict for at least one Y. A design p is 'admissible' if no other p' is better than it.

Godambe and Joshi (1965) showed that among all designs with $v = E_p(v_s) = n$ (a positive integer, fixed) using HTE of course, every design p_n is admissible. Writing $[x] \equiv$ greatest integer $\leq x$ (>0), $\theta = v - [v]$, noting the inequality

$$\theta(1 - \theta) \leq V_p(v_s) \leq (N - v)(v - 1)$$

and restricting Y to $\mathbb{R}_N^+$, Hanurav (1966) conjectured that 'only admissible sampling designs' are those for which v_s (with $p(s) > 0$) is either $[v]$ or $[v] + 1$ when v is not an integer or v_s. Joshi (1971b) proved this conjecture false. He showed that when $Y \in \mathbb{R}_N^+$, a necessary condition for admissibility of a design is the existence of at least one pair of samples $s_1, s_2 . \ni . p(s_1) > 0, p(s_2) > 0, s_1 \subset s_2$ (i.e. every unit of s_1 is one of s_2), $v_{s_1} \leqslant v_{s_2} + 2$ (we denote this condition as A_1). For example, any design giving positive probability to samples of at least two sizes m_1, m_2 with $m_2 \geqslant m_1 + 2$, these samples of sizes m_1, m_2 being with equal probabilities is 'inadmissible' (i.e. there is one better than it, i.e. dominating or beating it). If in the class of all designs there is a sub-class c such that given any one outside it there is a better one inside it is called a 'complete class of designs', the latter is a 'minimal complete' class of designs c_1, if no sub-class of it is complete. Hanurav (1968) attempted to find a c_1 class. Joshi (1971'), however, observed that 'the class of designs for which A_1 is false' is a 'complete class' of designs, but 'this class is not minimal complete' i.e. 'violation of A_1 is not sufficient to make a design admissible'. He proved the proposition, viz.

I. 'Every sampling design p for which every sample s_1 with $p(s_1) > 0$ includes at least one i not in any other s_2 with $p(s_2) > 0$ is admissible', in particular, 'uni-cluster designs' are 'admissible'. Using I and illustrating an admissible design for which $V_p(t)$ is not minimized, Joshi (1971a) proved falsity of Hanurav's conjecture.

The above discussion of properties of a design is restricted with HTE as the only estimator used. Chaudhuri (1970) called 'a class of designs $\mathscr{P}$ admissible' if '$\not\exists$ any design outside $\mathscr{P}$ admitting an HLUE (for Y) which is uniformly better than every HLUE based on any design in $\mathscr{P}$'. Requiring every design to involve a common value for $v = E_p(v_s)$ (arbitrarily chosen) he proved admissibility for the class of UCD's. From Hanurav (1966) and Chaudhuri (1970) it is also known that of two HTE's based on two different designs one can be uniformly better than the other only if the former is based on a greater effective size on an average.

In Section 2.3 we considered admissibility of estimators for a fixed design p (say) and of uniform admissibility of strategies for a class of designs. Because of this we may call them p-admissibility and p-uniform admissibility to give the impression that the results are dependent on a chosen p or a chosen class of p-designs. But Scott (1975) and Scott and Smith (1977) show that these concepts are really design-free. Let p, p_0 be designs with supports $\overline{\mathscr{S}_p}, \overline{\mathscr{S}_{p_0}}$ respectively, $\overline{\mathscr{S}_p} \subset \overline{\mathscr{S}_{p_0}}$. Then, $p_0(s) = 0 \Rightarrow p(s) = 0$ i.e. p is absolutely continuous with respect to p_0 (i.e. $p \ll p_0$). Then Scott (1975) and Scott and Smith (1977) show that (i) if e be any p_0-admissible estimator then it is p-admissible, $\forall p . \ni . p \ll p_0$, and (ii) if a strategy (p_0, e) is p_0-uniformly admissible then (p, e) is also so if $p \ll p_0$ and both p, p_0 are p_n designs or p_n' designs for an arbitrary but fixed n.

Lanke (1975) calls a design p' better than (or at least as good as) another design p if given any p-unbiased estimator for $\tau(Y)$ there exists a p'-unbiased estimator e' for $\tau(Y)$ such that (p', e') is better than (or at least as good) (p, e) as an unbiased strategy. If given a p, no better p' in this sense exists, then p is an admissible design. He calls a design p' an extension of a design p, (denoted

$p' \supseteq p$), if $\exists$ a function f which to every pair of samples (s, s') assigns a probability $f(s, s') \ni \Sigma_{s'} f(s, s') = p(s)$, $\forall s$, $\Sigma_s f(s, s') = p'(s')$, $\forall s'$, $f(s, s') > 0$ $\Rightarrow s \subseteq s'$. If $p \neq p'$ then p' is a strict extension of p, (written $p' \supset p$), the choice of f is not unique. Operationally, to create an extension, one may 'take a sample s with a p-design' and 'take a sample s' with a p'-design' such that $s \subseteq s'$. Thus, one may choose s by p, then take another sample from $I_N - s$ with a conditional probability $q(\cdot/s)$, say, adjoin this to s and call s' their union; the resulting design p' is an extension of an SRSWR in n draws for $n > 1$. Two results to note are that (1) 'If p' is an extension of p, then p' is at least as good as p for any $\tau(Y)$: the truth of the converse is unknown', (2) p' is better than p if the strict extension is achieved through a connecting function f such that $\forall s$ with $p'(s') > 0$, $\exists$ an i_0 with $\pi_{i_0} = \Sigma_{s \ni i_0} p(s) > 0$ and an $s \not\ni i_0 . \ni . f(s, s') > 0$—this is so provided $\tau(Y)$ involves all the Y_i's, $i = 1, \ldots, N$.

Let $r = \{j_1, \ldots, j_m\}$ be a set of m distinct labels in I_N. For any design p, the quantity $\pi(r) = \Sigma_{s \supset r} p(s)$ is called an mth order inclusion probability (we will write $\pi'(r)$ with p' for p). Lanke (1975) obtains a 'necessary condition' for 'p' to be at least as good as p' as that $\pi'(s) \geq \pi(s)$, $\forall s$, when $\tau(Y) = Y$. (In particular, p' is at least as good as p if $v' > v$.) This condition is not, however, sufficient. His results have certain practical implications. His example, in estimating Y, SRSWOR-design p'_e, say, is better than SRSWR-design p_e (say), if $n \geq 2$ (n being the number of draws in both); to every p_e-unbiased t for Y there corresponds a p'_e-unbiased $t' . \ni . V_{p'_e}(t') < V_{p_e}(t)$, if the RHS > 0, (if RHS = 0, so is the LHS). If p_e involves n draws and p'_e involves n' draws, then neither is at least as good as the other if $n > n' > 2$. Thus, $n > n' \Rightarrow p'_e$ cannot beat p_e. Using particular estimators t, t' however for p_e, p'_e, it is possible to check that $(p'_e, t') \succ (p_e, t)$ even if $n > n'$, (see Seth and Rao, 1964; Chaudhuri, 1968; Ramakrishnan, 1969; among others).

A relevant problem is the choice of an optimal sample size for a design. Usually, for various sampling strategies in practice the larger the sample-size the smaller is the variance of the estimator involved. But Chaudhuri (1977, 1978), Chaudhuri and Arnab (1978), Chaudhuri and Mukhopadhyay (1978) noted that this may not necessarily be true with the HTE and laid down conditions on designs when this may hold good. Sinha (1980) tackled this issue using Lanke's (1975) concept of extension of designs. Sengupta (1980) got certain results relevant to this area trying to link Chaudhuri and Mukhopadhyay (1978) and Lanke's (1975) approaches. But neither his nor Sinha's findings provide guidelines for serviceable courses of action in practice.

2.6. Necessary bestness and hyper admissibility

As UMV estimators seldom exist and complete classes of estimators are not adequately narrow, Ajganokar (1965) introduced the criterion of 'necessary bestness' (NB, say) for a serviceable estimator for Y among HLUE's. Noting the difference in variances of two such estimators as a quadratic form (QF) in Y_i's, he considers the signs of the principal minors of order r ($r = 1, 2, \ldots, N$) of this

QF to call an estimator of this class the 'necessary best of order r'(NB(r)). For $r = 1$, it is known [see Ajganokar, 1965; Hege, 1969; Ogus, 1969, among others] that HTE is the unique NB(1), called just NB, the 'necessary best' estimator for Y. Following the works of Rao and Singh (1973), Dharmadhikari (1969), Patel (1974), Gupta and Yogi (1980) and others it is known that an NB(r) estimator does not exist when $r > 1$. The NB(1) criterion is no longer accepted as a useful one, thanks to recommendations from Basu (1971), and Rao and Singh (1973).

Hanurav (1965, 1968) recommended an alternative of 'hyper-admissibility' (h-admissibility, for short) for UE's for Y. The criterion was however modified, amended and opposed by Basu (1971), Joshi (1971, 1972), Rao and Singh (1973), Lanke and Ramakrishnan (1974). Taking the parametric space as $\Omega = \{Y = (Y_1, \ldots, Y_N) | -\infty < Y_i < \infty, i = 1, \ldots, N\} = \mathbb{R}_N$, Hanurav considers its linear sub-spaces of the form $R(i_1, \ldots, i_k) = \{Y | Y_j \neq 0 \text{ for } j = i_1, \ldots, i_k, Y_i = 0, \forall i \neq i_1, \ldots, i_k, i_1 \neq i_2 \neq \ldots \neq i_k$ are elements of $I_N = \{1, \ldots, N\}\}$, for $k = 0, 1, \ldots, N-1$. These are called k-dimensional principal hypersurfaces (phs's, ph(k), etc. in short) of $\mathbb{R}_N$. The total number of phs's is $2^N - 1$. Let $R^0(i_1, \ldots, i_k)$ consist of Y's in $R(i_1, \ldots, i_k)$ that do not lie in any phs of a dimension less than k. This is called an 'interior' of $R(i_1, \ldots, i_k)$. Among estimators in sub-class C of UE's for $\tau(Y)$ one is called by Hanurav (1965) an h-admissible estimator in C if it is admissible in C when Y is restricted within $R(i_1, \ldots, i_k)$ whatever may be the choice of $(i_1, \ldots, i_k)$ for $0 \leq k \leq N-1$, and $i_1 \neq \ldots \neq i_k = 1, \ldots, N$. We will call it h_1-admissibile to distinguish it from an h_2-admissible one when in the definition $R(i_1, \ldots, i_k)$ is replaced by $R^0(i_1, \ldots, i_k)$ (see Hanurav, 1968). The two definitions are not equivalent, when either of them is meant we will write simply h-admissible to mean either. The criterion is motivated by the demand that a same estimator may be admissible when used for 'domain'-estimation, whatever may be the domain. This of course is a tall order and the criterion seems unreasonable. Also it yields the unrealistic result that the HTE is the unique (with certain exceptions to be indicated below) h-admissible UE for Y whatever may be the p^+ design employed. Rao and Singh (1973) emphasize the key role ph(1)'s play in yielding this result just as they do in yielding the HTE as the NBE. Thus they regard both the criteria equivalently useless and justifiably so. They further show that among all UE'S for $V_p(\bar{t})$ with p^{++} designs, v_1 is the unique h-admissible estimator—ph(1)'s again playing the key role. However, it is known that v_1 is not a very useful estimator especially as it often turns out negative and is often less efficient than v_2. Both the NB(r) and h-admissibility criteria are virtually abandoned now-a-days though they evoked an initial curiosity. Let us recall some of the mathematical niceties associated with h-admissibility as treated in the literature. Let $\overline{\mathscr{S}}$ be the support of a design p and $\mathscr{P}$ be the class of polynomial p-unbiased estimators for Y. Hanurav (1968) gives two theorems below:

THEOREM 1. *For any NUCD, p^+, HTE is the unique h_2-admissible estimator in $\mathscr{P}$.*

THEOREM 2. *For any UCD, p^+, an estimator t in $\mathscr{P}$ is h_2-admissible iff*

$t = t(s, Y) = k(s) + \bar{t}(s, Y)$, $k(s)$ is free of Y, $k(s_1) = k(s_2)$ if $s_1 \sim s_2$ and $\sum_{s \in \mathscr{S}} k(s) p(s) = 0$.

Joshi (1971) showed (i) Theorem 2 is false, (ii) Theorem 1 holds for any UE; (iii) the h_2-admissibility property does not depend on UC property of a design.

So, he gave the following 'revised criterion' (RC) on a design. Let $\overline{\mathscr{S}_i} = \{s \in \overline{\mathscr{S}} | s \ni i\}$, $\overline{\overline{\mathscr{S}_i}} = \{s \in \mathscr{S} | s \not\ni i\}$, giving $\overline{\mathscr{S}} = \overline{\mathscr{S}_i} + \overline{\overline{\mathscr{S}_i}}$, let $\exists$ an ordered sequence of integers $i_1, \ldots, i_k (i_1 < i_2 < \cdots < i_k)$ $\forall m$ with $1 \leq m \leq k - 1$,

$$\overline{\mathscr{S}} = \bigcup_{r=1}^{k} \overline{\mathscr{S}_{i_r}}$$

and

$$\overline{\overline{\mathscr{S}_{i_{m+1}}}} \cap \left(\bigcup_{r=1}^{m} \right) \overline{\mathscr{S}_{i_r}} \neq \emptyset$$

($\emptyset$ being the empty set). Then Joshi (1971) gives the theorem:

THEOREM 3. *For any p^+ subject to RC, HTE is uniquely h_2-admissible for Y among all UE's.*

Joshi (1971) showed that any NUCD assigning positive selection-probability to not less than three mutually non-equivalent samples, satisfies RC. So, Hanurav's (1968) Theorem 3 is valid only for UCD's assigning positive selection-probabilities to at most only two non-equivalent samples. Joshi (1971) further showed that any NUCD, p^+ giving positive values to not less than three mutually non-equivalent samples necessarily meets RC. So, Hanurav's Theorem 2 applies only to UCD's with at most two non-equivalent samples with positive selection-probabilities. For a UCD not meeting the RC, HTE will yet be the unique h_2-admissible estimator if an additional condition (ARC, say) be imposed on the class of estimators viz. that it be required to have continuity at the origin i.e. be continuous in Y_i at the point at which Y_i's for $i \in s$ vanish.

Joshi (1971, 1972) gave the following interesting results: Let a design $p = p^+$ be subject to the following re-revised conditions (RRC, say): it is a non-census design $\bar{p}_c$, its support is partitioned into two subsets $\overline{\mathscr{S}_i}(\neq \emptyset)$, $i = 1, 2$, I_N is partitioned into two subsets I', I'' (not necessarily non-empty) (i) all units in I' appear in each sample s of $\overline{\mathscr{S}_1}$, (ii) all units in I'' appear in each sample s in $\overline{\mathscr{S}_2}$. Then it follows that

(1) A design satisfies RRC iff it does not meet RC and is a $\bar{p}_c$.

(2) If RRC holds, $\exists$ an infinity of h_2-admissible estimators. When RC holds, HTE is uniquely h_2-admissible. Also, it is uniquely h_2-admissible for any $\bar{p}_c$. So, NSC's for unique h_2-admissibility of HTE are:

(A) A design satisfies RC or is a $\bar{p}_c$ design and equivalently;

(B) the design does not satisfy RRC.

The RRC is easier to check in practice than RC. Lanke and Ramakrishnan (1974) critically studied the relationship between h_i-admissibility ($i = 1, 2$) and concluded through the following seven theorems:

THEOREM 4. (i) h_2-admissibility $\Rightarrow$ h_1-admissibility,
(ii) *If all UE's in the class C (see the definition of h-admissibility) are required to vanish at the origin, h_2-admissibility $\Rightarrow$ h_1-admissibility,*
(iii) *If $V_p(t)$ is continuous, $\forall t$ in C, h_1-admissibility $\Leftrightarrow$ h_2-admissibility (i.e. $h_1 = h_2$ i.e. they are equivalent), h_1-admissibility $\Rightarrow$ admissibility (evidently, taking $k = N$), h_2-admissibility $\Rightarrow$ admissibility (contrary to Hanurav's (1968) claim).*

THEOREM 5. *If $p^+ \ni \cdot p^+(s) = 0$ when $s = I_N$ and if t is h_2-admissible in C ($\subset UE$ class), then it is admissible in C.*

THEOREM 6. *If t is h_2-admissible in C ($\subset UE$ class) and if C be convex (i.e. $t_1, t_2 \in C \Rightarrow \lambda t_1 + (1 - \lambda)t_2 \in C, \forall 0 < \lambda < 1$), then t is admissible in C.*

THEOREM 7. *For any p^+, HTE is h_1-admissible among UE's.*

THEOREM 8. *If t is h_1-admissible among UE(0)'s (viz. the unbiased estimators that vanish at the origin), then t is the HTE.*

THEOREM 9. *For an NUCD, with exactly two clusters if t is h_1-admissible among UE's then, $t \in UE(0)$.*

THEOREM 10. *For a p^+, either (1) HTE is the only h_1-admissible estimator among UE's or (2) every t among UE' is h_1-admissible among UE's.*

For an NUCD or UCD with at least three clusters (1) occurs for a UCD with exactly two clusters, (2) occurs for a UCD with only one cluster (i.e. for a p_c design), both (1) and (2) occur.

A question to which an answer is yet to be available is whether h_2-admissibility $\Rightarrow$ admissibility when (i) $p(I_N) > 0$, (ii) C is convex, (iii) the sample is unordered with only distinct units.

Rao and Singh (1973) considered a class G (containing the HLUE's) of UE's for Y of the from $t = t(s, Y) = \Sigma_{i \in s} t_i(s, y_i) \ni \cdot t_i(s, y_i)$ is a function of i, s and y_i, is free of other Y_j's, is zero when $y_i = 0$, $i \in s$ and $p(s) > 0$. They showed that among estimators in G, HTE is both NB and 'uniquely h_1-admissible' ph(1) again playing the decisive role to ensure both properties. They also considered a general class G_u of unbiased estimators for V_p(HTE) of the form

$$v(s, y) = \sum_{i \in s} v_i(s, y_i) + \sum\sum_{i \neq j \in s} v_{ij}(s, y_i, y_j)$$

such that $v_i(s, y_i)$ involves only s, y_i, is free of Y_j's ($j \neq i$), equals 0 when $y_i = 0$, $i \in s$, $p(s) > 0$, $v_{ij}(s, y_i, y_j)$ involves s, y_i, y_j, is free of Y_k's ($k \neq i, j$), equals 0 when at least one of y_i, y_j equals 0, with $i, j \in s$, $p(s) > 0$. They showed that v_i is the unique h_1-admissible estimator for V_p(HTE) in the G_u class (here also ph(1) plays a key role). They further point out that v_2 does not possess the h_1-admissibility property as an unbiased estimator for $V_p(\bar{t})$.

2.7. Further optimality results: Admissibility, uniform admissibility, strong admissibility, completeness, minimaxity, uniformly least variance etc.

Let v be an arbitrarily chosen average effective sample size for any p' design, based on such designs. Let

$\text{AE}(v)$ the class of all estimators for Y,
$\text{UE}(v)$ the class of all p'-unbiased estimator for Y,
$\text{HTE}(v)$ the HTE for Y,
$\text{HTS}(v)$ a strategy using $\text{HTE}(v)$,
$\text{HLUS}(v)$ the class of strategies with HLUE's for Y,
$\text{US}(v)$ strategies with UE'S for Y,
$\text{AS}(v)$ all strategies for estimating Y.

We replace (v) by $(\leq v)$ to mean that the average effective sample size $\leq v$.

By p_v we as usual mean designs with a fixed effective size v, with v of course a positive integer in this case. Ramakrishnan (1975) has, inter alia, the following results:

(i) $\text{UE}(v)$ is a 'complete' sub-class of $\text{AE}(v)$ iff $v = N$.

(ii) $(p'^+, \text{HTE}(v))$ is uniformly admissible among $\text{US}(v)$'s. No best $\text{US}(v)$ exists. These extend to the cases $(\leq v)$.

(iii) The class $(p'^+, \text{HTE}(v))$ is complete among all $\text{HLUS}(v)$'s iff $v = 1$ or N.

(By a complete class of strategies obviously we mean a sub-class C of strategies such that any strategy in C is better than any one outside C.)

By implication, with the UMV criterion, strategies not using HTE's cannot be excluded in practice.

Joshi's (1966) result that $(p_v^+, t^* = (N/n)\Sigma_{i \in s} y_i)$ is uniformly admissible among $\text{AS}(v)$'s, is extended as follows by Ramakrishnan (1975):

(iv) Let $I = [v]$, $\theta = v - I$ and $\bar{p}'_v$ be a p'_v-design $\ni V_{\bar{p}'_v}(n_s) = \theta(1 - \theta)$. Then, the strategy $(\bar{p}'_v, t^*)$ is uniformly admissible among $\text{AS}(v)$'s. This extends when (v) is replaced by $(\leq v)$.

(Joshi's (1966) result follows as a special case when $\theta = 0$.)

If $V_{p'_v}(n_s) > \theta(1 - \theta)$ for a design p', the strategy (p', t^*) may be inadmissible, for example, he shows that, in particular (SRSWR, t) is inadmissible among $\text{AS}(v)$, with average effective size in SRSWR as v. Denoting this design as SRSWR(v), he shows that (p_1, t^*) is better than (SRSWR(v), t^*), where p_1 is an SRSWR design of size I with probability $1 - \theta$ and of size $1 + I$ with probability θ.

After proving the non-existence of a best (UMV) strategy in $\text{HLUS}(v)$ class and showing that the 'complete' class of admissible strategies is wider than the $(p'^+, \text{HTE}(v))$ class in most situations, Ramakrishnan (1975) introduces a new criterion of 'strong uniform admissibility'—a property midway between 'admissibility' and 'hyper admissibility'—in an attempt to find a narrow enough class of good strategies. In the $\text{HLUS}(v)$ class no 'hyper-admissibile strategy' (i.e. strategy with p' and a hyperadmissible estimator) exists. So, he considers 'strongly uniformly admissible' strategies, namely, those which are uniformly admissible for every E_r (i.e. when Y is restricted to E_r) where $E_r = \bigcup_{i=1}^{\binom{N}{R}} R_i^r$ where R_i^r is the ith ph(r) of $\mathbb{R}_N$. In HLUS(v) class there does not exist a hyper-admissible strategy but a strongly uniformly admissible strategy exists. Ramakrishnan's (1975) main result on this criterion is that 'HTS(v) is the set of all strongly uniformly admissible strategies in HLUS(v) class. Thus, HTS(v) is complete in HLUS(v) with respect to strong uniform admissibility'.

A physical interpretation of strong uniform admissibility is that it ensures 'admissible domain-estimators' when the domain size is known but not the domain frame.

About admissibility Hanurav (1966) made a conjecture that any non-zero function of the minimal sufficient statistic $\hat{d}$ is an admissible estimator for its expectation. Dharmadhikari (1969) pointed out its falsity in case the estimand is the 'zero function'. He, however, proved the falsity also of a modified conjecture that if for a $\tau(Y)$ ∃ a UMVE, then any $t = t(d)$ is an admissible estimator for its expectation $\tau(Y)$.

T. P. Liu (1973, 1974) proves that v_1 is not only 'admissible' in the class of all unbiased estimators of $V_{p_+}(\bar{t})$ but that it is 'admissible' in the class of all unbiased estimators of the variances of all HLUE'S of Y. This property he describes as 'double admissibility'. Let us write $d = (s, y) = ((i_1, \ldots, i_n), (y_{i_1}, \ldots, y_{i_n})) = (i_1, y_{i_1}), \ldots, (i_n, y_{i_n}))$ as $d' = ((r_1, y_{r_1}), \ldots, (r_n, y_{r_n})) = ((r_1, \ldots, r_n), (y_{r_1}, \ldots, y_{r_n}))$ to indicate that in d', the labels $r_1, \ldots, r_n$ out of I_N are obtained on n successive draws from I_N and $y_{r_1}, \ldots, y_{r_n}$ are the y-values for the respective labels $r_1, \ldots, r_n$ which may be 'repeats' i.e. the same label occuring with various multiplicities. Obviously, $s = (i_1, \ldots, i_n)$ and $s' = (r_1, \ldots, r_n)$ are ancillary statistics, their probability distributions, assuming non-informative designs, are free of Y. So, sometimes, in particular when the labels are not available for utilization at the analysis stage or when they are believed not to bear any relationship to the y-values to be profitably utilized in data-analysis, estimators for $\tau(Y)$ are sought in terms of $y = (y_{i_1}, \ldots, y_{i_n})$ or $y' = (y_{r_1}, \ldots, y_{r_n})$ above. If the labels are not all distinct and $\hat{s} = \{j_1, \ldots, j_k\}$ is the set of the distinct units in s, $(1 \leq k \leq n)$, we may consider estimation using only the sequence $\hat{y} = (y_{i_1}, \ldots, y_{i_k})$ of values associated (and arranged in an arbitrary but fixed manner, say, in ascending order of values of $(j_1, \ldots, j_k)$) with s or equivalently through the order statistics, viz., $y_0 = (y_{0_1}, \ldots, y_{0_k})$, where $y_{0_1} \leq y_{0_2} \leq \ldots \leq y_{0_k}$ are the values $y_{j_1}, \ldots, y_{j_k}$.

In the class of HLUE'S for Y of the particular form $t_r = \Sigma_{i=1}^n c_{r_i} y_{r_i}$, based on an SRSWOR design p_n, the expansion estimator $N\bar{y}$ is the unique UMV estimator, as proved by Neyman (1934). Watson (1964) following Halmos (1946) proved that for an SRSWOR design p_n, among HLUE'S for R based on Y, the unique UMVUE is $N\bar{y}$. Royall (1968) showed that for the SRSWOR design p_n, among all p_n-unbiased estimators based on y_0, for Y, again $N\bar{y}$ is the unique UMVE, by observing that it is unbiased and that y_0 is a 'complete' statistic, implying that among all functions of y_0 a p_n-unbiased estimator for any $\tau(Y)$ is unique with probability one. If a varying sampling design p^+ is used, the above results have straight forward extensions. (See Särndal, 1972, 1976.)

Writing $\xi_{r_i} = y_{r_i}/\pi_{r_i}$, $i = 1, \ldots, k$; $\xi_{j_i} = y_{j_i}/\pi_{j_i}$, $i = 1, \ldots, k$, $\xi_{0_i} = y_{0_i}/\pi_{0_i}$,

$i = 1, \ldots, k$, (with obvious notations for π_{0i}'s), we have the UMV properties of HTE for Y among (i) HLUE's of the form $\Sigma_1^n c_r \xi_{r_i}$, (ii) HLUE's as function of $(\xi_{j_1}, \ldots, \xi_{j_k})$ and (iii) all p^+-unbiased estimators based on $(\xi'_{0_1}, \ldots, \xi'_{0_k})$, writing $\xi'_{0_1}, \ldots, \xi'_{0_k}$ as the order-statistics corresponding to $(\xi_{0_1}, \ldots, \xi_{0_k})$.

Kempthorne (1969) considers a 'permutation' model such that $Y = (Y_1, \ldots, Y_i, \ldots, Y_N)$ is regarded as a random realization of one of the $N!$ points obtainable on permuting the coordinates of a fixed but unknown vector of real numbers $Z = (Z_1, \ldots, Z_i, \ldots, Z_N)$. He proves that in the class of 'translation invariant' HLUE'S for Y based on an SRSWOR $N\bar{y}$ is the estimator with the minimum average (with respect to the above permutation modeling) design-variance. In case a varying probability design p^+ is employed, C. R. Rao (1971) applies a modified version of this permutation model to establish this minimum average (permutation model based) design-variance property of HTE. He does not even need 'translation-invariance' restriction (which obviously is not needed for Kempthorne's result in case of the special case of SRSWOR). In his modified permutation model $(\xi_1, \ldots, \xi_N)$ is a random realization of one of the $N!$ points obtainable on permuting the co-ordinates of a fixed but unknown vector $(R_1, \ldots, R_N)$ of reals, where $\xi_i = y_i/\pi_i$, $i = 1, \ldots, N$. (π_i's fixed for p^+) giving the vector $Y = (Y_1, \ldots, Y_i, \ldots, Y_N)$. Thompson (1971) removed the linearity restrictions from Rao's result. Rao and Bellhouse (1978) utilized the permutation model to get further optimality results we will briefly describe in Section 3.

Royall (1970), Joshi (1979) and Stenger (1979) obtain optimality results for 'label-free' estimators essentially as follows through a reformulation of permutation modelling.

Let $\mathcal{P}$ be the permutation operator $\ni \mathcal{P}I_N = \mathcal{P}(1, \ldots, i, \ldots, N) = (\mathcal{P}_1, \ldots, \mathcal{P}_i, \ldots, \mathcal{P}_N)$ where $(\mathcal{P}_1, \ldots, \mathcal{P}_i, \ldots, \mathcal{P}_N)$ is a permutation of I_N, $\mathcal{P}Y' = (Y_{\mathcal{P}_1}, \ldots, Y_{\mathcal{P}_i}, \ldots, Y_{\mathcal{P}_N})$. Let $\mathcal{P}^{-1}$ be the inverse permutation $\ni$. for a sample s, $\mathcal{P}^{-1}(s) = \{j | \mathcal{P}_j \in s\}$. Let $P(n) = \Sigma_{(s|v_s = n)} p(s)$ and for an estimator $t = t(s, Y)$ for Y, let

$$\bar{t} = \bar{t}(s, Y) = \sum_{\mathcal{P}} t(\mathcal{P}^{-1}(s), \mathcal{P}Y) \frac{p(\mathcal{P}^{-1}(s))}{p(v_s)v_s!(N - v_s)!}.$$

This $\bar{t}$ is symmetric (invariant) with respect to $\mathcal{P}$ although t may not be so. Royall (1970) observes that with convex loss functions for estimating a symmetric $\tau(Y)$ ($= \tau(\mathcal{P}Y) \forall \mathcal{P}$) (1) average (w.r.t. $\mathcal{P}$) risk of $\bar{t}$ (for any p) is less than that of t, (2) for SRSWOR, (p_n), (i) average (w.r.t. $\mathcal{P}$) risk of $\bar{t}$ is less than that of t and (ii) maximum (w.r.t. $\mathcal{P}$) risk of $\bar{t}$ is less than that of t, unless t depends on d only through the order statistic y_0 and/or t is itself symmetric. In particular, among HLUE's for $\bar{Y}$ based on SRSWOR designs p_n, $\bar{y}$ has the (i) minimum average variance and (ii) minimum 'maximum variance' averaging as well as maximization being with respect to permutation of the co-ordination of Y.

Let $\mathcal{P}s = (\mathcal{P}i | i \in s)$. If for any design p and estimator $t = t(s, Y)$ we construct new designs $\bar{p} = \bar{p}(s) = (1/N!) \Sigma_{\mathcal{P}} p(\mathcal{P}s) \forall s$ and $\bar{t} = \bar{t}(s, Y) = (1/N!) \Sigma_{\mathcal{P}} t(\mathcal{P}s,$

$\mathscr{P}Y$), then $\bar{p}$, $\bar{t}$ are respectively called 'symmetric' (i) design and (ii) estimator.

With a convex loss function and linearity as well as symmetry restrictions on estimators for $\bar{Y}$, every strategy $(\bar{p}, \bar{y})$ is 'minimax' (i.e. minimizes the maximum variance, maximization being over permutation of Y), with any symmetric design $\bar{p}$.

Joshi (1979), on the other hand, in justifying the use of discarding the label-part of the data d, assumes that Y is 'homogeneous' so that each permutation of its co-ordinates is 'equally likely' to constitute the parameter vector associated with I_N—the definitions, however, are not made sufficiently clear. From this he justifies (1) minimizing the 'maximum' risk and (2) averaging the risk—both over permutation of co-ordinates in Y, in order to derive optimal strategies. With convex loss functions, among design-unbiased strategies for estimating $\bar{Y}$, on fixing the maximum allowable sample size as say, M, (from cost and feasibility considerations), he claims the strategy (SRSWOR with p_M design, $\bar{y}$) is the optimal one in achieving minimum (1) 'average' as well as (2) 'maximum' (both over permutations) risks.

An important distinction between the uses of 'permutation modelling' in deriving optimal estimators and/or strategies recommended (or discussed) by the two groups of workers viz. (1) Kempthorne, C. R. Rao, Thompson and (2) Royall, Stenger, Joshi is that the former group assumes the label-part unavailable at analysis stage for utilization while the latter group discards it purposely when label is believed to be 'non-informative'. We believe, however, that the course of action recommended by the later group does not seem justifiable. Following a private discussion with J. N. K. Rao and perusal of an unpublished article by Joshi we are led to this view which is of course personal. Let us cite the following example to elaborate our stand.

Suppose three persons named a, b, c have respective weights A, B, C (in kg) and we intend to estimate $T = A + B + C$ choosing an SRSWOR of two of them and using an HLUE, viz., say, $t. \ni$. its values for the respective unordered samples, viz. $s = \{a, b\}$, $\{a, c\}$, $\{b, c\}$ are $t = C_{ab}A + C_{ba}B$, $C_{ac}A + C_{ca}C$, $C_{bc}B + C_{ab}C$ with coefficients to be chosen to make t unbiased. With this formulation, the Royall–Stenger–Joshi-approach yields no solution. To formulate in line with their approach let (a, b, c) be labelled (1, 2, 3) and (A, B, C) written as (Y_1, Y_2, Y_3), such that correspondences among (a, b, c) $\to$ (1, 2, 3) and $(A, B, C) \to (Y_1, Y_2, Y_3)$ are not known before the survey. Denoting by s a sample from I_3, and the permutation operator by $\mathscr{P}$, let, as before. $\mathscr{P}^{-1}(s) = (j | \mathscr{P}_j \in s)$, $\mathscr{P}_Y = (Y_{\mathscr{P}_1}, Y_{\mathscr{P}_2}, Y_{\mathscr{P}_3})$, $(\mathscr{P}Y)_i = i$th co-ordinate of $\mathscr{P}Y$. Let $p(s) = \frac{1}{3} = p(\mathscr{P}^{-1}(s))$, $\forall \mathscr{P}$ and an HLUE for $Y = \Sigma_1^3 Y_i$ be $t = t(s, Y) = \Sigma_{i \in s} b_i(s) Y_i$. $\ni . t(\mathscr{P}^{-1}(s), \mathscr{P}Y) = \Sigma_{i \in \mathscr{P}^{-1}(s)} b_i(\mathscr{P}^{-1}(s)) (\mathscr{P}Y)_i$.

Then, the Royall–Joshi–Stenger's optimal estimator will be $\bar{t} = \bar{t}(s, Y) = \frac{1}{6} \Sigma_{\mathscr{P}} t(\mathscr{P}^{-1}(s), \mathscr{P}Y)$.

Let us work out $\bar{t}$ explicitly using Table 1.

Then, $\bar{t} = \bar{t}(s, Y) = t(\mathscr{P}^{-1}(s), \mathscr{P}Y) = \frac{1}{6} \Sigma_{\mathscr{P}} t(\mathscr{P}^{-1}(s), \mathscr{P}Y) = \frac{3}{2}(y_1 + y_2)$, $\forall (\mathscr{P}^{-1}(s), \mathscr{P}Y)$.

Table 1

Persons	Labels assigned permuted	$\mathscr{P}Y$	$\mathscr{P}^{-1}(s)$	$t(\mathscr{P}^{-1}(s), \mathscr{P}Y)$
a b c	1 2 3	$Y_1\ Y_2\ Y_3$	1, 2	$b_1(12)Y_1 + b_2(12)Y_2$
a b c	1 3 2	$Y_2\ Y_3\ Y_1$	1, 3	$b_1(13)Y_1 + b_3(13)Y_2$
a b c	2 3 1	$Y_2\ Y_3\ Y_1$	1, 3	$b_1(13)Y_2 + b_3(13)Y_1$
a b c	2 1 3	$Y_2\ Y_1\ Y_3$	1, 2	$b_1(12)Y_2 + b_2(12)Y_1$
a b c	3 1 2	$Y_3\ Y_1\ Y_2$	2, 3	$b_2(23)Y_1 + b_3(23)Y_2$
a b c	3 2 1	$Y_3\ Y_2\ Y_1$	2, 3	$b_3(23)Y_2 + b_3(23)Y_1$

Now if the data-analyst is given just two values y_1, y_2 and even after the survey he is not aware to whom they relate then this estimate is allright. But Royall, Joshi and Stenger do not envisage this but permit the ascertainment of who are the individuals bearing these values, i.e. the entire datum $d = (s, y_i|i \in s)$ and not just $(y_i|i \in s)$ is available for analysis. If so, then $\bar{t}$ can be ascertained only if all the y_i's ($i = 1, 2, 3$) are known and it equals $Y = T$. Hence the objection to their approach. But this criticism does not apply to Kempthorne–Rao–Thompson's approach where labels are not available for utilization at the analysis stage.

Godambe (1960) considered the problem of choosing an appropriate design to minimize the maximum (with respect to possible values of Y_i's, all positive) value of V_p (HTE) and showed that among p_n designs, SRSWOR gives this required minimax design.

Aggarwal (1959) showed that with a 'square error loss' and SRSWOR design, $\bar{y}$ is the minimax estimator for $\bar{Y}$ provided $Y \in D = \{Y | \Sigma_i^N (Y_i - \bar{Y})^2 \leq A\}$ where A is a fixed but unknown real constant. He extended the result to stratified simple random sampling (see Aggarwal, 1966). He imposed no restriction on the class of estimators (e.g. unbiasedness is also not demanded). Also there is no restriction on sample-size for the design, only for each sample size probability should be same for each sample.

Joshi (1969) extended Aggarwal's (1959) result by showing that his result holds when D is replaced by any subset of $\mathbb{R}_N$ which is symmetric in the co-ordinates of $Y \in \mathbb{R}_N$. Joshi (1969) also shows that $\bar{y}$ is 'admissible' on $D_0 = \{Y | \Sigma (Y_i - \bar{Y})^2 = A\}$ and also 'minimax' in it. Also, $\bar{y}$ is admissible provided MSR($\bar{y}$) is bounded and y's are restricted to any subset of $\mathbb{R}_N$ symmetric in Y_i's ($i = 1, \ldots, N$). Bickel and Lehmann (1981), and Hodges and Lehmann (1982) published further minimax results.

Joshi (1977a) again shows that for any arbitrary p_n^+ design, in the class C_1 (say) of all p_n^+-unbiased estimators for Y of the form $t = t(\hat{s}, y) = f(\xi_{i_1}, \ldots, \xi_{i_n})$ where $\xi_{i_k} = y_{i_k}/\pi_{i_k}$ and $\hat{s} = \{i_1, \ldots, i_n\}$ with any arbitrary f, the HTE is the UMVE. Of course, as a corollary, for SRSWOR design $N\bar{y}$ then is the UMVE, which however is corroborated by the Halmos–Watson (1946, 1964) result. Godambe (1969), in defence of Godambe and Joshi's (1965) non-existence result concerning UMVUE in the entire class C of UE's comments that the above existence result holds iff labels are ignored. Särndal (1972) is not satisfied with this. He notes that if a UE is to have a uniformly less variance than $V(\text{HTE})$, then it must be in the $C - C_1$ class. Joshi (1977) elaborates as follows on the features of the C_1 class. His 'invariance condition' on a design-dependent estimator $t = t_p(s, y)$, say, is this that it must satisfy the condition of invariance under permutations of labels, namely, $t_p(s, Y) = t_p(\mathscr{P}s, \mathscr{P}Y)$ for two designs p, p', and for any $s = (i_1, \ldots, i_n)$, $\mathscr{P}s = (\mathscr{P}_{i_1}, \ldots, \mathscr{P}_{i_n})$, $p(s) = p'(\mathscr{P}s)$. For the SRSWOR design p_n, estimators of the C_1 class satisfy this 'invariance or symmetry condition'. The converse holds too and we may write $t_p(s, Y) = f(y_{i_1}, \ldots, y_{i_n})$, for $s = (i_1, \ldots, i_n)$. Thus, C_1 class is obtained by imposing on estimators the restriction of invariance under permutation of labels. So, 'invariance condition' means restrictions to the class of estimators invariant under permutation of label numbers' and Särndal (1972) means by 'ignoring labels' just this. The invariance condition here agrees with 'the principle of invariance as such' as the estimation problem is invariant under the group of transformations G which is the set of all permutations $\mathscr{P}$. The HTE and many other usual estimators satisfy this invariance condition.

Blackwell and Girshick (1954) showed the minimaxity of SRSWOR in the following manner. Let $\alpha(a, s|Y)$ be the probability of drawing a sample s and taking an action 'a' concerning some inference problem relating to Y. Then, $\Sigma_a \alpha(a, s|Y)$ is the probability $p(s)$ of drawing the sample s. He applied the principle of invariance with respect to a group of transformations including permutations so as to demand $\alpha(a, s|Y) = \alpha(a, \mathscr{P}s|\mathscr{P}Y)$, $\forall \mathscr{P}$. Assuming p_n designs, this leads to the solution $p_n(s) = 1/\binom{N}{n}$. Godambe (1965) does not approve of this solution. Because, he would write $\alpha(a, s|Y, I_N)$ instead of $\alpha(a, s|Y)$ and $p(s|I_N) = \Sigma_a \alpha(a, s|Y, I_N)$ would be the probability of choosing s from I_N. Invariance would then demand $\alpha(a, s|Y, I_N) = \alpha(a, \mathscr{P}s|\mathscr{P}Y, \mathscr{P}I_N) \Rightarrow p(s|I_N) = p(\mathscr{P}s|\mathscr{P}I_N)$. But this leads to no solution for p. We are inclined to concede this 'objection' from Godambe.

Chaudhuri (1969, 1971) obtained a number of minimax, admissibility and complete class results concerning sampling strategies with a game-theoretic approach. The statistician is treated as player II choosing certain features of the sampling strategies as his moves. His adversary, the player I viz. Nature, makes moves to maximize the resulting risks i.e. the pay-offs' by choosing the right 'orderings' of magnitudes of certain parameters like Y_i-values unknown to the statistician. The latter's aim is to choose his moves to minimize this maximum pay-off to protect against the worst predicament when his state of knowledge is poor. If he correctly guesses Nature's choice of the 'orderings' he would appropriately choose the orderings of corresponding quantities at his disposal to find a complete class of strategies. He mainly uses the following inequalities connecting real numbers:

$$(a_i \geq a_j) \Leftrightarrow (b_i \geq b_j), \forall i, j = 1, \ldots, N,$$

$$\Rightarrow \text{(i)} \sum_i^N a_i b_i \geq \frac{1}{N} (\sum a_i)(\sum b_i),$$

$$\Rightarrow \text{ (ii) } \sum_{i}^{N} a_i b_{N-i+1} \leq \sum_{i=1}^{N} a_i b_{j_i} \leq \sum_{1}^{N} a_i b_i,$$

where $(j_1, \ldots, j_N)$ is any permutation of I_N.

Inequalities become strict in obvious circumstances. Making the following assumptions A_1–A_4, he obtained, inter alia the results R_1–R_4, of course on examining that there exist practical situations when A_1–A_4 hold.

A_1: Nature chooses out of N fixed numbers $Y_1, \ldots, Y_N$ to assign at random one each to each element of I_N.

A_2: Corresponding to permutations of $Y = (Y_1, \ldots, Y_N)$ one may choose orderings of some quantities concerning functions of estimators and/or variate-values similarly or reversively with orderings of some other quantities.

A_3: Support of every design is finite.

A_4: $p(s) > 0 \Rightarrow n_s < \infty$ (labels in s not necessarily distinct).

R_1: The minimax value of the risks (with square error loss) for estimators (for $\bar{Y}$) free of p is realized for a design giving a constant selection-probability to each sample.

R_2: A class of minimax (as well as a complete class of) sampling strategies is characterized for SRSWOR, with square error loss, among HLUE's for Y, the minimax estimator is $N\bar{y}$ (maximization is over permutations), forming complete class is also with respect to permutation.

R_3: For any design, among HLUE's for Y, with square error loss the minimax estimator is HTE (maximization over 'ordering'), provided $Y_i > 0$, $\forall i$.

R_4: In using HTE each class of sampling designs with any ordering of inclusion-probabilities of first two orders is 'admissible'.

Scott and Smith (1975) and Scott (1977) noted minimax property of PPS method of sampling in using a specific estimator for an asymmetric estimand.

Let MY_i based on a single observation (i, Y_i) be used to estimate $\theta = \Sigma_i^N M_i Y_i$, where M_i's are known, $M = \Sigma M_i$. Scott and Smith (1975) intend to choose a selection-probability p_i so as to control the risk $R_p(\theta) = \Sigma p_i (MY_i - \theta)^2 = R_p(Y)_1$, say. Assuming $Y \in B = \{Y | 0 \leq Y_i \leq B, \forall i\}$, for a given B they want to choose p_i and $p_i^* . \ni . \sup_{Y \in B} R_{p^*}(Y) \leq \sup_{Y \in B} R_p(Y)$. Assuming the existence of a subset S_0 of $I_N . \ni . \Sigma_{i \in S_0} M_i = \frac{1}{2}M$, their minimax solution yields $p_i^* = M_i/M$, $\forall i$. Extension to PPSWR with n draws is straightforward. Further generalization is given by Scott (1977). Let a fixed estimator t be used to estimate an estimand $\tau(Y)$ with a risk $R_p(Y)$ when based on a design p using any general loss structure. The problem is to choose a design $p^* . \ni . \sup_Y R_{p^*}(Y) \leq \sup_Y R_p(Y)$.

Let $Y \in \Omega \Rightarrow \mathscr{P}Y \in \Omega$, $\forall \mathscr{P}$. Let $\{\mathscr{P}_1\}$ be a subset of the totality $\{\mathscr{P}\}$ of all permutations and $I' \subset I_n . \ni . \mathscr{P}_i = i$, $\forall i \notin I'$ $\forall \mathscr{P} \in \{\mathscr{P}_1\}$. Then, Scott (1977) gives the result:

Let for any $\mathscr{P} \in \{\mathscr{P}_1\}$, $\mathscr{P}_Y \in \Omega$ for $Y \in \Omega$, and the loss $\mathscr{L} = \mathscr{L}_t(s, Y) = \mathscr{L}_t(\mathscr{P}_s, \mathscr{P}_Y)$. Then, for any p, $\exists$ a $p^* . \ni .$

$$p^*(s) = p^*(\mathscr{P}_s) \ \forall \mathscr{P} \in \{\mathscr{P}_1\},$$

i.e. $p^*(s) = \Sigma_{\mathscr{P} \in \{\mathscr{P}_1\}} p(\mathscr{P}_s)/n(\{\mathscr{P}_1\})$, where $n(\{\mathscr{P}_1\})$ = cardinality of $\{\mathscr{P}_1\}$. In case $I' = I_{N'}$, the Blackwell–Girshick's result follows as a corollary. Cheng and Li (1983) consider finding minimax strategies based on varying probability sampling WOR but end up on showing approximate minimaxity of the RHC strategy. Stenger (1979) presents a few results on the choice of reasonable loss functions and related admissibility of estimators with restricted parameter spaces.

He calls a design p symmetric if $p(s)$ is constant for every s with a common cardinality $|s|$, say. Let p_n be the SRSWOR design with sample size n. Every symmetric design p^* is of the form $p^* = \Sigma_{n=1}^{N} \alpha_n p_n$ with $0 \leq \alpha_n$ for every $n = 1, \ldots, N$, $\Sigma_{n=1}^{N} \alpha_n = 1$. He calls an HLUE symmetric if it is of the form $t = t(s, Y) = \Sigma_{i \in s} b_{si} Y_i$, $\forall s, s'$ with $|s| . \ni . i \in s$, $i' \in s'$ $\Rightarrow$ $b_{si} = b_{s'i'}$. Let H be the class HLUSE (homogeneous linear unbiased symmetric estimators) for some parametric function $\tau(Y)$. For example, the sample mean of distinct units, viz. $\bar{y} = \bar{y}(s) = (1/|s|) \Sigma_{i \in s} Y_i$ belongs to the HLUSE class. Every estimator t in HLUSE class may be written $t = t(s, Y) = \beta_{|s|} \bar{y}(s)$ when β_n, $n = 1, \ldots, N$, are constants free of Y. Stenger considers loss functions $\mathscr{L}_1 = \mathscr{L}_1(t, \tau(Y)) = (t - \tau(Y))^2$, $\mathscr{L}_2 = \mathscr{L}_2(t, \tau(Y)) = g(\tau)(t - \tau(Y))^2$, $\mathscr{L}_3 = \mathscr{L}_3(t, \tau(Y)) = (t/\tau - \log_e t/\tau - \tau(Y))$ where $\tau = \tau(Y)$ is an estimand and $g(\cdot)$ a positive valued function. He considers usual definition of admissibility with respect to the corresponding risks involved and obtains the results:

Among HLUSE's for $\tau = \bar{Y}$, with loss $\mathscr{L}_1$:

I. For SRSWOR p_n's, $t = t(s, y) = \beta_{|s|} \bar{y}(s)$ is 'admissible', if $\beta_n \in [n/N, 1]$, provided $y \in \mathbb{R}_N^+$.

II. For a given $p^*(s) = \Sigma_{n=1}^{N} \alpha_n p_n(s)$ $\forall s$, $t = t(s, Y) = \beta_{|s|} \bar{y}(s)$ is admissible iff a $\delta \in [0, 1]$ with

$$\beta_j = \frac{\delta + (1-\delta)/N}{\delta + (1-\delta)/\delta}, \quad j = 1, \ldots, N.$$

with $\alpha_j > 0$, $\forall j$ provided $Y \in \mathbb{R}_N^+$.

III. For a design $p . \ni . p(s) = \Sigma_{n=1}^{N} \alpha_n p_n(s)$ with $\alpha_n < 1$, 'given', $n = 1, \ldots, N$, $t = t(s, Y) = (|s|/\Sigma \alpha_n) \bar{y}(s)$ is the HTE w.r.t. p and is 'inadmissible', provided $Y \in \mathbb{R}_N^+$.

IV. For a loss function $\mathscr{L}_3$ above and any arbitrary design $p^*(s) = \Sigma \alpha_n p_n(s)$, for any $Y \in R_N$, $t = t(s, Y) = \beta_{|s|} \bar{y}(s)$ among HLUSE's, is admissible iff $\beta_n = 1$, $\forall n = 1, \ldots, N$ if $\alpha_n > 0$ $\forall n$.

Another result due to him is:

V. $\mathscr{L}_3$ is the only loss function, such that $\bar{y}$ is the only admissible estimator for $\bar{Y}$ among estimators of the class $\{\lambda \bar{y} | \lambda > 0\}$, provided $Y \in \mathbb{R}_N^+$.

Stenger (1977) illustrated a 'sequential' cum 'informative', NUC design of choosing in successive draws with selection probabilities permitted to depend on variate-values for units chosen on earlier draws and observed and constructed an estimator which is UMV among HLUE's based on such an NUC design. Hence an exception to the non-existence result of Godambe (1955) later modified by Godambe (1965), Hege (1965) and Hanurav (1966). Obviously, the latter holds only with non-informative designs i.e. designs with selection-probabilties totally free of Y.

We have already seen that a complete class of estimators can be constructed by making an estimator (a) free of order and multiplicities of units in a sample. Stenger (1977) considers two more possibilities; (b) for units chosen in successive draws the estimator may depend on the set of distinct units along with respective values realized prior to each successive draw but ignoring the order and multiplicity of the previously drawn units, and (c) selection is without replacement.

Stenger (1977) constructs essentially complete classes of estimators combining (a) and (b) and also by combining (a), (b) and (c).

Following Murthy (1957), and Tikkiwal (1972), Bhargava (1978) applies the technique of 'unordering' to construct additional complete classes of estimators so as to improve on estimators based on sampling with varying probabilities with or without replacement. Unlike Basu (1958) and Pathak (1961, 1962) he does not necessarily base his estimators on the available minimal sufficient statistic $\hat{d}$.

3. Optimality under super-population modelling

From discussions in Section 2 very little emerged to provide useful guidance in choosing strategies worthy of application with a claim to superiority to others. So, to facilitate discrimination among them under various classes of situations at hand suggesting specific structural features of Y, let us now consider postulating various models under super-population set-ups.

Let M_1 be a model such that Y is postulated as a random vector with (model-) expectations (E_m), Variances (V_m), Covariances (C_m), $E_m(Y_i) = \alpha_i + \beta X_i$ (α_i known $\alpha = \Sigma_1^N \alpha_i$, $X_i > 0$ known with $X = \Sigma X_i = N$), $V_m(Y_i) = \sigma^2 X_i^2$, $C_m(Y_i, Y_j) = \rho\sigma^2 X_i X_j$, with β, σ^2 (>0), $\rho(-1/(N-1) \leq \rho \leq 1)$ as unknown reals. Then, for any linear design unbiased estimator LUE for $\bar{Y}$ of the form $t = b_s + \Sigma_{i \in s} b_{si} y_i$ (with b_s, b_{si} free of Y, $b_{si} = 0$ if $s \not\ni i$) based on a p_n design we have (see CSW, 1976)

$$E_m V_p(t) \geq (1-\rho)(1-fA)\sigma^2/n = E_m V_{p_{nx}}(\bar{t}(x)),$$

where $A = (1/N)\Sigma_1^N X_i^2$, $f = n/N$, p_{nx} is any p_n design with $\pi_i = \pi_i(X) = fX_i$ and

$$N\bar{t}(X) = \sum_{i \in s} \frac{y_i - \alpha_i}{\pi_{i(x)}} + \alpha,$$

viz. the GDE for Y. Thus, under M_1 among LUE's for $\bar{Y}$ based on p_n designs, the strategy $(p_n(X), \bar{t}(X))$ is optimal. Important limitations are (i) α_i's must be known, (ii) $V_m(Y_i) \propto E_m(Y_i - \alpha_i)^2$, (iii) constancy of pairwise (model-) correlations.

Let $h = h(s, Y)$ be a UE for 0, t be any UE for Y, $t_b = \Sigma_{i \in s} b_{si} y_i$ any HLUE for Y and $\bar{t}$ the HTE for Y.

Let M be any model for Y such that the expectation operators E_p, E_m may commute. They may not, e.g. if p involves x_i's, say $p(s) \propto \Sigma_{i \in s} x_i$ and x_i's may be random variates. In that case only $E_m E_p$ is meaningful but not $E_p E_m$.

Under M, writing $\Delta_m(t) = E_m(t - Y)$, we have (see Godambe and Thompson, 1977).

$$E_m V_p(t) = E_p V_m(t) + E_p \Delta_m^2(t) - V_m(Y).$$

If, in particular, for a model $M = M_2$ (say), $C_m(Y_i, Y_j) = 0$, $\forall i \neq j$, $E_m(Y_i) = \theta_i$ (with $\theta = \Sigma \theta_i$), $V_m(Y_i) = \sigma_i^2 > 0$, then

$$E_m V_p(t_b) \geq \sum \sigma_i^2(1/\pi_i - 1) + E_p \Delta_m^2(t_b),$$
$$E_m V_p(\bar{t}) = \sum \sigma_i^2(1/\pi_i - 1) + E_p \Delta_m^2(\bar{t}).$$

Writing $p_{n\theta}$ for a p_n design with $\pi_i \propto \theta_i$ (possible, if for example $\theta_i = \beta X_i$, $X_i > 0$ known and $\beta > 0$ unknown) we have (see Godambe, 1955)

$$E_m V_{p_{n\theta}}(t_b) \geq E_m V_{p_{n\theta}}(\bar{t}).$$

This optimality of $(p_{n\theta}, \bar{t})$ among strategies $(p_{n\theta}, t_b)$ is not very encouraging because $p_{n\theta}$ may not be an appropriate design to use t_b; some other design may combine well with t_b to produce a better strategy than $(p_{n\theta}, \bar{t})$. If, in particular, $\sigma_i \propto \theta_i$, say, when $\theta_i = \beta X_i$, $\sigma_i = \sigma X_i$, $\infty > \sigma > 0$, then writing $p_{n\theta\sigma}$ for a p_n design with $\pi_i \propto \theta_i \propto \sigma_i$, we have (see Godambe, 1955) $E_m V_p(t_b) \geq E_m V_{p_{n\theta\sigma}}(\bar{t})$. This optimality for the strategies $(p_{n\theta\sigma}, \bar{t})$ is also quite restrictive by the requirement $\pi_i \propto \theta_i \propto \sigma_i$. If M_2 is replaced by M_3 demanding Y_i, Y_j's to be (model-) 'independent' (not merely 'uncorrelated' as earlier) we have

$$E_m V_p(t) = E_p V_m(\bar{t}) + E_p \Delta_m^2(t) + E_p V_m(h) - V_m(Y)$$

and $E_m V_p(t) \geq E_m V_p(t^*)$, (see Ho, 1980) where $t^* = \Sigma_{i \in s}(Y_i - \theta_i)/\pi_i + \theta$.

This optimality of (p, t^*) is not very useful because t^* is not available in practice because θ_i should be unknown. However, under M_3, we have (see Godambe and Joshi, 1965)

(i) $E_m V_{p_{n\theta}}(t) \geq E_m V_{p_{n\theta}}(\bar{t})$,

and

(ii) $E_m V_{p_n}(t) \geq E_m V_{p_{n\theta\sigma}}(\bar{t}) = \sigma^2/n - \Sigma \sigma_i^2$

if $\pi_i \propto \theta_i \propto \sigma_i$.

Though one may not be equipped with θ_i, σ_i in practice to implement such an optimal strategy these exercises are extremely useful, as amply demonstrated by Godambe and Thompson (1977) and Godambe (1982a, b, 1983), in visualizing that $\bar{t} = \Sigma_{i \in s} y_i/\pi_i$ based on appropriate designs involving stratification in practice remains close to t^* for samples with high selection-probabilities and $E_m V_p(\bar{t})$ remains close to $E_m V_p(t^*)$ so that $\bar{t}$ is nearly optimal and robust against a wide class of alternative models. Comparable results also apply to the ratio estimator. Since this paper is on 'optimality' we abstain from further elaboration on such a 'near-optimality' concept valuable for robustness against inaccurate modelling. Godambe and Thompson (1973) consider a model, say, M_4, such that given certain numbers α_i's ($0 < \alpha_i < 1$, $\forall i$), $\Sigma_1^N \alpha_i = n$ (a positive integer) such that the random vector $(Y_1/\alpha_1, \ldots, Y_i/\alpha_i, \ldots, Y_N/\alpha_N)$ is supposed to have an exchangeable (symmetric) distribution. They consider a class $p_{n\alpha}$ of designs (as a sub-class of p_n designs) with $\pi_i = \alpha_i$, $\forall i = 1, \ldots, N$. They show that under M_4, $E_m V_{p_{n\alpha}}(\bar{t})$ has a common value for each set of α's as above and that $E_m V_{p_n}(t) \geq E_m V_{p_n}(\bar{t})$ for every unbiased estimator t for Y. They also demonstrate that without the 'fixed sample size' requirement on designs this optimality does not hold for the HTE. This model may not be called a super-population model and may just be called an extension of the permutation model considered in Section 2.

Rao and Bellhouse (1978) postulate that the true Y_i's are observable only as $\hat{Y}_i$'s with response errors (supposed random variables) during surveys. They assume X_i's are available without observational errors. Writing $\hat{r}_j = Y_j/X_j$ and $r_j = \hat{Y}_j/X_j$ they consider a model M_5, say, such that

$$\hat{r}_j = \bar{R} + f_i, \quad E_m(\bar{R}) = \bar{R}, \quad E_m E_r(f_j) = 0,$$
$$E_m E_r(f_j^2) = \delta_1(>0), \quad E_m E_r(f_j f_k) = \delta_2, \quad (j \neq k),$$

where $\bar{R} = (1/N) \Sigma_{j=1}^N r_j$, E_r is the expectation-operator with respect to the random response error distribution. The stipulation $E_m(\bar{R}) = \bar{R}$ may be justified through random permutation modelling (vide section 2) for the vector $R = (r_1, \ldots, r_j, \ldots, r_N)$. They consider LUE's for Y of the form $t_b = b_s + \Sigma_{j \in s} b_{sj} \hat{y}_j$ and show inter alia that under M_5, $E_m V_{p_n}(t_b) \geq E_m V_{p_{nx}}(\Sigma_{i \in s} \hat{Y}_i/\pi_i)$. Here compared to Godambe–Joshi's (1965) optimality of (p_{nx}, HTE) (i) 'independence' is not needed but (ii) there is the linearity restriction. It is similar to CSW's result without 'response error' and also to Godambe and Thompson's (1973) but the latter's 'exchangeability' is a stronger restriction. Most of the optimality results point to the efficacy of the HTE. But still it may not often deliver the goods. For example, let us consider a model, say, M_6, such that Y_i's are independently distributed with $E_m(Y_i) = \beta X_i$, $V_m(Y_i) = \sigma^2 X_i^g$, where $0 \leq g \leq 2$, (a model studied by many including Fairfield Smith (1938), Cochran (1946), Mahalanobis (1944), Brewer and Mellor (1977), etc.). Then, writing, under M_6, the model-expected variances of HTE, RHC estimator and ratio-estimator based on Midzuno (1952) scheme (each for Y, based on the same sample size n and using the same size-measures X_i's, using the IPPS scheme for HTE) as E_1,

E_2, E_3 respectively, it is known (see Chaudhuri and Arnab, 1979) that $E_3 < E_2 < E_1$ if $g > 1$, $E_1 > E_2 > E_3$ if $g < 1$, $E_1 = E_2 = E_3$ if $g = 1$. Thus, of course, with $g = 2$, (p_{nx}, HTE) is optimal but with $g < 1$ it is no longer good though with $g > 1$, it is better than at least the other two strategies just considered. But it is not known how it fares vis-a-vis other competitors when $g < 2$.

This conveys a message that it may not often pay to go for an elusive optimal, rather pragmatism may be a better guide sometimes.

In double sampling where for a first phase sample say, s', values on X_i's are ascertained (they are supposed unknown) and from a second phase sub-sample say, s, from it values are determined on Y_i's the problem is to estimate Y using the survey data and relation among Y_i's and X_i's. Postulating appropriate random permutation models and demanding pm-unbiasedness, Rao and Bellhouse (1978) derive an optimal estimator for a general fixed-size two-phase sample design. Their estimator however involves model-parameters and hence not usable. Chaudhuri and Adhikary (1983) formulate appropriately restrictive set-ups to show that most of the non-existence and optimality results in single-phase sampling carry over to double sampling. In their investigation a prominent figure is played by the estimator

$$t = \sum_{j \in s} \frac{y_j - x_j}{Q_j} + \sum_{j \in s'} \frac{x_j}{P_j},$$

where $P_j(Q_j)$'s are inclusion-probabilities for the first phase (over-all, in two phases) sampling. Chaudhuri (1980) applied the Rao–Bellhouse approach to derive an optimal strategy for estimating the 'current' total for a finite population on using samples drawn from it on the current and a previous occasion in an appropriate manner. His estimator also involves unknown model-parameters and hence not usable in practice. As a way out he suggests an alternative strategy which is not optimal but is expected to fare well in practice.

In case the y-variable relates to a confidential character, direct surveys may fail to elicit response from sample persons. Then using a suitable random device one may carry out a randomized enquiry generating responses

$$Z_i = Y_i \text{ with a probability } c \ (0 < c < 1),$$

$$= x_j \text{ with a probability } \frac{1-c}{M}, \quad j = 1, \ldots, M,$$

where x_j's are suitably chosen numbers purported to be close to Y_i's and covering all possible Y_i's and given to the respondents to choose from. Optimal strategies are then derived by Chaudhuri and Adhikary (1983) using estimators of the form

$$t = \sum_{i \in s} Z_i / \pi_i - \frac{1-c}{c} \frac{1}{M} \left(\sum_{i \in s} \frac{1}{\pi_i} \right) \sum_{j=1}^{M} x_j.$$

They also consider some variations of this also retaining optimal properties.

4. Likelihood approach

Given the data $d = (s, y) = ((i_1, y_{i_1}), \ldots, (i_n, y_{i_n}))$, based on a sample s chosen with probability $p(s)$, the likelihood for the underlying parameter Y is $L_d(Y) = p(s)I_d(Y)$ where $I_d(Y) = 1$ if Y is consistent with $d(=0$, else). This by itself admits no further inference about unobserved co-ordinates of Y. Also, given d, its possible further analysis for inference consistently with likelihood and conditionality principles must be independent of a design. This finding of Godambe (1966b) led to 'model-based' and 'Bayesian' theories of inference in finite populations, both irrespective of designs.

Following Basu (1969) if we postulate a prior density $q = q(Y)$ for Y, then this likelihood has an immediate use in yielding the posterior density viz.

$$q^* = q_d^*(Y) = \frac{q(Y)L_d(Y)}{\int_{Y_d} q(Y)L_d(Y)\,\Pi_1^N\,dY_i}, \quad Y = Y_d,$$

$$= 0, \qquad \text{otherwise.}$$

Here $Y_d = \{Y | Y \text{ is consistent with } d\}$.

Assuming 'non-informative designs', this q^* is just the restriction of q to Y_d. Any Bayesian inference based necessarily on $q_d^*(Y)$ is free of p. That is, design does not influence 'post-survey' Bayesian data-analysis. Using a square error loss, the Bayes estimator for Y minimizing the posterior risk is $t_B = \Sigma_s Y_i + \Sigma_{i \in s} E_{q^*}(Y_i)$, where E_{q^*} is the expectation operator with respect to the posterior q^*. To get this optimal (Bayesian) estimate no reference is needed to repeated sampling and there is no speculation about the sample-space but only about the parametric space Ω.

However Zacks (1969) suggests that one should choose an appropriate design p so as to minimize the design-average of the prior risk namely the quantity $\mathcal{R} = E_p E_q(Y - t(d))^2$, where E_q is the expectation-operator with respect to the prior q. Assuming $\mathcal{R} < \infty$, one may note

$$\mathcal{R} = \sum_s p(s) E_{q(Y)} E_{q^*}(Y - t(d))^2$$

$$\geq \sum_s p(s) E_{q(Y)} E_{q^*}(Y - t_B)^2 = \sum_s p(s) U(s), \text{ say.}$$

Here $E_{q(Y)}$ is the expectation operator over the marginal density of Y and $U(s) = E_{q(Y)} E_{q^*}(Y - t_B)^2$. Thus the optimal design is a purposive one namely, p_B (say), such that

$$p_B(s_m) = 1 \quad \text{if } U(s_m) = \min_s U(s),$$

$$p_B(s) = 0 \quad \text{for any } s \neq s_m,$$

s_m being the 'minimal' sample out of s's. Applicability of this optimal Bayesian strategy (p_B, t_B) is heavily dependent on the choice of the prior.

If sampling is sequential and informative, then a more appropriate purposive design is also available. Zacks (1969) has shown that this optimal purposive sequential scheme turns out a 'without replacement' selection scheme. With the above Bayesian approach design has its role prior to the survey but is irrelevant at post survey analysis. But Scott (1977) permits design-dependent Bayesian analysis when a design and a prior are of the forms $p(s|X)$, $q(Y|X)$ depending on X such that (i) X is available at the design stage but (ii) is 'not available' at the analysis stage. Then, a prior $r(X)$ for X may be postulated to derive the posterior for Y as

$$q_d^*(Y) = \frac{\int_x p(s|X)q(Y|X)r(X)I_d(Y) \Pi_{i=1}^N \, dX_i}{\int_{Y_d}\left[\int_x p(s|X)q(Y|X)r(X)I_d(Y) \Pi_1^N \, dX_i\right]\Pi_{i=1}^N \, dY_i}.$$

This is not free of design and so the Bayes estimate as well depends on the design in this case. But the optimal Bayes strategy may be derived in this case as earlier. But in either case the approach is strongly 'model-dependent' and investigations are needed for finding 'robust' procedures i.e. procedures that work well even if something other than what is postulated is really true.

If the detailed data d or $\hat{d}$ are not available for analysis then also design may influence analysis. For example, the posterior based on the order statistic y_0 is

$$q^*(Y|y_0) = \frac{\Sigma' \, p(s)q(Y)}{\int_{Y(y_0)} \Sigma' \, p(s)q(Y) \Pi_1^N \, dY_i}, \quad \text{for } Y \in Y(y_0)$$

$= 0$ elsewhere, $Y(y_0) = [Y|Y \text{ is consistent with } y_0]$,

Σ' is sum over samples of the type d with order statistic equal to y_0.
This is not free of design and hence the Bayes estimate should depend on design, but not so if the prior is exchangeable (symmetric) i.e. $q(Y_1, \ldots, Y_N) = q(Y_{i_1}, \ldots, Y_{i_N})$ for every permutation $(i_1, \ldots, i_N)$ of $(1, \ldots, N)$ and/or $Y(y_0)$ consists only of one point with positive $q(Y)$. Let $Y_0 = (Y_{(1)}, \ldots, Y_{(N)})$ with $Y_{(1)} \leq \cdots \leq Y_{(N)}$ for a permutation $((1), \ldots, (N))$ of $(1, \ldots, N)$. Any symmetric function of $Y_1, \ldots, Y_N$ is a function of Y_0. Let Ω_0 be the totality of Y_0's over variation in magnitudes of $Y_{(i)}$, $i = 1, \ldots, N$. Let $\Omega_0(y_0)$ be the totality of Y_0's consistent with y_0. Then the posterior of Y_0 given y_0 is $q_{y_0}^*(y_0) \propto \Sigma_{\Omega_d} \Sigma' \, p(s)q(Y)$, $\forall Y_0 \in \Omega_0(y_0)$ (and 0 outside).

This depends on design but is free of designs if q is exchangeable. So, with an exchangeable prior Bayesian inference about any symmetric $\tau(Y)$ is free of designs.

If a design be symmetric (with respect to permutation of labels) Bayesian inference on symmetric $\tau(Y)$ is not affected if the prior is replaced by any symmetric prior (see Ericson, 1977).

In fact if two of the three elements namely (1) design, (2) prior, (3) estimand, are symmetric, the Bayesian inference is not affected by the lack of symmetry in the third.

Godambe (1968), Godambe and Thompson (1971) and Chaudhuri (1977) postulated product-measure priors $q(Y) = \prod_1^N (Y_i)$, with q_i's as marginal for Y_i, $i = 1, \ldots, N$, in order to obtain unique (and hence optimal) estimators for Y, V, $V/\bar{Y}$ etc., using suitably 'invariant' (with respect to chosen 'translation invariant', 'scale invariant' and other groups of transformations), 'Cochran consistent' (i.e. an estimator coinciding with the 'estimand' it seeks to estimate when the sample coincides with the population) and 'Bayes sufficient' statistics. Chaudhuri (1977) draws upon the concept of invariantly sufficient statistics (see Hall, Wijsman and Ghosh, 1965).

If the posterior distribution of any parametric function $\tau(Y)$ given d with respect to a class of priors $q(Y)$ depends on d only through a statistic $t(d)$, then $t(d)$ is called a 'Bayesian sufficient' statistic for $\tau(Y)$.

The product-measure prior is not generally acceptable (see C. R. Rao, 1977) as it involves both Y_i's for $i \in s$ and $i \notin s$. But if it involves only Y_i's outside s, no 'link' among Y_i's for $i \in s$ and $i \notin s$ is allegedly there to exploit in getting an estimate for Y. But Godambe (1968, pp. 368–69; 1982, p. 405) and Godambe and Thompson (1973, p. 1216) showed a way to look at such a 'link' details we omit to save space. One such way was also shown by Ericson (1969) who postulates an exchangeable prior as a mixture of product measure densities of the form

$$q(Y) = \int_\alpha \prod_1^N q(Y_i)|\alpha) \, dF(\alpha)$$

where Y_i's are priorly independently, identically distributed random variates (iidrv) conditional on a certain vector parameter α with a postulated distribution function $F(\alpha)$. With this prior the posterior is of the form

$$q^*(Y|d) \propto q^*(Y|\hat{d}) \propto \int_\alpha \left(\prod_{i \in s} q(Y_i|\alpha) \, dF(\alpha|d) \right)$$

for $Y \in Y(y)$ and 0 elsewhere.

Here $F(\alpha|d)$ is the distribution function of α given d. For example, postulating a prior such that given $\alpha = (\xi, v)$, $(Y_i|\alpha)$'s iid normal $N(\xi, v)$ variates (ξ unknown, v known) with ξ having a conjugate normal $N(m', v')$ prior distribution with m' unknown but v' known so that Y is assigned an N-dimensional symmetric normal prior with a common mean m', variance $v + v'$ and pairwise common covariance v' for the co-ordinates and paired co-ordinates. Then, $\bar{Y}$ has the prior $N(m', v' + v/N)$ and the posterior $N(E(\bar{Y}|d), V(\bar{Y}|d))$ where

$$E(\bar{Y}|d) = \frac{1}{N}\left[n\bar{y} + (N-n)\frac{\bar{y}v' + m'v/n}{v' + v/n}\right],$$

$$V(\bar{Y}|d) = \frac{N-n}{N^2}\frac{v(Nv' + v)}{(nv' + v)}.$$

Ericson studied further considering other forms of priors deriving posteriors for $\bar{Y}$ given data in various forms. Royall (1968) and Hartley and Rao (1968, 1969) derive and utilize the likelihood as follows from the reduced data suppressing the label-part. Royall (1968) assumes (though strongly objected to by Kempthorne (1969)) that Y contains 'only' the distinct (say, $m \leq n$ in number) values observed in Y. He assumes an SRSWOR design. He supposes the distinct values $y'_1, \ldots, y'_m$ (say) observed in the sample with frequencies n_j ($j = 1, \ldots, m$), say, occur with unknown frequencies N_j ($\geq n_j$), $j = 1, \ldots, m$, in the population. His likelihood is

$$L = L(N_1, \ldots, N_m|y) = \frac{1}{\binom{N}{n}}\prod\binom{N_j}{n_j}.$$

Then, $Y = \sum_{j=1}^{m} N_j y'_j$ and its maximum likelihood estimator (MLE) or best supported estimator (BSE, as called by Royall) is $Y = \sum \hat{N}_j y'_j$ where $\hat{N}_j$'s are values for N_j's that maximize L (i.e. are the MLE's for N_j's).

Virtually equivalently, Hartley and Rao (1968, 1969) make a scale-load approach. They postulate a discrete parametric space. They assume each Y_i ($i = 1, \ldots, N$) as just one of a possible large number M ($\gg N$) of values, say, $Z_1, \ldots, Z_M$, which are all pre-assigned in any given context. Then, letting $\phi_{ij} = 1$ if $Y_i = Z_j$ for $i = 1, \ldots, N_j$, $j = 1, \ldots, M$, $Y = \sum_1^N Y_i = \sum_1^N \sum_1^M \phi_{ij} Z_j = \sum_{j=1}^M K_j Z_j$, where $K_j = \sum_{i=1}^N \phi_{ij}$. For an SRSWOR the likelihood based on ϕ_{ij}'s is $L = L(K_1, \ldots, K_M|Y) \propto \sum_{j=1}^M \binom{K_j}{k_j}$, where k_j = observed number of Z_j's among Y, $j = 1, \ldots, M$. Writing $\hat{K}_j$'s as the MLE's for the scaleloads K_j's of the discrete scale-points Z_j's the MLE is derived as $\hat{Y} = \sum_{j=1}^M \hat{K}_j Z_j$.

There is a lot of controversy over the efficacy or otherwise of these approaches essentially involving elimination of labels whether available or not at the analysis. Extensions to SRSWR and PPSWR sampling have been implemented by Hartley and Rao (1968) themselves. The case of more general sampling with arbitrary positive inclusion probability admitting HTE as the MLE has been considered fruitfully by Särndal (1976). Additional reference is CSW (1977). Interesting applications of Hartley-Rao's scale-load concept are reported in recent works of Meeden and Ghosh (1981) who demonstrate admissibility of a wide class of estimators for Y of the form

$$\sum_{i \in s} y_i + \sum_{i \notin s}\left\{(y_i/x_i)c_i \Big/ \sum_{i \in s} c_i\right\}\sum_{i \notin s} X_i.$$

5. Prediction approach

As developed through the initiative mainly of Brewer (1963), Scott and Smith (1969) and Royall (1970), a model-based, non-Bayesian, predictive approach also yields optimality results. Some of the highlights are briefly noted below. Given the observed data d based on a sample s chosen with probability $p(s)$, one may write $Y = \Sigma_s Y_i + \Sigma_{i \notin s} Y_i$. In the predictive approach one needs an estimate for Y close to its unknown value for the data at hand. No speculation about what else might be observed is encouraged. For this a link among Y_i's for $i \in s$ and $i \notin s$ is needed. This is tried by treating Y as a random vector. Consequently, one needs to predict $\Sigma_{i \notin s} Y_i$ to estimate Y, given $\Sigma_{i \in s} Y_i$ as observed. So, a model is postulated on Y. Royall (1970) considers a simple model treating Y_i's as independent random variables with (model-) expectations and variances as

$$E_m(Y_i) = \beta X_i \quad (X_i > 0, \text{ known}), \quad V_m(Y_i) = \sigma^2 f(X_i),$$

where f is some known function.
He then looks for a model-unbiased predictor $t = t(s, Y)$ for Y. $\ni . E_m(t - Y) = 0$, $\forall s$ with $p(s) > 0$. Among such t's he seeks one, say, t_0, for which

$$E_p M(t_0) = \sum_s p(s) E_m[t_0(s, Y) - Y]^2 \leq \sum_s p(s) E_m[t(s, Y) - Y]^2.$$

The optimal t_0 for any given p, can be worked out as one for which β in

$$E_m(Y|d) = \sum_{i \in s} Y_i + \beta \sum_{i \notin s} X_i$$

is replaced by its least squares estimate (LSE) say, $\hat{\beta}$. By the Gauss–Markov theorem, this LSE, which is the best linear unbiased estimate (BLUE), turns out to be

$$\hat{\beta} = \sum_{i \in s} Y_i X_i / f(X_i) \Big/ \sum_{i \in s} X_i^2 / f(X_i).$$

Thus, the optimal t_0 becomes

$$t_0 = \sum_{i \in s} Y_i + \hat{\beta} \sum_{i \notin s} X_i = t_R$$

(say, the Royall estimator).
The optimal design minimizing $\Sigma_s p(x) E_m[t_0 - Y]^2$ turns out a purposive one namely, a $p_0 . \ni . p_0(s_0) = 1 . \ni . E_m[t_0(s_0, Y) - Y]^2 = \min_s E_m[t_0(s, Y) - Y]^2$ and $p_0(s) = 0$, $\forall s \neq s_0$.
Chang (1983), incidentally, proved admissibility of the Royall (1970) estimator t_0 for Y.

Various results with this approach are available in the literature with more complicated models including polynomial regression curves with various variance functions. We noted in Section 3 that with $E_m(Y_i) = \beta X_i$, $V_m(Y_i) = \sigma^2 X_i^2$ and Y_i's as independent random variables, the strategy $(p_{nx} \cdot \text{HTE}) \equiv (p_{ns}, \bar{t} = (X/n) \sum_{i \in s} Y_i/X_i)$ is optimal in the sense of the value of $E_m E_{p_{nx}}(\bar{t} - Y)^2$ being optimally controlled. Here also, $E_m(\bar{t}) = \beta X = E_m(Y)$, $\forall s$, with $p_{nx}(s) > 0$, i.e. $\bar{t}$ is model-unbiased. But here the Royall estimator is

$$t_R = \sum_{i \in s} Y_i + \frac{1}{n} \sum_{i \in s} \frac{Y_i}{X_i} \left(\sum_{i \notin s} X_i \right)$$

and if we assume the X_i's to be non-stochastic, the operators E_p, E_m commute. So, we have

$$E_p E_m (t_R - Y)^2 \leqslant E_p E_m (\bar{t} - Y)^2 = E_m E_p (\bar{t} - Y)^2 = E_m V_p(\bar{t}).$$

This however does not imply any contradiction because t_R is not subject to $E_p(t_R) = Y$. Also, t_R is a predictor and the interpretations in the inferential values of t_R and $\bar{t}$ are different—$\bar{t}$ is assessed through repeated sampling but t_R is assessed on the basis of a sample at hand.

Efficacy of the predictive approach is dependent on tenability of the model. The strategy itself depends on model-parameters which are unknown in general. In the traditional approach with super-population modelling the strategies themselves are free of unknown parameters, only their performances are judged under postulated models. If models are true, certain comparisons are valid, else they may not be so. Yet inference is still possible through an adopted strategy through its interpretation in terms of repeated sampling. But in predictive approach model alone affords an inferential basis. If a model goes wrong, an adopted strategy merits considerations only if it has a property of 'robustness' in respect of model-bias and model-mean-square error. The question is if rather than the one postulated some other or a more general model is true, whether the chosen strategy continues to remain model-unbiased with unaffected model-mean square error. In maintaining robustness sampling design has an important role to play. Works of Royall and Herson (1973), Scott, Brewer and Ho (1978), and Mukhopadhyay (1977) are relevant in this context.

6. A summing up: Efficacy of an asymptotic theory

In the search for optimal sampling strategies several approaches are made. In order to facilitate taking a right stand this and the following section are appended.

Identifiability of the units through labelling gives a special status to inference problems with survey populations. A twin problem emerges—one of sample-selection and another of estimation. Neyman's traditional approach is decision-theoretic, envisaging speculation on the sample space. Godambe opens up problems showing non-existence of UMV estimators. Hence various optimality criteria e.g. admissibility, completeness, hyper-admissibility etc. are tried to yield acceptable strategies. This exercise helps weeding out some ones, admitting others but is not decisive. The Horvitz–Thompson estimator through πps sampling is recommended by some criteria but its variance often explodes if π_i's are very small for certain units. For the sake of comparing among strategies in an attempt to restrict choice to a specific few of them in diverse situations a superpopulation modelling is considered a way out to take account of structural specialities. This of course affords good insights on occasions but the modelling is often simplistic rather than realistic. A third approach is trying out classical methods of maximum likelihood estimation and UMV estimation through utilization of reduced data suppressing the label-part. This is justified only if labels are not available at analysis but not otherwise and/or if identifiability and hence the peculiar finite population character itself is not available for exploitation. A fourth approach is Bayesian. This is too model-dependent—estimation, appraisal of error in estimation etc. all depend on the choice of a prior which must both be realistic and analytically tractable. With complex survey sampling situations its success is doubtful, it is rather a non-starter in the practical field as yet. A fifth approach is a predictive one which is model-dependent. Its robustness is being studied by many workers now-a-days. But a modification of it as a 'model-based' rather than a 'model-dependent' approach is now fast gaining a popular approbation. Let us briefly illustrate this new trend.

For the model with independent Y_i variates with $E_m(Y_i) = \beta X_i$, $V_m(Y_i) = \sigma_i^2$, Royall's predictor for Y is $t_R = \Sigma_s Y_i + \hat{\beta}_{BLU} \Sigma_{i \notin s} X_i$, where

$$\hat{\beta}_{BLU} = \frac{\Sigma_{i \in s} Y_i X_i / \sigma_i^2}{\Sigma_{i \in s} X_i^2 / \sigma_i^2}$$

is the BLU estimator for β and is free of design but involves model parameter. Instead, modern researchers (e.g., Brewer, 1979; Särndal, 1980; Robinson and Särndal, 1981, etc.) prefer an estimator

$$t_w = \sum_s Y_i + \frac{\Sigma_{i \in s} W_i Y_i}{\Sigma_{i \in s} W_i X_i} \left(\sum_{i \notin s} X_i \right)$$

with weights to be suitably chosen so as to be (i) model-free but (ii) design-dependent rendering good properties to t_w in-terms of design affording it an interpretation through repeated sampling and hence bestowing on it robustness properties as well against any violation of models. This predictor is model-based because its form is suggested by the postulated model.

Earlier we had seen that an optimal (in terms of superpopulation modelling) p-unbiased estimator for Y is of the form $t^* = \Sigma_i Y_i / \pi_i + \beta(X - \Sigma X_i / \pi_i)$, (see the form $\Sigma_{i \in s}(Y_i - \mu_i)/\pi_i + M$, $M = \Sigma \mu_i$ with $\mu_i = E_m(Y_i) = \beta X_i$ in this case).

This suggests (see Särndal, 1980) the use (when β is unknown) of

$$t_f = \sum_{i \in s} \frac{Y_i}{\pi_i} + \frac{\sum_{i \in s} f_i Y_i}{\sum_{i \in s} f_i X_i}\left(X - \sum_{i \in s} \frac{X_i}{\pi_i}\right),$$

replacing β in t^* by $\sum_{i \in s} f_i Y_i / \sum_{i \in s} f_i X_i$, with f_i's as assignable weights. This is suggested by the model but is model-free. So, it is 'model-based' but not 'model-dependent'.

In order to examine properties of t_w, t_f etc. an asymptotic theory is needed. The weights w_i's, f_i's are to be so assigned that they may be asymptotically (1) 'design-unbiased' and/or (2) 'design-consistent' and subject to such requirement(s) they may have suitably controlled asymptotic values of design-averages of mean square errors.

Relevant asymptotic theories are studied by Brewer (1979), Särndal (1980), Robinson and Särndal (1981), Isaki and Fuller (1980) among others.

Brewer (1979) assumes N and n large and conceptualizes hypothetical reproduction of the same population k times along with independent sampling following the identical scheme each time so that the same estimator based on the pooled sample s_k (say) of size nk is used to estimate the total kY for the amalgamated population of size Nk, the sampling fraction n/N remaining constant. An asymptotic theory is then developed allowing $k \to \infty$. His asymptotic design-unbiasedness for t_w demands $\lim_{k \to \infty} E_p[t_w(s_k)/Nk] = \bar{Y}$.

He also gives optimal choice of w_i's and π_i's so as to control the magnitude of

$$\lim_{k \to \infty} E_p E_m[\{t_1(s_k) - kY\}^2/Nk]$$

by finding a lower bound for it and showing how the bound may be attained.

Chaudhuri (1983) compares asymptotic performances of t_w and t_f and points out inadequacy of optimality claimed for t_w by Brewer (1979).

Isaki and Fuller (1981), on the other hand, consider nested sequences of finite populations U_k of sizes N_k, $\ni$ $0 < N_1 < N_2 < \cdots < N_k < \cdots$, $U_1 \subset U_2 \subset \cdots \subset U_k \subset \ldots$, U_1 consisting of 1st N_1, U_2 of 1st N_2 units and so on and sequences of samples independently chosen from U_1, U_2, etc. using sequences of designs of respective sizes n_1, n_2, etc. with inclusion-probabilities π_{i1}, π_{i2}, etc. Denoting by $\bar{Y}_k$ the kth population mean, t_k an estimator for it and by p_k the design corresponding to the over-all sampling from $U_1, \ldots, U_k$ as above, they call t_k a design-consistent estimator for $\bar{Y}_k$ if, given $\varepsilon > 0$, $\lim_{k \to \infty} \text{Prob}\{|t_k - \bar{Y}_k| > \varepsilon\} = 0$, the probability being calculated with respect to p_k.

They also lay down rules for obtaining asymptotically optimal estimators with controlled magnitudes of $\lim_{k \to \infty} E_p E_m[(t_k - \bar{Y}_k)^2]$. A point worth noting with these asymptotic approaches is this that the emerging strategies save the tendencies marked with (IIPS design, HTE) to yield explosive magnitudes of MSE's in exceptional circumstances with a few 'too small π_i's'.

Finally we may mention Godambe's and Godambe–Thompson's (1971) fiducial approach of finding suitable pivotal quantities to estimate population parameters. We will not discuss the details as the theory is too complicated to yield practically useful results except noting that unlike Bayesian and predictive approaches this one is design-dependent at the bottom of its root.

Also we are inclined to mention Sampford's (1978) predictive estimation following the principle of internal congruence (IC). To explain it let us illustrate. Writing $I_i(s) = 1(0)$ if $i \in s$ ($i \notin s$), let

$$Y = \sum_{i \in s} Y_i + \sum_{i \notin s} Y_i = \sum_{i=1}^{N} Y_i I_i(s) + \sum_{1}^{N} Y_i(1 - I_i(s)).$$

Then,

$$E_p \sum_{1}^{N} Y_i(1 - I_i(s)) = \sum_{i}^{N} Y_i(1 - \pi_i) = \sum_{1}^{N} Z_i = Z,$$

say, where $Z_i = (1 - \pi_i)Y_i$.

Then, an estimator (predictor) for Y is, say,

$$\hat{Y} = \sum_s Y_i + \hat{Z}.$$

The criterion of IC demands that $\hat{Z}$ should be of the same form as $\hat{Y}$, only Y_i in it should be replaced by Z_i. Thus, if an HLUE for Y is $\hat{Y} = \Sigma_1^N Y_i b_{si} I_i(s)$, then $\hat{Z} = \Sigma Z_i b_{si} I_i(s)$ giving an HLICE for Y as

$$\hat{Y} = \sum_{i \in s} Y_i + \sum_{1}^{N} Z_i b_{si} I_i(s).$$

Sampford illustrates that many well-known estimators meet this IC criterion.

This seems a rather artificial criterion and does not yield any new optimal estimators.

Also we may mention Mukhopadhyay's (1978) work in finding Royall-type estimators for V using predictive approach and studying its efficiency properties.

In combining the virtues of the traditional and predictive approaches sometimes estimators are required to satisfy the pm-unbiasedness criterion viz. the property

$$E_p E_m(t - \tau(Y)) = 0.$$

Rao and Bellhouse (1978) studied optimality of certain strategies meeting this pm-unbiasedness and having minimal values for $E_p E_m(t - \tau(Y))^2$.

7. Roles of labels and randomization: controversies

A ticklish problem is to pin-point and justify the role of labels in finite population inference. This created a lot of controversies among theoreticians. It is really not easy to get over the generated heat to see the light of truth. Basu and Godambe insist on the use of labels giving different reasons. According to Basu, before the survey the labels are useful in relating the available auxiliary information unit-wise with the variable of interest to help the choice of an appropriate prior. They are influential on posterior analysis too through consideration of sufficiency. According to Godambe, identifiability of units through labelling is the characteristic that differentiates survey sampling theory from the general statistical theory. It is through the use of labels alone that one may implement schemes of sample selection leading to specification of sampling designs through which probability theory may be applied in making statistical inference about survey populations. It is 'identifiability' and 'labelling' that are at the root of opening up the 'finite population inference' problem. Identifiability nullifies the extension of the Gauss–Markov theory to finite populations in deriving best linear estimators. Even though the label-part of survey data is an ancillary statistic in case of non-informative designs, it is still an integral part of the minimal sufficient statistic in the Sample Survey model. So, one cannot leave aside the labels without sacrificing efficiency, in his opinion.

However, there are leading survey statisticians who recommend discarding labels (i) when they are believed to bear no relation to the respective variate-values and (ii) also purposely in order to obtain optimal strategies falling in line with the classical approach of viewing survey sampling problem nothing different from general statistical ones. Some of them not so committed are inclined to search for good strategies when labels are actually not available at the analysis stage because of considerations of confidentiality or even otherwise. Some theorists are most extremes and permit ignoring labels even at the design stage and thus visualize lack of even identifiability of units. They only assume but not ensure randomness in the selection process as in general statistical theory. Hartley and Rao (1968, 1969), Royall (1968), Kempthorne (1969), C. R. Rao (1971), Ericson (1969), Särndal (1972, 1976) are the main contributors among those who suggest optimal procedures for estimation from survey data even when appearing only as $(y_i | i \in s)$ or $(y_i/\pi_i | i \in s)$ with 'labels' non-available otherwise. A paradox is this that if labels are used throughout it is hard to come by positive optimality results that are useful in precctice, but if labels are not used at analysis optimal results are available according to alternative criteria. Thus, ignorance becomes a bliss in a sense. Of course, if labels are available, then their utilization is more profitable than their non-utilization as adequately illustrated by Godambe (1975). So, whatever the provocations from the conflicting schools a reasonable stand to take is (1) not to eschew labels when available and live with the deficiency in optimality for strategies and (2) to employ optimal procedures without using labels when unavailable. However it behoves us to observe that it is not possible to find a label-dependent estimator with which to beat uniformly any given label-free estimator or vice-versa, if the class of estimators is wide enough accommodating the biased ones too, as e.g. Joshi's result that $\bar{y}$ is admissible for $\bar{Y}$ in terms of MSE irrespective of any designs.

When Godambe (1966) presented the likelihood for Y given survey data based on a probability sample and noted its aridlooking form incapable of yielding any inference on the unobserved co-ordinates of Y and more importantly that this is so for any given data no matter how (i.e. with what probability) gathered (which he calls the 'problem of randomization') germs of further intricacies were bred. Godambe (1966b) himself looked for alternative principles and criteria for inference and courses of action in view of problems inherent in (1) Neyman's structure-free approach and (2) likelihood-conditionality principles relevant to 'labelled' finite populations. He applied principles of 'linear sufficiency', 'distribution-free sufficiency' and 'censoring' in order to find serviceable strategies based on unequal probability sampling (details omitted). Basu (1969, 1978) vehemently opposed the use of any kind of probability sampling and recommended Bayesian method as the only sensible way to analyse survey data. However, through the writings of C R. Rao, Ericson, Scott and others a point is made that forms of the likelihood for specific parametric functions and suitably reduced data can be so contrived as to make them capable of (1) yielding fruitful inference and (2) accommodating a positive role of probability sampling as a source of influence on the inference. Godambe's (1982a, b, 1983) works emphasize strongly the efficacy of 'probability sampling' in rendering robustness on strategies yielded through a model-based approach consistently with likelihood/Bayesian principles guarding them against inaccurate modelling. He shows how a proper sampling design yields a sample with a high probability for which an estimator is kept close to an optimal one under a specific model even in case of a model-failure retaining efficiency appreciably close to the optimal level.

However, T.M.F. Smith (1976) and others effectively suggest that inference on survey populations should be attempted not just through randomization but be dominated by structural features of the population in specific situations.

With the prediction approach randomization is unnecessary in deriving optimal results when a postulated model holds true in a usable form. But when model goes wrong, randomness gathers importance as a means of protecting its robustness both (i) against specific model-violations (see Royall and Herson, 1973; Scott, Brewer and Ho, 1978, etc.) and (ii) against all sorts of model-crackdown—see Godambe (1976). For the model-based predictive approach 'optimal model-dependent, design-free' predictors attain maximal efficiency levels when based on optimal purposive designs but are protected well against model-inaccuracies when based on balanced equal-probability samples with or without stratification or on over-balanced unequal probability samples, but this probability sampling is only an ad hoc or make-shift precautionary measure. But Godambe's (1955, 1982a,b, 1983) unified approach straightaway prescribes (1) 'probability sampling and postulating modelled structures for Y' in a 'joint' fashion in deriving 'design-cum-model-based' optimal strategies. Also he uses sampling designs as a tool to achieve 'criterion and efficiency robustness' allowing high selection-probability for a sample yielding a convenient estimator close to an optimal one under a model with a limited loss in efficiency despite a failure in modelling to a certain extent. The recently developing asymptotic theory at the hands of Brewer (1979), Isaki and Fuller (1980), Särndal (1980), etc. (see Section 6) also vindicates the wholesome effect of randomization on model-dependent strategies in revising them to rid them of the vulnerability to model-failures. In fact the asymptotic model-based approach is chiefly motivating current research trends in sampling. In this, pragmatism rather than optimality, has a crucial role to play. So, the importance of optimality needs to be assessed in a right perspective in the context of finite population theory.

References

Aggarwal, O. P. (1959). Bayes and minimax procedures in sampling from finite and infinite populations. I. *Ann. Math. Stat.* **30**, 206–218.
Aggarwal, O. P. (1966). Bayes and minimax procedures for estimating the arithmetic mean of a population with two-stage sampling. *Ann. Math. Stat.* **37**, 1186–1195.
Ajgaonkar, S, G. P. (1962). Some aspects of successive sampling. An unpublished Ph.D. thesis, Karnataka University, Dharwad, India.
Ajgaonkar, S, G. P. (1965). On a class of linear estimators in sampling with varying probabilities without replacement. *J. Amer. Stat. Ass.* **60**, 637–642.
Basu, D. (1958). On sampling with and without replacement. *Sankhyā* **20**, 287–294.
Basu, D. (1969). Role of the sufficiency and likelihood principles in sample survey theory. *Sankhyā, Ser. A* **31**, 441–454.
Basu, D. (1971). An essay on the logical foundations of survey sampling. Part one. In: V. P. Godambe and D. A. Sprott, eds., *Foundations of Statistical Inference*. Holt, Rinehart, Winston, Toronto, 203–242.
Basu, D. (1978). Relevance of randomization in data analysis (with discussion). In: N. E. Namboodiri, ed., *Survey Sampling and Measurement*, Academic Press, New York.
Basu, D. and Ghosh, J K. (1967). Sufficient statistics in sampling from a finite universe. *Proc. 36th Session, Int. Stat. Inst.*, 850–859.
Bhargava, N. K. (1978). On some applications of the technique of combined unordering. *Sankhyā, Ser. C* **40**, 74–83.
Bickel, P. J. and Lehmann, E. L. (1981). A minimax property of the sample mean in finite populations. *Ann. Stat.* **9**, 1119–1122.
Biyani, S. H. (1980). On inadmissibility of the Yates–Grundy variance estimator in unequal probability sampling. *J. Amer. Stat. Ass.* **75**, 709–712.
Blackwell, D. (1947). Conditional expectation and unbiased sequential estimation. *Ann. Math. Stat.* **18**, 105–110.
Blackwell, D. and Girshick, G. A. (1954). *Theory of Games and Statistical Decisions*. Wiley, New York.
Brewer, K. R. W. (1963). Ratio estimation and finite populations. Some results deducible from the assymption of an underlying stochastic process. *Aust. J. Stat.* **5**, 93–105.
Brewer, K. R. W. (1979). A class of robust sampling design for large-scale surveys. *J. Amer. Stat. Ass.* **74**, 911–915.
Brewer, K. R. W., Foreman, Mellor, R. W. and Trewin, D. J. (1977). Use of experimental design and population modelling in survey sampling. Paper presented at Int. Stat. Inst. Conference at Delhi.
Cassel, C. M., Särndal, C. E. and Wretman, J. H. (1976). Some results on generalized difference estimation and generalized regression estimation for finite populations. *Biometrika* **63**, 615–620.
Cassel, C. M., Särndal, C. E. and Wretman, J. H. (1977). *Foundations of Inference in Survey Sampling*. Wiley, New York.

Chang, H. J. (1981). On admissibility in finite population sampling. Tamkang, *J. Manag. Sci.* **2**, 51–62.

Chaudhuri, A. (1968). A comparison between sampling with and without replacement from finite populations. *Bull. Cal. Stat. Assoc.* **17**, 137–160.

Chaudhuri, A. (1969). Minimax solutions of some problems in sampling from a finite population. *Bull. Cal. Stat. Assoc.* **18**, 1–24.

Chaudhuri, A. (1970). Admissibility of the class of unicluster sampling designs. *Bull. Cal. Stat. Assoc.* **19**, 87–93.

Chaudhuri, A. (1971). Some problems in sampling from a finite population. Unpublished Ph.D. thesis. Calcutta University.

Chaudhuri, A. (1977). Some applications of the principles of Bayesian sufficiency and invariance to inference problems with finite populations. *Sankhyā C* **39**, 140–149.

Chaudhuri, A. (1977). On some problems of choosing the sample-size in estimating finite population totals. *Proc. Bull. Int. Stat. Inst.*, 116–119.

Chaudhuri, A. (1978). On estimating the variance of a finite population. *Metrika* **25**, 65–76.

Chaudhuri, A. (1978). On the choice of sample-size for a Horvitz/Thompson estimator. *J. Ind. Soc. Agri. Stat.* **30**, 35–42.

Chaudhuri, A. (1980). On optimal and related strategies for sampling on two occassions with varying probabilities. Tech. Report. Ind. Stat. Inst. Calcutta. Now in *J. Ind. Soc. Agri. Stat.* **37** (1) (1985) 45–53.

Chaudhuri, A. (1981). Non-negative unbiased variance estimators. In: D. Krewski, R. Platek and J. N. K. Rao, eds., *Current Topics in Survey Sampling*. Academic Press, New York, 317–328.

Chaudhuri, A. (1983). Choosing sampling strategies with an asymptotic model-based approach. Tech. Report. Ind. Stat. Inst. Calcutta.

Chaudhuri, A. and Arnab, R. (1978). On the role of sample-size in determining efficiency of Horvitz/Thompson estimators. *Sankhyā A* **40**, 104–109.

Chaudhuri, A. and Arnab, R. (1979). On the relative efficiencies of sampling strategies under a super-population model. *Sankhyā C* **41**, 40–43.

Chaudhuri, A. and Mukhopadhyay, P. (1978). A note on how to choose the sample-size for Horvitz/Thompson estimator. *Bull. Cal. Stat. Ass.* **27**, 149–154.

Chaudhuri, A. and Adhikary Arun Kumar (1983). Sampling strategies with randomized response trials—Their properties and relative efficiencies. Tech. Report. Ind. Stat. Inst. Calcutta.

Chaudhuri, A. and Adhikary Arun Kumar (1983a). On optimality of double sampling strategies with varying probabilities. *J. Stat. Plan. Inf.* **8**, 257–265.

Chaudhuri, A. and Adhikary Arun Kumar (1983b). On sampling strategies with randomized response about quantitative data. Tech. Report. Ind. Stat. Inst. Calcutta.

Cheng, C. S. and Li, K. C. (1983). Optimality criteria in survey sampling. Unpub. ms., Dept. Stat. Purdue Univ.

Cochran, W. G. (1946). Relative accuracy of systematic and stratified random samples for a certain class of populations. *Ann. Math. Stat.* **17**, 164–177.

Dharmadhikari, S. W. (1969). A note on some inadmissible estimates. *Sankhyā, Ser. A* **31**, 361–364.

Ericson, W. A. (1969). Subjective Bayesian models in sampling finite populations. *J. Roy. Stat. Soc. (B)* **31**, 195–224.

Ericson, W. A. (1970). On a class of uniformly admissible estimators of a finite populations total. *Ann. Math. Stat.* **41**, 1369–1372.

Ericson, W. A. (1974). A note on the non-existence of minimum variance unbiased estimators for a finite population total. *Sankhyā C* **36**, 181–184.

Ghosh, M. and Meeden, G. (1983). Estimation of the variance in finite population sampling. *Sankhyā B* **45**, 362–375.

Godambe, V. P. (1955). A unified theory of sampling from finite populations. *J. Roy. Stat. Soc. (B)* **17**, 269–278.

Godambe, V. P. (1960). An admissible estimate for any sampling design. *Sankhyā* **22**, 285–288.

Godambe, V. P. (1965). A review of the contributions towards a unified theory of sampling from finite populations. *Rev. Int. Stat. Inst.* **33**, 242–256.

Godambe, V. P. (1966). Bayes and empirical estimation in sampling finite populations (Abstract). *Ann. Math. Stat.* **37**, 552.

Godambe, V. P. (1966). A new approach to sampling from finite populations I, II, *J. R. Stat. Soc. (B)*, 310–328.

Godambe, V. P. (1968). Bayesian sufficiency in survey sampling. *Ann. Inst. Stat. Math.* **20**, 363–373.

Godambe, V. P. (1969). Admissibility and Bayes estimation in sampling finite populations – V. *Ann. Math. Stat.* **40**, 672–676.

Godambe, V. P. (1969). A fiducial argument with applications to survey sampling. *J. Roy. Stat. Soc. (B)* **31**, 246–260.

Godambe, V. P. (1969). Some aspects of the theoretical developments in survey sampling. In: *New Developments in Survey Sampling*. Wiley Inter-Science, 27–53.

Godambe, V. P. (1970). Foundations of survey sampling. *Ann. Stat.* **24**, 33–38.

Godambe, V. P. (1975). A reply to my critics. *Sankhyā C* **37**, 53–76.

Godambe, V. P. (1976). A historical perspective of the recent developments in the theory of sampling from actual populations. *Ind. Soc. Agri. Stat.* (an invited talk), 1–12.

Godambe, V. P. (1976a). Philosophy of survey-sampling practice. In: *Foundations of Probability Theory, Statistical Inference and Statistical Theories of Science, Vol. II*. Harper and Hooker, 103–123.

Godambe, V. P. (1982a). Estimation in survey sampling: Robustness and optimality. *J. Amer. Stat. Assoc.* **77**, 393–406.

Godambe, V. P. (1982b). Likelihood principle and randomization. In: G. Kallainpur, P. R. Krishnaiah and J. K. Ghosh, eds., *Statistics and Probability—Essays in Honour of C. R. Rao*. Academic Press, New York, 281–294.

Godambe, V. P. (1983). Survey sampling: Modelling, randomization and robustness–A unified view. *Proc. Amer. Stat. Assoc. Sec. on Survey Research Methods*, 26–29.

Godambe, V. P. and Joshi, V. M. (1965). Admissibility and Bayes estimation in sampling finite populations. *I. Ann. Math. Stat.* **36**, 1707–1722.

Godambe, V. P. and Thompson, M. E. (1971). The specification of prior knowledge by classes of prior distributions in survey sampling estimation. In: V. P. Goodambe and D. A. Sprott, eds., *Foundations of Statistical Inference*. Holt, Rinchart and Winston, Toronto, 243–354.

Godambe, V. P. and Thompson, M. E. (1971a). Bayes, fiducial and frequency aspects of statistical inference n regression analysis in survey sampling. *J. Roy. Stat. Soc. (B)* **33**, 361–390.

Godambe, V. P. and Thompson, M. E. (1973). Estimation in sampling theory with exchangeable prior distributions. *Ann. Stat.* **1**, 1212–1221.

Godambe, V. P. and Thompson, M. E. (1977). Robust near optimal estimation in survey practice. *Rev. Int. Stat. Inst.* 129–146.

Gupta, P. C. and Yogi, A. K. (1980). Necessary best estimators in certain classes of linear estimators for Ikeda–Sen sampling. *J. Ind. Soc. Agri. Stat.* **42**, 61–72.

Hájek, J. (1949). Representative sampling by the method of two phases (in Czech). *Stat. Obzor.* **29**, 384–395.

Hájek, J. (1959). Optimum strategy and other problems in probability sampling. *Casopis pro Pest. Mat.* **84**, 387–423.

Halmos, P. (1946). The theory of unbiased estimation. *Ann. Math. Stat.* **17**, 34–43.

Hansen, M. H. and Hurwitz, W. N. (1943). On the theory of sampling from finite populations. *Ann. Math. Stat.* **14**, 333–362.

Hanurav, T. V. (1962). An existence theorem in sampling. *Sankhyā A* **24**, 227–230.

Hanurav, T. V. (1965). Optimum sampling strategies and some related problems. Ph.D. thesis, Ind. Stat. Inst.

Hanurav, T. V. (1966). Some aspects of unified sampling theory. *Sankhyā A* **28**, 175–204.

Hanurav, T. V. (1968). Hyperadmissibility and optimum estimators for sampling finite populations. *Ann. Math. Stat.* **39**, 621–642.

Hartley, H. O. and Rao, J. N. K. (1968). A new estimation theory for sample surveys. *Biometrika* **55**, 547–557.

Hartley, H. O. and Rao, J. N. K. (1969). A new estimation theory for sample surveys, II. In: N. L. Johnson and H. Smiths, eds., *New Developments in Survey Sampling*. Wiley–Interscience, New York, 147–169.

Hartley, H. O. and Ross, A. (1954). Unbiased ratio estimators. *Nature* **174**, 270–271.
Hege, V. S. (1965). Sampling designs which admit uniformly minimum variance unbiased estimators. *Bull. Cal. Stat. Ass.* **14**, 160–162.
Hege, V. S. (1972). A note on uni-cluster sampling designs. *Bull. Cal. Stat. Assoc.* **21**, 77–82.
Hall, W. J., Wijsman, R. A. and Ghosh, J. K. (1965). The relationship between sufficiency and invariance with applications in sequential analysis. *Ann. Math. Stat.* **36**, 575–614.
Ho, E. W. H. (1980). Model-unbiasedness and the Horvitz–Thompson estimator in finite population sampling. *Aust. J. Stat.* **22**, 218–225.
Hodges, Jr., J. L. and Lehmann, E. L. (1982). G. Kallianpur, P. K. Krishnaiah and J. K. Ghosh, *Statistics and probability Essays in Honour of C. R. Rao*, North-Holland, Amsterdam, 325–327.
Horvitz, D. G. and Thompson, D. J. (1952). A generalization of sampling without replacement from a finite universe. *J. Amer. Stat. Ass.* 663–685.
Isaki, C. T. and Fuller, W. A. (1980). Survey design under super population model. In: D. Krewski, R. Platek and J. N. K. Rao, eds., *Current Topics in Survey Sampling*. Academic Press, New York, 199–226.
Joshi, V. M. (1965). Admissibility and Bayes estimation in sampling finite populations, II, III. *Ann. Math. Stat.* **36**, 1723–1742.
Joshi, V. M. (1966). Do. IV. *Ann. Math. Stat.* **37**, 1658–1670.
Joshi, V. M. (1967). Confidence intervals for the mean of a finite population. *Ann. Math. Stat.* **38**, 1180–1207.
Joshi, V. M. (1968). Admissibility of the sample mean is estimate of the mean of a finite population. *Ann. Math. Stat.* **39**, 606–620.
Joshi, V. M. (1969). Admissibility of estimates of the mean of a finite population. In: N. L. Johnson and H. Smith, Jr., eds., *New Developments in Survey Sampling*. Wiley–Interscience, New York, 188–212.
Joshi, V. M. (1970). Note on the admissibility of the Sen-Yakes-Grundy variance estimator and Murthy's estimator and its variance estimator for samples of size 2. *Sankhyā A* **32**, 431–438.
Joshi, V. M. (1971). Hyperadmissibility of estimators for finite populations. *Ann. Math. Stat.* **42**, 680–690.
Joshi, V. M. (1971a). Admissibility of arbitrary estimates and unbiased estimates of their expectations in a finite population. *Ann. Math. Stat.* **42**, 839–841.
Joshi, V. M. (1971b). A note on admissible sampling designs for a finite population. *Ann. Math. Stat.* **42**, 1425–1428.
Joshi, V. M. (1972). A note on hyperadmissibility of estimators for finite populations. *Ann. Math. Stat.* **43**, 1323–1328.
Joshi, V. M. (1977). Estimators of the mean of a finite population. *J. Ind. Stat. Ass.* **15**, 1–9.
Joshi, V. M. (1977a). A note on estimators in finite populations. *Ann. Stat.* **5**, 1051–1053.
Joshi, V. M. (1979). Joint admissibility of the sample mean as estimators of the means of finite populations. *Ann. Stat.* **7**, 995–1002.
Joshi, V. M. (1980). Joint admissibility of the sample means as estimators of he means of finite populations under a general loss function. *Sankhyā C*, 118–123.
Kempthorne, O. (1969). Some remarks on statistical inference in finite sampling. In: N. D. Johnson and H. Smith, Jr., eds., *Developments in Survey Sampling*. Wiley–Interscience, New York, 671–695.
Kempthorne, O. (1971). Probability, Statistics and Knowledge Business. In: V. P. Godambe and D. A. Sprott, eds., *Foundations of Statistical Inference*. Holt, Rinehart, Winston, Toronto, 482–499.
Koop, J. C. (1957). Contributions to the general theory of sampling finite populations without replacement and with unequal probabilities. Ph.D. thesis. N. Carolina Stat. Conf. Inst. Stat. No. 296.
Koop, J. C. (1963). On the aximas of sample formation and their bearing on the construction of linear estimators in sampling theory for finite unvierses, I, II, III. *Metrika* **7**, 81–114, 165–204.
Lahiri, D. B. (1951). A method of sample selection providing unbiased ratio estimators. *Bull. Int. Stat. Inst.* **33**, 133–140.
Lanke, J. (1973). On UMV-estimators in survey sampling. *Metrika* **20**, 196–202.
Lanke, J. (1975). Some contributions to the theory of survey sampling. Ph.D. thesis, University of Lund.

Lanke, J. and Ramakrishnan, M. K. (1974). Hyper-admissibility in survey sampling. *Ann. Stat.* **2**, 205–215.

Liu, T. P. (1973). Double admissibility in sampling from a finite population. *Bull. Inst. Math. C Acad. Sin.* **1**, 71–74.

Liu, T. P. (1974). Bayes estimation for the variance of a finite population. *Sankhyā C* **36**, 23–42.

Liu, T. P. (1974a). Bayes estimation for the variance of a finite population. *Matrika* **21**, 127–132.

Mahalanobis, P. C. (1944). On large-scale sample surveys. *Phil. Trans. Roy. Soc. Lond. (B)* **231**, 329–151.

Meeden, G. and Ghosh, M. (1981). Admissibility in finite problems. *Ann. Stat.* **9**, 846–852.

Midzuno, H. (1952). On the sampling system with probability proportional to sum of sizes. *Ann. Inst. Stat. Math.* **3**, 99–107.

Mukhopadhyay, P. (1977). Robust estimators of finite population total under certain linear regression models. *Sankhyā C* **39**, 71–87.

Mukhopadhyay, P. (1978). Estimating the variance of a finite population under a super population model. *Metrika* **25**, 115–122.

Murthy, M. N. (1957). Ordered and unordered estimators in sampling without replacement. *Sankhyā* **18**, 379–398.

Murthy, M. N. (1963). Generalized unbiased estimation in sampling from finite populations. *Sankhyā B* **25**, 245–262.

Neyman, J. (1934). On the two different aspects of the representative method. The method of stratified sampling and the method of purposive selection. *J. Roy. Stat. Soc.* **97**, 558–625.

Ogus, J. L. (1969). A note on the necessary best estimator. *J. Amer. Stat. Ass.* **64**, 1350–1352.

Patel, H. C. (1974). A note on necessary best estimator in the class of linear unbiased estimators. *Sankhyā C* **36**, 195–196.

Patel, H. C. and S. W. Dharmadhikary (1974). Some results on the minimal complete class of linear estimators in survey sampling. *Sankhyā C* **36**, 185–194.

Patel, H. C. and S. W. Dharmadhikary (1977). On linear invariant unbiased estimators in survey sampling. *Sankhyā C* **39**, 21–27.

Patel, H. C. and S. W. Dharmadhikary (1978). Admissibility of Murthy's and Midzuno's estimators within the class of linear unbiased estimators of finite populations totals. *Sankhyā C* **40**, 21–28.

Pathak, P. K. (1962). On simple random sampling with replacement. *Sankhyā A* **24**, 287–302.

Pathak, P. K. (1962a). On sampling units with unequal probabilities. *Sankhyā A* **24**, 315–326.

Raj, D. (1956). Some estimators in sampling with varying probabilties without replacement. *J. Amer. Stat. ss.* **51**, 269–284.

Ramakrishnan, M. K. (1969). Some results on the comparison of sampling with and without replacement. *Sankhyā A* **31**, 333–342.

Ramakrishnan, M. K. (1970). Optimum estimators and strategies in survey. Ph.D. thesis, Ind. Stat. Inst.

Ramakrishnan, M. K. (1973). An alternative proof of the admissibility of the Horvitz–Thompson estimator. *Ann. Stat.* **1**, 577–579.

Ramakrishnan, M. K. (1975). Choice of an optimum sampling strategy – I. *Ann. Stat.* **3**, 669–679.

Rao, C. R. (1945). Information and accuracy attainable in estimation of statistical parameters. *Bull. Cal. Math. Soc.* **37**, 81–91.

Rao, C. R. (1952). Some theorems on minimum variance unbiased estimations. *Sankhyā C* **12**, 27–42.

Rao, C. R. (1971). Some aspects of statistical inference in problems of sampling from finite populations. In: V. P. Godambe and D. A. Sprott, eds., *Foundations of Statistical Inference*. Holt, Rinehart and Winston, Toronto, 177–202.

Rao, C. R. (1977). Some problems of survey sampling. *Sankhyā C* **39**, 28–139.

Rao, J. N. K. (1975). On the foundations of survey sampling. In: J. N. Srivastava, ed., *A Survey of Statistical Design and Linear Models*. American Elsevier, New York, 489–525.

Rao, J. N. K. and Bellhouse, D. L. 1978). Optimal estimation of a finite population mean under generalized random permutation models. *J. Stat. Plan. and Inf.* **2**, 125–141.

Rao, J. N. K., Hartley, H. O. and Cochran, W. G. (1962). A simple procedure of unequal probability sampling without replacement. *J. Roy. Stat. Soc. (B)* **24**, 482–491.

Rao, J. N. K. and Singh, M. P. (1973). On the choice of estimator in survey sampling. *Aust. J. Stat.* **15**, 95–104.

Robinson, P. M. and Särndal, C. E. (1980). A symptotic properties of the generalized regression estimator in probability sampling. (unpublished ms).

Roy, J. and Chakravarti, I. M. (1960). Estimating the mean of a finite populations. *Ann. Math. Stat.* **31**, 392–398.

Royall, R. M. (1968). An old approach to finite population sampling theory. *J. Amer. Stat. Ass.* **63**, 1269–1279.

Royall, R. M. (1970). On finite population sampling theory under certain linear regression models. *Biometrika* **57**, 377–387.

Royall, R. M. (1971). Linear regression models in finite populations sampling theory. In: V. P. Godambe and D. A. Sprott, eds., *Foundation of Statistical Inference*. Holt, Rinehart and Winston, Toronto, 259–279.

Royall, R. M. and Herson, J. (1973). Robust estimation in finite populations, I, II. *J. Amer. Stat. Ass.* **68**, 80–893.

Sampford, M. R. (1978). Predictive estimation and internal congruency. In: H. A. David, ed., *Contributions to Survey Sampling and applied statistics*. Academic Press, New York.

Sankarnarayanan, K. (1980). A note on the admissibility of some non-negative quadratic estimators. *J. Roy. Stat. Soc. (B)* **42**, 387–389.

Särndal, C. E. (1972). Sample survey theory general statistical theory: Estimation of the population mean. *Rev. Int. Stat. Inst.* **40**, 1–12.

Särndal, C. E. (1976). On uniformly minimum variance estimation in finite populations. *Ann. Stat.* **4**, 993–997.

Särndal, C. E. (1980). On π-inverse weighting versus BLU weighting in probability sampling. *Biometrika* **67**, 693–650.

Scott, A. J. (1975). On admissibility and uniform admissibility in finite populations sampling. *Ann. Stat.* **2**, 489–491.

Scott, A. J. (1977). On the problem of randomization in survey sampling. *Sankhyā C* **39**, 1–9.

Scott, A. J. and Smith, T. M. F. (1969). A note on estimating secondary charactors in multivariate survey. *Sankhyā A* **31**, 497–498.

Scott, A. J. and Smith, T. M. F. (1969). Estimation in multistage surveys. *J. Amer. Stat. Assoc.* **64**, 830–840.

Scott, A. J. and Smith, T. M. F. (1975). Minimax design for sample surveys. *Biometrika* **62**, 353–357.

Scott, A. J., Brewer, K. R. W. and Ho, E. W. H. (1978). Finite population sampling and robust estimation. *J. Amer. Stat. Ass.* **73**, 359–361.

Sen, A. R. (1953). On the estimate of the variance in sampling with varying probabilities. *J. Ind. Soc. Agri. Stat.* **5**, 119–127.

Sen Gupta, S. (1980). Further studies on some strategies in sampling finite populations. Unpublished Ph.D. thesis, Calcutta University.

Sen Gupta, S. (1980). On the admissibility of the symmetrized Des Raj estimator for PPSWOR samples of size two. *Bull. Cal. Stat. Assoc.* **29**, 35–44.

Sen Gupta, S. (1982). Admissibility of the symmetrized Des Ray estimator for fixed size sampling designs of size two. *Bull. Cal. Stat. Assoc.* **31**, 201–205.

Sen Gupta, S. (1983). On the admissibility of Hartley-Ross estimator. To appear in *Sankhyā B*.

Sen Gupta, S. (1983). Admissibility of unbiased estimators in finite population sampling for samples of size at most two. *Bull. Cal. Stat. Assoc.* **32**, 91–102.

Sen Gupta, S. (1983). Optimality of a design unbiased strategy for estimating a finite population variance. To appear in *Sankhyā B*.

Seth, G. R. (1980). Some thoughts on statistical inference about finite populations. *J. Ind. Soc. Agri. Stat.* **32**, 25–46.

Seth, G. R. and J. N. K. Rao (1964). On the comparison between simple random sampling with and without replacement. *Sankhyā A* **26**, 85–86.

Sinha, B. K. (1976). On balanced sampling schemes. *Bull. Call. Stat. Ass.* **25**, 129–138.

Sinha, B. K. (1980). On the concept of admissible extensions of sampling designs and some related problems. Tech. Report. Ind. Stat. Inst. Calcutta.

Smith, T. M. F. (1976). The foundations of survey sampling. A review (with discussion). *J. Roy. Stat. Soc. (A)* **139**, 183–204.

Smith, H. F. (1938). An expirical law describing heterogeneity in the yields of agricultural crops. *J. Agri. Sc.* **28**, 1–23.

Stein, C. (1956). Inadmissibility of the usual estimator for the mean of a multi-variate normal distribution. *Proc. 3rd. Berk. Symp., Vol. 1*, 197–206.

Stenger, H. (1977). Sequential sampling from finite populations. *Sankhyā C* **39**, 10–20.

Stenger, H. (1979). A minimax approach to randomization and estimation in survey sampling. *Ann. Stat.* **7**, 395–399.

Stenger, H. (1981). Loss functions and admissible estimators in survey sampling. *Metrika*.

Thompson, M. E. (1971). Discussion of a paper by C. R. Rao. In: V. P. Godambe and D. A. Sprott, eds., *Foundations of statistical inference*. Holt, Rinehart and Winston, Toronto, 191–198.

Tikkiwal, B. D. (1972). On unordering of estimators in sampling with or without replacement and its impact on T-class of estimators. Unpublished manuscript, presented at IMS meeting, Montreal.

Watson, G. S. (1964). Estimation in finite populations. Unpublished report.

Yates, F. and Grundy, P. M. (1953). Selection without replacement from within strata with probability proportional to size. *J. Roy. Stat. Soc. (B)* **15**, 253–261.

Zacks, S. (1969). Bayes sequential designs for sampling finite populations. *J. Amer. Stat. Ass.* **64**, 1342–1349..

Simple Random Sampling

P. K. Pathak

0. Summary

The primary object of this article is to show that for estimating the population mean under three commonly used forms of simple random sampling, the sample mean based only on the distinct units possesses the following remarkable invariance property: It is admissible, unbiased, and asymptotically normally distributed if and only if the Erdös–Rényi–Hájek condition is satisfied. An important implication of this result is that from a practical point of view, the commonly used forms of simple random sampling are all asymptotically equally cost-efficient.

1. Introduction

In sample surveys, simple random sampling without replacement (SRSWOR) is commonly perceived as being more practical as well as more efficient than other forms of simple random sampling, such as simple random sampling with replacement (SRSWR), fixed cost simple random sampling (SRSFC), and inverse simple random sampling (SRSI). There is, however, no theoretically sound justification for its validity. Although over two decades ago, preliminary investigations of the relative efficiency of SRSWOR versus SRSWR were first undertaken by Basu (1958), Hájek (1959), Pathak (1962), Seth and Rao (1964) and others. They were largely inconclusive. A major weakness in these investigations might well have been the lack of satisfactory methods to deal with the first passage behavior of simple random sampling under realistic cost functions. This article focuses on the asymptotics of simple random sampling based on Hájek's imbedding methods, as put forth by him in his 1960 classic paper (Hájek, 1960). These imbedding methods furnish a complete answer concerning the asymptotic efficiency of the various forms of simple random sampling. In fact they show that under mild conditions as the cost of sampling gets larger and larger, the cost-adjusted relative efficiency of one form of simple random sampling versus another approaches one. Thus the various forms of simple random sampling, SRSWOR, SRSWR, SRSFC, etc., are all asymptotically equally cost-efficient.

A brief outline of Hájek's simple but powerful technique is as follows: Suppose that our object is to compare the asymptotic efficiencies of two estimators t_1 and t_2 (or more precisely two strategies of estimation). Also suppose that it is possible to imbed t_1 and t_2 into a common sampling scheme and thereby obtain a fairly accurate estimate of the quantity

$$H(t_1, t_2) = E(t_1 - t_2)^2/V(t_1). \tag{1.1}$$

In the sequel, we refer to $H(t_1, t_2)$ as Hájek's *measure of disparity* between t_1 and t_2.

Now suppose that $H(t_1, t_2)$ approaches zero asymptotically. It is easily seen that $\lim H(t_1, t_2) = 0$ implies that $\lim V(t_2)/V(t_1) = 1$, showing that both t_1 and t_2 are asymptotically equally efficient. A second and more important consequence of this result is the asymptotic equivalence of the limiting distributions of t_1 and t_2 in the following sense: Suppose that under a certain condition t_1 is asymptotically normally distributed with parameters $(Et_1, V(t_1))$. Then under the same set of conditions, t_2 is also normally distributed with parameters $(Et_2, V(t_2))$ and vice versa. This asymptotic equivalence follows from the identity:

$$\frac{t_2 - Et_2}{\sqrt{V(t_2)}} = \left[\frac{t_1 - Et_1}{\sqrt{V(t_1)}} + \frac{(t_2 - t_1)}{\sqrt{V(t_1)}}\right]\frac{\sqrt{V(t_1)}}{\sqrt{V(t_2)}} \tag{1.2}$$

and the observation that $\lim H(t_1, t_2) = 0$ implies $\lim V(t_1)/V(t_2) = 1$, as well as the convergence of $(t_2 - t_1)/\sqrt{V(t_1)}$ to zero in probability.

It is perhaps logical now to digress slightly and define the term asymptotics in our context. The standard practice in sample surveys is to formally work with a triangular array of populations, say $\pi = \{P_k: k \geq 1\}$ in which for each integer k, P_k is a population of size N_k from which a sample of size n_k is to be drawn. It is assumed that for each k, the population size N_k, the sample size n_k, the size of the complement of the sample, $N_k - n_k$, are all approaching infinity as k increases. Asymptotics in sample surveys is viewed as the investigation of the limiting properties of statistics of interest based on such a sequence of surveys as k increases. With this framework at the back of our minds, we now proceed to study three different forms of simple random sampling. For brevity, we however have chosen not to use the extra suffix 'k' in the discussion that follows.

Consider a population $P = (U_1, \ldots, U_j, \ldots, U_N)$ of N units in which the j-th unit $U_j = (j, Y_j, C_j)$, where j denotes its label, Y_j its unknown Y-characteristic value under study and C_j the unknown cost of ascertaining the value of Y_j. We assume that like Y_j, the true value of C_j is unknown and varies from unit to unit. In practice such an assumption is undoubtedly more realistic than the customary assumption that the cost of observing a unit is a known constant and does not vary from unit to unit. We assume that the cost characteristic has been suitably scaled so that $\overline{C} = N^{-1} \Sigma C_j = 1$. We also assume that there is a universal constant Δ such that $0 \leq C_j \leq \Delta$ for all j. We assume that the total cost, n, to be spent on sample selection is fixed in advance and is an integer. The noninteger case can be easily reduced to the integer case by replacing n by the greatest integer contained in n. We assume that $\min(n, N - n)$ is at least as large as 2Δ. This last condition is a technicality designed to ensure that the sample size is at least 2 and at most $(N - 2)$.

With this preliminary background, we now turn to the three most common forms of simple random sampling.

2. Simple random sampling without replacement (SRSWOR)

Under this scheme, units are drawn one-by-one with equal probabilities without replacement and sampling stopped when the total sample size reaches a preassigned number n. The object of the sampling scheme is to select n distinct population units in the sample. The actual cost of selecting such a sample is a random variable; the expected cost of sample selection is nevertheless n.

Under SRSWOR, the customary estimator of the population mean $\bar{Y} = N^{-1} \Sigma Y_j$ is the sample mean $\bar{y}_n = n^{-1} \Sigma y_i$, where y_i denotes the Y-value of the ith sample unit, $1 \leq i \leq n$. It is unbiased, and admissible (cf. Joshi, 1961, p. 1660). The unbiasedness of $\bar{y}_n$ is easily established by invoking the symmetric nature of SRSWOR which implies that $E\bar{y}_n = k\bar{Y}$. Letting $Y_j = 1$ for all j shows that $k = 1$.

An elementary proof of the variance formula for $\bar{y}_n$ can be given along similar lines by noting that the complement of an SRSWOR sample of size n is in itself an SRSWOR sample of size $N - n$. So if $\Sigma_s y_i$ and $\Sigma_{s\sim} y_i$ respectively denote totals of Y-values based on the sample s and its complement $s^\sim$, we have

$$\sum_s y_i = \sum Y_j - \sum_{s\sim} y_i \tag{2.1}$$

in which the population total ΣY_j is a constant. So $V(\Sigma_s y_i) = V(\Sigma_{s\sim} y_i)$. The symmetric nature of SRSWOR implies that $V(\Sigma_s y_i) = P_2(n)$, a quadratic function of the sample size n. Clearly $P_2(n) = 0$ when $n = N$, and $= N^{-1} \Sigma (Y_j - \bar{Y})^2$ when $n = 1$. Also $V(\Sigma_s y_i) = V(\Sigma_{s\sim} y_i)$ implies that $P_2(n) = P_2(N - n)$. These three constraints uniquely identify $P_2(n)$ to be $n(N - n) \Sigma_j (Y_j - \bar{Y})^2 / N(N - 1)$. Therefore

$$V(\bar{y}_n) = \frac{P_2(n)}{n^2} = \frac{(N-n)}{Nn} S^2, \tag{2.2}$$

where

$$S^2 = (N - 1)^{-1} \sum_j (Y_j - \bar{Y})^2.$$

An unbiased estimator of $V(\bar{y}_n)$ is given by

$$v(\bar{y}_n) = \frac{(N-n)}{nN} s^2 \tag{2.3}$$

where $s^2 = (n-1)^{-1} \Sigma (y_i - \bar{y}_n)^2$ denotes the sample variance. An elementary proof of the unbiasedness of $v(\bar{y}_n)$ is based on the following representation

$$s^2 = \frac{1}{2n(n-1)} \sum_h \sum_i (y_h - y_i)^2, \tag{2.4}$$

in which $1 \leq h \leq n$ and $1 \leq i \leq n$.

The symmetric nature of SRSWOR implies that $Es^2 = E(y_1 - y_2)^2/2$. Letting $n = N$ now yields the identity $ES^2 = E(y_1 - y_2)^2/2$, where $S^2 = (N-1)^{-1} \Sigma (Y_j - \bar{Y})^2$. Since S^2 is a parameter, $S^2 = ES^2$. So

$$Es^2 = E(y_1 - y_2)^2/2 = ES^2 = S^2.$$

Since we will be interested in the asymptotic efficiency of $\bar{y}_n$, it is useful at this point to recall the following elegant result.

THEOREM 2.1. *Suppose that* $\lim n(N-n)N = \infty$. *Then the sample mean* $\bar{y}_n$ *is asymptotically normally distributed with parameters* $(E\bar{y}_n, V(\bar{y}_n))$ *if and only if for each* $\varepsilon > 0$,

$$\lim \left[\sum_{\tau(\varepsilon)} (Y_j - \bar{Y})^2 \bigg/ \sum (Y_j - \bar{Y})^2 \right] = 0 \tag{2.5}$$

where

$$\tau(\varepsilon) = \left\{ j : (Y_j - \bar{Y})^2 > \varepsilon \, \frac{n}{N} \left(1 - \frac{n}{N}\right) \sum (Y_j - \bar{Y})^2 \right\}. \tag{2.6}$$

It is worth noting that 'lim' in (2.5) is meant to be taken with respect to a sequence of populations as explained in Section 1. Although this sequence has not been mentioned explicitly in the above discussion, it will, as earlier noted, always be there in the back of our minds.

This remarkable Lindeberg type condition for the asymptotic normality of the sample mean under SRSWOR is due to Erdös and Rényi (1959) and Hájek (1960), and furnishes a complete solution to the problem of asymptotic normality in SRSWOR. We shall refer to this condition in (22.5) as the Erdös–Rényi–Hájek condition. Hájek (1960) furnished an ingenious proof of this asymptotic normality by showing that SRSWOR is, in a certain sense, equivalent to the following binomial sampling scheme:

First perform N independent Bernoulli trials with a constant probability of success $P = n/N$. Let $i_1, i_2, \ldots, i_v$ denote the trials which result in a success, $1 \leq i_1 < i_2 < \cdots < i_v \leq N$. The observed sample then consists of the population units with labels $i_1, i_2, \ldots, i_v$.

Note than under Hájek's sampling scheme, the cost of sample selection and the number of distinct sample units are both random variables. However, the expected number of distinct units and the expected cost of sample selection both continue to be n for this sampling scheme as well. Since Hájek's scheme is the result of

N independent Bernoulli trials, he succeeded in showing that the problem of existence of a limiting distribution for the sample mean in SRSWOR can be reduced to a problem concerning the existence of a limiting distribution for sums of N independent random variables. Since conditions concerning the existence of a limiting distribution for sums of independent random variables are all well-known, Hájek's approach provided an elegant and complete solution to this problem. The crux of Hájek's approach is a remarkable imbedding lemma. To describe this lemma, consider the following sampling scheme. Let $s_v = \{u_1, \ldots, u_v\}$ denote the observed sample of size v obtained under Hájek's binomial sampling scheme. If $v \geq n$, then select an SRSWOR subsample s_n of size n from s_v treating s_v as a population in itself. On the other hand, if $v < n$, then draw an SRSWOR sample $s_{\tilde{n}}$ of size $(N - n)$ from $s_{\tilde{v}}$. Let s_n denote the complement of $s_{\tilde{n}}$. Then the marginal distribution of s_n is that of an SRSWOR sample of size n from the whole population. This imbedding of both SRSWOR and his binomial sampling scheme into a larger sampling scheme enabled Hájek to establish the following important result.

LEMMA 2.1. *Under the above sampling scheme, the following inequality holds.*

$$\frac{E(\bar{y}_n - \bar{y}_*)^2}{V(\bar{y}_n)} \leq \sqrt{\left(\frac{1}{n} + \frac{1}{(N-n)}\right)} \tag{2.7}$$

where $\bar{y}_n$ *denotes the sample mean based on an SRSWOR of size n, and*

$$\bar{y}_* = (v/n)\bar{y}_v + (1 - v/n)\bar{Y}$$

in which $\bar{y}_v$ denotes the sample mean based on the corresponding Hájek's binomial sampling scheme. Note that $\bar{y}_*$ can be thought of as an estimator of $\bar{Y}$ based on the binomial sampling scheme.

The implications of the above inequality are quite deep. As $n \to \infty$ and $(N - n) \to \infty$, it follows from (2.7) that Hájek's measure of disparity between $\bar{y}_n$ and $\bar{y}_*$ approaches zero asymptotically. Consequently $\lim[V(\bar{y}_n)/V(\bar{y}_*)] = 1$ so that $\bar{y}_n$ and $\bar{y}_*$ are asymptotically equally efficient. Besides, the respective conditions for the asymptotic normality of $\bar{y}_n$ and $\bar{y}_*$ must also be the same. Since $\bar{y}_*$ is based on a sum of independent Bernoulli trials, the Lindberg condition for its asymptotic normality can be easily worked out. The simplicity with which Hájek thus established the asymptotic normality of $\bar{y}_n$ in SRSWOR is truly remarkable. Unfortunately this simplicity and the potential of Hájek's imbedding approach in the study of asymptotics have been overlooked in the sampling literature.

To conclude this section we digress briefly and say a few words about the nature of the Erdös–Rényi–Hájek necessary and sufficient condition (2.5) for the asymptotic normality of the sample mean. This condition is a scale-free measure of the uniform summability (integrability) of the second moment. It can also be thought of as a scale-free measure of sum of squares due to the tails or the outliners. In broad terms if the tails of the population make a negligible contribution to the total sum of squares, then one can expect the sample mean in SRSWOR to be asymptotically normally distributed. It can be easily seen that (2.5) holds if for example (i) the Y-values are uniformly bounded by a certain multiple of the population standard deviation, or (ii) the kurtosis of the population is uniformly bounded, and so forth. Similarly in sampling from a binary population with zero–one values, (2.5) holds if (i) the sampling fraction n/N lies strictly between 0 and 1, or (ii) population proportion of 1's lies strictly between 0 and 1. None of these conditions are however necessary for (2.5). In case the sampling fraction lies strictly between 0 and 1, the condition (2.5) admits the simpler form

$$\lim [\max_{j \leq N}(Y_j - \bar{Y})^2 / \sum (Y_j - \bar{Y})^2] = 0.$$

This last condition is known as the Noether condition.

We turn now to simple random sampling with replacement.

3. Simple random sampling with replacement (SRSWR)

Under this sampling scheme units are drawn one-by-one with equal probabilities with replacement and sampling stopped when the total sample size reaches n. The observed sample can now be expressed as

$$s = (u_1, u_2, \ldots, u_n) \tag{3.1}$$

in which $u_1, u_2 \ldots$ are all i.i.d. variates with common distribution $P(u_1 = U_j) = N^{-1}$, $1 \leq j \leq N$.

Since SRSWR is carried out with replacement, generally not all of the n units in the sample are distinct. Let $T_0 = \{u_{(1)}, \ldots, u_{(v)}\}$ denote the set of v distinct sample units, $1 \leq v \leq n$. Then T_0 is known to be a minimal sufficient statistic (Basu and Ghosh, 1967; Pathak, 1976). Based on T_0, an admissible and unbiased estimator of the population mean is the following sample mean (Basu, 1958; Raj and Khamis, 1958):

$$\bar{y}_{(v)} = v^{-1} \sum_{(i)} y_{(i)}. \tag{3.2}$$

This estimator is however quite different from the classical estimator

$$\bar{y}_n = n^{-1} \sum_i y_i \tag{3.3}$$

which is the overall average of all the n units in the sample. In fact it is easy to show that $E(\bar{y}_n | T_0) = \bar{y}_{(v)}$. So by the Rao–Blackwell theorem it follows that $\bar{y}_{(v)}$ has a smaller variance.

Often while comparing SRSWOR and SRSWR for the purpose of estimating the population mean, one is tempted to compare $\bar{y}_n$ of SRSWOR with $\bar{y}_n$ of SRSWR and thus claim that SRSWOR is superior since $V(\bar{y}_n | \text{WOR}) < V(\bar{y}_n | \text{WR})$. Of course we all know now that such a comparison is totally erroneous for two reasons. First, since for samples of the same size, SRSWOR is more expensive than SRSWR, such a comparison makes no adjustment for unequal sampling costs. Second, the sample mean $\bar{y}_n$ in SRSWR is inadmissible. So we are comparing an inadmissible estimator in SRSWR with an admissible estimator in SRSWOR. For a valid comparison of these two schemes, we should really be comparing the sample mean based on distinct units in SRSWR with the sample mean in SRSWOR and that too only after making the necessary adjustments for the differing sampling costs under the two schemes. When such a cost-adjusted comparison is made, we find that SRSWR and SRSWOR are both asymptotically equally cost-efficient. This asymptotic comparison is based on the following results.

LEMMA 3.1 (Lanke, 1975). *Let v denote the number of distinct units in an SRSWR sample of size n drawn from a population of size N. Then*

$$N[1 - \exp(-n/N)] \leqslant Ev \leqslant 1 + N[1 - \exp(-n/N)] \qquad (3.4)$$

and, for $N \geqslant 3$,

$$V(v) \leqslant 4N \exp(-n/N)[1 - \exp(-n/N)]. \qquad (3.5)$$

Now consider the following sampling scheme. First draw an SRSWR sample of size n. Let s_v denote the set of v distinct units so drawn. Let $[Ev]$ denote the greatest integer contained in $1 + N(1 - \exp(-n/N))$. If $v \geqslant [Ev]$, draw an SRSWOR sample $s_{[Ev]}$ of size $[Ev]$ from s_v. If $v < [Ev]$, draw an SRSWOR subsample $s_{\widetilde{[Ev]}}$ of size $N - [Ev]$ from $s_{\widetilde{[Ev]}}$. Let $s_{[Ev]}$ be the complement of $s_{\widetilde{[Ev]}}$. Then it is easy to check that $s_{[Ev]}$ is an SRSWOR sample of size $[Ev]$ from the given population.

LEMMA 3.2. *Under the above sampling scheme, for $N \geqslant 3$,*

$$\frac{E\left[\left\{(\bar{y}_{[Ev]} - \bar{Y}) - \frac{v}{[Ev]}(\bar{y}_{(v)} - \bar{Y})\right\}^2\right]}{V(\bar{y}_{[Ev]})}$$

$$\leqslant \frac{c}{\sqrt{[N \exp(-n/N)\{1 - \exp(-n/N)\}]}} \qquad (3.6)$$

where c is a constant and $\bar{y}_{[Ev]}$ denotes the sample mean based on an SRSWOR sample $s_{[Ev]}$ of size $[Ev]$.

(For a proof see Pathak, 1982, Lemma 3.2, Eq. (3.9), p. 571.)

THEOREM 3.1. *Let* $n \to \infty$ *and* $N \to \infty$ *so that*

$$\lim N \exp(-n/N)[1 - \exp(-n/N)] = \infty. \tag{3.7}$$

Then $\bar{y}_{(v)}$ *is asymptotically normally distributed with parameters* $(\bar{Y}, V(\bar{y}_{[Ev]}))$ *if and only if*

$$\lim \sum_{Q(\varepsilon)} (Y_j - \bar{Y})^2 \Big/ \sum (Y_j - \bar{Y})^2 = 0 \tag{3.8}$$

where

$$Q(\varepsilon) = \{j: (Y_j - \bar{Y})^2 > \varepsilon^2 \exp(-n/N)\{1 - \exp(-n/N)\} \sum (Y_j - \bar{Y})^2\}.$$

Lemma 3.1 and (3.7) imply that $v/[Ev]$ converges to one in probability. Consequently under (3.7), $(\bar{y}_{(v)} - \bar{Y})$ and $\{v/[Ev]\}(\bar{y}_{(v)} - \bar{Y})$ have identical asymptotic distributions and conditions for their asymptotic normality are also the same. Equation (3.6) implies that $\{v/[Ev]\}(\bar{y}_{(v)} - \bar{Y})$ and $(\bar{y}_{[Ev]} - \bar{Y})$ are asymptotically equivalent in the same preceding sense. The main assertion of the theorem now follows from Theorem 2.1.

The implication of Theorem 3.1 is quite significant. It shows that if we compare SRSWR of size n with SRSWOR of equivalent size $[Ev] \sim N[1 - \exp(-n/N)]$, then the sample means based on distinct units under the two schemes are equally efficient. Thus if the cost of sampling is assumed proportional to the number of distinct units in the sample then SRSWR and SRSWOR are asymptotically equally cost-efficient.

It would be pertinent now to say a few words about the meaning of the condition in (3.7). First let us note that (3.7) holds if and only if $\lim N[1 - \exp(-n/N)] = \infty$ and $\lim N \exp(-n/N) = \infty$. Since $Ev \sim N[1 - \exp(-n/N)]$, the condition in (3.7) says that for asymptotics to go through the expected number of distinct units in the sample as well as in its complement must approach infinity. Also observe that when $\lim(n/N)$ exists, (3.7) is equivalent to either (i) $\lim(n/N) = \alpha$, $0 \leq \alpha < \infty$, or (ii) $\lim(n/N) = \infty$ and $\lim N \exp(-n/N) = \infty$.

4. Fixed cost simple random sampling (SRSFC)

Under SRSFC sample units are drawn sequentially with equal probabilities and without replacement (wor) until the cumulative cost of sampling first exceeds the preassigned amount n. This version of simple random sampling eliminates the randomness of the total cost of sampling and ensures the collection of maximum possible information at a given cost. Under this sampling scheme, we record the observed sample as:

$$s = (u_1, u_2, \ldots, u_v), \tag{4.1}$$

where $u_i = (y_i, c_i)$ is the ith sample unit with y_i being its Y-variate value and c_i its cost of selection. For reasons that will be clear from the discussion in the sequel, we define the stopping variable v as follows: $v = r$ if and only if $\Sigma_{i \leq r} c_i \leq n$ and $\Sigma_{i \leq r+1} c_i > n$. The random variable v is not a stopping variable in the strict sense since the event $\{v = r\}$ is not just a function of $u_1, \ldots, u_r$, but also depends on u_{r+1}. Nevertheless, the new definition does indeed provide an added bonus. Under this stopping rule, the conditional distribution of $u_1, \ldots, u_v$ given v is a symmetric function of $u_1, \ldots, u_v$. This property allows us to imbed SRSFC and SRSWOR into a larger sampling scheme. We are then able to get very good estimates of Hájek's disparity between these two schemes. These estimates furnish a complete solution to the problem of asymptotics in SRSFC. In what follows we shall refer to 'v' as a *symmetric stopping rule*.

In SRSFC, an unbiased estimator of the population mean $\bar{Y}$ is given by the sample mean $\bar{y}_v = (1/v) \Sigma_i y_i$. Similarly an unbiased estimator of the population variance $S^2 = (N-1)^{-1} \Sigma (Y_j - \bar{Y})^2$ is the corresponding sample variance $s_v^2 = (v-1)^{-1} \Sigma_i (y_i - \bar{y}_v)^2$. Thus although the sample size in SRSFC is a random variable, the customary estimators of the population mean and the population variance continue to remain unbiased estimators. Admissibility of $\bar{y}_v$ is also preserved. A more interesting result about SRSFC is regarding the estimation of the variance of the sample mean. We all know that in SRSWOR, the customary unbiased estimator of the variance of the sample mean is given by $(n^{-1} - N^{-1})s^2$ in which s^2 is the sample variance. Now even though in SRSFC the sample size v is random, this formula with n replaced by v continues to be an unbiased estimator of the variance of the sample mean despite the fact that the actual form of $V(\bar{y}_v)$ is quite complicated (Pathak, 1976). Perhaps the most intriguing aspect of SRSFC is that under mild restrictions the necessary and sufficient condition for the asymptotic normality of the sample mean under SRSFC is identical to that for the sample mean under SRSWOR.

We turn now to a few main results concerning the first passage behavior of SRSFC. These play a central role in the study of asymptotics. As noted earlier, for simplicity, we assume that the cost-characteristic has been suitable normalized so that $\bar{C} = N^{-1} \Sigma C_j = 1$. (Cf. Mitra and Pathak (1984) for proofs of these results.)

LEMMA 4.1. *Let v_n denote the sample size for SRSFC of cost n. Then*

$$|Ev_n - n| \leq 1 + \Delta \tag{4.2}$$

where Δ is such that $\max C_j \leq \Delta$.

In spirit the above lemma is much like Wald's identity for martingales. But unlike the proof of Wald's identity, the proof of this lemma relies heavily on the conditional itnerchangeability of $u_1, \ldots, u_v$ given v, and this is one of the underlying reasons for using a slightly nontraditional definition of the stopping rule in SRSFC.

LEMMA 4.2.

$$V(v_n) \leq 8\Delta n + un^2 \frac{S_c^2}{N} \tag{4.3}$$

where $S_c^2 = (N-1)^{-1} \Sigma (C_j - 1)^2$.

A consequence of the preceding lemmas is that v_n/Ev_n converges to one in probability. We can thus expect SRSFC of cost n to behave must like an SRSWOR of size $Ev_n = n$. We shall furnish a precise justification for this equivalence shortly. A second point worth discussing here is the invariant nature of SRSFC under complementation. It is well-known that the complement of an SRSWOR sample of size n is an SRSWOR sample of size $(N - n)$. Therefore, it is natural to ask if this property is also shared by SRSFC. In other words, does the complement of an SRSFC sample of cost n behave much like an SRSFC sample of cost $(N - n)$? This is not true in general. Indeed if it were true, we should then be able to establish a dual of Lemma 4.2 with n replaced by $(N - n)$ on the right side of (4.3). This can be done only if we assume that the cost-characteristic assumes strictly positive values.

LEMMA 4.3. *Suppose that the cost-characteristic assumes strictly positive values. Then*

$$V(v_n) \leq 20 \left[\Delta \min(n, N - n) + \frac{S_c^2}{N} (\min(n, N - n))^2 \right] \tag{4.4}$$

This lemma provides an important result for establishing asymptotic normality under SRSFC, as well as the asymptotic equivalence of SRSFC and SRSWOR when $n \to \infty$ and $(N - n) \to \infty$. The weakness of the lemma is that it is only valid under the added requirement that the cost-characteristic be strictly positive. There are, however, a number of applications in which this requirement is not met. The simplest example of this kind is inverse simple random sampling. Under this scheme units are drawn sequentially (wor) until a preassigned number n of units with a specified trait are included in the sample. This scheme is a special case of SRSFC when the cost is taken to be the indicator of the given trait, i.e., $C_j = 1$ if the unit possesses the trait and $= 0$ otherwise. Asymptotic normality in these schemes requires somewhat stronger constraints. It is worth adding here that one cannot treat inverse simple random sampling as a special case of the negative binomial distribution, doing so tantamounts to the assumption that $\lim(n/N) = 0$.

We turn now to the discussion of asymptotic normality in SRSFC. Let R_c denote an SRSFC sampling experiment of selecting v units at a total cost of n from the given population P, and R_s refer to an SRSWOR sampling experiment of selecting n units from P. Let $\bar{y}_v = (1/v) \Sigma_c y_i$ and $\bar{y}_n = (1/n) \Sigma_s y_i$ denote the sample means under SRSFC and SRSWOR respectively. The following random experiment provides a good estimate of Hájek's disparity between $\bar{y}_v$ and $\bar{y}_n$.

A random experiment

(1) First draw an SRSWOR sample of size n from P as follows: Let $u = (u_1, \ldots, u_N)$ be a random permutation of P. Let s_n denote the observed SRSWOR based on the first n coordinates of u and let $S(n)$ denote the total cost of selecting this sample.

(2) If $S(n) > n$, select an SRSFC subsample s_v of cost n from s_n, treating s_n as a population in its own right.

(3) If $S(n) \leq n$, select sequentially additional units from the remaining units in P until an SRSFC sample s_v of cost n has been selected.

(4) Repermute the v units in s_v at random. For notational simplicity, denote the repermuted sample by the same symbol s_v.

Under this experiment given s_n and that $s_v \subset s_n$ and $s_v \neq s_n$, s_v is also an SRSFC sample of cost n from s_n, while given s_v and that $s_n \subset s_v$, s_n is an SRSWOR sample of size n from s_v. This experiment imbeds both SRSWOR and SRSFC into a larger sampling scheme, thereby enabling us to establish the following key results.

LEMMA 4.4. *Under the given experiment,*

$$E(\bar{y}_v - \bar{y}_n)^2 = \tfrac{1}{2} E \left| \frac{1}{v} - \frac{1}{n} \right| (y_1 - y_2)^2 . \tag{4.5}$$

LEMMA 4.5. *Under the given experiment,*

$$H(\bar{y}_v, \bar{y}_n) = \frac{E(\bar{y}_v - \bar{y}_n)^2}{V(\bar{y}_n)} \leq 10\Delta[\Delta + \sqrt{V(v)}] \left[\frac{1}{n} + \frac{1}{(N-n)} \right] . \tag{4.6}$$

It is clear from the above lemma that Hájek's disparity between $\bar{y}_v$ of SRSFC and $\bar{y}_n$ of SRSWOR can be made to approach zero asymptotically provided as $n \to \infty$ and $(N - n) \to \infty$, $\lim V(v)/n^2 = 0$ and $\lim V(v)/(N - n)^2 = 0$. Of these two conditions, the first says that v/Ev must converge to one in the mean. Lemma 4.2 implies that all SRSFC satisfy this condition. The second condition is, however, more restrictive. Roughly speaking, it can be interpreted to mean that the complement of an SRSFC sample of size (cost) n must also behave like an SRSFC sample of size $(N - n)$. In other words SRSFC sample must display the same sort of duality which SRSWOR samples do, namely that the complement of an SRSWOR sample of size n is an SRSWOR sample of size $(N - n)$. This condition is more restrictive. For example, if we assume that the cost-characteristic is strictly positive, Lemma 4.3 holds. Then both $V(v)/n^2$ and $V(v)/(N - n)^2$ approach zero. As noted earlier, inverse simple random sampling is a special case of SRSFC if we allow the cost-characteristic to also assume zero values. For such SRSFC schemes $\lim H(\bar{y}_v, \bar{y}_n) \neq 0$ without added restrictions on the rate of growth of n and $(N - n)$. A simple condition that suffices is to require that as

$n \to \infty$ and $(N - n) \to \infty$, $\lim n/(N - n)^2 = 0$. It is easily seen that this would be true if for example the sampling fraction n/N is strictly bounded by one.

Thus through the preceding lemmas, we are able to establish the following main results.

THEOREM 4.1. *Let* $\lim n = \lim (N - n) = \infty$ *and suppose that the cost-characteristic is strictly positive and bounded. Then under SRSFC of cost n, the sample mean $\bar{y}_v$ is asymptotically normally distributed if and only if the Erdös–Rényi–Hájek condition of Theorem* 2.1 *holds.*

THEOREM 4.2. *Suppose that the cost-characteristic is nonnegative and bounded. Then the conclusion of the above theorem goes through if* $\lim n = \lim (N - n) = \infty$ *and* $\limsup (n/N) < 1$.

For completeness, it is worthwhile now to mention the connection between above results and those of Gordon (1983) and Holst (1973), both of whom obtain results very similar to those of Theorems 4.1 and 4.2. However, their techniques are quite different and involve intricate analyses with Fourier transforms and order-statistics of independent exponentially distributed waiting times. Gordon's results apply to the case: $0 < \Delta^{-1} \leq C_j \leq \Delta$, while Holst's to the case $0 < \Delta^{-1} \leq C_j \leq \Delta$ and $0 < \liminf (n/N) \leq \limsup (n/N) < 1$.

It should be clear from the foregoing discussion that asymptotic normality in SRSFC is no more restrictive than in SRSWOR. Besides as the cost of sampling gets larger and larger, the cost-adjusted relative efficiency of SRSFC versus SRSWOR approaches one. The preceding theorems provide a truly rigorous justification for results of this nature first anticipated by Basu (1958) and others. They show that purely from the viewpoint of cost-effectiveness, the three forms of simple random sampling, namely SRSWOR, SRSWR, and SRSFC, are all asymptotically equally cost-efficient.

Acknowledgement

This work was partially supported by the National Science Foundation Grants INT-8020450 and DMS-8703798.

References

Basu, D. (1958). On sampling with and without replacement. *Sankhyā* **20**, 287–294.
Basu, D. and Ghosh, J. K. (1967). Sufficient statistics in sampling from a finite universe. *Proc. 36th Session Int. Statist. Inst.* 850–859.
Erdös, P. and Rényi, A. (1959). On the central limit theorem for samples from afinite population. *Publ. Math. Inst. Hung. Acad. Sci. Ser. A* **4**, 49–61.
Gordon, L. (1983). Successive sampling in large finite populations. *Ann. Statist.* **11**, 702–706.
Hájek, J. (1959). Optimum strategy and other problems in probability sampling. *Casopis Pest. Mat.* **84**, 387–425.

Hájek, J. (1960). Limiting distributions in simple random sampling from a finite population. *Publ. Math. Inst. Hung. Acad. Sci. Ser. A* **5**, 361–374.

Holst, L. (1973). Some limit theorems with applications in sampling theory. *Ann. Statist.* **1**, 644–658.

Joshi, V. M. (1961). Admissibility and Bayes estimation in sampling finite populations – IV. *Ann. Math. Statist.* **37**, 1958–1970.

Lanke, J. (1975). *Some contributions to the theory of survey sampling*. Thesis, Department of Mathematics and Statistics, Univ. of Lund, Sweden.

Mitra, S. K. and Pathak, P. K. (1984). The nature of simple random sampling. *Ann. Statist.* **12**, 1536–1542.

Pathak, P. K. (1962). On simple random sampling with replacement. *Sankhyā A* **24**, 287–302.

Pathak, P. K. (1976). Unbiased estimation in fixed cost sequential sampling schemes. *Ann. Statist.* **4**, 1012–1017.

Pathak, P. K. (1982). Asymptotic normality of the average of distinct units in simple random sampling with replacement. In: G. Kallianpur et al., eds., *Essays in Honour of C. R. Rao*. North-Holland, Amsterdam, 567–573.

Raj, D. and Khamis, S. H. (1958). Some remarks on sampling with replacement. *Ann. Math. Statist.* **29**, 550–557.

Seth, G. R. and Rao, J. N. K. (1964). On the comparison of simple random sampling with and without replacement. *Sankhyā A* **26**, 85–86.

On Single Stage Unequal Probability Sampling

V. P. Godambe and M. E. Thompson

1. Introduction

Unequal probability sampling is a term used to describe a probability sampling design under which the population units have differing probabilities of inclusion. A widely used and accepted form of unequal probability sampling is stratified random sampling with allocation not proportional to the stratum sizes. The usual justification for this practice and other single stage unequal probability sampling procedures is increased efficiency in the point estimation of certain population quantities. That is, such procedures produce with high probability samples from which the population quantities can be estimated with precision, while at the same time retaining the advantage of liberal randomization, namely the neutralization of the effects of nuisance parameters in the model and the provision of design based support for estimates and inferences.

In this article we discuss uses and justifications for single stage unequal probability sampling. We begin by examining the role of randomization in inference. We illustrate this discussion with simple random and stratified random sampling. We then consider the more sophisticated (and controversial) unequal probability sampling procedures sometimes recommended for use with certain regression models for the characteristic of interest. There follows a discussion of monetary unit sampling as used in auditing. We conclude with a brief historical note.

2. Use of randomization

One primary purpose that random sampling serves is that of the elimination of nuisance parameters. This can be illustrated first by an example from the design of experiments. Suppose there are M plots and two treatments. Considering all possible assignments of treatments to plots provides 2^M possible yield vectors $[y_1(t_1), \ldots, y_M(t_M)]$, each t being 1 or 2. Consider a model under which $E(y_i(t)) = \mu + a_i + b_t$, where a_i and b_t are the plot and treatment effects respectively and E denotes expectation. Then the basic problem is to find a 'sampling scheme' by which to choose one of the 2^M possible yield vectors so that the estimation of the treatment contrast $b_2 - b_1$ can be done efficiently, or equivalently so that the elimination of the nuisance parameters $a_1, \ldots, a_M$ can be done effectively.

In an analogous example, which is closer to the sampling problem, suppose that a statistician can observe n out of N random variates y_i distributed independently and normally, with $Ey_i = \theta + \alpha_i$ and $\text{Var } y_i = \sigma^2$, $i = 1, \ldots, N$. The α_i are unknown nuisance parameters with $\Sigma_{i=1}^N \alpha_i = 0$. From n observations the statistician is to obtain a point estimate for θ. If the α_i were all zero it is clear that the best estimate under most optimality criteria would be

$$\hat{\theta}_0 = \frac{1}{n} \sum_{i \in s} y_i,$$

where s is the subset of $\{1, \ldots, N\}$ which the statistician has chosen to observe. If the α_i were non-zero but all known, the best estimate would be

$$\hat{\theta}_1 = \frac{1}{n} \sum_{i \in s} (y_i - \alpha_i).$$

Under this situation one can ask the following question:

Can the statistician design a method to choose a subset s of the variates so that the two estimates $\hat{\theta}_0$ and $\hat{\theta}_1$ would tend to agree for all reasonable or likely values of $\alpha_1, \ldots, \alpha_N$?

That is, we need a design which tends to produce 'balanced' samples, for which

$$\hat{\theta}_0 - \hat{\theta}_1 = \frac{1}{n} \sum_{i \in s} \alpha_i$$

is approximately 0. For these samples the point estimate $\hat{\theta}_0$ would be approximately unbiased for θ under the given probability model for the y_i.

Suppose we knew, for example, that α_i tended to be positive for small values of i and negative for large values of i. Then it would make sense to try to choose the sample so that both high and low values of i were always represented.

In the absence of any such knowledge about the way in which the α_i are produced or arranged, choosing s by simple random sampling would seem to give the best chance of providing samples which are approximately balanced in α_i much of the time. For under this design

$$E\left(\frac{1}{n} \sum_{i \in s} \alpha_i\right) = \frac{1}{N} \sum_{i=1}^N \alpha_i = 0$$

and

$$\text{Var}\left(\frac{1}{n} \sum_{i \in s} \alpha_i\right) = \frac{1}{n}\left(1 - \frac{n}{N}\right) \frac{1}{N-1} \sum_{i=1}^N \alpha_i^2;$$

the latter can be small if n is large and/or the α_i are small.

We can turn the preceding example into a typical sampling problem simply by supposing that instead of estimating θ we are trying to estimate

$$\bar{Y} = \frac{1}{N}\sum_{i=1}^{N} y_i,$$

given the y_i for i in the sample s. In fact, under the model

M1:   (i) $y_1, \ldots, y_N$ are distributed independently,
         (ii) $Ey_i = \theta + \alpha_i$, $i = 1, \ldots, N$,   where $\sum_1^N \alpha_i = 0$,
         (iii) Var $y_i = \sigma^2$, $i = 1, \ldots, N$,

these two problems are essentially equivalent for N large by the laws of large numbers. In the rest of this section, let us equate θ and $\bar{Y}$.

The second main purpose of randomization in sampling is in providing a frequency interpretation to expressions of inference such as interval estimates. To continue the example above, suppose that the α_i are assumed to be small and reasonably uniform but with no particular relationship to one another. As indicated, simple random sampling seems to be a suitable way of obtaining samples which are approximately balanced on α_i. Now according to the theory of simple random sampling, an approximate 95% confidence interval for $\bar{Y}$ (from a sample of size $n > 50$, assuming a reasonably symmetric y_i distribution) would be

$$\bar{y}_s \pm 1.96 \sqrt{\frac{1}{n} s_y^2} \tag{2.1}$$

where

$$\bar{y}_s = \frac{1}{n}\sum_{i \in s} y_i$$

and

$$s_y^2 = \frac{1}{n-1} \sum_{i \in s}(y_i - \bar{y}_s)^2.$$

(We ignore the finite population correction since N is large.) This is precisely the interval estimate which would be used for θ under the model

M2, which is M1 above with $\sum_{i=1}^N \alpha_i = 0$ replaced by the assumption that $\alpha_1, \ldots, \alpha_N$ are independent and normal with mean 0 and unknown variance σ_0^2.

Thus the same inference expression has validity under the model M2 (and in fact also under departures from the assumed normal distribution because of the central limit effect). Note that the model M2 is consistent with the prior notion that the α_i are roughly uniform in size but with unrelated values. Thus the simple random sampling design endows the (now model based) interval (2.1) with 'appropriate' long run frequency properties.

In summary, it may be pointed out that simple random sampling and some form of, say, judgement sampling can both lead to the same sample and the same y-values. However, under the model M1 with no distributional assumption on the α_i's, only with the first of these can the approximate balance of the sample and the approximate unbiasedness of $\hat{\theta}_0$ be quantitatively assessed. Only with the probability sampling design can the interval estimate (2.1), based on M2 which requires a distributional assumption on $\alpha_1, \ldots, \alpha_N$, be supported by a long run frequency interpretation.

3. Estimating finite population means and totals

Let us now turn full attention to the primary problem of finite population sampling theory, the estimation of totals and means.

Let $\mathscr{P} = \{1, \ldots, N\}$ be a finite population. Let y_i be the value of a variate y associated with unit i, so that the values for the population $\mathscr{P}$ form an array

$$y = (y_1, \ldots, y_N).$$

The population total may be denoted

$$Y = \sum_{i=1}^{N} y_i,$$

while the population mean is

$$\overline{Y} = Y/N.$$

If $\mathscr{S} = \{s\}$ is the collection of all samples s or subsets s of $\mathscr{P}$, a probability sampling design is a probability function p on $\mathscr{S}$. Thus $p(s) \geq 0$ for all s and $\sum_{s \in \mathscr{S}} p(s) = 1$. The observation from the sampling experiment is

$$\chi_s = \{(i, y_i); i \in s\},$$

and the sampling design p yields a distribution (indexed by y) for any function $e(\chi_s)$. The function or estimator $e(\chi_s)$ is called p-unbiased for a population function $\phi(y)$ if

$$Ee(\chi_s) = \phi(y)$$

for all y, where E denotes expectation with respect to p. Thus under simple random sampling, as we have seen above, since

$$E\bar{y}_s = \bar{Y},$$

$\bar{y}_s$ is a p-unbiased estimator of $\bar{Y}$.

Given a sampling design p, the inclusion probability for unit i is given by

$$\pi_i = \sum_{s:\, i \in s} p(s).$$

The design p yields unequal probability sampling if the π_i are not all equal.

The Horvitz–Thompson estimator

$$\hat{Y} = \sum_{i \in s} \frac{y_i}{\pi_i}$$

is easily seen to be an unbiased estimator of the total Y for any sampling design. It evidently gives greater 'weight' to units i in the sample for which the probability of inclusion is smaller. In fact the term y_i/π_i may be interpreted loosely as an estimate of a total over units in the population 'represented by' the individual i; a unit from a sparsely sampled part of the population would usually have to represent a relatively large number of other units.

4. Stratified random sampling

Let us now suppose that $\mathscr{P} = \mathscr{P}_1 \cup \mathscr{P}_2$, where $\mathscr{P}_1$ and $\mathscr{P}_2$ are two well delineated strata in the population with sizes N_1, N_2. Assume that for $i \in \mathscr{P}_j$ the y_i are independent and identically distributed with means $\theta_j + \alpha_i$ and variances σ_{0j}^2, $j = 1, 2$. Suppose also that

$$\sum_{i=1}^{N} \alpha_i = 0.$$

Otherwise the values of the α_i and θ_j are unknown. Call this model MST1.

Suppose a sample $s = s_1 \cup s_2$ is taken, where s_j is from $\mathscr{P}_j$ and has size n_j, $j = 1, 2$.

Clearly if the α_i were all 0, the 'best' estimator (or predictor) of Y in the sense of the model would be

$$\hat{Y}_{st0} = N_1 \bar{y}_1 + N_2 \bar{y}_2$$

where $\bar{y}_1$ and $\bar{y}_2$ are the sample means for s_1 and s_2 respectively. If the α_i were not 0 but known the best estimator or predictor would be

$$\hat{Y}_{st1} = N_1(\bar{y}_1 - \bar{\alpha}_1) + N_2(\bar{y}_2 - \bar{\alpha}_2)$$

where $\bar{\alpha}_j$ is the sample mean of α_i for s_j, $j = 1, 2$. Since in fact the α_i are unknown a stratified random sampling design could help to ensure approximate 'balance' so that the computable estimator $\hat{Y}_{st0}$ is close to the optimal estimator $\hat{Y}_{st1}$.

Coming now to the selection of n_1, n_2 we see that $\hat{Y}_{st0} - \hat{Y}_{st1}$ will be minimized if the sampling variance

$$\operatorname{Var}(N_1\bar{\alpha}_1 + N_2\bar{\alpha}_2) = \frac{N_1^2}{n_1}\left(1 - \frac{n_1}{N_1}\right)S_1^2(\boldsymbol{\alpha}) + \frac{N_2^2}{n_2}\left(1 - \frac{n_2}{N_2}\right)S_2^2(\boldsymbol{\alpha})$$

is minimized, where

$$S_j^2(\boldsymbol{\alpha}) = \frac{1}{N_j - 1} \sum_{i \in \mathscr{P}_j} \alpha_i^2, \quad j = 1, 2.$$

This occurs when

$$\frac{n_j}{N_j} \propto S_j(\boldsymbol{\alpha}) \tag{4.1}$$

(similar to Neyman allocation). In the happy situation where

$$S_j^2(\boldsymbol{\alpha}) \propto \sigma_{0j}^2 \tag{4.2}$$

it is easy to see that this same allocation guarantees also the minimization of

$$E(\hat{Y}_{st1} - Y)^2$$

under model MST1.

If (4.2) does not hold, then the best allocation must clearly be a compromise between (4.1) and the allocation

$$\frac{n_j}{N_j} \propto \sigma_{0j}^2.$$

In fact, suppose we introduce the model MST2 which is MST1 with '$\Sigma_{i=1}^{N} \alpha_i = 0$' replaced by added assumption that α_i are i.i.d. with mean 0 and variance $\sigma_{\alpha j}^2$ for i in $\mathscr{P}_j$, $j = 1, 2$. Then the optimal allocation for minimizing $E(\hat{Y}_{st0} - Y)^2$ is provided by

$$\frac{n_j}{N_j} \propto \sigma_{0j}^2 + \sigma_{\alpha j}^2.$$

Finally we note that under stratified random sampling, the probability of inclusion π_i for a unit i in stratum $\mathscr{P}_j$ is n_j/N_j. Thus, unless the $\sigma_{0j}^2 + \sigma_{\alpha j}^2$ in (4.3) are all equal, the best allocation under MST2 is an example of unequal probability sampling. It can evidently be used both to decrease the chance of model bias of $\hat{Y}_{st0}$ for the model MST1 having no distributional assumption on the α_i, and to increase the efficiency of this estimate under the model MST2 where the α_i are i.i.d. within strata.

At the same time, the randomization in the stratified random sampling design provides frequency support for intervals of the form

$$\hat{Y}_{st0} \pm Z_\alpha \sqrt{\sum_{j=1}^{2} \frac{N_j^2}{n_j}\left(1 - \frac{n_j}{N_j}\right) s_{jy}^2},$$

also justifiable under the second model (cf. the support of (2.1) by simple random sampling). Here s_{jy}^2 is the sample variance for subsample s_j and Z_α is the appropriate standard normal percentage point.

5. Sampling designs for regression models

The situation of the last section can obviously be generalized to the case of several strata. Let us turn now to what is in some ways a generalization of this situation where the intrinsic stratification may be as fine as the population itself. That is, we might suppose in a model MR1 that the y_i are independent or uncorrelated, with

$$Ey_i = \theta_i + \alpha_i, \quad \text{Var } y_i = \sigma_i^2.$$

We again assume that the α_i are unknown and $\sum_{i=1}^{N} \alpha_i = 0$. As was not the case in the previous example, but to make sense of the possibility of inferring from the sample to the population, we assume a relationship among the θ_i, namely that for some unknown $\theta > 0$, $\theta_i = \theta x_i$ where $x_i > 0$ is the value for i of a 'size' variate x which is known for each i. Thus

$$Ey_i = \theta x_i + \alpha_i, \quad \text{Var } y_i = \sigma_i^2.$$

Assuming N large, an approximately optimal predictor of Y if the α_i were all 0 would be

$$\hat{Y}_0 = \hat{\theta}_0 X$$

where

$$\hat{\theta}_0 = \frac{\sum_{i \in s} x_i y_i / \sigma_i^2}{\sum_{i \in s} x_i^2 / \sigma_i^2},$$

and $X = \sum_{i=1}^{N} x_i$. An approximately optimal predictor of Y if the α_i were known would be

$$\hat{Y}_1 = \hat{\theta}_1 X$$

where

$$\hat{\theta}_1 = \frac{\sum_{i \in s} x_i(y_i - \alpha_i)/\sigma_i^2}{\sum_{i \in s} x_i^2/\sigma_i^2}.$$

Thus

$$\hat{Y}_1 - \hat{Y}_0 = X(\hat{\theta}_1 - \hat{\theta}_0) = X \sum_{i \in s} \frac{x_i \alpha_i/\sigma_i^2}{\sum_{i \in s} x_i^2/\sigma_i^2}.$$

If we were in the happy situation where $\sigma_i^2 = \sigma^2 x_i^2$, we would have

$$\hat{Y}_1 - \hat{Y}_0 = X \left(\sum_{i \in s} \frac{\alpha_i}{x_i} \right) / n$$

and by choosing a sampling design with

$$\pi_i = n x_i / X$$

we would ensure that the sample is balanced at least on average in the sense that the sampling expectation

$$E(\hat{Y}_1 - \hat{Y}_0) = 0.$$

In the more general situation where $\sigma_i^2 \neq \sigma^2 x_i^2$, a design which had

$$\pi_i \propto \sigma_i^2 / x_i \tag{5.1}$$

and for which $\sum_{i \in s} x_i^2/\sigma_i^2$ varied little among possible samples would clearly help to ensure approximate balance, making

$$E\left[\left(\sum_{i \in s} x_i \alpha_i / \sigma_i^2 \right) \left(\sum_{i \in s} x_i^2 / \sigma_i^2 \right) \right] \cong \sum_{1}^{N} \alpha_i = 0.$$

If σ_i^2 is a function $\sigma^2(x_i)$, evidently sharp stratification by the variable σ_i^2/x_i^2, with allocation determined by (5.1), would give an approximate solution; in this allocation the sample size n_j from the stratum $\mathscr{P}_j$ would satisfy

$$n_j \propto \sum_{i \in \mathscr{P}_j} (\sigma_i^2/x_i^2) x_i, \quad j = 1, \ldots, k.$$

If σ_i^2/x_i^2 were a constant A_j for all $i \in \mathscr{P}_j$, then $n_j \propto A_j \sum_{i \in \mathscr{P}_j} x_i$ in this allocation.

Note that if sampling inclusion probabilities are π_i and the α_i are all 0, the sampling expectation of the model mean squared error of $\hat{Y}_0$ is minimized for

$$\pi_i \propto \sigma_i. \tag{5.2}$$

This 'optimal design' again ensures approximate balance on α_i only if $\sigma_i \propto x_i$, for then (5.2) is the same as (5.1). If we specify the model MR2 by adding the further assumption that $E\alpha_i = 0$, $\text{Var}\,\alpha_i \propto x_i^2$, then the model based interval estimate

$$\hat{Y}_0 \pm Z_\alpha \sqrt{\frac{N^2 X^2}{2} \frac{1}{n^2(n-1)} \sum_{\substack{i \neq j \\ i,j \in s}} \left(\frac{y_i}{x_i} - \frac{y_j}{x_j}\right)^2}$$

for Y is supported by the probability sampling design with $\pi_i \propto \sigma_i \propto x_i$, provided that N is large and the design is of a type which can be thought of as being implemented through almost identical independent draws.

6. Monetary unit sampling

Suppose the population $\mathscr{P} = \{1, \ldots, N\}$ consists of a set of N 'accounts', with each of which is associated a nominal monetary value x. Let us suppose for each i that the ith account, independently of the other accounts, has a positive error y_i (in the value of x_i) with probability $\rho(x_i)$. Given that the y_i is positive, suppose that it has density $q(y|x_i)$ with mean $\mu(x_i)$ and variance $\sigma^2(x_i)$. If we take y_i to be 0 when the x_i value is correct, then marginally the y_i are independent with

$$Ey_i = \rho(x_i)\mu(x_i) \qquad (6.1)$$

and

$$\begin{aligned}\text{Var}\,y_i &= \rho(x_i)(\sigma^2(x_i) + \mu^2(x_i)) - \rho^2(x_i)\mu^2(x_i) \\ &= \sigma^2(x_i)\rho(x_i) + \mu^2(x_i)\rho(x_i)(1 - \rho(x_i)).\end{aligned} \qquad (6.2)$$

If for example $\rho(x_i) \equiv \text{const.}\ \rho$, and $\mu(x_i) = \mu x_i$, $\sigma^2(x_i) = \sigma^2 x_i^g$, g near 1, we would have

$$Ey_i = \rho\mu x_i, \qquad \text{Var}\,y_i = \sigma^2 \rho x_i^g + \mu^2 \rho (1 - \rho) x_i^2.$$

The second term would tend to dominate for large x_i and the model would be close to MR2 of Section 5, under which it has been shown that the use of the point estimator

$$\hat{Y} = \frac{X}{n} \sum_{i \in s} \frac{y_i}{x_i} \qquad (6.3)$$

together with a design with inclusion probabilities $\pi_i \propto x_i$ is ideal. This interestingly enough corresponds to the practice of so-called 'dollar unit sampling' (more generally, monetary unit sampling) together with the 'tainted dollar estimate' (Teitlebaum, 1973) in the auditing literature. See also Smith (1979).

In this type of sampling, the elementary sampling units are taken to be the individual dollars or other monetary units which make up the book values of the 'accounts'. A systematic or simple random sample of n dollars is taken, and the corresponding accounts are examined for error. This amounts to sampling accounts with probability proportional to book value.

The superpopulation model implied in (6.1) and (6.2) was introduced by Cox and Snell (1979) in their paper on sampling for rare errors. In pointing out the connection with classical finite population theory, they noted the futility of confidence intervals based on estimating the sampling variance of $\hat{Y}$ and the usual normal approximations. In auditing applications, and to a lesser extent in other 'quality control' setups, it is hoped that errors will occur rarely, and in such a case the normal approximation will be poor.

Cox and Snell consider in detail two approaches to the problem of estimating Y. In the first (here slightly modified) they note that if a single account is drawn with probability proportional to size, so that $\pi_i = x_i/X$, then the account is in error with probability

$$\sum x_i \rho(x_i)/X,$$

which is EX_d/X where X_d is the total of the nominal values for those accounts in error. Thus the number of accounts in error in a sample of size n is approximately Poisson with mean nEX_d/X. This Poisson approximation, applied to the number of sample accounts in error, allows the derivation of an upper confidence limit for nEX_d/X. This can be transformed into a still more approximate limit for Y using the fact that

$$Y \approx \mu(EX_d)$$

when $E(y_i | y_i > 0) = \mu x_i$; the mean μ is estimated using prior experience if the sample number in error is 0 or positive but small.

The second approach uses parametric modelling. In the illustration $\rho(x)$ is taken to be a constant ϕ; and given $y > 0$, y/x is taken to be exponential with mean μ. A conjugate prior is postulated for ϕ and μ, and a posterior density for $\chi = \phi\mu$ is obtained. (In the exponential case the density depends on the data through $\hat{Y}$ of (6.3).) An approximate density for Y can be obtained through

$$Y \approx \phi\mu X.$$

Both approaches can evidently give sensible results, but they depend on fairly specific prior knowledge. The first approach to some extent exploits the sampling design, but discards the labels from the sample outcome; the second uses the estimator $\hat{Y}$ but not the sampling design in its derivation. It may be asked whether there is any practical way of approaching the problem which exploits the sampling design more fully, in a way which is consistent with a reasonable model for the y_i. The following outlines an approach in the spirit of Sections 1–5 of this paper.

Adopting further the notation of Cox and Snell, let $d_i = 1$ if the i-th account is in error and $d_i = 0$ otherwise. Assume that the cost of sampling is the same for every account, but that k_i is the 'cost' of error in the i-th account. For example, k_i could be proportional to y_i, or it could be constant over all accounts in error. The total cost of error will be

$$K_d = \sum k_i d_i ;$$

the object of sampling is to detect at minimal sampling cost a situation where K_d is too high. The estimator

$$\hat{K}_d = \sum_{i \in s} k_i d_i / \pi_i \qquad (6.4)$$

may be proposed for K_d. Note that if the cost of error $k_i \equiv 1$, then $K_d = N_d$, the number of accounts in error.

The estimator $\hat{K}_d$ of (6.4) has both model and sampling design validity if under the model $k_i d_i / \pi_i$ are approximately i.i.d. and if the sampling design approximates a sequence of independent draws with selection probabilities $p_i \approx \pi_i / n$. Thus for Cox and Snell's first approach to have both justifications, since $k_i \equiv 1$ in this case, the π_i and $\rho(x_i)$ should be constant as i varies.

If the conditions for both justifications are present, as when $\rho(x_i) \equiv$ constant ρ, the k_i / x_i are i.i.d., and $\pi_i \propto x_i$, then the problem becomes that of determining a rule for flagging or rejecting the population based on a high value of $\hat{K}_d$ together with other information in the sample. One possibility might involve post stratification by value of k_i / x_i (cf. Fienberg et al., 1977). Another possibility would be to use the sample to estimate $P(\hat{K}_d > c)$ for some predetermined c and to flag the population if this estimate is too low. To obtain a useful estimator of $P(\hat{K}_d > c)$ without introducing some kind of model would in general be impossible, but a model as explicit as an exponential one for k_i / x_i may not be required. For example, it may be possible to use sample estimates of skewness and kurtosis profitably, or to find a smooth estimate of the characteristic function of $k_i d_i / x_i$ and invert its n-th power to estimate $P(\hat{K}_d \leq c)$ (cf. Heiner and Paulson, 1983).

7. A historical note

Unequal probability sampling has an interesting history. Actually, as indicated earlier the concept of giving different individuals of the population different probabilities of inclusion in the sample is already tacit in what is called Neyman allocation for stratified sampling, though Neyman (1934) did not mention that aspect. Cochran (1942) questioned whether units with different areas should be given the same probabilities of selection. Assigning different probabilities to different units within a stratum seems to have been first implemented by Hansen and Hurwitz (1943), in the first stage of a two stage sample. This was clearly a step ahead of what was implied by Neyman's stratified sampling, though it served a different purpose: Hansen and Hurwitz (1943, 1949) used unequal probability sampling for primary units to implement equal or approximately equal probability of inclusion in the sample for all the 'elements' in a stratum. General schemes of drawing different individuals or 'elements' with varying probabilities of selection were developed a little later by Goodman and Kish (1950), Midzuno (1950), and Horvitz and Thompson (1952). Thus single stage unequal probability sampling was on its way. Many refinements were proposed by other authors subsequently.

Though earlier authors clearly suggested the use of unequal probability sampling for reducing the variance of the estimate, a general theory formalizing the involved concepts of linear estimator, sampling design, etc. was first put forward by Godambe (1955). In this theory, probabilistic models were linked with sampling designs to establish optimum estimaton. On the other hand completely model based estimation was developed by Brewer (1963), Ericson (1969), Royall (1971) and others.

The idea that randomization enables elimination of some kind of 'unwanted or unknown factors' from estimation is quite old. With the introduction of probabilistic models in survey sampling, these unknown factors were identified as the unknown *departures* from the assumed model. The optimum estimator would in general be different for different departures. Hence if different departures are denoted by different values of an additional (unwanted or *nuisance*) parameter, the optimum estimator would depend on the value of the nuisance parameter.

A new turn to the interpretation of sampling designs, particularly unequal probability sampling, was given by the recognition that if the sample is drawn with an appropriate sampling design the dependence of the optimal estimation on the nuisance parameters mentioned in the preceding paragraph can be eliminated or appreciably reduced. A related idea is to use the sampling design for frequency interpretation of 'inference statements' based only on the probabilistic models. These two aspects form the approach adopted in this article. Some related papers in this respect are following: Godambe and Thompson (1971, 1976, 1977), Thompson (1984), Godambe (1982a, 1982b).

References

Brewer, K. W. R. (1963). Ratio estimation and finite populations: Some results deducible from the assumption of an underlying stochastic process, *Aust. J. Statist.* **5**, 93–105.

Cochran, W. G. (1942). Sampling theory when the sampling units are of unequal sizes. *J. Amer. Statist. Assoc.* **37**, 199–212.

Cox, D. R. and Snell, E. J. (1979). On sampling and estimation of rare errors. *Biometrika* **66**, 125–132.

Fienberg, S. E., Neter, J. and Leitch, R. A. (1977). Estimating the total overstatement error in accounting populations. *J. Amer. Statist. Assoc.* **72**, 295–302.

Godambe, V. P. (1955). A unified theory of sampling from finite populations. *J. Roy. Statist. Soc. Ser. B* **17**, 269–278.

Godambe, V. P. (1982a). Likelihood principle and randomisation. In: G. Kallianpur, P. R. Krishnaiah and J. K. Ghosh, eds., *Essays in Honour of C.R. Rao.* North-Holland, Amsterdam, 281–294.

Godambe, V. P. (1982b). Estimation in survey sampling, robustness and optimality. *JASA* **77**, 393–403.

Godambe, V. P. and Thompson, M. E. (1971). Bayes, fiducial and frequency aspects of statistical inference in survey sampling. *J. Roy. Statist. Soc. Ser. B* **33**, 361–390.

Godambe, V. P. and Thompson, M. E. (1976). Philosophy of survey-sampling practice. In: *Foundations of Probability Theory, Statistical Inference and Statistical Theories of Science, Vol. II.* Reidel, Dordrecht, Holland, 103–123.

Godambe, V. P. and Thompson, M. E. (1977). Robust near optimal estimation in survey practice. *Bull. Int. Statist. Inst.* **XLVII**(3), 129–145.

Goodman, R. and Kish, L. (1950). Control beyond stratification, a technique in probability sampling. *J. Amer. Stat. Assoc.* **45**, 350–372.

Hansen, M. H. and Hurwitz, W. N. (1943). On the theory of sampling from finite populations. *Ann. Math. Statist.* **14**, 333–362.

Hansen, M. H. and Hurwitz, W. N. (1949). On the determination of optimum probabilities in sampling. *Ann. Math. Statist.* **20**, 426–432.

Heiner, K. and Paulson, A. S. (1983). Some new methods for setting confidence limits for auditing accounting populations. Paper presented at ORSA/TIMS Joint National Meeting, November, 1983.

Horvitz, D. G. and Thompson, D. J. (1952). A generalization of sampling without replacement from a finite universe. *J. Amer. Statist. Assoc.* **47**, 663–685.

Midzuno, H. (1950). An outline of the theory of sampling systems. *Ann. Inst. Statist. Math.* **1**, 149–156.

Smith, T. M. F. (1979). Statistical sampling in auditing: A statistician's viewpoint. *The Statistician* **28**, 267–280.

Teitlebaum, A. D. (1973). Dollar-unit sampling in auditing. Paper presented at Annual Meeting of American Statistical Association, December 1973.

Thompson, M. E. (1984). Model and design correspondence in finite population sampling. *J. Statist. Plan. Inf.* **10**, 323–334.

Systematic Sampling

D. R. Bellhouse

1. Introduction

Consider a population in which the units on which the measurements are made are linearly ordered in some way. The ordering may be natural, such as points in time or contiguous strips of land, or at the discretion of the sampler, such as cards in a file drawer or lists of names. Given the ordering, a systematic sampling scheme is one in which the sample units are equally spaced. Systematic samples are often drawn for their convenience over other sampling designs. In addition, systematic sampling is often more efficient, in terms of variance or mean square error, than other sampling designs.

Two reviews of the literature of systematic sampling have already been made: Buckland (1951) and Iachan (1982c). Also Cochran (1977, Ch. 8) and other sampling textbooks contain discussions of systematic sampling. This article attempts to provide a sequel to the results in Cochran (1977, Ch. 8). In doing so, the discussion will be more detailed than Iachan's (1982c) review article and will contain additional references and results.

2. Sampling theory based on the randomization alone

2.1. Systematic random sampling

Two types of populations from which a sample of size n is to be drawn may be enumerated:

(I) The population consists of a finite number N of units with measurements y_u attached to the uth unit, $u = 1, \ldots, N$.

(II) The population consists of a continuum of values on the interval $[0, N]$. A measurement y_u is made at the point u, $0 \leq u \leq N$.

For the purpose of efficiency comparisons, in the finite population (I) it is usually assumed that $N/n = k$ is an integer. In this case, a systematic sample is obtained by selecting a random start r between 1 and k and then selecting every

kth unit thereafter, i.e. $r, r + k, r + 2k, \ldots, r + (n - 1)k$. The population mean $\bar{Y} = \Sigma_{u=1}^{N} y_u/N$ is estimated by the sample mean. If the starting value r has been chosen, the sample mean is $\bar{y}_r = \Sigma_{j=0}^{n-1} y_{r+jk}/n$. The variance

$$V_p(\bar{y}_{sy}) = \sum_{r=1}^{k} (\bar{y}_r - \bar{Y})^2/k, \tag{2.1}$$

where V_p denotes the variance with respect to the sampling design. Cochran (1977) gives the appropriate formulae for finite population comparisons with stratified sampling with one unit per stratum and simple random sampling without replacement.

When N/n is not integral several options are available to obtain a systematic type sample. When choosing among the options available, several properties are desirable: the sample size is constant; the sample mean is unbiased for $\bar{Y}$; and the inclusion probability is the same for all units. These properties are all satisfied by the method of circular systematic sampling introduced by Lahiri in 1952 (see Murthy, 1967, p. 139). A random start r between 1 and N and a sampling interval k are chosen. The sample consists of the units $r + jk \pmod{N}$ for $j = 0, \ldots, n - 1$ and with unit N chosen when $r + jk \pmod{N} = 0$. The choice of the sampling interval is arbitrary. Most textbooks including Cochran (1977) suggest that k should be chosen as the integer closest to N/n or $[N/n + 1/2]$ where $[\cdot]$ denotes the 'greatest integer in'. No problem arises when $k = [N/n]$. However, when N/n is rounded up to the next integer, there may be situations in which the sample size is not the desired size n. Sudakar (1978) has given an example of this situation. When, for example, $N = 15$, $n = 6$ and $k = 3$, the sample with starting point $r = 3$ consists of the units numbered 3, 6, 9, 12 and 15. In response to this problem Sudakar (1978) found that for all starting points r and all sample sizes n, the sample contains n distinct units if and only if N and k are relatively prime. Sudakar (1978) says that this result is appropriate when n is not fixed in advance. The theorem suggests that a sampling interval k be chosen in advance with N and k relatively prime. Then any sample size, when decided upon, will be of the required size. However, depending on the choice of k this method will not necessarily give an even spread of the sample units across the population as in the case of systematic sampling when $N = nk$. Also, the sample size is usually fixed in advance by budgetary contraints. With the additional desire to have the sample units appear as close as possible to a systematic sample when $N = nk$, k should be close to N/n. Bellhouse (1984b) suggests that the sampling interval k be the greatest integer in N/n when $N/(n - 1)$ is integral and the integer nearest $N/(n - 1)$ is nonintegral.

For the infinite population set-up (II) the sampling interval $k = N/n$ can be any real number. In this situation a random start is chosen with uniform probability on the interval $[0, k]$ and the sample points $r, r + k, \ldots, r + (n - 1)k$ are chosen. The sample mean $\bar{y}_r = \Sigma_{j=0}^{n-1} y_{r+jk}/n$ has variance

$$\frac{1}{k} \int_0^k (\bar{y}_r - \bar{Y})^2 \, dr, \tag{2.2}$$

where $\bar{Y} = \int_0^N y_u \, du/N$ is the population mean.

2.2. Systematic sampling with probability proportional to size (PPS)

In the finite population set-up (I) suppose that it is desired to have the jth unit included with probability π_u, $u = 1, \ldots, N$. One way to obtain the sampling scheme with the desired inclusion probabilities is first to obtain the cumulative totals $T_i = \Sigma_{u=1}^{i} \pi_u$, $i = 1, \ldots, N$, with $T_N = n$ and $T_0 = 0$. Select a uniform random number r on the interval $[0, 1]$. The unit numbered i is included in the sample if $T_{i-1} \leqslant r + j < T_i$ for some $j = 0, 1, \ldots, n - 1$. A discussion of this sampling method, originally due to Madow (1949) and using size variables $x_u \propto \pi_u$, is given in Murthy (1967, Ch. 11e) and Cochran (1977, Ch. 9A.10), among others. Hartley (1966) obtained the variance of the Horvitz–Thompson estimator of $\bar{Y}$ for this method and provided two methods of variance estimation. Using the Horvitz–Thompson estimator he showed that the variance of the estimation for this scheme is smaller than the variance based on sampling with replacement if there is a negative intra-class correlation between units in the same sample. Pinciaro (1978) has provided a general algorithm to compute the joint inclusion probabilities for pps systematic sampling.

Hartley and Rao (1962) assumed that the N units have first been arranged in random order and then a pps systematic sample is taken. This will be referred to as randomized pps systematic sampling. From their equation 5.15 Hartley and Rao (1962) obtained the joint inclusion probability to $O(N^{-4})$

$$\pi_{ij} = n(n - 1)p_i p_j [1 + (p_i + p_j) - S_2 + 2(p_i + p_j)^2 \\ - 2p_i p_j - 3(p_i + p_j)S_2 + 3S_2^2 - 2S_3], \qquad (2.3)$$

where the inclusion probability $\pi_j = np_j$ and where $S_2 = \Sigma_{u=1}^{N} p_u^2$ and $S_3 = \Sigma_{u=1}^{N} p_u^3$. The determination of (2.3) is based on the assumption that $\pi_j \leqslant CN^{-1}$ for all j and N and $C_1 N^{-2} \leqslant S_{ij} \leqslant C_2 N^{-2}$ for all pairs (i, j) and N, where C, C_1, and C_2 are universal constants. The term S_{ij} is the mean square of the π_u's excluding units i and j.

The approximate evaluation of π_{ij} in randomized pps systematic sampling through (2.3) has been the subject of some controversy in the recent literature. Hájek (1981, pp. 115–116) claimed to give a counter-example to show that (2.3) is not the correct asymptotic expansion. This was further reported uncritically by Iachan (1982b, c). However, Joshi (1983) showed that Hájek's (1981) counter-example is invalid: it fails to satisfy the conditions assumed by Hartley and Rao (1962) in the derivation of (2.3). In Hájek's example, $S_{12}^2 = 0$. Iachan (1983) replied by questioning the general applicability of the condition $S_{ij}^2 \leqslant C_1 N^{-2}$ so that (2.3) is not applicable. This may not be a fair criticism. Hidiroglou and Gray (1975) in an empirical study have shown that (2.3) is fairly accurate, in the populations they studied, for $N \geqslant 10$ and $n = 2$. Moreover, when $\pi_u = n/N$ so that $S_{ij}^2 = 0$, (2.3) produces $\pi_{ij} = n(n - 1)[1 + (1/N) + (1/N^2)]/N^2$ which is $\pi_{ij} = n(n - 1)/[N(N - 1)]$ to $O((N^{-4})$, obtained by expanding $[1 + (1/N)]^{-1}$ in a Taylor series. However, in Hájek's (1981) example, with $\pi_1 = \pi_2 = 1/(N - 1)$ and $\pi_3 = \cdots = \pi_N = 2/(N - 1)$, $S_{12}^2 = 0$ and when N is odd, using Connor's (1966) method of calculation, $\pi_{ij} = 0$ so that the result in (2.3) does not apply. In a private communication Joshi showed that the condition $S_{ij}^2 > 0$ holds for a wide class of populations. Let $\pi_1 = \cdots = \pi_M = a/N$, where $a < 2$ is a fixed numbmer. Let $\pi_{M+1} = \cdots = \pi_N = b/(N-M)$, where b is determined so that $\Sigma_{u=1}^N \pi_u = 2$. Let $M/N \to k$. Provided that $k > 0$ then $S_{ij}^2 > 0$ so that (2.3) applies with $n = 2$. In Hájek's (1981) example $k = 0$ and $a = 1$. In summary, it appears that the asymptotic expansion has wide applicability. The approximation works in some cases when $S_{ij}^2 = 0$ and not in others. Further, there are many situations in which $S_{ij}^2 > 0$ so that (2.3) applies. However, the statement, given in Hartley and Rao (1962), that the assumption $S_{ij}^2 \geqslant C_1 N^{-2}$ can be circumvented is not necessarily true.

Connor (1966) provided a method to calculate π_{ij} exactly in randomized pps systematic sampling exactly. Using a modification of this method, Hidiroglou and Gray (1980) have given a FORTRAN program for the calculation of π_{ij}. Using (2.3) with $n = 2$, Des Raj (1965) studied the stability of the Yates–Grundy variance estimator and found it compared favourably to pps sampling with replacement.

Systematic sampling with probability proportional to size, either randomized or nonrandomized, has been used in a number of large scale surveys. Murthy (1967) mentions the use of nonrandomized pps systematic sampling in the Indian National Sample Survey. Des Raj (1964) describes a survey on employment in Greece in which 147 provinces were allocated to 22 strata with 2 provinces chosen by pps systematic sampling in each stratum. Systematic sampling is used extensively, for reasons of flexibility and ease of expansion of the sample, in the Canadian Labour Force Survey (Statistics Canada, 1976). Randomized pps systematic sampling is used in the rural sample at two stages of sampling. The nonrandomized version is used in the apartment sample to choose apartment buildings within strata.

3. Trends in the population

3.1. Linear trend

In some situations it is reasonable to assume that the finite population itself is a sample from an infinite superpopulation. A superpopulation model of the finite population (Type I) in which there is a linear trend may be expressed as

$$y_u = \alpha + \beta u + e_u,$$
$$E_m(e_u) = 0, \quad E_m(e_u^2) = \sigma^2, \quad E_m(e_u e_v) = 0.$$
(3.1)

In (3.1) the operator E_m denotes expectation with respect to the model.

With the assumption of a model the efficiency of an estimator, in this case the sample mean $\bar{y}$, can be measured in one of several ways. The criterion for efficiency used here is due to Cochran (1946): take the finite population variance of the estimator averaged over the model. For systematic sampling this is (2.1) averaged over (3.1) which yields

$$E_m V_p(\bar{y}_{sy}) = \beta^2(k^2 - 1)/12 + \bar{\sigma}^2, \qquad (3.2)$$

where $\bar{\sigma}^2 = \sigma^2(k-1)/(nk)$. The first term on the right of (3.2) is the variance component due to the linear trend; the second term is due to the random error. Two other commonly used sampling designs, stratified sampling and simple random sampling, may be compared to systematic sampling. The finite population variance for simple random sampling without replacement averaged over (3.1) yields

$$E_m V_p(\bar{y}_{ran}) = \beta^2(k-1)(nk+1)/12 + \bar{\sigma}^2. \qquad (3.3)$$

For stratified sampling it is assumed that the population consists of n strata comprised of the sets of units $\{1, \ldots, k\}$, $\{k+1, \ldots, 2k\}$, ... $\{(n-1)k+1, \ldots, nk\}$. A random sample of one unit is taken from each stratum. The finite population variance for stratified sampling averaged over model (3.1) is

$$E_m V_p(\bar{y}_{strat}) = \beta^2(k^2 - 1)/12n + \bar{\sigma}^2. \qquad (3.4)$$

It may be easily seen from (3.2), (3.3) and (3.4) that

$$E_m V_p(\bar{y}_{strat}) \leqslant E_m V_p(\bar{y}_{sy}) \leqslant E_m V_p(\bar{y}_{ran}). \qquad (3.5)$$

3.2. Removal of the linear trend as a component of variance

On examining (3.5) it might be concluded that a stratified sampling scheme is the most efficient method of sampling when a linear trend is present. However, there are several systematic-type sampling designs with $\bar{y}$ as the estimator of $\bar{Y}$ which eliminate the linear trend as a component of variance. Bellhouse and Rao (1975) have studied three of these designs, both theoretically and empirically, for various types of trends (other than linear) in the population. They found that the centrally located systematic sampling design due to Madow (1953) was the most promising in terms of efficiency. For this scheme the starting point $r = (k+1)/2$, k odd, is chosen with probability 1 so that the sample is $r, r+k, r+2k, \ldots, t+(n-1)k$ with $r = (k+1)/2$. When k is even, the starting points $r = k/2$ and $r = k/2 + 1$ are each chosen with probability 1/2.

In many surveys, especially multi-purpose surveys, it may be desirable to use systematic sampling for convenience or for its efficiency in the estimation of the population mean for some of the variables. It is then desirable to keep the systematic sampling design and to use a different estimator. Yates (1948) has suggested an estimator in systematic sampling which removes the linear trend as a component of variance. The estimator is obtained by taking the sample mean and then adding weights w and $-w$ to the first and last sample units respectively. The weight is obtained by setting the estimator equal to the finite population mean under a perfect linear trend. This yields $w = (2r - k - 1)/[2(n - 1)k]$ so that the estimator becomes

$$\bar{y}_r + \frac{2r - k - 1}{2(n - 1)k} (y_r - y_{r+(n-1)k}). \tag{3.6}$$

Bellhouse and Rao (1975) have studied the efficiency of (3.6) both theoretically and empirically under several trend models. They also generalized (3.6) to circular systematic sampling designs. In this situation the weights are applied to the largest and smallest unit numbers in the sample. There are two cases:

(i) When the last sample unit chosen, $r + (n - 1)k \leq N$ the estimator is

$$\bar{y}_r + \frac{2r + (n - 1)k - (N + 1)}{2(n - 1)k} (y_r - y_{r+(n-1)k}).$$

(ii) When $r + (n - 1)k > N$ the estimator is

$$\bar{y}_r + \frac{2r + (n - 1)k - (N + 1) - 2Nn_2/n}{2(N - k)} (y_{r+n_1 k - N} - y_{r+k(n_1-1)})$$

where n_1 is the number of sample units such that $r + jk \leq N$, $j = 0, \ldots, n - 1$, and $n_2 = n - n_1$.

3.3. Examples

Four examples of populations with trends are examined. All populations are mentioned in the empirical study of Bellhouse and Rao (1975). Populations 1 and

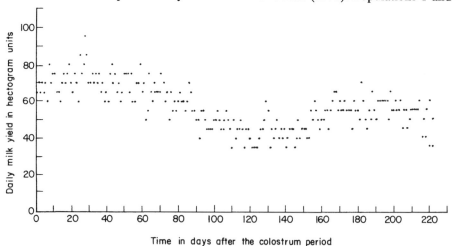

Fig. 1. Cow number 1.

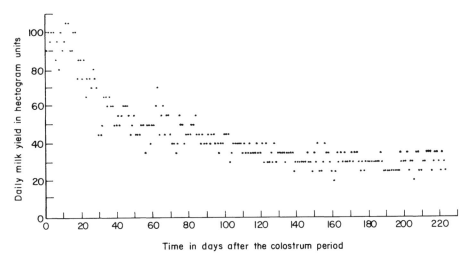

Fig. 2. Cow number 2.

Table 3.1
Mean square errors

Sampling interval k	Sample size n	Systematic sampling	Centrally located systematic sampling	Yates' end corrections
Population 1				
56	4	13.44	18.37	11.10
28	8	7.72	14.55	8.09
14	16	5.07	6.14	4.08
7	32	2.23	0.14	1.85
Population 2				
56	4	46.26	3.52	16.39
28	8	8.94	2.60	4.25
14	6	2.69	1.25	1.13
7	32	1.06	0.03	0.88
Population 3 (mse × 1000)				
32	4	191.4	32.6	124.5
16	8	89.4	85.4	80.6
8	16	34.1	67.2	29.8
4	32	8.5	13.8	14.1
Population 4 (mse × 1000)				
322	4	487.7	66.0	69.2
16	8	133.2	2.3	19.2
8	16	39.6	8.7	3.4
4	32	5.9	0.1	1.2

2 are the daily milk yields ($N = 224$) of two Hariana cows. Their scatter diagrams, given in Figures 1 and 2 show a linear trend with periodic variation for population 1 and a quadratic trend for population 2. Since the observations on the Hariana cows are taken at equally spaced points in time, it would be appropriate to consider fitting time series models as an alternative. Using the models described in Box and Jenkins (1976) both populations 1 and 2 fit integrated moving average processes with one order of differencing and one moving average parameter. Populations 3 and 4 are the cultivated area and the 1961 census population of $N = 128$ villages when arranged in increasing order of the geographical area and the 1951 census population respectively. Their scatter diagrams, given in Murthy (1967, p. 161) indicate a linear trend with heteroscedastic error variances for population 3 and a quadratic trend for population 4. Table 3.1 shows the variance or mean square error, at various sample sizes, of systematic sampling and centrally located systematic sampling using the sample mean as estimator and of systematic sampling using Yates' method of end corrections. Centrally located systematic sampling is generally the most efficient method but exhibits erratic behaviour, as n increases, for populations 3 and 4.

3.4. Periodic variation

To examine the efficiency of systematic sampling using the most general models of periodic variation it is useful to work with continuous populations (Type II) and to evaluate (2.2) under the assumed model. On assuming that the measurement y_u for $0 \leq u \leq N$ has period p, i.e. $y_u = y_{u+p}$, and that N is an integral multiple of p the period, then y_u has the Fourier series representation,

$$y_u = \sum_{v=-\infty}^{\infty} C_v \exp[2\pi i v u/p] \tag{3.7}$$

where $i = \sqrt{-1}$ and where C_v, $v = 0, \pm 1, \pm 2, \ldots$ are the Fourier coefficients. The population mean $\bar{Y} = \int_0^N y_u \, du/N = C_0$.

Mickey (1972) evaluated (2.2) under a model similar to (3.6); Bellhouse (1985) obtained an alternate derivation. Under model (3.6) with the additional assumption that the period is an integral multiple of the sampling interval, say $p = ak$, Bellhouse (1983b) obtained

$$V_p(\bar{y}_{sy}) = 2 \sum_{v=1}^{\infty} |C_{va}|^2 \tag{3.8}$$

for (2.2) where $|C_{va}|$ is the absolute value of the Fourier coefficient. The equivalent expression for simple random sampling is

$$V_p(\bar{y}_{\text{ran}}) = 2 \sum_{v=1}^{\infty} |C_v|^2/n. \tag{3.9}$$

The model used by Madow and Madow (1944) for periodic variation is notationally more complex and less general than (3.7). Their results are obtained as special cases in the set-up using (3.7). When the period and the sampling interval coincide, then systematic sampling is equivalent to simple random sampling with a sample of size one. This follows from (3.8) and (3.9) with $a = 1$. In the notation given here the model of Madow and Madow (1944) may be expressed as

$$y_u - C_0 = -[y_{u+p/2} - C_0] \tag{3.10}$$

for $0 \leq u \leq p/2$. When the period is twice the sampling interval ($a = 2$) the variance of the mean for systematic sampling is zero under assumption (3.10). This result follows as a special case of (3.8) since, on using the inverse Fourier transform $c_v = (1/p) \int_0^p y_u \exp[-2\pi i v u/p] \, du$, and assumption (3.10), $C_{2v} = 0$ for any integer v.

3.5. Random ordering—lack of a trend

When the finite population of N units are in random order, Madow and Madow (1944) showed that (2.1) averaged over the $N!$ permutations of the measurements $y_1, \ldots, y_N$ yielded $(k-1)S^2/(nk)$, the variance of the mean for simple random sampling without replacement, where $(N-1)S^2 = \sum_{u=1}^{N}(y_u - \bar{Y})^2$. A simplified algebraic proof of this result can be obtained by using the random permutation models of Rao (1975) and Rao and Bellhouse (1978). Assume that the finite population consists of N fixed numbers $z_1, \ldots, z_N$ and that the measurements $y_1, \ldots, y_N$ are obtained as a random permutation of $z_1, \ldots, z_N$. This yields $\Pr(y_u = z_i) = 1/N$ and $\Pr(y_u = z_i, y = z_j) = 1/[N(N-1)]$. The random permutation assumption implies the linear model

$$y_u = \bar{Y} + e_u, \tag{3.11}$$

where
$$E_m(e_u) = 0, \quad E_m(e_u^2) = \sigma^2, \quad E_m(e_u e_u) = -\sigma^2/(N-1),$$

$$\bar{Y} = \sum_{u=1}^{N} y_u/N = \sum_{u=1}^{N} z_u/N,$$

and
$$\sigma^2 = \sum_{u=1}^{N}(y_u - \bar{Y})^2/N = \sum_{u=1}^{N}(z_u - \bar{Z})^2/N.$$

Then it is straightforward to show that (2.1) averaged over model (3.11) is $(k-1)\sigma^2/(N-1) = (k-1)S^2/(nk)$. This result also follows as a special case of, among others, Theorem 2.1 in Rao and Bellhouse (1978), a theorem dealing with optimal estimation of $\bar{Y}$. The result of Sedransk (1969), showing that systematic sampling is equivalent to sampling with one unit per stratum under the random permutation model, also follows as a special case.

4. Autocorrelated populations

Cochran (1946), when he introduced super population models to the finite population sampling literature, assumed the autocorrelated model

$$E_m(y_u) = \mu, \quad E_m(y_u - \mu)^2 = \sigma^2,$$
$$E_m(y_{u+v} - \mu)(y_u - \mu) = \sigma^2 \rho(v), \quad (4.1)$$

where $\rho(v)$ is the correlation between units of distance v part. In many practical applications, it is reasonable to assume that $\rho(v)$ is positive, decreasing and convex, i.e.

$$\rho(v) \geq 0, \quad (4.2)$$
$$\Delta\rho(v) = \rho(v+1) - \rho(v) \leq 0, \quad (4.3)$$
$$\Delta^2\rho(v) = \rho(v+2) - 2\rho(v+1) - \rho(v) \geq 0. \quad (4.4)$$

Cochran (1977) has given three examples of autocorrelation functions which satisfy (4.2) to (4.4), namely linear, exponential and hyperbolic tangent. In addition, (4.2) to (4.4) are satisfied for any autoregressive process whose roots of the characteristic equation are real.

4.1. Small sample results

The variance of the sample mean for systematic sampling given by (2.1) when averaged over model (4.1) yields

$$\frac{\sigma^2}{n}\left(1 - \frac{1}{k}\right)\left\{1 - \frac{2}{kn(k-1)}\sum_{v=1}^{kn-1}(kn-v)\rho(v)\right.$$
$$\left. + \frac{2k}{n(k-1)}\sum_{n(k-1)}^{2k}\sum_{v=1}^{n-1}(n-v)\rho(kv)\right\}. \quad (4.5)$$

Hájek (1959), as a special case, has obtained the most general result relating to (4.5). This may be expressed as

THEOREM 4.1. *Among all sampling designs which have inclusion probabilities of the uth unit $\pi_u = n/N$ where $N = nk$, the minimum value of the finite population variance of the sample mean averaged over model (4.1) is (4.4), the variance for systematic sampling, provided that the correlation function is positive, decreasing and convex.*

Hájek's (1959) result is actually more general than Theorem 4.1. He allowed $\pi_u \propto x_u$, a size variable, and using the Horvitz–Thompson estimator he showed under assumptions (4.2) to (4.4) that pps systematic sampling was the optimal design. In the statement of his result, Lemma 9.2, Hájek (1959) explicitly gave only convexity (4.4) as a condition. However, conditions (4.2) and (4.3) are also necessary. Upon examining the last line of his proof involving an infinite sum, it may be noted that (4.2), (4.3) and (4.4) are all necessary to get convergence of the infinite series using the ratio test.

Theorem 4.1 contains a number of results in the literature as special cases. Cochran (1946) compared systematic sampling given by (4.5) to simple random and stratified sampling and Gautschi (1957) compared systematic sampling with a single random start given by (4.5) to systematic sampling with multiple random starts. Singh et al. (1968) and Murthy (1967, p. 165) have given two systematic type sampling designs which eliminate a linear trend as a component of the variance. In each case $\pi_u = n/N$ so that Hájek's (1959) result may be applied.

When the condition $\pi_u = n/N$ in Theorem 4.1 is relaxed, systematic sampling is no longer optimal. Madow (1953) showed that when $\rho(v)$ is decreasing, i.e. (4.3) is satisfied, the mean square error of $\bar{y}$ for centrally located systematic sampling averaged over (4.1) is smaller than (4.5).

Blight (1973) considered a different criterion for obtaining an optimal sampling design. In particular, his criterion was $V_m(\bar{Y}|y_u, u \in s)$ the variance under the model of the population mean $\bar{Y}$ given the observations y_u in the sample denoted by s. Under a first order autoregressive model with normally distributed errors,

$$y_u - \mu = \phi(y_{u-1} - \mu) + e_u, \tag{4.6}$$

Blight (1973) showed that centrally located systematic sampling minimized $V_m(\bar{Y}|y_u, u \in s)$ when $\phi > 0$. When $\phi < 0$, the optimal choice of sampling units is to choose two clusters of units at the beginning and the end of the sequence of population units. The autocorrelation function for (4.6), $\rho(v) = \phi^u$.

For continuous populations (II), Jones (1948) assumed the model

$$x_u = y_u + e_u,$$

where y_u follows the model equivalent to (4.1) in the continuous framework with $\rho(v) = e^{-\alpha v}$ and where e_u is an uncorrelated error measurement with zero expectation and variance λ^2. Jones (1948) used a mean square error with respect to the model as his minimization criterion. He showed that the optimal estimator of $\bar{Y} = \int_0^N y_u \, du/N$ in the class of linear model unbiased estimators of μ is proportional to the sample mean and that systematic sampling with a purposive start depending on the autocorrelation function is the optimal choice of sample points.

4.2. Asymptotic results

The first asymptotic result for finite populations (I) was obtained by Cochran (1946). He assumed in (4.1) that $\rho(v) = e^{-\alpha v}$. Then in (4.5) he let $n \to \infty$ and $k \to \infty$ such that $k\alpha = t$ is a constant. He obtained

$$\frac{\sigma^2}{n}\left[1 - \frac{2}{t} + \frac{2}{e^t - 1}\right] \tag{4.7}$$

as the limit of (4.5) and used this result to make numerical efficiency comparisons to simple random and stratified sampling. Iachan (1982) let $n \to \infty$ in (4.5) and obtained

$$\frac{\sigma^2}{n}\left(1 - \frac{1}{k}\right)\left\{1 - \frac{2}{k-1}\sum_{v=1}^{\infty}\rho(v) + \frac{2k}{k-1}\sum_{v=1}^{\infty}\rho(vk)\right\}. \quad (4.8)$$

The infinite population (II) equivalent to (4.5) is

$$\frac{\sigma^2}{n}\left\{1 - \frac{2}{nk}\int_0^N (N-v)\rho(v)\,dv + \frac{2}{n}\sum_{v=1}^{n-1}(n-v)\rho(vk)\right\}. \quad (4.9)$$

Quenouille (1949) obtained an asymptotic expression as $n \to \infty$ for (4.9) similar to (4.8), namely

$$\frac{\sigma^2}{n}\left\{1 - \frac{2}{k}\int_0^\infty \rho(v)\,dv + 2\sum_{v=1}^{\infty}\rho(vk)\right\}.$$

Iachan (1982a) went further in the finite population framework. He obtained limiting variances for systematic sampling with multiple random starts and then showed that the limiting variance decreases as the number of random starts decreases while keeping the total sample size the same for all schemes. This generalizes Gautschi's (1957) result concerning systematic sampling with multiple random starts. Because of the asymptotic argument Iachan (1982a) needed only to assume a decreasing correlogram (4.3) while Gautschi (1957) in the small sample framework, also required the convexity condition (4.4).

Jowett (1952) obtained a limiting form of (4.5) based on large k rather than a large sample size n. Although Jowett (1952) states only that $k \to \infty$ there is more to his asymptotic argument. He first assumes that the distance between the finite population units is w and that the correlation between the units depends on this distance as well as the difference in the unit number. In this set-up the variance of the sample mean is (4.5) with v replaced by vw. Jowett (1952) then assumes that $k \to \infty$ and $w \to 0$ but $wk \to d$ where d is the distance between the sampled units. Then the total length of the process, which is continuous in the limit, is nd. Jowett (1952) obtains

$$\frac{\sigma^2}{n}\sum_{u=-n}^{n}\left\{\left(1 - \frac{|ud|}{nd}\right)\rho(ud) - \int_{(u-1/2)d}^{(u+1/2)d}\frac{1}{d}\left(1 - \frac{|v|}{nd}\right)\rho(v)\,dv\right\} \quad (4.10)$$

as the limit, where the integrand in (4.10) is 0 for $|v| > nd$. Upon some rearrangement of terms it is seen that (4.9) and (4.10) are equal when $k = d$. Thus the result confirms a footnote of Quenouille (1949, p. 356) that 'in practice we can sample a continuous process only as if it were a discontinuous process with k large'. It appears that Quenouille (1949) arrived at the variance in the continuous framework by using the same asymptotic argument as Jowett (1952).

It is interesting to note the difference in the limiting arguments used by Iachan (1981, 1982a) and Jowett (1952). In both cases the population size becomes infinitely large. In Iachan's (1981, 1982a) case the population becomes infinitely large by adding more units to the end of the population; in Jowett's (1952) case the population becomes infinitely large by fixing an interval $[0, nd]$ and by packing more observations into this interval. Quenouille's (1949) asymptotic argument for large N is obtained by increasing the interval on which the population is defined $[0, N]$ to the whole real line.

Williams (1956) derived an asymptotic expression for the variance similar to one in Jowett (1952) but used weaker assumptions. Hannan (1962) obtained an asymptotic expression using the spectral density function rather than the autocorrelation function.

4.3. Relationship to experimental design

Suppose that the N finite population unts, arranged in some natural ordering, can be considered as N experimental units to which k treatments are to be applied with n applications of each treatment, $N = nk$. The model for x_u, the response of the uth experimental unit when the ith treatment is applied, is

$$x_u = \alpha_i + y_u, \tag{4.11}$$

where α_i is the ith treatment effect such that $\Sigma_{i=1}^k \alpha_i = 0$ and where y_u follows (4.1). In this model there are two sources of variation: one due to the randomization and resulting treatment, and one due to the assumption of model (4.1). The variation as a result of the randomization is similar to the variation due to sampling the finite population; the variation due to the model without the treatment application is equivalent to the super population model.

Consider any linear contrast $\Sigma_{i=1}^k h_i \bar{x}_i$ where $\Sigma_{i=1}^k h_i = 0$ and where $\bar{x}_i$ is the ith treatment mean. Given an application of the treatments, the usual variance of the treatment contrast is $\text{Var}_m(\Sigma_{i=1}^k h_i \bar{x}_i)$, where Var_m is the variance with respect to the model given by (4.10) and (4.1). Because of the autocorrelation, this variance depends upon the outcome of the allocation of the treatments to the experimental units. Now average this variance over the possible allocations of the treatments, i.e. the randomization distribution, with the restriction that the probability that treatment i is applied to experimental unit u is $1/k$. Bellhouse (1984a) has shown that a systematic application of the treatments minimizes $E_p \text{Var}_m(\Sigma_{i=1}^k h_i \bar{x}_i)$, the variance of the treatment contrast averaged over the application of the treatments. The design may be described as

$$1 \ 2 \ 3 \ldots (k-1) \ k \ 1 \ 2 \ 3 \ldots (k-1) \ k \ldots$$

with random allocation of the treatments to the integers $1, 2, \ldots, k$.

4.4. Systematic sampling of time series

A large subclass of models which give rise to model (4.1) are autoregressive moving average processes (ARMA) of order (p, q). These models are given by

$$\Phi(B)(y_u - \mu) = \Theta(B)e_u \tag{4.12}$$

$(u = 1, \ldots, N)$, where $\Phi(B) = 1 - \phi_1 B - \phi_2 B^2 - \cdots - \phi_p B^p$, $\Theta(B) = 1 - \theta_1 B - \theta_2 B^2 - \cdots - \theta_q B^q$ and B is the backwards shift operator (for example $B^2 y_u = y_{u-2}$). Brewer (1973) has shown that a systematic sample with interval k, of the process given by (4.12) can be represented as an ARMA (p, r) process where $r = [p + (q - p)/k]$ and $[\cdot]$ denotes 'the greatest integer in'. This result has been generalized by Wei (1981) to the non-stationary autoregressive integrated moving average process. An ARIMA model of order (p, d, q) is (4.12) with $\Phi(B)$ replaced by $\Phi(B)(1 - B)^d$. Wei (1981) showed that a systematic sample of this process can be represented by an ARIMA (p, d, r) process where $r = [(p + d) + (q - p - d)/k]$. For large k, Wei (1981) has shown that the sampled ARIMA (p, d, q) process may be represented by an IMA (d, l) process where $l = d - 1$. Earlier, Brewer (1973) had mistakenly given the limiting form for the ARMA (p, q) process. His incorrect result was based on the usual inspection of $r = [p + (q - p)/k]$ for large k. For $d = 0$, yielding an ARMA (p, q) process, the correct limiting process is white noise.

5. Variance estimation

There are two possibilities for variance estimation. One is to use only the systematic sample and attempt to estimate the variance from the single sample. The second method is to change the design either by taking additional systematic samples or additional sample or samples by a different design. The problem with the former method is that the variance estimate is usually biased; the problem with the latter method is that the convenience and possibly the efficiency of systematic sampling is lost.

5.1. Variance estimation from a single systematic sample

A single systematic sample is a cluster sample of one cluster so that there does not exist an estimator of the variance (2.1) which is unbiased with respect to the sampling design. In this situation estimates of the variance are usually based on an underlying model assumption. The variance estimate is then appropriate only when the assumed model reflects the true state of nature. This point appears to have been missed in an empirical study by Wolter and McCann (1977). They compared a number of variance estimators in seven artificial and six real populations. The variance estimates were calculated and evaluated whether or not the variance estimator was appropriate for the population under study.

For purposes of notational convenience the sample measurements

$y_r, y_{r+k}, \ldots, y_{r+(n-1)k}$ will be represented by $x_1, \ldots, x_n$, i.e. $x_j \equiv y_{r+(j-1)k}$ for $j = 1, \ldots, n$. The sample mean $\bar{y}_r \equiv \bar{x} = \Sigma_{j=1}^{n} x_j/n$.

When the population is in random order, model (3.11) applies. Then the estimate of variance, based on simple random sampling without replacement,

$$\frac{1}{n}\left(1 - \frac{1}{k}\right)\frac{1}{n-1}\sum_{j=1}^{n}(x_j - \bar{x})^2 \qquad (5.1)$$

is unbiased, with respect to model (3.11), for the variance (2.1), i.e. (5.1) averaged over model (3.11) is equal to (2.1) averaged over model (3.11). Estimator (5.1) appears in Cochran (1977, eq. 8.43).

When a linear trend is present model (3.1) may be considered. In this case, the estimator given by (3.6), Yates' end corrections estimator, is of interest and it is therefore of interest to estimate the mean square error of this estimator,

$$\sum_{r=1}^{k}(\bar{y}'_r - \bar{y})^2/k \qquad (5.2)$$

where $\bar{y}'_r$ is (3.6). The mean square error (5.2) averaged over model (3.1) yields

$$\left[\frac{k^2 - 1}{6(n-1)^2 k^2} + \frac{k-1}{nk}\right]\sigma^2. \qquad (5.3)$$

The variance estimate

$$\frac{k-1}{nk}\left[\frac{1}{n} + \frac{(2r-k-1)^2}{2(n-1)^2 k^2}\right]\sum_{j=1}^{n-2}\frac{(x_j - 2x_{j+1} + x_{j+2})^2}{6(n-2)} \qquad (5.4)$$

when averaged over the random starts $r = 1, \ldots, k$ and the model (3.1) yields (5.3). Hence (5.4) is unbiased for (5.2), the mean square error of Yates' end corrections estimator, with respect to a composite design-model expectation. When n is large, the term $(2r - k - 1)^2/[2(n-1)^2 k^2]$ may be dropped in (5.4). The variance estimator (5.4) is no longer unbiased when model (3.1) has heteroscedastic error variances, i.e. σ^2 in (3.1) is replaced by σ_u^2 ($u = 1, \ldots, N$). Estimator (5.4) also appears in Cochran (1977, eq. 8.45).

Several estimators of variance have been suggested when autocorrelation, given by model (4.1), is present in the population. Two estimators, given in Matérn (1960),

$$(\Delta^m x_1)^2 \bigg/ \binom{2m}{m}$$

and

$$[-x_1 + 2x_2 - 2x_3 + 2x_4 - \cdots + (-1)^m(2x_m - x_{m+1})]^2/(4m - 2), \qquad (5.5)$$

where $m \leq n$ and Δ^m is the mth finite difference operator, were shown to be biased under model (4.1) with $\rho(u) = \phi^u$. The estimator given in Cochran (1977, eq. 8.46), originally due to Yates (1949) is based on (5.5) with $m = 9$ and hence is also biased under model (4.1). Koop (1971) showed that the attempt to use replication techniques leads to biased results. He studied an estimator of variance based on splitting the sample in two half-samples. The most promising estimator for autocorrelated population is due to Cochran (1946). The estimator is based on the assumption of an autoregressive process of order one in the population, i.e. $\rho(u) = \phi^u$. Cochran's (1946) estimator, based on the asymptotic argument (4.7) is

$$s^2 \left\{ 1 + \frac{2}{k \ln \hat{\phi}} + \frac{2}{(\hat{\phi}^{-k} - 1)} \right\} / n,$$

where $s^2 = \Sigma_{j=1}^n (x_j - \bar{x})^2 / (n-1)$ and where $\hat{\phi}^k = [\Sigma_{j=1}^{n-1} (x_j - \bar{x})(x_{j+1} - \bar{x})]/[\Sigma_{j=1}^n (x_j - \bar{x})^2]$ is the estimated first order autocorrelation in the sample. An alternative estimator may be obtained by first order autocorrelation in the sample. An alternative estimator may be obtained by first setting $\rho(u) = \phi^u$ in (4.5) and then by replacing σ^2 by s^2 and ϕ by $\hat{\phi}$ in (4.5). Heilbron (1978), in an empirical study, has shown that these two estimators yield almost the same efficiency. Bellhouse and Sutradhar (1983), also in an empirical study, have shown that the second estimator, based on (4.5) with $\rho(u) = \phi^u$, performs very well against the alternative estimators given in Cochran (1977, eqs. 8.43, 8.45, and 8.46) when the population follows an autoregressive process of order one.

5.2. Variance estimation based on a change in the design

One way to obtain an unbiased estimate of variance is to choose more than one systematic sample. Gautschi's (1957) design is obtained by choosing l random starts without replacement from the k possible starts and systematic samples thereafter. Since this design yields a cluster sample of l clusters, an unbiased variance estimate may be obtained using the theory of cluster sampling with equal cluster sizes. However, this sampling design is less efficient than systematic sampling with a single radom start of equivalent sample size when the population is from an autocorrelated process following (4.1) to (4.4) or when a linear trend is present in the population. Also, some of the simplicity of systematic sampling is lost. Tornqvist (1963) suggested taking l independent systematic samples and using the technique of interpenetrating subsamples to obtain an estimate of variance. Since this method is based on sampling with replacement it will be less efficient than Gautschi's (1957) method no matter what the underlying model is.

Zinger (1963, 1964) suggested taking an additional simple random sample of size m without replacement from the $N - n$ remaining units remaining after the initial systematic sample has been taken. The idea behind the method is to retain the simplicity of systematic sampling and to make an additional expenditure to obtain an unbiased estimate of variance. Zinger found that an unbiased variance estimator based on the pooled estimate $(n\bar{y}_r + m\bar{y}_m)/(n + m)$, where $\bar{y}_r$ is the mean of the systematic sample with random start r and $\bar{y}_m$ is the mean of the random sample of size m, often yields negative estimates of variance. Zinger (1980) considered a class of estimators given by $(1 - \beta)\bar{y}_r + \beta\bar{y}_m$ where $0 < \beta < 1$. He found that when $\beta = 1/2$ he could obtain an estimate of variance which is always non-negative. Wu (1984) showed that a non-negative variance exists if and only if $\beta \geq (k - 1)/(2k)$ and gave a general form for the non-negative estimate of variance.

6. Spatial sampling

The theory of spatial sampling is concerned with two-dimensional populations such as fields or groups of contiguous quadrats. There are two approaches in the literature which have been used for spatial sampling: one originally due to Quenouille (1949); the other due to Zubrzycki (1958).

6.1. Quenouille's method of spatial sampling

Assume that the population consists of MN units, usually points or quadrats, arranged in M rows and N columns. There are three classes of sampling designs to choose mn sampling units: designs in which the sample units are aligned in both the row and column directions; designs in which the sample units are aligned in one direction only, say the rows, and unaligned in the column direction, and designs in which the sample units are unaligned in both directions. The three classes of sampling designs may be characterized in the following way. In sampling designs which have the sample units aligned in both directions, the number of sampled elements in any row of the population will be 0 or n and the number of sampled elements in any column of the population will be 0 or m. In sampling designs which have sample units aligned in the rows only the number of sample elements in any row of the population will be 0 or n and the number of sample elements in any column will be at most m. With the exception of simple random sampling without replacement of mn units from MN in the population, designs which have sample units unaligned in both directions are characterized by having at most n sample elements in any row and at most m elements in any column of the population.

To obtain a systematic sample in each type of alignment, assume further that $M = mk$ and $N = nl$ where m, n, k and l are integers. Let $r_1, \ldots, r_n$ be n integers between 1 and k and let $s_1, \ldots, s_m$ be m integers between 1 and l. Consider the set of points

$$S = \{(r_j + (i - 1)k, s_i + (j - 1)l): i = 1, \ldots, m; j = 1, \ldots, n\}.$$

An aligned systematic sample in both directions is obtained when, in S, $r_1 = \cdots = r_n = r$ with probability 1, $s_1 = \cdots = s_m = s$ with probability 1 and r and

s are chosen at random from the integers $1, \ldots, k$, and $1, \ldots, l$ respectively. A systematic sample which is aligned in the rows only is obtained when, in S, $r_1 = \cdots = r_n = r$ with probability 1 and when r is chosen at random from the integers $1, \ldots, k$ and $s_1, \ldots, s_m$ are chosen by random sampling with replacement from the integers $1, \ldots, l$. An unaligned systematic sample in both directions is obtained when, in S, $r_1, \ldots, r_n$ are chosen by random sampling with replacement from the integers $1, \ldots, k$ and $s_1, \ldots, s_m$ are chosen by random sampling with replacement from the integers $1, \ldots, l$. This latter method is known in the geographical literature as stratified unaligned systematic sampling, a term first used by Berry (1962).

Bellhouse (1981a) obtained the finite population variance for each of these systematic sampling designs as well as the associated stratified and random sampling designs. Under a two-dimensional linear trend model, the spatial analogue of (3.1), Bellhouse (1981a) showed that within each type of alignment stratified sampling was more efficient than systematic sampling which, in turn, was more efficient than random sampling. This corresponds to the one-dimensional results of Section 3.1. He also showed that unaligned designs were more efficient than aligned designs.

Koop (1976) has examined these spatial sampling methods in the estimation of cover-type areas on maps. He obtained the finite population variances for all the spatial sampling designs given above for an estimate of an area A_α on a map of total area A. However, only a limited number of efficiency comparisons could be made. Bellhouse (1981b) gave a model for the boundary of the area of interest and was able to show that unaligned stratified sampling is more efficient than aligned systematic sampling. The latter design was more efficient than aligned systematic sampling with multiple random starts.

6.2. Zubrzycki's method of spatial sampling

This approach, taken by Zubrzycki (1958), assumes a more general population structure. It is assumed only that the population consists of a number of non-overlapping domains which are congruent by translation. A systematic sample is obtained by choosing a point at random in one domain and then a point in each of the remaining domains by a translation which establishes the congruence of the domains. Other designs which could be considered are stratified sampling, in which a point is chosen at random in each domain, and random sampling, in which the points are chosen at random over the union of all the domains. With the addition of more population structure this approach was followed by Ripley (1981).

6.3. Spatial autocorrelation

Let y_a and y_b be any two measurements in the spatial population. The superpopulation model for spatial autocorrelation may be expressed as

$$E_m(y_a) = E_m(y_b) = \mu \,; \quad E_m(y_a - \mu)^2 = E_m(y_b - \mu)^2 = \sigma^2 \,;$$
$$E_m(y_a - \mu)(y_b - \mu) = \sigma^2 \rho(v, w) \,, \tag{6.1}$$

where the correlation function ρ depends on v, the absolute difference in the latitudes, and w, the absolute difference in the longitudes, for the points in plane associated with the measurements y_a and y_b. A further assumption often made is that

$$\rho(v, w) = \rho(d), \qquad (6.2)$$

where $d = (v^2 + w^2)^{1/2}$ is the Euclidean distance between the points.

Model (6.1) was used by Das (1950) and by Quenouille (1946) with $\rho(v, w) = \rho_1(v)\rho_2(w)$. Quenouille (1949) showed that unaligned sampling designs are more efficient than aligned sampling designs and that within each type of alignment the systematic sampling scheme was more efficient than either stratified or random sampling. Das (1950) also made some efficiency comparisons, but the results were restricted to three designs only. Bellhouse (1977) obtained the spatial analogue of Hajek's (1959) one-dimensional result. He showed that for correlation functions in (6.1) satisfying the restrictions $\rho(v, w) \geq 0$, $\Delta_v \rho(v, w) \leq 0$, $\Delta_w \rho(v, w) \leq 0$, $\Delta_v^2 \rho(v, w) \geq 0$, $\Delta_w^2 \rho(v, w) \geq 0$, $\Delta_v \Delta_w \rho(v, w) \geq 0$ and $\Delta_v^2 \Delta_w^2 \rho(v, w) \geq 0$, that within each type of alignment, a systematic sampling design is the most efficient sampling design in the class of designs with constant inclusion probability for each unit and the sample mean as the estimator. The operators Δ_v, Δ_v^2, Δ_w and Δ_w^2 are the first and second finite difference operators acting and v and w respectively. As in Hájek (1959), Theorem 4.1 above, Bellhouse (1977) gave only the restriction $\Delta_v^2 \Delta_w^2 \rho(v, w) \geq 0$. The remaining restrictions are, however, necessary to obtain convergence of the infinite sums in Bellhouse (1977, eq. 8).

Model (6.2) was studied by Zubrzycki (1958) and Hájek (1961). In the structure of section 6.2 Zubrzycki (1958) and Hájek (1961) both found that there is no simple relationship between systematic and stratified sampling, although Zubrzycki (1958) was able to show that stratified sampling is more efficient than random sampling. Dalenius et al. (1961) investigated the relationship further by obtaining optimal sampling schemes under model (6.2). The schemes consist of various triangular, rectangular and hexagonal lattices, the choice depending on the range of sampling point densities. Using the correlation structure $\rho(d) = e^{-\lambda d}$, Matérn (1960) earlier investigated the problem of the choice of sample points. For systematic sampling he found triangular lattices to be more efficient than square lattices, although the gains in efficiency were slight. For stratified sampling he found the opposite to be true—squares were better than triangles—and also found that hexagonal lattices were more efficient than either square or triangular lattices. In an empirical study of these sampling designs, he found that in systematic sampling, the square lattice design was usually slightly more efficient than the triangular design, the opposite of what the theory had predicted.

References

Bellhouse, D. R. (1977). Optimal designs for sampling in two dimensions. *Biometrika* **64**, 605–611.
Bellhouse, D. R. (1981a). Spatial sampling in the presence of a trend. *J. Statist. Plan. Inf.* **5**, 365–375.

Bellhouse, D. R. (1981b). Area estimation by point-counting techniques. *Biometrics* **37**, 303–312.
Bellhouse, D. R. (1984a). Optimal randomization for experiments in which autocorrelation is present. *Biometrika* **71**, 155–160.
Bellhouse, D. R. (1984b). On the choice of sampling interval in circular systematic sampling. *Sankhyā B*, 247–248.
Bellhouse, D. R. (1985). Systematic sampling of periodic functions. *Can. J. Statist.* **13**, 17–28.
Bellhouse, D. R. and Rao, J. N. K. (1975). Systematic sampling in the presence of a trend. *Biometrika* **62**, 694–697.
Berry, B. J. L. (1962). *Sampling, Coding, and Storing Flood Plain Data.* Agriculture Handbook No. 237, U.S. Dept. of Agriculture. U.S.G.P.O., Washington.
Blight, B. J. N. (1973). Sampling from an autocorrelated finite population. *Biometrika* **60**, 375–385.
Box, G. E. P. and Jenkins, G. M. (1976). *Time Series Analysis: Forecasting and Control.* 2nd Edn., Holden-Day, San Francisco.
Brewer, K. R. W. (1973). Some consequences of temporal aggregation and systematic sampling for ARMA and ARMAX models. *J. Econometrics* **1**, 133–154.
Buckland, W. R. (1951). A review of the literature of systematic sampling. *J. Roy. Statist. Soc. (B)* **13**, 208–215.
Cochran, W. G. (1946). Relative accuracy of systematic and random samples for a certain class of population. *Ann. Math. Statist.* **17**, 164–177.
Cochran, W. G. (1977). *Sampling Techniques.* 3rd Edn., Wiley, New York.
Connor, W. S. (1966). An exact formula for the probability that two specified sampling units will occur in a sample drawn with unequal probabilities and without replacement. *J. Amer. Statist. Assoc.* **61**, 384–390.
Das, A. C. (1950). Two-dimensional sampling and the associated stratified and random sampling. *Sankhyā* **10**, 95–108.
Dalenius, T., Hájek, J. and Zubrzycki, S. On plane sampling and related geometrical problems. *Proc. 4th Berkeley Symp.* **1**, 125–150.
Des Raj (1964). Systematic sampling with pps in a large scale survey. *J. Amer. Statist. Assoc.* **59**, 251–255.
Des Raj (1965). Variance estimation in randomized systematic sampling with probability proportional to size. *J. Amer. Statist. Assoc.* **60**, 278–284.
Gautschi, W. (1957). Some remarks on systematic sampling. *Ann. Math. Statist.* **28**, 385–394.
Hájek, J. (1959). Optimum strategy and other problems in probability sampling. *Cas. Pest. Mat.* **10**, 387–442.
Hájek, J. (1961). Concerning relative accuracy of stratified and systematic sampling in a plane. *Colloquium Mathematicum* **8**, 133–134.
Hájek, J. (1981). *Sampling From a Finite Population* Marcel Dekker, New York.
Hannan, E. J. (1962). Systematic sampling. *Biometrika* **49**, 281–283.
Hartley, H. O. (1966). Systematic sampling with unequal probabilities and without replacement. *J. Amer. Statist. Assoc.* **61**, 739–748.
Hartley, H. O. and Rao, J. N. K. (1962). Sampling with unequal probabilities and without replacement. *Ann. Math. Statist.* **33**, 350–374.
Heilbron, D. C. (1978). Comparison of estimators of the variance of systematic sampling. *Biometrika* **65**, 429–433.
Hidiroglou, M. A. and Gray, G. B. (1975). A computer algorithm for joint probabilities of selection. *Survey Methodology (Statistics Canada)* **1**, 99–108.
Hidiroglou, M. A. and Gray, G. B. (1980). Construction of joint probability of selection for systematic pps sampling. *Appl. Statist.* **29**, 107–112.
Iachan, R. (1981). An asymptotic theory of systematic sampling. Unpublished report.
Iachan, R. (1982a). Comparison of multiple random start systematic sampling schemes. Unpublished report.
Iachan, R. (1982b). Book review of J. Hájek, *Sampling From a Finite Population. J. Amer. Statist. Assoc.* **77**, 213.
Iachan, R. (1982c). Systematic sampling: A critical review. *Int. Statist. Rev.* **50**, 293–303.

Iachan, R. (1983). Reply to a letter to the editor by V. M. Joshi. *Amer. Statist.* **37**, 96.
Jones, A. E. (1948). Systematic sampling of continuous parameter populations. *Biometrika* **35**, 283–290.
Joshi, V. M. (1983). Letter to the editor. *Amer. Statist.* **37**, 96.
Jowett, G. H. (1952). The accuracy of systematic sampling from conveyor belts. *Appl. Statist.* **1**, 50–59.
Koop, J. C. (1971). On splitting a systematic sample for variance estimation. *Ann. Math. Statist.* **42**, 1084–1087.
Koop, J. C. (1976). Systematic sampling of two-dimensional surfaces and related problems. Research Triangle Institute Technical Report, NC.
Madow, L. H. and Madow, W. G. (1944). On the theory of systematic sampling. *Ann. Math. Statist.* **15**, 1–24.
Madow, W. G. (1953). On the theory of systematic sampling III: comparison of centred and random start systematic sampling. *Ann. Math. Statist.* **24**, 101–106.
Matérn, B. (1960). Spatial variation. *Medd. fr. Statens Skogsforsknings Institut.* **49**, 1–144.
Mickey, M. R. (1972). On the variance of the mean of systematically selected samples. In: T. A. Brancroft, ed., *Statistical Papers in Honor or George W. Snedecor*. Iowa State V.P., Ames, 227–243.
Murthy, M. N. (1967). *Sampling Theory and Methods*. Statistical Publishing Society, Calcutta.
Pinciaro, S. J. (1978). An algorithm for calculating joint inclusion probabilities under pps systematic sampling. *Proc. Surv. Res. Methods American Statistical Association*, 740.
Quenouille, M. H. (1949). Problems in plane sampling. *Ann. Math. Statist.* **20**, 335–375.
Rao, J. N. K. (1975). On the foundations of survey sampling. In: J. N. Srivastava, ed., *A Survey of Statistical Design and Linear Models*, North-Holland, Amsterdam, 489–505.
Rao, J. N. K. and Bellhouse, D. R. (1978). Optimal estimation of a finite population mean under generalized random permutation models. *J. Statist. Plan. Inf.* **2**, 125–141.
Ripley, B. D. (1981). *Spatial Statistics*. Wiley, New York.
Sedransk, J. (1969). Some elementary properties of systematic sampling. *Skand. Aktuar.* **52**, 39–47.
Singh, D., Jindal, K. K., and Garg, J. N. (1968). On modified systematic sampling. *Biometrika* **55**, 541–546.
Statistics Canada (1976). *Methodology of the Canadian Labour Force Survey*. Statistics Canada, Ottawa.
Sudakar, K. (1978). A note on circular systematic sampling. *Sankhyā C* **40**, 72–73.
Tornqvist, L. (1963). The theory of replicated systematic cluster sampling with random start. *Rev. Int. Statist. Inst.* **31**, 11–23.
Wei, W. W. S. (1981). Effect of systematic sampling on ARIMA models. *Comm. Statist. Theor. Meth.* **A10**, 2389–2398.
Williams, R. M. (1956). The variance of the mean of systematic samples. *Biometrika* **43**, 137–148.
Wolter, K. and McCann, S. (1977). Alternative estimators of variance for systematic sampling. *Proc. Social Statist. Sec.* American Statistical Association, Part II, 787–792.
Wu, C. F. (1984). Estimation in systematic sampling with supplementary observations. *Sankhyā B*, 306–315.
Yates, F. (1948). Systematic sampling. *Phil. Trans. Roy. Soc. (A)* **241**, 345–377.
Yates, F. (1949). *Sampling Methods for Censuses and Surveys*. 1st Edn., Griffin, London.
Zinger, A. (1963). Estimation de variances avec échantillonage systématique. *Revue de Statistique Appliquée* **11**, 89–97.
Zinger, A. (1964). Systematic sampling in forestry. *Biometrics* **20**, 553–565.
Zinger, A. (1980). Variance estimation in partially systematic sampling. *J. Amer. Statist. Assoc.* **75**, 206–211.
Zubrzycki, S. (1958). Remarks on random, stratified and systematic sampling in a plane. *Colloquium Mathematicum* **6**, 251–264.

Systematic Sampling with Illustrative Examples

M. N. Murthy and T. J. Rao

1. The basic procedure

Systematic sampling, in its simplest and commonly used form, is selection of every kth unit from a finite population of N units, k being an integer nearest to the inverse of the sampling fraction aimed at. Here k is termed the *sampling interval*, and if the first unit from 1 to k is selected at random, the resulting sample is said to be a systematic sample with a *random start*. The sample size is given by $n = (N/k)$, if N is an integral multiple of k with $N = nk$, and the sample size is a random variable taking the values $[N/k]$ and $[N/k] + 1$ with probabilities $[N/k] + 1 - N/k$ and $N/k - [N/k]$ respectively leading to the expected sample size (N/k), if N is not an integral multiple of k.

More specifically, the procedure consists of serially numbering the units in the population, $\{U_i\}$, $i = 1, 2, \ldots, N$, selecting a random start (r) from 1 to k, which is an integer nearest to the inverse of the sampling fraction, and then selecting the units

$$\{U_{r+jk}\}, \quad j = 0, 1, 2, \ldots, l_r - 1, \tag{1}$$

where l_r is an integer such that

$$(l_r - 1)k \leqslant N - r < l_r k \, .$$

In other words, starting from the rth unit, r being a random number from 1 to k, the sampling interval, every kth unit is selected till the entire population is covered in the process. This procedure is termed *linear systematic sampling*, as the selection is done in a linear direction, and abbreviated to lss. If N is a multiple of k with $N = nk$, then the systematic sample with random start consists of units $\{U_{r+jk}\}$, $j = 0, 1, 2, \ldots, (n-1)$.

It can be easily seen that in this sampling scheme, the arrangement of units has an important influence on its efficiency, as any change in the arrangement will change the composition of the sample possible through this scheme. For instance if $k \geqslant 2$, which is generally the case, no two consecutive units can occur in the same sample. Further, there is no possibility of the same unit being included more than once in a sample in this scheme, and the number of possible samples is only k, much less than the N_{C_n} samples possible in simple random sampling without replacement. This situation can be utilized to increase the efficiency of the scheme in estimating population parameters by adopting a suitable arrangement of the units in the population before selection. However, the noninclusion of some pairs of units in the samples makes it difficult to estimate the variance of the estimator based on a single sample unless some assumptions or approximations are resorted to.

Because of its simplicity and operational convenience systematic sampling is widely used in practice for selecting units, particularly in forest and land use surveys, in census work, in record sampling and in household and establishment surveys. For instance, it is operationally convenient to select every fifth household starting with a random household selected from the first five households, than select a 20% sample of households completely at random. Similarly, it is easier to select every tenth record from a mass of serially numbered records than selecting a 10% sample of records completely at random.

In the early years of the development of sampling methodology, systematic sampling was used in forest and land use surveys, where cruising along a path and observing volume of timber or cultivation in strips selected systematically, known as *strip sampling*, was found to be operationally feasible for collecting such data on a sample basis (Hasel, 1938; Osborne, 1942; Finney, 1948; Nair and Bhargava, 1951; Mokashi, 1954). The U.S.A. used systematic sampling in the 1940 population census for selecting a 1 in 20 sample of persons for getting information on supplementary questions, (Stephan, Deming and Hansen, 1940) and in India this technique was used to select and preserve a 2% sample of individual slips from the 1941 population census to enable tabulation of data on age and means of livelihood, as the financial and other resources available at the time were inadequate for tabulation of data on a 100% basis, (Lahiri, 1954a).

Following the theoretical framework developed by W. G. Madow and L. H. Madow (1944), Cochran (1946), L. H. Madow (1946) and W. G. Madow (1949, 1953), use of systematic sampling became common not only in forest and land use surveys, but also in household enquiries, record sampling, spatial sampling, industrial sampling and for obtaining quick tabulations, additional information and estimating non-sampling errors in census work.

Systematic sampling is simple to understand and carry out in practice. Training in the use of this technique can easily be imparted to lay staff and it can conveniently be used under normal conditions of large-scale operations. This technique can also be used when a sampling frame is not readily available, as a systematic sample can be selected by serially numbering the units while progressively listing them ensuring at least approximately a given sampling fraction. In some situations, such as strip sampling, even serial numbering may not be necessary at the listing stage, as serial numbering and sampling interval can be conceptually linked with counting, distance, time or some other measure. Thus systematic sampling is simpler, quicker, cheaper, operationally convenient and, with a suitable arrangement of units, more efficient than many other schemes. However, use of systematic sampling may result in certain types of bias in selecting units and these are considered in Section 6. Further, systematic sampling should be used with great care as a 'bad' arrangement of units makes this technique less efficient than other schemes.

A sample is said to be systematic if its members are selected according to some deterministic rule (Jones, 1948). In its generalized form, systematic sampling is forming clusters which are generally mutually exclusive but can be overlapping in special cases, following a certain pattern, and selecting one cluster at random or with some specified probability. Many variations of systematic sampling have been proposed by various authors and these are considered in Section 7.

Representations of linear systematic samples are given in Figure 1 and an enumeration of all the k possible systematic samples in lss is shown in Table 1.

✳–⊖– – – –✳–⊖– – – –✳–⊖– – – –✳–⊖– – – –✳–⊖– – –✳–⊖– – – –✳–⊖– – – –✳

Fig. 1. Showing linear systematic samples selected with sampling interval 7 and random starts 1 and 3 from a population of 50 units.

Table 1
Showing all the k possible systematic samples from a population of N units in linear systematic sampling with interval k

Random start	Sample observations					No. of units in sample
1	Y_1	Y_{1+k}	Y_{1+2k} $\cdots Y_{1+jk}$	$\cdots Y_{1+(l_1-1)k}$		l_1
2	Y_2	Y_{2+k}	$Y_{2+2k}\cdots Y_{2+jk}$	$\cdots Y_{2+(l_2-1)k}$		l_2
$\vdots$	$\vdots$	$\vdots$	$\vdots$	$\vdots$		$\vdots$
r	Y_r	Y_{r+k}	$Y_{r+2k}\cdots Y_{r+jk}$	$\cdots Y_{r+(l_r-1)k}$		l_r
$\vdots$	$\vdots$	$\vdots$	$\vdots$	$\vdots$		$\vdots$
k	Y_k	Y_{2k}	Y_{3k}	$\cdots Y_{(j+1)k}$	$\cdots Y_{l_k k}$	l_k

Note: $(l_r - 1)k \leq N - r < l_r k$, $l_r = [N/k] + 1$ for $r = 1$ to q and $l_r = [N/k]$ for $r = q + 1$ to k, where $q = N - [N/k]k$ and $\Sigma_{r=1}^{k} l_r = N$.

2. Estimation and sampling variance

If Y_i is the value of a characteristic y for the ith unit U_i, $i = 1, 2, \ldots, N$, it is generally of interst to estimate the population total $Y = \Sigma_{i=1}^{N} Y_i$ and the population mean $\bar{Y} = (Y/N)$ on the basis of a linear systematic sample selected with a random start. Let

$$\{Y_{r+jk}\}, \quad j = 0, 1, 2, \ldots, (l_r - 1), \tag{2}$$

be the sample observations, where l_r is as defined in (1). Then unbiased estimators of Y and $\overline{Y}$ are provided by

$$\hat{Y} = k \sum_{j=0}^{l_r-1} Y_{r+jk} = kl_r \overline{Y}_r \tag{3}$$

and

$$\hat{\overline{Y}} = \frac{k}{N} \sum_{j=0}^{l_r-1} Y_{r+jk} = \frac{kl_r}{N} \overline{Y}_r . \tag{4}$$

It is to be noted that the simple sample mean is not an unbiased estimator of the population mean in the general case of lss. However, if N is a multiple of k, then $l_r = n$ for all r and $kn = N$, in which case $\hat{Y} = N\overline{Y}_r$ and $\hat{\overline{Y}} = \overline{Y}_r$.

In what follows, we consider only the estimation of $\overline{Y}$, as the estimator of Y and its sampling variance can be obtained by multiplying the corresponding expressions for $\overline{Y}$ by N and N^2. Further, an estimator of a proportion P of units having a specified characteristic can also be obtained from the relevant expressions for $\overline{Y}$ by taking $Y_i = 1$ or 0, $i = 1, 2, \ldots, N$, depending on whether U_i has the specified characteristic or not. Similarly, an approximate expression for the variance of a ratio estimator $(\hat{\overline{Y}}/\hat{\overline{X}})\overline{X}$ of $\overline{Y}$ can be derived from that of $V(\hat{\overline{Y}})$ by applying the formula to the variable $Y_i - RX_i$, where R is the population ratio being estimated. Also an approximation to the variance of a product estimator $(\hat{\overline{Y}}\hat{\overline{X}})/\overline{X}$ can be obtained from $V(\hat{\overline{Y}})$ by applying the formula to the variable $Y_i + RX_i$.

The sampling variance of $\hat{\overline{Y}}$ given in (4) is given by

$$V(\hat{\overline{Y}}) = \frac{1}{k} \sum_{r=1}^{k} \left(\frac{kl_r}{N} \overline{Y}_r - \overline{Y} \right)^2 , \tag{5}$$

which can be written in the form

$$V(\hat{\overline{Y}}) = \frac{k}{N} \sigma^2 \left\{ 1 + \frac{\rho_c}{N} \sum_{r=1}^{k} l_r(l_r - 1) \right\}$$

$$+ \frac{k}{N^2} \overline{Y}^2 \left\{ \sum_{r=1}^{k} l_r^2 \left(2 \frac{\overline{Y}_r}{\overline{Y}} - 1 \right) - \frac{N^2}{k} \right\} , \tag{6}$$

where

$$\sigma^2 = \frac{1}{N} \sum_{i=1}^{N} (Y_i - \overline{Y})^2$$

and

$$\rho_c = \frac{2 \sum_{r=1}^{k} \sum_{j=0}^{l_r-1} \sum_{j'>j}^{l_r-1} (Y_{r+jk} - \overline{Y})(Y_{r+j'k} - \overline{Y})}{\sigma^2 \sum_{r=1}^{k} l_r(l_r - 1)} .$$

When N is a multiple of k, resulting in $l_r = n$ and $N = nk$, the variance in (5) and (6) simplify to

$$V(\hat{\bar{Y}}) = \frac{1}{k} \sum_{r=1}^{k} (\bar{Y}_r - \bar{Y})^2 \qquad (7)$$

and

$$V(\hat{\bar{Y}}) = \frac{\sigma^2}{n} \{1 + (n-1)\rho_c\}, \qquad (8)$$

where

$$\rho_c = \frac{2 \sum_{r=1}^{k} \sum_{j=0}^{n-1} \sum_{j'>j}^{n-1} (Y_{r+jk} - \bar{Y})(Y_{r+j'k} - \bar{Y})}{kn(n-1)\sigma^2},$$

since the additional term in (6) vanishes in this case. Here ρ_c stands for the intraclass correlation coefficient and measures the correlation between pairs of units in the systematic samples. From the expressions (6) and (8), it is clear that ρ_c, and hence the arrangement of units in the population, plays an important part in determining the sampling efficiency of systematic sampling. This situation can also be seen by considering the total variance as

$$\sigma^2 = \sigma_b^2 + \sigma_w^2, \quad \sigma_b^2 = \frac{1}{k} \sum_{r=1}^{k} (\bar{Y}_r - \bar{Y})^2$$

and

$$\sigma_w^2 = \frac{1}{N} \sum_{r=1}^{k} \sum_{j=0}^{n-1} (Y_{r+jk} - \bar{Y}_r)^2, \qquad (9)$$

where σ_b^2 is the variance between sample means and σ_w^2 is the variance within samples. Here σ_b^2 is the sampling variance given in (7) and (8) and

$$\sigma_w^2 = \sigma^2 - \sigma_b^2 = \frac{n-1}{n} \sigma^2 (1 - \rho_c). \qquad (10)$$

A high negative value of ρ_c, or equivalently a large value for σ_w^2 will make systematic sampling very efficient, and one way of ensuring this is to arrange the units in ascending or descending order of values of y or of a related variable, which will lead to systematic samples that are heterogeneous within themselves and homogeneous between themselves. However, it is to be noted that ρ_c and σ_b^2 are implicit functions of the sample size n and that the behaviour of the sampling variance with increase in sample size depends on the arrangement of the units and the variance need not necessarily be a decreasing function with n for a specific arrangement.

Comparing (8) and (9), ρ_c can be expressed in terms of σ_b^2/σ^2 or σ_w^2/σ^2 which are proportions of between and within sample variances to total variance, namely

$$\rho_c = \frac{1}{n-1}\left\{n\frac{\sigma_b^2}{\sigma^2} - 1\right\} \tag{11}$$

or

$$\rho_c = 1 - \frac{n}{n-1}\frac{\sigma_w^2}{\sigma^2}. \tag{12}$$

These forms are found convenient in computing ρ_c in some situations. Noting that the variance ratios σ_b^2/σ^2 and σ_w^2/σ^2 are non-negative and not more than unity, we get the limits of ρ_c as

$$-\frac{1}{n-1} \leqslant \rho_c \leqslant 1. \tag{13}$$

The dependence of the sampling variance on the arrangement of the units is explicitly brought out by expressing it in terms of circular serial correlation coefficients

$$V(\hat{\bar{Y}}) = \frac{\sigma^2}{n}\left\{1 + 2\sum_{d=1}^{m} \rho_{kd}\right\}, \tag{14}$$

where $m = n/2$ if n is even and $m = (n-1)/2$ if n is odd, and

$$\rho_{kd} = \frac{1}{N\sigma^2}\sum_{i=1}^{N}(Y_i - \bar{Y})(Y_{i+kd} - \bar{Y}),$$

with $r + dk$ taken as $r + dk - N$, when $r + dk > N$. From this expression also, it can be seen that the efficiency of systematic sampling for estimating Y is very high if the population values separated by k or multiples of k units are dissimilar, resulting in heterogeneous systematic samples and negative values of $\Sigma_{d=1}^{m} \rho_{kd}$.

In lss, unbiased estimation of variance of the estimator of population mean or total is not possible owing to the fact that not all the (N_{C_2}) pairs of population units get a chance to be included in the samples. In fact, the number of pairs included in the samples in lss is given by $\frac{1}{2}\Sigma_{r=1}^{k} l_r(l_r - 1)$ and the number of pairs not included in any systematic sample is $\frac{1}{2}(N^2 - \Sigma_{r=1}^{k} l_r^2)$, the corresponding figures in the particular case of N being a multiple of k are $\frac{1}{2}N(n-1)$ and $\frac{1}{2}N(N-n)$. Some variance estimators in lss are considered in Section 4 resorting to some assumptions and approximations.

3. Efficiency of systematic sampling

As mentioned in Sections 1 and 2, the sampling efficiency of systematic sampling depends on sample size and the intraclass correlation coefficient between units within samples, which in turn depends on the arrangement of units and is an implicit function of the sample size. In this Section the efficiency of systematic sampling is compared with those of simple random sampling (srs) with replacement and without replacement and stratified sampling and its behaviour is studied for certain types of arrangement of the units. For the sake of simplicity and clarity, we consider only the special case of N being a multiple of k with $l_r = n$ and $N = nk$ in this Section. However, the results discussed are usually applicable to the general case of the lss.

3.1. Comparison with simple random sampling

In selecting n units from a finite population of N units with equal probability with replacement (srswr), an unbiased estimator of $\overline{Y}$ and its variance are given by

$$\hat{\overline{Y}} = \bar{y} \quad \text{and} \quad V(\hat{\overline{Y}}) = V' = \frac{\sigma^2}{n}, \tag{15}$$

where $\bar{y}$ is the sample mean. Similarly, in simple random sampling without replacement (srs wor), an unbiased estimator of $\overline{Y}$ and its variance are given by

$$\hat{\overline{Y}} = \bar{y} \quad \text{and} \quad V(\hat{\overline{Y}}) = V'' = \frac{N-n}{N-1}\frac{\sigma^2}{n}. \tag{16}$$

Comparing the expression for the sampling variance V in the case of systematic sampling given in (8) and (9) with those of srswr and srs wor given in (15) and (16), we find that $V \leq V'$ if $\rho_c \leq 0$ and $V \leq V''$ if $\rho_c \leq -1/(N-1)$, and the efficiency of systematic sampling compared with those of srswr and srs wor are

and

$$E' = \frac{V'}{V} = \frac{1}{n}\frac{\sigma^2}{\sigma_b^2} = \frac{1}{1+(n-1)\rho_c} \tag{17}$$

$$E'' = \frac{V''}{V} = \frac{N-n}{N-1}\frac{1}{n}\frac{\sigma^2}{\sigma_b^2} = \frac{N-n}{N-1}\frac{1}{1+(n-1)\rho_c}. \tag{18}$$

Thus we see that the efficiency of systematic sampling compared to srs will be high when ρ_c is made highly negative by adopting a suitable arrangement of units before selection. If ρ_c is 0, then systematic sampling is as efficient as srswr, but is less efficient than srs wor. However, if ρ_c is positive and large, systematic sampling is highly inefficient compared to the other two schemes considered here.

3.2. Comparison with stratified sampling

Systematic sampling is basically grouping of the units into n groups of k units each ($N = nk$) and selecting one unit from each group in a systematic manner.

Thus it is comparable to stratified sampling with n strata and one unit per stratum. If $\{Y_{sj}\}$ is the value of the jth unit in the sth group or stratum $s = 1, 2, \ldots, n;\ j = 1, 2, \ldots, k$, then an unbiased estimator of $\bar{Y}$ based on a stratified simple random sample of one unit per stratum is given by the sample mean $(1/n) \sum_{s=1}^{n} y_{si}$, where y_{si} stands for the value of y for the sample observation in the sth stratum. The sampling variance is given by

$$V(\hat{\bar{Y}}_{st}) = V''' = \frac{1}{n^2} \sum_{s=1}^{n} V(y_{si}) = \frac{\sigma_{wst}^2}{n}, \qquad (19)$$

where

$$\sigma_{wst}^2 = \frac{1}{N} \sum_{s=1}^{n} \sum_{j=1}^{k} (Y_{sj} - \bar{Y})^2 .$$

With this notation, the sample observations in systematic sampling are given by $\{y_{si}\}$, $s = 1, 2, \ldots, n$, where the ith unit in each stratum is selected. An unbiased estimator of the population mean is given by $\hat{\bar{Y}} = \bar{y}_i$ and the variance V given in (8) can be rewritten in the form

$$V(\hat{\bar{Y}}) = V = \frac{\sigma_{wst}^2}{n}[1 + (n-1)\rho_c'], \qquad (20)$$

where

$$\rho_c' = \frac{2 \sum_{i=1}^{k} \sum_{s=1}^{n} \sum_{s'>s}^{n} (Y_{si} - \bar{Y}_s)(Y_{s'i} - \bar{Y}_{s'})}{kn(n-1)\sigma_{wst}^2} .$$

Here ρ_c' is the intraclass correlation coefficient between units in systematic samples with the deviations measured from the group or stratum means, whereas in ρ_c the deviations are measured from the overall population mean $\bar{Y}$.

By comparing (20) with (19), we can see that systematic sampling is more efficient than stratified sampling when ρ_c' is negative and that when ρ_c' is 0, the two schemes are equivalent. However systematic sampling is less efficient than stratified sampling when ρ_c' is positive and large. In fact, if ρ_c' is 0, then systematic sampling is comparable to stratified simple random sampling with proportional allocation. It is to be noted that when the units are arranged in increasing or decreasing order of the values of y or of a related characteristic, then systematic sampling is expected to be less efficient than stratified sampling with one unit per stratum, as in this case ρ_c' is likely to be positive due to the deviations of pairs of sample observations from the respective group or stratum means being positive or negative. In view of this, systematic sampling should not be considered in competition with stratified sampling, but as a complementary technique which can be used within strata to reduce variation when the strata are large with allocations greater than one unit.

The effect of stratification can broadly be achieved by putting together units having the same or similar characteristics in the arrangement before using systematic sampling, as this will ensure proportionate representation of the specified groups which can be domains of study. For instance, a serpentine serial numbering of area units will help in ensuring proper representation of different topographical features in an area sample selected systematically. Similarly, in selecting households systematically in sampled area units it is desirable to arrange them according to some dichotomy of the units such as farm households and non-farm households or some combinations of dichotomies such as farm households with less than 4 persons, farm households with 4 or more persons, non-farm households with 4 or more persons and non-farm households with less than 4 persons.

In industry, it is helpful to inspect products manufactured by sampling the units systematically over time to ensure proper representation of the entire production in a given period of time. An example of using systematic sampling over time to estimate the number of boats landing in a landing centre along a sea coast and the catch of marine fish is discussed in detail by Sukhatme, Panse and Sastry (1958).

However, it may be noted that in case of units arranged in increasing or decreasing order of the values of y or of a related characteristic, if the arrangement of units in alternative groups of k units are reversed before selection of a systematic sample, then the deviations of the sample observations from the respective group means in the neighbouring groups are expected to be of opposite signs leading to the possibility of systematic sampling being more efficient than stratified sampling.

3.3. Random arrangement of units

If the units in the population are arranged in a random order, then systematic sampling of n units from N units with k as interval ($nk = N$) is equivalent to srs wor, as in this case the number of possible samples taking into account all permutations of the N units is N_{C_n} and the probability of selecting any sample is $1/N_{C_n}$. Thus if we have reason to believe that the arrangement is random, then the results of simple random sampling can be applied to systematic sampling and we get the sample mean as an unbiased estimator of $\overline{Y}$ and its variance is given by (16). However, in practice, efforts are generally made to obtain a sampling frame with a suitable arrangement of units or to rearrange the units in a frame to make systematic sampling more efficient than srs. This situation is further considered in Sections 4 and 9.

3.4. Population with linear trend

To study the efficiency of systematic sampling in the presence of linear trend, we consider a hypothetical population where $Y_i = a + bi$, $i = 1, 2, \ldots, N$. Then

$$\overline{Y} = a + b\,\frac{N+1}{2} \quad \text{and} \quad \sigma^2 = b^2\,\frac{N^2 - 1}{12}.$$

In this case a systematic sample of n units drawn with k as the sampling interval and r as the random start has the following sample observations:

$$\{a + b(r + jk)\}, \quad j = 0, 1, 2, \ldots, (n-1), \tag{21}$$

in the special case $N = nk$. It can be seen that the sample mean

$$\overline{Y}_r = a + br + bk\frac{n-1}{2} \tag{22}$$

is an unbiased estimator of $\overline{Y}$ and its variance is given by

$$V(\overline{Y}_r) = V = b^2 \frac{k^2 - 1}{12}. \tag{23}$$

The variances of estimators of $\overline{Y}$ based on srswr, srs wor and stratified sampling with one unit per stratum are given by

$$V' = b^2 \frac{N^2 - 1}{12n}, \tag{24}$$

$$V'' = b^2 \frac{N-n}{N-1} \frac{N^2 - 1}{12n}, \tag{25}$$

$$V''' = b^2 \frac{k^2 - 1}{12n}, \tag{26}$$

showing that

$$V''' < V < V'' < V' \tag{27}$$

for $n \geq 2$. Thus, for this population, exhibiting a perfect linear trend, stratified sampling is the most efficient, followed by systematic sampling, srs wor and srswr in that order. In case of N being very large, the variances are related by the relation

$$V''' : V : V'' : V' = \frac{1}{n^2} : \frac{1}{n} : 1 : 1. \tag{28}$$

However, as mentioned in Section 3.2, if the arrangement of units in the alternate groups starting from the second group are reversed, then the sample mean for each sample is equal to $\overline{Y}$, showing that systematic sampling is more efficient than all the other three schemes.

When trend is present in the population, it may be desirable to weight the first and the last sample observations a little differently from the equal weights of $(1/n)$ given to other units, as this is expected to increase the efficiency of the estimator of Y. The suggested weights are

$$\frac{1}{n} + \frac{2r - k - 1}{2(n-1)k} \quad \text{and} \quad \frac{1}{n} - \frac{2r - k - 1}{2(n-1)k}, \tag{29}$$

which are determined so as to make the lss estimator of the mean for the population with perfect linear trend considered in this Section equal to $\overline{Y}$, giving rise to zero variance. When applied to populations exhibiting a rough trend, the adjusted estimator of $\overline{Y}$ is expected to have smaller variance, though it may be slightly biased. These corrections applied to the first and the last sample observations in a systematic sample are termed *end corrections* (Yates, 1948).

A representation of a population with linear trend and systematic samples drawn from it are shown in Figure 2.

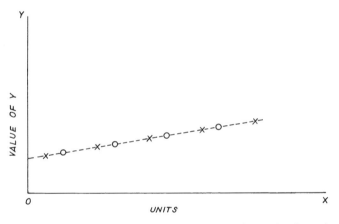

Fig. 2. Showing a population with linear trend and two systematic samples drawn from a population of 32 units with sampling interval 7 and random starts 3 (×) and 5 (○).

3.5. Periodic variation

If the population exhibits a periodic variation in that the values initially increase and then decrease and this pattern gets repeated in the arrangement in a regular manner, then the efficiency of systematic sampling depends very much on the value of k. For instance, if k is equal to the period of the cyclical pattern or a multiple of it, then there is considerable heterogeneity between samples leading to a large sampling variance. On the other hand, if k is taken as half of the period or odd multiples of it, then systematic samples are heterogeneous within themselves leading to high efficiency.

Periodic variations are likely to occur in certain natural populations such as land fertility, forest growth, events over time, etc. and in records such as payroll, census list of individuals arranged by households, etc. It is important to study an arrangement of units for any periodicity before using systematic sampling as a knowledge of the presence of a cyclical pattern and its periodicity would help in making systematic sampling more efficient by a proper choice of k or by suitable rearrangement of the units. This matter is again referred to in Section 6.

The efficiency of systematic sampling in autocorrelated populations is discussed later in Section 9, where superpopulation models are considered.

A representation of a population with a cyclical pattern and systematic samples drawn from it are shown in Figure 3.

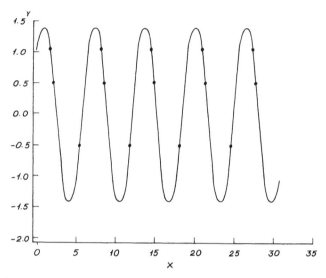

Fig. 3. Showing a cyclical population with a regular period and systematic samples drawn from it with sampling intervals equal to the period of the cycle and half the period.

4. Variance estimation

As mentioned in Section 1, it is generally not possible to estimate validly the sampling variance of the estimator of population mean based on a single systematic sample unless some assumptions or approximations are resorted to. For instance, if the arrangement of the units in the population is assumed to be random, then systematic sampling is equivalent to srs wor as mentioned in Section 3.3 and the sampling variance is given by

$$v_1(\hat{\bar{Y}}) = \frac{N-n}{Nn} s^2, \quad s^2 = \frac{1}{n-1} \sum_{i=1}^{n} (y_i - \bar{y})^2, \tag{30}$$

where $N = nk$. This variance estimator will be approximately valid if there is no perceptible trend or periodicity in the arrangement of units in the population. But in practice, units are arranged in a suitable manner to make systematic sampling more efficient than srs, in which case the variance estimator turns out to be a gross over estimate and hence the variance estimator (30) is not usually used. However, if systematic sampling is inefficient compared to srs, then $v_1(\hat{\bar{Y}})$ turns out to be an underestimate of the sampling variance.

Since systematic sampling can be considered as stratified simple random sampling with two units selected from each of $n/2$ strata, then an approximate variance estimator is given by

$$v_2(\hat{\bar{Y}}) = \frac{N-n}{Nn^2} \sum_{i=1}^{n/2} (y_{2i} - y_{2i-1})^2, \qquad (31)$$

where $\{Y_j\}$, $j = 1, 2, \ldots, n$, are the sample observations. An alternative variance estimator based on the same principle as (31), but taking into account successive differences is given by

$$v_3(\hat{\bar{Y}}) = \frac{N-n}{Nn} \frac{1}{2(n-1)} \sum_{i=1}^{n-1} (y_{i+1} - y_i)^2. \qquad (32)$$

Other variance estimators similar to (32) can be constructed by using *balanced differences*, such as

$$d_{3i} = \tfrac{1}{2} y_i - y_{i+1} + \tfrac{1}{2} y_{i+2} \qquad (33)$$

and

$$d_{5i} = \tfrac{1}{2} y_i - y_{i+1} + y_{i+2} - y_{i+3} + \tfrac{1}{2} y_{i+4} \qquad (34)$$

instead of successive differences and they are given by

$$v_{31}(\hat{\bar{Y}}) = \frac{N-n}{Nn} \frac{1}{1.5(n-2)} \sum_{i=1}^{n-2} d_{3i}^2 \qquad (35)$$

and

$$v_{32}(\hat{\bar{Y}}) = \frac{N-n}{Nn} \frac{1}{3.5(n-4)} \sum_{i=1}^{n-4} d_{5i}^2. \qquad (36)$$

The variance estimators in (35) and (36) are likely to be appropriate in case of systematic sampling estimator corrected by end corrections in the presence of linear trend and these are likely to be underestimates in general.

These variance estimators are based on the results obtained by Cochran (1946) and Yates (1948). These estimators are generally not unbiased in practical situations and hence should be used with considerable care. Efforts should be made to study the biases of the variance estimators in specific cases, as otherwise their use might result in misleading inferences based on the survey results. For instance, the variance estimators given here may be inappropriate if systematic sampling is made more efficient by reversing the order of units in alternate groups of the units mentioned earlier in Section 3.4.

To ensure high efficiency for the estimator based on systematic sampling benefitting from the 'good' arrangement, sometimes one systematic sample is drawn with a single random start and then to get an estimator of the sampling variance, this sample is split into two or more sub-samples systematically. For instance, units with odd serial numbers can be taken to form sub-sample A and those with even serial numbers to form sub-sample B, and assuming the two sub-samples are independently drawn, a variance estimator in this case is given by

$$v_4(\hat{\bar{Y}}) = \tfrac{1}{4}(\bar{y}_A - \bar{y}_B)^2 . \tag{37}$$

This variance estimator is biased, since

$$E[v_4(\hat{\bar{Y}})] = \left(1 - \frac{2\rho_0}{1 + \rho_0}\right) V(\hat{\bar{Y}}), \tag{38}$$

where ρ_0 is the correlation coefficient between the sub-sample means (Koop, 1971). If the value of ρ_0 is known from past experience, then an almost unbiased estimator of the sampling variance can be obtained as

$$v'_4(\hat{\bar{Y}}) = \frac{1 + \rho_0}{1 - \rho_0} v_4(\hat{\bar{Y}}) . \tag{39}$$

Instead of assuming that the two sub-samples are independently drawn, if we assume they are drawn with random starts selected without replacement, then the variance estimator can be taken as

$$v_{41}(\hat{\bar{Y}}) = \frac{k-1}{k} \tfrac{1}{4}(\bar{y}_A - \bar{y}_B)^2 . \tag{40}$$

In the general case of splitting a systematic sample into m sub-samples, the variance estimator given later in (42) can be used as an approximation to the variance.

None of the variance estimators based on a single linear systematic sample is unbiased due to the fact that not all the pairs of units get a chance of being included in the sample. However, this situation changes, when the sample is drawn in the form of m (≥ 2) systematic samples. For instance, if the sample is selected in the form of m independent systematic sub-samples with m random starts selected with replacement and with mk as the sampling interval when $N = nk$, then an unbiased variance estimator of the combined sample mean is given by

$$v_5(\hat{\bar{Y}}) = \frac{1}{m(m-1)} \sum_{i=1}^{m} (\bar{y}_i - \bar{y})^2 . \tag{41}$$

However, selection of sub-samples with independent random starts may lead to repeated sub-samples and to avoid this, the m sub-samples can be selected with

m random starts selected without replacement and in this case, the variance estimator is given by

$$v_{51}(\hat{\bar{Y}}) = \frac{k-1}{k} \frac{1}{m(m-1)} \sum_{i=1}^{m} (\bar{y}_i - \bar{y})^2, \tag{42}$$

where $\{\bar{y}_i\}$, $i = 1, 2, \ldots, m$, are sample means based on the m sub-samples. This technique known as *interpenetrating sub-samples* is generally used in practical sampling to estimate sampling variance unbiasedly (Mahalanobis, 1946). Though the variance estimator is unbiased, it is not likely to be precise if the number of sub-samples is small, and if m is large, then the combined estimator is likely to be less efficient than an estimator based on one systematic sample selected with one random start. Variance estimation in the context of super-population models is considered in Section 9.

5. Illustrative examples

The results discussed in the previous sections relating to the use of linear systematic sampling are illustrated in this section by applying them to two small populations. The populations considered are the following:
 A – a hypothetical population of 40 units exhibiting a steady linear trend in y; and
 B – data on number of seedlings in 80-feet bed.
Populations A and B are used to illustrate the efficiency of systematic sampling compared to srswr, srs wor and stratified sampling and to study the efficiency of some of the variance estimators given in Section 4. Population B is also con-

Table 2
A population of 40 units exhibiting a steady linear trend in the value of a variable y

Unit	y	Unit	y	Unit	y	Unit	y
(1)	(2)	(1)	(2)	(1)	(2)	(1)	(2)
1	0	11	10	21	23	31	41
2	1	12	11	22	25	32	43
3	2	13	12	23	29	33	46
4	3	14	12	24	30	34	50
5	4	15	13	25	32	35	52
6	5	16	14	26	33	36	53
7	7	17	15	27	35	37	57
8	7	18	17	28	38	38	59
9	8	19	20	29	39	39	62
10	9	20	22	30	40	40	63

Table 3
Number of seedlings in a 80-feet bed

1–10	11–20	21–30	31–40	41–50	51–60	61–70	71–80
(1)	(2)	(3)	(4)	(5)	(6)	(7)	(8)
26	16	27	37	4	36	20	21
28	9	20	14	5	20	21	26
11	22	25	14	11	43	15	16
16	26	39	24	9	27	14	13
7	17	24	18	25	20	13	11
22	39	25	17	16	21	9	19
44	21	18	14	13	18	25	27
26	14	44	38	22	19	17	29
31	40	55	36	18	24	7	31
26	30	39	29	9	30	30	29

sidered to study the behaviour of the sampling efficiency of systematic sampling compared to srs wor with increase in sample size.

To study the efficiency of systematic sampling compared to other sampling schemes, all the ten possible systematic samples of 4 units from Population A and all the ten possible systematic samples of 8 one-foot seed-beds from Population B are enumerated and the sampling variance is obtained as the variance of the estimates of population mean based on the possible samples. For this study the following four sampling schemes are considered:

(i) simple random sampling with replacement;
(ii) simple random sampling without replacement;
(iii) linear systematic sampling; and
(iv) stratified sampling with one unit per stratum.

The results of this study are given in Table 4.

Table 4
Variances of estimates of population means based on a sample of 4 units from 40 units of Population A and a sample of 8 one-foot seed-bed from Population B

Sampling scheme	Population A $n = 4$	Population B $n = 8$
Simple random sampling with replacement	90.76	12.86
Simple random sampling without replacement	83.78	11.72
Linear systematic sampling	23.48	16.54
Stratified sampling with one unit per stratum	6.65	9.52

From Table 4, it can be seen that stratified sampling is more efficient than all three schemes, but lss is more efficient than srs schemes only for Population A.

In case of sampling from Population A exhibiting a steady linear trend, the sampling variance for lss gets reduced to 0.26 when the end corrections mentioned in Section 3.4 are applied. Further, if the arrangement of units in sets 2 and 4 of Population A is reversed before selecting a systematic sample of 4 units, then the variance becomes 0.4.

For each of the ten possible systematic samples drawn from Populations A and B, the variance estimators v_1, v_2, v_3 and v_{41} given in Section 4 are computed and their expected values, variances, biases and mean square errors are shown in Table 5.

Table 5
Variance characteristics of four variance estimators computed from systematic samples of 4 units from Population A and of 8 units from Population B

Variance estimator characteristics	Variance estimators in systematic sampling			
	v_1	v_2	v_3	v_{41}
Population A				
Expected value	101.89	29.14	31.52	52.41
Sample bias	78.41	5.66	8.04	28.93
Sample variance	338.98	28.56	32.64	95.97
Mean square error	6487.11	60.60	97.28	932.91
Population B				
Expected value	11.10	9.91	10.41	23.37
Sampling bias	− 5.44	− 6.63	− 6.13	6.83
Sampling variance	24.31	24.05	29.38	451.59
Mean square error	53.90	68.01	66.96	498.24

From Table 5, it is clear that the variance estimator v_1, applicable only in the case of a random arrangement of the units in the population, is highly inefficient for Population A because of the presence of a steady linear trend, whereas it is reasonably efficient for Population B not having any specific trend. The variance estimators v_2 and v_3 have performed well for both Populations A and B, while v_{41} has a large bias for Population A and a large sampling variance for Population B. It is to be noted that all the four variance estimators considered here are biased and the bias of v_1 is particularly large for Population A. Thus in applying these variance estimators in practice, special studies are to be conducted to ensure that their biases and the sampling variances are not large enough to make the inference based on the results of systematic sampling misleading and ineffective.

To study the behaviour of the sampling efficiency of systematic sampling for Population B, all possible systematic samples of size 2, 4, 5, 8, 10, 20 and 40 are enumerated using one-foot bed as the sampling unit. The sampling variance of systematic sampling is computed as the variance of the sample means of the number of seedlings based on all the possible samples for each sample size. The variance of systematic sampling and its efficiency compared to srs wor measured as a variance ratio, E'' given in (18), is given in Table 6 along with the corresponding intraclass correlation coefficient.

Table 6
Sampling variance of systematic sampling and its efficiency expressed as a ratio of the variance of srs wor to that of systematic sampling together with ρ_c

Sample size	Sampling variance	Efficiency $E'' = v''/v$	Intraclass correlation ρ_c
2	43.37	1.171	−0.1571
4	23.04	1.074	−0.0349
5	19.87	0.983	−0.0086
8	16.54	0.709	−0.0409
10	11.17	0.816	−0.0095
16	3.45	1.511	−0.0309
20	0.654	5.977	−0.0459
40	0.0306	42.57	−0.0253

From Table 6, it is seen that the behaviour of the variance of systematic sampling and its efficiency compared to srs wor is rather irregular with increase in sample size. Though systematic sampling is more efficient for some sample sizes, there are also cases where it is less efficient than srs wor. It is to be noted that the efficiency of systematic sampling depends not only on the arrangement of the units but is also an implicit function of the sample size. The figures for ρ_c, defined in (8) and (11) or alternatively by

$$\rho_c = \frac{1}{n-1}\left\{\frac{N-n}{N-1}\frac{1}{E''} - 1\right\}, \tag{43}$$

also reflect a similar irregularity in its behaviour with increase in sample size. It may be noted that systematic sampling is highly efficient even when ρ_c is slightly negative provided the sample size is large making $(n-1\rho_c)$ nearly equal to -1. Similarly, even for a small positive value of ρ_c, systematic sampling is very inefficient when sample size is large making $(n-1)\rho_c$ a large positive value. Thus we see that in using systematic sampling, we should ensure that the arrangement of units adopted is efficient for the sample size being considered. This aspect requires special care when the arrangement of units exhibits some periodicity.

6. Bias in systematic sampling

Though it is possibe to obtain an unbiased estimator of the population mean based on a linear systematic sample selected with a random start from a sampling point of view, the estimator in some situations may turn out to be misleading because of the following two possibilities:

(i) deviations in practice in sample selection from the theoretically laid down procedure; and

(ii) high variability between samples due to presence of some pattern or patterns in the arrangement of the units in the population.

For instance, *line sampling*, where units are to be consecutively listed and a systematic sample is selected using a given random start and a sampling interval, requires that all the lines on a page are used up before starting listing on another page. However, in practice, for convenience, the operational procedure for sample selection is prescribed as taking all units listed on the first line of each page. Since the number of lines on a page is fixed and since there is a tendency on the part of enumerators to leave out the last few lines on a page if the space is not sufficient to list all the units in a group such as individuals of a household or family and to start listing the units of another group on the next page, a systematic sample of all the units listed on the first line of each page will lead to a higher sampling fraction than envisaged and to a higher proportion of units usually listed first in the groups under consideration such as the head of the household or family. This type of bias known as *line bias* has been discussed by Stephan, Deming and Hansen (1940) and Lahiri (1954a).

Since a systematic sample selected with a given random start and a specified sampling interval depends on the arrangement of the units in the population, if there is a conscious or unconscious bias in listing units such that specified types of units occur at approximately equal intervals, then under certain circumstances, some selected samples would have a large proportion of such units. This is a case of periodicity, which may lead to samples with high between variation and makes the estimates ineffective. Similarly, in strip sampling, if the crop growth shows regular fluctuations starting with thin growth in the beginning followed by thick growth, systematic sampling may result in samples with high between variation.

To avoid 'bias' in systematic sampling, it is important that sample selection is done according to an operational procedure closely following the theoretical procedure leading to unbiased estimation. To avoid high variation between systematic samples, the following steps may be considered:

   (i) to group the units into certain strata and select systematic samples within each stratum using independent random starts;
   (ii) to select a systematic sample in the form of two or more systematic samples with independent or linked random starts;
   (iii) to ensure that the sampling interval is not equal to the period or a multiple of it in case of a population exhibiting periodicity; and
   (iv) if the sampling interval cannot be changed in (iii), to rearrange the units in the population to break any periodicity in the original arrangement.

If in spite of the steps taken to improve the performance of systematic sampling, the selected sample does not reflect the population characteristics, then the possibility of using special estimation procedures needs to be considered.

7. Variations in systematic sampling

From the previous sections, it is clear that linear systematic sampling, though expected to be more efficient when the arrangement is 'good', does not admit of valid variance estimator without resorting to some assumptions and approximations. When N is not a multiple of k, lss results in a variable sample size and in this case the sample mean is not unbiased for the population mean. To obviate these difficulties and in some cases to increase the efficiency, a number of variations of systematic sampling have been proposed by many authors and some of these are considered in this Section.

7.1. Linear systematic sampling

This technique has been discussed in detail in the previous sections. It may be noted that when the sample size is not a submultiple of the population size, then the sample mean is biased for the population mean, but the bias is small and can be neglected if the sample size is large and in this case, the estimator, variance and variance estimator given for the case of $N = nk$ in Sections 2 and 4 can be used (Aoyama, 1951).

When N is not a multiple of n, then we can use the fractional interval k' ($= N/n$) without rounding it off to an integer. In this procedure, k' is expressed in its decimal form and the ith unit is selected if

$$(i - 1) < r + jk' \leq i, \quad j = 0, 1, 2, \ldots, (n - 1), \tag{44}$$

where r is a random number selected such that $r \leq k'$. For instance, if $N = 5$ and $n = 2$, then $k' = 2.5$ and in this case r is selected from 0.1 to 2.5 at random and the units are selected according to (44). The use of fractional interval is equivalent to association of n numbers with each unit and selecting a systematic sample from the Nn numbers using N as the sampling interval and a random start selected from 1 to N. For this procedure the sample mean is unbiased for the population mean and its variance is obtained as the between samples variation. In this case also, it is not possible to estimate the variance without resorting to some assumptions and approximations.

Since in large-scale sample surveys where systematic sampling is used in selecting households in sampled area units by enumerators, use of fractional interval is likely to create difficulties and this can be avoided by rounding off the fractional interval $k' = k + q$ ($q < 1$) to k or $k + 1$ with probabilities $k(1 - q)/k'$ and $q(k + 1)/k'$ respectively so that the expected value of the estimator $k' \Sigma Y_i$ is unbiased for the population total (Kish, 1965; Murthy, 1967).

7.2. Centrally located sample

Instead of selecting a random start from 1 to k, if we select the systematic sample with the start $(k + 1)/2$ when k is odd or any one of the two systematic samples with the starts $(k/2)$ and $(k/2) + 1$ when k is even, then we get a centrally located sample (cls). It can be seen that the estimate of the population mean based on a cls is biased, but the bias and the sampling variance are likely to be small for a population exhibiting a trend and this situation can also be expected in case of a population with periodicity when the sampling interval is a multiple of half the period is used. For the hypothetical population considered in Section 3.4, the cls estimate is equal to the population mean, making it the most efficient estimate in this case.

7.3. Circular systematic sampling

A simple modification of lss makes it possible to ensure a fixed sample size n and to make the sample mean unbiased for the population mean even in case of N being not a multiple of k. This technique consists in selecting a random start from 1 to N and selecting the units corresponding to $\{r + jk\}$, $j = 0, 1, 2, \ldots, (n - 1)$, where k is an integer nearest to (N/n) and $r + jk$ is taken as $r + jk - N$ whenever $r + jk > N$, and it is termed circular systematic sampling (css). In this case there are N possible samples and each unit occurs in n of these N samples, thus making the sample mean unbiased for the population mean. The sampling variance in this case is given by

$$V(\hat{\bar{Y}}) = \frac{1}{N} \sum_{r=1}^{N} (\bar{Y}_r - \bar{Y})^2 \qquad (45)$$

which can be rewritten as

$$V(\hat{\bar{Y}}) = \frac{\sigma^2}{n} \{1 + (n-1)\rho_c^*\}, \qquad (46)$$

where

$$\rho_c^* = \frac{2 \sum_{r=1}^{N} \sum_{j=0}^{n-1} \sum_{j'>j}^{n-1} (Y_{r+jk} - \bar{Y})(Y_{r+j'k} - \bar{Y})}{Nn(n-1)\sigma^2}$$

is the intraclass correlation coefficient between pairs of units in the N samples with the deviations measured from the population mean. Css reduces to lss when $N = nk$.

It may be noted that the sample mean is unbiased for the population mean for all vaues of k, though the spread of the sample and hence efficiency is better if k is taken as an integer nearest to (N/n). However, if repetition of the same unit in a sample is to be avoided, then it is desirable to take the sampling interval as $[N/n]$. It is shown that a necessary and sufficient condition for all samples in css to have distinct units is that N and k are relatively coprime (Sudakar, 1978). Css, in general, has the same difficulty as the lss in obtaining a valid variance estimator as not all the pairs of units get a chance of being selected. However, in a special case such as selecting $(N + 1)/2$ units with interval 2, it is possible to estimate the variance unbiasedly from a single css (Murthy, 1967).

Another advantage of the css is when the units are classified and arranged according to two dichotomies $(A, A'; B, B')$ giving rise to the arrangement $(AB, A'B, A'B', AB')$ ensuring proper representation not only for the four combined classes but also for the four single classes A, A', B and B' in css. When the population has a linear trend, then the end corrections for the end units in the css are given by $(1/n) + x$ and $(1/n) - x$, where

$$x = \frac{2r + (n-1)k - (N+1)}{2(n-1)k}, \tag{47}$$

which reduces to (29) when $N = nk$ (Bellhouse and Rao, 1975).

7.4. Balanced systematic sampling

If the units in the population are arranged in increasing or decreasing order of a variable y, then the optimum pairing of units in the sense of getting minimum variance for the sample mean based on one pair selected at random is given by pairs with units equidistant from the two ends, that is, the optimum pairing consists of the pairs

$$\{r, N - r + 1\}, \quad r = 1, 2, \ldots, N/2, \tag{48}$$

when N is even (Sethi, 1965). Thus if the sample size is 2, then pairing the units as in (48) and selecting one pair with equal probability results in an optimum sampling scheme where sample mean is more efficient than in any other pairing of units including lss. This procedure of balanced pairing can be applied to the balanced pairs to obtain samples of size 4, 8 and so on, using same or different auxiliary information for arrangement in successive pairings.

A population is first divided into $(n/2)$ (n even) groups of $2k$ units each ($nk = N$), then using the principle of optimum pairing or balancing, we select 2 units from each group in a systematic manner, choosing a pair in the first group with equal probability. Thus this technique consists in selecting a number r at random from 1 to k and selecting the units corresponding to

$$\{r + 2jk, 2(j+1)k - r + 1\}, \quad j = 0, 1, 2, \ldots, (n/2) - 1, \tag{49}$$

and this is termed *balanced systematic sampling* (bss) (Murthy, 1967). This procedure is expected to be more efficient than even the lss when the population exhibits a trend.

7.5. Modified systematic sampling

Following the principle of pairing using equidistant units, it is possible to modify the above scheme by selecting a number r at random from 1 to k and selecting the units corresponding to

$$\{r + jk, N - r - jk + 1\}, \quad j = 0, 1, 2, \ldots, (n/2) - 1, \tag{50}$$

when n is even, and this technique is termed *modified systematic sampling* (mss) and its performance under certain population models is considered by Singh,

Jindal and Garg (1968). Assuming the sample size to be even in bss and mss is not too restrictive in practice. The main difference between bss and mss is that in the former the balancing is done within each group of $2k$ units, whereas in the latter balancing is achieved for the population as a whole. However, it may be noted that though the pairing of units leads to optimum results, the formation of the entire sample using equidistant units may not necessarily be the optimum.

7.6. Multi-start systematic sampling

In Section 4, we have already considered the possibility of selecting a systematic sample in the form of m subsamples with m independent random starts to enable unbiased estimation of the variance of the combined estimator using (41). This procedure which is extensively used to estimate sampling variance in complex situations is termed interpenetrating subsamples (Mahalanobis, 1946), Tukey's independent samples (Deming, 1950) and replicated samples. If, however, the m random starts are selected without replacement to avoid repeating subsamples, then we get a multiple-start systematic sample (msss) and its properties have been studied by Gautschi (1957) and Shiue (1960). The variance estimator in this case is as given in (42).

7.7. Partially systematic sampling

Another apporach to getting an unbiased variance estimator in systematic sampling is to supplement the selected sample with a sample of m units selected with srs wor from the remaining $N - n$ units, and this scheme is termed partially systematic sampling (Hasel, 1938). Zinger (1980) has considered this scheme in detail and compared it with the msss. However, the variance estimator in this case is relatively more complex than in the msss. Wu (1981) has given further results in case of this scheme.

7.8. Balanced random sampling

Another scheme which enables unbiased estimation of variance in systematic sampling is to select a sample of $(n/2)$ units from the first $(N/2)$ units in the population with srs wor and to supplement it by a sample of $(n/2)$ units from the remaining $(N/2)$ units using the principle of balancing mentioned in Section 7.4. The sample in this case consists of units corresponding to

$$\{r_j, N + 1 - r_j\}, \quad j = 1, 2, \ldots, (n/2), \tag{51}$$

where r_j is the first sample of $(n/2)$ from the first $(N/2)$ units. This scheme is termed *balanced random sampling* (brs) (Padam Singh and Garg, 1979). The inclusion probabilities π_i and π_{ij} are given by

$$\pi_i = \frac{n}{N} \quad \text{for all } i,$$

$$\pi_{ij} = \frac{n}{N} \qquad \text{for } i + j = N + 1, \tag{52}$$

$$= \frac{n(n-2)}{N(N-2)} \quad \text{otherwise}.$$

We can obtain the Horvitz and Thompson (1952) estimator and the Yates and Grundy (1953) variance estimator by using the values of π_i and π_{ij}.

When N is even and n is odd, the technique consists of selecting a brs of $(n-1)$ units from N units and supplementing it with one unit selected at random from the remaining $N - n + 1$ units. If N is odd and n is even, one unit is selected from N units, then a brs of size $(n-2)$ is selected from $(N-1)$ units and then one unt is chosen from the remaining $(N - n + 1)$ units. When both N and n are odd, one unit is selected from N units and this is supplemented by a brs of size $(n-1)$ selected from the remaining $(N-1)$ units. The inclusion probabilities for these cases are given in the paper referred in this Section.

7.9. New systematic sampling

Yet another scheme which provides an unbiased variance estimator in systematic sampling is to select a sample of u consecutive units starting with a random number chosen from 1 to N and then selecting v units circular systematically with d as the sampling interval, $(u + v = n)$. The sample consists of units corresponding to

$$\{r + i, r + u - 1 + jd\}, \quad i = 0, 1, 2, \ldots, (u-1), \; j = 1, 2, \ldots, v. \tag{53}$$

This technique is termed *new systematic sampling* (nss) (Singh and Padam Singh, 1977). It has been shown by the authors of this scheme that a sufficient condition for all units in a sample to be distinct is $u + vd \leq N$ and that the inclusion probabilities for all the pairs of units will be non-zero if $d \leq u$ and $u + vd \geq (N/2) + 1$, which leads to unbiased estimation of the variance. However, this requires that the sample size should be at least $\sqrt{(2N+4)} - 1$. The efficiency of the technique under different population models has been discussed in detail in the paper already referred to in this Section.

7.10. An empirical study

An empirical study is conducted to study the behaviour of the sampling variance and the efficiency of some selected variations of systematic sampling with increasing sample size using the following three populations:

A – Population of 40 units showing a rising trend (Table 2);
B – Population of 128 villages (Murthy, 1967, pp. 127–130); and
C – Population of 80 factories arranged in increasing order of number of workers (Murthy, 1967, p. 228).

The characteristics considered are value of y for Population A, number of cultivators for Population B and fixed capital and output for Population C. For Population A and C, the study is done with the arrangement of the units as mentioned above and for Population B, the study is carried out both for the frame arrangement and arrangement in ascending order of number of households. The results of the study in the form of sampling variance for the estimation of the population mean and its efficiency compared to srs wor are given in Table 7 for different sample sizes.

From Table 7, it can be seen that the behaviour of sampling variances and efficiencies of different variations of systematic sampling is rather irregular, as they are implicit functions of the arrangement of the units and the sample size. Generally, when the arrangement is in ascending order of a related variable, the selected versions of systematic sampling have performed better than srs wor, whereas their performance is relatively worse in some cases when the frame

Table 7
Sampling variances and efficiencies (in brackets) of selected variations of systematic sampling compared to simple random sampling for different sample sizes

Sample size	Sampling variance and efficiency (in brackets) of scheme						
	srs wor	lss	bss	mss	brs	msss	
						$m = 2$	$m = 4$
(1)	(2)	(3)	(4)	(5)	(6)	(7)	(8)

A: Population of 40 units; arrangement: increasing order of y; variable: y

2	176.870	81.700	9.100	9.100	9.100	176.870	–
	(100)	(216)	(1943)	(1943)	(1943)	(100)	
4	83.781	23.475	0.400	2.362	4.311	38.700	83.781
	(100)	(357)	(20945)	(3546)	(1943)	(216)	(100)
8	37.236	6.225	0.019	0.781	1.916	10.433	17.200
	(100)	(598)	(198591)	(4766)	(1943)	(357)	(216)

B1: Population of 128 villages; arrangement: as in frame; variable: number of cultivators

2	162373	216702	149763	149763	149759	162374	–
	(100)	(75)	(108)	(108)	(108)	(100)	
4	79898	108318	99995	70316	73691	106631	79898
	(100)	(74)	(80)	(114)	(108)	(75)	(100)
8	38660	47793	37393	41783	35657	52412	51596
	(100)	(81)	(103)	(93)	(108)	(74)	(75)
16	18041	17125	21036	9405	16639	22303	24459
	(100)	(105)	(86)	(192)	(108)	(81)	(74)
32	7732	1722	2218	4943	7131	7339	9559
	(100)	(449)	(349)	(156)	(108)	(105)	(81)
64	2577	1547	128	4793	2377	574	2446
	(100)	(167)	(2008)	(54)	(108)	(449)	(105)

Table 7 (*Continued*)

Sample size	Sampling variance and efficiency (in brackets) of scheme						
	srs wor	lss	bss	mss	brs	msss	
						$m = 2$	$m = 4$
(1)	(2)	(3)	(4)	(5)	(6)	(7)	(8)

*B*2: Population of 128 villages; arrangement: increasing order of number of households; variable: number of cultivators

Sample size	srs wor	lss	bss	mss	brs	msss ($m=2$)	msss ($m=4$)
2	162375 (100)	138241 (117)	64883 (250)	64883 (250)	64881 (250)	162375 (100)	–
4	79898 (100)	40298 (198)	28859 (277)	25919 (308)	31926 (250)	68023 (117)	79899 (100)
8	38661 (100)	22664 (171)	16000 (242)	12238 (316)	15448 (250)	19499 (198)	32915 (117)
16	18042 (100)	10022 (180)	8529 (212)	4709 (383)	7209 (250)	10577 (171)	9100 (198)
32	7732 (100)	2429 (318)	3235 (239)	1407 (550)	3090 (250)	4295 (180)	4533 (171)
64	2577 (100)	1360 (190)	685 (376)	195 (1321)	1030 (250)	810 (318)	1432 (180)

*C*1: Population of 80 factories; arrangement: increasing order of number of workers; variable: fixed capital (in '00 $)

Sample size	srs wor	lss	bss	mss	brs	msss ($m=2$)	msss ($m=4$)
2	348587 (100)	226902 (154)	94663 (368)	94663 (368)	94660 (368)	348588 (100)	–
4	169825 (100)	58429 (291)	3577 (4747)	25387 (669)	46117 (368)	110542 (154)	169825 (100)
8	80443 (100)	15559 (517)	1705 (4717)	7032 (1144)	21845 (368)	27677 (291)	52362 (154)
16	35753 (100)	6035 (592)	169 (21170)	2706 (1321)	9709 (368)	6915 (517)	12301 (291)

*C*2: Population of 80 factories; arrangement: increasing order of number of persons; variable: output (in '00 $)

Sample size	srs wor	lss	bss	mss	brs	msss ($m=2$)	msss ($m=4$)
2	1642696 (100)	970074 (169)	34883 (4709)	34883 (4709)	34895 (4708)	1642696 (100)	–
4	800288 (100)	335913 (238)	9896 (8087)	7162 (11174)	17000 (4708)	472600 (169)	800288 (100)
8	379083 (100)	95473 (397)	4666 (8124)	2095 (18094)	8053 (4708)	159117 (238)	223863 (169)
16	168482 (100)	22052 (764)	1342 (12555)	844 (19957)	3579 (4708)	42432 (397)	70719 (238)

arrangement is used for Population B. It is evident that of all the schemes considered, bss and mss are very efficient when the arrangement of the units is favourable. The performance of bss has particularly been outstanding for Population A. Also when the arrangement is good, msss is more efficient when m, the number of subsamples, is 2 than when $m = 4$.

8. Systematic sampling with probability proportional to size

When the units in the population differ considerably in respect of a size measure x, positively correlated with the study variable y with the line of regression passing through the origin, then very efficient estimators of the population mean or total can be obtained by combining systematic sampling with probability proportional to size (pps) sampling. This is done by associating with each unit a number of numbers equal to its size measure and selecting a systematic sample using a systematic sample of the associated numbers. This technique consists of the following steps:

(i) cumulate the sizes of the units, $T_i = T_{i-1} + X_i$;
(ii) determine the sampling interval, $I = T/n \ (= X/n)$;
(iii) select a number r at random from 1 to I;
(iv) select unit U_i if $T_{i-1} < r + jk \leq T_i$, $j = 0, 1, 2, \ldots, (n-1)$.
(54)

This technique is termed pps-systematic sampling and was suggested by Madow (1949) and utilized in controlled selection by Goodman and Kish (1950). This technique is presently widely used in many sampling situations because of its simplicity, operational convenience and efficiency. In view of this technique, systematic sampling of units so far considered may be termed simple systematic sampling.

The operational procedure can be illustrated by considering sampling of 2 units from a population of 4 units, A, B, C and D with sizes 10, 20, 30 and 40. First the cumulative sizes 10, 30, 60 and 100 are obtained and the sampling interval in this case is $50 = (100/2)$. If the random start selected from 1 to 50 is 29, then units corresponding to the number 29 and 79, which are B and D, are selected in the sample.

If (X/n) is not an integer, the sampling interval can be taken as the integer nearest to (X/n) and to avoid the problem of variable sample size in this case, we may select the random start from 1 to X instead of from 1 to I and select units according to (54) proceeding circularly.

The population mean and total are unbiasedly estimated by

$$\hat{\bar{Y}} = \frac{1}{N}\frac{1}{n}\sum_{i=1}^{N} r_i \frac{y_i}{p_i} \quad \text{and} \quad \hat{Y} = \frac{1}{n}\sum_{i=1}^{N} r_i \frac{y_i}{p_i}, \qquad (55)$$

where $P_i = X_i/X$ and r_i is the number of times the ith unit is selected in the sample and it is a random variable taking the values $[nP_i]$ and $[nP_i] + 1$ with probabilities $[nP_i] + 1 - nP_i$ and $nP_i - [nP_i]$, so that $E(r_i) = nP_i$. Thus we see that if $X_i \leq I$, the ith unit either occurs once in the sample or does not occur. However, if $X_i > I$, the ith unit will get repeated in the sample. In such situations, we may consider the possibility of selecting those units with $X_i \geq I$ with certainty in the sample and choosing the remaining sample from the remaining units in the population using pps-systematic sampling. If the number of repetitions is small, it may be operationally convenient to select the sample from the whole population and weight the selected units according to (54).

By considering the ith unit as made up of X_i subunits, each having the value (Y_i/X_i), it can be seen that pps-systematic sampling is equivalent to simple systematic sampling of the subunits. This subunits approach makes it possible to apply the results derived so far for simple systematic sampling to the case of pps-systematic sampling. Thus a good arrangement of units for pps-systematic sampling is the one where the subunits are arranged in ascending or descending order of their values, which are $\{Y_i/X_i\}$. In applying pps-systematic sampling, we should ensure that any periodicity in (Y_i/X_i) does not lead to highly variable samples. It may be mentioned that some of the variations of simple systematic sampling such as bss considered in Section 7 can be applied to pps-systematic sampling with the help of the subunits approach to get variations of this technique such as balanced pps-systematic sampling.

Though the estimator in pps-systematic sampling is simple, the expressions for its variance and variance estimator are generally complex. Further, as in the case of simple systematic sampling, it is not possible to get valid variance estimators in pps-systematic sampling without resorting to some assumptions and approximations. A relatively simple method of estimating sampling variance in pps-systematic sampling is to resort to multiple random start pps-systematic sampling where the sample is drawn in the form of m subsamples of size $u\ (= n/m)$ each with m random starts selected with or without replacement from 1 to mI. The variance estimator is given by

$$v(\hat{Y}) = \frac{1}{m(m-1)} \sum_{i=1}^{m} (t_i - \bar{t})^2, \quad t_i = \frac{1}{u} \sum_{j=1}^{u} \frac{y_{ij}}{P_{ij}}, \quad \bar{t} = \frac{1}{m} \sum_{i=1}^{m} t_i. \quad (56)$$

The sampling variance can be found by determining the inclusion probabilities. When the arrangement of units in the population is random, Hartley and Rao (1962) have derived approximate expressions for the variance and variance estimator for pps-systematic sampling when N is moderately large and $nP_i < 1$ as

$$V(\hat{Y}) = \frac{1}{n} \sum_{i=1}^{N} \left(\frac{Y_i}{P_i} - Y\right)^2 P_i\{1 - (n-1)P_i\} \quad (57)$$

and

$$v(\hat{Y}) = \frac{1}{n^2(n-1)} \sum_{i=1}^{n} \sum_{j>i}^{n} \left\{1 - n(p_i + p_j) + n \sum_{k=1}^{N} P_k^2\right\} \left(\frac{y_i}{p_i} - \frac{y_j}{p_j}\right)^2, \quad (58)$$

where $\{y_i, p_i\}$, $i = 1, 2, \ldots, n$, are the sample observations and corresponding probabilities. From (57) it is clear that pps-systematic sampling is more efficient than pps with replacement sampling even with a random arrangement of the units. Thus if the arrangement of the units is 'good', then pps-systematic sampling can be expected to be considerably more efficient than pps sampling. Connor (1966) has derived exact expressions for the variance of pps-systematic sampling and this technique has been considered in detail, among others, by Grundy (1954), Murthy and Sethi (1965), Des Raj (1965), Hartley (1966), Murthy (1967), Hidiroglou and Gray (1975) and Isaki and Pinciaro (1977).

9. Superpopulation models and asymptotic results

9.1. Basic results and extensions

Theoretical comparison of systematic sampling with other schemes is in general difficult since the results depend on the particuilar arrangement of the units used in systematic sampling. Thus theoretical comparison of the efficiency of systematic sampling with other schemes was done in Section 3 assuming random arrangement of units and for a population exhibiting a linear trend. To enable such comparisons in more realistic cases of units arranged in ascending or descending orders of the values of the study variable or those of an auxiliary variable, the concept of a superpopulation is used where it is assumed that the population units themselves are drawn from a superpopulation with certain characteristics (Cochran, 1946). For instance, the situation considered in Section 3.3 is a case of the use of the concept of the superpopulation model where the concerned population with a particular arrangement of units is considered as a sample from the N possible populations.

Cochran (1946) considered the following model, where $\{Y_i\}$, $i = 1, \ldots, N$, may be considered as sample observations from a superpopulation such that

$$\mathscr{E}(Y_i) = \mu, \quad \mathscr{E}(Y_i - \mu)^2 = \sigma^2, \quad \mathscr{E}(Y_i - \mu)(Y_{i+u} - \mu) = \rho(u)\sigma^2, \tag{59}$$

with $\rho(u) \geq \rho(v) \geq 0$ whenever $u < v$, which model can be termed Cochran's model.

In using the concept of the superpopulation model, it is to be noted that the variance of an estimator is based only on one realisation of the population and that its expected variance over all possible realisations is used for comparison of efficiencies. Under the superpopulation model given in (59), the expected values of the sampling variances in simple random sampling, stratified sampling and systematic sampling given respectively in (16), (19) and (7) are given by

$$\mathscr{E}_{\text{srs}} = \mathscr{E}(V'') = \frac{\sigma^2}{n}\left(1 - \frac{1}{k}\right)\left[1 - \frac{2}{N(N-1)}\sum_{u=1}^{N-1}(N-u)\rho(u)\right], \tag{60}$$

$$\mathscr{E}_{st} = \mathscr{E}(V''') = \frac{\sigma^2}{n}\left(1 - \frac{1}{k}\right)\left[1 - \frac{2}{k(k-1)}\sum_{u=1}^{k-1}(k-u)\rho(u)\right], \qquad (61)$$

$$\mathscr{E}_{sy} = \mathscr{E}(V) = \frac{\sigma^2}{n}\left(1 - \frac{1}{k}\right)\left[1 - \frac{2}{N(k-1)}\sum_{u=1}^{N-1}(N-u)\rho(u)\right.$$

$$\left. + 2\frac{k}{n(k-1)}\sum_{u=1}^{n-1}(n-u)\rho(ku)\right]. \qquad (62)$$

Even comparisons between these expected variances are not possible unless some further assumptions are made on the population. For example, if $\rho(u) = 0$ for $u > k - 1$, then systematic sampling is more efficient than stratified sampling, while for a population for which $\rho(1) = \rho(2) = \cdots = \rho(k)$ and $\rho(u) = 0$, $u > k$, systematic sampling is less efficient than simple random sampling. Further, if

$$\rho(i) \geqslant \rho(i+1) \geqslant 0, \quad i = 1, 2, \ldots, N-1,$$
$$\delta_i^2 = \rho(i-1) + \rho(i+1) - 2\rho(i) \geqslant 0, \quad i = 2, 3, \ldots, N-2, \qquad (63)$$

then

$$\mathscr{E}_{sy} \leqslant \mathscr{E}_{st} \leqslant \mathscr{E}_{srs} \qquad (64)$$

for any sample size. Further, $\mathscr{E}_{sy} < \mathscr{E}_{st}$ unless $\delta_i^2 = 0$, $i = 2, 3, \ldots, N-2$.

For the special cases of linear and exponential correlograms, asymptotic formulae for the variances can be obtained. First, consider the linear case for which

$$\rho(u) = (L-u)/L, \quad u \leqslant L,$$
$$= 0, \quad u > L. \qquad (65)$$

For $L \geqslant N - 1$, $\mathscr{E}_{srs}/\mathscr{E}_{sy} \sim n$. For the exponential case where $\rho(u) = e^{-\lambda u}$, the corresponding $\mathscr{E}_{sy}$ decreases steadily and the relative efficiency compared to stratified and simple random sampling increases as n tends to infinity.

Simple generalisations of Cochran's results given here have been proposed and considered by various authors including Quenouille (1949), Jowett (1952) and Williams (1956).

In the case of multiple start systematic sampling considered in Section 7.6, the use of the superpopulation model helps in comparing the efficiencies of selecting one single combined systematic sample of n units with that of selecting s systematic samples of n/s units each. For a random population with a superpopulation model

$$\mathscr{E}(Y_i) = \mu, \quad \mathscr{E}(Y_i - \mu)^2 = \sigma^2, \quad \mathscr{E}(Y_i - \mu)(Y_j - \mu) = 0, \qquad (66)$$

the expected variances in the above two cases are the same. For populations with linear trend under the superpopulation model

$$\mathcal{E}(Y_i) = \alpha + \beta i, \quad \mathcal{E}(Y_i - \mu)^2 = \sigma^2, \quad \mathcal{E}(Y_i - \mu)(Y_j - \mu) = 0 \quad (67)$$

and for auto-correlated populations under the superpopulation model (59) satisfying the conditions (63), the single systematic sampling scheme is more efficient than the multiple start systematic sampling scheme. These results are due to Gautschi (1957).

Further work in this area has been reported by Hannan (1962), Blight (1973) and Iachan (1981a, b).

9.2. Optimality of systematic sampling

Hájek (1959) considered a slightly more general model than Cochran's, which incorporates the information on an auxiliary variable taking values $\{X_i\}$ on the units $\{U_i\}$, $i = 1, 2, \ldots, N$. The model is specified by

$$\mathcal{E}(Y_i) = \mu X_i, \quad \mathcal{E}(Y_i - \mu X_i) = \sigma^2 X_i^2$$

and

$$\mathcal{E}(Y_i - \mu X_i)(Y_j - \mu X_j) = \sigma^2 X_i X_j R|i - j|, \quad (68)$$

where R is a convex function of the distance between the units (U_i, U_j). Under this model, it is shown that systematic sampling scheme with (a) the sample size n_s is fixed, (b) the probability of inclusion of the ith unit, π_i, is proportional to the size measure X_i of that unit, and (c) the estimator is the Horvitz–Thompson estimator $\hat{Y}_{HT} = \Sigma_{i \in s} y_i/\pi_i$ has minimum variance.

9.3. Comparison of variations in systematic sampling procedures

In what follows we shall review some of the comparisons between the variations of systematic sampling discussed in Section 7, under suitable superpopulation models. Singh, Jindal and Garg (1968) have compared their modified systematic sampling (cf. Section 7.5) procedure with other schemes for different types of populations. For populations with a linear trend for which a suitable superpopulation model is assumed as

$$Y_i = \alpha + \beta i + e_i, \quad i = 1, 2, \ldots, N,$$
$$\mathcal{E}(e_i) = 0, \quad \mathcal{E}(e_i^2) = \sigma^2, \quad \mathcal{E}(e_i e_j) = 0, \quad i \neq j, \quad (69)$$

it is proved that

$$\mathcal{E}(V_{mss}) \leq \mathcal{E}(V_{sy}) \leq \mathcal{E}(V_{srs}), \quad (70)$$

where V_{mss} denotes the variance for the modified systematic sampling procedure. For populations with linear trend and periodic variation, a suitable model is taken as

$$Y = \alpha + \beta i + X_i + e_i \tag{71}$$

with e_i as in (69), and X_i is a periodic function of i with $2h$ as its period and $N = 2hQ$ (cf. Madow and Madow, 1944). In this case, when h is an odd multiple of half period, the contribution of linear trend to the variance of the modified systematic sampling is zero when $n/2$ is even and trivial for n large and $n/2$ odd. Under certain conditions, it is shown that their scheme is better for populations with parabolic trend which assume a model of the type

$$Y_i = \alpha + \beta i + \gamma i^2 + e_i. \tag{72}$$

Finally, for autocorrelated populations for which the assumed model is given in (59), Singh, Jindal and Garg (1968) have shown that $\mathscr{E}(V_{\text{mss}}) < \mathscr{E}(V_{\text{sy}})$ for upward correlograms, which do not favour their method. Under the model

$$Y_i = \lambda Y_{i-1} + e_i, \quad i = 2, 3, \ldots, N, \tag{73}$$

where the e_i's are uncorrelated and each is distributed as $N(0, \sigma^2)$, the centrally located systematic sampling plan is the optimum (Blight, 1973). Bellhouse and Rao (1975) have compared this optimum plan with other variations for different populations. They assume $N = nk$ and consider the plans

$$\hat{\bar{Y}}_1 = \bar{y}_i, \quad \text{based on the } i\text{th systematic sample}, \tag{74}$$

$$\hat{\bar{Y}}_2 = \bar{y}_i + \frac{2i - k - 1}{2(n-1)k} (y_i - y_{i+(n-1)k}), \tag{75}$$

using Yates' (1948) end corrections (cf. Sections 3.4);

$$\hat{\bar{Y}}_3 = \text{mean of the units selected by modified systematic sampling}, \tag{76}$$

$$\hat{\bar{Y}}_4 = \text{mean of the selected sample in balanced systematic sampling (cf. Section 7.4);} \tag{77}$$

and

$$\hat{\bar{Y}}_5 = \bar{y}_{(k+1)/2} \quad \text{for } k \text{ odd},$$
$$= \bar{y}_{k/2} \text{ or } \bar{y}_{k/2+1} \quad \text{for } k \text{ even}; \tag{78}$$

based on centrally located systematic sampling (Madow, 1953; also cf. Section 7.2).

Let M_i denote the Mean Squared Error corresponding to the estimator $\hat{\bar{Y}}_i$ described above, $i = 1, 2, \ldots, 5$. Bellhouse and Rao (1975) first considered populations with a linear trend specified by (69) and found that $\mathscr{E}(M_1)$ is the largest and there is not much to choose from the rest. For populations with linear trend and periodic variation the model given by (71) is chosen and in this case modified systematic sampling is preferred. Centrally located systematic sampling is found to be most efficient if k is odd for populations with parabolic trend which assume the model (72). For autocorrelated populations specified by (73), centrally located systematic sampling again beats the rest of the schemes even in the absence of trend.

For the method of balanced random sampling (brs) (cf. Section 7.8), when N and n are even, Padam Singh and Garg (1979) have shown that $V(\hat{\bar{Y}}_{\text{brs}}) = 0$ for populations with a linear trend $Y_i = \alpha + \beta i$ and when Y_i is an exact periodic function as in (71). For autocorrelated populations as in (73), no specific conclusion is reached. They have also given modifications when N and n are not even.

9.4. Variance estimation and superpopulation models

The problem of estimating the variance of the estimate of the population mean based on a single systematic sample has already been discussed in Section 4 and to some extent in Section 7. In this Sub-section, we shall only deal with superpopulation models and asymptotic theory in connection with the estimation of variance. Cochran (1946; 1977, Section 8.11), Quenouille (1949), Jowett (1952) and Williams (1956) have given unbiased estimators of variance of systematic sampling, V_{sy} under explicit assumptions on the superpopulation models. For example, Cochran (1977) considered the model

$$Y_i = \mu_i + e_i \quad \text{with } \mathscr{E}(e_i) = 0, \quad \mathscr{E}(e_i^2) = \sigma_i^2, \quad \mathscr{E}(e_i, e_j) = 0,$$

and obtained unbiased estimators $\hat{V}_{\text{sy}}$ in the sense that $\mathscr{E}E(\hat{V}_{\text{sy}}) = \mathscr{E}(V_{\text{sy}})$. Also formulae based on certain assumptions regarding the correlogram were given by Osborne (1942) and Cochran (1946). To illustrate, we consider the case of an exponential correlogram for which Cochran used the formula

$$\mathscr{E}_{\text{sy}} = \frac{\sigma^2}{n}\left(1 = \frac{2}{t} + \frac{2}{e^t - 1}\right)$$

for large k and n, for the estimation of variance from a single systematic sample. Here, an estimate of e^{-t} and of t is provided by the correlation between successive terms in the systematic sample. Also neglecting terms in $1/n$, the mean square within systematic sample is found to be an unbiased estimator of σ^2 and by substitution a consistent estimate of variance based on a single systematic sample is obtained.

On the other hand, Quenouille (1949) has suggested the estimation of variance by using sets of q systematic samples of sufficient length with randomly chosen starting points. Recalling the concept of random permutations used by Sedransk (1969), it may be noted that under the assumption of exchangeability, the method of collapsed strata can be used for estimating V_{sy}. It is also suggested that use of the msss with m systematic samples in each stratum is desirable to estimate the variance unbiasedly based on $(m-1)$ degrees of freedom.

For any superpopulation model for the population $(Y_1, Y_2, \ldots, Y_N)$ a natural estimator of V_{sy} is proposed by Heilbron (1978) which is given by the conditional expectation of V_{sy} given the sample values,

$$\mathscr{E}(V_{sy}| Y_r, Y_{r+k}, \ldots, Y_{r+(n-1)k}).$$

This has the property that it minimises the expected mean square error. The estimator is derived when the underlying model is Gaussian, with special reference to Markov serial correlation model with $\rho(u) = \rho^u$.

Iachan (1981a) has suggested the variance estimator

$$\hat{V}_{sy} = \left(1 - \frac{1}{k}\right)\frac{W}{n}, \qquad (79)$$

where W is the mean sum of squares within the jth sample based on the assumption that the population is in random order [i.e. $\rho(u) = 0$]. Under certain regularity conditions, he has shown that the confidence intervals

$$\overline{Y}_{sy} \pm Z_\alpha \sqrt{\hat{V}_{sy}} \qquad (80)$$

with confidence level $1 - 2\alpha$, where Z_α is the upper α-point of the standard Normal distribution, are asymptotically conservative, as $n \to \infty$ with $k = N/n$ fixed.

10. Applications and illustrations

In Section 1 of this chapter, we have mentioned that though the early applications of systematic sampling were in forestry and land use surveys and in census work, following the theoretical developments, the use of systematic sampling gained momentum and was applied in household enquiries, spatial sampling, geological and geographical sampling, industrial and business surveys as well and a wide variety of other fields. In this Section we shall briefly mention a few of these applications just to illustrate how important and useful the technique is, without going into details of some of the well-known applications. We have already mentioned some applications of systematic sampling method in the previous sections.

Systematic sampling has been used for the estimation of volume of timber in a forest and the number of seedlings in successive feet in a forest nursery bed taking into account periodicity. Bonnor (1972) has prepared an extensive bibliography on the applications of systematic sampling in Forestry. For a recent application of this technique in multi-stage (phase) sampling design, we refer to Tin and Thein (1977) and for the use of multiple random start systematic sampling schemes from a population of truckloads of trees from Jackson forest we refer to Iachan (1980).

Various applications of systematic sampling in geological mine sampling have been discussed by Journel and Huijbregts (1978).

Use of systematic sampling in land use and agricultural surveys is well known (Osborne, 1942; Mahalanobis, 1944; Yates, 1948). For large scale vegetation surveys, Lambert (1972) has discussed models for systematic sampling in sub-areas when the problem is the estimation of total abundance over the entire study area.

Nordskog and Crump (1948) have demonstrated the efficiency of systematic sampling of eggs based on 'interval days' when egg production of groups of poultry is noted on different days. An interesting illustrative example of estimation of catch of marine fish in India, using systematic sampling technique is available in a report of the Indian Council of Agricultural Research (ICAR, 1950) and in Sukhatme, Panse and Sastry (1958). Deming (1950) (cf. Jones, 1955) has used multiple random start systematic sampling for pole sampling and for sampling of other telephone property and for assessing the error, the technique of independent subsamples or replicated sampling was suggested following Tukey (also see Mahalanobis, 1946; Törnqvist, 1963). In Industrial Quality Control, systematic sampling is referred to sometimes as 'period sampling' (Cowden, 1957).

As mentioned earier, use of systematic sampling in population census was first discussed by Stephan, Deming and Hansen (1940) in the case of the 1940 U.S. Census where a further sample of 1 in 20 was chosen for supplementary information. Also, Shaul and Myburgh (1948) proposed sample survey of the African population of Southern Rhodesia by taking a systematic sample from lists and treating them as if they were random samples for analysis. Further examples of the application of this technique are provided by the 2% sample of persons selected in the 1941 Indian census for the tabulation of data on means of livelihood, the 1% sample of persons selected systematically from a 10% sample of 1950 Japanese census data for quick tabulation of important characteristics and the 25% systematic sample of households from the 1960 United States population for supplementary information.

Systematic sampling is increasingly used in large-scale household surveys. For instance, the Indian National Sample Survey (Lahiri, 1954b; Murthy, 1967; Bhattacharyya, 1981) and the World Fertility Survey described by Verma, Scott and O'Muircheartaigh (1980) are two examples of large-scale sample surveys where systematic sampling is widely used in sample selection at different stages of a multi-stage design, after suitably arranging the relevant units.

11. Some of the available computer programs

It was mentioned earlier that Hartley and Rao (1962) gave an asymptotic formula for the joint inclusion probability for pairs of units under unequal probability systematic sampling design. Connor (1966) derived an exact formula for the computation of this joint probability, π_{ij}. Hidiroglou and Gray (1980) give an algorithm for computing this, using Gray's version of Connor's formula which is easily programmable. Here, π_{ij} is written as a sum of conditional joint probabilities and these are averaged over all possible orderings of units in the population between units i and j. The computation of each conditional probability involves the identification of the units between i and j and the inclusion probabilities of these units.

Sasaki (1981) has proposed a generalized multidimensional systematic sampling scheme which consists of dividing the population into cells using data on d auxiliary variables with $p_1, p_2, \ldots, p_d$ partitions, where p's are prime numbers, not necessarily distinct and then selecting the cells systematically. The algorithm used for this purpose is called SYSCLUST. Several merits of this procedure have been discussed by Sasaki (1981).

12. Two-dimensional systematic sampling

Certain populations may consist of units having a natural arrangement or units arranged on a plane instead of a line. Application of systematic sampling from such populations is termed 'two-dimensional or plane systematic sampling'. For instance, if the population consists of a specified region divided into N square or rectangular grids of equal size and if a sample of n such grids is to be selected so as to get representation from different parts of the region in the sample, then it would be desirable to divide the region into square or rectangular cells of equal size such that each cell consists of $k = N/n$ grids and to select the grids occupying the rth place in each of the n cells, where r is randomly chosen from 1 to k.

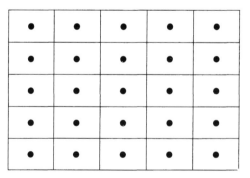

Fig. 4. A representation of a two-dimensional systematic sample.

Estimators of population total, mean and proportion, their variances and variance estimators in two-dimensional systematic sampling can be obtained proceeding on lines similar to those discussed in this paper for one-dimensional systematic sampling. Considerable work has been done in this field and reference may be made, among others, to Quenouille (1949), Das (1951), Patterson (1954), Koop (1976), Bellhouse (1977, 1981) and Ripley (1981).

Bibliographical note

The earliest developments in the field of systematic sampling were reviewed by Yates (1948) and Buckland (1951). The later developments during the last three decades have been reviewed by Iachan (1982). Many books on sampling such as those by Cochran (1977), Deming (1960), Hájek (1981), Konijn (1973), Murthy (1967) and Sukhatme and Sukhatme (1970) include a substantive chapter on systematic sampling. Some selected references, mainly referred to in the text, are given at the end of the chapter.

References

Aoyama, H. (1951). On practical systematic sampling, *AISM* 3, 57-64.
Bellhouse, D. R. and Rao, J. N. K. (1975). Systematic sampling in the presence of a trend. *Biometrika* 62, 694-697.
Bellhouse, D. R. (1977). Optimal designs for systematic sampling in two dimensions. *Biometrika* 64, 605-611.
Bellhouse, D. R. (1981). Spatial surveys in the presence of a trend. *JSPI* 5, 365-370.
Bhattacharyya, A. K. (1981). Evolution of the sample design of the Indian National Sample Survey. *Sankhyā C* 38, 9-25.
Blight, B. J. N. (1973). Sampling from an autocorrelated finite population. *Biometrika* 60, 375-385.
Bonnor, G. M. (1972). Forest sampling and inventories - A bibliography. Internal Report, Forest Management Institute, Ottawa.
Buckland, W. R. (1951). A review of the literature of systematic sampling. *JRSS Ser. B* 13, 208-215.
Cochran, W. G. (1946). Relative accuracy of systematic and stratified random samples for a certain class of populations. *AMS* 17, 164-177.
Cochran, W. G. (1977). *Sampling Techniques.* Wiley, New York, 3rd ed.
Connor, W. S. (1966). An exact formula for the probability that two specified sampling units will occur in a sample drawn with unequal probabilities and without replacement. *JASA* 61, 348-390.
Cowden, D. J. (1957). *Statistical Methods in Quality Control.* Prentice-Hall, NJ.
Das, A. C. (1951). Systematic sampling. *BISI* 33 (2), 119-132.
Deming, W. E. (1950). Some Theory of Sampling. Wiley, New York.
Deming, W. E. (1960). *Sample Design in Business Research.* Wiley, New York.
Des Raj (1965). Variance estimation in randomized systematic sampling with probability proportional to size. *JASA* 60, 278-284.
Finney, D. J. (1948). Volume estimation of standing timber by survey. *Forestry* 21, 179-203.
Gautschi, W. (1957). Some remarks on systematic sampling. *AMS* 28, 385-394.
Goodman, R. and Kish, L. (1950). Controlled selection - A technique in probability sampling. *JASA* 45, 350-372.
Grundy, P. M. (1954). A method of sampling with probability exactly proportional to size. *JRSS Ser. B* 16, 236-238.
Hájek, J. (1959). Optimum strategy and other problems in probability sampling. *Casopis pro Pestovani Matematiky* 84, 387-420.
Hájek, J. (1981). *Sampling from a Finite Population.* Marcel Dekker, New York.

Hannan, E. J. (1962). Systematic sampling. *Biometrika* **49**, 281–283.
Hartley, H. O. (1966). Systematic sampling with unequal probabilities and without replacement. *JASA* **61**, 739–748.
Hartley, H. O. and Rao, J. N. K. (1962). Sampling with unequal probabilities and without replacement. *AMS* **33**, 350–374.
Hasel, A. A. (1938). Sampling error in timber surveys. *Journal of Agricultural Research* **57**, 713–736.
Heilbron, D. C. (1978). Comparison of estimators of the variance of systematic sampling. *Biometrika* **65**, 429–433.
Hidiroglou, M. A. and Gray, G. B. (1975). A computer algorithm for joint probabilities of selection. *Survey Methodology (Statistics Canada)* **1**, 99–108.
Horvitz, D. G. and Thompson, D. J. (1952). A generalization of sampling without replacement from affinite universe. *JASA* **47**, 663–685.
Iachan, R. (1980). The weight scaling problem – Systematic sampling from a population of truckloads of trees. Univ. of Wisconsin Tech. Report No. 626.
Iachan, R. (1981a). An asymptotic theory of systematic sampling. Univ. of Wisconsin Tech. Report No. 632.
Iachan, R. (1981b). An asymptotic comparison between systematic sampling and simple random sampling. Univ. of Wisconsin Tech. Report No. 635.
Iachan, R. (1982). Systematic sampling – A critical review. *ISR* **50**, 293–303.
Indian Council of Agricultural Research, New Delhi (1950). Estimation of catch of marine fish in a sample of landing centres. Report on the pilot sample survey conducted on the Malabar coast for estimating the catch of marine fish (unpublished).
Isaki, C. T. and Pinciaro, S. J. (1977). Numerical comparison of some estimators of the variance under pps systematic sampling. In: *Proceedings of the Social Statistics Section, American Statistical Association*, **I**, 308–313.
Jones, A. E. (1948). Systematic sampling of continuous parameter populations. *Biometrika* **35**, 283–290.
Jones, H. L. (1955). The application of sampling procedures to business operations. *JASA* **50**, 763–776.
Journel, A. G. and Huijbregts, C. J. (1978). *Mining Geostatistics*. Academic Press, London.
Jowett, H. H. (1952). The accuracy of systematic sampling from conveyor belts. *Applied Statistics* **1**, 50–59.
Kish, L. (1965). *Survey Sampling*. Wiley, New York.
Konijn, H. S. (1973). *Statistical Theory of Sample Survey Design and Analysis*. North-Holland, Amsterdam.
Koop, J. C. (1971). On splitting a systematic sample for variance estimation. *AMS* **42**, 1084–1087.
Koop, J. C. (1976). Systematic sampling of two-dimensional surfaces and related problems. Research Triangle Institute, NC.
Lambert, J. M. (1972). Theoretical methods for large-scale vegetation survey. In: J. N. R. Jeffers, ed., *Mathematical Models in Ecology*. Oxford, Blackwell, 87–109.
Lahiri, D. B. (1954a). On the question of bias in systematic sampling. *Proceedings of the World Population Conference* **6**, 349–362.
Lahiri, D. B. (1954b). Technical report on some aspects of the development of the sample design. National Sample Survey Report No. 5, Government of India, Reprinted in *Sankhyā* **14**, 264–316.
Madow, L. H. (1946). Systematic sampling and its relation to other sampling designs. *JASA* **41**, 204–217.
Madow, W. G. (1949). On the theory of systematic sampling II. *AMS* **20**, 333–354.
Madow, W. G. (1953). On the theory of systematic sampling III. *AMS* **24**, 101–114.
Madow, W. G. and Madow, L. H. (1944). On the theory of systematic sampling. *AMS* **15**, 1–24.
Mahalanobis, P. C. (1944). On large-scale sample surveys. *Roy. Soc. Phil. Trans. Ser. B* **231**, 329–451.
Mahalanobis, P. C. (1946). Recent experiments in statistical sampling in the Indian Statistical Institute, *JRSS Ser. A* **109**, 325–378, reprinted in *Sankhyā* (1958), 1–68.
Mokashi, V. K. (1954). Efficiency of systematic sampling in forest sampling, *JISAS* **6**, 101–114.
Murthy, M. N. and Sethi, V. K. (1965). Self-weighting design at tabulation stage. *Sankhyā B* **27**, 201–210.

Murthy, M. N. (1967). *Sampling Theory and Methods*. Statistical Publishing Society, Calcutta, 2nd ed.

Nair, K. R. and Bhargava, R. P. (1951). Statistical sampling in timber surveys in India. Indian Forest Leaflet, No. 153, Forest Research Institute, Dehradun.

Nordskog, A. W. and Crump, S. L. (1948). Systematic and random sampling for estimating egg production in poultry. *Biometrics* 4, 223–233.

Osborne, J. G. (1942). Sampling errors in systematic and random surveys of covertype areas. *JASA* 37, 256–264.

Padam Singh and Garg, J. N. (1979). On balanced random sampling. *Sankhyā Ser. C* 41, 60–68.

Patterson, H. D. (1954). The errors of lattice sampling. *JRSS Ser. B* 16, 140–149.

Quenouille, M. H. (1949). Problems in plane sampling. *AMS* 20, 335–375.

Ripley, B. (1981). *Spatial Statistics*. Wiley, New York.

Sasaki, T. (1981). Multidimensional systematic sampling. *Journal Inform. Process.* 4, 79–88.

Sedransk, J. (1969). Some elementary properties of systematic sampling. *Skand. Aktuar.* 52, 39–47.

Sethi, V. K. (1965). On optimum pairing of units. *Sankhyā Ser. B* 27, 315–320.

Shaul, J. R. H. and Myburgh, C. A. L. (1948). A sample survey of the African population of Southern Rhodesia. *Pop. Studies* 2, 339–353.

Shiue, G. J. (1966). Systematic sampling with multiple random starts. *Forestry Science* 6, 142–150.

Singh, D., Jindal, K. K. and Garg, J. N. (1968). On modified systematic sampling. *Biometrika* 55, 541–546.

Singh, D. and Padam Singh (1977). New systematic sampling. *JSPI* 1, 163–177.

Stephan, F. F., Deming, W. E. and Hansen, M. H. (1940). The sampling procedure of the 1940 population census. *JASA* 35, 615–630.

Sudakar, K. (1978). A note on circular systematic sampling. *Sankhyā Ser. C* 40, 72–73.

Sukhatme, P. V., Panse, V. G. and Sastry, K. V. R. (1958). Sampling techniques of estimating the catch of sea-fish in India. *Biometrics* 14, 78–96.

Sukhatme, P. V. and Sukhatme, B. V. (1970). *Sampling Theory of Surveys with Applications*. FAO, Rome, 2nd ed.

Tin, M. and Thein, P. (1977). Bamboo inventory in Burma. *BISI* 47, 597–600.

Törnqvist, L. (1963). The theory of replicated systematic cluster sampling with random start. *RISI* 31, 11–23.

Verma, V., Scott, C. and O'Muircheartaigh, C. (1980). Sample designs and sampling errors for the World Fertility Survey. *JRSS Ser. A* 143, 431–473.

Williams, R. M. (1956). Variance of the mean of systematic samples. *Biometrika* 43, 137–148.

Wu, Chien-Fu (1980). Estimation in systematic sampling with supplementary observations. Univ. of Wisconsin Technical Report.

Yates, F. (1948). Systematic sampling. *Philosophical Transactions of the Royal Society Ser. A* 241, 345–347.

Yates, F. and Grundy, P. M. (1953). Selections without replacement from within strata with probability proportional to size. *JRSS Ser. B* 15, 253–261.

Zinger, A. (1980). Variance estimation in partially systematic sampling. *JASA* 75, 206–211.

Abbreviations used

AISM	Annals of Institute of Statistical Mathematics
AMS	Annals of Mathematical Statistics
BISI	Bulletin of the International Statistical Institute
ISR	International Statistical Review
JASA	Journal of the American Statistical Association
JISAS	Journal of the Indian Society of Agricultural Statistics
JRSS	Journal of the Royal Statistical Society
JSPI	Journal of Statistical Planning and Inference
RISI	Review of the International Statistical Institute

Sampling in Time

D. A. Binder and M. A. Hidiroglou

1. Introduction

It is common for survey organizations to estimate at regular intervals of time a population parameter such as a mean or total which varies with time. If there exists a relationship between the values attached to a unit of the population at two or more points of time, then it is possible to use the information contained in earlier samples to improve the current estimate of a population parameter. In order to use the earlier sample information (without resorting to time series models discussed later), it is necessary to sample the population such that the samples referring to different points in time have some elements in common. The term 'rotation sampling' refers to the organized process of eliminating some of the elements from the sample and adding new elements to the sample as time advances. This method of sampling is also referred to as 'sampling on successive occasions with partial replacement of units' (Yates, 1949; Patterson, 1950) and 'sampling for a time series' (Hansen, Hurwitz and Madow, 1953).

The level of overlap between occasions will depend on the objectives of the survey, which include measures of level and change, as well as operational constraints which suggest how long a unit remains in the sample before it is rotated out. In terms of objectives, measures of change are best achieved by keeping the same sample between two time occasions (providing there are no births and deaths in the population), whereas for estimating the average of totals over a number of occasions, it is best to draw a new sample on each occasion. In practice, however, we are usually interested in efficient estimation of change between occasions, level within each occasion, and average level over a number of occasions. Rotation sampling offers a way to balance out these different requirements.

There are two main approaches, in terms of estimation, for incorporating the information contained in rotating samples. These are: the classical approach and the time series approach. In the classical approach (e.g. Jessen, 1942; Patterson, 1950; Eckler, 1955; Rao and Graham, 1964; Tikkiwal, 1979; Wolter, 1979), the sequence of population parameters of interest $\{\theta_t\}$, usually the population means or totals, is considered to be a fixed quantity. The individual values, y_{ti}, attached to the ith unit at time t, $i = 1, 2, \ldots$; $t = 1, 2, \ldots$, are the variate values. These values are assumed to be related to the previous values $y_{t-1,i}$ of the same unit under some correlation structure.

The population parameters θ_t need not to be considered as fixed unknown quantities. They may be regarded as random quantities resulting from a time series process. Methods of estimation which make use of this philosophy are associated with the time series approach (e.g. Blight and Scott, 1973; Scott and Smith, 1974; Scott, Smith and Jones, 1977; Jones, 1980). This approach can be considered as being akin to a Bayesian framework. In this paper, we also discuss an empirical Bayes method which extends this theory.

The article is structured as follows. In Section 2, some of the commonly used rotation designs are reviewed. In Section 3, the estimation procedures associated with the classical approach are discussed. These include sampling on two and more occasions as well as multi-stage sampling. In Section 4, the effect of rotation bias on composite estimators is discussed. In Section 5, several time series approaches are studied which include statespace models and ARMA models. The statespace approach is illustrated by an example.

2. General approaches to designs with overlapping units

Rotation sampling may be executed in a variety of ways: one-level rotation sampling, semi one-level rotation sampling and multi-level rotation sampling. The one-level rotation sampling functions as follows. Assuming that the sample consists of n elements on all sampling occasions, $(1 - \mu)n$ of the elements, $0 \leq \mu \leq 1$, which occur in the sample at period t_{i-1} are retained for observation at period t_i. The remaining μn elements are replaced with an equal number of new ones. On each occasion, the enumerated sample reports only one period of data. An example of a survey which uses such a scheme is the Canadian Labour Force Survey, conducted monthly by Statistics Canada. It employs a rotation scheme that permits the replacement of one-sixth of the households in the sample each month. The sample is composed of six panels, with each panel remaining in the sample for six consecutive monthly enumerations before rotating out of the sample.

The semi one-level rotation, introduced by Hansen, Hurwitz, Nisselson and Steinberg (1955) for the Current Population Survey conducted monthly by the U.S. Bureau of the Census, is a variation on the one-level rotation scheme. For this scheme, units are in the sample for some number of consecutive occasions, are then not sampled for a specified number of consecutive occasions and then return into the sample for a number of occasions. A generalization of this scheme is given by Rao and Graham (1964). Let N and n be the respective population and sample sizes. A group of $n_2 \geq 1$ units stays in the sample for r occasions ($n = n_2 r$), leaves the sample for m occasions, comes back into the sample for another r occasions, then leaves the sample for m occasions, and so on. The repetition may stop after a fixed number of cycles. In the case of the Current

Population Survey, for any given month the sample is composed of eight subsamples or rotation groups. All the units belonging to a particular rotation group enter and drop out of the sample at the same time. A given rotation group stays in the sample for four consecutive months, leaves the sample during the next eight succeeding months, and then returns for another four consecutive months. It is then dropped from the sample entirely. Of the two new rotation groups that are enumerated each month, one is completely new and the other returns after an absence of eight months from the sample. This rotation pattern is illustrated in Figure 2.1, where 'x' denotes which subsamples are being surveyed in a given month. S_1, S_2, ... are samples each containing eight subsamples (1, 2, ..., 8).

Figure 2.1
Rotation pattern for U.S. Bureau of the Census's Current Population Survey

Year and Month		S_1: 1 2 3 4 5 6 7 8	S_2: 1 2 3 4 5 6 7 8	S_3: 1 2 3 4 5 6 7 8	S_4: 1 2 3 4 5 6 7 8
YY	Jan	x x x x · · · ·	· · · · x x x x	· · · · · · · ·	· · · · · · · ·
	Feb	· x x x x · · ·	· · · · · x x x	x · · · · · · ·	· · · · · · · ·
	Mar	· · x x x x · ·	· · · · · · x x	x x · · · · · ·	· · · · · · · ·
	Apr	· · · x x x x ·	· · · · · · · x	x x x · · · · ·	· · · · · · · ·
	May	· · · · x x x x	· · · · · · · ·	x x x x · · · ·	· · · · · · · ·
	Jun	· · · · · x x x	x · · · · · · ·	· x x x x · · ·	· · · · · · · ·
	Jul	· · · · · · x x	x x · · · · · ·	· · x x x x · ·	· · · · · · · ·
	Aug	· · · · · · · x	x x x · · · · ·	· · · x x x x ·	· · · · · · · ·
	Sep	· · · · · · · ·	x x x x · · · ·	· · · · x x x x	· · · · · · · ·
	Oct	· · · · · · · ·	· x x x x · · ·	· · · · · x x x	x · · · · · · ·
	Nov	· · · · · · · ·	· · x x x x · ·	· · · · · · x x	x x · · · · · ·
	Dec	· · · · · · · ·	· · · x x x x ·	· · · · · · · x	x x x · · · · ·
YY + 1	Jan	· · · · · · · ·	· · · · x x x x	· · · · · · · ·	x x x x · · · ·
	Feb	· · · · · · · ·	· · · · · x x x	x · · · · · · ·	· x x x x · · ·

The survey starts with eight subsamples S_1 (1 2 3 4) and S_2 (5 6 7 8). The next month, $S_1(1)$ is dropped and $S_1(5)$ is substituted and similarly for S_2.

This rotation scheme, known as the 4–8–4 scheme, has the following characteristics. In any given month, 75 percent of the subsamples are in common with the previous month. There is a 50 percent year-to-year sample overlap in any month with the same month of the previous year.

In the previous schemes, only the current occasion's data were collected during an enumeration. If information for previous occasions is also collected, then the rotation scheme is known as multi-level sampling, the most widely used of which is the two-level scheme. In this scheme a new set of elements is drawn from the population on each occasion and the associated sample values for the current occasion and preceding occasion are recorded. A generalization to higher level schemes is obvious.

The U.S. Census Bureau's monthly Retail Trade Survey (Wolter et al., 1976) incorporates a two-level design. In this survey, part of the sample is based on three rotating panels. One rotating panel reports in January, April, July, and October of each calendar year; the second in February, May, August, and November; and the third in March, June, September and December. During the monthly enumeration each member of the reporting panel is requested to report both current and previous months' retail sales.

3. The classical approach

3.1. Sampling on two occasions

The problem of sampling on two successive occasions with a partial replacement of sampling units was first considered by Jessen (1942) in the analysis of a survey which collected farm data. The survey was designed so that of the $n = 900$ sampling units employed in the 1938 phase of the survey, 450 were retained for further observation in 1939. An additional 450 different units were selected to bring the 1939 sample up to strength. The matched sample on the second occasion can therefore be regarded as a subsample of the original sample in the first occasion. The principles of double sampling can be applied to this situation.

Suppose that the samples on occasions t and $t-1$ are both of size n. Let the population mean and variance for the tth occasion be θ_t and σ_t^2, respectively. It is assumed that observations on different units are uncorrelated. Denote the unmatched and matched portions of the two samples between occasions be 'u' and 'm', and let $\bar{u}_{t-1,u}$ and $\bar{x}_{t-1,m}$ be the respective means for the unmatched and matched portions on occasion $t-1$. Correspondingly, $\bar{y}_{tm}$ and $\bar{y}_{tu}$ will denote the means for the unmatched and matched portions on occasion t. Assuming a linear relationship to hold between observations which are common to both occasions, an adjusted mean for occasion t is $\bar{y}'_{tm} = \bar{y}_{tm} + b(\bar{x}_{t-1} - \bar{x}_{t-1,m})$ where b is the least squares regression coefficient as computed from the matched sample results. The variance of $\bar{y}'_{tm}$ can be obtained by viewing n_{tm} (the number of matched units between occasions t and $t-1$) as a subsample of n. This variance, which is the result of double sampling, is w_{tm}, where $w_{tm}^{-1} = \sigma_t^2(1-\rho^2)/n_{tm} + \rho^2\sigma_t^2/n$, and ρ is the correlation between variate values realized by the same unit on consecutive occasions. Another estimate of the mean for occasion t is $\bar{y}'_{tu} = \bar{y}_{tu}$, the mean of unmatched units, with variance w_{tu} where $w_{tu}^{-1} = \sigma_t^2/n_{tu}$ (n_{tu} is the number of elements entering into the sample at occasion t). Jessen considered a linear combination of these two means given by

$$\hat{\theta}_t = \phi_t \bar{y}'_{tu} + (1-\phi_t)\bar{y}'_{tm}, \qquad (3.1)$$

where

$$\phi_t = w_{tu}/(w_{tu} + w_{tm}). \qquad (3.2)$$

The variance of $\hat{\theta}_t$ is given by

$$\text{Var}(\hat{\theta}_t) = \sigma_t^2 \frac{(n - n_{tu}\rho^2)}{n^2 - (n_{tu}\rho)^2} . \tag{3.3}$$

Assuming the cost of matched and unmatched units to be the same, Jessen derived the optimum match fraction which would minimize the above variance to be approximately

$$n_{tm}/n_{tu} = (1 - \rho^2)^{1/2} .$$

For his problem at hand, Jessen compared the estimator $\hat{\theta}_t$ in efficiency with the unweighted mean $\bar{y}_t$ of all 900 units. For fourteen items studied, 22 to 45 percent efficiency gains were achieved.

Hansen, Hurwitz and Madow (1953, p. 268) developed a minimum-variance linear unbiased estimator (MVLUE) for the mean on the second occasion when the overlapping sample design is the same as Jessen's (1942). The linear unbiased estimator is of the form

$$\hat{\theta}_t = a(\bar{x}_{t-1, u} - \bar{x}_{t-1, m}) + c\bar{y}_{tm} + (1 - c)\bar{y}_{tu} . \tag{3.4}$$

If it is to be a minimum variance linear unbiased estimator, it must be uncorrelated with every linear zero function (Rao, 1973, p. 317). Minimizing the variance of (3.4) yields the $\hat{\theta}_t$ given by (3.1), with minimum variance given by (3.3). Hence, the minimum variance linear unbiased estimation and regression approach yield identical results.

3.2. Sampling on more than two occasions

The extension of Jessen's results to the situation where the population mean of the characteristic is to be estimated on each of $t > 2$ occasions, was considered by Yates (1949). He specified that (a) a given fixed fraction μ, where $0 < \mu < 1$, of the units was to be replaced on each occasion, (b) the population variance on each occasion and the correlation ρ between the same sampling unit on successive occasions were stationary, (c) an exponential correlation pattern of the type $\rho, \rho^2, \rho^3, \ldots$, held between the same sampling units separated by one, two, three, $\ldots$, occasions, (d) the correlation coefficient ρ was assumed to be known. The composite estimator considered was

$$\hat{\theta}_t = \psi_t \bar{y}_{tu} + (1 - \psi_t)\{\bar{y}_{tm} + \rho(\hat{\theta}_{t-1} - \bar{x}_{t-1, m})\} . \tag{3.5}$$

The optimum value of the weight coefficient ψ_t which minimized the variance of $\hat{\theta}_t$ was given as a function of ρ, μ, and the total number of occasions on which sampling had taken place. Yates also discussed the estimation of change and the possibility of improving the composite estimator of the mean on the $(t - 1)$th occasion by using data provided on the tth occasion as auxiliary information.

Patterson (1950) provided the general theory for estimation of means in repeated surveys with overlapping units. He considered the following function of the observations for estimating the mean on the tth occasion.

$$\hat{\theta}_t = \sum_{j=1}^{t} \sum_{i=1}^{n_j} w_{ji} y_{ji}. \tag{3.6}$$

The condition of unbiasedness requires that

$$\sum_{i=1}^{n_j} w_{ij} = 1 \quad \text{for } j = t,$$
$$\phantom{\sum_{i=1}^{n_j} w_{ij}} = 0 \quad \text{for } j \neq t.$$

The minimum variance linear unbiased estimator of θ_t is obtained by minimizing the Lagrangian

$$\text{Var}\left\{\sum_{j=1}^{t}\sum_{i=1}^{n_j} w_{ji} y_{ji}\right\} - 2\sum_{j=1}^{t}\left\{\lambda_{tj}\sum_{i=1}^{n_j} w_{ij}\right\} \tag{3.7}$$

where λ_{tj} are the Lagrangian multipliers. This minimization led to a set of equations involving all values $w_{ji} y_{ji}$ satisfying the relationship

$$\text{Cov}(y_{ji}, \hat{\theta}_t) = \lambda_{tj}, \quad \text{for all } j \text{ and } i. \tag{3.8}$$

Using expression (3.8), Patterson showed that the minimum variance linear unbiased estimator of the population mean under the following model

$$y_{ti} - \theta_t = \rho(y_{t-1,i} - \theta_{t-1}) + \varepsilon_{ti}, \quad \text{Var}(\varepsilon_{ti}) = (1 - \rho^2)\sigma^2 \tag{3.9}$$

is of the form given by expression (3.5). Note that (3.9) is equivalent to Yates' (1949) assumption of an exponential correlation pattern.

The values of ψ_t which minimize $\text{Var}(\hat{\theta}_t)$ are given recursively by

$$1 - \psi_t = \frac{n_{tm} n_{t-1,u}}{n_t n_{t-1,u} - \rho^2 n_{tu}(n_{t-1,u} - \psi_{t-1} n_{tm})} \tag{3.10}$$

for $n_{tu} \neq 0$. The variance of the estimate $\hat{\theta}_t$ is

$$\text{Var}(\hat{\theta}_t) = \frac{\psi_t \sigma^2}{n_{tu}} \quad \text{if } n_{tu} \neq 0, \, n_{t-1,u} \neq 0,$$

$$= \sigma^2 \left\{\frac{(1-\rho^2)}{n_t} + \frac{\rho^2 \psi_{t-1}}{n_{t-1,u}}\right\} \quad \text{if } n_{tu} = 0, \, n_{t-1,u} \neq 0.$$

The iterations start with an initial trial value $\psi_1 = 1$.

Patterson also presented a minimum variance linear unbiased estimator of the change in level between occasions. This procedure uses the information from t periods to produce an efficient estimator of θ_{t-1}, given by

$$_t\hat{\theta}_{t-1} = \hat{\theta}_{t-1} - (\hat{\theta}_t - \bar{y}_{tu})\rho\psi_{t-1}\frac{n_{tu}}{n_{t-1,u}}. \tag{3.11}$$

The corresponding estimator of change is

$$\Delta\hat{\theta}_t = \hat{\theta}_t - \hat{\theta}_{t-1} + \rho\psi_{t-1}(\hat{\theta}_t - \bar{y}_{tu}). \tag{3.12}$$

The gains in efficiency produced by using these estimators have been extensively examined by Patterson (1950) and Cochran (1977). Matching 75% of the sample on successive occasions gives an appreciable gain in efficiency for estimating change, while for estimating level a 50% match of the sample is the optimum, assuming the survey has run for enough occasions to support the asymptotic assumption. Patterson (1950) notes that a 75% matched sample has little effect on the variance of the mean compared to the optimum 50%.

Eckler (1955) extended Patterson's method to two and three-level rotation sampling. Iterative solutions for the minimum-variance estimates of the current mean were developed for these two schemes. The problem of improving past estimates by incorporating more recent data was discussed for the two-level case. For the two-level rotation sampling with $n_t = n$ for all t, the proposed minimum variance linear unbiased estimator of θ_t is of the form

$$\hat{\theta}_t = \bar{y}_{tm} - \psi_t \bar{y}_{t-1,m} + \psi_t \hat{\theta}_{t-1}, \tag{3.13}$$

where $\bar{y}_{tm}$ and $\bar{y}_{t-1,m}$ are the means of the new reporting elements at time t for periods t and $t-1$ respectively. Minimizing the variance of (3.13) yields

$$\psi_t = \frac{\rho}{2 - \psi_{t-1}\rho}. \tag{3.14}$$

The variance of $\hat{\theta}_t$ is given by

$$\mathrm{Var}(\hat{\theta}_t) = \frac{\sigma^2}{n}(1 - \psi_t\rho). \tag{3.15}$$

The results for the rotation sampling obtained thus far were derived under the assumption that the population of units is infinite.

Rao and Graham (1964), Tikkiwal (1979), Prabhu-Ajgaonkar (1962), Singh (1968) and others extended Patterson's theory to other designs and incorporated finite population assumptions. Prabhu-Ajgaonkar and Tikkiwal also extended the theory of rotation sampling to the case of an arbitrary correlation pattern among occasions. They showed that

$$\hat{\theta}_t = \psi_t \bar{y}_{tu} + (1 - \psi_t)\{\bar{y}_{tm} + \beta_t(\hat{\theta}_{t-1} - \bar{y}_{t-1,m})\} \tag{3.16}$$

is the general form of the minimum variance linear unbiased estimator (MVLUE) under an arbitrary correlation pattern, provided the common units between any two occasions, two or more time points apart, form a sub-sample of the new units on the earlier occasion. When this condition does not hold, they argued that although the estimator is no longer MVLUE it is still convenient to use it without much loss of efficiency.

The MVLUE requires

$$\beta_t = \rho_{t,t-1} \frac{\sigma_t}{\sigma_{t-1}},$$

$$\frac{\psi_t}{1 - \psi_t} = \frac{n_{tu}}{n_{tm}}(1 - \rho_{t,t-1}^2) + \rho_{t-1,t}^2 \frac{n_{tu}}{n_{t-1,u}} \psi_{t-1},$$

where $\rho_{t,t-1}$ is the correlation between occasions $t-1$ and t and σ_t^2 is the variance of the population for occasion t.

As mentioned in Section 2, Rao and Graham (1964) extended the 4–8–4 scheme developed by Hansen, Hurwitz, Nisselson and Steinberg (1955) to a k cycle r–m–r, scheme. This means that a group of units stays in the sample for r occasions, leaves the samples for m occasions, returns for another r occasions, and so on. If a unit returns to the samples after having dropped $(k-1)$ previous times from the sample, the unit is said to be in its kth cycle. Rao and Graham (1964) studied the properties of the estimator of the population mean given by

$$\hat{\theta}_t = K(\hat{\theta}_{t-1} + \bar{y}_{tm} - \bar{x}_{t-1,m}) + (1 - K)\bar{y}_t \tag{3.17}$$

where $0 \leqslant K \leqslant 1$ and $\bar{y}_t$ is the sample mean at time t, assuming sampling from a finite population. They provided the variance of their composite estimator under the general situation and tabulated the gain in efficiency of the estimator for optimum K over the efficiency of the classical sample mean estimator for different r, ρ and m values. For measuring the mean, the gain in efficiency of $\hat{\theta}_t$, for fixed ρ and optimum K, is maximum for $r = 2$ and $m = \infty$. However, for measuring change $\Delta\theta_t = \theta_t - \theta_{t-1}$, the gain in efficiency of $\Delta\hat{\theta}_t = \hat{\theta}_t - \hat{\theta}_{t-1}$ over $\Delta\bar{y}_t = \bar{y}_t - \bar{y}_{t-1}$ increases with r. Thus, if both current occasion and change are of interest, the use of an r other than 2 may be preferred.

The extension from simple random sampling to multi-stage designs has been considered by a number of authors. Singh (1968) examined in detail the estimation of the mean on the second occasion when two samples are considered, assuming an exponential correlation pattern and a two-stage design on each occasion. Agarwal and Tikkiwal (1975) subsequently expanded the theory and Singh's (1968) results follow as a special case.

3.3. Minimum variance unbiased estimators and approximation to the optimum estimate

Repeated surveys can be broken down into elementary components determined by the pattern of the overlap. An estimate which does not make use of the survey data for any time period except that period to which the estimate refers is called an 'elementary estimate'. Gurney and Daly (1965) generalized the theory previously given by using the elementary estimate concept. Elementary estimates can be incorporated into a linear model framework which uses the correlation structure between the elementary estimates in order to produce minimum variance linear unbiased estimators (MVLUE). Gurney and Daly solved the general problem of MVLUE using Hilbert space theory. Smith (1978) and Wolter (1979) later provided simpler developments of their results.

To this end, let $\{\theta_t\}$ be the sequence of population parameters of interest. Assuming that y_{it} is the ith elementary estimate at time t ($i = 1, \ldots, n_t$; $t = 1, \ldots, T$), let

$$y_{it} = \theta_t + e_{it}$$

where e_{it} is the sampling error which has expectation zero and known correlation structure. In vector form, the above model may be written as

$$Y = X\Theta + e \tag{3.18}$$

where

$$Y = (y_{11}, \ldots, y_{1n_1}, \ldots, y_{T1}, \ldots, y_{Tn_T})',$$
$$e = (e_{11}, \ldots, e_{1n_1}, \ldots, e_{T1}, \ldots, e_{Tn_T})',$$
$$\Theta = (\theta_1, \theta_2, \ldots, \theta_T)',$$

and X is a design matrix of 0's and 1's.

Assuming that $E(e) = 0$ and $E(ee') = S$, the MVLUE of θ is given by

$$\hat{\Theta} = (X'S^{-1}X)^{-1}X'S^{-1}Y \tag{3.19}$$

and the covariance matrix of $\hat{\Theta}$ is $(X'S^{-1}X)^{-1}$. Furthermore, the MVLUE of any linear combination $\lambda'\Theta$ is the same linear combination $\lambda'\hat{\Theta}$.

Examples of the use of elementary estimates to form MVLUE have been given by Gurney and Daly (1965), Smith (1978) and Wolter (1979).

One of the difficulties in implementing (3.19) in practice is that the computation of MVLUE estimators requires the inversion of the covariance matrix S which can become large if many historical preliminary estimates are brought into the computations.

Composite estimation as introduced by Hansen, Hurwitz and Madow (1953) and investigated by Rao and Graham (1964) is an alternative to MVLUE. Composite estimation has been shown to be both computationally and statistically efficient by Gurney and Daly (1965) for the Current Population Survey employing the 4-8-4 semi one-level rotation scheme and by Wolter (1979) for the Census Bureau Retail Trade Survey employing a two-level rotation scheme. For a two level rotation scheme, the composite estimator of θ_t is

$$\hat{\theta}_t = K \hat{\theta}_{t-1} + (y_{2,t} - K y_{1,t-1}) \tag{3.20}$$

where $y_{2,t}$ and $y_{1,t-1}$ are unbiased elementary simple estimates of the population parameters θ_t and θ_{t-1}.

This estimator is exactly like the one used in the Current Population Survey, given by Hansen, Hurwitz, Nisselson and Steinberg (1955). The optimal value of K is

$$K = \frac{1 - \sqrt{1 - \rho^2}}{\rho}$$

and the variance of $\hat{\theta}_t$ is

$$\mathrm{Var}(\hat{\theta}_t) = \sigma^2 \sqrt{1 - \rho^2}.$$

Wolter (1979) studied the properties of composite estimation for a two-level rotation scheme and applied it to the panel design used by the U.S. Census Bureau's monthly Retail Trade Survey. Wolter considered two composite estimators of the totals: a preliminary composite estimator and a final composite estimator. The preliminary composite estimator is given by (3.20). In this case, $y_{2,t}$ and $y_{1,t-1}$ are unbiased elementary estimates of the population totals θ_t and θ_{t-1} obtained from the panel reporting in month t. Assuming a given covariance structure between the elementary estimates provided by the different panels, the coefficient K is obtained using a numerical method which minimizes the variance of $\hat{\theta}_t$. The final composite estimator is given by

$$\theta^*_{t-1} = (1 - L) y_{1,t-1} + L \hat{\theta}_{t-1}$$

where L is obtained by minimizing the variance of θ^*_{t-1}.

The composite estimators of month-to-month change and year-to-year change are given by

$$\Delta \hat{\theta}_t = \hat{\theta}_t - \theta^*_{t-1}$$

and

$$\delta \hat{\theta}_t = \hat{\theta}_t - \theta^*_{t-12},$$

respectively. The values of K and L which minimize the variances of $\hat{\theta}_t$, θ_t^*, $\Delta\hat{\theta}_t$ and $\delta\hat{\theta}_t$ differ from one another. This suggests that different coefficients should be used for each of these estimates whenever permitted by the survey conditions. If different coefficients cannot be used in the various estimators, then compromises need to be made. In the case of the Census Bureau's Retail Trade Survey, the values $K = 0.75$ and $L = 0.80$ are used as compromises. Wolter's investigation also showed that the composite estimators $\hat{\theta}_t$ and θ_{t-1}^* are approximately 95 percent efficient relative to the MVLU estimator for characteristics that have either high or low correlation structures.

In general, the composite estimator of θ_t is of the form

$$\hat{\theta}_t = K(\hat{\theta}_{t-1} + d_{t,t-1}) + (1 - K)\bar{y}_{\cdot,t} \tag{3.21}$$

where $\bar{y}_{\cdot,t}$ is the mean of the elementary estimates for month t and $d_{t,t-1}$ is the mean difference of elementary estimates common to months t and $t-1$. Gurney and Daly (1965) showed how the simple composite estimator given by (3.21) can be generalized so that the resulting estimator has a variance closer to the MVLUE variance.

Assuming that for a given occasion t there are l elementary estimates and that b of these elementary estimates are associated with panels entering the sample for the first time, Gurney and Daly showed that if more weight was given to those b elementary estimates than to be the remaining $l - b$ remaining elementary estimates, then the coefficients associated with the elementary estimates would be closer to those obtained using the MVLUE approach. The general form of this estimator, known as the 'AK-composite estimator' is

$$\tilde{\theta}_t = K(\tilde{\theta}_{t-1} + d_{t,t-1}) + \frac{1}{l}\left\{1 - K + A\right) \sum_{j=1}^{b} y_{ij,t}$$

$$+ \left(1 - K - \frac{l}{l-b}A\right) \sum_{j=b+1}^{l} y_{ij,t}\right\}, \tag{3.22}$$

where $(i_1, t), \ldots, (i_b, t)$ denote panels on their first visit on occasion t and (i_{b+1}, t), $\ldots, (i_l, t)$ denote panels which have been in the sample prior to occasion t.

The simple composite and the AK-composite estimators will have a smaller variance than the simple estimator which uses only the current occasion's information. However, as will be seen, this advantage can be offset by a rotation bias.

4. Rotation bias

Rotation bias can arise in periodic surveys used to collect data on such characteristics as health, consumer expenditures and labor force participation. The distinguishing feature of such surveys is that for many characteristics, estimates from the different panels relating to the same time period may not have the same expected value. The number of times respondents have been exposed to the survey seems to affect the data. A list of possible hypotheses about the causes of this bias contains several factors relating to the methods of data collection.

The effect of rotation bias on the estimates of unemployment rates based on the Current Population Survey was first discussed by Hansen, Hurwitz, Nisselson and Steinberg (1955). Gurney and Daly (1965) set up a simple model to study the effect of rotation bias on a number of estimators for the CPS. They compared the bias and root mean square error for the simple estimator, the MVLUE estimator, the simple composite estimator and the AK-composite estimator. Bailar (1975, 1979) subsequently studied the effect of rotation group bias on the estimates of level and change using labor force data from the Current Population Survey covering the years 1968–1979. Bailar (1975) showed that the rotation group bias has a different effect on the ratio estimate than it does on the composite estimate for estimates of level. For some characteristics, the mean square error of the composite estimate was larger than that of the ratio estimate, even though the variance of the composite estimate was smaller. In the CPS, the ratio estimator is obtained as a result of a four-step procedure which takes into account probabilities of selection, adjusts for non-response, uses a ratio estimation procedure from the non-self representing primary sample units and finally adjusts the sample estimates of population for a number of age-sex-race groups. The simple composite estimator studied was of the form given by (3.21) with $K = 0.5$; $\bar{y}_{\cdot,t}$ is the estimate for month t based on all eight rotation groups; $d_{t,t-1}$ is the mean difference between the current month and the previous month based on the six rotation groups common to month t and $t - 1$.

In terms of our previously defined notation, the 'i' associated with the elementary estimate y_{it} will denote the number of months ($i = 1, 2, \ldots, 8$) a rotation group has been in sample for reference month t. Let the population parameter being estimated be denoted by θ_t. Let a_{it} be the rotation group bias for month t associated with the rotation group in its ith month in the sample, so that $a_{it} = E(y_{it}) - \theta_t$.

Assuming that the biases for the rotation groups of a given month in sample are constant over months, i.e., $a_{it} = a_i$ for all t, Bailar (1975) showed that for the CPS

$$E(\bar{y}_{\cdot,t}) = \theta_t + \frac{1}{8} \sum_{i=1}^{8} a_i$$

and that for a sufficiently large sample size

$$E(\tilde{\theta}_t) = \theta_t + \frac{1}{8} \sum_{i=1}^{8} a_i + \frac{1}{5}[(a_4 + a_8) - (a_1 + a_5)] .$$

Thus, unless $\sum_{i=1}^{8} a_i = 0$, the ratio estimator $\bar{y}_{\cdot,t}$ is biased. Furthermore, the simple composite estimator is unbiased only if

$$\sum_{i=1}^{8} a_i + \tfrac{4}{3}[(a_4 + a_8) - (a_1 + a_5)] = 0.$$

Upon examining CPS data covering 1970–1972 for selected labour characteristics (i.e.: civilian labour force by sex, employment status), Bailar (1975) found that although the ratio estimator was negatively biased (based on the assumption of more correct re-interview data), the simple composite estimator was even more negatively biased, resulting in even larger losses of efficiency (in terms of mean square error). For estimates of month-to-month change, the simple composite estimator seemed better for several characteristics since it had smaller estimated variance on the average. Bailar (1979) subsequently re-examined the CPS data covering 1968–1979 for similar labor characteristics and concluded that the use of the simple composite estimator would result in a small percentage of understatement in the estimate of level, corresponding to a fluctuating, sometimes large, percentage of understatement of overestimate in the estimate of change.

Huang and Ernst (1981) studied the variance and bias aspects of the AK composite estimator versus the simple composite estimator examined by Bailar (1975, 1979) for the CPS. They investigated these properties using the 4–8–4 rotation pattern as well as an alternative 3–9–3 rotation pattern for selected labor characteristics covering the period September 1976 to December 1977. The form of the AK-estimator for the 4–8–4 scheme is

$$\tilde{\theta}_t = K(\tilde{\theta}_{t-1} + d_{t,t-1}) + \tfrac{1}{8}\bigg\{(1 - K + A)(y_{1t} + y_{5t}) + \bigg(1 - K - \frac{A}{3}\bigg)\bigg(\sum_{i=1}^{8} y_{it} - y_{1t} - y_{5t}\bigg)\bigg\}.$$

This estimator assigns more weight to rotation groups which have been in the sample for the first and fifth time, and less weight to the remaining rotation groups. Assuming a constant variance and covariance for all observations at all time periods, Huang and Ernst concluded the following. For each characteristic studied, the optimum AK-composite estimator had greater efficiency than the simple composite estimator for monthly level, month-to-month change, and annual average for both rotation patterns. Assuming a 4–8–4 rotation pattern and no bias for the average of the ratio estimates based on the eight rotation group, the bias of the AK-composite estimator is smaller than the bias for the simple composite estimator.

Kumar and Lee (1983) conducted a study similar to Huang and Ernst (1981) for the Canadian Labour Force Survey. The rotation group bias was examined for a number of labor force characteristics. It was found to be significant in the first-rotation group for the following characteristics: In Labour Force, Employed and Employed Non-Agriculture. The optimal-AK composite estimator was more efficient in terms of mean square error and had smaller bias than the corresponding optimal simple composite estimator. However, for the labor force characteristics where rotation group bias was significant, these estimators had a relative efficiency (measured as ratios of mean square error) smaller than 10% with respect to the currently used simple post-stratified estimator. Kumar and Lee subsequently obtained better composite estimators by minimizing the mean square error as opposed to the variance. The resulting estimator had gains in efficiency ranging 0 to 22% over the simple post-stratified estimator.

5. Time series approaches

5.1. Time series adjustments to survey estimates

In the previous sections, we considered only the classical approach to estimating a characteristic on repeated occasions. There, the mean (or total) was assumed fixed and unknown, without any assumptions on its behaviour over time. However, as Smith (1978) pointed out:

'The fact that econometricians and sociometricians use government records (assumed accurate) as inputs in their stochastic models shows that it is quite common to treat θ_t as a random quantity rather than a fixed quantity'.

Blight and Scott (1973) were one of the first authors to exploit the assumption of a stochastic model for the population means over time. Their set-up was similar to Pattersons's (1950) exponential correlation pattern except that they assumed a first order autoregressive process for the population means. The complete model is:

$$\theta_t - \mu = \alpha(\theta_{t-1} - \mu) + \varepsilon_t, \qquad \text{Var}(\varepsilon_t) = \sigma^2, \qquad (5.1a)$$

$$\bar{y}_{tm} - \theta_t = \rho(\bar{x}_{t-1,m} - \theta_{t-1}) + e_{1t}, \quad \text{Var}(e_{1t}) = w'_t, \qquad (5.1b)$$

$$\bar{y}_{tu} - \theta_t = e_{2t}, \qquad \text{Var}(e_{2t}) = w''_t, \qquad (5.1c)$$

where $|\alpha| < 1$.

Expression (5.1a) describes the behaviour of the finite population means over time, whereas expressions (5.1b) and (5.1c) describe the mechanism which generates the sampling error arising from sampling the finite population. The $\bar{y}$'s and $\bar{x}$'s are as defined in Section 3.1. The random terms $[\varepsilon_t, e_{1t}, e_{2t}]$ are assumed to be independent normal with zero mean. The parameters α, μ, ρ, σ^2, w'_t and w''_t are all assumed known.

Blight and Scott (1973) gave a recursive formulation for the Bayes estimator, $\hat{\theta}_{t|t} = E(\theta_t|Y_t)$, where Y_t is the vector of all the observations up to time t. Their recursions are equivalent to those given by the Kalman filter to be described in Section 5.2. The Bayes estimator is also the minimum mean square error predictor of θ_t.

Blight and Scott (1973) also used the model to derive the Bayes estimator of $\theta_t - \theta_{t-1}$, the change between occasions. Noting that w'_t and w''_t are functions of the sizes of the overlapping and non-overlapping samples, Blight and Scott derived the proportion of overlap to minimize the posterior variance of $\theta_t - \theta_{t-1}$.

One of Scott and Smith's (1974) considerations was the case of non-overlapping samples, where the sampling errors are assumed to be independent between occasions. In this case, the theory of Sections 2 and 3 is of little value, since the minimum variance linear unbiased estimate of θ_t is simply the survey estimate for data on the current occasion, t, only. The model, conditional on Y_{t-1}, may be written as:

$$\theta_t = \hat{\theta}_{t|t-1} + \varepsilon_t, \quad \text{Var}(\varepsilon_t) = v_{tt|t-1}, \tag{5.2a}$$

$$y_t = \theta_t + e_t, \quad \text{Var}(e_t) = S_t^2, \tag{5.2b}$$

where $\{\varepsilon_t, e_t\}$ are assumed to be independent normal with zero mean. The following recursions are then obtained. Denoting $\hat{\theta}_{t|t} = E(\theta_t | Y_t)$ and $v_{tt|t-1} = \text{Var}(\theta_t | Y_{t-1})$, we have

$$\hat{\theta}_{t|t} = (1 - \pi_t) y_t + \pi_t \hat{\theta}_{t|t-1}, \tag{5.3b}$$

where

$$\pi_t = S_t^2 / (v_{tt|t-1} + S_t^2). \tag{5.3a}$$

The posterior variance of θ_t is

$$\text{Var}(\theta_t | Y_t) = v_{tt|t} = (1 - \pi_t) S_t^2. \tag{5.3c}$$

Note the gain in efficiency over the survey estimate y_t, given by $\text{Var}(\theta_t | Y_t)/\text{Var}(e_t)$, is $1 - \pi_t$ from (5.3c). Therefore, greatest gains are made when π_t is close to unity. This occurs when $v_{tt|t-1}$ is small relative to S_t^2. Scott and Smith (1974) suggested that the one-step predictor of θ_{t+1} at time t and its variance, given by $\hat{\theta}_{t+1|t}$ and $v_{t+1,t+1|t}$ respectively, can be obtained from the assumed underlying time series model on $\{\theta_t\}$.

Jones (1980) generalized these approaches, as well as Gurney and Daly's (1965) linear model using elementary estimates, into a single framework. Letting Θ be the vector of θ_t's, and Y the vector of elementary estimates, the model can be written as:

$$\Theta = \mu + \varepsilon, \quad E(\varepsilon\varepsilon') = V, \tag{5.4a}$$

$$Y = X\Theta + e, \quad E(ee') = S, \quad E(\varepsilon e') = 0, \tag{5.4b}$$

where μ denotes the vector of mean values for $\Theta = (\theta_1, \theta_2, \ldots)'$.

The minimum mean squared error predictor (also the Bayes estimator under multivariate normality) is given by

$$E(\Theta | Y) = \mu + (X' S^{-1} X + V^{-1})^{-1} X' S^{-1} (Y - X\mu), \tag{5.5a}$$

$$\mathrm{Var}(\Theta|Y) = (X'S^{-1}X + V^{-1})^{-1}. \tag{5.5b}$$

For example, if $V = \sigma^2 R$, where R is a correlation matrix, then as σ^2 increases to infinity, this formulation captures the minimum variance linear unbiased estimator discussed in Section 3.3.

The practicality of this general result is limited for a variety of reasons. As with minimum variance linear unbiased estimates, the application of (5.5a, b) can involve inverting large matrices. Even with modern matrix inversion methods, the inverse can become unstable. In Sections 5.2 and 5.3, we consider simplifications which could arise by using statespace models and Kalman filters. These require some structure for the matrix S. Without such structure, optimal estimators may be difficult to compute.

Another problem is specifying the matrix V in (5.4). Using time series models, this can often be reduced to a parameter estimation problem. However, many standard time series estimation methods are not appropriate, due to the presence of (possibly correlated) survey errors. We discuss this in Section 5.4.

5.2. Statespace models

We give here a formulation of a statespace model, which will prove useful for implementing Jones (1980) result in (5.5). To this end, let $\{z_t\}$ be a series of r-dimensional state vectors. In general these are not directly observable. We will denote by $\{\theta_t\}$ the scalar valued mean values for the population over time. These are related to the z_t's. In the presence of survey error, the θ_t's are also not directly observable. Instead, what we do observe at time t is y_t, a vector of elementary estimates. Associated with each of these, there is a vector of survey errors denoted by e_t. It will be assumed that this survey error can also be related to the state vectors $\{z_t\}$. The assumed model is:

$$z_0 \sim N(0, \Sigma_0), \tag{5.6a}$$

$$z_t = F_t z_{t-1} + G_t \varepsilon_t, \tag{5.6b}$$

where F_t is $r \times r$, G_t is $r \times R_t$ and ε_t is an $R_t \times 1$ random $N(0, V_t)$ vector. Note that both F_t and G_t are transition matrices composed of fixed constants. The model formulation continues with

$$\theta_t = \mu_t + h'z_t, \tag{5.6c}$$

$$e_t = K_t z_t, \tag{5.6d}$$

$$y_t = x_t \theta_t + e_t = x_t \mu_t + A_t z_t, \tag{5.6e}$$

where $A_t = x_t h' + K_t$. The matrix K_t in (5.6d) is normally a diagonal matrix to allow the survey error to be a stationary model up to a multiplicative factor which varies with time. The vector x_t in (5.6e) is usually a column vector of 1's to reflect that y_t is a vector of elementary estimates of θ_t at time t. The vector h' in (5.6c) is a row vector so that (5.6c) would be the measurement equation of a model with no survey error. The transition equation is represented by expression (5.6b) and the measurement equation by expression (5.6e).

To exemplify this formulation, we consider the simple case where $\{\theta_t\}$ is the first order autoregressive process: $\theta_{t+1} = \alpha\theta_t + \varepsilon_{t+1}$, $\{e_t\}$ is the first order moving average process: $e_{t+1} = \eta_{t+1} - \beta\eta_t$ and $y_t = \theta_t + e_t$. We let the state vector z_t be $z_t' = (\theta_t, e_t, \xi_t)$.

Initial conditions for stationarity are that z_0 is trivariate normal with mean zero and covariance matrix

$$\begin{bmatrix} \sigma_\varepsilon^2/(1-\alpha^2) & 0 & 0 \\ 0 & \sigma_\eta^2(1+\beta^2) & -\sigma_\eta^2\beta \\ 0 & -\sigma_\eta^2\beta & \sigma_\eta^2\beta^2 \end{bmatrix}. \tag{5.7a}$$

The transition equations corresponding to (5.6b) may be written:

$$\theta_{t+1} = \alpha\theta_t + \varepsilon_{t+1}, \tag{5.7b}$$

$$e_{t+1} = \xi_t + \eta_{t+1}, \tag{5.7c}$$

$$\xi_{t+1} = -\beta\eta_{t+1}. \tag{5.7d}$$

The measurement equation corresponding to (5.6e) is $y_t = \theta_t + e_t$.

Now, for the general case, we denote by Y_t the vector of all observations up to time t and we let

$$\hat{z}_{t|i} = E(z_t|Y_i), \tag{5.8a}$$

$$P_{t|i} = \text{Var}(z_t|Y_i), \tag{5.8b}$$

$$\hat{\theta}_{t|i} = E(\theta_t|Y_i) = \mu_t + h'\hat{z}_{t|i}, \tag{5.8c}$$

$$u_{t|i} = \text{Var}(\theta_t|Y_i) = h'P_{t|i}h, \tag{5.8d}$$

$$\hat{y}_{t|i} = E(y_t|Y_i) = x_t\mu_t + A_t\hat{z}_{t|i}, \tag{5.8e}$$

$$W_{t|i} = \text{Var}(y_t|Y_i) = A_tP_{t|i}A_t'. \tag{5.8f}$$

In the above, the index i usually has value t or $t-1$.

A recursion for $\hat{z}_{t|t-1}$, $\hat{z}_{t|t}$, $P_{t|t-1}$, $P_{t|t}$ for $t = 1, 2, \ldots, T$ is given as:

$$\hat{z}_{1|0} = 0, \tag{5.9a}$$

$$P_{1|0} = F_1\Sigma_0 F_1' + G_1 V_1 G_1', \tag{5.9b}$$

$$\hat{z}_{t|t} = \hat{z}_{t|t-1} + P_{t|t-1}A_t' W_{t|t-1}^{-1}(y_t - \hat{y}_{t|t-1}), \tag{5.9c}$$

$$\hat{z}_{t+1|t} = F_{t+1}\hat{z}_{t|t}, \tag{5.9d}$$

$$P_{t|t} = P_{t|t-1} - P_{t|t-1}A_t' W_{t|t-1}^{-1} A_t P_{t|t-1}, \tag{5.9e}$$

$$P_{t+1|t} = F_{t+1}P_{t|t}F_{t+1}' + G_{t+1}V_{t+1}G_{t+1}'. \tag{5.9f}$$

This is the Kalman filter; see Kalman (1960) and Anderson and Moore (1979) for general results. We note that the only matrix which requires direct inversion is $W_{t|t-1}$ which is generally of small order. (The order of the matrix is the length of the vector of elementary estimates at time t.)

For example, for the case above where θ_t is a first order autoregressive process and e_t is a first order moving average process, we have the following results:

$$W_{t|t-1} = [1 \quad 1 \quad 0] P_{t|t-1} \begin{bmatrix} 1 \\ 1 \\ 0 \end{bmatrix}, \tag{5.10a}$$

$P_{1|0}$ is given by (5.7a),

$$\begin{bmatrix} \hat{\theta}_{t|t} \\ \hat{e}_{t|t} \\ \hat{\xi}_{t|t} \end{bmatrix} = \begin{bmatrix} \hat{\theta}_{t|t-1} \\ \hat{e}_{t|t-1} \\ \hat{\xi}_{t|t-1} \end{bmatrix} + P_{t|t-1} \begin{bmatrix} 1 \\ 1 \\ 0 \end{bmatrix} (y_t - \hat{\theta}_{t|t-1})/W_{t|t-1}, \tag{5.10b}$$

$$\hat{\theta}_{t+1|t} = \alpha \hat{\theta}_{t|t-1}, \tag{5.10c}$$

$$\hat{e}_{t+1|t} = \hat{\xi}_{t|t-1} = 0, \tag{5.10d}$$

$$P_{t|t} = P_{t|t-1} - P_{t|t-1} \begin{bmatrix} 1 & 1 & 0 \\ 1 & 1 & 0 \\ 0 & 0 & 0 \end{bmatrix} P_{t|t-1}/W_{t|t-1}, \tag{5.10e}$$

$$P_{t+1|t} = \begin{bmatrix} 1 & 0 & 0 \\ 0 & 0 & 1 \\ 0 & 0 & 0 \end{bmatrix} P_{t|t} \begin{bmatrix} 1 & 0 & 0 \\ 0 & 0 & 0 \\ 0 & 1 & 0 \end{bmatrix} + \begin{bmatrix} \sigma_\varepsilon^2 & 0 & 0 \\ 0 & \sigma_\eta^2 & -\beta\sigma_\eta^2 \\ 0 & -\beta\sigma_\eta^2 & \beta^2\sigma_\eta^2 \end{bmatrix}.$$

$$\tag{5.10f}$$

From (5.8c) and (5.9c), we can obtain $\hat{\theta}_{t|t}$, which is the estimate of θ_t given data Y_t. The variance is given by $u_{t|t}$ in (5.8d).

Sometimes we also wish to smooth the previous data in light of the later observations, especially for estimating change. Backwards recursions are available for performing this; see for example, Harvey (1984). For $\hat{z}_{t|T}$ and $P_{t|T}$, we start with $\hat{z}_{T|T}$ and $P_{T|T}$ and use:

$$\hat{z}_{t|T} = \hat{z}_{t|t} + P_{t|t}F'_{t+1|t}P^{-1}_{t+1|t}(\hat{z}_{t+1|T} - \hat{z}_{t+1|t}), \tag{5.11a}$$

$$P_{t|T} = P_{t|t} - P_{t|t}F'_{t+1|t}P^{-1}_{t+1|t}(P_{t+1|t} - P_{t+1|T})P^{-1}_{t+1|t}F_{t+1}P_{t|t}. \tag{5.11b}$$

Expression (5.8c) and (5.8d) can then be used to obtain $\hat{\theta}_{t|T}$ and $u_{t|T} = \mathrm{Var}(\theta_t | Y_t)$. Note that the matrix requiring inversion, $P_{t+1|t}$ is of order r, the dimensionality of the state vector. This is generally fixed, unless there are missing values or irregularly spaced observations.

5.3. ARMA models

A general extension of the example given by (5.7a) to (5.7d) where θ_t was AR(1) and e_t was MA(1) can be described as follows. The general statespace formulation, given by Harvey and Phillips (1979) for the classical ARMA(p, q) model is:

$$\theta_t - \alpha_1 \theta_{t-1} - \cdots - \alpha_p \theta_{t-p} = \varepsilon_t - \beta_1 \varepsilon_{t-1} - \cdots - \beta_1 \varepsilon_{t-q}. \tag{5.12}$$

In the classical setup, the θ_t's are directly observable.

Letting $r = \max(p, q + 1)$,

$$F_t = F = \begin{bmatrix} \alpha_1 & 1 & 0 \ldots 0 \\ \alpha_2 & 0 & 1 \ldots 0 \\ \vdots & \vdots & \vdots \\ \alpha_{r-1} & 0 & 0 \ldots 1 \\ \alpha_r & 0 & 0 \ldots 0 \end{bmatrix}, \quad G_t = G = \begin{bmatrix} 1 \\ -\beta_1 \\ -\beta_2 \\ \vdots \\ -\beta_r \end{bmatrix} \tag{5.13}$$

$V_t = \sigma_\varepsilon^2$, $h' = (1, 0, \ldots, 0)$, $\mu_t = 0$, $K_t = (0, 0, \ldots, 0)$, $x_t = 1$, we have the statespace formulation given by (5.7). A necessary condition for stationary is $F\Sigma_0 F' + \sigma^2 GG' = \Sigma_0$.

Once we incorporate survey errors in the set of observations, the process is more complex. Suppose instead that the true mean process is an ARMA(p, q) model given by (5.12). Suppose also that for each point in time there is a single estimate with sampling error, e_t, which is a stationary ARMA(p^*, q^*) process up to a multiplicative constant; $e_t = k_t e_t^*$, where

$$e_t^* - \alpha_1^* e_{t-1}^* - \cdots - \alpha_{p^*}^* e_{t-p^*}^* = \eta_t - \beta_1^* \eta_{t-1} - \cdots - \beta_q^* \eta_{t-q^*}. \tag{5.14}$$

For example, a pure moving average process, ARMA(0, q^*) may be reasonable for rotating panel surveys.

The complete model for $\{\theta_t, e_t, y_t\}$ can then be formulated as a statespace model given by (5.6), where the state vector includes components from both the $\{\theta_t\}$ process and the $\{e_t\}$ process.

These models can be extended within the statespace framework given by (5.6). A regression component can be added by letting $\mu_t = c_t' \gamma$. The model for the survey errors given by (5.14) can be extended to a vector of errors corresponding to the elementary estimates. The scalars $\{\alpha_1^*, \ldots, \alpha_{p^*}^*, \beta_1^*, \ldots, \beta_{q^*}^*\}$ could be matrices and the matrices F^* and G^* would be analogously defined. Missing time points can also be incorporated by extending ε_t to a vector including all the ε's corresponding to the missing time points.

We see then that this statespace formulation can be used effectively for a wide variety of problems in the context of repeated surveys. The computations using the Kalman filters (5.9) and (5.11) are easily performed.

An outstanding problem, though, is the estimation of the matrices F_t, G_t and V_t. (Often Σ_0 is defined in terms of these other matrices by imposing a condition such as stationarity.)

It will be assumed that the parameters associated with the survey errors, $\{e_t\}$, can be estimated using the usual design-based methods. In Section 5.4 we discuss the estimation of the other parameters. Whereas the methods described in this section can be considered as Bayesian, when we estimate the parameters of the underlying time series for the θ's the procedure becomes empirical Bayes.

5.4. Parameter estimation

The estimation of the parameters of an ARMA(p, q) model in the presence of survey error has received only scant attention in the literature. The methods suggested here not only have applications in the sampling theory context but also in the estimation of time series parameters when the input series is subject to survey error. Under the assumption that the sequence of e_t's is a stationary process, Scott, Smith and Jones (1977) and Jones (1980) suggested using the estimated autocovariance function of the observations $\{y_t\}$ to model the observation process. Since $y_t = \theta_t + e_t$ and the autocovariances of the e_t's can be estimated using design-based methods, the autocovariances of the θ_t's can be estimated by subtraction using: $\text{Cov}(\theta_t, \theta_{t-1}) = \text{Cov}(y_t, y_{t-1}) - \text{Cov}(e_t, e_{t-1})$.

Miazaki (1985) considered the model where θ_t is ARMA(p, 0) and e_t is ARMA(0, q^*), the parameters of the e_t process being estimated directly from the survey. Recognizing that this results in the observations $\{y_t\}$ being a ARMA(p, $p + q^*$), Miazaki proposed estimating the parameters using a restricted maximum likelihood approach.

However, by formulating the model in statespace form, direct likelihood methods are available for a large class of problems. Using the notation of (5.8) the exact log-likelihood function may be written as:

$$l = -\tfrac{1}{2}\sum_{t=1}^{T} \log|W_{t|t-1}| - \tfrac{1}{2}\sum_{t=1}^{T} (y_t - \hat{y}_{t|t-1})' W_{t|t-1}^{-1}(y_t - \hat{y}_{t|t-1}).$$
(5.15)

Maximizing this log-likelihood function with espect to $\alpha_1, \ldots, \alpha_p, \beta_1, \ldots, \beta_q, \sigma^2$ and possibly regression coefficients $\gamma_1, \ldots, \gamma_s$ (when $\mu_t = c_t' \gamma$ for known c_t) involves finding first and second derivatives of (5.15) with respect to these parameters. Therefore, it is sufficient to obtain the first and second derivations of $\hat{y}_{t|t-1}$ and $W_{t|t-1}$ with respect to the unknown parameters $\{\alpha_1, \ldots, \alpha_p, \beta_1, \ldots, \beta_q, \sigma^2, \gamma_1, \ldots, \gamma_s\}$. These can be obtained recursively using (5.9) and (5.8e, f).

In order to obtain the approximate (asymptotic) variance of the estimated parameters, we consider the Fisher information matrix $J = [J_{ij}]$ where $J_{ij} = -E\{\partial^2 l/\partial \lambda_i \partial \lambda_j\}$ for parameters λ_i and λ_j. Under model (5.6), the vector of observations, Y, is multivariate normal with mean $X\mu$ and some variance matrix which we denote as U. Supposing $\mu = C\Gamma$ we have the log-likelihood function

$$l = -\tfrac{1}{2}\log|U| - \tfrac{1}{2}(Y - D\Gamma)' U^{-1}(Y - D\Gamma)$$
(5.16)

where $D = XC$.

Therefore the Fisher information matrix can be obtained using:

$$E\left\{\frac{\partial^2 l}{(\partial \Gamma)(\partial \Gamma)'}\right\} = D' U^{-1} D,$$
(5.17a)

$$E\left\{-\frac{\partial^2 l}{\partial \Gamma \partial \lambda_i}\right\} = 0,$$
(5.17b)

$$E\left\{-\frac{\partial^2 l}{\partial \lambda_i \partial \lambda_j}\right\} = \left\{\tfrac{1}{2}\mathrm{tr}\left(\frac{\partial U}{\partial \lambda_i}\right) U^{-1} \left(\frac{\partial U}{\partial \lambda_j}\right) U^{-1}\right\}.$$
(5.17c)

The values for $U = [u_{ij}]$ can be obtained numerically from the model using the following recursion:

$$\mathrm{Var}(z_0) = \Sigma_0,$$ (5.18a)

$$\mathrm{Var}(z_t) = F_t[\mathrm{Var}(z_{t-1})]F_t' + G_t V_t G_t',$$ (5.18b)

$$\mathrm{Cov}(z_{t+i}, z_t) = F_{t+i}[\mathrm{Cov}(z_{t+i-1}, z_t)], \quad i > 0,$$ (5.18c)

$$u_{t+i,t} = \mathrm{Cov}(y_{t+i}, y_t) = A_{t+i}[\mathrm{Cov}(z_{t+i}, z_t)]A_t', \quad i \geq 0,$$ (5.18d)

where A_t is given in (5.6). These same recursions can be used to derive the values for $\partial U/\partial \lambda$.

However, (5.17) also contains terms involving U^{-1}. If matrix inversion is to be avoided, we note from (5.16) that the (i, j)th element of U^{-1} is given by $-2\partial^2 l/\partial y_i \partial y_j$. Using expression (5.15) for the form of the likelihood, we can obtain U^{-1} numerically by employing the recursion implied by (5.8e) and (5.9c, d).

Once all the parameters have been estimated, the smoothed values for θ, given by $\hat{\theta}_{t|T}$ from (5.8c), will be equivalent to Jones' (1980) formulation in (4.5a), where μ and V are replaced by $\hat{\mu}$ and $\hat{V}$, respectively.

We now consider the variance of $\hat{\theta}_{t|T}$. Letting $\mu = C\Gamma$, we have

$$\hat{\Theta} = C\hat{\Gamma} + (X'S^{-1}X + \hat{V}^{-1})^{-1}X'S^{-1}(Y - D\hat{\Gamma}) \quad (5.19a)$$

where

$$\hat{\Gamma} = (D'\hat{U}^{-1}D)^{-1}D'\hat{U}^{-1}Y. \quad (5.19b)$$

This may be written as

$$\hat{\Theta} = \hat{B}Y \quad (5.20)$$

where $\hat{B}$ is a function of the maximum likelihood estimates of α, β and σ^2. The variance of $\hat{\Theta}$ can be estimated by $\hat{B}\hat{U}\hat{B}'$, assuming that $\hat{B} - E(\hat{B})$ is small relative to $Y - D\Gamma$.

5.5. An example

To exemplify this time series technique we consider a small data series based on the Canadian Travel Survey. In Table 5.1 we provided (1) y_t, the original estimates for the number of overnight person-trips from the province of New Brunswick to the province of Quebec, (2) $\sigma^2\{y_t - \theta_t\}$, the estimated variances for $y_t - \theta_t$, (3) $\hat{\theta}_{t|T}$, the smoothed estimates based on the Kalman filter using the estimated model parameters and (4) $\sigma^2\{\hat{\theta}_{t|T}\}$ the variances of the smoothed estimates. We note that no survey was conducted for certain time points. These points are easily handled in the state space formulation, although the details have been left out here. It is possible to obtain smoothed estimated for these points, but these would be purely model-based and would be more susceptible to model misspecification.

The survey errors $\{e_t\}$ were assumed to be independent. The fitted model was:

$$\theta_t = \gamma_0 + \gamma_1(t - 12.5) + \gamma_2 Q_{1t} + \gamma_3 Q_{2t} + \gamma_4 Q_{3t} + \varepsilon_t, \quad (5.21a)$$

$$y_t = \theta_t + e_t, \quad (5.21b)$$

where ε_t is ARMA(2, 0) with parameters α_1, α_2 and σ^2.

$$Q_{it} = 1 \quad \text{if observation is in quarter } i; i = 1, 2, 3,$$
$$= -1 \quad \text{if observation is in quarter 4},$$
$$= 0 \quad \text{otherwise.}$$

In Table 5.2 we give the estimated parameters under the assumed model. We also give the parameter estimates when the sampling error is ignored (assumed to be zero).

Table 5.1
New Brunswick to Quebec overnight person trips (in 1000's)

t	Year	Quarter	y_t	$\sigma^2\{y_t - \theta_t\}$	$\hat{\theta}_{t|T}$	$\sigma^2\{\hat{\theta}_{t|T}\}$
1	79	1	21.2	27.1	23.1	15.8
2		2	60.6	97.5	45.3	30.3
3		3	63.2	102.6	58.5	32.4
4		4	20.6	8.7	21.5	7.4
5	80	1	20.0	25.2	23.4	13.5
6		2	44.9	22.6	41.5	15.2
7		3	43.4	64.8	50.1	24.2
8		4	35.1	50.0	27.0	19.3
9	81	1	26.4	11.8	25.4	8.3
10		2	–	–	–	–
11		3	70.5	117.1	50.0	24.1
12		4	–	–	–	–
13	82	1	21.8	28.0	23.0	13.8
14		2	17.9	22.0	24.0	13.7
15		3	37.4	18.0	42.3	12.5
16		4	27.0	36.4	26.9	17.5
17	83	1	–	–	–	–
18		2	–	–	–	–
19		3	–	–	–	–
20		4	–	–	–	–
21	84	1	19.0	7.9	17.7	6.8
22		2	23.5	10.2	23.8	8.8
23		3	48.9	25.0	45.1	17.2
24		4	16.9	6.9	17.3	6.1

Table 5.2
Parameter estimates

	Accounting for sampling error		Ignoring sampling error	
Regression parameters				
Intercept : γ_0	31.45	(0.74)[a]	32.81	(0.57)
Linear : γ_1	– 0.60	(5.67)	– 0.77	(4.16)
Quarter 1: γ_2	– 11.00	(0.44)	– 13.08	(0.32)
Quarter 2: γ_3	0.42	(0.30)	1.53	(0.27)
Quarter 3: γ_4	18.18	(0.40)	19.46	(0.31)
Autoregressive parameters				
Lag 1 : α_1	– 0.141	(0.42)	– 0.118	(0.28)
Lag 2 : α_2	– 0.624	(0.34)	– 0.194	(0.27)
Variance : σ^2	17.75		84.71	

[a] Bracketed figures are estimated standard errors.

First we note in Table 5.2 that the estimates of the autoregressive component parameters are substantially different when the sampling error is ignored. This should serve as a warning to those doing time series modelling when sampling errors cannot be assumed to be small. It is important to incorporate the sampling error variances in the model.

On the other hand, the estimates for the regression coefficients do not change substantially. This is because the estimate for $\hat{\varGamma}$ given in (5.19b) is still unbiased even when $\hat{U}$ is incorrectly specified.

In Table 5.1, the smoothed estimates $\{\hat{\theta}_{t|T}\}$ are all well within the variability to be expected from the survey sampling error variances. The largest jumps are at observations $t = 2$ and $t = 11$. These two estimates change from 60.6 and 70.5 to 45.3 and 50.0, respectively. Although the sampling errors for these two estimates are substantial, so that the adjusted values are within reason, an economic analysis would be necessary to determine whether these higher than expected original estimates can be explained.

In general, the technique seems to give reasonable values for the smoothed estimates. A criticism of the method may be the heavy reliance on some underlying model. However, if the sampling error variance is small relative to the estimated model variance, the smoothed estimates will be close to the original values. In this sense, the procedure is design-consistent.

Acknowledgements

The authors are grateful to Peter Dick for providing us with an advance copy of his M. Sc. Thesis at the University of Guelph: *Estimation in Repeated Surveys*, which contained some of the background material. The authors are also indebted to Professor J. E. Graham and Professor J. N. K. Rao of Carleton University for constructive comments and suggestions.

References

Agarwal, C. L. and Tikkiwal, B. D. (1975). Two-stage sampling on successive occasions. *Proceedings of the 62nd session of Indian Science Congress Association*. Part III, p. 31 (abstract).

Anderson, B. D. O. and Moore, J. B. (1979). *Optimal Filtering*. Prentice-Hall, Englewood Cliffs, NJ.

Bailar, B. A. (1975). The effects of rotation group bias on estimates from panel surveys. *Journal of the American Statistical Association* **70**, 23–29.

Bailar, B. A. (1978). Rotation sample biases and their effects on estimates of change. In: N. Krishan Namboodini, ed., *Survey Sampling and Measurement*. Academic Press, New York, 385–407.

Blight, B. J. N. and Scott, A. J. (1973). A stochastic model for repeated surveys. *Journal of the Royal Statistical Society, Series B* **35**, 61–68.

Cochran, W. G. (1977). *Sampling Techniques*. Third Edition, Wiley, Toronto.

Eckler, A. R. (1955). Rotation sampling. *Annals of Mathematical Statistics* **26**, 664–685.

Gurney, M. and Daly, J. F. (1965). A multivariate approach to estimation in periodic sample surveys. *Proceedings of the American Statistical Association, Section on Survey on Research Methods*, 247–257.

Hansen, M. H., Hurwitz, W. N. and Madow, W. G. (1953). *Sample Survey Methods and Theory, Volume 2*. Wiley, New York.

Hansen, M. H., Hurwitz, W. N., Nisselson, H. and Steinberg, J. (1955). The redesign of the Census Current Population Survey. *Journal of the American Statistical Association* **50**, 701–719.

Hanson, R. H. (1978). The Current Population Survey-Design and Methodology. Technical Paper 40, Bureau of the Census, Washington, DC.

Harvey, A. C. (1984). A unified view of statistical forecasting procedures. *Journal of Forecasting* **3**, 245–275.

Harvey, A. C. and Phillips, G. D. A. (1979). Maximum likelihood estimation of regression models with autoregressive-moving average disturbances. *Biometrika* **66**, 49–58.

Huang, E. T. and Ernst, L. R. (1981). Comparison of an alternative estimator to the current composite estimator of CPS. *Proceedings of the American Statistical Association, Section on Survey Research Methods*, 303–308.

Jessen, R. J. (1942). Statistical investigation of a farm survey for obtaining farm facts. *Iowa Agricultural Station Research Bulletin* **304**, 54–59.

Jones, R. G. (1980). Best linear unbiased estimators for repeated surveys. *Journal of the Royal Statistical Society, Series B* **42**, 221–226.

Kalman, R. E. (1960). A new approach to linear filtering and prediction problems. *Transactions ASME Journal of Basic Engineering* **82**.

Kumar, S. and Lee, H. (1983). Evaluation of composite estimation for the Canadian labour force survey. *Survey Methodology* **9**, 1–24.

Miazaki, E. S. (1985). Estimation for time series subject to the error of rotation sampling. Ph.D. Thesis, Iowa State University, Ames, IA.

Patterson, H. D. (1950). Sampling on successive occasions with partial replacement of units. *Journal of the Royal Statistical Society, Series B* **12**, 241–255.

Prabhu-Ajgaonkar, S. G. (1967). The theory of univariate sampling on successive occasions under the general correlation patterns. *Australian Journal of Statistics* **18**, 56–63.

Rao, C. R. (1973). *Linear Statistical Inference and its Applications*. Second Edition, Wiley, Toronto.

Rao, J. N. K. and Graham, J. E. (1964). Rotation designs for sampling on repeated occasions. *Journal of the American Statistical Association* **50**, 492–509.

Scott, A. J., Smith, T. M. F. and Jones, R. G. (1974). Analysis of repeated surveys using time series methods. *Journal of the American Statistical Association* **69**, 674–678.

Scott, A. J. and Smith, T. M. F. (1977). The application of time series methods to the analysis of repeated surveys. *International Statistical Review* **45**, 13–28.

Singh, D. (1968). Estimates in successive sampling using multi-stage design. *Journal of the American Statistical Association* **63**, 99–112.

Smith, T. M. F. (1978). Principles and problems in the analysis of repeated surveys. In: N. Krishan Namboodini, ed., *Survey Sampling and Measurement*, Academic Press, New York, 201–206.

Tikkiwal, B. D. (1979). Successive sampling—A review. *Proceedings of the 42nd Session of the International Statistical Institute held in Manila. Book 2*, 367–384.

Wolter, K. M., Isaki, C. T., Sturdevant, T., Hansen, N. and Mayes, F. (1976). Sample selection and estimation aspects of the Census Bureau's monthly business surveys. *Proceedings of the American Statistical Association, Business and Economic Statistics Section*, 99–109.

Wolter, K. M. (1979). Composite estimation in finite populations. *Journal of the American Statistical Association* **74**, 604–613.

Yates, F. (1949). *Sampling Methods for Censuses and Surveys*. First Edition. Charles Griffin, London.

Bayesian Inference in Finite Populations

William A. Ericson

Introduction

The appearance of V. P. Godambe's pioneering paper (Godambe, 1955), in which he showed the non-existence of minimum variance unbiased estimators for a labelled finite population total, resulted in renewed interest and research in this important area of statistics. This renewed research activity saw the application to the finite population setting of virtually all known approaches to inference and some new ones as well. Thus we have seen the Bayes, the pseudo-Bayes, the empirical Bayes, the predictive, the superpopulation, the pseudo-likelihood, the fiducial and other approaches applied to finite populations. Concepts like hyperadmissibility, model unbiasedness etc., have been introduced. Many of these approaches, in attempts to obtain more informative forms of inference, incorporate prior information or distributions in one guise or another. Smith (1976, 1984) presents a review of much of this work.

The present author (Ericson, 1969a) introduced an openly subjectivistic Bayesian approach to problems of inference and sample design for finite populations. Convinced that no new principles of inference were needed and that the Bayesian model was straightforward, simple and natural, he and some others have pursued this point of view. In addition, the Bayesian model seems to provide one simple way of handling the major applied problems of non-response, response error and response bias. The present paper reviews some of the results obtained from this Bayesian approach.

2. The basic model

We assume the now usual finite population model, i.e., a population of N distinguishable units labelled by the integers $1, 2, \ldots, N$. Let $\eta = \{1, 2, \ldots, N\}$ be the label set and $Y = (Y_1, Y_2, \ldots, Y_N)$, where Y_i is the unknown value (possibly a vector) of some characteristic possessed by the ith population unit. Inference concerns Y or, more realistically, some function, g, of Y. Here we will be mainly concerned with inference about $T = g(Y) = \sum_{i=1}^{N} Y_i$ or $\mu_y = T/N$. In addition, we assume that there is a known vector $X_i = (X_{i1}, X_{i2}, \ldots, X_{ip})'$ of concomitant variables available for each $i \in \eta$. Let X be the N by p matrix of these variables.

Note that in this model the label $i \in \eta$ is just a convenient coding of some actual label which may contain information about Y_i. For example, the label may consist of a person's name and address, the serial number of an automobile, the names of the streets defining a city block etc. There is some flexibility in the model as to which information about Y_i is carried by $i \in \eta$ and which by X_i.

We define a sample by the pair (s, y_s) where $s = \{i_1, i_2, \ldots, i_n\} \subseteq \eta$ is the set of distinct unit labels comprising the sample and $y_s = (y_{i_1}, y_{i_2}, \ldots, y_{i_n})$ and y_{i_j} is that which is observed for sample unit i_j.

If one assumes no nonresponse and no response bias or error then $y_{i_j} = Y_{i_j}$ if $i_j \in s$. If, on the other hand, one wishes to incorporate the possibility of missing data—item and/or unit nonresponse—and non-sampling errors, then the model may be extended to include these. This may be accomplished by allowing y_{i_j} to assume null values or by taking y_{i_j} to be equal to Y_{i_j} plus some unknown bias plus some error. Such extended models have been considered by Albert (1983), Ericson (1983b), Little (1982) and Rubin (1976, 1977). For much of the present paper we will assume the simpler model where $y_{i_j} = Y_{i_j}$ if $i_j \in s$.

A sample design is defined by S, the set of all subsets of η, together with $p_X(s)$, a discrete probability measure on S; i.e., $p_X(s) \geq 0$ and $\Sigma_{s \in S} p_X(s) = 1$. Also, for any s in S let $n(s)$ be the sample size, the number of elements in s. Note that we have eliminated sampling designs with replacement; but, as is now well known, there is no loss of generality in doing so since for any design (s, y_s) is a minimal sufficient statistic. Note too that the sample design may depend on the values of the concomitant variables, X, where $X = (X_1, X_2, \ldots, X_N)$.

For notational convenience, for any sample (s, y_s) define the matrix operator $S(Y) = (Y_{i_1}, Y_{i_2}, \ldots, Y_{i_n})$ and its complement $\bar{S}(Y) = (Y_{j_1}, Y_{j_2}, \ldots, Y_{j_{N-n}})$ for $j_k \in \eta - s$. For definteness assume that $i_1 < i_2 < \cdots < i_n$ and $j_1 < j_2 < \cdots < j_{N-n}$. With this notation the likelihood function of Y is given by

$$l(Y|(s, y_s)) = \begin{cases} p_X(s)k & \text{for } Y|S(Y) = y_s, \\ 0 & \text{otherwise}, \end{cases} \tag{2.1}$$

where $k > 0$ is any constant. For any given prior distribution of Y, denoted by $p(Y|X)$ (the prior may be dependent on the known X), the posterior of Y is given by

$$p(Y|(s, y_s)) = \begin{cases} \dfrac{p(Y|X)}{p_{S(Y)}(y_s|X)} & \text{for } Y|S(Y) = y_s, \\ 0 & \text{otherwise}, \end{cases} \tag{2.2}$$

where $p_{S(Y)}(y_s|X)$ is the marginal prior on $S(Y)$.

The likelihood function in (2.1) has disturbed a number of workers in the field since it is but one step removed from the most uninformative likelihood function, viz., one that is constant for all Y, and also since it does not depend upon the design by which (s, y_s) was observed. This observation, we believe, has led to a number of consequences.

First, it has led some to more feverently espouse methods of inference which are not compatible with the likelihood principle. Next, it has led many to conclude that from a Bayesian point of view the sample design is irrelevant and that randomization plays no role. Finally, it has lead some to adopt the modelling or superpopulation approach in which Y is viewed as arising from a random sample of N units from some superpopulation having a probability density function given by $f(Y_i | \theta)$ where θ may be either known or some unknown superpopulation parameter.

The point of view adopted here is that, despite the relatively uninformative nature of the likelihood, by specification of reasonable priors, $p(Y|X)$, one often obtains useful inferences (posterior distributions) on various functions, $g(Y)$. And in cases where such prior distributions are suitably diffuse or uninformative the resulting Bayes estimators or their limits are in agreement with those obtained by the traditional sampling distribution viewpoint. In addition, it can be shown that, under appropriate assumptions regarding one's prior distribution, various of the traditional sampling designs (for example, stratified and multistage samples) are optimal from a Bayesian decision-theoretic view.

The superpopulation concept seems to be nothing but a partial specification of a prior distribution. The 'superpopulation' is a subjective artifact of no practical interest per se. It is true that one might, at times, specify an exchangeable prior distribution on Y via

$$p(Y|X) = \int_\theta \prod_{i=1}^N f(Y_i | X, \theta) \, dF(\theta | X)$$

—a mixture of independent, identically distributed random variables conditional on θ. The view here is that the specification of a superpopulation via $\prod_{i=1}^N f(Y_i | X, \theta)$ is rather subjective and should be part of a prior distribution and *not* be combined with $p_X(s)$ in order to yield a more informative 'likelihood' than that of (2.1).

3. Some basic results

Specification of the N-dimensional prior distribution as required in (2.2) may seem, at first glance, difficult or even impossible. This task, however, is considerably alleviated and rendered feasible in several different ways. One is by the application of notions of exchangeability, partitioning of η, modelling prior knowledge of the relation between Y and X, use of randomization, etc. These will be elucidated in subsequent sections. Another is via the application of various studied methods of eliciting or assessing prior information and distributions. Among the relevant papers here are those by Case and Keats (1982), Crosby (1981), Hogarth (1975), Kadane (1980), Kadane et al. (1980), Savage (1971) and Winkler (1967). A third approach is by the use of linear Bayes estimators which only rely on the specification of low moments of the prior distribution. Many of the results indicated in subsequent sections are based on this approach.

Linear Bayes estimators have been independently discovered, rediscovered and studied by many researchers. Among these are: Diaconis and Ylvisaker (1979), Ericson (1969c, 1970, 1975, 1983a, 1983b), Fienberg (1980), Goel and DeGroot (1980), Goldstein (1979, 1981b), Hartigan (1969), Jewell (1974a, 1974b, 1975) and Smouse (1984). The basic result here says roughly that if one chooses to approximate a posterior mean by a 'best' linear function of the observed data then that estimator is the same as that resulting from a multivariate normal distribution model. One useful form of this result is the following (Ericson, 1983a):

RESULT 3.1. *Suppose that X_1 ($n_1 \times 1$) and X_2 ($n_2 \times 1$) are any jointly distributed random vectors having means μ_1 and μ_2 and covariance matrix*

$$\Sigma = \begin{bmatrix} \Sigma_{11} & \Sigma_{12} \\ \Sigma_{21} & \Sigma_{22} \end{bmatrix}. \tag{3.1}$$

If

$$E(X_1|X_2) = PX_2 + \gamma \tag{3.2}$$

for some P ($n_1 \times n_2$) and γ ($n_1 \times 1$) not depending on X_2, or if P and γ are chosen to minimize

$$E_{X_2} \| E(X_1|X_2) - PX_2 - \gamma \|^2, \tag{3.3}$$

then

$$PX_2 + \gamma = \Sigma_{12} \Sigma_{22}^{-1} X_2 + (\mu_1 - \Sigma_{12} \Sigma_{22}^{-1} \mu_2). \tag{3.4}$$

Furthermore,

$$E_{X_2} \mathrm{Var}(X_1|X_2) - (\Sigma_{11} - (\Sigma_{12} \Sigma_{22}^{-1} \Sigma_{21})) \leq 0, \tag{3.5}$$

that is, the matrix on the left in (3.5) is negative semi-definite, with equality holding if (3.2) holds.

Note that the above expressions for the conditional expectation of X_1 given X_2 and for the expectation of the conditional variance in the case that (3.2) holds are exactly what one would obtain in the case that X_1 and X_2 had a multivariate *normal* distribution. Also, if (3.2) fails to hold then the 'best' linear approximation to $E(X_1|X_2)$ and an upper bound to $E\mathrm{Var}(X_1|X_2)$ are still given by (3.4) and (3.5) respectively. Note, also that (3.2) holds for a number of joint distributions, see for example, Diaconis and Ylvisaker (1979) and Ericson (1970). However, under mild added conditions, (3.2) is equivalent to multivariate normality, see Goel and DeGroot (1980).

Another impact of Result 3.1 is that some results of various researchers under the assumption of multivariate normality are immediately generalizable. Examples of this are contained in the work of Leamer (1978) and Royall and Pfeffermann (1982).

The main results of this paper are based upon this result. It may be used to characterize the posterior mean of Y, T and μ in many cases, as well as the prior expectation of the posterior covariance matrix of Y. The latter is useful in determining optimal sample designs.

One illuminating and useful special case of Result 3.1, Ericson (1976), is given by

COROLLARY 3.1. *Suppose $\xi\,(=X_1)$ and $X\,(=X_2)$ are any $n+1$ jointly distributed random variables such that $E(\xi)=m$, $\mathrm{Var}(\xi)=v(\xi)<\infty$, $V(X)$, the variance–covariance matrix of X, is positive definite, and $E(X_i|\xi)=\xi$, for $i=1,2,\ldots,n$. If either*

$$E(\xi|X) = Xa + b, \quad a = (a_1, \ldots, a_n)' \tag{3.6}$$

or a and b are chosen to minimize

$$E_X[E(\xi|X) - Xa - b]^2, \tag{3.7}$$

then

$$Xa + b = \frac{E(\xi)EV(\hat{\xi}|\xi) + \hat{\xi}v(\xi)}{V(\xi) + EV(\hat{\xi}|\xi)}, \tag{3.8}$$

where $\hat{\xi} = \hat{\xi}(X) = \mathbf{1}[V(X)]^{-1}X'/\mathbf{1}[V(X)]^{-1}\mathbf{1}'$, $(\mathbf{1}=(1,1,\ldots,1))$ is the usual BLUE or weighted least squares estimator of ξ with respect to either the variance–covariance matrix $V(X)$ or $E_\xi V(X|\xi)$. Also,

$$E_X V(\xi|X) \leq v(\xi)[1 - v(\xi)\mathbf{1}(V(X))^{-1}\mathbf{1}'] = v(\xi)\left[\frac{EV(\hat{\xi}|\xi)}{v(\xi) + EV(\hat{\xi}|\xi)}\right], \tag{3.9}$$

with equality if (3.6) holds.

Thus in this case the linear Bayes estimator of ξ is simply a weighted average of a traditional or classical estimator and the prior mean, with weights proportional to natural measures of the precisions of each. Also, the nicely interpretable quantity on the right side of (3.9) is a natural quantity to minimize for sample design purposes. Applications of these results are given in the next sections.

4. Exchangeability and simple random sampling

Traditionally, one of the most basic sample designs is that of simple random samples (without replacement and of size n). Here, every subset of n of the N

population elements has the same probability of constituting the sample or, in the notation of Section 1, the design consists of the pair $(S, p_X(s))$, where

$$p_X(s) = \begin{cases} \dfrac{1}{\binom{N}{n}} & \text{if } n(s) = n, \\ 0 & \text{otherwise}. \end{cases} \qquad (4.1)$$

The root of this design is that the N population units are viewed as indistinguishable, in the sense that one would just as soon base inference on any subset of n units as any other subset of n units. Here there is no apparent information about Y contained in the unit labels or at least none that the sample designer is willing or able to specify. The actual sample is then selected using randomization to provide protection against unforseen biases and to provide an 'objectivistic' sampling distribution of various statistics. The subjectivist would interpret this as meaning that the variate values were essentially exchangeable in the sense of de Finetti (1937).

4.1. Exchangeability

The ties between simple random sampling results concerning inference about the finite population mean, μ, and those obtained under a subjectivistic Bayesian viewpoint with an exchangeable prior on Y are extremely close. We now assume that there are no concomitant variables—no X—and, for simplicity only, that the Y_i's are scalars.

First, the random variables $Y_1, Y_2, \ldots, Y_N$ are said to be *exchangeable* whenever the joint probability of each of the $N!$ permutations of the Y_i's is the same. Exchangeability also results in having the distribution of any subset of r $(1 \leq r \leq N)$ of the Y_i's be the same as that for any other subset of r of the Y_i's. Now suppose that a subjectivist assesses a prior on Y under which the Y_i's are exchangeable with mean, $E(Y_i) = m$, variance, $V(Y_i) = v$, and covariance, $\text{Cov}(Y_i, Y_j) = c$, for all i and $j \neq i$. Let μ be the mean of Y_i in the finite population and let $\sigma^2 = \sum_{i=1}^{N} (Y_i - \mu)^2 / N$ be the finite population variance. Finally, let $\overline{Y}$ be the mean (sample of Y_i for *any* subset of n of the N population units. Using the fact that the Y_i's remain exchangeable when conditioned on any symmetric function of them, one can easily demonstrate the following, Ericson (1969b):

RESULT 4.1. *If the random variables* $Y_1, Y_2, \ldots, Y_N$ *are exchangeable with means, variances and covariances as defined above, then*

$$E(\mu) = E(\overline{Y}) = m, \quad V(\mu) = [v + (N-1)c]/N,$$
$$E(\sigma^2) = [(N-1)(v-c)]/N, \quad V(\overline{Y}) = [v + (n-1)c]/n, \qquad (4.2)$$
$$E(\overline{Y} | \mu, \sigma^2) = \mu,$$

and

$$V(\bar{Y}|\mu, \sigma^2) = \frac{(N-n)\sigma^2}{(N-1)n}. \qquad (4.3)$$

Note that these properties are those of the prior distribution only, no sampling design has been specified. The conditional mean and variance of $\bar{Y}$, (4.2) and (4.3), are in the exact form as for a simple random sample design. The analogy is even closer, for it may also be shown that under an exchangeable prior on Y, given the collection of the N population variate values, Y_i's, *but not the units to which they are attached*, the probability that *any prespecified* subset of n of the population elements will assume the values given by any collection of n of the N population values is precisely the same as the objective probability that the subset was selected by simple random sampling, namely $1/\binom{N}{n}$.

4.2. Bayes linear estimator

Now suppose that one has an exchangeable prior on Y, of the sort described above, selects a sample, s, of some arbitrary subset of n of the N population units and observes $Y_i = y_i$ for $i \in s$. One then has met all the conditions to apply Corollary 3.1, by taking $\mu = \xi$ and $(s, y_s) = x$, a realization of X. One could also apply Result 3.1 directly to Y, partitioned into $S(Y)$ and $\bar{S}(Y)$ corresponding to the sampled and unsampled Y_i's. In either case the posterior mean of μ or the Bayes linear estimate of μ, denoted by $\tilde{\mu}$, is by (3.8) given by

$$\tilde{\mu} = \frac{\bar{y}V(\mu) + mEV(\bar{Y}|\mu, \sigma^2)}{V(\mu) + EV(\bar{Y}|\mu, \sigma^2)}, \qquad (4.4)$$

where

$$EV(\bar{Y}|\mu, \sigma^2) = \frac{(N-n)E(\sigma^2)}{(N-1)n}. \qquad (4.5)$$

Also, from (3.9), one has

$$E_{(s, y)}V(\mu|(s, y_s)) \leq V(\mu) \frac{EV(\bar{Y}|\mu, \sigma^2)}{V(\mu) + EV(\bar{Y}|\mu, \sigma^2)}. \qquad (4.6)$$

Note that in these expressions if one had only vague prior information regarding μ, in the sense that $V(\mu)$ grows large with $E(\sigma^2)$ fixed then the linear Bayes estimator will be nearly $\bar{Y}$ and the precision measure on the right of (4.6) is $(N-n)E(\sigma^2)/(N-1)n$ in tune with traditional simple random sampling results.

4.3. The role of randomization

It is immediately evident, by symmetry, that if one's prior knowledge regarding the Y_i's is truly reflected by an exchangeable prior distribution on Y then, economics aside, there are no a priori grounds for preferring a sample consisting of any particular subset of n of the N units over any other subset of the same size.

Although randomization and random samples are not required formally by the Bayesian view put forth here it seems that there are several compelling reasons for using randomization. One is the 'lazy Bayesian' argument given in Ericson (1969b)—viz., drawing a random sample of size n is equivalent to taking any convenient sample of n units from a random permutation of the N population units and, no matter what prior distribution one had on the original population Y_i's, one's prior distribution on the randomly permuted Y values must be a member of the class of prior distributions which reflect the fact that they are exchangeable and thus randomization may often be used to justify using a prior distribution in this class. Next, the argument, put forth in Royall and Pfeffermann (1982), to the effect that random samples are approximately balanced samples and hence are approximately model robust is also compelling. Finally, it must be realized that most surveys serve multiple users having diverse views of inference. While the subjective Bayesian approach may be quite useful in the formal use of prior information in designing a sample, the sample itself, (s, y_s) (or summaries of properties of it), should be published in order to be fully useful to others. Consequently, randomization, while costing the subjectivist little, may buy him much in terms of the utility of his sample to others.

5. Stratified sampling

In many real applications the prior information about Y_i contained in the label, i, and/or in the value, X_i, of the concomitant variables may be used to partition η in such a way that within *each* element of the partition the Y_i's are roughly exchangeable a priori. Such partitioning, for example, may be by geographic region, sex and race, farm size, etc. This partitioning for approximate exchangeability is strictly akin to the traditional stratification of the population aimed at achieving within stratum homogeneity. For obvious reasons this partitioning process will be termed *stratification* and the elements of the partition *strata*.

5.1. Basic definitions and model

Let η_k, $k = 1, 2, \ldots, K$, be the K strata (a partition of η). Let N_k be the number of units in η_k, Y_{ik} be the value of the variable of interest, Y, for the ith unit in the kth stratum and $Y_k = (Y_{1k}, Y_{2k}, \ldots, Y_{N_k k})$. Also let μ_k and σ_k^2 be the mean and variance of the Y_{ik}'s in the kth stratum, i.e.,

$$\mu_k = \sum_{i=1}^{N_k} Y_{ik} \quad \text{and} \quad \sigma_k^2 = \sum_{i=1}^{N_k} (Y_{ik} - \mu_k)^2 / N_k$$

and let μ be the overall finite population mean given by

$$\mu = \sum_{k=1}^{K} N_k \mu_k / N.$$

Now, suppose that a priori within each stratum the Y_{ik}'s are exchangeable with means m_k, variances v_k and covariances c_k. Finally, assume that one's prior information is such that Y_{ik}'s in different strata are uncorrelated, i.e., $\text{Cov}(Y_{ik}, Y_{hj}) = 0$, for $k \neq j$. A sample, s, of n observations is obtained with n_k being in the kth stratum, $0 \leq n_k \leq N_k$, and for $ik \in s$ one observes $Y_{ik} = y_{ik}$. Here, again we denote this sample by (s, y_s). Let $\overline{Y}_k$ be the mean of the sample observations in the kth stratum.

5.2. Bayes linear estimator

Under the model above, it is a straightforward generalization of the results of the preceding section to show that the linear Bayes estimator of μ is given by

$$\tilde{\mu} = \sum_{k=1}^{K} \left[\frac{N_k (\overline{Y}_k V(\mu_k) + m_k EV(\overline{Y}_k | \mu_k, \sigma_k^2))}{N[V(\mu) + EV(\overline{Y}_k | \mu_k, \sigma_k^2)]} \right]. \tag{5.1}$$

Also,

$$E_{(s, y)} V(\mu | (s, y_s)) \leq \sum_{k=1}^{K} \frac{N_k^2 V(\mu_k)}{N^2} \frac{EV(\overline{Y}_k | \mu_k, \sigma_k^2)}{V(\mu_k) + EV(\overline{V}_k | \mu_k, \sigma_k^2)}, \tag{5.2}$$

where $EV(\overline{Y}_k | \mu_k, \sigma_k^2)$ and $V(\mu_k)$ are as given in (4.5) and above (4.2) by adding the subscript, k.

Here, too, if one had only vague prior knowledge of the μ_k's in the sense that $V(\mu_k)$ gets large with $E(\sigma_k^2)$ fixed, then the posterior mean of μ (or its best linear approximation) and the precision measure in (5.2) are given by $\sum_{k=1}^{K} N_k \overline{Y}_k / N$ and

$$\sum_{k=1}^{K} \left[\frac{(N_k - n_k) E(\sigma_k^2)}{(N_k - 1) n_k} \right]$$

respectively, in agreement with traditional sampling distribution theory.

5.3. Optimum stratified design

As pointed out in Section 4.3, from a subjectivistic Bayesian viewpoint one is indifferent as to which n_k units are sampled from the kth stratum because of the within stratum exchangeability. However, the arguments for random selection of the n_k persist. A subjectivist is not necessarily indifferent among different allocations of sampling effort to the K strata because of differential prior information about the Y_{ik}'s and sampling costs.

If one were interested in estimation of μ with a quadratic loss function then it is well known that the Bayes estimator is $E(\mu | (s, y_s))$ and the Bayes risk is

$EV(\mu|(s, y_s))$. Then a natural criterion for choosing a sample design (allocation) would be to adopt that design defined by $\boldsymbol{n} = (n_1, n_2, \ldots, n_k)$ which minimizes the Bayes risk subject to a budget constraint of the sort $\sum_{k=1}^{K} d_k n_k \leq C$; where d_k is the per unit cost of an observation in the kth stratum and C is a given sampling budget. Alternatively, one might add a sampling cost component to the loss function, $L(\tilde{\mu}, \mu)$, by defining it as

$$L(\tilde{\mu}, \mu) = \gamma(\tilde{\mu} - \mu)^2 + \sum_{k=1}^{K} d_k n_k, \tag{5.3}$$

where γ represents the trade off between squared estimation errors and sampling costs. In this case it is easily shown that the optimum strategy is that $\boldsymbol{n}$ which minimizes $EV(\mu|(s, y_s)) + \sum_{k=1}^{K} d_k n_k$. Now, if the linearity condition, (3.6), here that

$$E(\mu_k|(s_k, y_k)) = \sum_{i=1}^{n_k} a_{ik} y_{ik} + b_k,$$

holds in each stratum then the optimum designs involve either the minimization of the right hand side of (5.2) subject to a budget constraint or the minimization of γ times it plus a linear function of the n_k's. If the linearity condition fails in one or more strata then the same minimization seems quite reasonable in that it provides an upper bound to the function of interest.

Both of the minimization problems described above have been solved in detail in Ericson (1969a); some multivariate generalizations appear in Sekkappan (1981). Some of the simpler basic results are given below. First, let the right hand side of the inequality in (5.2) be denoted by $r(\boldsymbol{n})$. Then it is easily seen that

$$r(\boldsymbol{n}) = \sum_{k=1}^{K} (N_k - n_k) W_k/(n_k + n'_k), \tag{5.4}$$

where $W_k = n'_k[v_k + (N_k - 1)c_k]/N^2$ and $n'_k = (v_k - c_k)/c_k$. Note from above (4.2) that since $V(\mu_k) > 0$, we have $c_k > -v_k/(N_k - 1)$, which in most applications is tantamount to having $c_k > 0$, which we now assume. The first result deals with finding that $\boldsymbol{n}$, call it $\boldsymbol{n}^0$, which minimizes $r(\boldsymbol{n})$ subject to the constraints that $0 \leq n_k \leq N_k$ and $\sum_{k=1}^{K} d_k n_k \leq C$. To this end, let

$$u_k = [W_k(N_k + n'_k)/d_k]^{1/2}, \quad k = 1, \ldots, K.$$

Now $r(\boldsymbol{n})$ can be shown to be a convex function of $\boldsymbol{n}$ and by application of the Kuhn–Tucker theorem the following can be shown:

RESULT 5.1. *If* $C < \sum_{k=1}^{K} d_k N_k$ *and if* $\min_k(u_k/n'_k) > \max_k(u_k/(N_k + n'_k))$ *then for all budgets, C, in the interval*

$$\frac{\sum_{k=1}^{K} u_k d_k}{\min_k(u_k/n'_k)} - \sum_{k=1}^{K} n'_k d_k < C < \frac{\sum_{k=1}^{K} u_k d_k}{\max_k(u_k/(N_k + n'_k))} - \sum_{k=1}^{K} n'_k d_k, \quad (5.5)$$

the optimum allocation is given by $0 < n_k^o < N_k$ where

$$n_k^o = u_k \left[\frac{C + \sum_{k=1}^{K} d_k n'_k}{\sum_{k=1}^{K} d_k u_k} \right] - n'_k. \quad (5.6)$$

A relation between this allocation and the well-known Neyman allocation may be seen by the following argument. If $V(\mu_k)$ gets large with $E(\sigma_k^2)$ fixed it follows that v_k and c_k are going to infinity with $v_k - c_k$ fixed; thus $n'_k = (v_k - c_k)/c_k$ is approaching zero. Then by (5.5) for all C in an interval of the form $0 < C < C^* < \sum_{k=1}^{K} d_k N_k$ the optimum allocation is given approximately by (5.6) or by

$$n_k^o = u_k \left[\frac{C}{\sum_{k=1}^{K} d_k u_k} \right]. \quad (5.7)$$

Thus, in this limiting case, the n_k^o's are proportional to the u_k's. Now note that as $n'_k \to 0$ via $v_k \to \infty$, $c_k \to \infty$ with $v_k - c_k \to N_k E(\sigma_k^2)/(N_k - 1)$, fixed, then

$$u_k \to \frac{N_k}{N} \left[\frac{N_k E(\sigma_k^2)}{(N_k - 1) d_k} \right]^{1/2}. \quad (5.8)$$

Thus, in this *diffuse* prior case, the optimum sample size in any stratum is proportional to the stratum size and a measure of the standard deviation within the stratum and inversely proportional to the square root of the per unit sampling cost.

In Ericson (1969b), the author has given an algorithm for obtaining the optimum allocation for all values of C, not just those in the interval given by (5.5). What this amounts to, essentially, is that if for any C the optimum n_k^o given in (5.6) is negative, take no observations in that stratum and recompute n_k^o, via (5.6) deleting that stratum. Also if any n_k^o exceeds N_k then census that stratum, replace C by $C - N_k d_k$ delete that stratum from consideration and recompute the other n_k^o via (5.6). In the same paper it is also shown how to obtain the optimal design using the full loss function of (5.3).

6. Two stage sampling

There are also subjective Bayesian analogs to cluster sampling and two (and higher) stage sampling schemes. To illustrate these we consider one simple special case of equal primary sampling unit sizes; the case of unequal sizes can also be easily handled. We adopt the same notation as in Section 5, except that now Y_{ik} refers to the ith unit in the kth primary sampling unit (psu). The traditional two-stage sampling design selects a random sample of psu's and then a random sample of units within the selected psu's. We assume that there are K psu's and M units within each psu. Thus $k = 1, 2, \ldots, K$; $i = 1, 2, \ldots, M$ and $N = MK$. Note that cluster sampling seems to differ from two-stage sampling only by taking a census in the sampled psu's and calling the psu's *clusters*.

Here μ_k and σ_k^2 are the *unknown* population mean and variance within the kth psu. Further, let $\mu = \Sigma_{k=1}^{K} \mu_k/K$ be the overall population mean and let

$$\sigma^2 = \sum_{k=1}^{K} \sum_{i=1}^{M} (Y_{ik} - \mu)^2/N = \sigma_b^2 + \sigma_w^2, \tag{6.1}$$

where

$$\sigma_b^2 = \sum_{k=1}^{K} (\mu_k - \mu)^2/K \quad \text{and} \quad \sigma_w^2 = \sum_{k=1}^{K} \sigma_k^2/K. \tag{6.2}$$

6.1. A class of prior distributions

The main result of this section, providing a Bayesian interpretation of some standard sampling theory results, have been obtained (Ericson, 1975) under a fairly general class of prior distributions. Related work is given in Bellhouse et al. (1977), Mohd (1976), Scott and Smith (1969) and Sundberg (1983). The class of priors we assume here is defined as follows:

A1: Conditional on μ_k and σ_k^2, the Y_{ik}'s are exchangeable for $i = 1, 2, \ldots, M$ and also, for $k \neq h$, Y_{ik} and Y_{jh} are independent for all i and j.

A2: For $k = 1, 2, \ldots, K$ the ordered pairs (μ_k, σ_k^2) are exchangeable random two-tuples such that the following moments exist:

$$E(\mu_k) = m, \quad V(\mu_k) = v, \quad \text{Cov}(\mu_k, \mu_k') = c \quad \text{and} \quad E(\sigma_k^2) = \phi, \tag{6.3}$$

where $|c| < v/(K - 1)$. This latter condition ensures the positive definiteness of the covariance matrix of the psu means.

Now let $\bar{Y}$ be the mean of any subset of n_k units in the kth psu. The following is then easily shown:

RESULT 6.1. *If a prior distribution on Y satisfies A1 above, then*

$$E(Y_{ik}|\mu_k, \sigma_k^2) = E(\bar{Y}_{ik}|\mu_k, \sigma_k^2) = \mu_k, \tag{6.4}$$

$$V(Y_{ik}|\mu_k, \sigma_k^2) = \sigma_k^2, \tag{6.5}$$

$$\text{Cov}(Y_{ik}, Y_{jk}|\mu_k, \sigma_k^2) = -\sigma_k^2/(M-1) \tag{6.6}$$

and

$$V(\bar{Y}_k|\mu_k, \sigma_k^2) = \frac{(M - n_k)\sigma_k^2}{(M - 1)n_k}. \tag{6.7}$$

Also one can show

RESULT 6.2. *If a prior distribution on Y satisfies A1 and A2 above, then*

$$E(\mu) = E(\overline{Y}_k) = E(Y_{ik}) = m, \quad V(\mu) = (v + (K-1)c)/K, \quad (6.8a)$$

$$E(\sigma_w^2) = \phi, \quad E(\sigma_b^2) = (K-1)(v-c)/K, \quad (6.8b)$$

$$V(V_{ik}) = \phi + v, \quad \operatorname{Cov}(Y_{ik}, Y_{jk}) = v - \phi/(M-1) \text{ for } j \neq i, \quad (6.8c)$$

$$\operatorname{Cov}(Y_{ik}, Y_{jh}) = \operatorname{Cov}(\overline{Y}_k, \overline{Y}_h) = c \quad \text{for } k \neq h, \quad (6.8d)$$

$$V(\overline{Y}_k) = v + \frac{(M-n_k)\phi}{(M-1)n_k}, \quad E(\mu_k | \mu) = E(\overline{Y}_k) = \mu, \quad (6.8e)$$

$$E_\mu V(\overline{Y}_k | \mu) = E(\sigma_k^2) + \frac{(M-n_k)E(\sigma_w^2)}{(M-1)n_k} \quad (6.8f)$$

and

$$E_\mu \operatorname{Cov}(\overline{Y}_k, \overline{Y}_h | \mu) = -E(\sigma_b^2)/(K-1). \quad (6.8g)$$

Note that to establish this prior on N-space one need only specify four parameters—m, v, c, and ϕ.

6.2. *The Bayes linear estimator*

Assume now that one selects a sample of r psu's $(1 \leq r \leq K)$ and then selects n_k observations from the ith sampled psu $(1 \leq i \leq M)$. One may now apply Corollary 3.1 to the joint distribution of μ, and the $\overline{Y}_k$'s. By doing this it can be shown that the posterior mean of μ of the best linear approximation to it is given by

$$\tilde{\mu} = \frac{mEV(\hat{\mu}|\mu) + \hat{\mu}V(\mu)}{V(\mu) + EV(\hat{\mu}|\mu)}, \quad (6.9)$$

where $\hat{\mu}$ is the BLUE or WLSE of μ given here by

$$\hat{\mu} = \sum_{i=1}^{r} v_i \overline{Y}_i \Big/ \sum_{i=1}^{r} v_i, \quad (6.10)$$

where

$$v_i = \left[\frac{KE(\sigma_b^2)}{(K-1)} + \frac{(M-n_i)E(\sigma_w^2)}{(M-1)n_i} \right]^{-1}, \quad (6.11)$$

and

$$EV(\hat{\mu}|\mu) = \frac{E(\sigma_b^2)}{(K-1)} \left[\frac{K \sum_{i=1}^{r} v_i^2 - (\sum_{i=1}^{r} v_i)^2}{(\sum_{i=1}^{r} v_i)^2} \right]$$

$$+ \sum_{i=1}^{r} v_i^2 \left[\frac{(M-n_i)E(\sigma_w^2)}{(M-1)n_i (\sum_{i=1}^{r} v_i)^2} \right]. \quad (6.12)$$

It also follows from (3.9) that

$$EV(\mu|(s, Y_s)) \leq V(\mu)\left[1 - \frac{V(\mu)\sum_{i=1}^{r} v_i}{1 + c\sum_{i=1}^{r} v_i}\right]. \quad (6.13)$$

Here again, as one's prior information gets weak about μ the posterior mean of μ approaches the weighted least squares estimator, $\hat{\mu}$. Also, for optimal design purposes, the quantity on the right hand side of the inequality in (6.13) is a natural quantity to minimize. The basic result is indicated in the next section.

6.3. Optimum two-stage design

Suppose now that it costs d_1 to sample a psu and d_2 to sample any observation within any selected psu. Suppose also that a budget of C is available for sampling. We define the optimal design by r^0 and n_i^0, $i = 1, 2, \ldots, r^0$, which minimizes the right hand side of (6.13) subject to the constraints that $1 \leq r^0 \leq K$, $1 \leq n_i^0 \leq M$ for $i = 1, 2, \ldots, r^0$ and

$$r^0 d_1 + d_2 \sum_{i=1}^{r^0} n_i^0 \leq C. \quad (6.14)$$

The solution to this minimization problem is given by the following two basic results:

RESULT 6.3. *If* $(M-1)(v-c) \leq \phi$, *the optimal design is given by utilizing the entire budget in censusing as few psu's as possible. More specifically, the design is given by*

$$r^0 = \begin{cases} r^* \equiv [C/(d_1 + Md_2)] & \text{if } C < (r^* + 1)d_1 + (r^*M + 1)d_2, \\ r^* + 1, & \text{otherwise}, \end{cases} \quad (6.15)$$

$$n_i^0 = M, \quad i = 1, 2, \ldots, r^0, \quad (6.16)$$

and

$$n_{r+1}^0 = [(C - (r_0 + 1)d_1 - r^0 M d_2)/d_2], \quad (6.17)$$

where in these expressions $[\cdot]$ denotes the greatest integer in function.

Note that the condition that $(M-1)(v-c) \leq \phi$ is equivalent to

$$E(\sigma_b^2) \leq \frac{(K-1)E(\sigma_w^2)}{K(M-1)}, \quad (6.18)$$

which provides an interpretation of the solution in Result 6.3 and agrees with the traditional sample survey two-stage design result by simply replacing σ_b^2 and σ_w^2

by their prior expectations. See, for example Cochran (1963, Chapter 10). The solution to the design problem is completed by

RESULT 6.4. *If* $(M - 1)(v - c) > \phi$ *the optimal design takes some* r^0 *psu's and makes the* n_i^o*'s as nearly equal as possible.*

Thus in the case that $(m - 1)(v - c) > \phi$ the optimum design simply consists of finding an optimum r^0 and $n_i^o = n^o$. After a bit of manipulation, it can be shown that these values are those which minimize

$$V^* = \frac{(K - r)E(\sigma_b^2)}{(K - 1)r} + \frac{(M - n)E(\sigma_w^2)}{(M - 1)nr} . \qquad (6.19)$$

The expression in (6.19) is well-known, for if the expectation operators are removed then it is precisely equal to the sampling variance of the standard estimator of μ from a two-stage sample design. This means that for any prior distribution on Y the optimum Bayes sample design is identical to the classical optimum design where the subjectivist's prior expectations, $E(\sigma_b^2)$ and $E(\sigma_w^2)$, are substituted for the sampling theorist's σ_b^2 and σ_w^2. Specifically, n^0 and r^0 may be found by applying the results given in Cochran (1963, Chapter 10), for example, and making the suitable identifications between (6.19) above and his equation (10.8).

Again the optimum Bayesian design simply specifies the best number of psu's to sample and the best sample size within a psu, and, by the exchangeability assumption, which particular psu's and which units within the psu's is of no consequence. But for reasons discussed previously a subjectivist would select them randomly.

7. Ratio and regression estimation

In the present section we assume that one's prior knowledge about Y may be expressed in the form of a general linear model, i.e., we assume that

$$Y = X\beta + \varepsilon , \qquad (7.1)$$

where $\beta = (\beta_1, \beta_2, \ldots, \beta_p)'$ and $\varepsilon = (\varepsilon_1, \varepsilon_2, \ldots, \varepsilon_N)'$ are not known for sure. Here X is the known matrix of concomitant variables. We assume that conditional on β, ε has some N-dimensional prior distribution having mean $\mathbf{0}$ and covariance matrix Σ, not depending on β, and that β is assigned a prior distribution with mean $\bar{\beta}$ and covariance matrix V_β. Thus the class, C, of prior distributions, $p(Y|X)$, assumed here is that class having mean $X\bar{\beta}$ and covariance matrix given by $XV_\beta X' + \Sigma$.

7.1. Some basic results

Result 3.1 may be immediately applied to obtain properties of the posterior distribution of a finite population mean or total under the class of prior distributions defined above. To do this first assume that for any given sample, (s, Y_s), the elements of Y, X, and Σ have been permuted and partitioned so that

$$Y = \begin{bmatrix} Y_s \\ Y_r \end{bmatrix}, \quad E(Y|X) = \begin{bmatrix} X_s \\ X_r \end{bmatrix}\beta, \tag{7.2}$$

and

$$\text{Var}(Y|X) = \begin{bmatrix} \Sigma_{ss} + X_s V_\beta X'_s & \Sigma_{sr} + X_s V_\beta X'_r \\ \Sigma_{rs} + X_r V_\beta X'_s & \Sigma_{rr} + X_r V_\beta X'_r \end{bmatrix}, \tag{7.3}$$

where $Y_s = S(Y)$ and $Y_r = \bar{S}(Y)$.

Obviously, given the sample, (s, y_s), inference about Y is equivalent to inference about Y_r.

Applying Result 3.1, taking $X_1 = Y_r$ and $X_2 = Y_s$ one imediately has

RESULT 7.1. *For any prior distribution, $p(Y|X) \in C$, and any sample (s, Y_s),*

$$E(Y_r|Y_s) = V_{rs} V_{ss}^{-1} y_s + (X_r - V_{rs} V_{ss}^{-1} X_s)\bar{\beta}, \tag{7.4}$$

at least approximately, and

$$E_{Y_s} \text{Var}(Y_r|Y_s) \leq V_{rr} - V_{rs} V_{ss}^{-1} V_{sr}, \tag{7.5}$$

where

$$V_{ij} = (\Sigma_{ij} + X_i V_\beta X'_j) \quad \text{for } i, j = r, s. \tag{7.6}$$

Some special cases of Result 7.1 are interesting and usseful. These are given by

COROLLARY 7.1. *If Σ is diagonal then*

$$E(Y_r|Y_s) = X_r(V_\beta X'_s V_{ss}^{-1} Y_s + (I - V_\beta X'_s V_{ss}^{-1} X_s)\bar{\beta}). \tag{7.7}$$

COROLLARY 7.2. *If one's prior distribution is such that, conditional on β, the Y_i's are independent or Σ is diagonal and $E(Y_r|Y_s, \beta) = E(Y_r|\beta) = X_r\beta$, for all Y_r and Y_s then*

$$E(Y_r|Y_s) = X_r E(\beta|Y_s) \tag{7.8}$$

where

$$E(\beta|Y_s) = V_\beta X'_s V_{ss}^{-1} Y_s + (I - V_\beta X'_s V_{ss}^{-1} X_s)\bar{\beta}. \tag{7.9}$$

The expression in (7.9) can also be obtained directly by assuming that $E(\beta|Y_s) = PY_s + \gamma$, for some P ($p \times n$) and γ ($p \times 1$), or by choosing P and γ to minimize $E_{Y_s} \| E(\beta|Y_s) - PY_s - \gamma \|^2$. This follows as simply another application of Result 3.1. Also, the expression in (7.9) has the form of a weighted average of the prior mean, $\bar{\beta}$, and a weighted least squares estimator, $\hat{\beta}$, with weights proportional to $\text{Var}(\beta)$ and $E \text{Var}(\hat{\beta}|\beta)$ respectively. This weighted average form is given explicitly in

COROLLARY 7.3. *Under the conditions of Corollary 7.2*

$$E(\beta|Y_s) = V_\beta [E \text{Var}(\hat{\beta}|\beta) + V_\beta]^{-1} \hat{\beta}$$
$$+ E \text{Var}(\hat{\beta}|\beta) [E \text{Var}(\hat{\beta}|\beta) + V_\beta]^{-1} \bar{\beta}, \qquad (7.10)$$

where

$$\hat{\beta} = (X_s' V_{ss}^{-1} X_s)^{-1} X_s' V_{ss}^{-1} Y_s. \qquad (7.11)$$

Note that in order to compute the posterior mean of Y_r, the sum of the unsampled Y_i's (or the best linear approximation to it), involves the inversion of the $n \times n$ matrix, V_{ss}. This may be done analytically if V_{ss} assumes various paterned forms, and, in this connection, the following result is useful.

RESULT 7.2. *If $V = \Sigma + XV_\beta X'$, where X is $n \times p$, then*

$$V^{-1} = \Sigma^{-1} - \Sigma^{-1} XV_\beta (I + X' \Sigma^{-1} XV_\beta)^{-1} X' \Sigma^{-1}, \qquad (7.12)$$

provided these inverses exist.

This result may be easily verified by direct computation; it also appears in Graybill (1969). Note that if Σ^{-1} is known, then using (7.12) to find V^{-1} involves inversion of a $p \times p$ matrix instead of an $n \times n$ matrix.

In the next sections we examine special cases and show that various standard estimators are limits of posterior means.

7.2. *Conditional exchangeability with one concomitant variable*

In the present section we will be concerned only with the case that $p = 2$ and where one's prior distribution on the Y_i's is such that, conditional on $\beta^* = (\alpha, \beta)'$, there exist numbers, f_i, $i = 1, 2, \ldots, N$, such that the $(Y_i - \alpha - \beta X_i)/f_i$'s are exchangeable with mean zero, variance σ^2 and covariances $\rho\sigma^2$. In addition, β^* is assigned a joint prior distribution with mean $(\bar{\alpha}, \bar{\beta})'$ and covariance matrix

$$V_\beta = \begin{bmatrix} v_\alpha & v_{\alpha\beta} \\ \hline v_{\alpha\beta} & v_\beta \end{bmatrix}.$$

In such a case the prior distribution of Y has mean $X\beta^*$, where the first column of X consists of 1's and the second of the values of the X_i's, and covariance matrix

$$\text{Var}(Y|X) = F \Sigma F + XV_\beta X',$$

where F is diagonal with entries f_i and Σ has σ^2 on the diagonal and $\rho\sigma^2$ elsewhere. We assume that $\text{Var}(Y|X)$ is partitioned as in (3.1), so that $V_{ss} = F_s \Sigma_{ss} F_s + X_s V_\beta X_s'$, etc. We also consider here inference about the sum of the Y_i's for i's not in the sample, and let $Y_r = \Sigma_{i \notin s} Y_i$; posterior inference about μ follows immediately since $\mu = (Y_r + n\bar{y}_s)/N$. By application of Results 3.1 and 7.2 and some lengthy calculations the following can be shown (Ericson, 1976):

$$E(Y_r | (s, y_s)) = (1/D) [((N-n)V_1 + X_r V_2)d + \sigma^2(1-\rho)vV_3$$
$$+ (\rho F/v)V_4] + C_\alpha \bar{\alpha} + C_\beta \bar{\beta} \tag{7.13}$$

where

$$D = \sigma^4(1-\rho)^2 v^2 + \sigma^2(1-\rho)v(v_\alpha T_{11} + v_\beta T_{22} + 2v_{\alpha\beta}T_{12}) + dE,$$

$$V_1 = T_{13}T_{22} - T_{12}T_{24}, \qquad V_2 = T_{11}T_{22} - T_{13}T_{12},$$

$$X_r = \sum_{i \notin s} X_i, \qquad F = \sum_{i \notin s} f_i, \qquad d = v_\alpha v_\beta - v_{\alpha\beta}^2, \qquad v = (1 + (n-1)\rho),$$

$$T_{11} = vS_2 - \rho S_1^2, \qquad T_{22} = vS_5 - \rho S_3^2, \qquad T_{12} = vS_4 - \rho S_1 S_3,$$

$$T_{13} = vS_6 - \rho S_1 S_7, \qquad T_{24} = vS_8 - \rho S_3 S_7,$$

$$S_1 = \sum_{i \in s} f_i^{-1}, \qquad S_2 = \sum_{i \in s} f_i^{-2}, \qquad S_3 = \sum_{i \in s} x_i f_i^{-1},$$

$$S_4 = \sum_{i \in s} x_i f_i^{-2}, \qquad S_5 = \sum_{i \in s} x_i^2 f_i^{-2}, \qquad S_6 = \sum_{i \in s} f_i^{-2} y_i,$$

$$S_7 = \sum_{i \in s} f_i^{-1} y_i, \qquad S_8 = \sum_{i \in s} f_i^{-2} x_i y_i,$$

$$V_3 = (N-n)T_{13}v_\alpha + X_r T_{24} v_\beta + (X_r T_{13} + (N-n)T_{24})v_{\alpha\beta},$$

$$V_4 = DS_7 - \sigma^2(1-\rho)vV_5 - dvV_6,$$

$$V_5 = S_1 T_{13} v_\alpha + S_3 T_{24} v_\beta + (S_3 T_{13} - S_1 T_{24})v_{\alpha\beta},$$

$$V_6 = T_{13}(S_1 S_5 - S_3 S_4) + T_{24}(S_2 S_3 - S_1 S_4),$$

and

$$E = T_{11}T_{22} - T_{12}^2.$$

The coefficients C_α and C_β of $\bar{\alpha}$ and $\bar{\beta}$ are complex as the first term in (7.13), however they do possess the same form, i.e., if we let $i = \alpha, \beta$ then

$$C_i = (1/D) [\sigma^4(1-\rho)^2 v^2 B_{i1}^2 + B_{i1} \sigma^2(1-\rho)v(v_i^{-1}B_{i2} + v_{\alpha\beta}T_{12})$$
$$- \sigma^2(1-\rho)vB_{i1}^{-1}(v_i^{-1}T_{12} + v_{\alpha\beta}B_{i2}^{-1})$$
$$- (pF/v) [\sigma^4(1-\rho)^2 v^2 B_{i3} + \sigma^2(1-\rho)v^2(v_i^{-1}B_{i4}^{-1} - v_{\alpha\beta}B_{i4})$$
$$+ d(EB_{i3} - v(T_{12}B_{i4} + B_{i2}^{-1}B_{i4}^{-1}))],$$

where the values of the B_{ij}'s are given in Table 1 and the negative one exponent means reversing the roles of α and β, i.e. $v_\alpha^{-1} = v_\beta$, $B_{\beta 1}^{-1} = B_{\alpha 1}$, etc.

In the following subsections we examine special cases of (7.13).

Table 1

i	B_{i1}	B_{i2}	B_{i3}	B_{i4}
α	$(N-n)$	T_{22}	S_1	$S_1 S_5 - S_3 S_4$
β	X_r	T_{11}	S_3	$S_2 S_3 - S_1 S_4$

7.3. Uncorrelated 'errors', $\rho = 0$

In the case that $\rho = 0$ the expression (7.13) simplifies considerably. In fact, in that special case

$$E(Y_r | y_s) = (1/D)[dA_1 + A_2 + A_3 \bar{\alpha} + A_4 \bar{\beta}], \qquad (7.14)$$

where

$$D = \sigma^4 + \sigma^2(v_\alpha S_2 + v_\beta S_5 + 2v_{\alpha\beta} S_4) + d(S_2 S_5 - S_4^2),$$

$$A_1 = (N-n)(S_5 S_6 - S_4 S_8) + X_r(S_2 S_8 - S_4 S_6),$$

$$A_2 = \sigma^2[S_6(v_\alpha(N-n) + X_r v_{\alpha\beta}) + S_8(v_{\alpha\beta}(N-n) + X_r v_\beta)],$$

$$A_3 = \sigma^2[(N-n)(\sigma^2 + v_\beta S_5 + v_{\alpha\beta} S_4) - X_r(v_{\alpha\beta} S_2 + v_\beta S_4)],$$

$$A_4 = \sigma^2[X_r(\sigma^2 + v_\alpha S_2 + v_{\alpha\beta} S_4) - (N-n)(v_\alpha S_4 + v_{\alpha\beta} S_5)],$$

the S_i's and d being as defined above.

There are many special subcases that one might examine here; we examine one, leaving others as subcases of other special cases. The one examined here is that of regression through the origin, i.e., suppose that one is a priori sure that $\alpha = 0$. Then by taking $\bar{\alpha} = 0$, $v_\alpha = 0$ and hence $v_{\alpha\beta} = 0$ in (7.14) one finds that the posterior mean of Y_r is simply

$$E(Y_r | y_s) = X_r \left\{ \frac{v_\beta \frac{\sum x_i y_i f_i^{-2}}{\sum x_i^2 f_i^{-2}} + \frac{\sigma^2}{\sum x_i^2 f_i^{-1}} \bar{\beta}}{v_\beta + \frac{\sigma^2}{\sum x_i^2 f_i^{-1}}} \right\}, \qquad (7.15)$$

which is X_r times a simple weighted average of $\bar{\beta}$ and $\hat{\beta} = \sum x_i y_i f_i^{-2} / \sum x_i^2 f_i^{-2}$, in precisely the form (7.10).

If, in addition, $f_i = 1$ and one has little prior knowledge about β, $(v_\beta \to \infty)$, then

$$E(Y_r | y_s) \to \left[\frac{\sum_{i \in s} x_i y_i}{\sum_{i \in s} x_i^2} \right] X_r, \qquad (7.16)$$

the usual sampling theory regression estimator. This is discussed further in Section 7.4.

7.4. Diffuse priors on β

Suppose now that one's prior knowledge about β is rather vague reflected by large values of v_α and v_β. If we let $v_\alpha \to \infty$ in such a way that $v_{\alpha\beta}^2/(v_\alpha v_\beta) = r^2$, a constant; then it is relatively easy to see that by dividing both numerator and denominator of the terms in (7.13) by $v_\alpha v_\beta$, $D/v_\alpha v_\beta \to E(1-r^2)$, $C_\alpha/v_\alpha v_\beta \to 0$, $C_\beta/v_\alpha v_\beta \to 0$ and the first term in the numerator of (7.13) approaches $[(N-n)V_1 + X_r V_2](1-r^2)$ and thus that

$$E(Y_r|y_s) \to \frac{(N-n)(T_{13}T_{22} - T_{12}T_{24}) + X_r(T_{11}T_{24} - T_{13}T_{12})}{T_{11}T_{22} - T_{12}^2}$$

$$= (N-n)\hat{\alpha} + X_r \hat{\beta}, \tag{7.17}$$

say, where $\hat{\alpha}$ and $\hat{\beta}$ are implicitly defined as the coefficients of $(N-n)$ and X_r in (7.17). It is also easy to see that $(\hat{\alpha}, \hat{\beta})$ are just the classical weighted least squares estimators of (α, β), i.e.,

$$(\hat{\alpha}, \hat{\beta})' = (X_s' \Sigma^{*-1} X_s)^{-1} X_s' \Sigma^{*-1} y_s, \quad \text{where } \Sigma^* = F_s \Sigma_{ss} F_s.$$

Certain subcases are of interest, for example if $f_i = 1$, $i = 1, 2, \ldots, N$, then (7.17) reduces to the usual regression estimator, i.e.

$$E(Y_r|y_s) = (N-n)(\bar{y} - \bar{x}\hat{\beta}) = X_r \hat{\beta}, \tag{7.18}$$

where

$$\hat{\beta} = \sum[(x_i - \bar{x})(y_i - \bar{y})]/\sum(x_i - \bar{x})^2.$$

Note also that the other limiting case, $\sigma^2 \to \infty$ with ρ fixed in (7.13) leads to the limit

$$E(Y_r|y_s) \to (N-n)\bar{\alpha} + X_r \bar{\beta} = E(Y_r),$$

as one might expect.

7.5. 'Regression' through the origin

The case where one is a priori certain that $\alpha = 0$ leads to several classical survey estimators as being approximate posterior means under the class of priors C. If one takes $\bar{\alpha} = 0$, $v_\alpha = 0$ and $v_{\alpha\beta} = 0$ in (7.13) one finds that

$$E(Y_r|y_s) = \left[\frac{1}{(\sigma^2(1-\rho)v + v_\beta T_{22})}\right][X_r T_{24} v_\beta + \rho F(\sigma^2(1-\rho)S_7$$
$$+ v_\beta(S_5 S_7 - S_3 S_8)) + (\sigma^2(1-\rho)vX_r - \rho F(\sigma^2(1-\rho)S_3)\bar{\beta}], \tag{7.19}$$

and letting $V_\beta \to \infty$ in this expression one finds that, under this model, with vague prior knowledge about β,

$$E(Y_r|y_s) = \frac{X_r(vS_8 - \rho S_3 S_7) + (S_5 S_7 - S_3 S_8)\rho(N-n)}{vS_5 - \rho S_3^2}, \quad (7.20)$$

and here $(vS_8 - \rho S_3 S_7)/(vS_5 - \rho S_3^2) = \hat{\beta}$ is the classical weighted least squares estimator of β. So if additionally one assumes that $\rho = 0$ then $E(Y_r|y_s) = X_r \hat{\beta}$. Also, if $f_i = 1$, as well as $\rho = 0$, then (7.20) reduces to the standard regression estimator given in (7.16).

If one takes $f_i = x_i^{1/2}$, for $i = 1, 2, \ldots, N$, in (7.20) one finds that

$$E(Y_r|y_s) = \left[\frac{1}{v \sum x_i - \rho(\sum x_i^{1/2})^2}\right][X_r \sum y_i - \rho(\sum x_i^{1/2} \sum (y_i/x_i^{1/2}))$$
$$+ (\sum x_i \sum (y_i/x_i^{1/2}) - \sum x_i^{1/2} \sum y_i)\rho \sum_{i \notin s} x_i^{1/2}]$$

and if additionally, $\rho = 0$ then

$$E(Y_r|y_s) = \frac{\bar{y}_s}{\bar{x}_s} X_r,$$

the usual ratio estimator.

Finally, if one takes $f_i = x_i$ in (7.20) one finds that

$$E(Y_r|y_s) = \sum_{i \in s}\left(\frac{y_i}{x_i}\right) X_r/n, \quad (7.21)$$

the usual average of ratios estimator. Note that (7.21) holds regardless of the value of ρ, unlike the previous two special cases. Note also that these last three special cases of standard survey estimators being, at least approximately, posterior means were previously shown by Ericson (1969) to hold under the much more stringent assumption of a normal prior distribution.

8. Response error and bias

As mentioned in the introduction, the subjective Bayesian veiw adopted here seems to provide the only natural way to incorporate non-sampling biasses and errors into the model. In the present section we present some results in this direction. We concentrate here on response bias and error. The approach is a generalization of the original treatment by Schlaifer (1959). The problems of nonresponse have been considered by Albert (1983), Basu and Pereira (1982),

Ericson (1967), Little (1982), Rubin (1976, 1977) and Singh and Sedransk (1978).

The Bayesian approach adopted here incorporates prior information (perhaps only vague) concerning the bias and errors. Such information would certainly be based on the sampler's previous experience, including sampling experiments designed to measure such biases and errors. In addition, we consider the case where biased and error prone (tainted) observations are available, as well as pure or untainted observations, the latter presumably at a higher per unit cost. One example of such a situation would be the case where for a subsample a randomized response technique can be used (to obtain relatively untainted observations) and for the balance of the sample the usual questionnaire or interview is administered. Other examples include situations where for a subsample original records are examined while for the balance of the sample personal recollection is used; or, indeed any situation where a combination of some reliable source of information is used for part of the sample and some less reliable source is used for the balance. In such cases we obtain an optimum allocation of sampling effort among the two sources of data. It turns out that, up to a certain point, the tainted observations may be useful and informative.

Finally, we consider the case where sample information on the bias is obtainable. Here in the first subsample tainted observations are gotten and in the second both pure and tainted are obtained. The optimum allocation of sampling effort among these two categories is obtained.

8.1. The model and prior distribution

Suppose now that if $i \in s$ then one observes

$$x_i = Y_i + \beta + \varepsilon_i, \tag{8.1}$$

where β is some unknown constant or average response bias and the ε_i's are unobservable response errors.

The observed sample is now assumed to be given by (s, x_s) where $x_s = (x_{i_1}, x_{i_2}, \ldots, x_{i_n})'$ for $i_j \in s$. It is assumed that there is some prior information regarding β and the ε_i's which is incorporated into a joint prior distribution of (Y, β, ε) where $\varepsilon = (\varepsilon_1, \varepsilon_2, \ldots, \varepsilon_N)'$, perhaps by specifying

$$f(Y, \beta, \varepsilon) = f(\beta, \varepsilon | Y) f(Y). \tag{8.2}$$

From such a prior distribution, using (2.2), it is easy (again in principle) to obtain $f(Y, (s, x_s))$ and $f((s, x_s))$ and thus the desired posterior distribution of Y:

$$f(Y | (s, x_s)) = \frac{f(Y, (s, x_s))}{f((s, x_s))}. \tag{8.3}$$

Under the model in (8.1) suppose that one has a symmetric prior on X in the sense that $E(X_i) = m$, $V(X_i) = v$ and $\text{Cov}(X_i, X_j) = c$ for $i = 1, 2, \ldots, N$, $i \neq j$.

Suppose also that the joint prior on (β, ε) conditional on Y has moments

$$E(\beta|Y) = \bar{\beta}, \qquad V(\beta|Y) = v_\beta, \qquad E(\varepsilon_i|Y) = 0, \qquad V(\varepsilon_i|Y) = v_\varepsilon,$$

and the ε_i's are uncorrelated as are the ε_i's and β.

Now suppose a sample, s, of n units is selected and (s, x_s) is observed. Let Y_s be the unknown vector of Y_i's for $i \in s$ and let Y_r, $((N-n) \times 1)$, be similarly defined for the unsampled units. By (8.1) and the properties of the prior given above, it is easy to see that

$$E(X_i) = m + \bar{\beta}, \quad i = 1, 2, \ldots, n, \qquad E(Y_j) = m, \quad j = 1, 2, \ldots, N, \quad (8.4)$$

$$\text{Var}(X_i) = v + v_\beta + v_\varepsilon, \quad i = 1, 2, \ldots, n, \tag{8.5}$$

$$\text{Cov}(X_i, X_j) = c + v_\beta, \quad i \neq j = 1, 2, \ldots, n,$$

$$\text{Cov}(X_i, Y_i) = v, \quad i = 1, 2, \ldots, n,$$

and
$$\text{Cov}(X_i, Y_j) = c, \quad i \neq j, \quad i = 1, 2, \ldots, n, \quad j = 1, 2, \ldots, N. \tag{8.6}$$

8.2. The Bayes linear estimator

Now, with the mean vector and the covariance matrix of $(Y_r', Y_s', X_s')'$ having the components given in (8.4)–(8.6), one can apply Result 3.1 taking $X_1 = (Y_r', Y_s')'$ and $X_2 = X_s$. Doing this and letting $\mu = \sum_{i=1}^{N} Y_i/N$, one finds, after some algebra, the result that the posterior mean of μ, the finite population mean, is given by

$$E(\mu|(s, x_s)) = \frac{\text{Var}(\mu)(\bar{x}_s - \bar{\beta}) + E_\mu \text{Var}(\bar{X}_s|\mu)m}{\text{Var}(\bar{Y}) + v_\beta + v_\varepsilon/n}, \tag{8.7}$$

at least approximately, and where

$$\text{Var}(\mu) = \frac{v + (N-1)c}{N}, \qquad \text{Var}(\bar{Y}) = \frac{v + (n-1)c}{n}, \tag{8.8}$$

$$E_\mu \text{Var}(\bar{X}_s|\mu) = (N-n)(v-c)/Nn + v_\varepsilon/n + v_\beta, \tag{8.9}$$

$$\bar{X}_s = \sum_{i \in s} X_i/n \quad \text{and} \quad \bar{Y}_s = \sum_{i \in s} Y_i/n. \tag{8.10}$$

This is, of course, just the usual weighted average of the prior mean of μ and a simple 'unbiased' estimator of μ with weights inversely proportional to the prior variance of μ and the prior expectation of the sampling variance of the estimator. In addition, Result 3.1 says that

$$E_{x_s} \text{Var}(\mu|(s, x_s)) \leqslant \frac{\text{Var}(\mu) E_\mu \text{Var}(\bar{x}_s|\mu)}{\text{Var}(\bar{Y}_s) + v_\varepsilon/n + v_\beta}. \tag{8.11}$$

Note that, as expected, even if one censused the population the Bayes estimator of μ remains a weighted average of m and $\bar{x}_s - \bar{\beta}$, and that the upper bound to $E_{x_s} \text{Var}(\mu|(s, x_s))$ is never less than

$$\text{Var}(\mu) \times \frac{v_\beta + v_\varepsilon/N}{\text{Var}(\mu) + v_\beta + v_\varepsilon/N}. \tag{8.12}$$

Note also that, in (8.9), $v - c = E(\sigma^2)$, where $\sigma^2 = \sum_{i=1}^{N} (x_i - \mu)^2/(N-1)$, the finite population variance. Note the change in the definition of σ^2 from that above Result 4.1; this is merely for convenience.

8.3. On the value of tainted observations

Suppose now that one can draw a sample, s_1, and observe n_1 tainted observations, x_{s_1}, per (8.1) and that, at a higher per unit cost, one may also select a second, non-overlapping, sample, s_2, and observe n_2 untainted observations, Y_{s_2}. Presumably the second sample would obtain the true Y_i's from a different source or by a different method of collection. Under such a model what should be the allocation of sampling effort among s_1 and s_2 or the optimum choice of n_1 and n_2?

Let Y_r be the $N - n_1 - n_2 \times 1$ vector of the variable of interest for the unsampled units; Y_{s_i} be the $n_i \times 1$ vector of such values for the units in s_i, $i = 1, 2$; and let x_{s_1} be the $n_1 \times 1$ vector of the observed tainted observations. Suppose again that one's prior distribution is as in Section 8.2. One may again apply Result 3.1, now taking $X_1 = (Y_r', Y_{s_1}')'$ and $X_2 = (Y_{s_2}', x_{s_1}')'$. After some algebraic manipulation, one can show that under this model

$$E(\mu|(s_1, x_{s_1}), (s_2, Y_{s_2})) = \frac{w_1(\bar{x}_{s_1} - \bar{\beta}) + w_2 \bar{Y}_{s_2} + w_3 m}{w_1 + w_2 + w_3}, \tag{8.13}$$

at least approximately, and where

$$w_1 = (v - c) \text{Var}(\mu) n_1, \tag{8.14}$$

$$w_2 = (v - c + v_\varepsilon + n_1 v_\beta) n_2 \tag{8.15}$$

and

$$w_s = \frac{E(\sigma^2)}{N} ((N - n_1 - n_2) E(\sigma^2) + (N - n_2)(v_\varepsilon + n_1 v_\beta)), \tag{8.16}$$

where again $v - c = E(\sigma^2)$. Also,

$$E_{x_{s_1}, Y_{s_2}} \text{Var}(\mu|(s_1, x_{s_1}), (s_2, Y_{s_2})) \leq \frac{(v - c)}{N} \text{Var}(\mu) W, \tag{8.17}$$

where

$$W = \frac{(v - c)(N - n_1 - n_2) + (v_\varepsilon + n_1 v_\beta)(N - n_2)}{(v + (n_2 - 1)c)(v_\varepsilon + n_1 v_\beta + v - c) + n_1 c(v - c)}.$$

Now suppose that each tainted observation in s_1 costs c_1 and each untainted one in s_2 costs c_2 ($c_1 < c_2$), and that a total budget of C is available for sampling. A reasonable choice of n_1 and n_2 would then be those values, n_1^o and n_2^o which minimize $g(n_1, n_2)$, the upper bound (given as the right side of the inequality of (8.17)), to the prior expectation of the posterior variance of μ, subject to the constraints that $c_1 n_1 + c_2 n_2 \leq C$, $n_1 \geq 0$, $n_2 \geq 0$, and $n_1 + n_2 \leq N$.

One then has the following

RESULT 8.1. (a) $g(n_1, n_2)$ *is decreasing in* n_1 *and* n_2, *so that* $n_2^o = (C - n_1^o c_1)/c_2$.
(b) $g(n_1, n_2)$ *is a convex function of* n_1 *and* n_2 *when* $c > 0$.
(c) $g(n_1, n_2)$ *is minimized subject to the constraints above by taking*

$$n_1^o = \frac{1}{v_\beta}\left[\left(\frac{c_2(v-c)(v-c+v_\varepsilon)}{c_1}\right)^{1/2} - (v - c + v_\varepsilon)\right] \tag{8.18}$$

provided that the right hand side of (8.18) *is between zero and* $\min[N, C/c_1]$, *otherwise the minimum is attained by taking* n_1^o *to be the nearest endpoint of this interval.*

Note that from (8.18) the optimum number of tainted observations to select increases as c_2 increases or as c_1 or v_β decreases. The effect of $(v - c)$ and v_ε is more complex. By examining the derivative of n_1^o, (8.18), with respect to $(v - c)$ it is straightforward to see that if $c_2/c_1 \geq v_\beta$ then n_1^o is increasing with $(v - c)$ and is decreasing otherwise. Similarly, if $v_\varepsilon \leq (v - c)(c_2 - 4v_\beta^2 c_1)/4v_\beta^2 c_1$, then n_1^o is increasing in v_ε, and is decreasing otherwise. Note, too, that n_1^o is independent of C, the sampling budget, that is, if the budget is increased then it should be spent in buying more untained observations.

8.4. Sample information on the bias

Suppose now that one can obtain both tainted and pure observations on a subsample of units in order to learn about the magnitude of the bias. More specifically, assume that for $i \in s_1 \subset \eta$ one obtains tainted observations, $x_i = Y_i + \beta + \varepsilon_i$, and for a second sample, $s_2 \subset \eta$, $(s_1 \cap s_2 = \emptyset)$, one observes both Y_i and x_i. The two sample sizes will be denoted by n_1 and n_2. Let x_1, x_2 and Y_2 denote the column vectors of these sample observations, let Y_1 be the unobserved pure values for the units in s_1 and let Y_r be the $n_r = N - n_1 - n_2 \times 1$ vector of the true values for the remaining unsampled units. We also will assume the same class of prior distributions on Y_i, β and ε_i as in Section 8.2.

Using (8.4)–(8.6) one can easily write down the prior mean and covariance matrix of $Y \equiv (\beta, Y_1', Y_r', x_1', x_2', X_2')'$.

If one were to apply Result 3.1 directly in this case to evaluate or approximate $E(\beta, Y_1', Y_r' | x_1', x_2', Y_2')$ by the expression in (3.4), then it would require the inversion of the $n_1 + 2n_2 \times n_1 + 2n_2$ matrix, Σ_{22}. Fortunately this is not necessary since one is only interested here in the posterior distribution of β, $Y_1 = \Sigma_{i \in s_1} Y_i$, and $Y_r = \Sigma_{i \in s_r} Y_i$, where $s_r = \eta - s_1 - s_2$. The posterior distribution of this triple can be shown to depend on the sample only through x_1, x_2, and Y_2, the sums of the tainted and pure observations in the two subsamples. More specifically, one has

RESULT 8.2. *The posterior mean of* (β, Y_1, Y_r) *given* x_1, x_2, *and* Y_2 *depends on the data only through the sums* $x_1 = \Sigma_{i \in s_1} x_i$, $x_2 = \Sigma_{i \in s_2} x_i$, *and* $Y_2 = \Sigma_{i \in s_2} Y_i$.

It thus suffices to consider the 6×1 vector $Y = (\beta, Y_1, Y_r, x_1, x_2, Y_2)'$, which has mean

$$E(Y) = (\bar{\beta}, n_1 m, n_r m, n_1(\bar{\beta} + m), n_2(\bar{\beta} + m), n_2 m)'$$

and covariance matrix

$$V = \begin{bmatrix} V_{11} & V_{12} \\ V_{21} & V_{22} \end{bmatrix}, \tag{8.19}$$

where

$$V_{11} = \begin{bmatrix} v_\beta & 0 & 0 \\ 0 & n_1 f_1 & n_1 n_r c \\ 0 & n_1 n_r c & n_r f_r \end{bmatrix}, \tag{8.20}$$

$$V_{12} = V'_{21} = \begin{bmatrix} n_1 v_\beta & n_2 v_\beta & 0 \\ n_1 f_1 & n_1 n_2 c & n_1 n_2 c \\ n_1 n_r c & n_2 n_r c & n_2 n_r c \end{bmatrix}, \tag{8.21}$$

$$V_{22} = \begin{bmatrix} n_1(f_2 + g_2) & n_1 n_2(c + v_\beta) & n_1 n_2 c \\ n_1 n_2(c + v_\beta) & n_2(f_2 + g_2) & n_2 f_2 \\ n_1 n_2 c & n_2 f_2 & n_2 f_2 \end{bmatrix}, \tag{8.22}$$

and $f_i = v + (n_i - 1)c$, $g_i = v_g + n_i v_\beta$ for $i = 1, 2, r$.

8.5. The posterior mean of β

The matrix V_{22} is, with some algebraic effort, invertible and $V_{12} V_{22}^{-1}$ is calculable. Given this, the posterior means of β, Y_1 and Y_r are obtainable using (3.4). In particular, letting (s, y) denote the sample data x_1, x_2, Y_2, it may be seen that

$$E(\beta|(s, y)) = \frac{a_1\bar{\beta} + a_2(\bar{x}_2 - \bar{Y}_2) + a_3[a_4(\bar{x}_1 - m) - a_5(\bar{x}_2 - m)]}{a_1 + a_2 + a_3},$$
(8.23)

where

$$a_1 = v_\varepsilon[sf_{12} + v_\varepsilon f_2], \qquad a_2 = n_2 v_\beta(v - c + v_\varepsilon)f_{12},$$
$$a_3 = n_1 v_\varepsilon v_\beta s, \qquad a_4 = f_2/s, \qquad a_5 = n_2 c/s,$$

and $s = v - c$ and $f_{12} = v + (n_1 + n_2 - 1)c$. From this expression it is easily seen that, as $n_1 \to 0$,

$$E(\beta|(s, y)) \to \frac{(v_\varepsilon/n_2)\bar{\beta} + v_\beta(\bar{y}_2 - \bar{Y}_2)}{(v_\varepsilon/n_2) + v_\beta},$$

as one would expect.

8.6. The posterior mean of μ

Turning now to the finite population mean, μ, and noting that

$$E(\mu|(s, y)) = [n_2\bar{x}_2 + E(Y_1 + Y_r|(s, y))]/N,$$

after considerable computation, one finds that

$$E(\mu|(s, y)) = [w_1 m + NV(\mu)\{a_1\bar{Y}_2 + a_2(\bar{x}_1 - \bar{\beta}) - a_3(\bar{x}_2 - \bar{\beta})\}]/Nh,$$
(8.24)

where

$$w_1 = s[(n_1 + n_r)v_\varepsilon g_{12} + n_r s g_2],$$
$$V(\mu) = (v + (N - 1)c)/N,$$
$$a_1 = n_2 g_{12}(v - c + v_\varepsilon), \qquad a_2 = n_1 g_2 s, \qquad a_3 = n_1 n_2 v_\beta s,$$

and

$$h = g_2 s f_{12} + f_2 v_\varepsilon g_{12}.$$

Two special cases of this result are of interest. First note that if $n_2 = 0$, then the expression above agrees precisely with that in (8.7) taking $n = n_1$. If $n_1 = 0$ then it is easy to see that the expression above reduces to

$$E(\mu|(s, y)) = \frac{V(\mu)\bar{Y}_2 + EV(\bar{Y}_2|\mu, \sigma^2)m}{V(\mu) + EV(\bar{Y}_2|\mu, \sigma^2)},$$

where $EV(\bar{Y}_2|\mu, \sigma^2) = (N - n_2)E(\sigma^2)/Nn_2$, another expected form (Ericson, 1969b). Thus for inference regarding μ obtaining sample information on the bias is of no use unless other biassed observations are also obtained. Certain limits of $E(\mu|(s, y))$ are also of interest; but will not be pursued here.

8.7. Optimum sample design

The matrix $V_{11} - V_{12}V_{22}^{-1}V_{21}$ in (3.5) may also be evaluated. Doing so, one can show that

$$EV(\beta|(s, y)) \leq \frac{v_\beta v_\varepsilon [sf_{12} + v_\varepsilon f_2]}{g_2 sf_{12} + f_2 v_\varepsilon g_{12}} \tag{8.25}$$

and

$$EV(\mu|(s, y)) \leq \frac{V(\mu)s[sn_r g_2 + g_{12} v_\varepsilon (n_1 + n_r)]}{N[g_2 sf_{12} + f_2 g_{12} v_\varepsilon]} . \tag{8.26}$$

As one check on the computation involved in obtaining these expressions, various special cases yield expected results. For example, if one takes $n_2 = 0$ in (8.26) one obtains the same expression as in (8.11) and taking $n_1 = 0$ in (8.26) one finds that

$$EV(\mu|(s, y)) \leq V(\mu) \frac{EV(\overline{Y}_2|\mu, \sigma^2)}{V(\mu) + EV(\overline{Y}_2|\mu, \sigma^2)} .$$

Other limiting cases yield interesting results which are in agreement with earlier findings.

Turning to the question of sample design, it is again reasonable to choose a design, (n_1, n_2) which minimizes the upper bound, (8.26), to the prior expectation of the posterior variance of μ, subject to the constraints that $c_1 n_1 + c_2 n_2 \leq C$, $n_i \geq 0$, $n_1 + n_2 \leq N$ with $c_2 > c_1 > 0$. One then has the following result.

RESULT 8.3. *Let $D = C/c_1$, $d = c_2/c_1$ and $m = \min[N, C/c_2]$. The upper bound in (8.26) is decreasing in both n_1 and n_2 and is minimized when $n_1 = n_1^o = D - dn_2$ and $n_2 = n_2^o$, where*
(a) *if $s > (d - 1)v_\varepsilon$ then*

$$n_2^o = \begin{cases} 0 & \text{if } n_2(+) < 0, \\ n_2(+) & \text{if } 0 \leq n_2(+) \leq m, \\ m & \text{if } n_2(+) > m, \end{cases} \tag{8.27}$$

where $n_2(+)$ is given by

$$n_2 = n_2(\pm) \equiv \frac{-v_\varepsilon(d-1)(v - c + v_\varepsilon + Dv_\beta) \pm [v_\varepsilon(d-1)s]^{1/2}(dv_\varepsilon + Dv_\beta)}{v_\beta(d-1)(s - (d-1)v_\varepsilon)} .$$

$$\tag{8.28}$$

(b) *If* $s = (d - 1)v_\varepsilon$ *then* n_2^o *is as in* (8.27); *but where now* $n_2(+)$ *is given by*

$$n_2 = n_2(+) \equiv \frac{v_\varepsilon[(v_\varepsilon + Dv_\beta)^2 - v_\varepsilon(d - 1)s]}{2v_\varepsilon v_\beta(d - 1)(s + v_\varepsilon + Dv_\beta)}. \qquad (8.29)$$

(c) *If* $s < (d - 1)v_\varepsilon$ *then*

$$n_2^o = \begin{cases} m & \text{if } n_2(+) \geq m, \\ n_2(+) & \text{if } n_2(-) \geq m \text{ and } n_2(+) \geq 0, \\ 0 & \text{if } n_2(+) < 0 \text{ and } n_2(-) \geq m, \\ 0 \text{ or } m & \text{if } n_2(+) < 0 \text{ and } n_2(-) < m, \\ n_2(+) \text{ or } m & \text{otherwise}, \end{cases}$$

where, in the last two subcases (8.26) *must be evaluated for each of the two possibilities and the minimizing value chosen.*

Note that in the unlikely case that $0 \leq d \leq 1$, it can be shown that $n_2^o = C/c_2$, as one would expect. Also as d gets large case (c) of the result will hold and $n_2(+)$ will be negative and near zero, while $n_2(-)$ will be small and positive. Hence the fourth subcase of (c) will obtain. By comparing (8.26) at $(C/c_1, 0)$ with that at $(0, C/c_2)$ it may be seen (most easily for N large) that $n_2^o \to 0$.

In most applications the optimal n_2^o will be given by $n_2(+)$ in part (a) of the result, i.e.,

$$n_2^o = \frac{[s/(v_\varepsilon(d - 1))]^{1/2}(dv_\varepsilon + DV_\beta) - (s + v_\varepsilon + Dv_\beta)}{(v_\beta/v_\varepsilon)(s - (d - 1)v_\varepsilon)}. \qquad (8.31)$$

In such a case it is easily seen that n_2^o is increasing in D, as is n_1^o. Thus the bigger the sampling budget the more of each type of observation should be selected, unlike the case in the model of Section 4. Finally how does n_2^o change with increasing prior fuzziness about the unknown bias? By examining the derivative of (8.31) with respect to v_β it is easily seen that n_2^o is increasing in v_β provided that $d^2/(d - 1) < (s + v_\varepsilon)^2/sv_\varepsilon$ that is, provided that the relative cost of the (Y_i, y_i), $i \in s_2$ is not too high.

9. Discussion

The results given in the previous sections may be combined in many different ways to provide estimators and appropriate optimum sample designs under a variety of other classes of prior distributions. Examples of this include: The combination of the results of Sections 5 and 6 to provide results in situations where the population psu's are themselves stratified on the basis of certain prior information. Or, one may have data on concomitant variables in some or all of the K strata which may be linearly related to the unknown stratum Y_k's; here the results of Sections 5 and 7 may be combined in various ways. One may also have different prior knowledge about response bias and error for subsets of the population units, e.g., whites and non-whites, and may find that applying the results of Section 8 separately in each subset (strata) provides a satisfactory model. There are many such possibilities.

The classes of prior distributions examined in this exposition may also be extended in a variety of ways. For example, the Y_{ik}'s in different strata need not be independent a priori. The psu's in Section 6 need not be of the same sizes. One may find it convenient to assess a prior distribution in three stages by some heirarchical grouping of the ultimate sample units—leading to some three stage sample design as optimal in the Bayes sense. Here, again the possibilities are endless.

Note that, with the sometimes exception of exchangeability, *all* of the results indicated in the preceding sections required one to specify *only* the means, variances and covariances of one's prior distribution. This minimal specification of a prior distribution suffices for the derivation of a Bayesian posterior mean, or its best linear approximation, *and* an optimal sampling design. In addition, if the specification of these low prior moments reflected little real prior information then the Bayes results are equivalent to those of the traditional sampling distribution point of view. In this context, two comments seem relevant. The first is that, not unexpectedly, the basic estimators and designs of the sampling distribution view have a subjectivistic interpretation only for broad general classes of prior distributions reflecting considerable uncertainty. Perhaps this is as it should be in a number of practical, multipurpose applications. The second is that with the minimal prior specification here, the Bayesian gets sort of a minimal output in terms of the posterior. Clearly, a more detailed specification of a prior distribution will yield a much more informative posterior distribution. Examples of this are given in Ericson (1969b), among others.

The selection of topics presented in the earlier sections is not meant to be a complete summary of the work in applying Bayesian methods to survey sampling. The selection clearly reflects the author's own research interests; but also was made to indicate how the very basic building blocks, estimators and designs, of traditional sampling theory arise as natural consequences or limits of Bayes procedures under appropriate prior distributions. For example, we have not discussed the problem of nonresponse, an increasingly important practical problem, a natural one to attack from a Bayesian view, although we have indicated the main relevant references at the beginning of Section 8. Another topic, which has been treated from a Bayesian view, which we have not discussed is the randomized response technique. Relevant papers here include Spurrier and Padgett (1980) and Winkler and Franklin (1979). Another omitted area is that of estimation of subdomain means and totals, see for example Laake (1979). Another important class of omitted topics concerns Bayesian nonparametrics and inference about percentiles. One, somewhat unsatisfactory, approach here is that based upon the Dirichlet or Dirichlet process prior distributions, see Ericson (1969b) and Binder (1982). More promising approaches appear in Hill (1968), Lane and Sudderth (1978) and in forthcoming work by Lenk (1984).

The main practical applications of the Bayesian approach to survey sampling seem to be in accounting and auditing, quality control and in estimation of population sizes (capture re-capture situations). In the accounting area some of the relevant work is given in Andrews and Smith (1983), Crosby (1981), and Moors (1983). In quality control see, for example, Case and Keats (1982) and for population size problems see Castledine (1981) and Hunter and Griffiths (1978). To date textbook writers, with the exception of Sudman (1976), have strictly adhered to the standard sampling theory approach.

So far, the bulk of the research in survey or finite population sampling which has emanated from the (1955) Godambe paper has concentrated on the equivalent problems of inference to population means, totals and proportions. Some work has been carried out on variance estimation, e.g., Mukhopadhyay (1978), Royall and Cumberland (1978), Zacks (1981) and Zacks and Solomon (1981). It seems that much work remains to be done for inference about $g(Y)$'s other than totals.

References

Albert, J. H. (1983). A Bayes treatment of non-response when sampling from a dichotomous population. In: K. W. Heiner, R. S. Sacher and J. W. Wilkinson, eds., *Computer Science and Statistics: Proceedings of the 14th Symposium on the Interface*. Springer, Berlin, 298–301.

Andrews, R. W. and Smith, T. M. F. (1983). Pseudo-Bayes and Bayes approaches to auditing. *The Statistician* 32, 124–126.

Basu, D. and Pereira, A. deB. (1982). On the Bayes analysis of categorical data, the problem of non-response. *J. Stat. Plann. Inference* 6, 345–362.

Bellhouse, D. R., Thompson, M. E. and Godambe, V. P. (1977). Two-stage sampling with exchangeable prior distributions. *Biometrika* 64, 97–103.

Binder, D. A. (1982). Non-parametric Bayesian models for samples from finite populations. *J. Roy. Statist. Soc. B* 44, 388–393.

Case, Kenneth E. and Keats, J. B. (1982). On the selection of prior distributions in Bayes acceptance sampling. *J. Qual. Tech.* 14, 10–18.

Castledine, B. J. (1981). A Bayesian analysis of multiple recapture sampling from a closed population. *Biometrika* 68, 197–210.

Chaudhuri, A. (1977). Some applications of the principle of Bayesian sufficiency and invariance to inference problems with finite populations. *Sankhya C* 39(3), 140–149.

Cochran, W. G. (1963). *Sampling Techniques*. 2nd edition, John Wiley and Sons, New York.

Crosby, M. A. (1981). Bayesian statistics in auditing: A comparison of probability elicitation techniques. *Accounting Rev.* 56, 355–365.

Cumberland, W. G. and Royall, R. M. (1981). Prediction models and unequal probability sampling. *J. Roy. Statist. Soc. B* 43, 353–367.

deFinetti, B. (1937). La prévision, ses lois logiques, ses sources subjectives. Annales de l'Institut Henri Poincaré 7, 1–68. Appearing in English translation with new notes in: H. E. Kyburg and H. E. Smokler, eds., *Studies in Subjective Probability*. Wiley, New York, 1964.

Diaconis, P. and Ylvisaker, D. (1979). Conjugate priors for exponential families. *Ann. Statist.* 7, 269–281.

Ericson, W. (1967). Optimal sample design with non-response. *J. Amer. Statist. Assoc.* 62, 63–78.

Ericson, W. (1969a). Subjective Bayesian models in sampling finite populations: stratification. In: N. L. Johnson and H. Smith, Jr., eds., *New Developments in Survey Sampling*. J. Wiley, New York, 326–357.
Ericson, W. (1969b). Subjective Bayesian models in sampling finite populations (with discussion). *J. Roy. Statist. Soc. B* **31**, 195–233.
Ericson, W. (1969c). A note on the posterior mean of a population mean. *J. Roy. Statist. Soc. B* **31**, 32–334.
Ericson, W. (1970). On the posterior mean and variance of a population mean. *J. Amer. Statist. Assn.* **65**, 649–652.
Ericson, W. (1975). A Bayesian approach to two-stage sampling Technical Report AFFDL-TR-75-145, *Air Force Dynamics Laboratory*, Wright-Patterson AFB, Ohio, 1–25.
Ericson, W. (1983a). A Bayesian approach to regression estimation in finite populations. Technical Report #120, Department of Statistics, University of Michigan, 1–21.
Ericson, W. (1983b). Response bias and error in sampling finite populations. Technical Report #122, Department of Statistics, University of Michigan.
Fienberg, S. E. (1980). Linear and quasi linear Bayes estimators. In: A. Zellner, ed., *Bayesian Analysis in Econometrics and Statistics*. North-Holland, New York.
Godambe, V. P. (1955). A unified theory of sampling from finite populations. *J. Roy. Statist. Soc. B* **17**, 267–278.
Goel, P. K. and DeGroot, M. H. (1980). Only normal distributions have linear posterior expectations in linear regression. *J. Amer. Statist. Assoc.* **75**, 895–900.
Goldstein, M. (1979). The variance modified linear Bayes estimator. *J. Roy. Statist. Soc. B* **41**, 96–100.
Goldstein, M. (1981a). A Bayes criterion for sample size. *Ann. Statist.* **9**, 670–672.
Goldstein, M. (1981b). Revising previsions: a geometric approach (with discussion). *J. Roy. Statist. Soc. B* **43**, 105–130.
Graybill, F. A. (1969). *Introduction to matrices with applications in statistics*. Wadsworth Publishing Co., Belmont, CA.
Hartigan, J. A. (1969). Linear Bayes methods. *J. Roy. Statist. Soc. B.* **31**, 446–454.
Hill, B. M. (1968). Posterior distribution of percentiles: Bayes theorem for sampling from a population. *J. Amer. Statist. Assoc.* **63**, 677–691.
Hogarth, R. M. (1975). Cognitive processes and the assessment of subjective probability distributions. *J. Amer. Statist. Assoc.* **70**, 271–289.
Hunter, A. J. and Griffiths, H. J. (1978). Bayesian approach to estimation of insect population size. *Technometrics* **20**, 231–234.
Jewell, W. S. (1974a). Credible means are exact Bayesian for exponential family response models. *Astin Bulletin* **8**, 77–90.
Jewell, W. S. (1974b). Exact multidimensional credibility. *Mitt. Vereini. Schweiz. Versicherungsmath.* **74**(2), 193–204.
Jewell, W. S. (1975). Regularity conditions for exact credibility. *Astin Bulletin* **8**, 336–341.
Kadane, J. B. (1980). Predictive and structural models for eliciting prior distributions. In: A. Zellner, ed., *Bayesian Analysis in Econometrics and Statistics*. North-Holland, New York.
Kadane, J. B., Dickey, J. M., Winkler, R., Smith, W. S. and Peters, S. C. (1980). Interactive elicitation of opinion for a normal linear model. *J. Amer. Statist. Assoc.* **75**, 845–854.
Khan, M. Z. (1976). Optimum allocation in Bayesian stratified two phase sampling when there are m attributes. *Metrika* **23**, 211–220.
Kolehmarnen, O. (1981). Bayesian models in estimating the total of a finite population: Towards a general theory. *Scand. J. Statist.* **8**, 27–32.
Laake, P. (1979). A prediction approach to subdomain estimation in finite populations. *J. Amer. Statist. Assoc.* **74**, 355–358.
Lane, D. A. and Sudderth, W. D. (1978). Diffuse models for sampling and predictive inference. *Ann. Statist.* **6**, 1318–1336.
Leamer, E. E. C. (1978). Specification searches: Ad hoc inference with nonexperimental data. John Wiley, New York.

Lenk, P. J. (1984). Bayesian nonparametric predictive distributions. Ph.D. Thesis, Department of Statistics, The University of Michigan. v + 261 pp.

Little, R. J. A. (1982). Models for non-response in sample surveys. *J. Amer. Statist. Assoc.* **77**, 237–250.

Mohd, Z. K. (1976). Optimum allocation in Bayes stratified two phase samples. *J. Indian Statist. Assoc.* **14**, 65–74.

Moors, J. J. A. (1983). Bayes estimation in sampling for audits. *The Statistician* **32**, 281–288.

Mukhopadhyay, P. (1978). Estimating the variance of a finite population under a superpopulation model. *Metrika* **25**, 115–122.

Pfeffermann, D. and Nathan, G. (1981). Regression analysis of data from a cluster sample. *J. Amer. Statist. Assoc.* **76**, 681–689.

Royall, R. M. and Cumberland, W. G. (1978). Variance estimation in finite population sampling. *J. Amer. Statist. Assoc.* **78**, 351–358.

Royall, R. M. and Pfeffermann, D. (1982). Balanced samples and robust Bayesian inference in finite population sampling. *Biometrika* **69**, 401–409.

Rubin, D. B. (1977). Formalizing subjective notions about the effect of nonrespondents in sample surveys. *J. Amer. Statist. Assoc.* **72**, 538–543.

Rubin, D. B. (1976). Inference and missing data. *Biometrika* **63**, 581–590.

Rubin, D. B. (1978). Bayesian inference for causal effects—The role of randomization. *Ann. Statist.* **6**, 34–58.

Savage, L. J. (1971). Elicitation of personal probabilities and expectations. *J. Amer. Statist. Assoc.* **66**, 783–801.

Schlaifer, R. (1959). Probability and statistics for business decisions. McGraw-Hill, New York.

Scott, A. J., Brewer, K. R. W. and Ho, E. W. H. (1978). Finite population sampling and robust estimation. *J. Amer. Statist. Assoc.* **73**, 359–361.

Scott, A. J. and Smith, T. M. F. (1969). Estimation in multistage surveys. *J. Amer. Statist. Assoc.* **64**, 830–840.

Scott, A. J. and Smith, T. M. F. (1973). Survey design, symmetry and posterior distributions. *J. Roy. Statist. Soc. B* **35**, 57–60.

Sekkappan, R. M. (1981). Generalization of Ericson's subjective Bayesian models in sampling finite populations for k characteristics. *J. Ind. Statist. Assoc.* **19**, 159–164.

Sekkappan, R. M. (1981). Subjective Bayes multivariate stratified sampling for finite populations. *Metrika* **28**, 123–132.

Singh, Bahadur and Sedransk, J. (1978). Sample size selection in regression analysis when there is nonresponse. *J. Amer. Statist. Assoc.* **73**, 362–365.

Smith, T. M. F. (1976). The foundations of survey sampling: a review. *J. Roy. Statist. Soc. A* **139**, 1–9.

Smith, T. M. F. (1984). Sample surveys, present position and potential developments: some personal views. *J. Roy. Statist. Soc. A* **147**, 208–221.

Smouse, E. P. (1982). Bayesian estimation of a finite population total using auxiliary information in the presence of non-response. *J. Amer. Statist. Assoc.* **77**, 97–102.

Smouse, E. P. (1984). A note on Bayesian least squares inference for finite population models. *J. Amer. Statist. Assoc.* **79**, 390–392.

Spurrier, J. D. and Padgett, W. J. (1980). The application of Bayes techniques in randomized response. *Socio. Meth.*, 533–544.

Sudman, S. (1976). *Applied Sampling*. Academic Press, New York.

Sugden, R. A. (1979). Inference on symmetric functions of exchangeable populations. *J. Roy. Statist. Soc. B* **41**, 269–273.

Sugden, R. A. and Smith, T. M. F. (1984). Ignorable and informative designs in survey sampling inference. *Biometrika* **71**, 495–506.

Sundberg, R. (1983). The prediction approach and randomized population type models for finite population inference for two stage samples. *Scand. J. Statist.* **10**, 223–238.

Thompson, M. E. (1978). Stratified sampling with exchangeable prior distributions. *Ann. Statist.* **6**, 1168–1169.

Winkler, R. L. (1967). The assessment of prior distributions in Bayesian analysis. *J. Amer. Statist. Assoc.* **62**, 776–800.

Winkler, R. L. and Franklin, LeRoy A. (1979). Warner's randomized response model: a Bayesian approach. *J. Amer. Statist. Assoc.* **74**, 207–214.

Zacks, S. (1981). Bayes equivariant estimators of the variance of a finite population for exponential priors. *Comm. Statist. A* **10**, 427–437.

Zacks, S. and Solomon, H. (1981). Bayes and equivariant estimators of the variance of a finite population: Part I, simple random sampling. *Comm. Statist. A* **10**, 407–426.

Inference Based on Data from Complex Sample Designs

Gad Nathan

1. Introduction

When data are obtained from a sample survey of complex design (e.g. clustered or stratified), standard methods are available for the point estimation of sample functions of the population variables, both for linear functions, such as means and totals, and for non-linear functions, such as ratios. In addition, consistent estimators of the sample variances of these estimators are available. These point estimators and estimators of their sample variances, in conjunction with a sampling theory variant of the Central Limit Theorem, provide all that is necessary for descriptive or enumerative purposes of sample surveys. However, the increasing requirements for analytical inference on the basis of data from surveys of complex sample design are not met by classical sampling theory methods. On the other hand, classical methods and models of inference, such as regression, analysis of variance and qualitative data analysis are, in general, based on the assumption of simple random sampling, rarely realized in the practice of sample surveys. The standard methods of analysis, which are based on this assumption, have been widely propagated in the form of computer packages, but can rarely be applied directly to data from a complex sample survey (see e.g. Cohen and Gridley, 1981).

The duality of complex sample designs and of hypothetical models required for inference naturally leads to the consideration of some combination of the 'design-based' approach of classical sampling theory and the 'model-based' approach of classical inference. In order to do this systematically, we introduce the following concepts and notation, basically following Cassel, Sarndal and Wretman (1977).

Both approaches start out with a finite population of distinguishable (labelled) units, $U = \{U_1, U_2, \ldots, U_N\}$. The size, N, of the population is assumed known. The 'design-based' approach considers fixed (possibly vectorial) values $y^T = (y_1, \ldots, y_N)$, associated with the population units, and a parametric function, $G(y)$, which is the object of inference. In the case of analytical inference, $G(y)$ will in general be a non-linear function, such as the finite population regression coefficients

$$B = (X_N^T X_N)^{-1} X_N^T y, \tag{1.1}$$

where X_N is an $N \times r$ matrix of values of r auxiliary variables known for each population unit. The sample selected, $s = \{U_{i_1}, \ldots, U_{i_{n(s)}}\}$, is an (unordered) set of population units, where $n(s)$ is the sample size. The sample design is defined as a probability measure, $P(\cdot)$ (the p-distribution), on the set of all subsets of U. Denote by S the random variable which takes the value s with probability $P(s)$. The observed sample data consist of the fixed values for units in the sample and their associated labels

$$d = \{(y_k, k); k \in s\} \tag{1.2}$$

and are to be regarded, according to the design-based approach, as a realization of the random variable

$$\tilde{d} = \{(y_k, k); k \in S\}. \tag{1.3}$$

Sample statistics

$$\tilde{t} = \tilde{t}(\tilde{d}) = \tilde{t}\{(y_k, k); k \in S\} \tag{1.4}$$

are considered as estimators of $G(y)$ and their properties are investigated with respect to the sample distribution $P(\cdot)$, only. Thus, $\tilde{t}(\tilde{d})$ is a *p-unbiased* (or *design-unbiased*) estimator of $G(y)$ if

$$E_p[\tilde{t}(\tilde{d})] = \sum_s P(s) t(d) = G(y), \tag{1.5}$$

for all $y \in \mathbb{R}^N$, where

$$t(d) = t\{(y_k, k); k \in s\}, \tag{1.6}$$

is the realization of $\tilde{t}(\tilde{d})$ for $S = s$.

The model-based approach considers the vector y as the realization of an N-variate random variable $Y^T = (Y_1, \ldots, Y_N)$ under a super-population or model distribution (the ξ-distribution). The distribution of Y is assumed to be defined by a probability density function, $f_Y(y; \theta)$, with an unknown (possibly vectorial) parameter θ, which belongs to a known parameter set Θ. The object of analytical inference according to the model-based approach is then the model parameter θ. For instance, under the super-population model

$$Y = X\beta + \varepsilon; \quad \varepsilon \sim N(0, \sigma^2 I_n), \tag{1.7}$$

β is the parameter of interest for inference. The data d—equation (1.2)—are now to be regarded as a realization of the random variable

$$D = \{(Y_k, k); k \in s\} \tag{1.8}$$

for a fixed sample of units, s. Sample statistics

$$T = T(D),$$

are considered as estimates of θ and their properties are investigated with respect to the model-distribution (ξ-distribution)—$f_Y(y; \theta)$. Thus $T(D)$ is a ξ-unbiased (or model-unbiased) estimator of θ if

$$E_\xi[T(D)] = \int_{\mathbb{R}^N} t(d) \, dF_Y(y; \theta) = \theta \qquad (1.9)$$

for any sample s and for any value of θ, where $F_Y(y; \theta)$ is the cumulative distribution function of Y and $t(d)$ is the realization of $T(D)$ for $Y = y$.

It should be noted that model-based inference can also be used with respect to a finite population parameter function $G(Y)$, in which case $T(D)$ is used as a predictor of $G(Y)$, but that design-based inference can obviously not be used with respect to a model parameter θ (see Hartley and Sielken, 1975).

In the controversy between the advocates of design-based inference and those of model-based inference (or model-free and design-free inference, respectively) (see, e.g., Brewer and Mellor, 1973; Smith, 1976; Sarndal, 1978) extreme positions have been taken. Thus Kish and Frankel (1974) consider finite population parameters as the only relevant objects of inference and their inference is purely design-based with no recourse to super-population models. Fienberg (1980), on the other hand, considers only inference relating to the parameters of a super-population model as relevant. Examples of such inference can be found in Konijn (1962), Fuller (1975), Thomsen (1978) and Pfeffermann and Nathan (1981).

Serious objections have been raised with respect to each of these extreme positions. Model-based inference relies heavily on assumptions about the form of the model-distribution which are difficult to verify and may not be robust to departures from the model assumptions.

On the other hand, design-based inference relates only to finite population parameters, $G(y)$, which have little descriptive value in themselves and are usually only 'copies' of theoretical model parameters. For instance, a finite population correlation coefficient is a useful measure of the relationship between two variables only if the relationship between them is approximately linear.

The two approaches are, however, not completely irreconcilable. For instance, although the conceptual difference between the finite population parameter $G(y)$ and the model parameter θ may be important, in many cases the quantitative difference may be extremely small. Thus, since the finite population parameter, B (1.1), is a consistent estimator of the super-population parameter, β, under the model (1.7), their values will in practice be close because of the large size, N, of the population. Furthermore, there will often be close correspondence between the population structure as defined by the super-population model and the sample design, which is usually based on known aspects of the population structure. This is true if information on the population structure—such as a natural cluster structure—which is incorporated into the model, is both available and utilized for the sample design. However, in other cases, especially when inference is of a secondary analysis type, or in multipurpose surveys on items of secondary importance, the variables of interest for inference may differ from those for which the descriptive-enumerative survey was originally designed, resulting in different parameters for inference.

Finally, a joint model- and design-based approach may be taken. This considers the data: $\tilde{D} = \{(Y_k, k); k \in S\}$ and statistics based on them—$\tilde{T} = \tilde{T}(\tilde{D})$—as depending jointly on the random variable Y and on the random variable S. $\tilde{T}$ then estimates θ or predicts $G(y)$ on the basis of the joint $p\xi$-distribution of Y and S.

We consider only the case of noninformative designs, in the sense that the sample design does not depend on the realizations, y, of Y, or, in other words, that the random variables Y and S are independent. This implies that the likelihood of θ, given $\tilde{D} = d$, can be decomposed into

$$f_{\tilde{D}}(d; \theta) = g_{Y_s}(y_s; \theta) P(s), \tag{1.10}$$

where $g_{Y_s}(y_s; \theta)$ is the marginal probability density function of $Y_s = \{Y_k; k \in s\}$. Pure model-based inference for θ is then based on the conditional likelihood $g_{Y_s}(y_s; \theta)$, given the sample of units s. If the sample size $n(s)$ is fixed and the distribution is exchangeable, then $g_{Y_s}(y_s; \theta)$ depends only on the values y_s and inference derived from it will not depend on the sample s (i.e. will be 'label-independent'). In other cases, however, inference will depend on the sample, s, and use of the joint $p\xi$-distribution for inference eliminates this dependence of inference on the particular sample selected. Similarly, pure design-based inference on $G(y)$ can be considered as being based on the conditional distribution of $\tilde{D}$ given the values y_s. The use of the joint $p\xi$-distribution then eliminates the dependence of inference on the particular values of y_s. This is basically the approach underlying the original introduction of super-population models (Cochran, 1946) by considering the ξ-expectation of the design variance.

In the following we shall consider inference both on finite population parameters, $G(y)$, and on super-population parameters, θ, either via the joint $p\xi$-distribution or via the marginal p-distribution or the marginal ξ-distribution, where appropriate. The parameter of interest for inference and the approach used will either be clear for the context or, in other cases, can be considered as interchangeable, at least when, as is often the case, the correspondence between the model and the design and between the relevant parameters leads to similar inferences.

The correspondence between model and design cannot, however, be taken for granted and the initial task of the data analyst is to find and use a model which reflects the sample design used to obtain the data, if appropriate. Thus, if a regression model is to be fitted to data from a stratified sample, the possibility of a different regression (different constants and, possibly, different slopes) for each stratum must be carefully checked. The exploratory analysis will usually be based on simple descriptive measures or on graphical displays. However, the results of this descriptive analysis must be interpreted with care to take into account the sample design. For example, in checking a linear regression model, many small residuals displaying systematic deviations with large sample weights may be considerably more important than a few large residuals with small sample weights. One possibility to overcome the effect of sample design at the exploratory analysis stage is to base the analysis on a self-weighting sub-sample of the data. A useful diagnostic tool in regression is the comparison of weighted regression estimators (with sample weights) and the unweighted estimators. DuMouchel and Duncan (1983) propose a test of the difference to help in choosing a model.

Once an appropriate model and parameters for inference have been determined, there still remains the question of the type of inference required. While point estimation and interval inference are appropriate both for finite population parameters and for super-population parameters, tests of hypotheses are rarely appropriate for finite population parameters. For example, the equality of two finite population domain means cannot be seriously hypothesized, whereas the test of this hypothesis with respect to the super-population expected values is both relevant and frequently carried out in practice. In any case, although classical inference concentrates on hypothesis testing, often with simple hypotheses as the null ones, interval inference (or inference about a composite hypothesis) might, in many cases, be preferable for the analysis of data coming from complex sample surveys.

In the following, in section two, some general methods of inference on the basis of data from complex sample designs are discussed, while in sections three, four and five, specific methods for linear models and regression, for categorical data analysis and for other types of analysis are presented.

2.1 General methods

2.1. Multivariate analysis methods

The multivariate analysis of categorical data by weighted least squares (Grizzle, Starmer and Koch, 1969) has been generalized to apply to inference on a general linear model on the basis of complex sample data (Koch, Freeman and Freeman, 1970; Freeman et al., 1976; Shah, Holt and Folsom, 1977; Koch, Stokes and Brock, 1980). Consider a vector of p sample estimators, $F^T = (F_1, \ldots, F_p)$, on whose expected values we wish to infer. The estimators could, for instance, be domain means or categorical frequencies. We assume only the assymptotic multivariate normality for F:

$$\sqrt{n}\,[F - E(F)] \rightrightarrows N[0, V_F\} \tag{2.1}$$

and the availability of a consistent estimator $\hat{V}_F$, of the covariance matrix of F.

A linear model for the expected values of F is hypothesised:

$$E(F) = X\beta, \tag{2.2}$$

where X is a known $p \times r$ design matrix of full rank $r(<p)$ and β is a vector of r unknown parameters. The weighted least squares estimator of β under the model is then

$$\hat{\beta} = (X^T \hat{V}_F^{-1} X)^{-1} X^T \hat{V}_F^{-1} F, \tag{2.3}$$

which has an asymptotic multivariate normal distribution with mean β and a covariance matrix which can be consistently estimated by

$$\hat{V}_\beta = (n X^T \hat{V}_F^{-1} X)^{-1}. \tag{2.4}$$

Thus the linear hypothesis (2.2) can be tested by the generalized Wald statistic

$$X_W^2 = n(F - X\hat{\beta})^T \hat{V}_F^{-1} (F - X\hat{\beta}), \tag{2.5}$$

which, under (2.2), is asymptotically distributed chi-squared with $p - r$ degrees of freedom. Furthermore, point and interval estimators of linear functions $l'\beta$ can be obtained via the asymptotic normality of $l'\hat{\beta}$. Linear hypotheses, such as

$$H_0: C\beta = \theta_0, \tag{2.6}$$

where C is a full rank $q \times r$ matrix, can be tested by the use of the statistic

$$X_C^2 = n(C\hat{\beta} - \theta_0)^T [C(X^T \hat{V}_F^{-1} X)^{-1} C^T]^{-1} (C\hat{\beta} - \theta_0), \tag{2.7}$$

which is approximately distributed chi-squared with q degrees of freedom under H_0. This holds also without the linear assumption (2.2), when F are already themselves consistent estimators of β, with a consistent estimator of the covariance matrix $\hat{V}_F/n$, and can be obtained by replacing X by the identity matrix in (2.7). Furthermore, even in cases where the hypothesis of interest is not linear, a first-order Taylor series linear approximation, yielding a form as in (2.6), can be used, so that (2.7) remains a valid asymptotically chi-squared distributed test statistic (see, e.g., Nathan, 1972; Shuster and Downing, 1976).

Note that all the above can be applied equally well both to model-based inference, where β is a vector of super-population parameters and all distributions and moments refer to the ξ-distribution, and to design-based inference, where β is a vector of finite population parameters and the p-distribution is used throughout.

2.2. Estimation and approximation of the covariance matrix

The major problem associated with the above approach involves the difficulty in obtaining a consistent estimator, $\hat{V}_F$, of the covariance matrix, in particular when F is not linear in the sample observations y_s. The basic methods available, surveyed by Rao (1975), are as follows:

(a) Linearization or Taylor expansion methods (Tepping, 1968): the covariances between first-order Taylor approximations for the non-linear statistics are computed via standard methods (e.g. the classical approximation to the sample variance of a ratio estimator).

(b) Balanced repeated replication (McCarthy, 1969): statistics based on half-samples, which are selected so as to ensure an orthogonal balanced set, are computed and the empirical covariances of these statistics are used as the appropriate estimator.

(c) Jackknife (Miller, 1974): the sample is first split into subsamples each of which reflects the original complex design. Statistics based on the sample data without one of the subsamples are computed and the empirical covariances of these statistics serve as covariance estimators.

While the standard statistical package computer programmes (BMDP, SPSS) do not in general deal with the complex sample design situation, several special-purpose programmes for covariance estimation have been developed for use with complex sample designs: SUPERCARP and MINICARP (Hidiroglou, Fuller and Hickman, 1980; Hidoroglou, 1981) are based on the direct estimation of covariances at each stage of a multi-stage stratified cluster sample; SUDAAN (Shah, 1978) is based on Taylor expansion; and OSIRIS IV-PSALMS (Lepkowski, 1982) is based primarily on Balanced Repeated Replications. For regression coefficients, SURREGR, which can be accessed via SAS, can be used in conjunction with GENCAT (Makuc, 1981).

Theoretical comparisons of the three methods of covariance estimation by Krewski and Rao (1981) and empirical comparisons by Kish and Frankel (1974) and by Richards and Freeman (1980) indicate that their performance is very similar in many cases.

Finally, it should be noted that when there is a large number of statistics for which variance–covariance estimators are required, the stability and approximate unbiasedness of estimators based on small numbers of observations have to be carefully checked. The asymptotic properties of the test statistics also have to be ascertained for the specific complex sample design at hand.

The difficulty in obtaining stable estimates of large scale covariance matrices has led to attempts to use simplified approximations based on structural assumptions. This can be done directly on the covariance matrix itself under a design-based approach. For instance, for a stratified cluster sample, Nathan (1973) approximates expressions for the covariances of functions of contingency table frequencies under the simplified assumption that within-stratum intraclass correlations are the same for all cells. Lepkowski and Landis (1980) propose four different more sophisticated models for the design effects of linear contrasts of subclass proportions.

Direct modelling of the population structure using a super-population approach can also lead to simplified covariance estimation (Altham, 1976; Cohen, 1976). Brier (1980) assumes a Dirichlet distribution for population proportions which is equivalent to cluster sampling with equal cell design effects. Pfeffermann and Nathan (1981) assume a random effects model for regression coefficients specific to each cluster. This leads to a covariance matrix for which methods of estimating its components are available. Fuller and Battese (1973) and Holt, Richardson and Mitchell (1980) use random effects models for the observations themselves in a clustered structure, thereby reducing considerably the number of parameters to be estimated in order to determine the covariance matrix. Further examples may be found in Tomberlin (1979) and Imrey, Sobel and Francis (1980).

2.3. Modifications of standard tests

The widely used standard computer packages provide a variety of test procedures based on simple random sampling. An obvious approach to the problem of inference under complex sample designs is to attempt to modify, in a simple way, the standard tests to take the complex design into account. This can be regarded as a natural extension of the use of design effects (deffs) as multiplicative factors to correct variance estimators based on simple random sampling, so as to estimate the variances of estimators under complex sampling.

Consider, for instance, the linear model (2.2) and the linear hypothesis on the parameters, β (2.6). Had the statistics F been obtained under simple random sampling, the standard large sample Wald statistic for testing the hypothesis (2.6) would be, similarly to (2.7):

$$X_0^2 = n(C\hat{\beta} - \theta_0)^T [C(X^T \hat{V}_0^{-1} X)^{-1} C^T]^{-1} (C\hat{\beta} - \theta_0), \qquad (2.8)$$

where $\hat{V}_0/n$ is a consistent estimator of the covariance matrix V_0/n of F under simple random sampling. X_0^2 would have an asymptotic chi-square distribution with q degrees of freedom under (2.6), if sampling were simple random. From standard results on the distribution of quadratic forms of multinormal variables it is easily shown that the asymptotic distribution of X_0^2 for the complex sample design is given by a weighted sum of independent chi-squared variables with one degree of freedom each:

$$X_0^2 \rightsquigarrow \sum_{i=1}^{q} \lambda_i Z_i^2, \qquad (2.9)$$

where Z_i $(i = 1, \ldots, q)$ are independent standard normal variables and $\lambda_1 \geq \lambda_2 \geq \cdots \geq \lambda_q > 0$ are the eigenvalues of

$$D = [C(X^T V_0^{-1} X)^{-1} C^T]^{-1} [C(X^T V_F^{-1} X)^{-1} C^T]. \qquad (2.10)$$

The largest eigenvalue, λ_1, can be shown to equal the maximal design effect of linear combinations of $\hat{\theta}$. If an estimate of λ_1, or an upper bound for it,

$\lambda_0 \geq \lambda_1$, is available, then obviously the use of the statistic X_0^2/λ_0 results in a conservative test of the hypothesis (2.6), which is exact only if $\lambda_0 = \lambda_1 = \cdots = \lambda_q$. An alternative is to use $X_0^2/\bar{\lambda}$, where $q\bar{\lambda} = \Sigma_{i=1}^q \lambda_i = \text{tr}(D)$. In some cases $\bar{\lambda}$ is a weighted average of simple design effects. For instance, if under simple random sampling the $\hat{\theta}_i$'s are uncorrelated so that their covariance matrix $C(X^T V_0^{-1} X) C^T$ is diagonal, then $\bar{\lambda}$ is the average of the design effects of $\hat{\theta}_i$:

$$\bar{\lambda} = \frac{1}{q} \sum_{i=1}^{q} V_P(\hat{\theta}_i)/V_{\text{srs}}(\hat{\theta}_i), \tag{2.11}$$

where $V_P(\hat{\theta})$ is the variance of $\hat{\theta}_i$ under the complex sample design, P, and $V_{\text{srs}}(\hat{\theta}_i)$ its variance under simple random sampling.

Corrections of standard test statistics based on generalized design effects or on average design effects have been used by Cowan and Binder (1978), Felligi (1980), Rao and Scott (1981), Scott and Holt (1982) and Fay (1982). Although the modifications above are formulated in terms of design effects under a design-based approach, similar relationships can be obtained under a model-based approach (Cohen, 1976; Campbell, 1977).

3. Regression and linear models

The widespread use of regression based on sample survey data requires a careful assessment of the use of standard techniques and their validity in the case of complex sample design. The approach taken to inference (design-based or model-based) and the determination of the parameters of interest and of the assumptions about the underlying model are of extreme importance.

Under a pure design-based approach, the finite population regression coefficient vector, B (1.1), is an obvious choice for the parameter of interest. However, the interest in B can only be justified by some belief in the underlying homoscedastic model

$$E(Y|X_N) = X_N B; \quad V(Y|X_N) = \sigma^2 I_N. \tag{3.1}$$

The OLS estimate for (3.1),

$$\hat{\beta} = (X_s^T X_s)^{-1} X_s^T y_s, \tag{3.2}$$

where X_s is the $n \times r$ matrix of rows of X_N which relate to be sampled units, is obviously not, in general, a design-consistent estimator of B. If we denote the sample inclusion probabilities by $\pi_i = P(U_i \in s)$ and the weighting matrix by $W_s = \text{diag}(\pi_1^{-1}, \ldots, \pi_n^{-1})$, where for convenience we assume $s = \{1, \ldots, n\}$, then the design-weighted (Horvitz–Thomson) estimator

$$\hat{B}_W = (X_s^T W_s X_s)^{-1} X_s^T W_s y_s \tag{3.3}$$

has a small design-bias, but is a design-consistent estimator of **B**. The estimators $\hat{\beta}$ and $\hat{B}_W$, coincide only for self-weighing designs, i.e. when $\pi_i = n/N$, $i = 1, \ldots, N$. For most sample designs used in practice the sampling variance of $\hat{B}_W$ cannot be estimated simply by standard computer packages and one of the variance estimating techniques mentioned above — linearization, balanced repeated replication or jackknife — has to be used. Applications of these techniques may be found in Kish and Frankel (1974), Jonrup and Rennermalm (1976), Shah, Holt and Folsom (1977) and Holt and Scott (1981) and the SUPERCARP programme (Hidiroglou, Fuller and Hickman, 1980) provides correct estimators of the design-variances.

Under the model-based approach, the relationship between the sample design and the explanatory variables X_N must be considered. If, indeed, the linear model (3.1) holds and X_N includes all the variables on which the sample design depends, then classical regression analysis, as if the sample were simple random, is valid under the model-based approach. Thus $\hat{\beta}$ (3.2) is the model-unbiased and efficient estimator of **β**. However, $\hat{B}_W$ is also model-unbiased and has the further advantage of being design-consistent as well. In the case of a heteroscedastic model

$$E(Y|X_N) = X_N\beta; \quad V(Y|X_N) = \sigma^2 V, \tag{3.4}$$

both $\hat{\beta}$ and $\hat{B}_W$ remain model-unbiased and $\hat{B}_W$ remains design-consistent. However, the suitability of estimating **B** is doubtful under (3.4) and the finite population parameter

$$B^* = (X_N^T V^{-1} X_N)^{-1} X_N^T V^{-1} y \tag{3.5}$$

would be a more suitable parameter for inference. While $\hat{\beta}$ and $\hat{B}_W$ remain model-umbiased also for B^* under (3.4), although neither is efficient, $\hat{B}_W$ is not, in general, a design-consistent estimator of B^*. If V is diagonal and $\pi_i \propto V_{ii}$, i.e. inclusion probabilities are proportional to the variances, then $\hat{B}_W$ coincides with the generalized OLS estimator

$$\hat{\beta}^* = (X_s^T V_s^{-1} X_s)^{-1} X_s^T V_s^{-1} y_s, \tag{3.6}$$

where V_s is the appropriate $n \times n$ submatrix of V_N, and is both model-efficient and design-consistent, under (3.3).

Standard programmes ordinarily compute the OLS estimator, $\hat{\beta}$, and can often also compute the generalized OLS estimator, $\hat{\beta}^*$, together with unbiased estimators of their model-variances — $\hat{\sigma}^2(X_s^T X_s)^{-1}$ and $\hat{\sigma}^2(X_s^T V_s^{-1} X_s)^{-1}$ under (3.1) and (3.4), respectively. The design-weighted estimator, $\hat{B}_W$, can also be obtained by the weighted regression options of standard programmes (e.g. BMDP) by using the weights $1/\pi_i$. Alternatively, $\hat{B}_W$ can be obtained by unweighted regression on the transformed variables $Y_i/\sqrt{\pi_i}$ and $X_i/\sqrt{\pi_i}$ (but *not* on the weighted variables Y_i/π_i and X_i/π). However, it must be emphasized that under both alternatives *the reported variances and covariances are incorrect* (both as estimators of design-variances and as estimators of model-variances). Thus *the standard significance tests* (e.g. *F* tests) *based on them are invalid and can result in grossly misleading conclusions.*

The weighted regression programmes, with weights $1/\pi_i$, report the estimator of the variance-covariance matrix as $\hat{\sigma}^2(X_s^T W_s X_s)^{-1}$, whereas the model-variance of $\hat{B}_W$, under the homoscedastic model (3.1), is easily seen to be

$$V(\hat{\beta}_W | X_s) = \sigma^2 (X_s^T W_s X_s)^{-1} X_s^T W_s^2 X_s (X_s^T W_s X_s)^{-1}. \tag{3.7}$$

Thus, the reported variance estimator will be model-unbiased, under (3.1), only for self-weighting designs (i.e. $W_s \propto I_n$) and under the heteroscedastic model (3.3) only if V is diagonal and the inclusion probabilities are proportional to the variances (i.e. $W_s \propto V_s^{-1}$).

However, the estimator of the multiple regression coefficient obtained from weighted regression

$$\hat{R}^2 = \frac{(y_s - x_s \hat{B}_W)^T W_s (X_s - X_s \hat{B}_W)}{(y_s - \bar{y}_s 1_n)^T W_s (y_s - \bar{y}_s 1_n)} \tag{3.8}$$

where $\bar{y}_s = (\Sigma_s y_i / \pi_i)/(\Sigma_s \pi_i^{-1})$, is a design-consistent estimator of the population multiple correlation coefficient

$$R^2 = \frac{(y - X_N B)^T (y - X_N B)}{(y - \bar{y} 1_N)^T (y - \bar{y} 1_N)} \tag{3.9}$$

where $\bar{y} = (1/N) 1_N^T y$. Thus the multiple correlation coefficient, (3.8), reported by weighted regression programmes, can be used for descriptive design-based inference on the goodness of fit of the model.

When all the variables on which the sample design depends are in the model of interest, the model may have to reflect a complex population structure. For instance, different regression relationships may have to be assumed for different strata or for different PSU's in a two-stage clustered design. In this case, the vector of regression coefficients, β, may be too large as an object of inference and interest may center on a simple average of the regression coefficients (Konijn, 1962); on a weighted average of the coefficients (Pfeffermann and Nathan, 1981); or on their expected value, under some prior distribution (Porter, 1973).

In other cases, especially when secondary analysis is being carried out, the design variables cannot be considered as part of an explanatory model. Thus, let $X_N = (X_{N1}; X_{N2})$ where X_{N1} is the $N \cdot r_1$ matrix of values of the explanatory variables of interest and X_{N2} is the $N \cdot r_2$ matrix of values of variables on which the sample design is based, but which are of no interest for inference. Similarly, let $\beta^T = (\beta_{1.2}^T; \beta_{2.1}^T)$. Then the model of interest is

$$E(Y|X_{N1}) = X_{N1}\beta_1,\qquad(3.10)$$

rather than

$$E(Y|X_{N1}, X_{N2}) = X_{N1}\beta_{1.2} + X_{N2}\beta_{2.1} = X_N\beta.\qquad(3.11)$$

If the design variables included in (3.11) but not in (3.10), X_{N2}, are conditionally correlated with Y (given X_{N1}), then the standard OLS estimator of β_1 is not model-consistent. Bishop (1977) gives expressions for its bias in a certain case and Nathan and Holt (1980) and Holt and Smith (1979) propose modified weighted and unweighted estimators of β_1 which are consistent. Holt, Smith and Winter (1980) give an example of the application of these estimators.

4. Categorical data analysis

For many sample surveys, especially in the social sciences, the variables involved are categorical. The application of a wide range of categorical data analysis techniques, such as chi-square tests for goodness of fit, for homogeneity and for independence, and the use of log-linear models for sample survey data is standard practice. However, in many cases of complex sample design the results of standard analysis may be misleading. The basic methods, described in section two, for taking into account the complex sample design, have been applied to categorical data analysis and compared with standard tests.

The simplest case involves the classification of the population by a single categorical variable. If the population is classified into k categories with probabilities (under the model-based approach) or relative frequencies (under the design-based approach), $p_1, \ldots, p_k$, let $p^T = (p_1, \ldots, p_{k-1})$ and assume that a consistent survey estimator, $\hat{p}^T = (\hat{p}_1, \ldots, \hat{p}_{k-1})$, of p is available. We assume asymptotic multivariate normality of $\hat{p}^T$:

$$\sqrt{n}(\hat{p} - p) \rightarrow N(O, V)\qquad(4.1)$$

and wish to test the goodness of fit of the estimate, $\hat{p}$, to a known distribution defined by $p_0^T = (p_{01}, \ldots, p_{0k-1})$, i.e.:

$$H_0: p = p_0.\qquad(4.2)$$

If a consistent estimator, $\hat{V}$, of V is available, the methods of section 2.1 can be applied and the generalized Wald statistic:

$$X_W^2 = n(\hat{p} - p_0)^T V^{-1}(\hat{p} - p_0),\qquad(4.3)$$

which is distributed asymptotically chi-squared with $k-1$ degrees of freedom, under (4.2), can be used to test H_0.

The estimation of V can be carried out by the methods outlined in Section 2.2. More general hypotheses, such as (2.2), where $F = F(\hat{p})$ are known functions of $\hat{p}$, can similarly be tested by X_W^2, as defined in (2.5), where $\hat{V}_F$, the variance matrix of $F(\hat{p})$, is obtained from $\hat{V}$ by Taylor linearization. Examples of applications may be found in Freeman et al. (1976), Shah, Holt and Folsom (1977), Koch, Stokes and Brock (1980), Lepkowski and Landis (1980) and Imrey, Koch and Stokes (1982). A general computer programme has been developed within the OSIRIS IV system to implement this method (Lepkowski, Bromberg and Landis, 1981).

In many cases, especially when goodness of fit tests have to be carried out for a large number of categorical variables, or in secondary analysis of published reports, appropriate modification of standard tests according to the methods of Section 2.3, may be necessary. For this case the standard X^2 statistic is:

$$X_0^2 = n \sum_{i=1}^{k} (\hat{p}_i - p_{0i})^2/p_{0i} = n(\hat{p} - p_0)^T P_0^{-1}(\hat{p} - p_0) \qquad (4.4)$$

where $P_0 = \text{diag}(p_0) - p_0 p_0^T$. Rao and Scott (1981) show, by application of the general results of Section 2.2 (with $X = C = I_{k-1}$; $\hat{\beta} = \hat{p}$; $\theta_0 = p_0$ and $V_F = P_0/n$), that X_0^2 has the asymptotic distribution of a weighted sum of $k - 1$ independent chi-squared variables with one degree of reedom each:

$$X_0^2 \approx \sum_{i=1}^{n-1} \lambda_i Z_i^2; \quad Z_i \sim N(0, 1) \text{ i.i.d.}, \qquad (4.5)$$

where $\lambda_1 \geq \lambda_2 \geq \cdots \geq \lambda_{k-1}$ are the eigenvalues of $D = P_0^{-1} V$. The largest eigenvalue λ_1 can be shown to equal the maximal design effect over all linear combinations of the p_i's:

$$\lambda_1 = \sup_l [V_P(l^T \hat{p})/V_{\text{srs}}(l^T \hat{p})], \qquad (4.6)$$

where V_p denotes variance under the complex sample design and V_{srs} variance under simple random sampling. Thus, for stratified random sampling with proportional allocation $\lambda_1 \leq 1$, so that X_0^2 will provide a conservative test of (4.2). However, for multi-stage cluster sampling, the use of X_0^2 may result in a test with considerably larger size than the nominal one—see Holt, Scott and Ewings (1980) for empirical evidence on the basis of the analysis of large-scale U.K. sample surveys. The eigenvalues $\lambda_1, \ldots, \lambda_{k-1}$ can be estimated consistently by $\hat{\lambda}_1, \ldots, \hat{\lambda}_{k-1}$, the eigenvalues of $\hat{D} = \hat{P}^{-1} \hat{V}$ where $\hat{P} = \text{diag}(\hat{p}) - \hat{p}^T \hat{p}$ and $\hat{V}$ is a consistent estimator of V. If a consistent estimator of V were available, then X_W^2 (4.3) could be used directly. However, in many cases when a consistent estimator of V is not available, estimates of the design effects in the individual cells, $\hat{d}_i = \hat{V}_{ii}/[p_i(1 - \hat{p}_i)]$, may be available. It can be easily seen that a weighted average of these estimated design effects:

$$\hat{\lambda} = \text{tr}(\hat{D})/(k-1) = \sum_{i=1}^{k} (1-p_i)\hat{d}_i/(k-1) \tag{4.7}$$

is a consistent estimator of $E(X_0^2)/(k-1)$. On the basis of this result, Rao and Scott (1981) propose the use of the modified statistic $X_0^2/\hat{\lambda}$ as approximately distributed chi-squared with $k-1$ degrees of freedom under (4.2). Empirical results of Hidiroglou and Rao (1981) for the Canada Health Surveys and of Holt, Scott and Ewings (1980) for U.K. surveys show that a close approximation to nominal levels of significance is achieved by $X_0^2/\hat{\lambda}$. Fellegi (1980) has proposed the modification $X_0^2/\hat{d}$, where $\hat{d} = \sum_{i=1}^{k} \hat{d}_i/k$ is the sample average of the design effects.

Direct modelling for p under a two-stage cluster design has been proposed by Cohen (1976) and Altham (1976) with respect to their moment structure and by Brier (1980) by a Dirichlet-multinomial distribution. Their models imply a constant design effect, so that $\lambda_1 = \lambda_2 = \cdots = \lambda_{k-1} = \lambda$, which can be consistently estimated and used to modify X_0^2.

Similar methods can be used for multivariate categorical analysis. For the case of testing for independence in a two-way $r \times c$ contengency table the hypothesis of independence can be formulated as:

$$H_0: F_{ij}(p) \equiv p_{ij} - p_{i+}p_{+j} = 0, \quad i = 1, \ldots, r-1; j = 1, \ldots, c-1, \tag{4.8}$$

where p_{ij} is the population probability of cell (i,j), p_{i+} and p_{+j} are the marginal probabilities and $p^T = (p_{11}, \ldots, p_{r-1,c-1})$. The generalized Wald statistic then becomes:

$$X_{WI}^2 = nF(\hat{p})^T \hat{V}_F^{-1} F(\hat{p}), \tag{4.9}$$

where $F(\hat{p})^T = [F_{11}(\hat{p}), \ldots, F_{r-1,c-1}(\hat{p})]$ and $\hat{V}_F/n$ is a consistent estimator of the covariance matrix of $F(\hat{p})$. Under (4.8), X_{WI}^2 is asymptotically distributed chi-squared with $(r-1)(c-1)$ degrees of freedom. For specific designs, Garza-Hernandez and McCarthy (1962), Nathan (1969, 1975), Shuster and Downing (1976) and Fellegi (1980) have developed approximations for $\hat{V}_F$ for use in (4.9). Fay (1982) has developed a general computer programme, CPLX, which provides tests of more general log-linear models, using the generalized Wald statistic in conjunction with estimation of $\hat{V}_F$ by replication.

Similarly to the case of goodness of fit, a modification of the chi-square statistic:

$$X_I^2 = n \sum_{i=1}^{r} \sum_{j=1}^{c} (\hat{p}_{ij} - \hat{p}_{i+}\hat{p}_{+j})^2/(\hat{p}_{i+}\hat{p}_{+j}), \tag{4.10}$$

can be used for testing independence when a consistent estimator of the full covariance matrix is not available. Rao and Scott (1981) show that, under H_0, X_I^2 has the asymptotic distribution of a weighted sum of $b = (r-1)(c-1)$ independent chi-squared variables, with one degree of freesom each:

$$X_I^2 \approx \sum_{i=1}^{r-1} \sum_{j=1}^{c-1} \delta_{ij} Z_{ij}^2 ; \quad Z_{ij} \sim N(0, 1) \text{ i.d.d.}, \tag{4.11}$$

where δ_{ij} are the eigenvalues of $D_F = (P_r^{-1} \otimes P_c^{-1}) V_F$, $P_r = \text{diag}(p_r) - p_r p_r^T$, $P_c = \text{diag}(p_c) - p_c p_c^T$, $p_r^T = (p_{1+}, \ldots, p_{r-1,+})$ and $p_c^T = (p_{+1}, \ldots, p_{+,c-1})$. The values of δ_{ij} can be considered as generalized deffs of the statistics $F_{ij}(\hat{p})$. An approximation to a chi-squared distribution with b degrees of freedom is then obtained by the modified statistic $X_{0I}^2 / \hat{\delta}$ where $\hat{\delta}$ is the weighted average of the estimated design effects of $F_{ij}(\hat{p})$:

$$\hat{\delta} = \sum_{1}^{r} \sum_{1}^{c} (1 - \hat{p}_{i+})(1 - \hat{p}_{+j}) \hat{\delta}_{ij} / b,$$

$$\hat{\delta}_{ij} = \hat{V}_{ij} / [\hat{p}_{i+} \hat{p}_{+j} (1 - \hat{p}_{i+})(1 - \hat{p}_{+j})], \tag{4.12}$$

where $\hat{V}_{ij}$ is a consistent estimator of the variance of $F_{ij}(\hat{p})$ under H_0. Rao and Scott (1984) have shown that $\hat{\delta}$ can be evaluated in terms of the estimated design effects, $\hat{d}_{ij}$, of the cell probability estimators, $\hat{p}_{ij}$, and the estimated design effects, $\hat{d}_i(r)$ and $\hat{d}_j(c)$, of the marginal probability estimators $\hat{P}_{i+}$ and $\hat{P}_{+j}$, respectively, by:

$$b\hat{\delta} = \sum_i \sum_j (1 - \hat{p}_{i+})(1 - \hat{p}_{+j}) \hat{d}_{ij}$$

$$- \sum_i (1 - \hat{p}_{i+}) \hat{d}_i(r) - \sum_j (1 - \hat{p}_{+j}) \hat{d}_j(r). \tag{4.13}$$

Alternative modifications based only on the estimated design effects of the cell probabilities have been proposed by Fellegi (1980)—$X_0^2 / \bar{d}$, where $\bar{d} = \Sigma_1^r \Sigma_1^c \hat{d}_{ij}$—and by Rao and Scott (1981)—$X_0^2 / \hat{\lambda}$, where $\hat{\lambda} = \Sigma_1^r \Sigma_1^c (1 - \hat{p}_{ij}) \hat{d}_{ij} / (rc - 1)$. Empirical comparisons by Holt, Scott and Ewings (1980) and by Hidoroglou and Rao (1981) show that $X_0^2 / \hat{\delta}$ is a good approximation, whereas $X_0^2 / \bar{d}$ and $X_0^2 / \hat{\lambda}$ give conservative tests. A conservative test is also obtained by a further alternative proposed by Holt, Scott, and Ewings (1980), based only on the estimated design effects of the marginal probability estimators—$X_0^2 / \bar{d}_m$ where $\bar{d}_m = \max[\Sigma_1^r \hat{d}_i(r)/r, \Sigma_1^c \hat{d}_j(c)/c]$.

The modifications above have been generalized for multiway tables to the testing of any general direct log-linear model, i.e. one in which the model admits explicit solutions in terms of marginal tables (Rao and Scott, 1984). They show that for any such model the appropriate weighted average design effect $\hat{\delta}$ can be evaluated in terms only of the estimated design effects of the cell probability estimators and those of the marginal table probabilities on which the direct estimators are based. For instance, for testing conditional independence, $p_{ijk} = p_{i+k} p_{+jk}/p_{++k}$, in a three-way table only the estimates $\hat{p}_{ijk}$, $\hat{p}_{i+k}$, $\hat{p}_{+jk}$, $\hat{p}_{++k}$ and estimates of their design effects are required for calculating $\hat{\delta}$. Since similar results are not available for indirect models (e.g. for testing hypothesis of no three-way interaction in a three-way table), they suggest using the value of $\hat{\delta}$ for a 'close' direct model as an approximation.

Finally, it should be noted that if all the design variables (e.g. stratification variables) are of intrinsic interest and are included in the log-linear model together with their interactions, the application of standard analysis is valid. However, design variables are very often of no interest as explanatory variables and in this case the methods outlined above are required.

5. Other methods of analysis

While linear regression, linear models, log-linear models, tests of goodness of fits and tests of independence, dealt with above, are the prevalent methods of analysis of sample survey data, many other methods of analysis are available and are used in fact for the analysis of data from complex sample design. Among these are multivariate analysis methods, such as principal component analysis, factor analysis, cluster analysis and discriminant analysis; path analysis; logistic regression; non-metric methods; and non-parametric methods.

The modifications required to take into account complex sample design can be developed for some of these techniques, from the principles outlined in section two and from the results of sections three and four. For instance, since path analysis is based on partial regression coefficients, the methods of section three can be directly applied. Similarly logistic regression with categorical explanatory variables can be derived from appropriate log-linear models. Insofar as the models involved are direct models, the methods of the previous section can be applied. However, hardly any examples of such applications have so far been reported.

The methods of multivariate analysis are primarily based on correlation coefficients and some results on the effects of sample design on the estimation of correlation coefficients are available. Kish and Frankel (1974) investigate the estimation of variance of the weighted design-unbiased estimator, (3.8), of the correlation coefficient, under a design-based approach. They compare, by simulation, the effects of the use of linearization, of balanced repeated replication and of jackknife estimation of the variances. Bebbington and Smith (1977), also under a design-based approach, investigate the effect of complex sample design on the bias of the simple random sampling estimator of the correlation coefficient and on the resulting biases of estimators of the principle components and latent roots of the correlation matrix. They show that the biases can be considerable even for self-weighting designs, except for proportional stratified sampling. Empirical results reinforce these conclusions and furthermore indicate serious downward biases in the estimates of variance.

Holt (1977) and Holt, Richardson and Mitchel (1980) take a model-based approach to the effect of complex sample design on the analysis of correlations. They consider a superpopulation random effects model for a clustered structure of the population and use standard model-unbiased estimators of the model parameters and estimates of their variances and covariances. The results are applied to principle component analysis and to cluster analysis of a national attainment survey.

Molina-Cuevas (1982) investigates the effect of complex sample design on measures of association in contingency tables based on cross-ratios and derives their asymptotic distributions and approximations based on generalized design effects. An empirical example based on data from a general household survey demonstrates their application.

In other areas Lepkowski and Landis (1980) have studied design effects for logits and Woodruff (1952) and Sedransk and Meyer (1978) give confidence intervals for quantiles on the basis of data from complex sample designs.

References

Altham, P. M. E. (1976). Discrete variable analysis for individuals grouped into families. *Biometrika* **63**, 263–269.
Bebbington, A. C. and Smith, T. M. F. (1977). The effect of survey design on multivariate analysis. In: C. A. O'Muircheartaigh and C. Payne, eds., *The Analysis of Survey Data, Vol. 2, Model Fitting*, Wiley, New York, 175–192.
Bishop, J. (1977). Estimation when the sampling ratio is a linear function of the dependent variable. *Proc. Soc. Statist. Sect., Amer. Statist. Assoc.*, 848–853.
Brewer, K. R. and Mellor, R. W. (1973). The effect of sample structure on analytical surveys. *Austral. J. Statist.* **15**, 145–152.
Brier, S. E. (1980). Analysis of contingency tables under cluster sampling. *Biometrika* **67**, 91–596.
Campbell, C. (1977). Properties of ordinary and weighted least squares estimators for two stage samples. *Proc. Soc. Statist. Sect., Amer. Statist. Assoc.*, 800–805.
Cassel, C.-M., Sarndal, C.-E. and Wretman, J. H. (1977). *Foundations of Inference in Survey Sampling*. Wiley, New York.
Cochran, W. G. (1946). Relative accuracy of systematic and stratified samples for a certain class of populations. *Ann. Math. Statist.* **17**, 164–172.
Cohen, J. E. (1976). The distribution of the chi-squared statistic under clustered sampling. *J. Amer. Statist. Assoc.* **71**, 665–670.
Cohen, S. B. and Gridley, G. (1981). Present limitations in the availability of statistical packages for the analysis of complex survey data. *Proc. Statist. Comp. Sect., Amer. Statist. Assoc.*, 20–24.
Cowan, J. and Binder, D. A. (1978). The effect of a two-stage sample design on tests of independence. *Survey Methodology (Canada)* **4**(1), 16–29.
DuMouchel, W. H. and Duncan, G. J. (1981). Using sample survey weights in multiple regression analyses of stratified samples. *J. Amer. Statist. Assoc.* **78**, 735–543.
Fay, R. E. (1982). Contingency table analysis for complex sample designs, *Proc. Sect. Survey Res. Meth., Amer. Statist. Assoc.*, 44–53.
Fellegi, I. P. (1980). Approximate tests of independence and goodness of fit based on stratified multistage samples. *J. Amer. Statist. Assoc.* **75**, 261–268.
Fienberg, S. E. (1980). The measurement of crime victimization: prospects for panel analysis of a panel survey. *The Statistician* **29**, 313–350.
Fuller, W. A. (1975). Regression analysis for sample survey. *Sankhya C* **37**, 117–132.

Fuller, W. A. and Battese, G. E. (1973). Transformations for estimation of linear models with nested-error structure. *J. Amer. Statist. Assoc.* **68**, 626–632.

Freeman, D. H. Jr., Freeman, J., Brock, D. B. and Koch, G. G. (1976). Strategies in the multivariate analysis of data from complex surveys II: An application to the United States National Health Interview Survey. *Internat. Statist. Rev.* **44**, 317–330.

Garza-Hernandez, T. and McCarthy, P. J. (1962). A test of homogeneity for a stratified sample. *Proc. Soc. Statist. Sect., Amer. Statist. Assoc.*, 200–202.

Grizzle, J. E., Starmer, C. F. and Koch, G. G. (1969). Analysis of categorical data by linear models. *Biometrics* **25**, 489–504.

Hartley, H. O. and Sielken, R. L. (1975). A superpopulation viewpoint for finite population sampling. *Biometrics* **31**, 411–422.

Hidiroglou, M. A. (1981). Computerization of complex survey estimates. *Proc. Statist. Comp. Sect., Amer. Statist. Assoc.*, 1–7.

Hidiroglou, M. A., Fuller, W. A. and Hickman, R. D. (1980). *Super Carp: Sixth Edition*, Statistical Laboratory Survey Section, Iowa State University, Ames, Ia.

Hidiroglou, M. A. and Rao, J. N. K. (1981). Chi-square tests for the analysis of categorical data from the Canada Health Survey, *Bul. Inst. Internat. Statist.* **49**, 699–718.

Holt, D. (1977). Correlation analysis using survey data. *Bull. Internat. Statist. Inst.* **47**(4), 228–231.

Holt, D. (1982). The use of logistic models for the analysis of survey data. *Paper presented at Israel Statistical Association International Meeting on Analysis of Sample Survey Data and on Sequential Analysis, Jerusalem.*

Holt, D., Richardson, S. C. and Mitchell, P. W. (1980). *The analysis of correlations in complex survey data.* Unpublished.

Holt, D. and Scott, A. J. (1981). Regression analysis using survey data. *The Statistician* **30**, 169–178.

Holt, D., Scott, A. J., and Ewings, P. O. (1980). Chi-squared tests with survey data. *J. Roy. Statist. Soc. Ser. A* **143**, 302–330.

Holt, D. and Smith, T. M. F. (1979). Regression analysis of data from complex surveys. *Royal Statistical Society Conference, Oxford.*

Holt, D., Smith, T. M. F. and Winter, P. O. (1980). Regression analysis of data from complex surveys. *J. Roy. Statist. Soc. Ser. A* **143**, 474–487.

Hu, T. W. and Stromsdorfer, E. W. (1970). A problem of weighting bias in estimating the weighted regression model. *Proc. Bus. Econ. Statist. Sect., Amer. Statist. Assoc.*, 513–516.

Imrey, P., Sobel, E. and Francis, M. (1980). Modeling contingency tables from complex surveys. *Proc. Sect. Survey Res. Meth., Amer. Statist. Assoc.*, 213–217.

Imrey, P. B., Koch, G. G. and Stokes, M. E. (1982). Categorical data analysis: some reflections on the log-linear model and logistic regression, Part II: data analysis. *Internat. Statist. Rev.* **50**, 35–63.

Jonrup, H. and Rennermalm, B. (1976). Regression analysis in samples from finite populations. *Scand. J. Statist.* **3**, 33–37.

Kaplan, B., Francis, I., and Sedransk, J. (1979). A comparison of methods and programs for computing variances of estimators from complex sample surveys. *Proc. Sect. Survey Res. Meth., Amer. Statist. Assoc.*, 97–100.

Kish, L. and Frankel, M. R. (1970). Balanced repeated replication for standard errors. *J. Amer. Statist. Assoc.* **65**, 1071–1094.

Kish, L. and Frankel, M. R. (1974). Inference from complex samples (with discussion). *J. Roy. Statist. Soc. Ser. B* **36**, 1–37.

Koch, G. G., Freeman, D. H., Jr. and Freeman, J. L. (1975). Strategies in the multivariate analysis of data from complex surveys. *Internat. Statist. Rev.* **43**, 59–78.

Koch, G. G., Stokes, M. E. and Brock, D. (1980). Applications of weighted least squares methods for fitting variational models to health survey data. *Proc. Sect. Survey Res. Meth., Amer. Statist. Assoc.*, 218–223.

Konijn, H. S. (1962). Regression analysis for sample surveys. *J. Amer. Statist. Assoc.* **57**, 590–606.

Krewski, D. and Rao, J. N. K. (1981). Inference from stratified samples: properties of the linearization, jackknife and balanced repeated replication methods. *Ann. Statist.* **9**, 1010–1019.

Lepkowski, J. M. (1982). The use of Osiris IV to analyze complex sample survey data. *Proc. Sect. Survey Res. Meth., Amer. Statist. Assoc.*, 38–45.

Lepkowski, J. M. and Landis, J. R. (1980). Design effects for linear contrasts of proportions and logits. *Proc. Sect. Survey Meth., Amer. Statist. Assoc.*, 224–229.

Lepkowski, J. M., Bromberg, J. A. and Landis, J. R. (1981). A program for the analysis of multivariate categorical data from complex sample surveys. *Proc. Statist. Comp. Sect., Amer. Statist. Assoc.*, 8–15.

Makuc, D. (1981). Interfacing SURREGR and GENCAT to analyze complex survey data. *Proc. Statist. Comp. Sect., Amer. Statist. Assoc.*, 16–19.

McCarthy, P. J. (1969). Pseudo-replication: half-samples. *Internat. Statist. Rev.* **37**, 239–264.

Miller, R. G. (1974). The jackknife – A review. *Biometrika* **61**, 1–15.

Molina-Cuevas, E. A. (1982). The effect of sample design on the measurement of associations. *Paper presented at Israel Statistical Association International Meeting on Analysis of Sample Data and on Sequential Analysis, Jerusalem.*

Nathan, G. (1969). Tests of independence in contingency tables from stratified samples. In: N. L. Johnson and H. Smith, eds., *New Developments in Survey Sampling*. Wiley, New York, 578–600.

Nathan, C. (1972). On the asymptotic power of tests for independence in contingency tables from stratified samples. *J. Amer. Statist. Assoc.* **67**, 917–920.

Nathan, G. (1973). *Approximate tests of independence in contingency tables from stratified proportional samples.* National Center for Health Statistics, Vital and Health Statistics Series 2, No. 53, Washington, DC.

Nathan, G. (1975). Tests of independence in contingency tables from stratified proportional samples. *Sankhya C* **37**, 77–87. [Corrigendum: *Sankhya C* **40** (1978) 190.]

Nathan, G. (1981). Notes on inference based on data from complex sample designs. *Survey Methodology (Canada)* **7**, 109–129.

Nathan, G. and Holt, D. (1980). The effect of survey design on regression analysis. *J. Roy. Statist. Soc. Ser. B* **42**, 377–386.

Pfeffermann, D. and Nathan, G. (1981). Regression analysis of data from complex samples. *J. Amer. Statist. Assoc.* **76**, 681–689.

Porter, R. M. (1973). On the use of survey sample weights in the linear model. *Annals of Economic and Social Measurement* **2**, 141–158.

Rao, J. N. K. (1975). Analytic studies of sample survey data. *Survey Methodology (Canada)* **1** (Supplementary Issue).

Rao, J. N. K. and Scott, A. J. (1981). The analysis of categorical data from complex surveys: chi-squared tests for goodness of fit and independence in two-way tables. *J. Amer. Statist. Assoc.* **76**, 221–230.

Rao, J. N. K. and Scott, A. J. (1984). On chi-square tests for multiway contingency tables with cell proportions estimated from survey data. *Ann. Stat.* **12**, 46–60.

Richards, V. and Freeman, D. H., Jr. (1980). A comparison of replicated and pseudo-replicated covariance matrix estimators for the analysis of contingency tables. *Proc. Sec. Survey Res. Meth., Amer. Statist. Assoc.*, 209–211.

Sarndal, S. E. (1978). Design-based and model-based inference in survey sampling. *Scand. J. Statist.* **5**, 27–52.

Sedransk, S. and Meyer, J. (1978). Confidence intervals for the quantiles of a finite population: simple random and stratified simple random sampling. *J. Roy. Statist. Soc. Ser. B* **40**, 239–252.

Scott, A. and Holt, D. (1982). The effect of two-stage sampling on ordinary least squares methods. *J. Amer. Statist. Assoc.* **77**, 848–854.

Scott, A. and Rao, J. N. K. (1981). Chi-squared tests for contingency tables with proportions estimated from survey data. In: D. Krewski, J. N. K. Rao and R. Platek, eds., *Current Topics in Survey Sampling*. Academic Press, New York, 247–266.

Shah, B. V. (1978). SUDAAN: survey data analysis software. *Proc. Statist. Comp. Sect., Amer. Statist. Assoc.*, 146–151.

Shah, B. V., Holt, M. M. and Folsom, R. E. (1977). Inference about regression models from sample survey data. *Bull. Internat. Statist. Inst.* **47**(3), 43–57.

Shuster, J. J. and Downing, D. J. (1976). Two-way contingency tables for complex sampling schemes. *Biometrika* **63**, 271–278.

Skinner, C. J. (1982). Design effects in two-stage sampling as effects of model misspecification. Paper presented at Israel Statistical Association International Meeting on Analysis of Sample Survey Data and on Sequential Analysis, Jerusalem.

Smith, T. M. F. (1976). The foundations of survey sampling: A review (with discussion). *J. Roy. Statist. Soc. Ser. A* **139**, 183–195.

Smith, T. M. F. (1981). Regression analysis for complex surveys. In: D. Krewski, J. N. K. Rao and R. Platek, eds., *Current Topics in Survey Sampling*. Academic Press, New York, 267–292.

Smith, T. M. F., Skinner, C. S. and Holmes, D. J. (1982). The effect of sample design on principal component analysis. Paper presented at Israel Statistical Association International Meeting on Analysis of Sample Survey Data and on Sequential Analysis, Jerusalem.

Tepping, B. J. (1968). The estimation of variance in complex surveys. *Proc. Soc. Statist. Sect., Amer. Statist. Assoc.*, 411–414.

Thomsen, I. (1978). Design and estimation problems when estimating a regression coefficient from survey data. *Metrika* **25**, 27–35.

Tomberlin, T. J. (1979). The analysis of contingency tables of data from complex samples. *Proc. Sect. Survey Res. Meth., Amer. Statist. Assoc.*, 152–157.

Woodruff, R. S. (1952). Confidence intervals for medians and other position measures. *J. Amer. Statist. Assoc.* **47**, 635–646.

Inference for Finite Population Quantiles

J. Sedransk and Philip J. Smith

Introduction

Since skewed distributions are very common when sampling from finite populations, it is useful and important to make inferences about finite population quantiles as well as traditional measures such as means. Increasingly, this need is being recognized. For instance, at the U.S. Bureau of the Census, there are plans to make inferences about medians and other quantiles for the Survey of Income Program Participation and National Prisoner Survey.

In this paper we describe currently available methodology for point estimation and confidence interval formation for finite population quantiles. Letting $Y_{(1)} < \cdots < Y_{(N)}$ denote the ordered characteristic values of a finite population of size N and $y_{(1)} < \cdots < y_{(n)}$ denote the ordered values obtained from a probability sample of size n, a confidence interval for the (t/N)th quantile, $Y_{(t)}$, is given by

$$[y_{(k)}, y_{(r)}] \tag{1}$$

where $1 \leq k \leq r$ and $k \leq t$. Because

$$\Pr\{y_{(k)} \leq Y_{(t)}\} = \Pr\{y_{(k)} \leq Y_{(t)}, y_{(r)} \geq Y_{(t)}\} + \Pr\{y_{(k)} \leq Y_{(t)}, y_{(r)} < Y_{(t)}\}$$

$$= \Pr\{y_{(k)} \leq Y_{(t)} \leq y_{(r)}\} + \Pr\{y_{(r)} \leq Y_{(t-1)}\}$$

the confidence coefficient associated with (1) is

$$\Pr\{y_{(k)} \leq Y_{(t)} \leq y_{(r)}\} = \Pr\{y_{(k)} \leq Y_{(t)}\} - \Pr\{y_{(r)} \leq Y_{(t-1)}\}. \tag{2}$$

For the case where $y_{(1)}, \ldots, y_{(n)}$ are the order statistics from a simple random sample, the confidence coefficient of (1) is

$$\left\{\sum_{i=k}^{n}\binom{t}{i}\binom{N-t}{n-i} - \sum_{i=r}^{n}\binom{t-1}{i}\binom{N-t+1}{n-i}\right\} \Big/ \binom{N}{n}$$

$$= \left\{\sum_{i=k}^{r-1}\binom{t}{i}\binom{N-t}{n-i} + \binom{t-1}{r-1}\binom{N-t}{n-r}\right\} \Big/ \binom{N}{n}. \tag{3}$$

Although (3) is satisfying because it is exact and concise, its utility is limited since very few surveys employ a simple random sampling design. Unfortunately, for more complicated survey designs determining reasonable confidence intervals and approximating their associated confidence levels is much more complicated.

In Section 2, methods for constructing frequentist confidence intervals are given for more complicated designs. Subsection 2.1 presents a general method (Woodruff, 1952) that may be applied to any sampling design. The results of a numerical investigation of properties of modifications of Woodruff's procedure are given by Haskell and Sedransk (1980).

Subsection 2.2 presents several confidence interval methods for stratified random sampling. Assuming proportional allocation, McCarthy (1965) gives a simple confidence interval and an exact lower bound for the associated confidence coefficient. Sedransk and Meyer (1978) present several alternative confidence intervals, and derive exact confidence coefficients for situations where (a) there are a small number of strata, and (b) there is prior information about the relationships among the distributions of Y over the strata. Smith and Sedransk (1983) consider the simple confidence interval investigated by McCarthy (1965) and show how to derive lower bounds for the confidence coefficient under any sample size allocation.

In Ph.D. dissertations Chapman (1970) and Blesseos (1976) give methods for determining confidence coefficients for confidence intervals associated with two stage cluster sampling designs. Each of these authors considers a confidence interval whose terminals depend upon the order statistics from the entire sample. Chapman (1970) provides a lower bound for the confidence coefficient that may also serve as an approximation to the exact value, while Blesseos (1976) outlines methods to extend procedures in Sedransk and Meyer (1978) to cluster sampling. These results are presented in Subsection 2.3.

Section 3 discusses point estimation for finite population quantiles. We sketch the very general results given by Loynes (1966) who emphasizes simple random sampling, but briefly treats stratified random sampling. Also, we summarize work by Gross (1980) who investigates point and variance estimation for stratified random sampling, equal probability single stage cluster sampling, and stratified single stage cluster sampling.

Section 4 includes synopses of work by Hill (1968), Ericson (1969), and Binder (1982) who give Bayesian inferential methods for finite population quantiles. Each of these authors emphasizes simple random sampling, but extensions to stratified random sampling can be made.

2. Confidence interval methods for complex sample designs

2.1. General methods

Woodruff (1952) describes a large sample procedure for determining confidence intervals for a finite population median and other finite population quantiles. His method accounts for most of the possible complexities in complex sample survey designs.

To describe Woodruff's general procedure, suppose a confidence interval is required for the quantile $Y_{(t)}$. For a sample of size n selected from a finite population of size N using a probability sampling design, let w_i denote the sampling weight of the ith ordered sample unit, $y_{(i)}$. Woodruff's procedure may be described by the following steps:

i. For an arbitrary value of y the finite population cumulative distribution function (cdf), F, is estimated by

$$\hat{F}(y) = \begin{cases} 0 & \text{if } y < y_{(1)}, \\ \sum_{i=1}^{n} d_i w_i / \sum_{i=1}^{n} w_i & \text{if } y_{(1)} \leq y \leq y_{(n)}, \\ 1 & \text{if } y > y_{(n)}, \end{cases}$$

where $d_i = 1$ if $y_{(1)} \leq y$ and $d_i = 0$ otherwise.

ii. The quantile $Y_{(t)}$ is estimated by $\hat{Y}_{(t)} = \hat{F}^{-1}(t/N)$. Alternatively, one may obtain an estimate of $Y_{(t)}$ from a smoothed estimate of F.

iii. The standard deviation of $\hat{F}(Y_{(t)})$ is estimated by $\hat{\sigma}_{\hat{F}(Y_{(t)})}$. This estimator accounts for the original sampling design.

iv. Assuming that

$$U_t = \{\hat{F}(Y_{(t)}) - (t/N)\} / \hat{\sigma}_{\hat{F}(Y_{(t)})}$$

has a standard normal distribution, a confidence interval with confidence coefficient $1 - \alpha$ for $\hat{F}(Y_{(t)})$ is

$$(t/N) - z_{\alpha/2} \hat{\sigma}_{\hat{F}(Y_{(t)})} \leq \hat{F}(Y_{(t)}) \leq (t/N) + z_{\alpha/2} \hat{\sigma}_{\hat{F}(Y_{(t)})}.$$

where $z_{\alpha/2}$ denotes the upper $1 - (\alpha/2)$ percentage point of the standard normal distribution.

v. Finally, the confidence interval for $Y_{(t)}$ is

$$\hat{F}^{-1}\{(t/N) \pm z_{\alpha/2} \hat{\sigma}_{\hat{F}(Y_{(t)})}\}.$$

To illustrate Woodruff's method consider a stratified random sampling design. As before, let $Y_{(1)} < Y_{(2)} < \cdots < Y_{(N)}$ denote the ordered variate values of a finite population of size N. Assume that the finite population has been divided into L strata with N_i elements being members of stratum i. Simple random samples of sizes n_i are selected independently from the L strata, and the order statistics in the combined sample are denoted by $y_{(1)}, \ldots, y_{(n)}$ where $n = \sum_{i=1}^{L} n_i$. It is desired to form a $100(1 - \alpha)\%$ confidence interval for $Y_{(t)}$, the Pth quantile ($P = t/N$) of the finite population.

Next, define the empirical cumulative distribution function, $\hat{F}(y)$, by

$$\hat{F}(y) = \begin{cases} 0 & \text{if } y < y_{(1)}, \\ \sum_{i=1}^{L} W_i\{m_i(y)\}/n_i & \text{if } y_{(1)} \leq y \leq y_{(n)}, \\ 1 & \text{if } y_{(n)} < y, \end{cases}$$

where $W_i = N_i/N$, and $m_i(y)$ denotes the number of observations in the sample from stratum i with $Y \leq y$.

To obtain a nominal $100(1 - \alpha)\%$ confidence interval for $Y_{(t)}$ first estimate the quantile by $\hat{Y}_{(t)} = \hat{F}^{-1}(P)$. Next, the sample proportion, p_i, of elements in stratum i with $Y \leq \hat{Y}_{(t)}$ is evaluated: $p_i = m_i(\hat{Y}_{(t)})/n_i$. An estimator of the variance of $\hat{F}(Y_{(t)})$ is

$$\hat{\sigma}_p^2 = \sum_{i=1}^{L} W_i^2 \left(1 - \frac{n_i}{N_i}\right) \frac{p_i(1-p_i)}{n_i - 1}.$$

Defining $\xi = \min\{P + (z_{\alpha/2})\hat{\sigma}_p, 1\}$ and $\gamma = \max\{P - (z_{\alpha/2})\hat{\sigma}_p, 0\}$, the nominal $100(1 - \alpha)\%$ confidence interval for $Y_{(t)}$ is

$$\hat{F}^{-1}(\gamma) \leq Y_{(t)} \leq \hat{F}^{-1}(\xi). \tag{4}$$

As noted previously, the basic assumption in the Woodruff approach is that $U_t \sim N(0, 1)$ where

$$U_t = \{\hat{F}(Y_{(t)}) - (t/N)\}/\hat{\sigma}_p.$$

However, from studies conducted by Smith and Sedransk it is clear that when sample sizes are small it is difficult to estimate the standard deviation of $\hat{F}(Y_{(t)})$ well. In this case it is uncertain whether the distribution of U_t may be satisfactorily approximated by the standard normal distribution.

Haskell and Sedransk (1980) have devised several modifications of Woodruff's method to try to overcome this difficulty. In the first modification Haskell and Sedransk use (4) as the confidence interval, but consider alternative estimation of $\sigma_p^2 = \text{Var}[\hat{F}(Y_{(t)})]$ specified by two choices of p_i

$$p_i(1) = m_i\{\hat{Y}_{(t)}\}/n_i \quad \text{and} \quad p_i(2) = [m_i\{\hat{Y}_{(t)}\} + 0.5]/[n_i + 1],$$

and two choices of $\hat{\sigma}_p^2$

$$\hat{\sigma}_p^2(1) = \sum_{i=1}^{L} W_i^2 \left(1 - \frac{n_i}{N_i}\right) \frac{p_i(1-p_i)}{n_i}$$

and

$$\hat{\sigma}_p^2(2) = \sum_{i=1}^{L} W_i^2 \left(1 - \frac{n_i}{N_i}\right) \frac{p_i}{n_i}.$$

That is, the *four* combinations specified by these choices of p_i and $\hat{\sigma}_p^2$ have been investigated. In simulation studies Haskell and Sedransk based their evaluation of the four alternatives on two criteria: (a) closeness of the actual confidence level (as estimated from the simulations) to the nominal level $(1 - \alpha)$; and (b) the expected length of the confidence interval. In their investigation, Haskell and Sedransk found that $p_i(1)$ and $\hat{\sigma}_p^2(2)$ performed better than the alternatives.

As a second modification to Woodruff's method, Sedransk and Haskell approximated $\hat{F}(Y_{(t)})$ by a beta distribution with parameters (a, b). Since $E[\hat{F}(Y_{(t)})] = a/(a + b)$ and $\text{Var}[\hat{F}(Y_{(t)})] = ab/(a + b)^2(a + b + 1)$, the parameters (a, b) of the beta distribution are selected so that $P = a/(a + b)$ and $\hat{\sigma}_p^2 = ab/(a + b)^2(a + b + 1)$. Then the approximate $100(1 - \alpha)\%$ confidence interval for $Y_{(t)}$ is

$$\hat{F}^{-1}(\gamma^*) \leq Y_{(t)} \leq \hat{F}^{-1}(\xi^*)$$

where

$$\xi^* = \beta^{-1}_{1-(\alpha/2)}(a, b), \qquad \gamma^* = \beta^{-1}_{\alpha/2}(a, b)$$

and β^{-1} denotes the inverse of the beta cumulative distribution function. Simulation studies indicate that use of the beta distribution to approximate $\hat{F}(Y_{(t)})$ yields confidence intervals with actual confidence coefficients which are close to the nominal levels.

2.2. Stratified random sampling

Assuming stratified random sampling, methods for constructing confidence intervals for finite population quantiles and determining their associated confidence coefficients are reviewed in this section.

Letting $y_{(k)}$ denote the kth order statistic of sampled values obtained from a stratified random sample of size n, McCarthy (1965) has proposed symmetric confidence intervals

$$[y_{(k)}, y_{(n-k+1)}] \tag{5}$$

for the finite population median, denoted, for simplicity, as $Y_{(N/2)}$. McCarthy's conclusions are important when strata sample sizes are small and proportional sample size allocation has been used. Also, his conclusions may be particularly useful in analytical investigations in which primary interest is on a 'domain of study' where the domain sample sizes are small in some strata.

McCarthy has derived two special results for proportional allocation: (a) a confidence interval with an exact confidence coefficient for $L = 2$ strata of equal size, and (b) a lower bound for the confidence coefficient for the confidence interval (5).

To summarize (a), assume that the population has been divided into two strata of equal size. Let a random observation from the ith stratum be denoted by Y_i with cdf F_i ($i = 1, 2$). Then

$$F(Y_{(N/2)}) = 0.5F_1(Y_{(N/2)}) + 0.5F_2(Y_{(N/2)}) = 0.5$$

and
$$F_1(Y_{(N/2)}) = 1 - F_2(Y_{(N/2)})$$

where $F(\cdot)$ is the cdf for the entire population.

Let n_i independent observations $y_i = y_{i1}, \ldots, y_{in_i}$ be drawn from the ith stratum. Then the number of sample elements in y_1 with $Y_1 < Y_{(N/2)}$, $n_Y(Y_1 < Y_{(N/2)})$, has a binomial distribution with parameters n_1 and $F_1(Y_{(N/2)})$; and the number of sample elements in y_2 with $Y_2 > Y_{(N/2)}$, $n_Y(Y_2 > Y_{(N/2)})$, has a binomial distribution with parameters n_2 and $1 - F_2(Y_{(N/2)}) = F_1(Y_{(N/2)})$. For any arbitrary finite population value, Y, the sample results can be presented in the following 2×2 contingency table:

Stratum 1	$n_Y(Y_1 < Y)$	$n_Y(Y_1 \geq Y)$	n_1
Stratum 2	$n_Y(Y_2 > Y)$	$n_Y(Y_2 \leq Y)$	n_2

For a fixed value of Y, Fisher's exact test can be used to test the hypothesis that the distributions of the binomial variables $n_Y(Y_1 < Y)$ and $n_Y(Y_2 > Y)$ have a common parameter against a two-tailed alternative. The set of all Y for which this hypothesis is accepted defines a confidence interval for the median.

McCarthy's second result is a lower bound for the confidence coefficient for the symmetric confidence interval (5). We extend this development to the more general case where a lower bound for the confidence coefficient is desired for a (not necessarily symmetric) confidence interval $[y_{(k)}, y_{(r)}]$ for the (t/N)th population quantile, $Y_{(t)}$. Let $P = t/N$ and assume that the population has been divided into $L \geq 2$ strata, the weights associated with the strata being denoted by $W_1, \ldots, W_L$, with $\Sigma_j^L W_j = 1$. Let the cdf in the jth stratum, F_j, evaluated at $Y_{(t)}$ be denoted by $F_j(Y_{(t)}) = P_j$. Then

$$\sum W_j P_j = P.$$

If the sample is proportionally allocated among the strata, i.e., $n_j = W_j n$, then

$$\sum n_j P_j = \sum n W_j P_j = nP.$$

Letting S_j denote the number of observations in a random sample of size n_j from stratum j that are less than the quantile $Y_{(t)}$. S_j is a binomial random variable with $E(S_j) = n_j P_j$. Letting S denote the number of observations in the combined sample having values of the variable less than $Y_{(t)}$, $S = \Sigma_j^L S_j$ and

$$E(S) = \sum n_j P_j = nP.$$

The probability that the interval, $[y_{(k)}, y_{(r)}]$, covers $Y_{(t)}$ is

$$\sum_{s=k}^{r-1} \Pr(S = s) = \Pr(k \leq S \leq r - 1).$$

Because the distribution of S is a convolution of binomial variables, it is cumbersome to assess this coverage probability exactly. However, Hoeffding's (1956) theorem may be used to obtain a lower bound for the coverage probability:

If S is the number of successes in n independent trials where P_i is the probability of success in the ith trial, and $E(S) = nP$ is fixed, and if b and c are two integers such that

$$0 \leq b \leq nP \leq c \leq n$$

then

$$\sum_{k=b}^{c} \binom{n}{k} P^k (1-P)^{n-k} \leq P(b \leq S \leq c) \leq 1.$$

Both bounds are attained. The lower bound is attained only if $P_1 = \cdots = P_n = P$ unless $b = 0$ and $c = n$.

In our application, there are n_j independent trials for which the probability of success (i.e., that an observation is less than $Y_{(t)}$) is P_j, $j = 1, \ldots, L$. Since $E(S) = \Sigma_j^L n_j P_j$ is fixed at nP, by applying Hoeffding's theorem the lower bound for the probability of coverage of $Y_{(t)}$ by the kth and rth order statistics is

$$\Pr(k \leq S \leq r-1) \geq \sum_{i=k}^{r-1} \binom{n}{i} P^i (1-P)^{n-i} \qquad (6)$$

for $1 \leq k \leq (n/2)$. The lower bound is attained only if $Y_{(t)}$ is the (t/N)th quantile of each individual stratum, as well as the (t/N)th quantile of the overall population.

It should be noted that the lower bound is the probability that $[y_{(k)}, y_{(r)}]$ covers $Y_{(t)}$ when a sample of size n is obtained by random sampling. However, if the allocation of the sample to the L strata is not proportional it is no longer necessarily true that a pair of order statistics from the combined stratified sample has greater probability of covering the population quantile than does the corresponding pair of order statistics from a random sample of the same total size. For the case where the quantile of interest is the median, McCarthy gives an example which indicates that the result may be robust with respect to deviations from proportional allocation and from equality of the strata medians. However, there has been no systematic investigation of the effects of such deviations.

Finally, McCarthy observes that the gain in the coverage probabilities achieved by using symmetric order statistics from a stratified random sample is achieved at the expense of increasing the expected length of the confidence intervals.

Sedransk and Meyer (1978) have provided alternative methods for constructing confidence intervals for finite population quantiles and give methods for determining their confidence coefficients exactly. These methods include use of the interval (1) as well as alternatives based on the empirical cdf. It is assumed that simple random samples (not necessarily proportional to stratum sizes) are selected independently from strata not necessarily of equal size. Their methods are most appropriate when there are a small number of strata and when there is prior information about the relationships among the distributions of Y over the strata.

To illustrate these methods, it is assumed that the population consists of two strata with N_i elements being members of stratum i ($i = 1, 2$). Let $Y_{i(j)}$ denote the jth ordered variate value in stratum i ($j = 1, 2, \ldots, N_i$). Simple random samples of sizes n_1 and n_2 are selected independently from the two strata, and the order statistics within stratum 1, within stratum 2, and in the combined sample are $y_{1(1)}, \ldots, y_{1(n_1)}$, $y_{2(1)}, \ldots, y_{2(n_2)}$, and $y_{(1)}, \ldots, y_{(n_1 + n_2)}$, respectively. Sedransk and Meyer give three different methods of forming a confidence interval for $Y_{(t)}$. The first ('combined') method uses the interval $[y_{(k)}, y_{(r)}]$. The second ('cdf') method uses the estimated finite population cdf to obtain an interval based on the combined order statistics. The third ('separate') method forms the confidence interval from the order statistics of each stratum; i.e., the confidence interval is of the type $[y_{1(k)}, y_{2(r)}]$ or $[y_{2(r)}, y_{1(k)}]$. Each method is designed for a specific purpose. For example, for a stratified random sample where the sampling fractions in the two strata are highly disparate, e.g., $(n_1/N_1) \gg (n_2/N_2)$, a wide confidence interval of the combined type, $[y_{(k)}, y_{(r)}]$, may be necessary to obtain a desired confidence coefficient. To attempt to eliminate this problem, the cdf method was developed.

To illustrate how the confidence coefficients are derived, we sketch some of the steps for the combined method.

If it were known that $Y_{(t)} = Y_{i(j)}$ then

$$\Pr(y_{(k)} \leq Y_{(t)} | Y_{(t)} = Y_{i(j)}) = \sum_{h=k}^{\min(n, t)} \Pr(A_h | Y_{(t)} = Y_{i(j)}),$$

where A_h is the event 'exactly h observations in the combined sample have values of Y with $Y \leq Y_{(t)}$'. It may be seen that

$$\Pr(A_h | Y_{(t)} = Y_{1(j)})$$

$$= \sum_m \binom{j}{m}\binom{N_1 - j}{n_1 - m}\binom{t - j}{h - m}\binom{N_2 - (t - j)}{n_2 - (h - m)} \bigg/ \binom{N_1}{n_1}\binom{N_2}{n_2} \quad (7)$$

where

$$\max(0, h - n_2, j + n_1 - N_1, h + j - t) \leq m$$
$$\leq \min(h, j, n_1, N_2 - t + j - n_2 + h).$$

$\Pr(A_h | Y_{(t)} = Y_{2(j')})$ is derived similarly and may be obtained by interchanging both N_1 and N_2, and n_1 and n_2 in (7).

When the same finite population is sampled repeatedly (over time), it is reasonable to utilize the unconditional probability (2) as the confidence coefficient to be associated with $[y_{(k)}, y_{(r)}]$. In this case

$\Pr(y_{(k)} \leq Y_{(t)})$

$$= \sum_{h=k}^{\min(n,t)} \sum_{j=\max(1,t-N_2)}^{\min(t,N_1)} \Pr(A_h | Y_{(t)} = Y_{1(j)}) \Pr(Y_{(t)} = Y_{1(j)})$$

$$+ \sum_{h=k}^{\min(n,t)} \sum_{j=\max(1,t-N_1)}^{\min(t,N_2)} \Pr(A_h | Y_{(t)} = Y_{2(j)}) \Pr(Y_{(t)} = Y_{2(j)}).$$

Use of (2) will also be appropriate for many single surveys. In many circumstances there is considerable prior information[1] about the $\binom{N}{N_1}$ configurations of the variate values $Y_{1(1)}, \ldots, Y_{1(N_1)}, Y_{2(1)}, \ldots, Y_{2(N_2)}$. Then it is reasonable to utilize such prior information to assess $\Pr(Y_{(t)} = Y_{i(j)})$ for $j = 1, \ldots, N_i$ and $i = 1, 2$. In such cases (2) is a reasonable measure of the confidence to be associated with the interval $[y_{(k)}, y_{(r)}]$.

To determine $\Pr(Y_{(t)} = Y_{1(j)})$ and $\Pr(Y_{(t)} = Y_{2(j)})$ an assumption about the stratification must be made. The simplest assumption is that among the $N = N_1 + N_2$ ordered variate values in the population, each of the $(N!/N_1!N_2!)$ permutations of the N_1 ordered variate values from stratum 1, together with the N_2 ordered variate values from stratum 2, is equally likely. This assumption of 'random' stratification may be appropriate where strata are formed primarily for administrative convenience. For example, administrative districts may be used as separate strata even though there is little difference in the distribution of Y among the strata.

Thus, assuming 'random' stratification,

$$\Pr(Y_{(t)} = Y_{1(j)}) = \binom{t-1}{j-1}\binom{N-t}{N_1-j} \bigg/ \binom{N}{N_1} \qquad (8)$$

for $\max(1, t - N_2) \leq j \leq \min(t, N_1)$. $\Pr(Y_{(t)} = Y_{2(j)})$ may be derived similarly and can be obtained from (8) by interchanging N_1 and N_2.

However, if random stratification is postulated it can be shown that the resulting confidence coefficient will be identical to the confidence coefficient for $[y_{(k)}, y_{(r)}]$ obtained as if the data were obtained via simple random sampling. This result can be proved for any number of strata.

---

[1] This condition typically obtains where the sampling units are institutions (e.g. hospitals, farms, banks, colleges) subject to periodic censuses or periodic reporting requirements.

Alternatively, one may assume that the stratification is 'ordered'. In many surveys the stratification variable, X, is closely related to the variable under study, Y. For example, X may be the value of Y at some previous time. Then, the usual stratification consists of placing those units with the smallest values of X in stratum 1, those with the next smallest values of X in stratum 2, etc. Knowledge about the relation between Y and X and the method of stratification may enable the investigator to assert a priori that $Y_{1(s)} < Y_{2(u)}$. (For instance, in some situations it may be reasonable to assume that the median among the variate values in stratum 1 is less than the median among the variate values in stratum 2.) Then, it is assumed that each permutation of the $N_1 + N_2$ variate values $Y_{1(1)}, \ldots, Y_{1(N_1)}, Y_{2(1)}, \ldots, Y_{2(N_2)}$ consistent with the requirement that $Y_{1(s)} < Y_{2(u)}$ is equally likely to occur. For this case Sedransk and Meyer give explicit expressions for

$$\Pr(Y_{(t)} = Y_{i(j)} | Y_{1(s)} < Y_{2(u)}), \quad i = 1, 2.$$

Using these expressions the exact confidence coefficient associated with the interval $[y_{(k)}, y_{(r)}]$ may be computed and is conditional on the ordering $Y_{1(s)} < Y_{2(u)}$.

From the numerical examples that Sedransk and Meyer have considered it appears that

$$P_{kr} = \Pr(y_{(k)} \leq Y_{(t)} \leq y_{(r)} | Y_{1(s)} < Y_{2(u)})$$

is not sensitive to moderate departures from $Y_{1(s)} < Y_{2(u)}$. For example, with $N_1 = N_2 = 10$ and $n_1 + n_2 = 6$, the values of P_{kr} are quite similar for $(s, u) = (5, 5)$ and $(8, 4)$.

Finally, when contrasted in a simulation study with the alternative methods, Sedransk and Meyer (1978) recommend the combined method.

Smith and Sedransk (1983) obtain lower bounds for the confidence coefficient associated with the confidence interval $[y_{(k)}, y_{(r)}]$ studied by McCarthy (1965) and Sedransk and Meyer (1978). Smith and Sedransk assume a stratified random sample with L strata and arbitrary sample sizes. They give expressions for the lower bound for the two cases corresponding to presence and absence of prior information regarding the 'ordering' of the strata variate values (e.g., $Y_{1(s)} < Y_{2(u)}$).

For the case when there is no available prior information about the ordering of strata variate values, $[Y_{i(j)}: j = 1, \ldots, N_i, i = 1, \ldots, L]$, one may consider use of a lower bound, P^*_{krt}, for $\Pr[y_{(k)} \leq Y_{(t)} \leq y_{(r)}]$ as a measure of the confidence to be associated with the interval $[y_{(k)}, y_{(r)}]$. The lower bound is defined as

$$P^*_{krt} = \min_C [\Pr\{y_{(k)} \leq Y_{(t)} \leq y_{(r)} | C\}]$$

for all admissible configurations, C, among the $(N!/\Pi N_i!)$ arrangements of the strata variate values. When the total number of strata, L, is small, evaluation of $\min_C [\Pr(y_{(k)} \leq Y_{(t)} \leq y_{(r)} | C)]$ poses no difficulties. For example, consider $L = 2$, and define

$$P_{krt}(i, j) = \Pr\{y_{(k)} \leq Y_{(t)} \leq y_{(r)} | Y_{(t)} = Y_{i(j)}\}.$$

Then, $P^*_{krt} = \min_i \min_j \{P_{krt}(i, j)\}$

where $i = 1, \ldots, L$, $\max(1, t - N + N_i) \leq j \leq \min(t, N_i)$, and $P_{krt}(i, j)$ can be obtained from (7).

However, for surveys with more strata ($L \geq 4$) the computational effort required to evaluate P^*_{krt} increases considerably. Thus, an easily computed and good approximation for P^*_{krt} has been developed.

For the second case, where there is prior information regarding the ordering of strata variable values, an alternative strategy is available for obtaining a lower bound which accounts for the additional information. Consider the special case of $L = 2$ strata and assume that this information enables the investigator to make assertions such as $Y_{1(s)} < Y_{2(u)}$ (e.g., the median of the variate values in stratum 1 is less than the median in stratum 2). For each stratum such knowledge restricts the possible values of j for which $Y_{(t)}$ may equal $Y_{i(j)}$.

Consequently, for fixed (k, r), t and (s, u), a restricted lower bound for $\Pr[y_{(k)} \leq Y_{(t)} \leq y_{(r)}]$ is given by

$$P^*_{krt, su} = \min\{P_{krt}(1, j_1), P_{krt}(2, j_2)\}$$

where $P_{krt}(i, j_i) = \min_{j \in D_i} P_{krt}(i, j)$, $i = 1, 2$, and D_i denotes the admissible range of j in stratum i for which $Y_{(t)}$ may equal $Y_{i(j)}$.

When it is appropriate to assert that $Y_{1(s)} < Y_{2(u)}$, it is always advantageous to use $P^*_{krt, su}$ rather than the unrestricted lower bound, P^*_{krt}, since $P^*_{krt, su} \geq P^*_{krt}$. Also, use of $P^*_{krt, su}$ may permit one to find a confidence interval with a guaranteed confidence coefficient and smaller length than if one ignored the prior information. Further details and numerical illustrations are given in Smith and Sedransk (1983) and Smith (1979).

2.3. Two-stage cluster sampling

In Chapman's (1970) unpublished thesis, two stage cluster sampling is assumed, and a confidence interval for $Y_{(N/2)}$, the finite population median, is given. This interval has as its terminals symmetric order statistics from the combined cluster sample. The main objective of Chapman (1970) is to approximate the confidence coefficient associated with such a confidence interval.

Several simplifying assumptions are made. First, the random variable of interest is assumed to have a continuous distribution over the entire population and also within each cluster. Second, the sample design is a random sample of m clusters drawn from the (infinite) population, and a random sample of t elements is selected from each of the clusters drawn. These assumptions, which exclude the possibility of unequal sized clusters and unequal cluster sample sizes, are necessary to permit a meaningful approach to the problem.

Because of the continuity assumptions the population median is well defined. Also, for each cluster, the probability, P, that a randomly selected element has a value less than $Y_{(N/2)}$ is well defined. The random variable, P, defined on the population of clusters, plays a central role in the analysis. Clearly,

$$0.5 = \Pr(Y < Y_{(N/2)}) = \int_0^1 \Pr(Y < Y_{(N/2)} | P = p) \, dF(p) = E(P). \tag{9}$$

For the two-stage design, Chapman considers the confidence interval for $Y_{(N/2)}$, $(y_{(k)}, y_{(r+1)})$, where $y_{(j)}$ denotes the jth order statistic from the entire sample of size $n = mt$. Letting R_i denote the number of sampled elements in cluster i with values less than $Y_{(N/2)}$, and $R^* = \Sigma_{i=1}^m R_i$,

$$\gamma = \Pr\{y_{(k)} < Y_{(N/2)} < y_{(r+1)}\} = \Pr(k \leq R^* \leq r). \tag{10}$$

Thus, to evaluate the confidence coefficient, γ, the distribution of R^* must be determined. Now, $\{R_i: i = 1, \ldots, m\}$ are independent, identically distributed random variables with distribution

$$\Pr\{R = r\} = \int_0^1 \binom{t}{r} p^r (1-p)^{t-r} f(p) \, dp. \tag{11}$$

If the distribution of P is taken to be beta,

$$f_\beta(p | r', n') \propto p^{r'-1} (1-p)^{n'-r'-1}, \tag{12}$$

then each R_i has the beta-binomial distribution with parameters r', n' and t.

Exact confidence coefficients can, in principle, be obtained from (10), (11) and (12). However, an alternative approach provides a lower bound for $\Pr(k \leq R^* \leq r)$, and, thus, for the desired confidence coefficient, γ. This lower bound, which is easily computed, may also serve as an *approximation* for γ.

Letting P_i denote the value of P for the ith selected cluster, define $\bar{P} = (\Sigma_{i=1}^m P_i / m)$. Then using results from Hoeffding (1956), Chapman proves the following theorem (subject to minor caveats):

THEOREM. *Let $(y_{(k)}, y_{(r+1)})$ be a confidence interval for $Y_{(N/2)}$ from the combined sample of size $n = mt$. Assuming that $\bar{P}$ has the beta distribution in (12),*

$$\gamma \geq \gamma' = \sum_{i=k}^r f_{\beta b}(i | r', n', n) \tag{13}$$

where $f_{\beta b}$ is the beta-binomial distribution obtained from (11) (n replacing t) and (12).

Now, since $E(P) = 0.5$ (formula (9)), $E(\bar{P}) = 0.5$ which implies that $r'/n' = 0.5$ (see (12)). Thus, the distribution assumed for $\bar{P}$ in the Theorem is

$$f_\beta(\bar{P}|r') \propto \bar{P}^{r'-1}(1-\bar{P})^{r'-1} \qquad (14)$$

which is symmetric. Since P is the mean of m quantities, Chapman conjectures that (14) may be an appropriate assumption for practical situations. One may assign the value of r' in (14) by specifying a value for $\text{Var}(\bar{P})$ and using $r'/n' = 0.5$. Since $\bar{P}$ is the mean of m independent observations on P, $\text{Var}(\bar{P}) = \text{Var}(P)/m$. Thus, the remaining questions are: (a) the appropriateness of assuming a beta distribution for $\bar{P}$, and (b) the assessment of $\text{Var}(P)$. In an empirical study, Chapman shows that his conjecture (a) is strongly supported and that the beta distribution for $\bar{P}$ in (14) will be appropriate in many cases. The second main question, (b), that of evaluating $\text{Var}(P)$, is addressed after finding an expression for the density function of P. A limited investigation of the quality of using γ' to approximate the exact confidence coefficient, γ, was carried out. Using γ' appears to be very satisfactory, especially for small values of the intraclass correlation coefficient.

In another Ph.D. dissertation, Blesseos (1976) has also considered the problem of determining confidence coefficients for confidence intervals for two stage cluster sampling. In this development the finite population is assumed to consist of N units each with a distinct Y value. Also, each unit belongs to one of C clusters and the ith cluster consists of M_i units.

A sample of size n is obtained via a two stage design. The confidence interval for $Y_{(t)}$ is formed from the order statistics of the combined sample; i.e., given by (1). Specifically, Blesseos obtains expressions for the confidence coefficient of $[y_{(k)}, y_{(r)}]$ for eight different two stage cluster sampling conditions: In the first stage the clusters are either selected by

A. simple random sampling, or by

B. probability proportional to size.

In the second stage, both simple random sampling and complete enumeration of units is considered. For simple random sampling, three allocation strategies are considered. Either

1. $m_i < M_i$ units are selected, or
2. m units are selected in each cluster, or
3. the subsample size is proportional to the cluster size.

To illustrate how the confidence coefficient is obtained Blesseos (1976) considers $C = 3$. From (2), only $\Pr\{y_{(k)} \leq Y_{(t)}\}$ must be obtained. Letting

E_i denote the event that cluster i is in the sample,
D_{t_i} denote the event that exactly t_i elements in cluster i have values of $Y \leq Y_{(t)}$,
Q be the number of elements in the combined sample with values of $Y \leq Y_{(t)}$, and
n be the total sample size,

$$\Pr\{y_{(k)} \le Y_{(t)}\} = \sum_{q=k}^{\min(t,n)} \Pr(Q = q)$$

$$= \sum_{q} \sum_{i<j<k} \sum_{s} \Pr(Q = q | E_i E_j E_k D_{t_i} D_{t_j} D_{t_k})$$

$$\times \Pr(D_{t_i} D_{t_j} D_{t_k} | E_i E_j E_k) \Pr(E_i E_j E_k)$$

where the range of summation on s is $\{t_i, t_j, t_k: 0 \le t_i + t_j + t_k \le t\}$. Blesseos gives explicit expressions for the individual components of $\Pr\{y_{(k)} \le Y_{(t)}\}$ above.

3. Point estimation for finite population quantiles

In this section we review the literature concerned with point estimation of population quantiles when there are complex sample designs. Loynes (1966) investigates point estimation of the quantiles of a distribution F assuming, initially, only that F is continuous, and that a random sample has been selected. Then the assumption of symmetry of F is added. Finally, there is a brief, much less successful, treatment of stratified random sampling.

Let $Y = (Y_1, \ldots, Y_n)$ denote the sample values from a random sample of size n from the unknown continuous cdf, $F(y)$. Let $T \equiv T(Y)$ denote the estimator of u_α, the α-quantile of F. The basic assumption,

$$\Pr\{T \le y | F\} = \Psi_T\{F(y)\},$$

is shown to be a 'possible expression of the (somewhat vague) requirement that the estimation procedure be distribution-free, because, for example, it follows that for any F with $F(y) = \alpha$

$$\Pr\{T \le y | F\} = \Psi_T(\alpha).'$$

Now, let $y_{(i)}$ denote the ith order statistic. Then, it is proved that if T corresponds to a distribution-free procedure to estimate quantiles, then $T = y_{(i)}$ with probability $p_i(Y)$ ($\Sigma\, p_i(Y) = 1$). If T is assumed to be permutation-invariant and the $p_i(Y)$ are continuous, then for T to be distribution-free it is necessary and sufficient that $p_i(Y) = p_i$, independently of Y. Such estimators are termed 0-statistics (i.e., randomized order statistics).

Loynes then exhibits, in Theorem 2, the admissible 0-statistics corresponding to the risk function $a(u, \Theta, \hat{\Theta})$ where

$$a(u, \Theta, \hat{\Theta}) = \Pr\{\hat{\Theta}(Y) \le u | \Theta\} \quad \text{for } u < \Theta,$$
$$= \Pr\{\hat{\Theta}(Y) \ge u | \Theta\} \quad \text{for } u > \Theta.$$

It is said that the estimator $\hat{\Theta}$ is at least as good as $\hat{\Theta}'$ if $a(u, \Theta, \hat{\Theta}) \le a(u, \Theta, \hat{\Theta}')$ for all Θ and $u \ne \Theta$ (and $\hat{\Theta}$ is better than $\hat{\Theta}'$ if there is also strict inequality for some Θ and $u \ne \Theta$). Then $\hat{\Theta}$ is admissible if there exists no estimator better than $\hat{\Theta}$.

The problem of choice among the admissible estimators is considered briefly. One method of selecting a single estimator is discussed. This leads to the 'natural' estimator of u_α, $y_{(r)}$ where $r = [n\alpha]$ or $r = [n\alpha] + 1$. This discussion is followed by consideration of the case where F is symmetric about some unknown point Θ.

Loynes explicitly considers stratified random sampling with two strata having known relative sizes W_1 and W_2 ($W_1 + W_2 = 1$). Assuming that the strata cdf's F_1 and F_2 are continuous, the cdf for the entire population is $H = W_1 F_1 + W_2 F_2$. Letting Y_1 and Y_2 denote independent simple random samples of sizes n_1 and n_2 from the two strata, let $T \equiv T(Y_1, Y_2)$ denote a (possibly randomized) estimator of u_α, the α-quantile of H. The estimator T is said to generate a distribution-free procedure if

$$\Pr\{T \le z | F_1, F_2\} = \Psi_T\{H(z)\}.$$

If continuity is required it is shown that $T = y_{1(i)}$ with probability $p_{1i}(\rho)$ and $T = y_{2(j)}$ with probability $p_{2j}(\rho)$ where ρ indicates which of the $\binom{n_1 + n_2}{n_2}$ possible patterns of Y_1 and Y_2 has occurred. However, Loynes has not been able to identify the acceptable values of the $p_{1i}(\rho)$ and $p_{2j}(\rho)$. Thus, he is unable to proceed to determine the admissible estimators.

Finally, Loynes considers a large sample solution where the requirement that the estimation procedure be distribution-free has also been dropped. Since u_α is the solution of $W_1 F_1(y) + W_2 F_2(y) = \alpha$, a natural estimator is the (approximate) solution of $W_1 \hat{F}_1(y) + W_2 \hat{F}_2(y) = \alpha$ where $\hat{F}_1$ and $\hat{F}_2$ are the sample cdf's. Letting T be the smallest value of y for which $W_1 \hat{F}_1(y) + W_2 \hat{F}_2(y) \ge \alpha$, the asymptotic ($n_1, n_2 \to \infty$ with $n_1/n_2 = \rho/\sigma$ where ρ and $\sigma = 1 - \rho$ are fixed) distribution of T is normal with mean u_α and variance

$$\left\{\frac{W_1^2}{\rho} F_1(1 - F_1) + \frac{W_2^2}{\sigma} F_2(1 - F_2)\right\} / n_2(W_1 f_1 + W_2 f_2)^2$$

provided $W_1 f_1 + W_2 f_2 > 0$, and (f_1, f_2, F_1, F_2) are evaluated at u_α. See Gross (1980), discussed below, for an analogous treatment of this problem.

Gross (1980) considers estimation of the finite population median, $Y_{(N/2)}$, when stratified random sampling, equal probability, single stage cluster sampling and stratified, single stage cluster sampling are employed. In each case, a point estimator, $\hat{Y}_{(N/2)}$, of $Y_{(N/2)}$, an asymptotic expression for $\text{Var}(\hat{Y}_{(N/2)})$ and a procedure to estimate $\text{Var}(\hat{Y}_{(N/2)})$ are given. Generally, estimators are reported to have asymptotic normal distributions. However, exact conditions and proofs of statements concerning the asymptotic distributions are not given.

Assuming stratified random sampling, a simple random sample of size n_i is selected from the N_i units in stratum i ($i = 1, 2, \ldots, L$). Denote by F_i and $\hat{F}_i$, the population and sample cumulative distribution functions in stratum i, and

$W_i = N_i/N$. Then the finite population cdf

$$F = \sum_{i=1}^{L} W_i F_i$$

is consistently estimated by

$$\hat{F} = \sum_{i=1}^{L} W_i \hat{F}_i.$$

To compute $\hat{Y}_{(N/2)}$) all of the L samples are pooled and ordered with each observation from stratum i being assigned the weight $\phi_i = W_i/n_i$. The weights of the ascending sequence are accumulated until 0.5 is first crossed; $\hat{Y}_{(N/2)}$ may be defined as the first observation, $y_{(k)}$, encountered after this crossing. Alternatively, one may take $\hat{Y}_{(N/2)} = 0.5(y_{(k)} + y_{(k-1)})$.

Asymptotically, under some (unspecified) conditions, $\hat{Y}_{(N/2)} \sim N(Y_{(N/2)}, \sigma^2)$ with

$$\sigma^2 = \{Nf^2(Y_{(N/2)})\}^{-1} \times \sum_{i=1}^{L} W_i(N_i - n_i)n_i^{-1} F_i(Y_{(N/2)})(1 - F_i(Y_{(N/2)})) \quad (15)$$

where f denotes the overall population density.

One may easily show that the allocation of a fixed total sample of size n to minimize σ^2 is given by

$$n_i \propto N_i \{F_i(Y_{(N/2)})\}^{1/2} \{1 - F_i(Y_{(N/2)})\}^{1/2}.$$

From this expression for the optimal allocation it is easily seen that strata having their medians far from $Y_{(N/2)}$ will have smaller values of n_i than those having their medians close to $Y_{(N/2)}$.

An estimator of $\text{Var}(\hat{Y}_{(N/2)})$ is developed by first finding an (complicated) expression for $\text{Var}(\hat{Y}_{(N/2)})$ and then replacing F_i by $\hat{F}_i$ in $\text{Var}(\hat{Y}_{(N/2)})$. Specifically, letting $Y_{(t)}$ denote the tth ordered *population* value, Gross shows that

$$p_t = \Pr(\hat{Y}_{(N/2)} > Y_{(t)}) = \sum \prod_{i=1}^{L} \binom{N_i F_i(Y_{(t)})}{t_i} \binom{N_i(1 - F_i(Y_{(t)}))}{n_i - t_i} \bigg/ \binom{N_i}{n_i} \quad (16)$$

where the sum in (16) is over all $(t_1, \ldots, t_L)$ such that $\sum_{i=1}^{L} \phi_i t_i < 0.5$ with $\phi_i = W_i/n_i$.

Since $\Pr(\hat{Y}_{(N/2)} = Y_{(t)}) = p_{t-1} - p_t$,

$$E(\hat{Y}_{(N/2)}^r) = \sum_{t=1}^{N} (p_{t-1} - p_t) Y_{(t)}^r. \quad (17)$$

Using (17), the estimator of variance of $\hat{Y}_{(N/2)}$ is given by

$$\text{var}(\hat{Y}_{(N/2)}) = \sum_{t=1}^{n} (\hat{p}_{t-1} - \hat{p}_t) y_{(t)}^2 - \left\{ \sum_{t=1}^{n} (\hat{p}_{t-1} - \hat{p}_t) y_{(t)} \right\}^2 \quad (18)$$

where, as usual, $y_{(t)}$ denotes the tth ordered observation in the combined sample of size n, and $\hat{p}_t$ is obtained from (16) by replacing $F_i(Y_{(t)})$ with $\hat{F}_i(\hat{Y}_{(t)})$.

Gross next considers single-stage cluster sampling with n clusters selected by simple random sampling from the population of N clusters. Cluster j is assumed to have M_j elements all of which are sampled. The usual ratio estimator of the finite population cdf is

$$\hat{F} = \sum_{j=1}^{n} M_j F_j \bigg/ \sum_{j=1}^{n} M_j$$

where F_j is the cdf in cluster j. Since there is no subsampling, no special weighting is required (as in stratified random sampling). Hence, the sample median, $\hat{Y}_{(N/2)}$, is a consistent estimator of the population median, $Y_{(N/2)}$, if $Y_{(N/2)}$ is unique. Under some (unspecified) conditions, asymptotically $\hat{Y}_{(N/2)} \sim N(Y_{(N/2)}, \sigma_c^2)$ where

$$\sigma_c^2 = (N-n)(Nn)^{-1}(N-1)^{-1} \sum_{j=1}^{N} M_j^2$$
$$\times \{F_j(Y_{(N/2)}) - F(Y_{(N/2)})\}^2 \{\overline{N} f(Y_{(N/2)})\}^{-2}$$

where F is the population cdf, $\overline{N} = \Sigma_{j=1}^{N} M_j/N$ and $f(Y_{(N/2)}) > 0$ is the asymptotic density evaluated at $Y_{(N/2)}$, assumed to exist when $N \to \infty$.

There are extensions to stratified, single-stage cluster sampling using the same techniques as indicated above. Two cases are considered: the total number of elementary units in each stratum is (a) known, or (b) unknown prior to sampling. Since straightforward variance estimation cannot be carried out as in stratified random sampling, a bootstrap method is suggested.

4. Bayesian methods

In this section an account of Bayesian inferential methods for finite population quantiles is given.

Ericson (1969) finds the posterior distribution of a finite population quantile when a single multinomial model can be assumed to generate the finite population of interest, and there is extreme prior vagueness about the parameters of the multinomial distribution.

It is postulated that Y can assume one of the finite set of values, $Y_{(1)} < Y_{(2)} < \cdots < Y_{(k)}$. Further, define

$$\Pr(Y = Y_{(j)}|\boldsymbol{p}) = p_j, \quad j = 1, \ldots, k, \quad \sum_{j=1}^{k} p_j = 1, \quad (19)$$

where $p = (p_1, \ldots, p_{k-1})$. The values of Y for the finite population of size N, $Y_1, \ldots, Y_N$, are assumed to be i.i.d. with the distribution (19). Because of the discrete model generating the finite population, it is most convenient to proceed by considering the vector $N = (N_1, \ldots, N_{k-1})$ where N_j denotes the unknown number of the N finite population elements having $Y = Y_{(j)}$. Then

$$p'(N) = \int_p \frac{\Gamma(N+1)}{\prod_{j=1}^{k-1} \Gamma(N_j+1)\Gamma(N - \Sigma_{j=1}^{k-1} N_j + 1)} \prod_{j=1}^{k-1} p_j^{N_j}$$

$$\times \left(1 - \sum_{j=1}^{k-1} p_j\right)^{N - \Sigma_{j=1}^{k-1} N_j} f'(p) \, dp \qquad (20)$$

where $f'(p)$ is a prior density on p, and the integral in (20) is over the simplex

$$\left\{ p \mid 0 \leq p_j \leq 1, \sum_{j=1}^{k-1} p_j \leq 1 \right\}.$$

Ericson takes as the prior on p the $k - 1$ dimensional Dirichlet distribution with density

$$f'(p) = \frac{\Gamma(\varepsilon)}{\prod_{j=1}^{k-1} \Gamma(\varepsilon_j) \Gamma(\varepsilon - \Sigma_{j=1}^{k-1} \varepsilon_j)} \prod_{j=1}^{k-1} p_j^{\varepsilon_j - 1} \left(1 - \sum_{j=1}^{k-1} p_j\right)^{\varepsilon - \Sigma_{j=1}^{k-1} \varepsilon_j - 1} \qquad (21)$$

where $\varepsilon_j > 0$ and $\varepsilon = \Sigma_{j=1}^{k} \varepsilon_j$. Ericson states that such a Dirichlet prior is only tenable as a representation of extreme prior vagueness, obtained by taking the ε_j (and ε) to be very small. Using (20) and (21), one obtains the Dirichlet-multinomial (prior) distribution of N,

$$p'(N) = \frac{\Gamma(N+1) \prod_{j=1}^{k-1} \Gamma(N_j + \varepsilon_j) \Gamma(N - \Sigma_{j=1}^{k-1} N_j + \varepsilon - \Sigma_{j=1}^{k-1} \varepsilon_j) \Gamma(\varepsilon)}{\prod_{j=1}^{k-1} \Gamma(N_j+1) \Gamma(N - \Sigma_{j=1}^{k-1} N_j + 1) \Gamma(N+\varepsilon) \prod_{j=1}^{k-1} \Gamma(\varepsilon_j) \Gamma(\varepsilon - \Sigma_{j=1}^{k-1} \varepsilon_j)} \qquad (22)$$

for $N_j = 0, 1, \ldots, N$ and $\Sigma_{j=1}^{k-1} N_j \leq N$.

Now, letting Θ_P denote the Pth finite population quantile, $\Theta_P = Y_{(j)}$ if j is the smallest integer for which $\Sigma_{i=1}^{j} N_i/N \geq P$; i.e., $\Theta_p \leq Y_{(j)}$ whenever $\Sigma_{i=1}^{j} N_i \geq NP$. Thus, the (prior) cdf can be obtained from

$$p'(\Theta_p \leq Y_{(j)}) = \begin{cases} \sum_{c_j = [NP]}^{N} \binom{N}{c_j} \frac{\Gamma(\varepsilon) \Gamma(c_j + \Sigma_{i=1}^{j} \varepsilon_i) \Gamma(N + \varepsilon - c_j - \Sigma_{i=1}^{j} \varepsilon_i)}{\Gamma(\Sigma_{i=1}^{j} \varepsilon_i) \Gamma(\varepsilon - \Sigma_{i=1}^{j} \varepsilon_i) \Gamma(N + \varepsilon)}, \\ \qquad \qquad \qquad \qquad \qquad j = 1, \ldots, k-1, \\ 1, \qquad \qquad \qquad \qquad \qquad j = k, \end{cases} \qquad (23)$$

where $[NP]$ denotes the smallest integer not less than NP. The result in (23) follows since, given p, $\Sigma_{i=1}^{j} N_i$ is distributed binomialle $(N, \Sigma_{i=1}^{j} P_i)$ while $\Sigma_{i=1}^{j} P_i$ has a beta distribution. Thus, the terms in the sum in (23) are terms of the beta-binomial distribution.

For posterior inference, recall that the $Y_1, \ldots, Y_N$ are assumed to be i.i.d. with the distribution (19). Letting n_j ($j = 1, \ldots, k$) denote the number of the n sampled units having $Y = Y_{(j)}$ and using (21), the posterior distribution of p is Dirichlet as in (21), but with $(\varepsilon_j + n_j)$ replacing ε_j. Now define $M_j = N_j - n_j$, the number of nonsampled units with $Y = Y_{(j)}$, and let $M = N - n$. Then it is easily seen that, given $\boldsymbol{n} = (n_1, \ldots, n_{k-1})$, $\boldsymbol{M} = (M_1, \ldots, M_k)$ has the Dirichlet-multinomial distribution as in (22) with $(M, M, n_j + \varepsilon_j, n + \varepsilon)$ replacing $(N, N, \varepsilon_j, \varepsilon)$. One may then obtain the posterior distribution of N by substitution.

To obtain the posterior distribution of Θ_p, note that $\Theta_p \leq Y_{(j)}$ whenever $\Sigma_{i=1}^{j}(M_i + n_i) \geq NP$. Thus, letting double primes indicate posterior distributions,

$$p''(\Theta_p \leq Y_{(j)}) = p''\left(\sum_{i=1}^{j} M_i \geq NP - \sum_{i=1}^{j} n_i\right).$$

The posterior distribution of $\Sigma_{i=1}^{j} M_i$ is beta-binomial, and it follows that

$$p''(\Theta_p \leq Y_{(j)}) = \begin{cases} 0, & [NP] - \sum_{i=1}^{j} n_i > N - n, \\ & j = 1, \ldots, k, \\ \sum_{u=[NP]-\Sigma_{i=1}^{j} n_i}^{N-n} p''\left(\sum_{i=1}^{j} M_i = u\right), & 0 < [NP] - \sum_{i=1}^{j} n_i \leq N - n, \\ & j = 1, \ldots, k-1, \\ 1, & j = k \text{ and/or} \\ & [NP] - \sum_{i=1}^{j} n_i \leq 0, \end{cases} \quad (24)$$

where

$$p''\left(\sum_{i=1}^{j} M_i = u\right) = \binom{N-n}{u}$$

$$\times \frac{\Gamma(n + \varepsilon)\Gamma\{u + \Sigma_{i=1}^{j}(n_i + \varepsilon_i)\}\Gamma\{N + \varepsilon - u - \Sigma_{i=1}^{j}(n_i + \varepsilon_i)\}}{\Gamma\{\Sigma_{i=1}^{j}(n_i + \varepsilon_i)\}\Gamma\{n + \varepsilon - \Sigma_{i=1}^{j}(n_i + \varepsilon_i)\}\Gamma(N + \varepsilon)},$$

$$u = 0, 1, \ldots, N - n.$$

Finally, Ericson exhibits a parallel between inferences based upon (24), and the usual frequentist confidence intervals for quantiles. He assumes that all n sample observations are distinct (i.e., $n_j = 0$ or $n_j = 1$), and takes $\varepsilon_j = 0$, $j = 1, \ldots, k$.

Letting $y_{(i)}$ denote the ith order statistic, and assuming that N is large relative to n,

$$p''(y_{(i)} < \Theta_p \leq y_{(j)}) \cdot \sum_{v=i}^{j-1} \binom{n-1}{v} P^v(1-P)^{n-1-v} \qquad (25)$$

where the right side of (25) is, approximately, the confidence coefficient attached to the confidence interval, $(y_{(i)}, y_{(j)})$, for Θ_p.

Binder (1982) extends Ericson's (1969) results for the multinomial model for generating the finite population. Binder notes that 'one drawback of this [i.e., Ericson's] approach is that the discrete values of the superpopulation distribution are assumed to be a subset of some *known* countable set'. Binder's extension is to allow the discrete values to take any real values by using (non-parametric) Dirichlet process priors. Unfortunately, if P is a random probability distribution arising from a Dirichlet process, then, with probability one, P is a discrete distribution. Thus, many of Binder's results are analogous to ones found in Ericson (1969). However, for the case of a single population, Binder finds the asymptotic posterior distribution of a finite population quantile.

Assuming stratified populations, Binder gives exact and asymptotic expressions for the posterior distribution of a finite population quantile. Brief comparisons with frequentist methods are also made.

Hill (1968), assuming a sample of size n selected without replacement from a finite population of N units, wishes to find the posterior distribution of ξ_p, the Pth percentile of the *unsampled* units in the population.

Let $M = N - n$ denote the number of non-sampled units in the population, and $Y_{(1)} < Y_{(2)} < \cdots < Y_{(k)}$ the k distinct values of the variable of interest, Y. Denoting by L_i the number of units with $Y = Y_{(i)}$, $L_i \geq 1$ and $\sum_{i=1}^{k} L_i = N$. The data, d, consists of n_i units with the value $y_{(i)}$,

$$y_{(1)} < y_{(2)} < \cdots < y_{(m)}, \quad n_i > 0, \quad \sum_{i=1}^{m} n_i = n.$$

Finally, let J_i be the population rank of $y_{(i)}$ among $Y = (Y_{(1)}, \ldots, Y_{(k)})$.

Treating k and N as unknown, and assuming sampling is without replacement

$$\Pr(d, J = j | Y, L = l, k, N)$$

$$= \begin{cases} \left\{ \binom{N}{n} \right\}^{-1} \prod_{i=1}^{m} \binom{l_{j_i}}{n_i} & \text{if } Y_{(j_i)} = y_{(i)}, \quad i = 1, \ldots, m, \\ = 0 & \text{otherwise}, \end{cases} \qquad (26)$$

where d denotes 'data'. Taking the expectation of (26) over the distribution of Y given L, k, N,

$$\Pr(d, J = j | L = l, k, N)$$
$$= \left\{\binom{N}{n}\right\}^{-1} \prod_{i=1}^{m} \binom{l_{j_i}}{n_i} \Pr\{Y_{(j_i)} = y_{(i)} : i = 1, \ldots, m | L = l, k, N\}. \quad (27)$$

Now,

$$\Pr(d, J = j, L = l | k, N) = \Pr(d, J = j | L = l, k, N) \Pr(L = l | k, N) \quad (28)$$

where the first term on the right side of (28) is given by (27). Since

$$\Pr(d, J = j, L = l | k, N) = \Pr(J = j, L = l | k, N, d) \Pr(d | k, N),$$

knowledge of (28), together with a prior distribution for k and N, permits evaluation of the posterior distribution of J and L. Then, one may evaluate the proportions of unsampled units with values between and at the order statistics of the *given sample*, and, finally, the posterior distribution of ξ_p.

In this article, Hill considers the consequences of taking

$$\Pr(L = l | k, N) = \left\{\binom{N-1}{k-1}\right\}^{-1}, \quad l_i \geq 1, \quad \sum_{i=1}^{k} l_i = N, \quad (29)$$

and

$$\Pr\{Y_{(j_i)} = y_{(i)} : i = 1, \ldots, m | L, k, N\} \quad (30)$$

to be a function *only* of $y_{(1)}, \ldots, y_{(m)}$. The assumption in (29) is that all compositions of the population with the specified k and N are regarded as equally likely. Hill shows that there do not exist countably additive distributions of Y given L, k and N such that (30) holds, even if (30) is weakened to permit dependence upon k. However, he regards both (29) and (30) to be adequate approximations in appropriate situations.

Preliminary steps in the determination of the posterior distribution of ξ_p are: (a) finding $\Pr(d | k, N)$ by summing (28) over j and l and using (29) and (30); and (b) using (28), (29), (30) and the expression for $\Pr(d | k, N)$, to show that

$$\Pr\{J = j, L = l | k, N, d\} = \prod_{i=1}^{m} \binom{l_{j_i}}{n_i} \left\{\binom{k}{m}\binom{N+m-1}{n+k-1}\right\}^{-1}. \quad (31)$$

Now, $\xi_p \leq y_{(i)}$ if, and only if, the proportion of *unsampled* units in the population with values less than or equal to $y_{(i)}$ is at least P. Then,

$$\Pr\{\xi_p \leq y_{(i)} | k, N, d\} = \Pr\{L^{(i)} \geq n^{(i)} + MP | k, N, d\}$$

where $n^{(i)} = \Sigma_{k=1}^{i} n_k$ and $L^{(i)} = \Sigma_{k=1}^{J_i} L_k$. Using (31) and a combinatorial identity it is shown that for z a nonnegative integer,

$$\Pr(L^{(i)} = n^{(i)} + z | k, N, d)$$

$$= \left\{ \binom{k}{m} \binom{N+m-1}{n+k-1} \right\}^{-1} \sum_{t=i}^{k-m+i} \left\{ \binom{t-1}{i-1} \binom{k-t}{m-i} \right.$$

$$\left. \times \binom{n^{(i)} + i + z - 1}{n^{(i)} + t - 1} \binom{N - n^{(i)} + m - i - 1 - z}{n - n^{(i)} + k - t - 1} \right\}. \qquad (32)$$

Letting $z_0 = MP$, for $i = 1, \ldots, m$,

$$\Pr\{\xi_p \leq y_{(i)} | k, N, d\} = \Sigma_{z=z_0}^{M} \Pr(L^{(i)} = n^{(i)} + z | k, N, d). \qquad (33)$$

If (k, N) are not known, averaging (32) and (33) over $\Pr(k, N | d)$ (see Section 3 of Hill's paper) gives the corresponding marginal probabilities.

Finally, Hill suggests methodology for the situation when (33) does not provide a sufficiently detailed description of the posterior distribution of ξ_p.

Acknowledgement

The authors wish to thank the Economic and Social Research Council of Great Britain for partial financial support under grant HR 7152/1.

References

Binder, D. (1982). Non-parametric Bayesian models for samples from finite populations. *Journal of the Royal Statistical Society Series B* **44**(3), 388–393.
Blesseos, N. (1976). Distribution-free confidence intervals for quantiles in stratified and cluster sampling. Ph.D. dissertation, Ball State University, Muncie, IN.
Chapman, D. W. (1970). Cluster sampling and approximate distribution-free confidence intervals. Ph.D. dissertation, Cornell University, Ithaca, NY.
Ericson, W. (1969). Subjective Bayesian models in finite populations, *Journal of the Royal Statistical Society Series B* **31**(2), 195–233.
Gross, S. (1980). Median estimation in sample surveys, In: *Proceedings of the Section on Survey Research Methods*, American Statistical Association, 181–184.
Haskell, J. and Sedransk, J. (1980). Confidence intervals for quantiles and tolerance intervals of finite populations, Unpublished Technical Report, SUNY at Albany, Department of Mathematics and Statistics, Albany, NY.
Hill, B. (1968). Posterior distribution of percentiles: Bayes' theorem for sampling from a population, *Journal of the American Statistical Association* **63**, 677–691.
Hoeffding, W. (1956). On the distribution of the number of successes in independent trials. *Annals of Mathematical Statistics* **27**, 713–721.

Loynes, R. M. (1966). Some aspects of the estimation of quantiles, *Journal of the Royal Statistical Society Series B* **28**(3), 497–512.

McCarthy, P. J. (1965). Stratified sampling and distribution-free confidence intervals for a median, *Journal of the American Statistical Association* **60**, 772–783.

Sedransk, J. and Meyer, J. (1978). Confidence intervals for the quantiles of a finite population: Simple random and stratified random sampling, *Journal of the Royal Statistical Society Series B* **40**(2), 239–252.

Smith, P. J. and Sedransk, J. (1983). Lower bounds for confidence coefficients for confidence intervals for finite population quantiles, *Communications in Statistics, Theory and Methods* **12**(12), 1329–1344.

Smith, P. J. (1979). Bayesian approaches to two fisheries research problems; and lower bounds for confidence coefficients of confidence intervals for finite population quantiles. Ph.D. dissertation. State University of New York at Buffalo, NY.

Woodruff, R. S. (1952). Confidence intervals for medians and other position measures, *Journal of the American Statistical Association* **47**, 635–646.

Asymptotics in Finite Population Sampling

Pranab Kumar Sen

1. Introduction

The theory of (objective or probabilistic) sampling from a *finite population* plays a fundamental role in statistical inference in sample surveys. Indeed, in practice, one mostly encounters a set or collection of a *finite* number (say, N) of objects or *units* comprising a *population*, and, on the basis of a subset of these units, called a *sample* (drawn in an objective manner), the task is to draw (valid) statistical conclusions on (various characteristics of) the population. The population size N, though finite, needs not be small, and the sample size, say n, though presumably less than N, needs not be very small compared to N (i.e., the *sampling fraction* n/N needs not be very small). In survey sampling, the *sampling frame* defines the units and the size of the population unambiguously. It also reconstructs a population having an uncountable (or infinite) number of natural units interms of a finite population by redefining suitable *sampling units*. Thus, given the sampling frame and units, one may like to draw inference on the population through an objective sampling scheme. In some other cases, though the units are clearly defined, the size of the population may not be known in advance, and one may therefore like to estimate the population size (along with its other characteristics), through objective sampling schemes. In either case, usually sampling is made *without replacement* (generally, leading to relatively smaller margins of sampling fluctations), though, for sampling *with replacement*, the theory is relatively simpler in form. Again, in either with or without replacement schemes, the different units in the population may all have the common probability for inclusion in the sample (leading to *equal probability* or *simple random sampling*), or they may be stratified into some subsets, for each of which simple random sampling may be adopted. In the extreme case, the units in the population, depending on their sizes or some other characteristics, may have possibly different probabilities for inclusion in the sample (i.e., *varying probability sampling*). There may be other variations in a sampling scheme (such as *double sampling, interpenetrating sampling, successive sampling*, etc.). However, all of these cases are characterized by an objective procedure defined by a probability law governing the sampling distribution of suitable statistics based on the sample units.

When N is small, such a sampling distribution (of a statistic) may mostly be studied by direct enumeration of all possible cases. However, as N increases this enumerational process generally becomes prohibitively laborious. On the other hand, in survey sampling and in other practical situations N is usually large and n/N may not be very small. In such a case, there may be a profound need to examine the generally anticipated and applicable large sample approximations for the sampling distributions (and related probability inequalities) with a view to prescribing them in actual applications. Our main interest is centered in these asymptotics in finite population sampling. Naturally, the asymptotic theory depends on the sampling design and, for diverse schemes, diverse techniques have been employed to achieve the general goals. It is intended to provide here a general account of these techniques along with the related asymptotic theory. In line with the general objectives of the *Handbook*, mostly, the derivations will be replaced by motivations, and emphasis will be laid on the applications oriented theory only.

In Section 2, we start with the asymptotics in simple random sampling (with and without replacement) schemes. For general U-statistics, containing the sample mean and variances as special cases, asymptotic normality and related results are presented there. Some asymptotics on probability inequalities in *simple random sampling* (SRS) are then considered in Section 3. Asymptotics on *jackknifing* in finite population sampling (SRS) are presented in Section 4. *Capture–mark–recapture* (CMR) techniques and asymptotic results on the estimation of the size of a finite population are considered in Section 5. Asymptotic *results on sampling with varying probabilities* (along with the allied *coupon collector problem*) are presented in Section 6. In this context, some limit theorems arising in the *occupancy problem* are also treated briefly, The concluding section deals with *successive sub-sampling with varying probabilities*, and the relevant asymptotic theory is discussed there.

2. Asymptotics in SRS

Let the N units with values $a_1, \ldots, a_N$ constitute the finite population. In a sample of size n ($\leq N$), drawn without replacement, the observation vector $X_n = (X_1, \ldots, X_n)$ is a (random) subset of $A_N = (a_1, \ldots, a_N)$, governed by the basic probability law

$$P\{X_1 = a_{i_1}, \ldots, X_n = a_{i_n}\} = N^{-[n]} \tag{2.1}$$

for every $1 \leq i_1 \neq \cdots \neq i_n \leq N$, where $N^{-[n]} = (N^{[n]})^{-1}$ and $N^{[n]} = N \cdots (N - n + 1)$ for $n \leq N$ ($N^{[0]} = 1$). Based on X_n, we may be interested in the estimation of the *population mean*

$$\bar{A}_N = N^{-1} \sum_{i=1}^{N} a_i \tag{2.2}$$

and the *population variance*

$$\sigma_N^2 = (N-1)^{-1} \sum_{i=1}^{N} [a_i - \bar{a}_N]^2, \qquad (2.3)$$

among other characteristics of the population. The optimal sample estimators (viz., Nandi and Sen, 1963) are given by

$$\bar{X}_n = n^{-1} \sum_{i=1}^{n} X_i \quad \text{and} \quad s_n^2 = (n-1)^{-1} \sum_{i=1}^{n} (X_i - \bar{X}_n)^2, \qquad (2.4)$$

respectively. Both these estimators are special cases of U-statistics, which may be introduced as follows.

For a symmetric *kernel* $g(X_1, \ldots, X_m)$ of *degree* $m \ (\geq 1)$, we define a (population) *parameter*

$$\theta_N = \theta(A_N) = N^{-[m]} \sum_{1 \leq i_1 \neq \cdots \neq i_m \leq N} g(a_{i_1}, \ldots, a_{i_m}), \qquad (2.5)$$

and the corresponding sample function, viz.,

$$U_n = n^{-[m]} \sum_{1 \leq i_1 \neq \cdots \neq i_m \leq n} g(X_{i_1}, \ldots, X_{i_m}) \qquad (2.6)$$

termed a *U-statistic*, is an optimal unbiased estimator of θ_N. We may note that for $m=1$, $g(x)=x$, $U_n = \bar{X}_n$, $\theta_N = \bar{A}_N$ and for $m=2$, $g(x,y) = \frac{1}{2}(x-y)^2$, $U_n = s_n^2$, $\theta_N = \sigma_N^2$. In fact, as $g(\cdot)$ is assumed to be symmetric in its $m \ (\geq 1)$ arguments, in (2.5) and (2.6), we may take $1 \leq i_1 < \cdots < i_m \leq N$ (and $1 \leq i_1 < \cdots < i_m \leq n$) and replace $N^{-[m]}$ (and $n^{-[m]}$) by $\binom{N}{m}^{-1}$ (and $\binom{n}{m}^{-1}$), where $\binom{p}{q} = p!/q!(p-q)!$). Note that the X_i are not independent random variables (r.v.). Nevertheless, they are symmetric dependent r.v.'s. For any given A_N, the (exact) sampling distribution of U_n may be obtained by direct enumeration of all possible $\binom{N}{n}$ samples of size n from A_N. Obviously, this process becomes prohibitively laborious for large N (and n). As such, there is a genuine need to provide suitable approximations to the large sample distribution when N and n are both large, though $\alpha = n/N$ (the sampling fraction) needs not be very small. In this context, the permutational central limit theorems (PCLT) play a vital role. For the particular case of $\bar{X}_n$, Madow (1948) initiated the use of PCLPT in finite population sampling and, since then, this has been an active area of fruitful research. For general U_n, the asymptotic normality result in SRS has been studied by Nandi and Sen (1963), with further generalizations due to Sen (1972), Krewski (1978), Majumdar and Sen (1978), and others.

In an asymptotic setup we coinceive of a sequence $\{A_N\}$ of populations and a sequence $\{n\}$ of sample sizes such that, as N increases,

$$\alpha_N = n/N \to \alpha, \quad 0 \leq \alpha < 1. \qquad (2.7)$$

Though theoretically the asymptotic theory is justified for N indefinitely large, in practice the asymptotic approximations workout quite well for N even moderately large.

For every N and h, $0 \leq h \leq m$, let

$$g_h^{(N)}(a_{i_1}, \ldots, a_{i_h}) = (N-h)^{-[m-h]} \sum_{Nh}^{*} g(a_{i_1}, \ldots, a_{i_m}) \tag{2.8}$$

where the summation $\sum_{Nh}^{*}$ extends over all distinct $i_{h+1}, \ldots, i_m$ over the set $\{1, \ldots, N\} \setminus \{i_1, \ldots, i_h\}$. Note that $g_0^{(N)} = \theta_N$ and $g_m^{(N)} \equiv g$. Then let

$$\bar{\zeta}_{h,N} = N^{-[h]} \sum_{1 \leq i_1 < \cdots < i_h \leq N} \{g_h^{(N)}(a_{i_1}, \ldots, a_{i_h})\}^2 - \theta_N^2, \tag{2.9}$$

for $h = 0, 1, \ldots, m$ (where $\bar{\zeta}_{0,N} = 0$). For the asymptotic theory, we assume that

$$\liminf_{N \to \infty} \bar{\zeta}_{1,N} > 0 \quad \text{and} \quad \limsup_{N \to \infty} \bar{\zeta}_{m,N} < \infty. \tag{2.10}$$

Then, as in Nandi and Sen (1963), we have, for every $n \geq m$,

$$\text{Var}(U_n) = E[U_n - \theta_N]^2$$

$$= m^2(n^{-1} - N^{-1})\bar{\zeta}_{1,N} + O((n^{-1} - N^{-1})^2). \tag{2.11}$$

Let us now assume that as N increases,

$$\left\{ \max_{1 \leq i \leq N} \left\{ g_1^{(N)}(a_i) - \frac{1}{N} \sum_{i=1}^{N} g_1^{(N)}(a_i) \right\}^2 \right\} / \{N\bar{\zeta}_{1,N}\} \to 0. \tag{2.12}$$

Also, let $[k]$ denote the largest integer $\leq k$, and let

$$Y_N(t) = Y_N(N^{-1}[Nt]) = [Nt](U_{[Nt]} - \theta_N)/\{m(N\bar{\zeta}_{1,N})^{1/2}\}, \quad t \in [0,1], \tag{2.13}$$

where for $t < m/N$, we let $Y_N(t) = 0$. Then $Y_N = \{Y_N(t), 0 \leq t \leq 1\}$ is well defined for every $N (\geq m)$. Finally, let $W^\circ = \{W^\circ(t), 0 \leq t \leq 1\}$ be a *Brownian bridge*, i.e., $W^\circ(t)$ is Gaussian with $EW^\circ(t) = 0$, $\forall 0 \leq t \leq 1$ and $EW^\circ(s)W^\circ(t) = s \wedge t - st$, $\forall s, t \in [0,1]$, where $a \wedge b = \min(a,b)$. Then, we have the following general result, discussed in detail in Sen (1981, Section 3.5):

Under (2.10) and (2.12), as N increases, the stochastic processes Y_N converges in distribution to W°.

An immediate corollary to this basic result is the following:

Under (2.10) and (2.12), as n increases, satisfying (2.7),

$$n^{1/2}(U_n - \theta_N)/\{m\bar{\zeta}_{1,N}^{1/2}\} \sim \mathcal{N}(0, 1-\alpha). \tag{2.14}$$

It is also interesting to note that this weak convergance (of $\{Y_N\}$ to W°) provides a very simple proof for the asymptotic normality result when n is itself a positive integer-valued r.v. Suppose that $\{v_N\}$ is a sequence of non-negative integer valued r.v.'s, such that as N increases,

$$N^{-1}v_N \xrightarrow{P} \beta, \quad 0 < \beta < 1. \tag{2.15}$$

Then, under (2.10) and (2.12),

$$N^{1/2}(U_{v_N} - \theta_N)/\{m\bar{\zeta}_{1,N}^{1/2}\} \sim \mathcal{N}(0, \beta^{-1} - 1). \tag{2.16}$$

This last result is useful in the case where size n is determined by some other considerations, so that it may be stochastic in nature.

Note that for $m = 1$ and $g(x) = x$, $\bar{\zeta}_{1,N} = (1 - N^{-1})\sigma_N^2$. In general, $\bar{\zeta}_{1,N}$ is an (estimable) parameter. Knowledge of $\bar{\zeta}_{1,N}$ is useful in providing a confidence interval for θ_N, based on (2.14) or (2.15). The following estimator, due to Nandi and Sen (1963), is a variant form of the usual *jackknifed estimator*. For each i ($= 1, \ldots, n$), let $U_{n-1}^{(i)}$ be the U-statistic based on $(X_1, \ldots, X_{i-1}, X_{i-1}, \ldots, X_n)$, and let

$$U_{n,i} = nU_n - (n-1)U_{n-1}^{(i)}, \quad i = 1, \ldots, n. \tag{2.17}$$

Let then

$$\tilde{S}_n^2 = (n-1)^{-1} \sum_{i=1}^n (U_{n,i} - U_n)^2. \tag{2.18}$$

Note that, for $m = 1$ and $g(x) = x$, $U_{n,i} = X_i$, $1 \leq i \leq n$, so that $\tilde{s}_n^2 = 2_n^2$, defined by (2.4), while in general we have (cf. Nandi and Sen, 1963), under (2.10) and (2.12),

$$|\tilde{S}_n^2 - m^2\bar{\zeta}_{1,N}| \xrightarrow{P} 0, \quad \text{as } n \text{ increases}. \tag{2.19}$$

The asymptotic results considered here extend easily to the case of more than one U-statistic. Further, we have so far studied the case of sampling without replacement. In the case of sampling with replacement we have $X_1, \ldots, X_n$ independent and identically distributed (i.i.d.) r.v., where

$$P\{X_1 = a_i\} = N^{-1} \quad \text{for } i = 1, \ldots, N. \tag{2.20}$$

As such, the classical central limit theorems and weak convergence results for U-statistics (viz., Hoeffding, 1948; Miller and Sen, 1972) repain applicable. In particular, in such a case, for (2.14), $1 - \alpha$ has to be replaced by 1, and, in (2.16), $\beta(1 - \beta)$ has to be replaced by β alone; (2.19) follows from Sen (1960a).

In simple random sampling with replacement (SRSWR), the classical estimators of the population mean and variance, considered in (2.4), are not optimal in the sense that there are other estimators (based on the distinct units in the sample) which have smaller variance (or risk with a convex loss function). In a SRSWR(N, n), let v_n be the number of distinct units (so that $1 \leq v_n \leq n$), and let $\bar{X}_{(v_n)}$ and $s^2_{(v_n)}$ be the sample mean and variance based on these distinct units. Then, it is known that $\bar{X}_{(v_n)}$ has a smaller risk than $\bar{X}_n$ and a similar result holds for the variance estimators, although $s^2_{(v_n)}$ is not an unbiased estimator of σ^2 (but the same can be made unbiased by introducing a multiplicative factor $c(n, v_n)$). The better performance characteristics of these estimators are mainly due to the Basu (1958) sufficiency of the number of distinct units in SRSWR, and a more comprehensive account of this is given in Chapter 11 (written by S. K. Mitra and P. K. Pathak). We may also refer to Sinha and Sen (1987) for some deeper results in this direction. We like to conclude this section with some discussions on the asymptotics in SRSWR based on the distinct units in the sample.

First, we may note that, in a SRSWR(N, n),

$$P\{v_n = k\} = N^{-n}\binom{N}{k}(\Delta^k 0^n) \quad \text{for } k = 1, \ldots, n, \tag{2.21}$$

where $\Delta^k a^q = (a+q)^k - \binom{k}{1}(a+q-)^k + \cdots + (-1)^k \binom{k}{k}a^q$ for $a \geq 0$, $q \geq 0$ and $k \geq 0$. As such, it is easy to verify that

$$E(v_n) = N\{1 - (1 - N^{-1})^n\} \quad \text{for every } n \geq 1, \tag{2.22}$$

$$\lim_{N \to \infty} (n/N) = \alpha, \quad 0 < \alpha < \infty \quad \Rightarrow \quad \lim_{N \to \infty} \{N^{-1}E(v_n)\} = 1 - e^{-\alpha}, \tag{2.23}$$

$$E(v_n^{-1}) = N^{-n} \sum_{k=1}^{N} (N - k + 1)^{n-1}$$

$$\Rightarrow \lim_{N \to \infty} \{NE(v_n^{-1})\} = (1 - e^{-\alpha})^{-1}. \tag{2.24}$$

If we put $Z_n = v_n/E(v_n)$, then from (2.22), (2.23) and (2.24), we obtain that both EZ_n and $E(Z_n^{-1})$ converge to 1 as n increases. On the other hand, Z_n is a positive valued r.v., and we know that for every positive x, $(x + x^{-1})/2 \geq 1$, where the strict equality holds only for $x = 1$. Thus, noting that $[EZ_n + EZ_n^{-1}]/2 \to 1$ as n increases, we immediately conclude from the above inequality that Z_n converges to 1, in probability, as $n \to \infty$. Consequently, we have for a SRSWR(N, n),

$$n/N \to \alpha, \quad 0 < \alpha < \infty \quad \Rightarrow \quad n^{-1} v_n \xrightarrow{P} (1 - e^{-\alpha})/\alpha. \tag{2.25}$$

In this context, we may note that

$$(1 - e^{-\alpha})/\alpha < 1 \quad \text{for every } \alpha > 0, \tag{2.26}$$

where for small values of α, the left hand side of (2.26) is close to 1. Further, we may note that by (2.21), the distribution of v_n is independent of the population units $\{a_1, \ldots, a_N\}$, and hence, given $v_n = k$ (≥ 1), the probability distribution of the distinct units (say, $X'_1, \ldots, X'_k$) is the same as in the SRSWR(N, k). Thus, given $v_n = k$, we are in a position to adapt all the asymptotic results for the classical situation in SRSWR(N, k). Finally, (2.25) ensures that $n^{-1} v_n \xrightarrow{P} (1 - e^{-\alpha})/\alpha$ as n (or N) increases with $n/N \to \alpha$ (>0). This last condition is parallel to (2.15), so that we can again use the central limit theorem for random sample sizes (and its ramifications for U-statistics, discussed in Miller and Sen, 1972), and conclude that the asymptotic results (on the distribution of U-statistics) for SRSWR(N, n) all extend smoothly for the distinct sample unit based U-statistics provided we replace n/N ($\simeq \alpha$) by $1 - (1 - N^{-1})^n$ ($\simeq 1 - e^{-\alpha}$). In view of (2.26), we may conclude that unless α is very small, the use of the distinct units only in SRSWR(N, n) generally leads to some increase in the efficiency of the estimators.

3. Some probability and moment inequalities for SRS

In SRS with replacement, the sample observations are independent and identically distributed random variables, so that the usual probability and moment inequalities holding for sampling from an infinite population also remain valid in this case. On the other hand, in SRS without replacement, the sample observations are no longer independent (but, exchangeable) random variables. In (2.11) and (2.14), we have observed that the dependence in SRS without replacement leads to a smaller variance for the sample mean or, in general, for U-statistics. This feature is generally shared by a general class of statistics and the related inequality is termed the *Hoeffding inequality*.

Let $X_1, \ldots, X_n$ be a sample in SRS without replacement from a finite population, and let $Y_1, \ldots, Y_n$ be a sample in SRS with replacement from the same population. Then, for any convex and continuous function $\phi(x)$, we have

$$E\phi(X_1 + \cdots + X_n) \leq E\phi(Y_1 + \cdots + Y_n). \tag{3.1}$$

We may refer to Hoeffding (1963) for a simple proof of (3.1). Rosen (1967) has extended the inequality in (3.1) for certain function other than convex, continuous $\phi(\cdot)$ and also for more general *symmetric sampling plans* which include the SRS with replacement as a particular case. A more general result in this direction is due to Karlin (1974). In his setup, $\phi(\cdot)$ needs to be a function of the sum of the X_i (or Y_i). Let $\phi(x_1, \ldots, x_n)$ be a function, symmetric in its n arguments, such that

$$\phi(a, a, x_3, \ldots, x_n) + \phi(b, b, x_3, \ldots, x_n) \geq 2\phi(a, b, x_3, \ldots, x_n), \tag{3.2}$$

for all $a, b, x_3, \ldots, x_n$. For all such $\phi(\cdot)$ and any symmetric sampling plan $\mathcal{S}$,

$$E\phi(X_1, \ldots, X_n) \leq E\phi(Y_1, \ldots, Y_n), \tag{3.3}$$

where the X_i are in SRS without replacement and the Y_i in the sampling plan $\mathscr{S}$. There are more general inequalities of this nature in Karlin etc. (1974), and they should be of considerable theoretical interest. In this context, we may define a random replacement sampling scheme (RRSS), $R(p_1, \ldots, p_{n-1})$, by a simple random sampling of size n from a population of size N when at stage i, the probability of replacing the observed unit is p_i, for $i = 1, \ldots, n-1$. If all the p_i are equal to 1, we ave the SRS with replacement, while if all the p_i are equal to 0, we have the SRS without replacement. Karlin (1974) conjectured that for $N \geq n$ and for the class of functions ϕ, satisfying (3.2),

$$E_{R(p_1, \ldots, p_{n-1})} \phi(X_1, \ldots, X_n) \leq E_{R(p_1', \ldots, p_{n-1}')} \phi(X_1, \ldots, X_n)$$

if and only if $p_i \leq p_i'$ for every $i = 1, \ldots, n-1$. \hfill (3.3a)

Though this inequality is true for the particular sampling schemes in (3.3) (i.e., SRS without and with replacements), there remains the question of its validity for general RRSS. Very recently, Bhandari (1984) and Kraft and Schaefer (1984) have shown that the 'only if' part of (3.3a) is false, and the falsity of the 'if' part is due to Schaefer (1985). However, when one confines oneself to certain class of non-negative Schur-convex functions which are less general than the ones in (3.2), then (3.3a) and some other related inequalities hold. We may refer to Bhandari (1985) for some of these related works. In passing, we may remark that the basic condition in (3.2) may not hold in general for functions of U-statistics. To illustrate this point, let us consider the special case of the sample variance when $n = 2$. Here, $U_2 = (X_1 - X_2)^2/2$, so that $EU_2 = \sigma_N^2$. Then for $\phi(x_1, x_2) = (\frac{1}{2}(x_1 - x_2)^2 - \sigma_N^2)^2$ we have $\phi(a, a) = \phi(b, b) = \sigma_N^4$ and $\phi(a, b) = (\frac{1}{2}(a - b)^2 - \sigma_N^2)^2$. Thus, whenever $\frac{1}{2}(a - b)^2 > 2\sigma_N^2$, $2\phi(a, b) > \phi(a, a) + \phi(b, b)$. Since $E[\frac{1}{2}(X_1 - X_2)^2] = \sigma_N^2$, in general $\frac{1}{2}(X_1 - X_2)^2$ exceeds $2\sigma_N^2$ with a positive probability, and hence (3.2) does not hold. The same picture holds for $n \geq 2$. Nevertheless, for convex functions of U-statistics, we have some simple moment inequalities (due to Hoeffding, 1963). Consider (as in Section 2) a kernel $g(X_1, \ldots, X_m)$ of degree m (≥ 1), and for every $n \geq m$, define $k_n = [n/m]$ and let

$$U_n^* = k_n^{-1} \sum_{i=1}^{k_n} g(X_{(i-1)m+1}, \ldots, X_{im}). \tag{3.4}$$

Let $\mathscr{F}_n$ be the sigma-field generated by the ordered collection of $X_1, \ldots, X_n$, so that we have $U_n = E[U_n^* | \mathscr{F}_n]$ and as a result, by the Jensen inequality, for any convex function ϕ (for which the expectation in (3.5) exists),

$$E[\phi(U_n)] \leq E[\phi(U_n^*)]. \tag{3.5}$$

On the other hand, for U_n^* the inequality in (3.1) is directly applicable, so that we have, for every continuous and convex ϕ,

$$E[\phi(U_n)] \leq E\left[\phi\left(\sum_{i=1}^{k_n} g(Y_{(i-1)m+1}, \ldots, Y_{im})/k_n\right)\right] \quad \forall n \geq m. \quad (3.6)$$

where the Y_i are defined as in (3.1). For the right hand side of (3.6), the usual moment inequalities are applicable, and these are therefore adaptable for U_n in SRS without replacement too.

In SRS without replacement the reverse martingale property of U-statistics (and hence sample means) has been established by Sen (1970), and this enables one to derive other probability inequalities, which will be briefly discussed here. In passing we may also note that by virtue of this reverse martingale property, for any convex ϕ, $\{\phi(U_n), m \leq n \leq N\}$ has the reverse sub-martingale property, so that suitable moment inequalities may also be based on this fact.

We define $V_n = \text{Var}(U_n)$ as in (2.11) and let $V_n^* = V_n - V_{n+1}$, for $n \geq m$. Also, let $\{c_k; k \geq m\}$ be a nondecreasing sequence of positive numbers. Then, we have the following (cf. Sen, 1970):

Whenever, for some $r \geq 1$, $E|U_n - \theta_N|^r$ exists (for $n \geq m$), for every $t > 0$ and $m \leq n' \leq N$,

$$P\{\max_{n \leq k \leq n'} c_k |U_k - \theta n| \geq t\}$$

$$\leq t^{-r}\{c_n^r E|U_n - \theta_N|^r + \sum_{i=n+1}^{n'} (c_i^r - c_{i-1}^r) E|U_i - \theta_N|^r\}, \quad (3.7)$$

so that, in particular, we have

$$P\{\max_{n \leq k \leq n'} c_k |U_k - \theta_N| \geq t\} \leq t^{-2}\{c_n^2 V_n + \sum_{i=n}^{n'-1} c_i^2 V_i^*\} \quad (3.8)$$

and

$$P\{\max_{n \leq k \leq N} |U_k - \theta_N| \geq t\} \leq t^{-2} V_n \quad \forall n \geq m. \quad (3.9)$$

For the particular case of sample means (or sums) i.e., kernels of degree 1, some related inequalities have also been studied by Serfling (1974). In this context, the following inequality (due to Sen, 1979b) is worth mentioning.

Let $\{d_{Ni}; 1 \leq i \leq N, N \geq 1\}$ be a triangular array of real numbers satisfying the normalizing constraints

$$\sum_{i=1}^{N} d_{Ni} = 0 \quad \text{and} \quad \sum_{i=1}^{N} d_{Ni}^2 = 1. \quad (3.10)$$

Also, let $q = \{q(t): 0 < t < 1\}$ be a continuous, nonnegative, U-shaped and square integrable function inside $I = [0, 1]$. Finally, let $Q = (Q_1, \ldots, Q_N)$ take on each permutation of $(1, \ldots, N)$ with the common probability $(N!)^{-1}$. Then

$$P\left\{\max_{1 \leq k \leq N-1} q(k/N) \left|\sum_{i=1}^{k} d_{NQ_i}\right| \geq 1\right\} \leq \int_0^1 q^2(t)\, dt. \quad (3.11)$$

Clearly, in SRS without replacement (3.11) may be used to provide a simultaneous (in k, $1 \leq k \leq N$) confidence band for θ_N by choosing q in an appropriate way. For a related inequality (exploiting the 4th moment but not the inherent martingale structure) we may refer to Hájek and Šidák (1967, p. 185):

$$P\left\{\max_{1 \leq k \leq n} \left|\sum_{i=1}^{k} d_{NQ_i}\right| \geq t\right\} \leq (n/N)[\max_{1 \leq i \leq N} d_{Ni}^2 + 3n/N]$$
$$\times t^{-4}(1 - n/N)^{-3}(1 + \varepsilon_N), \qquad (3.12)$$

where $\varepsilon_N \to 0$ as $N \to \infty$.

Generally, (3.7) with $r = 4$ (and $m = 1$) provides a better bound than (3.12). We may obtain even better bounds by exploiting the weak convergence results in Section 2, for large values of N. As in Sen (1972), we consider the case of general U-statistics (with the same notations as in Section 2), so that the case of sample means (or sums) can be obtained as a particular one. Note that by virtue of the weak convergence result, stated after (2.13), we have, for every $t > 0$ and n, $n/N \leq \alpha$ ($0 < \alpha \leq 1$),

$$\lim_{N \to \infty} P\{\max_{m \leq k \leq n} k|U_k - \theta_N| \geq tm[N\bar{\zeta}_{1,N}]^{1/2}\}$$
$$= P\{\sup_{0 \leq u \leq \alpha} |W^\circ(u)| \geq t\}, \qquad (3.13)$$

where $W^\circ = \{W^\circ(t), 0 \leq t \leq 1\}$ is a Brownian bridge. Noting that $W^\circ(s/(s+1)) = (s+1)^{-1}W(s)$, $s \geq 0$, where $W = \{W(t), t \geq 0\}$ is a standard Brownian motion process on $[0, \infty]$, we may rewrite the right hand side of (3.13) as

$$P\{\sup_{0 \leq u \leq \alpha/(1-\alpha)} |(u+1)^{-1}W(u)| \geq t\}. \qquad (3.14)$$

An upper bound for (3.14) is given by

$$P\{\max_{0 \leq u < \infty} |(u+1)^{-1}W(u)| \geq t\} = 2 \sum_{k=1}^{\infty} (-1)^{k+1} \exp(-2k^2 t^2). \qquad (3.15)$$

For small values of α (as is usually the case encountered in practice), we may get a better bound:

$$P\{\sup_{0 \leq u \leq \alpha/(1-\alpha)} |(1+u)^{-1}W(u)| \geq t\} \leq P\{\sup_{0 \leq u \leq \alpha/(1-\alpha)} |W(u)| \leq t\}$$
$$\leq 4P\{W(\alpha/(1-\alpha)) \geq t\} = 4[1 - \Phi(t(1/\alpha - 1)^{1/2})], \qquad (3.16)$$

where $\Phi(\cdot)$ is the standard normal d.f. In particular, for kernels of degree 1, (3.16) may be compared to (3.12), and, as $1 - \Phi(x)$ converges to 0 exponentially, as $x \to \infty$, usually (3.16) performs much better than (3.12). The same conclusion holds for the comparison between (3.8) and (3.16) (or (3.9) and (3.15)). We conclude

4. Jackknifing in finite population sampling

In SRS or other sampling plans, regression or other estimators, 'jackknifing' was mainly introduced to serve a dual purpose: To reduce the bias of estimators (which are typically of the non-linear form) and to provide an efficient (and asymptotically normally distributed) estimator of the sampling variance of the (jackknifed) estimator. In the same setup as in Section 2, for a general estimator $T_n = T(X_1, \ldots, X_n)$ (containing U_n as a special case), we may define the *pseudo values* $T_{n,i} = nT_n - (n-1)T_{n-1}^{(i)}$, $i = 1, \ldots, n$, as in (2.17). Then the jackknifed estimator is defined by

$$T_n^* = n^{-1}(T_{n,1} + \cdots + T_{n,n}), \tag{4.1}$$

and the (Tukey form of the) jackknifed variance estimator is given (as in (2.18)) by

$$S_n^2 = (n-1)^{-1} \sum_{i=1}^{n} (T_{n,i} - T_n^*)^2. \tag{4.2}$$

To motivate the jackknifed estimator, we may start with a possibly biased estimator T_n for which we may have

$$ET_n = \theta_N + n^{-1}a_1(N) + n^{-2}a_2(N) + \cdots, \tag{4.3}$$

where the $a_j(N)$ are real numbers depending possibly on the population size N and the set A_N. Using $n-1$ for n in (4.3) for each $T_{n-1}^{(i)}$ and (4.1), we obtain that under (4.3),

$$ET_n^* = \theta_n - a_2(N)/n(n-1) + \cdots = \theta_N + O(n^{-2}). \tag{4.4}$$

Thus, the bias of T_n is reduced from $O(n^{-1})$ to that of $O(n^{-2})$ for T_n^*. In addition to this important feature of 'bias reduction', the variance estimator S_n^2 also plays a very important tole in drawing statistical conclusions on θ_N. Since in this chapter, we are primarily concerned with the *asymptotics* in finite population sampling, we shall mainly restrict ourselves to the discussion of the large sample properties of T_n^* and S_n^2; hopefully, in some other chapter(s), there will be complementary discussions on other aspects of jackknifing.

Keeping in mind the ratio, regression and other estimators (which are all expressible as functions of some U-statistics), we conceive a general estimator T_n of the form $T_n = h(U_n)$ where $h(\cdot)$ is a smooth function and U_n is a vector of

U-statistics, defined as in Section 2. Also, we keep in mind the conditional (permutational) distribution generated by the $n!$ equally likely permutations of $X_1, \ldots, X_n$ among themselves, and define $\mathscr{F}_n$ as in after (3.4). Then it follows from the basic results in Majumdar and Sen (1978) that

$$T_n^* = T_n + (n-1)E[(T_n - T_{n-1})|\mathscr{F}_n] \quad \forall n > m, \tag{4.5}$$

$$S_n^2 = n(n-1)\operatorname{Var}[(T_n - T_{n-1})|\mathscr{F}_n] \quad \forall n > m. \tag{4.6}$$

Thus, for both the jackknifed estimator T_n^* and the variance estimator S_n^2, the inherent permutational distributional structure provides the access for the necessary modifications. This theoretical justification for jackknifing has been elaborately studied in Sen (1977).

To fix the notations, we let $\mu_N = EU_n$, and, in addition to (2.10) (in a matrix setup), we assume that

$$\sup_N E \| g(X_1, \ldots, X_m) \|^4 < \infty, \quad m = \max(m_1, \ldots, m_p), \tag{4.7}$$

where $g(\cdot)$ stands for the vector of kernels of degrees $m_1, \ldots, m_p$, respectively. Further, we assume that $h(u)$ has bounded second order (partial) derivatives (with respect to u) in some neighbourhood of μ_N and $h(\mu_N)$ is finite. Finally, let us define

$$\sigma_{Nn}^2 = E[(T_n^* - \theta_N)^2], \quad n \geq n_0, \quad \text{where } n_0 \, (\geq m) \text{ is finite}, \tag{4.8}$$

and assume that there exists a sequence $\{\sigma_N^2\}$ of positive numbers, such that

$$n \sigma_{Nn}^2 - [(N-n)/(N-1)] \sigma_N^2 \to 0 \quad \text{as } n \text{ increases},$$

$$\text{where } \varliminf \sigma_N^2 > 0. \tag{4.9}$$

Now, parallel to that in (2.13), we consider a stochastic process $Y_N = \{Y_N(t); 0 \leq t \leq 1\}$ by letting

$$Y_N(t) = Y_N(N^{-1}[Nt]) = [Nt](T_{[Nt]}^* - \theta_N)/\sigma_N, \quad \frac{m}{N} \leq t \leq 1, \tag{4.10}$$

where, for $t < m/N$, we complete the definition of $Y_N(t)$, by letting $Y_N(t) = 0$. Further, as in after (2.13), we define a Brownian bridge $W° = \{W°(t); 0 \leq t \leq 1\}$. Then we have the following result (viz., Majumdar and Sen, 1978):

> For the jackknifed estimator, under the assumed regularity conditions, Y_N converges in distribution (or law) to $W°$, as N increases. (4.11)

Further, under the same regularity conditions, $S_n^2 - \sigma_N^2$ strongly converges to 0 as

n increases; this strong convergence is in the sense that for every ε (>0) and δ (>0), there exists a positive integer $n_0 = n_0(\varepsilon, \delta)$, such that

$$P\{\max_{n_0 \leq n \leq N} |S_n^2 - \sigma_N^2| > \varepsilon\} < \delta, \quad N \geq n_0. \tag{4.12}$$

This strong convergence result enables us to replace in (4.10) σ_N by $S_{[Nt]}$, for every $t > 0$. Thus, if we denote such a studentized process by $Y_N^*(t)$, $t > 0$, then we conclude that, for every $\eta > 0$,

$$\{Y_N^*(t); t \in [\eta, 1]\} \text{ converges in law to } \{W^\circ(t); t \in [\eta, 1]\}$$
$$\text{as } N \text{ increases.} \tag{4.13}$$

In particular, it follows that, for any (fixed) α, $0 < \alpha \leq 1$, if $n/N \to \alpha$ as N increases, then

$$n^{1/2}(T_n^* - \theta_N)/S_n \text{ is asymptotically normal } (0, 1 - \alpha). \tag{4.14}$$

Further, if v_N is a nonnegative integer valued random variable such that $N^{-1} v_N$ converges in probability to α ($0 < \alpha \leq 1$), then

$$N^{1/2}(T_{v_N}^* - \theta_N)/S_{v_N} \text{ is asymptotically normal } (0, (1 - \alpha)/\alpha). \tag{4.15}$$

The last two results are very useful in setting up a confidence interval for the parameter θ_N or to test for a null hypothesis $H_0: \theta_N = \theta_0$ (specified). As a simple illustration, consider a typical *ratio-estimator* of the form

$$T_n = U_n^{(1)}/U_n^{(2)}, \quad U_n^{(j)} = n^{-1} \sum_{i=1}^n g_j(X_i), \quad j = 1, 2. \tag{4.16}$$

where the functions $g_1(\cdot)$ and $g_2(\cdot)$ may be of quite general form. In fact, we may even consider some U-statistics for $U_n^{(1)}$ and $U_n^{(2)}$ (of degrees ≥ 1). In such a setting, T_n is not generally an unbiased estimator of the population parameter $\theta_N = \mu_N^{(1)}/\mu_N^{(2)}$, though the $U_n^{(j)}$ may unbiasedly estimate the $\mu_N^{(j)}$, $j = 1, 2$. Typically, the bias of T_n is of the form in (4.3), and hence, jackknifing reduces the bias to the order n^{-2}. Further, here $h(a, b) = a/b$, so that

$$(\partial^2/\partial a^2) h(a, b) = 0, \quad (\partial^2/\partial a \partial b) h(a, b) = -b^{-2}$$

and

$$(\partial^2/\partial b^2) h(a, b) = 2b^{-2} h(a, b).$$

Consequently, whenever $\mu_N^{(2)}$ is strictly positive and finite, for finite θ_N, the regularity conditions are all satisfied, and hence (4.11) through (4.15) hold. For some specific cases, we may refer to Majumdat and Sen (1978) and Krewski (1978). The basic advantage of using (4.11), (4.12) and (4.13), instead of (4.14) or (4.15), is that these asymptotics are readily adoptable for sequential testing and estimation procedures. Further, the asymptotic inequalities discussed in the preceeding section also remain applicable for the jackknifed estimators. In particular, (3.13) through (3.16) also hold when we replace the U-statistics and their variances by the T_k^* and the Tukey estimator of their variances. Finally, the results are easily extendable to the case where the T_n are q-vectors, for some $q \geq 1$. In that case, instead of (4.11) or (4.13), we would have a *tied-down Brownian sheet* approximation (in law) and, instead of (4.12), we would have the strong convergence of the matrix of jackknifed variance–covariances. For (4.14) and (4.14), we would have an analogous result involving a multivariate normal distribution.

In the discussions so far, we have mainly confined ourselves to SRS (with or without replacement). Jackknifing is potentially useable in other sampling schemes as well, although the reverse martingale structure underlying the jackknifed versions may not generally hold in unequal probability sampling schemes. Moreover, the asymptotics on jackknifing considered here may not be totally appropriate for small or moderate sample sizes, particularly, in such unequal probability sampling plans. In this respect, the Tukey estimator of the variance is more vulnerable than the original jackknifed estimator, and in sample surveys, this variance estimation is of vital importance. In this respect, we may refer to the chapter on variance estimation (due to J. N. K. Rao) in this volume, while a significant amount of research work in this area is being conducted now. Related bootstrap methods are also under active consideration.

5. Estimation of population size: Asymptotics

The estimation of the total size of a population (of mibile individuals, such as the number of fish in a lake etc.) is of great importance in a variety of biological, environmental and ecological studies. Of the methods available for obtaining information about the size of such populations, the ones based on *capture, marking, release and recapture* (CMRR) of individuals, originated by Petersen (1896), have been extansively studied and adapted in practice. The Petersen method is a two-sample experiment and amounts to marking (or tagging) a sample of a given number of individuals from a closed population of unknown size (N) and then returning it into the population. The proportion of marked individuals appearing in the second sample estimates the proportion marked in the population, providing in turn, the estimate of the population size N. Schnabel (1938) considered a multi-sample extension of the Petersen method, where each sample captured commencing from the second is examined for marked members and then every member of the sample is given another mark before being returned to the population. For this method, the computations are simple, successive estimates enable the field worker to see his method as the work progresses and the method can be adapted for a wide range of capture conditions.

For the statistical formulation of the CMRR procedure, we use the following notations. Let

N = total population size (finite and unknown),
k = number of samples $\quad (k \geq 2)$,
n_i = size of the ith sample, $\quad i \geq 1$,
m_i = number of marked individuals in n_i, $\quad i = 1, \ldots, k$,
$u_i = n_i - m_i$, $\quad i = 1, \ldots, k$,
M_i = number of marked individuals in the population just before the ith sample is drawn (i.e., $M_i = \Sigma_{j=1}^{i-1} u_j$), $i = 1, \ldots, k$. Conventionally, we let $M_1 = u_1 = 0$ and $M_{k+1} = M_k + n_k - m_k = \Sigma_{j=1}^{k}(n_j - m_j)$. Now, the conditional distribution of m_i, given M_i and n_i, is given by

$$L_N^{(i)}(m_i|M_i, n_i) = \binom{M_i}{m_i}\binom{N - M_i}{n_i - m_i} \bigg/ \binom{N}{n_i}, \quad i = 2, \ldots, k, \tag{5.1}$$

so that the (*partial*) likelihood function is

$$L_N(n_1, \ldots, n_k) = \prod_{i=2}^{k} L_N^{(i)} = \prod_{i=2}^{k} \left\{ \binom{M_i}{m_i}\binom{N - M_i}{n - m_i} \bigg/ \binom{N}{n_i} \right\}. \tag{5.2}$$

Note that

$$L_N/L_{N-1} = N^{-(k-1)} \left\{ \prod_{i=2}^{k} (N - n_i) \right\} (N - M_2)/(N - M_{k+1}),$$

$$= N^{-(k-1)} \left\{ \prod_{i=1}^{k} (N - n_i) \right\} (N - M_{k+1}),$$

so that

$$L_N/L_{N-1} \gtreqless 1 \quad \text{according as} \quad (1 - N^{-1}M_{k+1}) \lesseqgtr \prod_{j=1}^{k}(1 - N^{-1}n_j). \tag{5.3}$$

Now, (5.3) provides the solution for the *maximum likelihood estimator* (MLE) of N. For the Petersen scheme (i.e., $k = 2$), (5.3) reduces to

$$L_N/L_{N-1} \gtreqless 1 \quad \text{according as} \quad N \lesseqgtr n_1 n_2/m_2, \tag{5.4}$$

so that $[n_1 n_2/m_2] = \hat{N}_2$ is the MLE of N. For $k \geq 3$, in general, (5.3) needs an iterative solution for locating MLE of N. Note that based on $L_N^{(i)}$, the MLE of N is given by $\hat{N}_i = [n_i M_i/m_i]$ for $i = 2, \ldots, k$. It is of natural interest to study the relationship between the MLE $\hat{N}$ (from (5.3)) and the $\hat{N}_j, j = 2, \ldots, k$, when $k \geq 3$. Before doing so we may note that, by virtue of (5.1),

$$P(m_i = 0|M_i, n_i) = \binom{N - M_i}{n_i} \bigg/ \binom{N}{n_i} > 0 \quad \text{for every } i = 2, \ldots, k,$$

so that the MLE $\hat{N}_i$ do not have finite moments of any positive order. To eliminate this drawback, we may proceed as in Chapman (1951) and consider the modified MLE

$$\check{N}_i = (n_i + 1)(M_i + 1)/(m_i + 1) - 1 \quad \text{for } i = 2, \ldots, k. \tag{5.5}$$

Asymptotically (as $N \to \infty$), both $\hat{N}_i$ and $\check{N}_i$ behave identically and hence this modification is well recommended. Using the normal approximation to the hypergeometric distribution, one readily obtains from (5.5) that

$$N^{-1/2}(\check{N}_2 - N) \text{ is asymptotically normal } (0, \gamma^2(\alpha_1, \alpha_2)), \tag{5.6}$$

whenever for same $0 < \alpha_1, \alpha_2 \leq 1$, $n_1/N \to \alpha_1$ and $n_2/N \to \alpha_2$ as $N \to \infty$, where

$$\gamma^2(a, b) = (1 - a)(1 - b)/ab \geq [(2 - a - b)/(a + b)^2], \quad 0 < a, b \leq 1, \tag{5.7}$$

and where the equality sign in (5.7) holds when $a = b$.

For the case of $k \geq 3$, a little more delicate treatment is needed for the study of the asymptotic properties of the MLE's as well as their interrelations. Using some martingale characterizations, such asymptotic studies have been made by Sen and Sen (1981) and Sen (1982a, b). First, it follows from Sen and Sen (1981) that a very close approximation N^* to the actual MLE $\hat{N}$ [in (5.3)] is given by the solution

$$N^* = \left[\sum_{s=2}^{k} \hat{N}_s m_s / (N^* - M_s)(N^* - n_s) \right] \bigg/ \left[\sum_{s=2}^{k} m_s / (N^* - M_s)(N^* - n_s) \right], \tag{5.8}$$

where the MLE $\hat{N}_i$ are defined as in after (5.4). Two other approximations, listed in Seber (1973), are given by

$$\tilde{N} = \left[\sum_{s=2}^{k} \hat{N}_s m_s / (\tilde{N} - M_s) \right] \bigg/ \left[\sum_{s=2}^{k} m_s / (\tilde{N} - M_s) \right], \tag{5.9}$$

and

$$\tilde{\tilde{N}} = \left[\sum_{s=2}^{k} \hat{N}_s m_s \right] \bigg/ \left[\sum_{s=2}^{k} m_2 \right]. \tag{5.10}$$

$\tilde{N}$ works out well when the n_j, $2 \leq j \leq k$ are all equal or $N^{-1} n_j$ are all small, while, (5.10) is quite suitable, when in addition, the $N^{-1} M_i$, $i = 2, \ldots, k$ are all small. For both (5.8) and (5.9), an iterative solution works out very well, and has been discussed is Sen and Sen (1981). Schumacher and Eschmeyer (1943) considered another estimator (pertaining to the same scheme):

$$\dot{N} = \left[\sum_{s=2}^{k} \hat{N}_s m_s M_s \right] \bigg/ \left[\sum_{s=2}^{k} m_s M_s \right], \tag{5.11}$$

which is also an weighted average of the Petersen estimators. Sen and Sen (1981) considered an alternative estimator

$$\hat{N} = \left[\sum_{s=2}^{k} \hat{N}_s(m_s/M_s)\right] \Big/ \left[\sum_{s=s}^{k} (m_s/M_s)\right]. \tag{5.12}$$

If we let

$$n_i = N\alpha_i \quad (0 < \alpha_i \leq 1) \quad \text{and} \quad \beta_i = \prod_{j=1}^{i}(1-\alpha_j), \quad i = 1, \ldots, k, \tag{5.13}$$

then, for the MLE $\hat{N}$ as well as the approximate MLE N^*, we have (viz., Sen and Sen, 1981)

$$N^{-1/2}(N^* - N) \text{ asymptotically normal } (0, \sigma^{*2}), \tag{5.13}$$

$$\sigma^{*2} = \left[\sum_{s=2}^{k} \alpha_s(1-\beta_{s-1})/\beta_s\right]^{-1}. \tag{5.14}$$

Parallel results for the other estimators are

$$N^{-1/2}(\tilde{N} - N) \text{ asymptotically normal } (0, \tilde{\sigma}^2),$$

$$\tilde{\sigma}^2 = \left[\sum_{s=2}^{k} \alpha_s(1-\alpha_s)(1-\beta_{s-1})/\beta_{s-1}\right] \left[\sum_{s=2}^{k} \alpha_s(1-\beta_{s-1})/\beta_{s-1}\right]^{-1}; \tag{5.15}$$

$$N^{-1/2}(\tilde{\tilde{N}} - N) \text{ asymptotically normal } (0, \tilde{\tilde{\sigma}}^2),$$

$$\tilde{\tilde{\sigma}}^2 = \left[\sum_{s=2}^{k} \beta_s^* \alpha_s(1-\alpha_s)(1-\beta_{s-1})\beta_{s-1}\right] \Big/ \left[\sum_{s=2}^{k} \alpha_s(1-\beta_{s-1})\right]^2,$$

$$\beta_s^* = 1 + \sum_{j=s+1}^{k} \alpha_j, \quad 2 = 2, \ldots, k-1, \quad \beta_k^* = 1; \tag{5.16}$$

$$N^{-1/2}(\dot{N} - N) \text{ asymptotically normal } (0, \dot{\sigma}^2),$$

$$\dot{\sigma}^2 = \left[\sum_{s=2}^{k}\left\{\left(\gamma_s + \sum_{j>s}\alpha_j\gamma_j\right)^2 \alpha_s(1-\alpha_s)\gamma_s(1-\gamma_s)\right\}\right] \Big/ \left[\sum_{s=2}^{k}\alpha_s\gamma_s^2\right]^2,$$

$$\gamma_s = 1 - \beta_{s-1} \quad \text{for } s = 2, \ldots, k; \tag{5.17}$$

$$N^{-1/2}(\ddot{N} - N) \text{ asymptotically normal } (0, \ddot{\sigma}^2),$$

$$\ddot{\sigma}^2 = \left(\sum_{s=2}^{k}\alpha_s\right)^{-1}\left[\sum_{s=2}^{k}\left(\gamma_s^{-1} + \sum_{j>s}\alpha_j/\gamma_j\right)^2\alpha_s(1-\alpha_s)\gamma_s(1-\gamma_s)\right]. \tag{5.18}$$

It follows from Sen and Sen (1981) that $\tilde{\sigma}$, $\tilde{\tilde{\sigma}}$, $\dot{\sigma}$, $\ddot{\sigma}$ are all greater than or equal to σ^*, where $\tilde{\sigma} = \sigma^*$ iff the α_s $(2 \leq s \leq k)$ are equal, while in the other three cases an approximate equality sign holds when the α_s are all small. Numerical comparison of these asymptotic variances reveals that, over the entire domain of variation of the α_s, none of the estimators $\tilde{N}$, $\hat{N}$ and $\dot{N}$ is uniformly better than the others. In (5.1) and (5.2), we have considered the so called sampling without replacement scheme. If we draw the 2nd, ..., kth samples with replacement, we need to replace the hypergeometric distributions in (5.1)–(5.2) by the corresponding binomial distributions, and this will lead to some simplifications in the formulae for the asymptotic variances.

In many situations when the n_j are very small compared to N, the m_j are also very small (may even be equal to 0 with a positive probability). This may push up the variability of the estimators considered earlier. For this reason, often an *inverse sampling* scheme is recommended. In this setup, at the sth stage, the sample units are drawn one by one until a preassigned number m_s of the marked units appear, so that the sample size n_s is a random variable while m_s is fixed in advance for $s = 2, \ldots, k$. For this inverse sampling scheme, parallel to (5.1), we have

$$L_N^{(i)}(n_i | M_i, m_i) = \binom{N}{n_i - 1}^{-1} \binom{M_i}{m_i - 1} \binom{N - M_i}{n_i - m_i}$$
$$\times \{(M_i - m_i + 1)/(N - n_i + 1)\}$$
$$= \left\{ m_i \binom{M_i}{m_i} \binom{N - M_i}{n_i - m_i} \right\} \Big/ \left\{ n_i \binom{N}{n_i} \right\}, \quad i = 2, \ldots, k,$$
(5.19)

and (5.2) can be modified accordingly. Note that (5.3) and (5.4) are not affected, so that the MLE remains the same. It follows from Bailey (1951) that $\check{N}_i = (M_i + 1)n_i/m_i - 1$, $i = 2, \ldots, k$, are unbiased estimators of N. Note that the exact variance of $\check{N}_2$ is equal to $(n_1 - m_2 + 1)(N + 1)(N - n_1)/m_2(n_1 + 2)$, so that on letting $m_2 = \alpha_1^* \cdot \alpha_1 N$ we have, parallel to (5.6)–(5.7), that

$$N^{-1/2}(\check{N}_2 - N) \text{ is asymptotically normal } (0, (1 - \alpha_1)(1 - \alpha_1^*)/\alpha_1 \alpha_1^*).$$
(5.20)

Note that, in (5.20), α_1^* plays the same role as α_2 in (5.6)–(5.7). With a similar modification for the other m_s the results considered earlier for the direct sampling scheme all go through for the inverse sampling scheme also (when N is large); the main advantage of this inverse sampling scheme is that the estimates have finite moments of positive orders, although the amount of sampling (i.e., $n_2 + \cdots + n_k$) is not predetermined (but is a random variable).

Inverse sampling schemes are the precursors of *sequential sampling tagging* considered by Chapman (1952), Goodman (1953), Darroch (1958) and others. Darling and Robbins (1967) and Samuel (1968) have studied some related problems on *stopping times* arising in sequential sampling tagging for the estimation of the population size N, and the asymptotic theory plays a vital role in this context. Lack of stochastic independence of the random variables at successive stages of drawing and nonstationarity of their marginal distributions call for a nonstandard approach for rigorous study of the asymptotic properties of the MLE of N in a multi-stage or sequential sampling procedure. Using a suitable martingale characterization, this asymptotic theory has been developed in Sen (1982a, b), and is presented below.

Individuals are drawn randomly one by one, marked and released before the next drawing is made. Let M_k be the number of *marked individuals* in the population just before the kth drawal, for $k \geq 1$. Thus, $M_0 = M_1 = 0$, $M_2 = 1$, $M_{k+1} \geq M_k$ for every $k \geq 1$, and $M_{k+1} = M_k + 1 - X_k$, $k \geq 1$, where, for every k (≥ 1), X_k is equal to 1 or 0 according as the kth drawal yields a marked individual or not. Now, the conditional probability function for X_k given $X_1, \ldots, X_{k-1}$ is

$$f_k(X_k | X_1, \ldots, X_{k-1}) = N^{-1} M_k^{X_k} (N - M_k)^{1 - X_k}, \quad k > 1,$$

so that at the nth stage, the (partial) likelihood function is given by

$$L_n(N) = \prod_{k=2}^{n} f_k(X_k | X_1, \ldots, X_{k-1}) = N^{-(n-1)} \prod_{k=2}^{n} \{M_k^{X_k} (N - M_k)^{1 - X_k}\}. \tag{5.21}$$

Note that

$$(\partial / \partial N) \log L_n(N) = \sum_{k=2}^{n} (1 - X_k)/(N - M_k) - (n - 1)/N. \tag{5.22}$$

The summands in (5.22) are neither independent nor identically distributed random variables. Nevertheless, they lead to a simple martingale-difference structure, with the asymptotic theory built in. The MLE $\hat{N}_{Sn}$ of N, based on $L_n(N)$, is a solution of (5.22) (equated to 0), and one is interested in the asymptotic behaviour of the partial sequence $\{\hat{N}_{Sn}; n \leq n^*\}$, where n^* is large, and the sequence is suitably normalized. For this purpose, we define

$$Z_n^*(N) = \sum_{k=2}^{n} (N - M_k)^{-1}(n - 1), \quad n \geq 2, \quad Z_0^*(N) = Z_1^*(N) = 0, \tag{5.23}$$

and, for every N, n, such that $n = [N\alpha]$, for some $\alpha > 0$, we define

$$n(t) = \max\{k: Z_k^*(N) \leq t Z_n^*(N)\}, \quad 0 \leq t \leq 1. \tag{5.24}$$

Then, for each (N, n) and every ε, $0 < \varepsilon < 1$, we may consider a stochastic process $W_{Nn}^{*\varepsilon} = \{W_{Nn}^*(t), \varepsilon \leq t \leq 1\}$, by letting

$$W_{Nn}^*(t) = N^{-1/2}(\hat{N}_{Sn(t)} - N)(e^\alpha - \alpha - 1)^{1/2}, \quad \varepsilon \leq t \leq 1. \tag{5.25}$$

Further, defining the standard Brownian motion process $W = \{W(t); t \in [0, 1]\}$ as in before (3.14), we let $W^{*\varepsilon} = \{W^*(t) = t^{-1}W(t), \varepsilon \leq t \leq 1\}$. Then, we have the following: For every ε, $0 < \varepsilon < 1$, as N increases,

$$W_{Nn}^{*\varepsilon} \text{ converges in law to } W^{*\varepsilon}, \text{ whenever } n = [N\alpha] \text{ for some } \alpha > 0. \quad (5.26)$$

A direct consequence of (5.26) is that whenever $n = [N\alpha]$, for some $\alpha > 0$,

$$N^{-1/2}(\hat{N}_{Sn} - N) \text{ is asymptotically normal } (0, (e^\alpha - \alpha - 1)^{-1}). \quad (5.27)$$

Further, if $\{v_n\}$ is any sequence of positive integer valued random variables such that $n^{-1}v_n \to 1$ in probability as n increases, then we have

$$N^{-1/2}(\hat{N}_{Sv_n} - N) \text{ is asymptotically normal } (0, (e^\alpha - \alpha - 1)^{-1}). \quad (5.28)$$

We are now in a position to compare (5.6) and (5.27), where we put $\alpha = \alpha_1 + \alpha_2$. By virtue of (5.7), the asymptotic variance in (5.6) is a minimum when $\alpha_1 = \alpha_2$. Comparing this minimum value with (5.27), we conclude that the asymptotic relative efficiency (ARE) of the two-sample Petersen estimator (for $\alpha_1 = \alpha_2$) with respect to the sequential estimator is given by

$$E(P, S) = \alpha^2/\{(2 - \alpha)^2 (e^\alpha - \alpha - 1)\}. \quad (5.29)$$

As α goes to 0, (5.29) converges to $\frac{1}{2}$, so that for small values of α, the Petersen estimator is about 50% efficient compared to $\hat{N}_{Sn}$. On the other hand, as α increases, $E(P, S)$ also increases, and in fact, for $\alpha \geq 0.7657$, (5.29) exceeds 1 and it can be quite large when α is close to 2. However, in all practical situations, α is generally quite small and hence the sequential estimator can be recommended with full confidence. From the operational point of view, often sequential schemes are not very practical, and hence the Petersen estimator may be used.

In the contest of sequential estimation of the total size of a finite population, the following *urn model* arises typically. Suppose that an urn contains an unknown number N of white balls and no others. We repeatedly draw a ball at random, observe its colour and replace it by a black ball, so that before each draw, there are N balls in the urn. Let W_n be the number of white balls observed in the first n drawals. Note that W_k is nondecreasing in k, $W_k \leq k$, for every $k \geq 1$ and $W_0 = 0$, $W_1 = 1$. For every $c > 0$, consider a *stopping variable*

$$t_c = \inf\{n: n \geq (c + 1)W_n\}. \quad (5.30)$$

Note that t_c can take on only the values $[(c + 1)k]$ for $k = 1, 2, \ldots$, and $W_{t_c} = m$ whenever $t_c = [m(c + 1)]$. In this situation, one is not interested in the study of the asymptotic properties of the MLE $\hat{N}_{St_c}$, but also of the standardized form of the stopping variable t_c. Samuel (1968) made some conjectures, and general results in this direction are due to Sen (1982b).

We may note that $W_n = W_{n-1} + w_n$, where w_n is equal to 1 or 0 according as the ball appearing at the nth draw is white or not for $n \geq 1$; $W_0 = w_0 = 0$. For every K $(0 < K < \infty)$ and N, we consider a stochastic process $Z_N = \{Z_N(t), t \in [0, K]\}$ by letting

$$Z_N(t) = N^{-1/2}(W_{[Nt]} - N(1 - (1 - N^{-1})^{[Nt]})), \quad t \in [0, K], \quad (5.31)$$

where $[s]$ denotes the largest integer contained in s. Also, let $Z = \{Z(t), t \in [0, K]\}$ be a Gaussian process with 0 drift and covariance function

$$EZ(s)Z(t) = e^{-t}\{1 - (1 + s)e^{-s}\} \quad \text{for } 0 \leq s \leq t \leq K. \quad (5.32)$$

Then, as N increases, Z_N converges in law to Z. Since $(1 - N^{-1})^n$ is close to $e^{-n/N}$, this suggests that a convenient estimator of N (based on n draws) is given by the solution (N_n^*) of the equation $W_n = N(1 - e^{-n/N})$, and the asymptotic properties of this estimator can then be studied by incorporating the convergence of Z_N to Z (in law). For the associated stopping time, we now define

$$I^* = [a, b] \quad \text{where } 0 < a < b < 1. \quad (5.33)$$

Also, for every $m \in I^*$, we define t_m^* as the solution of the equation

$$m = (1 - e^{-t_m^*})/t_m^*, \quad m \in I^*. \quad (5.34)$$

Note that $mt_m^* \leq 1$ for every $m \in [0, 1]$, and $t_1^* = 0$ and t_m^* monotonically goes to ∞ as m moves from 1 to 0. For every N, we consider a stochastic process $Y_N = \{Y_N(m), m \in I^*\}$ by letting

$$\tau_{Nm} = \inf\{n (\geq 1): mn \geq W_n\} \text{ and}$$
$$Y_N(m) = N^{-1/2}(\tau_{Nm} - Nt_m^*), \quad m \in I^*. \quad (5.35)$$

Note that $W_n (= W_{Nn})$ depends on N as well and m plays the role of $(1 + c)^{-1}$ in (5.30). Let then $Y = \{Y(m), m \in I^*\}$ be a Gaussian process on I^* with 0 drift and covariance function

$$EY(m)Y(m') = e^{-t_m^*}\{1 - (1 + t_{m'}^*)e^{-t_{m'}^*}\}/\{(m - e^{-t_m^*})(m' - e^{-t_{m'}^*})\},$$
$$m \geq m'. \quad (5.36)$$

Then, as N increases, Y_N converges in law to Y. This convergence result, in turn, provides the asymptotic normality of $Y_N(m)$ for every fixed m $(\in I^*)$ as well as for any sequence $\{m_n\}$ of positive random variables for which $m_n \to m$ $(\in I^*)$, in probability, as n increases. For m very close to 1 (i.e., c in (5.30) very close to 0), Poisson approximations for τ_{Nm}, suggested by Samuel (1968), works out well.

In the two or multi-sample capture–recapture model and, more critically, in the sequential tagiing scheme, there are certain basic assumptions which may not always match the practical applications. For example, effects of migration need to be taken into account when sampling is conducted over a period of time and new individuals may enter into the scheme as well as some existing ones may exit. Also, the catchability of an individual in the tagging scheme may depend on some other characteristics. Moreover, once caught, an individual may develop some trap-shyness or trap-addictions, so that at the subsequent stage(s), the capture-probabilities are affected. Farm (1971) considered the asymptotic normality in a capture-recapture problem when catchability is affected by the tagging procedure. Seber (1973) studied the robustness of CMRR procedures against possible departures from these basic homogeneity assumptions. For some CMRR models allowing some relaxations of these basic assumptions, large sample theory has been neatly developed in Rosen (1979). The basic problem with this development is that there are so many unknown parameters involved in the final structure that from the statistical inferential point of view, there is little encouragement in their possible adoptions.

For the estimation of the characteristics of rare animals in sampling on repeated occasions, the CMR technique may be fruitfully incorporated to provide better estimators. This has recently been studied by Sen and Sen (1986). In this setup, the population size (N) and the population mean (on a characteristic) are both unknown, as is usually the case with mobile populations which are rare. An initial sample of size n_1 is selected; these units are marked, measured and released, and allowed to mix with the rest of the population. Asqsuming a closed population, a second sample is drawn without replacement in the usual inverse sampling scheme, so that the sample size n_2 (random) is just enough to yield a prescribed number ($m < n_1$) of previously marked animals. Note that on these m marked units, we have the measurements on both the occasions, while on the $u = n_2 - m$ unmarked units, we have only the measurements at the second occasion. The hypergeometric law in (5.19) (for $i = 2$) applies to the second sample size n_2 and this also provides a suitable estimator of N, as has already been discussed after (5.19). While an estimator of the mean characteristic (on the second occasion) may easily be obtained from the n_2 measurements on the second occasion, in view of the fact that this scheme ignores the information contained in the earlier measurements on the matched units as well as the units in the first sample, a better estimator may be obtained by incorporating this additional information in the estimation rule. This has been treated in an asymptotic setup in Sen and Sen (1986). Separate estimates of the population mean (on the second occasion) may be obtained from the matched units and the unmatched ones. These estimators are generally correlated. On the top of that for the m matched units, granted the measurements on both the occasions, one may use the classical regression method to provide a 'regression estimator' of the mean on the second occasion which would generally have a smaller sampling variance than the marginal estimator ignoring the information on the measurements on the first occasion. The classical weighted least squares method may then be used to combine these two (correlated) estimators resulting in a smaller sampling variance. However, in this weighted least squares method the weights depend on the variance–covariances of the individual estimators which in turn depend on the unknown population size and other parameters. Hence, a variant form of this method is used wherein estimates of N and the other parameters are used to determine these weights. It has been shown by Sen and Sen (1986) that in an asymptotic setup, inverse sampling without replacement (at the second stage) is generally more efficient than simple random sampling based on the second sample only (when the expected sample sizes are comparable). The relative efficiency picture (in the asymptotic case) under diverse setups has also been studied by them. Since N is not known, if one wants to estimate the population total (on the second occasion), instead of the mean, the resulting estimator being the product of two random variables will have generally higher variance. In fact when the sampling fractions $\alpha = n_1/N$ and $\beta = m/n_1$ are small (viz., $(\alpha\beta)^{-1}$ is large), there may be a considerable increase in the sampling variance of the estimator of the population total when N is unknown (and is estimated by the CMR technique). Of course, in the estimation of the population total, when N is large, a more appropriate measure of the sampling variability is given by the usual coefficient of variation, and in the light of this measure, the picture looks quite good.

In the above problem, we are assuming that the population is closed, so that the size remains the same on both occasions. However, the characteristics may change (in measurement) from one occasion to the other. This is typically the case where the units may grow over time. In such a case, the CMR technique may not only be used (as in before) to provide the estimate of the mean characteristic on the second occasion, but also of the mean change of the characteristic over the time-period spanned by the two occasions. Though a similar asymptotic theory works out here, one may simultaneously consider the *change* and the *mean value at the second occasion*, and instead of the sampling variance, consider the corresponding generalized variance (in the light of which the asymptotic relative efficiency pictures may be studied on parallel lines).

6. Sampling with varying probabilities: Asymptotics

Hansen and Hurwitz (1943) initiated the use of unequal selection probabilities leading to more efficient estimators of the population total. If N and n stand for the number of units in the population and sample, respectively, and if $Y_1, \ldots, Y_N$ and $y_1, \ldots, y_n$ denote the values of these units in the population and sample respectively, then, one may consider the following sampling with replacement scheme. Let $P = (P_1, \ldots, P_N)$ be positive numbers which are normalized in such a way that $P_1 + \cdots + P_N = 1$. Typically, one may consider a measure S_i of the size of the ith unit in the population and set $P_i = S_i/(\Sigma_{i=1}^{N} S_i)$ for $i = 1, \ldots, N$. Now, corresponding to the sample entries $y_1, \ldots, y_n$, the associated P's are denoted by $p_1, \ldots, p_n$, respectively. Here, sampling is made with replacement and the jth unit in the population is drawn with the probability P_j for $j = 1, \ldots, N$. Then, the Hansen–Hurwitz estimator of the population total $Y = Y_1 + \cdots + Y_N$

is

$$\hat{Y}_{\text{HH}} = n^{-1}(y_1/p_1 + \cdots + y_n/p_n). \tag{6.1}$$

This estimator is unbiased and its sampling variance is given by

$$\text{Var}(\hat{Y}_{\text{HH}}) = n^{-1}\left\{\sum_{i=1}^{N} Y_i^2/P_i - Y^2\right\}$$

$$= (2n)^{-1} \sum_{1 \leq i \neq j \leq N} P_i P_j \{Y_i/P_i - Y_j/P_j\}^2. \tag{6.2}$$

We may further note that

$$S_{n\text{HH}}^2 = [n(n-1)]^{-1} \sum_{i=1}^{n} (y_i/p_i - \hat{Y}_{\text{HH}})^2$$

$$= [2n^2(n-1)]^{-1} \sum_{1 \leq i \neq j \leq N} \{y_i/p_i - y_j/p_j\}^2 \tag{6.3}$$

is an unbiased estimator of $\text{Var}(\hat{Y}_{\text{HH}})$. Since sampling is made with replacement and the y_i/p_i are independent with mean Y and variance $\sum_{i=1}^{N} Y_i^2/p_i - Y^2$ ($= \sigma_{N\text{HH}}^2$, say), standard large sample theory is adoptable to verify that as n increases,

$$nS_{n\text{HH}}^2/\sigma_{N\text{HH}}^2 \text{ converges to one, in probability}, \tag{6.4}$$

$$n^{1/2}(\hat{Y}_{\text{HH}} - Y) \text{ is asymptotically normal } (0, \sigma_{N\text{HH}}^2), \tag{6.5}$$

so that, by (6.4) and (6.5),

$$n^{1/2}(\hat{Y}_{\text{HH}} - Y)/S_{n\text{HH}} \text{ is asymptotically normal } (0, 1). \tag{6.6}$$

The situation becomes quite different when sampling is made without replacement. On one hand, one has generally more efficient estimators; on the other hand, the exact theory becomes so complicated that one is naturally inclined to rely mostly on the asymptotics. To encompass diverse sampling plans (without replacements), we identify the population with the set $\mathcal{N} = \{1, \ldots, N\}$ of natural integers and denote the sample by s. A sampling design may then be defined by the probabilities $p(s)$, $s \in \mathcal{S}$, associated with all possible samples. In particular, we let

$$\pi_i = P\{i \in s\} = \sum_{\{\text{all } s \text{ containing } i\}} p(s), \quad i = 1, \ldots, N. \tag{6.7}$$

These are termed the *first order inclusion probabilities*. Similarly, the *second order inclusion probabilities* are defined as

$$\pi_{ij} = P\{i, j \in s\} = \sum_{\{\text{all } s \text{ containing } (i,j)\}} p(s), \quad i \neq j = 1, \ldots, N. \tag{6.8}$$

The classical *Horvitz–Thompson* (1952) *estimator* of the population total Y is then expressible as

$$\hat{Y}_{HT} = \sum_{i \in s} (Y_i/\pi_i). \tag{6.9}$$

Various properties of this estimator are discussed in some other chapters, and, hence, we shall not repeat the discussion here. We shall mainly concentrate on the asymptotic theory. The sampling variance of this unbiased estimator of Y is

$$\text{Var}(\hat{Y}_{HT}) = \sum_{i=1}^{N} (\pi_i^{-1} - 1) Y_i^2 + \sum_{1 \leq i \neq j \leq N} (\pi_{ij}/\pi_i \pi_j - 1) Y_i Y_j. \tag{6.10}$$

When the number of units (n) in the sample s is fixed, an alternative expression for the variance in (6.10), due to Sen (1953) and Yates and Grundy (1953), is

$$\sum_{1 \leq i < j \leq N} (\pi_i \pi_j - \pi_{ij})(Y_i/\pi_i - Y_j/\pi_j)^2. \tag{6.11}$$

It is clear that if the Y_i are all (exactly or closely) proportional to the corresponding π_i, then (6.11) is (exactly or closely) equal to 0; this point advocates the choice of the π_i as proportional to the size of the units, and on that count, *'probability proportional to size'* (pps) sampling is quite a reasonable option.

Now, in the context of sampling (without replacement) with varying probabilities, various sampling designs have been considered by various workers. Among these, *rejective sampling* may be defined as in Hájek (1964) as sampling with replacement with drawing probabilities $\alpha_1, \ldots, \alpha_N$ at each draw, conditioned on the requirement that all drawn units are distinct. The α_i are positive numbers adding upto 1. As soon as one obtains a replication, one rejects the whole partially built up sample and starts completely new. In this scheme, the inclusion probabilities π_i can be computed, as in Hájek (1964), in terms of the α_i. A related sampling plan, known as the *Samford–Durbin sampling*, is defined in a similar manner, where the first unit in the sample is drawn from the population with the probabilities $\alpha_i^{(1)} = n^{-1} \pi_i$, $i = 1, \ldots, N$, and in the subsequent $(n-1)$ draws, one considers the drawing probabilities as $\alpha_i^{(*)} = \alpha \pi_i (1 - \pi_i)^{-1}$, $i = 1, \ldots, N$, where α is so selected that $\sum_{i=1}^{N} \alpha_i^{(*)} = 1$. Here also, a sample is accepted only if all selected units are distinct. For both these schemes, a rejection of the accumulating sample is made when at any intermediate tage, a repetition occurs. In a *successive sampling plan* one draws units one by one with drawing probabilities $P_1, \ldots, P_N$, and, if a replication occurs at any draw, that particular one is rejected, and the drawing is continued in this manner until one has the prefixed number n of distinct units in the sample. Following Rosen (1972, 1974), let $I_1, \ldots, I_n$ be the indices in the (random) order in which they appear in the sample of size n, and let

$$\Delta(r, n) = \text{Probability that item } r \text{ is included in the sample of size } n$$
$$\text{drawn according to the successive sampling plan,} \quad (6.12)$$

for $r = 1, \ldots, N$. For this scheme, the Horvitz–Thompson estimator in (6.9) reduces to the following:

$$\hat{Y}_{HT} = \sum_{i=1}^{n} Y_{I_i}/\Delta(I_i, n). \quad (6.13)$$

In passing, we may remark that if $P_1 = \cdots = P_N = N^{-1}$ the $\Delta(r, n)$ all reduces to n/N, so that (6.13) is given by $N(n^{-1} \sum_{i=1}^{n} Y_{I_i})$ and hence relates to the usual equal probability sampling scheme (without replacement). In the more general case where the P_i are not all equal, the $\Delta(r, n)$ can be obtained as in Rosen (1972) in terms of a set of inclusion probabilities. However, these expressions (as given below) are generally quite complicated and call for asymptotic considerations. For every n (≥ 1) and $r_1, \ldots, r_n$, $1 \leq r_1 \neq \cdots \neq r_n \leq N$, let

$$P(r_1, \ldots, r_n) = P_{r_1} \times \left\{ \prod_{k=2}^{n} P_{r_k} \left[1 - \sum_{j=1}^{k-1} P_{r_j} \right]^{-1} \right\}. \quad (6.14)$$

Then,

$$\Delta(r, n) = \sum_{j=1}^{n} \left\{ \sum_{(j)} P(r_1, \ldots, r_n) \right\}, \quad (6.15)$$

where the summation $\sum_{(j)}$ extends over all permutations of $(r_1, \ldots, r_n)$ over $(1, \ldots, N)$, subject to the constraint that $r_j = r$, $j = 1, \ldots, n$, for $r = 1, \ldots, N$.

The varying probability structure and the complications underlying the $\Delta(r, n)$ in (6.14)–(6.15) introduce certain complications in the study of the asymptotic distribution theory of the Horvitz–Thompson estimator (or other estimators available in the literature). Rosen (1970, 1972) considered an alternative approach (through the *coupon collector's problem*) and provided some deeper results in this context. To illustrate this approach, we first consider a coupon collector problem. Let

$$\Omega_N = \{(a_{N1}, P_{N1}), \ldots, (a_{NN}, P_{NN})\}, \quad N \geq 1, \quad (6.16)$$

be a sequence of coupon collector's situations, where the a_{Nj} and P_{Nj} are real numbers, the P_{Nj} are positive and $\sum_{j=1}^{N} P_{Nj} = 1$, $\forall N \geq 1$. Consider also a (double) sequence $\{J_{Nk}, k \geq 1\}$ of (row-wise) independent and identically distributed random variables, where, for each N (≥ 1), $k \geq 1$,

$$P\{J_{Nk} = s\} = P_{Ns} \quad \text{for } s = 1, \ldots, N. \quad (6.17)$$

Let then, for $k \geq 1$,

$$X_{Nnk} = \begin{cases} a_{NJ_{Nk}}/\Delta(J_{Nk}, n) & \text{if } J_{Nk} \notin \{J_{N1}, \ldots, J_{Nk-1}\}, \\ 0, & \text{otherwise}; \end{cases} \quad (6.18)$$

$$v_{Nm} = \inf\{n: \text{number of distinct } J_{N1}, \ldots, J_{Nn} = m\}, \quad m \geq 1. \quad (6.19)$$

Note that for each $N (\geq 1)$ the v_{Nm} are positive integer-valued random variables. Then, as Rosen (1970, 1972) has shown on identifying $\{Y_1, \ldots, Y_N; P_1, \ldots, P_N\}$ with Ω_N in (6.16),

$$Y_{HT} \stackrel{\mathcal{D}}{=} \sum_{k=1}^{v_{Nn}} X_{Nnk} = Z_{nv_{Nn}}, \quad \text{say}, \quad (6.20)$$

where $\stackrel{\mathcal{D}}{=}$ stands for equality in distributions. Now, for a coupon collector situation Ω_N in (6.16), $B_{Nn} = \sum_{k=1}^{n} a_{NJ_{Nk}}$ is termed the *Bonus sum after n coupons*, for $n \geq 1$. Thus, corresponding to the siruation Ω_N in (6.16), if we consider another situation $\Omega_{Nn}^* = \{(a_{N1}/\Delta(1, n), P_{N1}), \ldots, (a_{NN}/\Delta(N, n), P_{NN})\}$ and, as before, identify (a_{Nj}, P_{Nj}) with (Y_j, P_j), $j = 1, \ldots, N$, then, for a given n, $Z_{nv_{Nn}}$ in (6.20) is the bonus sum after v_{Nn} coupons in the collector's situation Ω_{Nn}^*. Thus, the asymptotic normality of (randomly stopped) bonus sums (for the reduced coupon collector's situation) provides the same result for the Horvitz–Thompson estimator. A similar treatment holds for many other related estimators in successive sampling with varying probabilities (without replacement). Towards this goal, we may note as in Rosen (1972), under some regularity conditions on the a_{Ni} as well as the P_{Ni}, as N increases,

$$\Delta(s, n) = 1 - \exp\{-P_{Ns}t(n)\} + o(N^{-1/2}), \quad s = 1, \ldots, N, \quad (6.21)$$

where the function $t(\cdot) = \{t(x), x \geq 0\}$, is defined implicitly by

$$N - x = \sum_{k=1}^{N} \exp(-t(x)P_{Nk}), \quad x \geq 0, \quad (6.22)$$

(and therefore, depends on $P_{N1}, \ldots, P_{NN}$). Given this asymptotic relation, we may write for every $s (= 1, \ldots, N)$,

$$a_{Ns}^* = a_{Ns}/\Delta(s, n) = (1 - \exp\{-P_{Ns}t(n)\})^{-1}a_{Ns} + o(N^{-1/2}). \quad (6.23)$$

It also follows from Rosen (1970) that under the same regularity conditions,

$$v_{Nn}/t(n) \text{ converges in probability to one}, \quad (6.24)$$

whenever n/N is bounded away from 0 and 1. Consequently, if we define the bonus sum for the reduced coupon collector's situation by $B_{Nnk}^* = \sum_{i=1}^{k} a_{NJ_{Ni}}^*$, $k \geq 1$, then, we need to verify that

(i) the normalized version of $B^*_{Nnt(n)}$ is asymptotically normal, and

(ii) $n^{-1/2} \max\{|B^*_{Nnk} - B^*_{Nnt(n)}|: |k/t(n) - 1| \leq \delta\}$ converges in probability to 0 (the later condition is known in the literature as the Anscombe (1952) 'uniform continuity in probability' condition). A stronger result, which ensures both (i) and (ii), relates to the weak convergence of the partial sequence $\{(B^*_{Nnk} - EB^*_{Nnk})/\{\text{var}(B^*_{Nnt(n)})\}^{1/2}; k \leq t(n)\}$ which has been established by Sen (1979a) through a martingale approach. For simplicity of presentation, we consider the case of the original coupon collector's situation (and the same result continues to hold for the reduced situation too). Let us denote

$$\phi_{Nn} = \sum_{s=1}^{N} a_{Ns}[1 - \exp\{-nP_{Ns}\}], \quad n \geq 0, \qquad (6.25)$$

$$d^2_{Nn} = \sum_{s=1}^{N} a^2_{Ns} \exp\{-nP_{Ns}\}[1 - \exp\{-nP_{Ns}\}]$$

$$- \left(\sum_{s=1}^{N} a_{Ns} P_{Ns} \exp\{-nP_{Ns}\}\right)^2, \quad n \geq 0. \qquad (6.26)$$

Then, under the usual (Rosen-) regularity conditions, it follows that

$$d^2_{Nn} = O_e(n) \quad \text{whenever } n/N \text{ is bounded away from 0 and } \infty. \qquad (6.27)$$

Further, under the same regularity conditions,

$$(B_{Nn} - \phi_{Nn})/d_{Nn} \quad \text{is asymptotically normal } (0, 1). \qquad (6.28)$$

In fact, if we consider any finite number, q, say, of the sample sizes, i.e., $n_1, \ldots, n_q$, where the n_j all satisfy the condition that $0 < n_j/N < \infty$ for $j = 1, \ldots, n_q$, then (6.28) readily extends to the multinormal case. Further, if we define $W_N = \{W_N(t), t \in T\}$, where $T = [0, K]$ for some finite K, and

$$W_N(t) = N^{-1/2}(B_{N[Nt]} - \phi_{N[Nt]}), \quad 0 \leq t \leq K, \qquad (6.29)$$

then it follows from Sen (1979a) that W_N converges in law to a Gaussian function on T, and this ensures the *tightness* of W_N as well (so that the Anscombe condition holds). In passing, we may remark that if the a_{Ns} are all nonnegative, the bonus sum B_{Nn} is then nondecreasing in n, so that we may define

$$U_N(t) = \min\{k: B_{Nk} \geq t\} \quad \text{for every } t \geq 0. \qquad (6.30)$$

Then, $U_N(t)$ is termed the *waiting time to obtain the bonus sum t in the coupon collector's situation* Ω_N. Note that, by definition,

$$P\{U_N(t) > x\} = P\{B_{N[x]} < t\} \quad \text{for all } x, t > 0. \qquad (6.31)$$

Therefore, the asymptotic distribution of the normalized version of the waiting time can readily be obtained from (6.28) and, moreover, the weak convergence result on W_N also yields a parallel result for a similar stochastic process constructed from the $U_N(t)$.

Note that, by (6.7), (6.8), (6.14) and (6.15), $\pi_r = \Delta(r, n)$ for every $r = 1, \ldots, N$, while for every $r \neq s \ (= 1, \ldots, N)$,

$$\pi_{rs} = \sum_{1 \leq i \neq j \leq n} \left\{ \sum_{(ij)} P(r_1, \ldots, r_n) \right\}, \tag{6.32}$$

where the summation $\Sigma_{(ij)}$ extends over all permutations of $(r_1, \ldots, r_n)$ over $(1, \ldots, N)$ subject to the constraints that $r_i = r$ and $r_j = s$, for $i \neq j = 1, \ldots, n$. Further, the expression for the variance in (6.10) can be rewritten as

$$\sum_{i=1}^{N} \sum_{j=1}^{N} \gamma_{ij} Y_i Y_j / (\pi_i \pi_j) - Y^2 \quad \text{where } Y = \sum_{i=1}^{N} Y_i, \tag{6.33}$$

so that the expressions for the π_{ij} and π_i may be incorporated to evaluate (6.33). This, however, is quite complicated (in view of (6.14), (6.15) and (6.32)), and therefore, we proceed to obtain simpler expressions. We way note that for the reduced coupon collector's situation Ω_N^* and for n replaced by $t(n)$ we have, parallel to (6.26),

$$d_{Nt(n)}^{*2} = \sum_{s=1}^{N} a_{Ns}^{*2} \exp\{-t(n)P_{Ns}\} [1 - \exp\{-t(n)P_{Ns}\}]$$

$$- t(n) \left[\sum_{s=1}^{N} a_{Ns}^* P_{Ns} \exp\{-t(n)P_{Ns}\} \right]^2$$

$$= \sum_{s=1}^{N} Y_s^2 \exp\{-t(n)P_{Ns}\} / [1 - \exp\{-t(n)P_{Ns}\}]$$

$$- t(n) \left[\sum_{s=1}^{N} Y_s P_{Ns} \exp\{-t(n)P_{Ns}\} / (1 - \exp\{-t(n)P_{Ns}\}) \right]^2, \tag{6.34}$$

where we may note that the P_{Ns} are all specified numbers, so that, by (6.22), $t(n)$ is a known quantity. As a result, we obtain that as N increases and n/N is bounded away from 0 (and is finite too),

$$(\hat{Y}_{HT} - Y)/d_{Nt(n)}^* \text{ is asymptotically normal } (0, 1). \tag{6.35}$$

Further, if we take the sample observations as $y_j \ (= Y_{I_{Nj}}), j = 1, \ldots, n$, and denote by $P_{NI_{Nj}} = p_{Nj}, j = , \ldots, n$, we may set

$$U_{Nn}^{(1)} = n^{-1} \sum_{j=1}^{n} Y_j^2 \exp\{-t(n)p_{Nj}\} / [1 - \exp\{-t(n)p_{Nj}\}]^2, \tag{6.36}$$

$$U_{Nn}^{(2)} = n^{-1} \sum_{j=1}^{n} y_j p_{Nj} \exp\{-t(n)p_{Nj}\}/[1 - \exp\{-t(n)p_{Nj}\}]^2, \quad (6.37)$$

$$V_{Nn} = U_{Nn}^{(1)} - t(n)[U_{Nn}^{(2)}]^2. \quad (6.38)$$

Then it follows that, as n increases, $V_{Nn/d_{Nt(n)}^{*2}}$ converges in probability to 1, so that, in (6.35), $d_{Nt(n)}^{*}$ may be replaced by $V_{Nn}^{1/2}$.

Besides the sampling strategies considered so far, there are some others, considered elsewhere. Among these, mention should be made of one special approach proposed by Rao, Hartley and Cochran (1962). They considered a simple procedure of unequal probability sampling (without replacement) leading to an estimator having a smaller variance than in the case of sampling with replacement. Moreover, their procedure provides an unbiased sample estimator of variance that is always positive. Both single-stage and two-stage designs were considered by them. In the single-stage design, let p_t be the probability of drawing the tth unit in the first draw from the whole population for $t = 1, \ldots, N$. They suggested that the population of N units be first divided at random into n groups of sizes $N_1, \ldots, N_n$, respectively, where $N = N_1 + \cdots + N_n$. Within each group, a sample of size one is drawn with probabilities proportional to p_t (for t belonging to the set of indices in the ith group), and this is done independently for each of the n groups. Thus, if the tth unit falls in group i, the actual probability that it will be selected is p_t/π_i where π_i is the sum over all values of p_t for which t belongs to the set of indices in the ith group. If $y_1, \ldots, y_n$ denote the sampled units from the n groups, then the estimator of the population total is $\hat{Y}_n = \pi_1 y_1/q_1 + \cdots + \pi_n y_n/q_n$, where $q_1, \ldots, q_n$ refers to the particular values of the p_t for these chosen units. $\hat{Y}_n$ is an unbiased estimator of the population total, and the sampling variance of $\hat{Y}_n$ is given by

$$V(\hat{Y}) = [N(N-1)]^{-1} \left(\sum_{i=1}^{n} N_i^2 - N \right) \left(\sum_{t=1}^{N} Y_t^2/p_t - Y^2 \right),$$

where Y_t denotes the value associated with the tth unit in the population (and $Y = Y_1 + \cdots + Y_n$). It is clear from the above that this variance is a minimum when all the N_i are equal (i.e., $N/n = R$ is a positive integer and $N_1 = \cdots = N_n = R$). Thus, in actual practice, these N_i should be taken as close to each other as possible. An estimator of $V(\hat{Y})$ is given by

$$V(\hat{Y}) = \left(N^2 - \sum_{i=1}^{n} N_i^2 \right)^{-1} \left(\sum_{i=1}^{n} N_i^2 - N \right) \left(\sum_{i=1}^{n} \pi_i(y_i/q_i - \hat{Y})^2 \right).$$

In the case of the two-stage design, the tth primary unit ($t = 1, \ldots, N$) is composed of M_t second-stage units (sub-units), so that following the selection of n primary units as in the single-stage dasign, for the tth primary unit selected, one draws a sample of m_t subunits without replacements and with equal probabilities from the M_t subunits. Again, sampling is done independently for the different groups. The estimator of the population total Y is given by

$$\sum_{i=1}^{n}\left\{\pi_i(M_i/m_i)\left(\sum_{j=1}^{m_i}Y_{ij}\right)/q_i\right\}.$$

Parallel expressions for the sampling variance of this estimator and an estimator of this sampling variance have also been provided by Rao, Hartley and Cochran (1962). Thus, in either design, the procedure has the advantage of exact variance formulae (and their estimators) for any population size N and sample size n. When n, the number of groups, is large and the groupings are made randomly (as has been prescribed by them), the asymptotic theory (developed earlier) remains applicable under quite general regularity conditions. However, this random division of the N units into n groups may introduce some uncontrolled feature, although from the efficiency point of view it leads to improved estimators. Hartley and Rao (1962) have also considered an alternative sampling scheme with unequal probabilities and without replacements. The N units in the population are listed in a random order and their x_i (sizes) are cumulated; a systematic selection of n elements from a 'random start' is then made on the cumulation. They were able to provide an asymptotic variance formula for their estimator. Comparing the two procedures by Rao et al. (1962) and Hartley and Rao (1962), we see that the former enjoys the advantahe of exact variance formula for any population size, while the later assumes N to be large; but, in terms of the sampling variance, the former may lead to an estimator with a slightly larger variance than the later (in many situations). In the same vein, Krewski and Rao (1981) considered general stratified multistage design relating to a sequence $\{\Pi_L\}$ of finite populations (with L strata in Π_L) in which the primary sampling units are selected with replacement and in which *independent subsamples* are taken within those primary sampling units selected more than once. The asymptotic normality of both linear and nonlinear statistics are studied under the assumption that $L\to\infty$, and in the same setup, the consistency of the variance estimators obtained by using the linearization, jackknifing and balanced repeated replication methods is established. Because of the independence of the subsamples, standard asymptotic theory, discussed before, remains applicable in this context too. Systematic procedured (random or ordered) for sampling with varying probabilities were also considered by Madow (1949) and Hartley (1966), among others. Some of these procedures are discussed in detail in some other chapters of this volume, and hence, we shall not elaborate on their related asymptotics. Besides the systematic procedures, there are other procedures due to Narain (1951), Midzuno (1952), Yates and Grundy (1953) and Sen (1953), among others. Most of these procedures work out well for small values of n (viz., for $n\leq 4$), and as n increases, these procedures become prohibitively cumbrous. We may refer to Brewer and Hanif (1983) for some detailed discussions of these procedures when n may not be large. However, as regards the asymptotic theory is concerned, a lot of work remains to be accomplished.

We conclude this section with some discussions on certain occupancy (and sequential occupancy) problems which are closely related to the coupon collector's problem, treated earlier, and hence, have important roles in the asymptotic theory of finite population sampling. Suppose that balls are thrown independently of each other into N cells such that each ball has the probability p_{Nk} of falling into the kth cell, for $k = 1, \ldots, N$, where $p_{N1} + \cdots + p_{NN} = 1$. Let Q_{Nn} denote the number of empty cells after n throws and let W_{Na} denote the throw for which for the first time exactly a cells remain empty for all $a = 0, 1, \ldots, N - 1$. The classical occupancy problem, treated in David and Barton (1962), Feller (1968) and others, relate to the case where all the p_{Ni} are equal (to N^{-1}), and one wants to study the limiting behaviour of Q_{Nn} when both N and n are large and certain other regularity conditions hold. Similarly, the classical sequential occupancy problem (same as the Coupon collector's problem) relates to the distribution of the waiting time W_{Na} are some general regularity conditions. In the asymptotic case, we allow both n and a to depend on N and write them as n_N and a_N, respectively. The limiting distributions depend very much on the nature of these n_N and a_N when N is large. In the special case of a bounded number of empty cells, we have the following result due to Sevastyanov (1972):

Suppose that the $\{p_{Ni}\}$ satisfy the following conditions:

(a) $\max\{(1 - p_{Ni})^n : 1 \leq i \leq N\} \to 0$ as $N \to \infty$, \hfill (6.39)

(b) $E(Q_{Nn}) = \sum_{i=1}^{N} (1 - p_{Ni})^n \to m \ (<\infty)$ as $N \to \infty$. \hfill (6.40)

Then, for every positive integer k,

$$P\{Q_{Nn} = k\} \to e^{-m}/k! \quad \text{as } N \to \infty, \quad (6.41)$$

that is, Q_{Nn} has asymptotically the Poisson law with parameter m.

It may noted that (6.39) entails that $n/N \to \infty$ as $N \to \infty$. In the context of finite population sampling (and elsewhere), often, one makes the assumption that for every N and some real number c^0,

$$Np_{Ni} \leq c^0 < \infty \quad \text{for all } i = 1, \ldots, N. \quad (6.42)$$

In such a case we have a different Poisson law when n^2/N is bounded: The following result is due to Chistyakov (1964).

If $n_N^2(\sum_{i=1}^{N} p_{Ni}^2)/2 \to m \ (<\infty)$ and (6.42) holds, then $n - Q_{Nn}$ is asymptotically Poisson with parameter m (when $n_N \to \infty$).

On the other hand, when $n_N^2/N \to \infty$ as $N \to \infty$, we may have asymptotic normality under additional regularity conditions. The following result is due to Chistyakov (1967):

If $\log(N^{-1}n_N)$ is bounded (implying that $N^{-1}n_N \not\to 0$) and (6.42) holds, then $N - Q_{Nn}$ is asymptotically normal with mean $\sum_{i=1}^{N} \exp(-n_N p_{Ni})$ and variance

$$\sigma_{Nn}^2 = \sum_{i=1}^{N} \exp(-n_N p_{Ni})\{1 - \exp(-n_N p_{Ni})\}$$
$$- n_N \left(\sum_{i=1}^{N} \exp(-n_N p_{Ni}) \right)^2 \qquad (6.43)$$

A related result due to Holst (1971) deals with the situation where $N^{-1}n_N \to 0$ but $N^{-1}n_N^2 \to \infty$:

Suppose that (6.42) holds and $N^{-1}n_N^2 \to \infty$ (but $N^{-1}n_N \to 0$) as $N \to \infty$. Then $N - Q_{Nn}$ is asymptotically normal with mean $\Sigma_{i=1}^{N} \exp(-n_N p_{Ni})$ and variance $(n_N^2 \Sigma_{i=1}^{N} p_{Ni}^2/2)^{1/2}$.

Holst (1977) has considered a related result:

Suppose that

$$0 < c_0 \leq N p_{Ni} \leq c^0 < \infty \quad \text{for all } i \text{ and } n, \qquad (6.44)$$

$$N^{-1}n_N \to \infty \qquad (6.45)$$

and

$$f(n_N) = EQ_{Nn} = \sum_{i=1}^{N} (1 - p_{Ni})^n N \to \infty. \qquad (6.46)$$

Then, when $n_N \to \infty$,

$$(Q_{Nn} - f(n_N))/(f(n_N))^{1/2} \sim \mathcal{N}(0, 1), \qquad (6.47)$$

and in (6.47) it is also possible to replace $f(n_N)$ by $g(n_N)$, where

$$g(n_N) = \sum_{i=1}^{N} \exp(-n_N p_{Nk}). \qquad (6.48)$$

Note that $N - Q_{Nn}$ is the number of occupied cells after n throws, so that $N - Q_{Nn} \leq n$ for every $n \geq 1$. In sampling with varying probabilities (and with replacement), $N - Q_{Nn}$ stands for the number of distinct units in a sample of size n and hence the asymptotic laws stated above relate to the limiting distribution of the number of distinct units in, with replacement, varying probability sampling schemes. In particular, if all the p_{Ni} are equal to N^{-1}, (6.42) and (6.44) both hold, so that the Poisson law in (6.41) holds whenever $N(1 - N^{-1})^n$ converges to some finite (and positive) m. On the other hand, the asymptotic normality result in 6.43 (or (6.47)) holds when $N^{-1}n_N$ does not converge to 0 (or (6.45) and $N(1 - N^{-1})^n \to \infty$ hold). These limit theorems are also nicely related to the asymptotic theory considered in Section 5.

Let us next consider some parallel results for the waiting time W_{Na} where the nature of a ($= a_N$) determines the form of the limiting distributions. The following results are due to Holst (1977):

Under (6.44) and for every fixed integer a, when $N \to \infty$,

$$2\left(\sum_{i=1}^{N} (1-p_{Ni})^{W_{Na}}\right) \quad \text{asymptotically has the chi square distribution with } 2(a+1) \text{ degrees of freedom}, \quad (6.49)$$

and

$$2\left(\sum_{i=1}^{N} \exp(-W_{Na} p_{Ni})\right) \quad \text{has the same limiting distribution as in (6.49)}.$$

Next, to consider the case where $a = a_N$ may increase with N, we denote by t_a the unique solution of the equation

$$a = g(t_a) = \sum_{i=1}^{N} \exp(-t_a p_{Ni}). \quad (6.50)$$

If $a_N \to \infty$ but $N^{-1} a_N \to 0$ as $N \to \infty$ and (6.44) holds, then

$$a_N^{-1/2} (W_{Na_N} - t_{a_N}) \left(\sum_{i=1}^{N} p_{Ni} \exp(-t_{a_N} p_{Ni})\right) \sim \mathcal{N}(0, 1). \quad (6.51)$$

Further, we refer to the following related result due to Holst (1971):
If $(N - a_N)^2 / N \to \infty$ but $N^{-1} a_N \to 1$ as $N \to \infty$, and (6.42) holds, then W_{Na_N} is asymptotically normal with t_{a_N} and variance $t_N^2 (\sum_{i=1}^{N} p_{Ni}^2)/2$.

Note that in (6.51) a_N is small compared to N while in the last limit theorem, $N - a_N$ increases at a faster rate than $N^{1/2}$, while a_N/N converges to 1 as $N \to \infty$. Finally, we may consider the case where $N^{-1} a_N$ does not converge to 0 or 1 as $N \to \infty$.

If $N^{-1} a_N$ is bounded away from 0 and 1, then W_{Na_N} is asymptotically normal with mean t_{a_N} and variance $t^2 (\sum_{i=1}^{N} p_{Ni}^2)/\{2(\sum_{i=1}^{N} \exp(-t_{a_N} p_{Ni})^2\}$, where t_a is defined by (6.50).

Holst (1971) has also a Poisson limit law for $W_{Na_N} - (N - a_N)$ when

$$(N - a_N)^2 \left(\sum_{i=1}^{N} p_{Ni}^2\right)/2 \to m \ (<\infty) \quad \text{as } N \to \infty. \quad (6.52)$$

These limit theorems on the occupancy problems are also of considerable importance from the combinatorial limit theorems point of view.

7. Successive sub-sampling with varying probabilities: Asymptotics

Sub-sampling or multi-stage sampling is often adopted in practice and has a great variety of applications in survey sampling. These are elaborated in some other chapters of this volume. Typically, we may consider a finite population of N units with variate values $a_{N1}, \ldots, a_{NN}$, respectively. Consider a successive sampling scheme where items are sampled one after the other (without replacement) in such a way that at each draw the probability of drawing item s is proportional to a number p_{Ns} if item s has not already appeared in the earlier drawals, for $s = 1, \ldots, N$, where $P_{N1}, \ldots, P_{NN}$ are a set of positive numbers, adding up to 1. We like to consider a multi-stage extension of this sampling scheme. Here, each of the N items in the population (called the *primary units*) is composed of a number of smaller units (*sub-units*), and may be more economic to select first a sample of n primary units, and then to use sub-samples of sub-units in each of these selected primary units. Suppose that the sth primary unit has M_s sub-units with variate values b_{sj}, $j = 1, \ldots, M_s$, so that $a_{Ns} = b_{s1} + \cdots + b_{sM_s}$ for $s = 1, \ldots, N$. For each s, we conceive of a set $\{P_{sj}^0, 1 \leq j \leq M_s\}$ of positive numbers (such that $\Sigma_{j=1}^{M_s} P_{sj}^0 = 1$) and consider a successive sampling scheme (without replacement), where m_s (out of M_s) sub-units are chosen. Then, as in (6.13), an estimator of a_{Ns} can be framed for each of the n selected primary units. Finally, these estimates can be combined as in (6.13) to yield the estimator of the total $A_N = a_{N1} + \cdots + a_{NN}$. The procedure can be extended to the multi-stage case in a similar way. This scheme may be termed the *successive sub-sampling with varying probabilities* (*without replacement*) or SSSVPWR. To study the asymptotic theory, first we may note that a Horvitz–Thompson estimator of a_{Ns} is

$$\hat{a}_{Ns} = \sum_{j=1}^{M_s} \omega_{sj}^* b_{sj} / \Delta_s^*(j, m_s), \qquad (7.1)$$

where the b_{sj} are defined as before, ω_{sj}^* is equal to 1 or 0 according as the jth sub-unit in the sth primary unit belongs to the sub-sample of size m_s or not, $j = 1, \ldots, M_s$ and $\Delta_s^*(j, m_s)$ is the probability that the jth sub-unit belongs to the sub-sample of m_s sub-units from the sth primary unit, $1 \leq j \leq M_s$, $s = 1, \ldots, N$. Combining (6.13) and (7.1), we may consider the natural estimator

$$\hat{A}_{N(\text{HT})} = \sum_{s=1}^{N} \omega_{Ns} \hat{a}_{Ns} / \Delta(s, n)$$

$$= \sum_{s=1}^{N} \sum_{j=1}^{M_s} \omega_{Ns} \omega_{sj}^* b_{sj} / [\Delta(s, n) \Delta_s^*(j, m_s)], \qquad (7.2)$$

where ω_{Ns} is equal to 1 or 0 according as the sth primary unit is in the sample of n primary units from the population $s = 1, \ldots, N$, and the inclusion probabilities $\Delta(s, n)$ are defined as in (6.12). Note that for each (selected) primary unit s, for the estimator $\hat{a}_{Ns}$ in (7.1), one may use the theory discussed in Section 6. This, however, leads to a multitude of stopping numbers and thereby introduces complications in a direct extension of the Rosen approach to SSVPWR. A more simple approach (based on some martingale constructions) has been worked out in Sen (1980), and we may present the basic asymptotic theory as follows.

Our primary interest is to present the asymptotic theory of the estimator $\hat{A}_{N(\text{HT})}$

in (7.2). In this context, as in earlier sections, we allow N to increase. As $N \to \infty$, we assume that n, the primary sample size, also increases, in such a way that n/N is bounded away from 0 and ∞, while the m_s (i.e., the sub-sample sizes) for the selected primary units may or may not be large. For this situation, the asymptotic theory rests heavily on the structure of the primary unit sampling and we may also allow the sampling scheme for the sub-units to be rather arbitrary (not necessarily a SSVPWR), while we assume that the primary units are sampled in accordance with a SSVPWR scheme. A second situation may arise where the number of primary units (i.e., N) is fixed or divided in to a fixed number of strata, and within each strata of secondary units is drawn according to SSVWPR scheme. This situation, however, is congruent to the stratified sampling scheme under SSVPWR, for which the theory in Section 6 extends readily. Hence, we shall not enter into the detailed discussions on this second scheme.

With the notations introduced before, we now set

$$a^0_{Ns} = E(\hat{a}_{Ns}) \quad \text{and} \quad \sigma^2_{Ns} = \text{Var}(\hat{a}_{Ns}) \quad \text{for } s = 1, \ldots, N; \tag{7.3}$$

$$A^0_N = \sum_{s=1}^N a^0_{Ns} = E(\hat{A}_{N(\text{HT})}). \tag{7.4}$$

In order that $A^0_N = A_N$, it is therefore preferred to have unbiased estimators at the sub-unit stage, so that $a^0_{Ns} = a_{Ns}$ for every s. Otherwise, the bias may not be negligible. Also, for every N, we consider a nondecreasing function $t_N = \{t_N(x): 0 \leq x \leq N\}$ by letting

$$N - x = \sum_{s=1}^N \exp\{-P_{Ns} t_N(x)\}, \quad x \in (0, N). \tag{7.5}$$

Let then

$$\delta^2_{Nn} = \sum_{s=1}^N [a^0_{Ns}]^2 \exp\{-P_{NS} t_N(n)\} [1 - \exp\{-P_{Ns} t_n(n)\}]^{-1}$$

$$+ \sum_{s=1}^N \sigma^2_{Ns} [1 - \exp\{-P_{Ns} t_N(n)\}]^{-1}$$

$$- t_N(n) \left[\sum_{s=1}^N a^0_{Ns} P_{Ns} \exp\{-P_{Ns} t_N(n)\}/(1 - \exp\{-P_{Ns} t_N(n)\}) \right]^2 \tag{7.6}$$

Finally, we assume that the sub-unit estimators $\hat{a}_{Ns}$ satisfy a Lindeberg-type condition, namely that, for every $\eta > 0$,

$$\max_{1 \leq s \leq N} E[(\hat{a}_{Ns} - a^0_{Ns})^2 I(|\hat{a}_{Ns} - a^0_{Ns}| > \eta N^{1/2})] \to 0 \quad \text{as } N \to \infty. \tag{7.7}$$

The other regularity conditions are, of course, the compatibility of the probabilities $P_{N1}, \ldots, P_{NN}$ and the sizes $a_{N1}, \ldots, a_{NN}$ (in the sense that for each sequence the ratio of the maximum to the minimum entry is asymptotically finite. Then, we have the following:

$$(\hat{A}_{N(\text{HT})} - A_N^0)/\delta_{Nn} \text{ is asymptotically normal } (0, 1). \tag{7.8}$$

Actually, parallel to (6.29), we may consider a stochastic process $\xi_N = \{\xi_N(t); c < t < 1\}$ (where $c > 0$), by letting $\xi_N(t) = N^{-1/2}(\hat{A}_{N(\text{HT})}^{(t)} - A_N^0)$, where $\hat{A}_{N(\text{HT})}^{(t)}$ is the estimator in (7.2) based on the sample size $n = [Nt]$ (for the primary sample), $t \in [c, 1]$. Then, the process ξ_N converges in law to a Gaussian function on $[c, 1]$. The proofs of these results are based on some asymptotic theory for an *extended coupon collector's problem*, where in (6.16) through (6.19), the real (non-stochastic) elements a_{Ns} are replaced by suitable random variables X_{Ns}, $s = 1, \ldots, N$. For details of these developments, we may refer to Sen (1980).

Note that in the above development, apart from the uniform integrability condition in (7.7), we have not imposed any restriction on the estimates $\hat{a}_{Ns}$. Thus we are allowed to make the sub-sample sizes m_s arbitrary, subject to the condition that (7.7) holds. In this context, we may note that if these m_s are also chosen to be large, then the σ_{Ns}^2, defined by (7.3), will be small, so that in (7.6) the second sum on the right hand side will be of smaller order of magnitude (compared to the first sum), and hence, in (7.8), δ_{Nn} may be replaced by d_{Nn}^*, defined by (6.34), where the Y_s are to be replaced by the a_{Ns}^0. In this limiting case, we therefore observe that sub-sampling does not lead to any significant increase of the variance (compared to SSVPWR for the primary units; although in many practical problems, sub-sampling is more suitable, because it does not presuppose the knowledge of the values of the primary units $\{a_{Ns}\}$ and a complete census for these may be much more expensive than the estimates $\{\hat{a}_{Ns}\}$ based on a handful of sub-units.

So far, we have considered sampling without replacement. In SSVP sampling with replacement, the theory of successive VP sampling with replacement, discussed in the beginning of Section 6, readily extands. In (6.1), instead of the primary units y_j, we need to use their estimates $\hat{y}_j$, derived from the respective sub-samples. As in (7.6), this will result in an increased variability due to the individual variances of the second-stage estimators. However, with replacement, the strategy yields simplifications in the treatment of the relevant asymptotic theory, and (6.5) and (6.6) both extend to this sub-sampling scheme without any difficulty.

In conclusion, we may remark that in finite population sampling, the usual treatment for the asymptotic theory (valid for independent random variables) may not be directly applicable. But, in most of these situations, by appeal to either some appropriate permutation structures (for equal probability sampling) or to some martingale theory (for VP sampling as well), the asymptotic theory has been established under quite general regularity conditions. These provide theoretical justifications of the asymptotic normality of different estimators (under diverse sampling schemes) when the sample size(s) may or may not be non-stochastic. In particular, for optimal allocation based on pilot data, often, we end up with sample sizes being random (positive integer valued) variables. In such a case, the asymptotic results on the stochastics referred to earlier are useful. These results are also useful for quasi-sequential or repeated significance testing problems in finite population sampling.

References

Anscombe, F. J. (1952). Large sample theory of sequential estimation. *Proc. Cambridge Phil. Soc.* **48**, 600–607.

Arnab, R. (1979). On strategies of sampling finite populations on successive occasions with varying probabilities. *Sankhyā C* **41**, 140–154.

Asok, C. and Sukhatme, B. V. (1976). On samford's procedure of unequal probability sampling without replacement. *J. Amer. Statist. Assoc.* **71**, 912–918.

Basu, D. (1958). On sampling with and without replacement. *Sankhyā A* **20**, 287–294.

Baum, L. E. and Billingsley, P. (1965). Asymptotic distributions for the coupon collector's problem. *Ann. Math. Statist.* **36**, 1835–1839.

Bhandari, S. K. (1984). On a conjecture of Karlin in sampling theory. Tech. Report No. 30/84, Indian Statist. Inst., Calcutta.

Bhandari, S. K. (1985). On some families of Schur-convex functions those satisfy Karlin's conjecture in random replacement schemes. Tech. Report No. 24/85, Indian Statis. Inst., Calcutta.

Brewer, K. R. W. and Hanif, M. (1983). *Sampling with Unequal Probabilities*, Lecture Notes in Statistics 15. Springer, New York.

Chapman, D. G. (1951). Some properties of the hypergeometric distribution with applications to zoological census. *Univ. Calif. Publ. Statist.* **1**, 131–160.

Chapman, D. G. (1952). Inverse, multiple and sequential sample censuses. *Biometrics* **8**, 286–306.

Chaudhuri, A. (1975). Some properties of estimators based on sampling schemes with varying probabilities. *Austral. J. Statist.* **17**, 22–28.

Chaudhuri, A. (1976). On the choice of sampling strategies. *Calcutta Statist. Asso. Bull.* **25**, 119–128.

Chistyakov, V. P. (1964). On the calculation of the power of the test of empty boxes. *Theor. Probability Appl.* **9**, 648–653.

Chistyakov, V. P. (1967). Discrete limit distributions in the problem of balls falling in cells with arbitrary probabilities. *Math. Notes Acad. Sc. USSR* **1,**, 6–11.

Cochran, W. G. (1977). *Sampling Techniques*. Wiley, New York, 3rd ed.

Cormack, R. M. (1968). The statistics of capture–recapture methods. *Oceanogr. Mar. Biol. Ann. Rev.* **6**, 455–506.

Darling, D. A. and Robbins, H. (1967). Finding the size of a finite population. *Ann. Math. Statist.* **38**, 1392–1398.

Darroch, J. N. (1958). The multiple recapture census, I: estimation of closed population. *Biometrika* **45**, 343–359.

Das, A. C. (1951). On two phase sampling and sampling with varying probabilities. *Bull. Internat. Statist. Inst.* **33**, 105–112.

David, F. N. and Barton, D. E. (1962). *Combinatorial Chance*. Charles Griffin, London.

Durbin, J. (1953). Some results in sampling when units are selected with unequal probabilities. *J. Roy. Statist. Soc. Ser. B* **15**, 262–269.

Farm, A. (1971). Asymptotic normality in a capture–recapture problem when catchability is affected by the tagging procedure. Tech. Rep. TRITA-MAT-1971-10, Roy. Inst. Tech., Stockholm, Sweden.

Feller, W. (1968). *An Introduction to Probability Theory and its Applications*, Wiley, New York, 3rd ed.

Goodman, L. A. (1953). Sequential sampling tagging for population size problems. *Ann. Math. Statist.* **24**, 56–69.

Hájek, J. (1959). Optimum strategy and other problems in probability sampling. *Časopis Pro Pěstování Matematiky* **84**, 387–423.

Hájek, J. (1964). Asymptotic theory of rejective sampling with varying probabilities from a finite population. *Ann. Math. Statist.* **35**, 1491–1523.

Hájek, J. (1974). Asymptotic theories of sampling with varying probabilities without replacement. In: J. Hájek, ed., *Proc. Prague Confer. Asymp. Statist.*, 127–138.

Hájek, J. (1981). *Sampling From a Finite Population*. Dekker, New York.

Hájek, J. and Šidák, Z. (1967). *Theory of Rank Tests*. Academic Press, New York.

Hanif, M. and Brewer, K. R. W. (1980). On unequal probability sampling without replacement: A review. *Internat. Statist. Rev.* **48**, 317–335.

Hansen, M. H. and Hurwitz, W. N. (1943). On the theory of sampling from a finite population. *Ann. Math. Statist.* **14**, 333–362.

Hansen, M. H., Hurwitz, W. N. and Madow, W. G. (1953). *Sample Survey Methods and Theory*. Wiley, New York.

Hartley, H. O. (1966). Systematic sampling with unequal probability and without replacement. *J. Amer. Statist. Asso.* **61**, 739–748.

Hartley, E. O. and Rao, J. N. K. (1962). Sampling with unequal probabilities and without replacement. *Ann. Math. Statist.* **33**, 350–374.

Hoeffding, W. (1948). On a class of statistics with asymptotically normal distribution. *Ann. Math. Statist.* **19**, 293–325.

Hoeffding, W. (1963). Probability inequalities for sums of bounded random variables. *J. Amer. Statist. Asso.* **58**, 13–30.

Holst, L. (1971). Limit for some occupancy and sequential occupancy problems. *Ann. Math. Statist.* **42**, 1671–1680.

Holst, L. (1977). Some asymptotic results for occupancy problems. *Ann. Probability* **5**, 1028–1035.

Horvitz, D. G. and Thompson, D. J. (1952). A generalization of sampling without replacement from a finite universe. *J. Amer. Statist. Asso.* **47**, 663–685.

Karlin, S. (1974). Inequalities for symmetric sampling plans, I. *Ann. Statist.* **2**, 1065–1094.

Kolchin, V. F. and Chistyakov, V. P. (1974). Combinatorial problems of probability theory. *J. Soviet Math.* **4**, 217–243.

Kraft, O. and Schaefer, M. (1984). On Karlin's conjecture for random replacement sampling plans. *Ann. Statist.* **12**, 1248–1535.

Krewski, D. (1978). Jackknifing U-statistics in finite populations. *Comm. Statist. Theor. Meth. A* **7**, 1–12.

Krewski, D. and Rao, J. N. K. (1981). Inference from stratified samples: Properties of the linearization, jackknife and balanced repeated replication methods. *Ann. Statist.* **9**, 1010–1019.

Lahiri, D. B. (1951). A method for selection providing unbiased estimates. *Bull. Internat. Statist. Inst.* **33**, 133–140.

Madow, W. G. (1949). On the theory of systematic sampling, II. *Ann. Math. Statist.* **20**, 333–354.

Majumdar, H. and Sen, P. K. (1978). Invariance principles for jackknifing U-statistics for finite population sampling and some applications. *Commun. Statist. Theor. Meth. A* **7**, 1007–1025.

Midzuno, H. (1952). On the sampling system with probability proportionate to sum of sizes. *Ann. Inst. Statist. Math.* **3**, 99–107.

Miller, R. G., Jr. and Sen, P. K. (1972). Weak convergence of U-statistics and von Mises' differentiable statistical functions. *Ann. Math. Statist.* **43**, 31–41.

Murthy, M. N. (1967). *Sampling Theory and Methods*. Statist. Pub. Soc., Calcutta.

Nandi, H. K. and Sen, P. K. (1963). On the properties of U-statistics when the observations are not independent. Part Two: Unbiased estimation of the parameters of a finite population. *Calcutta Statist. Asso. Bull.* **12**, 125–148.

Narain, R. D. (1951). On sampling without replacement with varying probabilities. *J. Indian Soc. Agricul. Statist.* **3**, 169–175.

Pathak, P. K. (1967a). Asymptotic efficiency of Des Raj's strategy, I. *Sankhyā A* **29**, 283–298.

Pathak, P. K. (1967b). Asymptotic efficiency of Des Raj's strategy, II. *Sankhyā A* **29**, 299–304.

Petersen, C. G. J. (1896). The yearly immigration of young plaice into the Limfjord from the German see. *Rep. Danish. Biol. Sta.* **6**, 1–48.

Raj, Des (1956). Some estimators in sampling with varying probabilities without replacement. *J. Amer. Statist. Asso.* **51**, 269–284.

Rao, C. R. (1977). Some problems of sample surveys. *Sankhyā C* **39**, 128–139.

Rao, J. N. K. (1961). On sampling with varying probabilities and with replacement in sunsampling designs. *J. Indian Soc. Agri. Soc.* **13**, 211–217.

Rao, J. N. K. (1966). On the comparison of sampling with and without replacement. *Rev. Internat. Statist. Inst.* **34**, 125–138.

Rao, J. N. K., Hartley, H. O. and Cochran, W. G. (1962). On a simple procedure of unequal probability sampling without replacement. *J. Roy. Statist. Soc. Ser. B* **24**, 482–491.

Renyi, A. (1962). Three new proofs and a generalization of a theorem of Irving Weiss. *Publ. Math. Inst. Hungar. Acad. Sci.* **7**, 203–214.

Rosen, B. (1967). On an inequality of Hoeffding. *Ann. Math. Statist.* **38**, 382–392.

Rosen, B. (1970). On the coupon collector's time. *Ann. Math. Statist.* **41**, 1952–1969.

Rosen, B. (1972a). Asymptotic theory for successive sampling with varying probabilities without replacement, I. *Ann. Math. Statist.* **43**, 373–397.

Rosen, B. (1972b). Asymptotic theory for successive sampling with varying probabilities without replacement, II. *Ann. Math. Statist.* **43**, 748–776.

Rosen, B. (1974). Asymptotic theory for Des Raj estimators. In: J. Hájek, ed., *Proc. Prague Confer. Asymptotic Statist.*, 313–330.

Rosen, B. (1979). A tool for derivation of asymptotic results for sampling and occupancy problems, illustrated by its application to some general capture–recapture sampling problems. In: P. Mandl and M. Hušková, ed., *Proc. 2nd Prague Confer. Asymptotic Statist.*, 67–80.

Samuel, E. (1968). Sequential maximum likelihood estimation of the size of a population. *Ann. Math. Statist.* **39**, 1057–1068.

Samuel-Cagn, E. (1974). Asymptotic distributions for occupancy and waiting time problems with positive probability of falling through the cells. *Ann. Probability* **2**, 515–521.

Schaefer, M. (1985). A counter-example to Karlin's conjecture for random replacement sampling plans. Tech. Report, Tech. Univ. Aachen, W. Germany.

Schnabel, Z. E. (1938). The estimation of the total fish population of a lake. *Amer. Math. Monthly* **45**, 348–352.

Seber, G. A. F. (1973). *The Estimation of Animal Abundance*. Griffin, London.

Sen, A. R. (1953). On the estimate of the variance in sampling with varying probabilities. *J. Indian Soc. Agri. Statist.* **5**, 119–127.

Sen, A. R. and Sen, P. K. (1981). Schnabel type estimators for closed populations with multiple markings. *Sankhyā B* **43**, 68–80.

Sen, A. R. and Sen, P. K. (1986). Estimation of the characteristics of rare animals based on inverse sampling at the second occasion. In: *V. M. Joshi Festschrift*, Reidel, Dordrecht, Vol. 1, 309–321.

Sen, P. K. (1960a). On some convergence properties of U-statistics. *Calcutta Ststist. Asso. Bull.* **10**, 1–19.

Sen, P. K. (1960b). On the estimation of population size by the capture–recapture method. *Calcutta Statist. Asso. Bull.* **10**, 91–110.

Sen, P. K. (1970). The Hajek–Renyi inequality for sampling from a finite population. *Sankhyā A* **32**, 181–188.

Sen, P. K. (1972). Finite population sampling and weak convergence to a Brownian bridge. *Sankhyā A* **34**, 85–90.

Sen, P. K. (1977). Some invariance principles relating to jackknifing and their role in sequential analysis. *Ann. Statist.* **5**, 315–329.

Sen, P. K. (1979a). Invariance principles for the coupon collector's problem: A martingale approach. *Ann. Statist.* **7**, 372–380.

Sen, P. K. (1979b). Weak convergence of some quantile processen arising in progressively censored tests. *Ann. Statist.* **7**, 414–431.

Sen, P. K. (1980). Limit theorems for an extended coupon collector's problem and for successive sub-sampling with varying probabilities. *Calcutta Statist. Asso. Bull.* **29**, 113–132.

Sen, P. K. (1981). *Sequential Nonparametrics: Invariance Principles and Statistical Inference*. Wiley, New York.

Sen, P. K. (1982a). On the asymptotic normality in sequential sampling tagging. *Sankhā A* **44**, 352–363.

Sen, P. K. (1982b). A renewal theorem for an urn model. *Ann. Probability* **10**, 838–843.
Sen, P. K. (1983). On permutational central limit theorems for general multivariate linear rank statistics. *Sankhyā A***45**, 141–149.
Serfling, R. J. (1974). Probability inequalities for the sum in sampling without replacement. *Ann. Statist.* **2**, 39–48.
Sevastyanov, B. A. (1972). Poisson limit law for a scheme of sums of dependent random variables. *Theor. Probability Appl.* **17**, 695–699.
Sinha, B. K. and Sen, P. K. (1987). On averaging over distinct units in sampling with replacement. *Sankhyā A* (submitted).
Yates, F. and Grundy, P. M. (1953). Selection without replacement from within strata with probability proportional to size. *J. Roy. Statist. Soc. Ser. B* **15**, 253–261.

The Technique of Replicated or Interpenetrating Samples

J. C. Koop

1. Introductory review

The technique of interpenetrating samples, in its several forms, is due to Mahalanobis, and its evolution to the form eventually used in the Indian National Sample Surveys (Lahiri, 1954, 1957a, b, 1964; Murthy, 1963, 1964; Murthy and Roy, 1975) is described in his early papers (Mahalanobis 1940a, b, 1944, 1945, 1946a, b) where he gives an account of his[1] experimental techniques and the theory for his jute and rice acreage surveys in Bengal and Bihar initiated in 1937 and 1943 respectively. He attributes the phrase 'interpenetrating net-works of samples', which he often used, to R. A. Fisher (Mahalanobis, 1946b).

The technique is founded on the principles of randomization, replication and control which are also the cardinal principles of *The Design of Experiments*. In sample surveys control is exercised by way of stratification and clustering. It can be said that the technique exploits the principle of replication much more than any other sampling method. In this connection see also Fisher (1956, p. 6; 1961, p. 110). The United Nations Subcommission on Statistical Sampling (1949) has strongly recommended its use and has also suggested the alternative term 'replicated sampling'.

The main purpose for the use of the technique was (and is) to control and reduce errors other than those of random sampling. These errors, collectively termed non-sampling errors, can be broadly classified into three types:

(i) *Coverage errors*. These are omission and/or duplication of sampling units that are members of the sample. Inclusion of originally unselected units is also an error under this category.

(ii) *Ascertainment errors*. These are a compound of measurement and response errors in the variate values for each of the characteristics of the ultimate sampling

---

[1] Mahalanobis was much influenced by the early work of Hubback (1927) who made sustained attempts to be as objective as possible in his crop-cutting experiments to determine rice yields in Bihar and Orissa by sampling methods. According to Fisher (Mahalanobis, 1946b) these were the earliest crop-cutting experiments anywhere in the world and also influenced his work in Rothamsted.

units of the selected sample occuring at the stage of data collection, editing, coding, data entry (either on punch cards or into a computer).

(iii) *Errors introduced by imputation for missing values and/or at the final stage of estimation.* Thus sub-type may include errors in weights or raising factors associated with the ultimate sampling unit and therefore with its corresponding variate values.

The causes for these errors are many. In summary they lie in defective operational definitions of the variate values under enquiry, poor training and supervision of investigators and/or those processing the data, lack of communication skills, and defective compliance with instructions (including mistakes or deliberate lies for recorded variate values).

Deming (1944) has classified errors in surveys into thirteen types. In the context of current sample survey practice his paper is still very relevant. The foregoing broad classification into three types follows that of Lahiri (1957a).

Thus even a complete census is not free from non-sampling errors. Thus also, estimates based on sample survey data are subject to both sampling and non-sampling errors. In the sequel it will appear that non-sampling errors can only be controlled by controlling the survey conditions unlike errors of random sampling, which for surveys can be controlled by increasing the total effective sample size.

In its most general form the technique of interpenetrating samples consists in drawing two or more independent samples from a stratified universe using the same sampling procedure for each sample. Statistical independence between samples is ensured by the entire replacement of each sample before the drawing of the next; the technique evolved into the independent form in the first three rounds of the Indian National Sample Survey which took place in the period October 1950 to November 1951. In its non-independent forms, used before 1950, the samples are linked. Because of its elegant statistical properties, we shall be concerned only with the form of the technique where the samples are independent. Later however, mention will be made of other non-independent forms.

When each independent sample is surveyed by its own team of investigators and processed and tabulated by its own unit, then control can be maintained at each of the component operations involved. For example if one estimate at the tabulation stage differs considerably from the others, then the sample data giving rise to that estimate can be examined to determine the causes; duplication of units may be the source of the problem. Remedial action can then be taken to correct the estimate. For large-scale surveys where hundreds of estimates for different characteristics are involved, such checks enhance the accuracy and utility of the final estimates.

The flexibility of the technique, in the following respects, may be noted.

If advance estimates are needed, one or two samples of the set may be used. With a single sample, subsampling would be necessary to achieve the same end; this may not always be technically convenient.

When the extent of funding is uncertain a survey based on the technique may proceed up to a point allowed by the resources, using one, two, or three samples of the original set as appropriate. In a similar situation a single sample would remain incomplete. The validity of its results would be in doubt, and when used would be subject to serious qualifications.

One important consequence of the technique is simplicity in the estimation of variance regardless of the complexity of the form of the estimate, e.g. a combined ratio estimate based on a stratified multi-stage sample, or the same for the domains of study cutting across strata.

If the survey is free from error, an unbiased estimator of the variance of the mean of the estimates from the replicated samples is given by the usual classical formula.

When there are errors from sources indicated under (ii) and (iii), for the case of linear estimators the estimator of total variance includes practically all components of variance and covariance due to ascertainment errors, excepting those for covariances between the same units appearing in different samples. This property is only very weakly possessed by the estimator of variance based on a *single sample*, which for unequal probability sampling designs may yield a *negative estimate*.

However, apart from all these qualifications, to a layman the extent of agreement between independent estimates (like agreement between witnesses in a court of law) is often more convincing as a sign of accuracy than a figure for total variance.

The property of mutual statistical independence possessed by the samples provides the logical basis for making probability statements about universe parameters by classical non-parametric arguments.

The technique of interpenetrating samples also provides the basis for studying the differential effect of varying methods of data collection, investigator training, editing, processing and tabulation procedures if appropriate experimental designs are superimposed on the samples, each treated as an experimental unit.

Whatever the nature of the sampling procedure the analysis of variance applies to the set of estimates (whether linear or nonlinear) computed from each of the samples subjected to experimental treatment. For examples, see Mahalanobis and Lahiri (1961), Som and Mukherjee (1962) and Som (1965). Because the sampling procedure is the same for each of the estimates involved in the analysis, the mean square for error is actually an estimate of the pure sampling variance. Even if F-tests are not applied, a comparison of the mean square for any effect under consideration with that of the mean square for error gives a measure of the degree of intensity of the effect.

Besides the uses described in the foregoing account the technique readily provides the basis for the method of fractile graphical analysis (Mahalanobis, 1960) useful for the visual and geometrical assessment and control of ascertainment errors, investigation of a functional relationship between any two measurable characteristics, and comparison of such relationships over time in a direct manner.

Incomplete coverage of a probability sample due to refusals or non-response is quite usual in large-scale surveys. With a single sample assumptions have to be made about the nature of the nonresponding part of the universe for each of the characteristics under study in order to introduce imputation procedures; this results in biases. On the other hand with interpenetrating samples, the incomplete samples can be compared with the complete ones (a) to determine if their inclusion would introduce non-negligible biases in the estimates, and (b) to set up, for each of the characteristics involved, sutiable imputation procedures as part of the respective estimation procedures. Thus with the technique of interpenetrating samples, the damage resulting from incomplete coverage is confined to the particular samples, and the sample data itself provides the information for its assessment and correction.

The use of the technique involves more cost than that of a single sample of equivalent size. This is because the samples are differentiated by different investigating, processing and tabulation units. However, Mahalanobis (1950) himself has pointed out that the extra cost is due mainly to travelling time by investigating parties and comes to about 8% for crop surveys and is much less for socio-economic surveys. Considering the advantages outlined in the foregoing account, the additional cost is worthwhile. Certainly its use would be redundant if non-sampling errors did not exist, or are uniformly negligible for every characteristic under study in a large-scale survey. But it has yet to be demonstrated that this is so, and that other methods can accomplish equally well. See Lahiri (1975a, pp. 149, 150; 1964, pp. 201–203).

Outside India the technique of interpenetrating samples, in its independent form has been used in the Phillipines (Lieberman, 1958), Ghana (Kpedekpo and Chacko, 1972) and Canada (Statistics Canada, 1973), in the first two countries for demographic studies, and in the third for the quality check of the 1971 census of agriculture.

The methods used by Shaul and Myburgh (1948, 1949, 1951), Shaul (1952), Deming (1956) and Jones (1956) have much in common with Mahalanobis' technique except for lack of statistical independence between samples.

Shaul and Myburgh, in their population surveys in Central Africa, used two or three systematic samples per stratum with random starts (Madow and Madow, 1944), and termed them 'inter-penetrating samples'.

Jones' replicates are also systematic samples with random starts.

Those of Deming may be described as stratified samples with one unit per stratum selected without replacement. Each of Deming's strata, which he calls zones, has the same number of sampling units. In his own work the technique has been used in a wide variety of economic and social investigations.

Flores (1957) has used Deming's method in Mexico in demographic, income and cost of living surveys.

Mehok (1963) studied the occupational structure of the Society of Jesus around the world using ten replicated samples drawn by Deming's method. Prabhu-Ajgaonkar (1969) also used the same technique to study the phonemic and graphemic frequencies of the Marathi language.

As with most pioneering theories and methods, Mahalanobis' technique has not escaped criticism and controversy, particularly in the evolutionary period before 1950. This reaction is inevitable human nature being what it is. However, in a book meant largely for practitioners, the discussion of this matter is obviously inappropriate. The papers indicated by asterisks, which are part of the bibliography captioned as works not cited in the text may be consulted by those who are interested. Viewed in historical perspective they appear to have provided desirable stimuli for the refinement of the technique.

In the next section the theory needed will be given. Applications of the theory with data from sample surveys carried out in Canada and India are considered in Section 3. In Section 4 the use of the analysis of variance, with data resulting from the superimposition of a simple experimental design on a set of interpenetrating samples, to study the differential effects of auspices, and the quality of work of investigating teams, is discussed. The chapter ends with summary comments on the technique in Section 5.

2. Theoretical basis of the technique

The technique of interpenetrating samples explicitly recognizes non-sampling errors. This is in sharp contrast to standard sample survey theory where such errors are ignored.

Our purpose is first to present, with an economy of model assumptions, a brief unified estimation theory which fully considers ascertainment errors described in Section 1. Needless to elaborate, this theory is also necessary to see if there are best estimators.

Results of practical importance following this theory, given in Sections 2.3 and 2.4, show that whatever the sampling design, variance estimation with the method of replicated or interpenetrating samples takes into account practically almost all types of non-sampling errors much more adequately than with the use of standard methods based on single samples which result in biased estimates.

2.1. Definitions, notation and explanations

There is a finite *universe* U consisting of N different identifiable units where u_i is the ith unit, i.e.

$$U = \{u_i : i = 1, 2, \ldots, N\} \tag{2.1.1}$$

with a set of N corresponding vectors of l real-valued components, $(x, y, z, \ldots)$ for each unit, i.e.

$$\{(x_i, y_i, z_i, \ldots) : i = 1, 2, \ldots, N\}. \tag{2.1.2}$$

The values given by (2.1.2) are *true* values, and they may be regarded as parameters.

The unit $u_i \in U$ is an *ultimate sampling unit*; all such units of U may be clustered in a hierarchy of units. Furthermore, the set of largest units in the hierarchy may be divided into strata. However, for the sake of generality we do not specify the internal structure of U.

The *frame F* identifies the units of U and for the purpose of this theory it is assumed to be perfect, i.e. there are no missing or overlapping units.

We have a *physical randomization procedure R*, e.g. tested random numbers, for selecting samples from U through the frame F and according to a set of *rules H*, specified in appropriate detail, for drawing the respective units of the hierarchy, either with or without replacemetn in each stratum. We also have a *probability system P* that defines the respective selection probabilities of the units in each stratum and which in turn is implemented by R.

In general the *sampling procedure* for selecting samples from U is defined by the combination (P, R). This of course directly implies that there is a unique sampling procedure for each stratum and we assume that for each stratum it is such that every stratum-unit is found in at least one stratum sample. The collection of such procedures for each stratum constitutes (P, R). Note that the principles of randomization, replication and control underlie (P, R).

When we apply (P, R) a sample of distinct ultimate sampling units

$$s = \{u_i, u_j, \ldots, u_m\}, \qquad (2.1.3)$$

for short a *distinct sample s*, that is found to contain $n(s)$ distinct units, will be eventually realized with probability $p(s) > 0$. We are saying that a distinct sample s will eventually be realized, because antecedent to s there is the realization of a sequence of units or a particular sample (Koop, 1979, p. 255), where order (in any of its senses) and multiplicity of units can be recognized; detailed explanations on this point have been omitted for the sake of brevity. Remark 5 of Koop (1979) summarizes this viewpoint.

The number $n(s)$, $s \in \mathscr{S}$, *is called the effective sample size.*

The entire collection of different distinct samples that can be realized by the repeated application of the sampling procedure (P, R) is designated $\mathscr{S} = \{s: s \subset U\}$.

We may also regard p as a function defined on $\mathscr{S}$ such that $\Sigma_{s \in \mathscr{S}} p(s) = 1$. In this context many writers now call $p(s)$ a *sampling design*.

The technique for selecting k *independant interpenetrating samples* is defined by the following operations:

(1) Apply the sampling procedure (P, R) to select a sample s in the sense of (2.1.3); call this sample s_1.

(2) Replace s_1. In practice replacement is technically achieved by regarding every sampling unit in the hierarchy of units constituting U as being eligible for selection again.

(3) Repeat operations (1) and (2) $k - 1$ times to realize $s_2, s_3, \ldots, s_k$.
Thus

$$\{s_1, s_2, \ldots, s_k\} \qquad (2.1.4)$$

constitutes a set of k independent interpenetrating samples.

In practice two to five samples are drawn, but for the purpose of experimental studies as many as twelve samples are known to have been drawn; of course costs set an upper limit to the total sample size.

The following points may be noted:

(a) Statistical independence between samples is ensured by replacement.

(b) It is possible for the k samples to be identical. However, this possibility will be extremely rare. Between samples, common units in the hierarchy of units are more likely, but their occurrence will decrease with increasing stratification and stages in sampling.

(c) In general the effective[2] sample size $n(s)$ will vary with s and so also $p(s)$.

With a large-scale survey the aim is usually to describe the universe in terms of totals, means and ratios of the values given by (2.1.2). Estimates of totals and sometimes means, are linear in the variates involved. An estimate of the mean per unit may be a ratio estimate.

For the purpose of an estimation theory it is sufficient to consider the estimation of the total for just one value, i.e.

$$T = \sum_{i \in U} x_i, \qquad (2.1.5)$$

which is an estimand linear in the x-values. Later, in Section 2.5, the estimation of ratios will be considered.

Consider the situation when a team of investigators collects data from the units of s with a given questionnaire and according to a prescribed set of instructions.

Denote by G the entire collection of circumstances and working conditions faced by the team of investigators and all associated personnal who later scrutinize, edit and process the data.

Regardless of G the attributes of the u_i, defined by (2.1.1), in the context of any *real* finite universe, are what they are. Therefore, the values or parameters defined by (2.1.2) are in no way conditioned by, or dependent on, G.

Under the conditions G let X_i be the variate value of u_i that is a member of s; note that this implies that X_i is a value subject to all the ascertainment errors described under (ii) of Section 1; in the sense of these explanations X_i is an observation on u_i.

Then the error of ascertainment in observing u_i is

$$X_i - x_i,$$

which in most real situations will remain unknown. Conceptually if $X_i = x_i$, then

---

[2] Many statisticians writing on survey estimation theory assume too frequently that $n(s) = n$, a constant integer for $s \in \mathcal{S}$. In the real world of survey practice this is only true for simple random sampling, stratified random sampling and special types of multistage sampling procedures, all of them *without exception* carried out without replacement at every stage. When there is replacement of units the possibility of repeated units exists so that the notion of a constant sample size is no longer applicable.

of course there is no error. But there is no way of knowing that this will be so except in a situation when a correct or true record for u_i exists.

On repeated observations on u_i, *all of which are hypothetical except the one actually made,* let

$$E(X_i|G) = x_{is}, \quad i \in s, \ s \in \mathcal{S}, \tag{2.1.6}$$

$$V(X_i|G) = V_{is}, \quad i \in s, \ s \in \mathcal{S}, \tag{2.1.7}$$

and

$$\text{Cov}(X_i, X_j|G) = C_{ijs}, \quad i \neq j \in s, \ s \in \mathcal{S}. \tag{2.1.8}$$

Note that the parameters x_{is}, V_{is} and C_{ijs} are conditioned by G through s. If G should change then it is reasonable to assume that they may all change.

Note also that there are conditional probability distributions underlying (2.1.6) and (2.1.8) all of which depend on G.

We may now say that the *ideal conditions* of the survey, G_0, are achieved when $x_{is} = x_i$, the true value for $i \in U$, and $V_{is} = C_{ijs} = 0$, for $i \in s$, $i \neq j \in s$ and $s \in \mathcal{S}$. Then the survey is in *perfect control* and the only source of error is the sampling error. Obviously G_0 cannot be attained.

In practice attempts are made to control G in order to reduce the non-sampling errors as much as possible for all estimates under consideration.

Finally we define the first and second order inclusion probabilities as

$$P_i = \sum_{s \supset i} p(s), \quad i \in U, \tag{2.1.9}$$

and

$$P_{ij} = \sum_{s \supset i \neq j} p(s), \quad i \neq j \in U, \tag{2.1.10}$$

respectively.

2.2. Brief estimation theory

To estimate T, as given by (2.1.5), we have the following sample data

$$\{s, X_i : i \in s\}. \tag{2.2.1}$$

Because T is a homogeneous linear functon we define a very general estimator that is linear in the X-values as

$$T(s) = \sum_{i \in s} \beta_{si} X_i, \tag{2.2.2}$$

where β_{si} is a coefficient to be applied to X_i whenever samples s is realized and $u_i \in s$. Thus the class of estimators linear in X is given by, say

$$\mathcal{B} = \left\{ T(s) \colon T(s) = \sum_{i \in s} \beta_{si} X_i, \ s \in \mathcal{S} \right\}. \tag{2.2.3}$$

With respect to (2.2.3) note that the number of β-coefficients involved in each possible $T(s)$ is $n(s)$, so that in all, say $\mathcal{M} = \Sigma_{s \in \mathcal{S}} n(s)$ coefficients are involved in the entire class $\mathcal{B}$.

Under G_0 it is easy to verify that $T(s)$ is an unbiased estimator of T if

$$\sum_{s \supset i} p(s)\beta_{si} = 1, \quad i = 1, 2, \ldots, N. \tag{2.2.4}$$

Under G however, $T(s)$ is a biased estimator of T, and this bias is given by

$$B_0 = \sum_{s \in \mathcal{S}} p(s) \sum_{i \in s} \beta_{si}(x_{is} - x_i). \tag{2.2.5}$$

Note that there is an unlimited number of solutions for the β-coefficients of $\mathcal{B}$, totalling in all $\mathcal{M}$, that can be obtained by the solution of the N simultaneous equations given by (2.2.4). Therefore, there will be an unlimited number of estimators of the form given by (2.2.2) for the class $\mathcal{B}$.

We shall show that *in the class $\mathcal{B}$ an estimator linear in the X-values with the least mean square error does not exist in a sense which will be specified.*

The variance of $T(s)$ is given by

$$V\{T(s)\} = V\left[E\left\{\sum_{i \in s} \beta_{st} X_i | s\right\}\right] + E\left[V\left\{\sum_{i \in s} \beta_{si} X_i | s\right\}\right]$$

$$= \sum_{s \in \mathcal{S}} p(s) \left\{\sum_{i \in s} \beta_{si}^2 x_{is} + \sum_{i \neq j \in s} \beta_{si} \beta_{sj} x_{is} x_{js}\right\}$$

$$- \left\{\sum_{s \in \mathcal{S}} p(s) \sum_{i \in s} \beta_{si} x_{is}\right\}^2$$

$$+ \sum_{s \in \mathcal{S}} p(s) \left\{\sum_{i \in s} \beta_{si}^2 V_{is} + \sum_{i \neq j \in s} \beta_{si} \beta_{sj} C_{ijs}\right\}. \tag{2.2.6}$$

Thus the mean square error of $T(s)$, given by

$$\mathrm{MSE}\{T(s)\} = V\{T(s)\} + B_0^2, \tag{2.2.7}$$

is a quadratic form in the β's involving the N unknown parameters $x_1, x_2, \ldots, x_N$ and all those parameters defined by (2.1.6), (2.1.7) and (2.1.8), and is succintly expressed as

$$\mathrm{MSE}\{T(s)\} = \beta H \beta', \tag{2.2.8}$$

where β is a row vector of $\mathcal{M}$ components made up of the β_{si}'s of each $T(s)$ taken estimator by estimator, and H is the $\mathcal{M} \times \mathcal{M}$ matrix of this quadratic form whose elements can be identified when the expressions given by (2.2.5) and (2.2.6) are substituted in (2.2.7).

We shall minimize (2.2.8) subject to the N restrictions given by (2.2.4). For this we set up the augmented function

$$F = \beta H \beta' + 2 \sum_{i=1}^{N} \lambda_i \left(\sum_{s \supset i} p(s) \beta_{si} - 1 \right)$$

$$= (\beta, \lambda) \begin{pmatrix} H & B \\ B' & O \end{pmatrix} \begin{pmatrix} \beta' \\ \lambda' \end{pmatrix} - 2(\beta, \lambda) \begin{pmatrix} o' \\ e' \end{pmatrix}, \qquad (2.2.9)$$

where $\lambda = (\lambda_1, \lambda_2, \ldots, \lambda_N)$ is a row vector of N Lagragian multipliers, B the $\mathcal{M} \times N$ matrix of the bilinear form

$$\beta B \lambda' = \sum_{i=1}^{N} \lambda_i \sum_{s \supset i} p(s) \beta_{si},$$

O a null $N \times N$ matrix, o a null row vector of $\mathcal{M}$ components, and finally e a unit row vector of N components.

Setting $\partial F / \partial (\beta, \lambda) = \mathcal{O}'$, in which $\mathcal{O}$ is a null row vector of $\mathcal{M} + N$ components, we find

$$\begin{pmatrix} H & B \\ B' & O \end{pmatrix} \begin{pmatrix} \beta' \\ \lambda' \end{pmatrix} = \begin{pmatrix} o' \\ e' \end{pmatrix}. \qquad (2.2.10)$$

Solving for β we have

$$\beta = e(B'H^{-1}B)^{-1} B' H^{-1}. \qquad (2.2.11)$$

The β-value given by (2.2.11), in a *strictly mathematical context*, minimizes MSE$\{T(s)\}$ and we find

$$\text{MSE}\{T(s)\} = \beta H \beta' \geq e(B'H^{-1}B)^{-1} e', \qquad (2.2.12)$$

equality being attained if and only if β is as given (2.2.11).

In a context that is statistically relevant β *cannot* be determined because we do not know the values of the parameters involved, given by (2.1.6), (2.1.7) and (2.1.8).

Therefore β is intractable. In the sense of this intractability we may say that an estimator linear in the X-values with the least mean square error does not exist, a proposition which is also true for all characteristics. This proposition is of course an extension of Godambe's (1955) proposition considered for the case when the variates are free from ascertainment errors.

Therefore to obtain an estimator that can be useful in the context of any given sampling design we choose β-values satisfying (2.2.4) that are tractable. One solution which is fairly general is obtained by putting $\beta_{si} = \beta_i$ for $s \supset i$ and $i \in U$.

Then

$$\beta_i = 1/P_i, \quad i = 1, 2, \ldots, N. \qquad (2.2.13)$$

Note that if the sampling design is specified in detail we can obtain solutions to (2.2.4) that are different from (2.2.13). Thus (2.2.13) yields the biased estimator

$$\hat{T} = \sum_{i \in s} X_i/P_i. \qquad (2.1.14)$$

In its error-free form the estimator was discovered by Horvitz and Thompson (1952). A similar estimator was independently found by Narain (1951).

The variance of $\hat{T}$ is given by

$$V(\hat{T}) = \sum_{i=1}^{N} \{(1 - P_i)/P_i\} x_i^2 + \sum_{i=1}^{N} \sum_{j(\neq i)=1}^{N} \{(P_{ij} - P_i P_j)/P_i P_j\} x_{i\cdot} x_{j\cdot}$$

$$+ \sum_{i=1}^{N} (\overline{V}_i/P_i) + \sum_{i=1}^{N} \sum_{j(\neq i)=1}^{N} (P_{ij} \overline{C}_{ij}/P_i P_j)$$

$$+ \sum_{i=1}^{N} (V_{i\cdot}/P_i) + \sum_{i=1}^{N} \sum_{j(\neq i)=1}^{N} (P_{ij} C_{ij\cdot}/P_i P_j), \qquad (2.2.15)$$

where

$$x_{i\cdot} = \sum_{s \supset i} \{p(s) x_{is}/P_i\}, \quad i \in U,$$

$$V_{i\cdot} = \sum_{s \supset i} \{p(s) V_{is}/P_i\}, \quad i \in U,$$

$$C_{ij\cdot} = \sum_{s \supset i \neq j} \{p(s) C_{ijs}/P_{ij}\}, \quad i \neq j \in U,$$

$$\overline{V}_i = \sum_{s \supset i} \{p(s) (x_{is} - x_{i\cdot})^2/P_i\}, \quad i \in U,$$

and

$$\overline{C}_{ij} = \sum_{s \supset i \neq j} \{p(s) (x_{is} - x_{i\cdot})(x_{js} - x_{j\cdot})/P_{ij}\}, \quad i \neq j \in U,$$

which is obtained from (2.2.6), noting also that the β-values are given by (2.2.13).

If the conditions of a particular survey, say G', warrant simplifying assumptions such as $x_{is} = \hat{x}_i$ and $V_{is} = \hat{V}_i$ for $i \in s$, $s \in \mathcal{S}$ and $C_{ijs} = \hat{C}_{ij}$ for $i \neq j \in s$, and $s \in \mathcal{S}$, then (2.2.15) reduces to

$$V_1(\hat{T}) = \sum_{i=1}^{N} \{(1 - P_i)/P_i\} \hat{x}_i^2 + \sum_{i=1}^{N} \sum_{j(\neq i)=1}^{N} \{(P_{ij} - P_i P_j)/P_i P_j\} \hat{x}_i \hat{x}_j$$

$$+ \sum_{i=1}^{N} (\hat{V}_i/P_i) + \sum_{i=1}^{N} \sum_{j(\neq i)=1}^{N} (P_{ij} \hat{C}_{ij}/P_i P_j). \qquad (2.2.16)$$

The first two terms of (2.2.15) constitute the *sampling variance*. The third and fourth terms are known as the *interactive variance* and the *interactive covariance* respectively and the fifth and sixth terms are known as the *response variance* and *response covariance* respectively; note that these four terms are really parametric functions of the underlying mean values, variances and covariances, that account for the ascertainment errors. The entire expression for $V(\hat{T})$ may be called the *total error variance*.

Theoretically whether or not the sampling variance dominates the four remaining parts will depend on the conditions G. However, it is believed that under most survey conditions, for most characteristics, the sampling variance is much smaller than the group of four variances and covariances; of course under G_0 these four vanish leaving only the sampling variance.

Thus there is a need for techniques of investigation for reducing the effect of the four non-sampling variances and covariances.

From (2.2.5) and (2.2.13) it turns out that even under the conditions G' the bias B_0 takes the value

$$B_1 = \sum_{i=1}^{N} (\hat{x}_i - x_i). \tag{2.2.17}$$

This expression can be formally estimated as $\Sigma_{i \in s}(X_i - x_i)/P_i$, but cannot be evaluated as a numerical value unless assumptions are made about the x_i's. The practical upshot of this situation is that we are left to deal only with $V(\hat{T})$ for assessing the accuracy of $\hat{T}$.

A simple example at this stage may be helpful in understanding the implications of (2.2.15).

EXAMPLE. For the case of simple random sampling when the n units are drawn without replacement the inclusion probabilities are

$$P_i = n/N, \quad i = 1, 2, \ldots, N,$$

and

$$P_{ij} = \{n(n-1)\}/\{N(N-1)\} \quad \text{for all } i \neq j,$$

and we find

$$\hat{T}' = (N/n) \sum_{i \in s} X_i, \tag{2.2.18}$$

with variance

$$V(\hat{T}') = N^2 \left[\frac{1}{n} \frac{\Sigma_{i=1}^{N}(x_{i\cdot} - \bar{X})^2}{N-1} \left(1 - \frac{n}{N}\right) \right.$$

$$+ \frac{1}{n} \frac{\Sigma_{i=1}^{N} \Sigma_{s \supset i}(x_{is} - x_{i\cdot})^2 / \binom{N-1}{n-1}}{N}$$

$$+ \left(\frac{n-1}{n}\right) \frac{\Sigma_{i=1}^{N} \Sigma_{j(\neq i)=1}^{N} \Sigma_{s \supset i \neq j}(x_{is} - x_{i\cdot})(x_{js} - x_{j\cdot})/\binom{N-2}{n-1}}{N(N-1)}$$

$$+ \frac{1}{n} \frac{\Sigma_{i=1}^{N} \Sigma_{s \supset i} V_{is}/\binom{N-1}{n-1}}{N}$$

$$+ \left(\frac{n-1}{n}\right) \frac{\Sigma_{i=1}^{N} \Sigma_{j(\neq i)=1}^{N} \Sigma_{s \supset i \neq j} C_{ijs}/\binom{N-2}{n-2}}{N(N-1)} \Bigg], \qquad (2.2.19)$$

where $\bar{X} = \Sigma_1^N x_i/N$. The first expression will be recognized as the sampling variance. The remaining parametric functions are the interactive variance, interactive covariance, response variance and response covariance respectively in the order written. The formula speaks for itself. With a complete census the sampling variance vanishes, leaving the four parametric functions, which under the simpler conditions G' reduces to

$$\sum_{i=1}^{N} \hat{V}_i + \sum_{i=1}^{N} \sum_{j(\neq i)=1}^{N} \hat{C}_{ij}. \qquad (2.2.20)$$

Theoretically, this value can be very large when the census operations are not in good control.

So far, errors introduced by missing values and imputation and all others described under (iii) of Section 1 have not been considered. Regardless of whether or not such errors are random variables, if they are incorporated as part of the formulation in (2.1.6), (2.1.7) and (2.1.8) as component quantities in X_i, i.e. as an extension of ascertainment error, then the estimation theory given in the foregoing paragraphs remains unchanged. If it turns out that they are not random variables, then their effects will be manifested in the bias B_0.

The expanded version of the foregoing theory is found in Koop (1974). The statements at (2.1.6), (2.1.7) and (2.1.8) are not restrictive and are motivated by the work of Hansen, Hurwitz and Bershad (1961).

Earlier, various aspects of the problem of non-sampling errors were considered by Hansen et al. (1951) and Sukhatme and Seth (1952). It should be noted that the mathematical theory of ascertainment errors has a lengthy background going back to the early work of astronomers and geodetic surveyors, and in this century to Karl Pearson (1902).

2.3. Estimation of variance with a single sample

For estimating the variance of $V(\hat{T})$ under G_0, when it is assumed that the X_i's are free from error, the usual estimator for any sampling design is

$$v(\hat{T}) = \sum_{i \in s} \{(1 - P_i) X_i^2/P_i^2\} + \sum_{i \neq j \in s} \{(P_{ij} - P_i P_j) X_i X_j / P_i P_j P_{ij}\}. \qquad (2.3.1)$$

Let us assess the effect of using the estimator under the conditions G. We find after some algebra that

$$E\{v(\hat{T})\} - V(\hat{T}) = -\left\{\sum_{i=1}^{N} \overline{V}_i + \sum_{i=1}^{N}\sum_{j(\neq i)=1}^{N} \overline{C}_{ij}\right.$$

$$\left.+ \sum_{i=1}^{N} V_{i\cdot} + \sum_{i=1}^{N}\sum_{j(\neq i)=1}^{N} C_{ij\cdot}\right\}. \quad (2.3.2)$$

Note that a very large number of parameters underlie the bias function. Although to some extent its magnitude is dependent on the sampling design, it can also be reduced by exercising control over errors described under (ii) and (iii) of Section 1 or what amounts to controlling the total survey conditions G.

Thus with a single sample the total error variance is underestimated or overestimated according as

$$\sum_{i=1}^{N} (\overline{V}_i + V_{i\cdot}) \gtrless -\sum_{i=1}^{N}\sum_{j(\neq i)=1}^{N} (\overline{C}_{ij} + C_{ij\cdot}). \quad (2.3.3)$$

Under simplified conditions G',

$$E\{v(\hat{T})\} - V(\hat{T}) = -\left\{\sum_{i=1}^{N} \hat{V}_i + \sum_{i=1}^{N}\sum_{j(\neq i)=1}^{N} \hat{C}_{ij}\right\}$$

$$= -V\left\{\sum_{i=1}^{N} X_i\right\}, \quad (2.3.4)$$

so that the total error variance is always *underestimated*. Note that $V\{\Sigma_1^N X_i\}$ is independent of the sampling design. Thus under these conditions, with a single sample the bias of underestimation can *only* be controlled by controlling the survey conditions.

2.4. Estimation of variance with interpenetrating samples

By the operations (1), (2) and (3) specified in Section 2.1 we have a set of k interpenetrating samples $\{s_1, s_2, \ldots, s_k\}$. On the question of the occurrence of identical samples or common units it is important to recall the points under (b) of Section 2.1. Each s is surveyed independently by its own team of investigators. The sample data from each is also processed separately and independently. Thus the conditions G apply to all samples so that (2.1.6), (2.1.7) and (2.1.8) hold.

Additionally it would be realistic to assume that the covariance between the observation on a unit of s_r and the observation on a unit of s_w is zero, unless the pair of units in question are identical, in which case we let

$$\text{Cov}(X_i, X_i' \mid s_r, s_w, G) = C_{i s_r i s_w}, \quad s_r, s_w = 1, 2, \ldots, \mathscr{C}(\mathscr{S}) \quad (2.4.1)$$

where X_i and X'_i are the observed values of the common unit u_i obtained by the respective teams investigating s_r and s_w, and where $\mathscr{C}(\mathscr{S})$ is the cardinality of $\mathscr{S}$, for which the new symbols r and w index all possible distinct samples from 1 to $\mathscr{C}(\mathscr{S})$. In relation to any real situation C_{is_r, is_w} is likely to be positive.

Let $T(s_1), T(s_2), \ldots, T(s_k)$ be the estimators constructed according to (2.2.2). The β-coefficients for them are chosen from any of the possible solutions of the N equations given by (2.2.4).

Under G, a biased estimator of T is

$$\tilde{T} = \sum_{r=1}^{k} T(s_r)/k, \qquad (2.4.2)$$

with variance given by

$$V(\tilde{T}) = \frac{1}{k^2}\left[\sum_{r=1}^{k} V\{T(s_r)\} + \sum_{r=1}^{k}\sum_{w(\neq r)=1}^{k} \operatorname{Cov}\{T(s_r), T(s_w)\}\right]. \qquad (2.4.3)$$

The formula for $V\{T(s_r)\}$ is given by (2.2.6), and its special form by (2.2.15), and is the *same* for $r = 1, 2, \ldots, \mathscr{C}(\mathscr{S})$ because of the uniformity of selection procedure as defined at (1), (2) and (3) of Section 2.1; let us denote it by V.

It can be shown for r or $w = 1, 2, \ldots, \mathscr{C}(\mathscr{S})$, that

$$\operatorname{Cov}\{T(s_r), T(s_w)\} = \sum_{r=1}^{\mathscr{C}(\mathscr{S})}\sum_{w=1}^{\mathscr{C}(\mathscr{S})} p(s_r)p(s_w) \sum_{i \in s_r \cap s_w} C_{is_r, is_w} \equiv C,$$

where the summation $\Sigma_{i \in s_r \cap s_w}$ applies only when, for any pair (s_r, s_w), whether or not $r = w$, the intersection $s_r \cap s_w$ yields a non-null or overlapping set of ultimate sampling units. Recall that the occurrence of overlapping samples, varying in the *extent of overlap* from identical pairs down to pairs with only one common unit, *will decrease with increasing stratification and stages in sampling*; experience shows that this is usually the case with large-scale surveys where the s_r are multi-stage stratified samples.

Thus we find that (2.4.3) reduces to a simple formula

$$V(\tilde{T}) = \frac{1}{k}\{V + (k-1)C\}. \qquad (2.4.4)$$

Furthermore we may set $C = \rho V$, for which ρ may be termed the *overlap correlation* in the sense explained, so that

$$V(\tilde{T}) = \frac{V}{k}\{1 + (k-1)\rho\}. \qquad (2.4.5)$$

Note that $\rho > -1/(k-1)$, and in view of the foregoing explanations it is possibly a small positive value close to zero in most practical situations.

The usual formula for estimating $V(\tilde{T})$ is

$$v(\tilde{T}) = \sum_{r=1}^{k} \{T(s_r) - \tilde{T}\}^2 / \{k(k-1)\}. \tag{2.4.6}$$

Whatever the sampling design, under the conditions G_0, (2.4.6) provides an unbiased estimate of the sampling variance. Under G it can be shown that

$$E\{v(\tilde{T})\} = \{V(1-\rho)\}/k, \tag{2.4.7}$$

so that with (2.4.5) we find

$$E\{v(\tilde{T})\} - V(\tilde{T}) = -\rho V = -C. \tag{2.4.8}$$

The bias in $v(\tilde{T})$ is likely to be small, as already explained, and it stems from those components of covariance induced by overlapping samples which may occur only infrequently. *Thus the total error variance $V(\tilde{T})$ is estimated almost without bias by $v(\tilde{T})$.* However, note that (2.4.8) holds regardless of the degree of success (or failure) in the exercise of control over all ascertainment errors.

In practice (2.4.6), or its special forms which recognize stratification, is used ignoring the effect of the overlap correlation. When stratification is taken into account more degrees of freedom are available to increase the precision of the estimate of variance.

2.5. *Estimation of ratios*

In the context of the unified formulation given in Section 2.1, consider the estimation of the *true* universe ratio or estimand

$$R = \sum_{i=1}^{N} y_i \Big/ \sum_{i=1}^{N} x_i, \tag{2.5.1}$$

with data from k interpenetrating samples $s_1, s_2, \ldots, s_k$ collected under the conditions G and given by

$$\{s_r, X_i, Y_i : i \in s_r\}, \quad r = 1, 2, \ldots, k. \tag{2.5.2}$$

The Y-observations like X's are subject to ascertainment errors whose variation and covariation with other Y's, and with corresponding X's, may be modelled in the same way as indicated at (2.1.6), (2.1.7) and (2.1.8).

It has already been shown in Section 2.2 that no best estimator for estimating $\Sigma_1^N x$, and for that matter $\Sigma_1^N y$, exists. We conjecture that none also exists for R.

With data from (2.5.2) let $x(s_r)$ and $y(s_r)$, $r = 1, 2, \ldots, k$, be estimates of $\Sigma_1^N x$ and $\Sigma_1^N y$, each of the form given by (2.2.14), which is fairly general.

Two types of well known ratio estimators are the separate ratio estimator given by

$$R' = \frac{1}{k} \sum_{r=1}^{k} \{y(s_r)/x(s_r)\} \equiv \frac{1}{k} \sum_{r=1}^{k} R'_r, \qquad (2.5.3)$$

and the combined ratio estimator given by

$$R'' = \left\{ \sum_{r=1}^{k} y(s_r)/k \right\} \bigg/ \left\{ \sum_{r=1}^{k} x(s_r)/k \right\} \equiv \frac{\bar{y}}{\bar{x}}. \qquad (2.5.4)$$

Under G, using the Hartley–Ross (1954) inequality it can be shown that

$$\frac{|E(R'') - R_G|}{\sqrt{V(R'')}} < \frac{1}{\sqrt{k}} \frac{\sqrt{V\{x(s_r)\}}}{E\{x(s_r)\}} \quad \text{and} \quad \frac{|E(R') - R_G|}{\sqrt{V(R')}} < \frac{\sqrt{V\{x(s_r)\}}}{E\{x(s_r)\}},$$

where $R_G = E\{y(s_r)\}/E\{x(s_r)\}$. Note that, under G_0, $R_G = R$ the true ratio. Hence with the hope of avoiding as much bias as possible the estimator R'' is to be preferred.

Murthy and Nanjamma (1959) have shown that $V(R')$ and $V(R'')$ are approximately the same under the ideal conditions G_0 when all observations are free from error. This approximation is likely to hold under G.

As a practical rule, R'' may therefore be used to estimate R taking $V(R')$ as its variance.

Ignoring the overlap correlations involved, an almost unbiased estimator of $V(R')$ is given by

$$v(R') = \sum_{r=1}^{k} (R'_r - R')^2 / \{k(k-1)\} \qquad (2.5.5)$$

and this value may be taken as $v(R'')$. Note that under the conditions G_0 it is an unbiased estimator of $V(R')$.

An estimator of $V(R'')$ can also be derived in its own right.

If s_r is recognized as a stratified sample from L strata and $x(s_r)$ and $y(s_r)$, $r = 1, 2, \ldots, k$, are expressed in terms of their respective stratum estimates x_{rl} and y_{rl}, $l = 1, 2, \ldots, L$, then $R'' = \bar{y}/\bar{x}$ can be rewritten as

$$R'' = \left\{ \sum_{l=1}^{L} \sum_{r=1}^{k} y_{rl}/k \right\} \bigg/ \left\{ \sum_{l=1}^{L} \sum_{r=1}^{k} x_{rl}/k \right\} = \sum_{l=1}^{L} \bar{y}_l \bigg/ \sum_{l=1}^{L} \bar{x}_l, \qquad (2.5.6)$$

where $\bar{y}_l$ and $\bar{x}_l$ are the means of the k samples in stratum l.

From the binomial series expansion of (2.5.6) and on the assumption that

$$\left| \sum_{l=1}^{L} \{\bar{x}_l - E(\bar{x}_l)\} / \sum_{l=1}^{L} E(\bar{x}_l) \right| < 1,$$

an approximate expression for $V(R'')$ can be derived, an estimate of which, given by Murthy (1963), is

$$v(R'') = \frac{1}{\bar{x}^2 k(k-1)} \sum_{l=1}^{L} \left[\sum_{r=1}^{k} (y_{rl} - \bar{y}_l)^2 - 2R'' \sum_{r=1}^{k} (y_{rl} - \bar{y}_l)(x_{rl} - \bar{x}_l) \right.$$

$$\left. + R''^2 \sum_{r=1}^{k} (x_{rl} - \bar{x}_l)^2 \right], \qquad (2.5.7)$$

where the overlap correlations are ignored. Note that there are more degrees of freedom for estimating $V(R'')$ than for $V(R')$.

Without the technique of interpenetrating samples the estimators given by (2.5.5) and (2.5.7) would not have been possible. Both take into account errors of ascertainment in the observations that are part of the numerator and denominator components of the respective ratios. Possibly the effect of the functional form of a ratio estimate is to dampen the joint effect of these errors in the estimated variances $v(R')$ and $v(R'')$. We shall recur to this question in Section 4 in respect to the former estimate.

While on the subject of ratios it can be said that the variance of a price index number, which is basically a ratio involving weights, prices and quantities, can be estimated by the use of the technique (Koop, 1960). Furthermore the recognized structure of an index number, with its groups of sub-indexes, allows for its computation in as many forms as are validly permitted by the combinatorial structure of the replication process. This important feature provides the basis for variance estimation.

2.6. McCarthy's method of variance estimation

For a universe with L strata, McCarthy (1966) noticed that for the case of two units drawn independently from each stratum, thus resulting in samples s_1 and s_2, a set of 2^{L-1} complement pairs of combined ratio estimates are available for estimating the variance of the arithmetic mean of the estimates from any complement pair.

He chose the orthogonal subset (because of its property of balance) consisting of a number of pairs of estimates equal to four times the integral part of $(L+3)/4$, i.e. say $L_0 = 4[(L+3)/4]$. These L_0 pairs are identified by the rows of the $L_0 \times L_0$ matrix of + and − signs, constructed by the method of Plackett and Burman (1946), after the deletion of any set of $L_0 - L$ columns. The columns of any such matrix are orthogonal to one another, and hence the property of balance.

The L plus or minus signs of a row of the resulting $L_0 \times L$ matrix now provide a convention for the choice of the (x, y)-pair of estimates either from sample s_1 or from sample s_2 for the computation of a combined ratio estimator and its complement, a $+$ sign for a stratum indicating choice from s, and a $-$ sign indicating choice from s_2. Thus L_0 complement pairs (R_1'', R_2'') are formed.

Note that this method can also be applied for computing complement pairs of estimates of other forms, e.g. correlation and regression coefficients with which McCarthy (1969) was later concerned.

The average of the L_0 estimated variances, each of the form $(R_1'' - R_2'')^2/4$ yields another estimate of $V(R'')$.

By this procedure computational work is reduced to a manageable level and at the same time the reliability of the variance estimate is enhanced.

Of course if L is small the entire set of complement pairs may be used. Alternatively, if L is very large, the strata can be collapsed into a smaller number of groups, say even six (thereby reducing the number of complement pairs to 32), to enable the use of all.

Ascertainment errors and their ramifications were not part of McCarthy's (1966, 1969) logical framework. Obviously his method has roots in Mahalanobis' technique.

2.7. Probability statements

The considerations which follow are in the framework of Section 2.1 except that the model assumptions at (2.1.6), (2.1.7), (2.1.8) and (2.4.1) are deleted. Thus the conditional probabilities implicit in them no longer apply. However, it is still realistic to postulate that the conditions G continue to hold without them.

Any set of k estimates from the independent samples, $s_1, s_2, \ldots, s_k$ for any given estimand are thus strictly independent in the context of the probability system P, despite the possibility that common units between the samples may induce similar observational records.

Let A be a parameter, or estimand of interest, that is a function of the true values given by (2.1.2). For examples, it may be a universe total $\Sigma_1^N x_i$ or a correlation coefficient

$$\left\{\sum_1^N (x_i - \mu_x)(y_i - \mu_y)\right\} \bigg/ \sqrt{\sum_1^N (x_i - \mu_x)^2 \sum_1^N (y_i - \mu_y)^2}.$$

Let $a(s_1), a(s_2), \ldots, a(s_k)$ be estimates of A based on relevant data from $s_1, s_2, \ldots, s_k$ respectively, collected under the conditions G. The set of possible estimates is $\{a(s): s \in \mathcal{S}\}$. We specify that A is such that $a(s)_{\min} < A < a(s)_{\max}$. Then the sampling distribution of $a(s)$ generated by (P, R) is given by

$$\{(a(s), p(s)): s \in \mathcal{S} \text{ and } a(s)_{\min} < A < a(s)_{\max}\}, \qquad (2.7.1)$$

and $a(s_\theta)$, $\theta = 1, 2, \ldots, k$, are estimates from (2.7.1).

Let

$$\bar{a} = \sum_{\theta=1}^{k} a(s_\theta)/k,$$

and

$$s_{\bar{a}}^2 = \sum_{\theta=1}^{k} \{a(s_\theta) - \bar{a}\}^2/\{k(k-1)\}.$$

Then with the following assumptions:
(i) $E\{a(s)\} = A$,
(ii) $(\bar{a} - A)/s_{\bar{a}}$ is distributed at t with $k - 1$ degrees of freedom,

(2.7.2)

we have the usual confidence interval $(\bar{a} - t_\alpha s_{\bar{a}}, \bar{a} + t_\alpha s_{\bar{a}})$ for A where t_α is the t-value with $k - 1$ degrees of freedom corresponding to confidence coefficient $1 - 2\alpha$.

Assumption (i) at (2.7.2) holds under G_0, if $a(s)$ is a linear form, with appropriate weights or coefficients, and A is correspondingly linear, but it may not hold under G, unless the accertainment errors completely cancel each other when its mathematical expectation is taken. However, it is unlikely to hold even under G_0 if $a(s)$ is a regression coefficient or even a ratio of linear forms.

Assumption (ii) is crucial. It may hold to a good approximation if $a(s)$ is a linear form and if $n(s)$ is very large, but under G the outcome is less certain. Whether under G or G_0 we know from standard theory that it is very unlikely to hold if $a(s)$ is non-linear.

It is essentially due to the foregoing reasons that non-parametric methods for inferring about A were proposed. Under the conditions G or G_0 if the median of the sampling distribution (2.7.1) is not appreciably different from A, then

$$P\{a_a < A < a_k\} = 1 - (\tfrac{1}{2})^{k-1}, \tag{2.7.3}$$

where a_1 and a_k are the least and greatest estimates in the set $\{a(s_1), a(s_2), \ldots, a(s_k)\}$. Furthermore when $k \geq 4$, if a_2 and a_{k-1} are the second smallest and second largest estimates respectively in the set, then also

$$P\{a_2 < A < a_{k-1}\} = 1 - (k+1)(\tfrac{1}{2})^{k-1}. \tag{2.7.4}$$

These probability statements, derived by binomial arguments, are due to Savur (1937). Although Student (1908) did not make generalized probability statements he was aware of the problem for the special case of two observations.

Koop (1979) has shown that the distributions of estimates from survey data, of whatever functional form, have kurtosis coefficients whose values are quite different from 3 under the conditions G_0 because of restrictions on randomization. It can also be shown that the skewness coefficients can be different from 0. The sample size $n(s)$, although implicit in the formulas for those coefficients, does not appear to effect these conclusions. Under the conditions G it is problematical

3. Applications

In this section two sample surveys carried out by the use of the technique of replicated or interpenetrating samples will be described.

For each the procedure of estimation and inference will be briefly discussed.

3.1. Forage crops survey in Alberta

In August 1969 a sample survey,[3] having as its prime objective the estimation of total acreage of forage crops (composed of various defined varieties of clover and grasses) in the area of the Peace River stretching northwards from Grande Prairie to Fort Vermillion, was conducted by the Agriculture Division of Statistics Canada in cooperation with the Agriculture Department of the Province of Alberta. The main reason for the survey was that precise estimates for these forage crops had not been available before.

The frame consisted of large-scale photographic maps of the area in question.

This area was divided into rectangular segments each measuring 2 miles × 1 mile, and eventually yielded a universe of 8500 segments which were then grouped into 22 contiguous strata. The segments are the first-stage sampling units.

The resources for field work allowed a total sample size of 448 segments.

Two independent stratified random samples each consisting of 224 segments were selected. In each stratum for each sample the segments were selected by simple random sampling without replacement.

The composition of each sample, designated as A and B, was as follows:

Stratum	1	2	3	4	5	6	7	8	9	10	11
N_h	105	151	15	83	302	349	552	314	389	13	29
n_h	5	7	2	4	15	16	25	16	18	2	3

Stratum	12	13	14	15	16	17	18	19	20	21	22	Total
N_h	595	656	444	510	434	543	630	711	962	379	334	8500
n_h	11	12	8	9	8	10	12	13	17	7	4	224

---

[3] I thank I. P. Fellegi and W. L. Porteous of Statistics Canada for permission, graciously given, to reproduce the results of the survey and also J. M. Gray, Y. S. Hwang, T. Yabut and Gordon Henderson with whom I was privileged to be associated in this survey as an advisor.

In this table N_h is the number of segments in stratum h and n_h is the sample size in that stratum for each sample.

A training session was held prior to the field enumeration to explain the objectives of the survey, the survey manual, the rules for ascertaining the acreages of each of the crops found in a segment from the farmer, and the method of aligned two-dimensional systematic sampling of 16 points in a random half of the segment, to be used if the farmer could not be contacted.

One point is to be located systematically in each of the grids of a selected half-segment consisting of four rows of grids, each row with four square grids of equal size.

The method of aligned systematic sampling was chosen because of its simplicity and convenience.

The systematic alignment of points is obtained as follows. A starting point is fixed by a pair of random co-ordinates in the north-west corner grid of the selected random half-segment which is either the top or the bottom half. In each of the remaining 15 grids, the same pair of random co-ordinates with the north-west corner point as origin fixes the position of the random point. Thus the crop identified at a selected point by an enumerator in the field will contribute 80 acres to the estimate for that particular segment. Note that the point-selection and the half-segment selection procedures constitute the third and second stages of sampling respectively.

The following circumstances of the field survey are noteworthy.

(a) The samples A and B, though clearly identified in office records, were not distinguished in the field.

(b) Three teams, each with two enumerators, completed the work in 25 days. During the survey they resided in their respective assigned areas. They had acquired some valuable experience from a similar survey conducted in August 1968 in the same area.

(c) There was no missing data.

The data was processed with a Wang computer fitted with an electric typewriter for typing out the computed estimates.

Let y_{hi} be the acreage of a given crop in the ith segment of the hth stratum, $i = 1, 2, \ldots, N_h$, $h = 1, 2, \ldots, 22$. The estimand is

$$T = \sum_{h=1}^{22} \sum_{i=1}^{N_h} y_{hi}. \tag{3.1.1}$$

With data from samples A or B an estimate of T is given by

$$\hat{T}_\alpha = \sum_{h=1}^{22} N_h \bar{y}_{h\alpha} \equiv \sum_{h=1}^{22} \hat{T}_{h\alpha}, \quad \alpha = A, B, \tag{3.1.2}$$

where $\bar{y}_{h\alpha}$ is the mean of the estimated segment acreages in stratum h with data from sample α, $\alpha = A, B$. Then a combined estimate of T is

$$T' = \tfrac{1}{2}(\hat{T}_A + \hat{T}_B), \tag{3.1.3}$$

with estimated total variance

$$v(T') = \sum_{h=1}^{22} (\hat{T}_{hA} - \hat{T}_{hB})^2/4, \qquad (3.1.4)$$

which includes estimates of components of variance and covariance due to errors of ascertainment if model assumptions about these errors are made as in Section 2.

If the observations are free from errors, T' and $v(T')$ are unbiased estimates of T and $V(T')$ respectively.

Estimates for various crops and land use were computed. The following table shows estimates for forage crops, the crop of chief interest to the sponsoring departments, by strata and by sample.

Table 1
Forage crops acreage estimates to nearest 100 acres

Stratum	Sample A	Sample B
1	17	84
2	621	475
3	0	12
4	0	0
5	266	395
6	297	192
7	291	380
8	196	196
9	372	251
10	3	0
11	0	0
12	0	22
13	306	590
14	44	222
15	91	0
16	109	152
17	500	586
18	168	210
19	153	131
20	226	45
21	0	65
22	0	0
Total	3660	4008

First and foremost, the rough agreement between the estimates of totals from samples A and B is remarkable considering that the sample size for each is only 2.6% of the total number of area segments, and far less in terms of 'points'.

An estimate of T according to (3.1.3) is $T' = (3660 + 4008)/2 = 3834$, with estimated variance, according to (3.1.4), $v(T') = 61\,442.5 = (247.9)^2$, yielding a coefficient of variation of 6.47%.

Since there is one degree of freedom for estimating the variance in each stratum, $v(T')$ may be taken to have 22 degrees of freedom. If the assumptions at (2.7.2), specialized for the case when $a(s)$ represents T', hold then the confidence interval for T with confidence coefficient 0.95 is

$$(3834 - 2.074 \times 247.9,\ 3834 + 2.074 \times 247.9)$$

where 2.074 is the t-value for 22 degrees of freedom. Thus the interval is (3320, 4348).

Note that a quick estimate of the standard error of T' is $|\hat{T}_A - \hat{T}_B|/2$ which comes to $(4008 - 3660)/2 = 174$. It is smaller than the former estimate. However, a confidence interval for T with the same confidence coefficient and with one degree of freedom would obviously yield a much wider interval.

If distributional assumptions are avoided except that the median of the distribution of possible T''s, all subject to error, is T, then the probability that the interval (3660, 4008) covers T is 0.5. In the next sub-section this problem will be considered further with more expansive data.

The nature of the estimator $\hat{T}_\alpha$, $\alpha = A, B$ and its variance function needs to be explained. We shall consider the problem corresponding to the situation when the farmers cannot be contacted in all selected segments of a stratum.

Dropping the sample and stratification identification subscripts, and assuming that there are no ascertainment errors, an unbiased estimate of the area under a specified crop is

$$\hat{T} = \frac{N}{n} \sum_{i=1}^{N} A \frac{2 \sum_{f=1}^{16} g_{idf}}{16} = \frac{N}{8n} \sum_{i=1}^{n} \sum_{f=1}^{16} g_{idf}, \qquad (3.1.5)$$

where g_{idf} is the characteristic random variable of grid f in half-segment d of segment i ($f = 1, 2, \ldots, 16$, $d = 1, 2$, $i = 1, 2, \ldots, n$) which assumes the value 1 with probability p_{idf} if the point falls on the crop and zero otherwise with probability $1 - p_{idf}$, and where A is the area of a half-segment consisting of 16 square grids.

Recalling that A is one square mile in area we have set $A = 1$ in (3.1.5). The estimate $\Sigma_f g_{idf}/16$ is an unbiased estimate of the proportion of the half-segment under the crop, and the multiplier 2 is the raising factor required to obtain an unbiased estimate of the area of the crop for the whole segment.

By the theorems expressing expected value and variance in terms of conditional expectations and conditional variances (Tschuprow, 1928) it can be shown that $E(\hat{T}) = \Sigma_1^N y_i$ and

$$V(\hat{T}) = N^2 \left[\frac{1}{n} \frac{\sum_{i=1}^{N}(y_i - \bar{\bar{y}})^2}{N-1} \left(1 - \frac{n}{N}\right) + \frac{1}{nN} \sum_{i=1}^{N} (y_{i1} - y_{i2})^2 \right.$$

$$+ \frac{1}{128nN} \sum_{i=1}^{N} \sum_{d=1}^{2} \left\{ \sum_{f=1}^{16} p_{idf}(1 - p_{idf}) \right.$$

$$\left. \left. + \sum_{f \neq f'} \text{Cov}(g_{idf}, g_{idf'}) \right\} \right], \qquad (3.1.6)$$

where

y_i = area of the crop in segment i,

y_{id} = area of the crop in half-segment d, $d = 1, 2$,

and

$\bar{\bar{y}} = \sum_{1}^{N} y_i/N$, the crop area per segment in the stratum.

Note that y_{i1} and y_{i2} are at most one square mile.

The covariance, $\text{Cov}(g_{idf}, g_{idf'})$, between any two grids f and f' of a given half-segment accounts for the covariation induced by the systematic alignment of the points vertically and horizontally with respect to the key random point located in the north-west corner grid. It can be evaluated geometrically, given the map of the crop in the half-segment (Koop, 1976). The entire covariance function $\Sigma_{i,d,f \neq f'} \text{Cov}(g_{idf}, g_{idf'})$, as also the preceeding variance function, depends on the areal distribution of the crop in the half-segments of the stratum at that period of time. If formal names are needed they may be called topographic covariance and topographic variance which are naturally time-specific.

The second variance function measures the variation due to sub-sampling one half-segment in each selected segment. The first will be recognized as the variance for simple random sampling.

If there are erros in locating the random points in the field, then crops may be misclassified if the points in question, with respect to their true points, indicate crops different from the ones on which the respective true points fall. Thus there will be errors in acreage estimates. With model assumptions similar to (2.1.6), (2.1.7) and (2.1.8), the variance function for $\hat{T}$ will be composed of four components similar to those of (3.1.6), augmented by four parametric functions similar to the last four of (2.2.15). Thus the total error variance is quite involved, in part due to systematic sampling.

The foregoing considerations indicate what the estimator of total error variance, given by (3.1.4), has accomplished. With a single stratified sample of 448 segments, at best there would be an approximate estimate of variance with no assurance that the total error variance is satisfactorily estimated.

3.2. The National Sample Survey of India: Eleventh and Twelfth rounds

In its eleventh and twelfth rounds, during the period August 1956 to August 1957, the National Sample Survey (1962), in addition to other enquiries, collected information on wages, employment, income, expenditure and indebtedness of agricultural labor households in the entire rural area of India, excepting the outlying islands in the Bay of Bengal and the Indian Ocean.

Each round of the survey was of six months duration. The entire field survey was organized in 12 sub-rounds, with equal work loads for each. Each sub-round was to be completed in a period of one month. The field work was completed as planned without any break between the two rounds.

It may be said that this approach was necessary for the purpose of obtaining monthly records, which otherwise would have been liable to, perhaps, more recording or reporting errors for longer memory periods.

The entire rural area of the 19 States of India (before their reorganization into 14 States which took effect on November 1, 1956) was divided into 72 strata, each with a rural population of approximately the same size except for a few.

The first-stage sampling units in a stratum were villages. The households of a village engaged in agricultural labor were the second-stage sampling units.

The sampling procedure was as follows. A certain number of villages were selected from each stratum with probability proportional to the population size of the village and with replacement. From each selected village agricultural labor households were selected systematically from a list compiled at the village. There was a rule prescribed for the selection of between 4 to 8 agricultural labor households; in practice it turned out that about 6 housholds were selected per village.

From each stratum four interpenetrating samples (of villages) were selected, the overall first-stage sample size being a multiple of 48, depending on the size of the rural population. This procedure of allocation ensured that at least one village was selected for each of the sub-rounds to represent the rural area of the stratum.

In all 3696 villages were selected from the 72 strata. However, of this number 56 villages, i.e. 1.54 percent, could not be surveyed for some reason or other. Substitutions for 48 villages were made at random from the relevant strata. No substitution for the remaining 8 sample villages was possible because they constituted the entire stratum for two monthly sub-rounds. Thus one stratum provided no data for two out of the 12 monthly sub-rounds.

To summarize, data was obtained from 20 720 agricultural labor households in 3688 sample villages constituting the four independent interpenetrating samples, each sample having 922 villages.

The average characteristics of agricultural labor households were of interest to the Government of India. These are in the nature of ratios, e.g. total net income during the year divided by the total number of households of a given State. As the result of the reorganization of States, some of their boundaries crossed stratum boundaries. Hence estimates for States, or other defined areas, are really estimates for domains of study in the terminology of Yates (1949, 1981). Thus estimates computed were ratio estimates for two-stage stratified sampling, with an estimate of a total for the characteristic of interest in the numerator, and a similar estimate in the denominator for the total number of agricultural labor households in the domain of study. The formulas for the numerator and denominator estimates, appropriate to this sampling design, are well known and given on page 166 of the report under reference.

As an example a table of estimates showing the average annual net income (in rupees) per agricultural labor household is given by regions, consisting of the

Table 2
Average annual total net income (rupees) per agricultural labour household by sample

Region	Sample				
	1	2	3	4	combined
(1)	(2)	(3)	(4)	(5)	(6)
1. Uttar Pradesh	369.66	344.55	344.52	339.87	354.74
2. Madhya Pradesh	326.98	356.44	417.81	320.66	350.86
3. Central Zone	352.90	360.22	370.56	332.19	352.88
4. Bihar	349.50	396.59	384.69	393.18	380.43
5. West Bengal	558.89	579.33	560.93	496.00	547.06
6. Orissa	347.62	271.76	299.70	260.12	296.49
7. Assam, Manipur and Tripura	765.22	755.26	984.85	506.54	777.52
8. Eastern Zone	420.09	438.98	432.39	401.22	423.67
9. Andhra Pradesh	324.14	336.26	409.50	357.16	355.08
10. Madras	284.92	269.47	278.86	244.07	269.25
11. Kerala	394.05	340.57	443.89	419.92	401.30
12. Southern Zone	316.26	309.22	360.34	315.72	325.08
13. Bombay	421.67	435.49	415.61	400.70	418.05
14. Mysore	388.60	398.96	408.91	354.50	389.73
15. Western Zone	411.30	422.71	413.02	386.92	408.87
16. Rajasthan	311.02	308.08	312.28	435.90	342.95
17. Punjab, Delhi and Himachal Pradesh	893.68	697.47	696.29	693.27	740.93
18. Jammu and Kashmir	225.83	392.96	437.94	476.77	356.63
19. Northern Zone	684.57	573.29	600.49	615.12	618.92
20. all India	383.43	385.51	404.79	366.98	385.38

reorganized States, and by sample. This table (Table 2) is essentially Table (E.1.11) of the report with a few minor deletions, considered appropriate for the purpose of this illustration. The combined estimates shown in column (6) of the table are computed as combined ratio estimates.

For a State or Zone the probability that the true average lies between the least and greatest estimates is $1 - (\frac{1}{2})^3 = 0.875$.

Even under the assumption that there are no ascertainment errors the variance functions for such ratio estimates are approximations. With the recognition of error, and consequently model assumptions similar to (2.1.6), (2.1.7) and (2.1.8) for the numerator and denominator variates, the approximations will be further complicated by the addition of 12 parametric functions. Eight of the parametric functions stem from the numerator and denominator variates, and four from the covariation between the respective numerator and denominator variates. Certainly such expressions will be formidable.

However, with the technique of independent replication the reader is reminded that the variance of the mean of any set of four sample estimates in the table is given simply by (2.5.5). For the State of Bombay the mean is 418.37 with estimated variance of $51.99 = (7.21)^2$; thus the estimate has a coefficient of variation of 1.7%. This estimate of variance may also be taken as the estimate of variance of the combined ratio estimate 418.05, given in the sixth column of the table. Note that it is an almost unbiased estimate of the total error variance implied by the model assumptions. Both these ratio estimates suffer from negligible biases, the former having a bias of 0.44 and the latter 0.11, according to formulas given by Murthy and Nanjamma (1959, p. 384).

With a single sample of equivalent size, estimation of variance, regardless of the degree of error in the observations, would be almost impracticable; this difficulty is in part compounded by the systematic selection of households.

4. Analysis of variance to study methods of investigation

In the eighth round of the National Sample Survey (1961) of India, conducted in the period July 1954 to April 1955, an opportunity to organize the entire field work in 19 of its States to enable a study of the differences between investigating teams under the administrative control of two separate types of agencies, viz. the Directorate of National Sample Survey and the respective State organizations (usually an agricultural department), presented itself.

For this round of the survey 345 strata, which did not cross State boundaries and with each having about the same size of agricultural population, had each been allocated 12 independent interpenetrating samples, with the exception of a few strata which, in some States, had more. In those States with more than 12 samples per stratum, only 12 in each stratum were selected. Thus for the purpose of the study there were 12 independent samples per stratum in each State.

Each of the 12 samples in each stratum was a village selected at the first-stage with probability proportional to population (or geographical area in the absence of population figures) and with replacement. Within a State, 4 villages selected in each stratum were allocated to the Directorate of National Sample Survey, the Central agency, and 8 villages selected in each stratum to the State agency. Within each selected village, the agricultural households were sub-stratified into three groups and households were selected systematically; this constituted the second-stage of sampling in each stratum. Thus 12 independent estimates for any given characteristic were possible for each stratum or each State.

In each State the organization of field work was as follows. Each agency controlled two parties of investigators. While the Central agency assigned two samples to each of its parties the State agency assigned four samples to each of its parties. Further, each agency processed its own data separately and independently.

Thus for any characteristic of interest 12 independent estimates would be available. The characteristic of interest chosen for each State was the area owned per household. Accordingly, 12 independent estimates of the area owned per household were computed for each of the 19 States. For each State the analysis of variance was applied to the 12 estimates. Note that the experimental units are the 12 samples, *not* their individual observations.

An estimate of the *true* ratio R_0 for each State, based on any of the samples, is of the following form:

$$R = \frac{\text{Linear estimate of area owned by agricultural households}}{\text{Linear estimate of the number of agricultural households}}.$$

Let R_{apk} be such an estimate from data of the kth sample of party p of agency a where $a = 1, 2$, $p = 1, 2$, and $k = 1, 2, \ldots, n_a$ for which $n_1 = 2$ if $a = 1$, indicating the Central agency, and $n_a = 4$ if $a = 2$, indicating the State agency. Further corresponding to R_{apk}, let $\hat{R}_{apk}$ be the *true* ratio estimate that would have resulted in the absence of influences of 'effects' introduced by agency auspices or party differences within agencies, for all relevant a, p and k. It will be assumed that $\hat{R}_{apk}$ is *virtually free from bias* because for any sample it is based on a fairly large number of observed households, varying from 69 to 89 observed households, in three small States, to 871 in the largest State, according to Table (2.3) of the 1961 report.

Then for any given State

$$E(\hat{R}_{apk} - R_0) = 0, \quad (4.1)$$

and

$$V(\hat{R}_{apk}) = V, \quad (4.2)$$

for all a, p and k, because the 12 independent samples are drawn by the same sampling procedure. Hence the basic conditions required for a valid and

Table 3

Source of variation	Sum of squares	Degrees of freedom
Between agencies	$2 \sum_{a=1}^{2} \sum_{k=1}^{n_a} (R_{a..} - R_{...})^2 \equiv \text{SSA}$	1
Between parties within agency	$\sum_{a=1}^{2} n_a \sum_{p=1}^{2} (R_{ap.} - R_{a..})^2 \equiv \text{SSP}$	2
Error	$\sum_{a=1}^{2} \sum_{p=1}^{2} \sum_{k=1}^{n_a} (R_{apk} - R_{ap.})^2 \equiv \text{SSE}$	8
Total	$\sum_{a,p,k} (R_{apk} - R_{...})^2$	11

meaningful analysis of variance are satisfied. Symbolically we have the analysis of variance, shown in Table 3.

In this analysis

$$R_{...} = \sum_{a, p, k} R_{apk}/12, \quad R_{a..} = \sum_{p=1}^{2} R_{ap.}/2 \quad \text{for } a = 1, 2$$

and

$$R_{ap.} = \sum_{k=1}^{n_a} R_{apk}/n_a \quad \text{for } a = 1, 2 \text{ and } p = 1, 2.$$

If circumstances are such that agency effects and party differences are entirely absent, then $R_{apk} = \hat{R}_{apk}$ for all a, p and k, and consequently it can be shown that

$$E(\text{SSA}) = E(\text{SSP}/2) = E(\text{SSE}/8) = V. \tag{4.3}$$

Thus even if F-tests are not applied the ratios of mean squares, viz. 8 SSA/SSE and 4 SSP/SSE are useful indicators of differences between agencies, and between parties within agency, respectively.

We reproduce the results of the analysis for each State given as Table (12) by Mahalanobis and Lahiri (1961) in our Table 4. Under the assumption that the F-distribution holds, the two authors noticed that there were significant differences between parties within agency for the States of Uttar Pradesh, Assam and Mysore. Agency differences were also significant for Uttar Pradash and Jammu and Kashmir.

The effect of the functional form of the ratio estimate R, which can possibly dampen the effect of errors of ascertainment in its numerator and denominator, to the extent that such errors are almost uniform increments of their true values, may also be a consideration relevant to the question of the significance of differences. Indeed this consideration appears to be implicit in the discussion presented by the two authors, particularly in their paragraphs 66 and 67.

If model assumptions similar to (2.1.6), (2.1.7) and (2.1.8) are used the resulting expressions for the expected values of the mean squares will be extremely complicated simply because the estimates under analysis are ratios of linear forms.

Furthermore, if linear estimates had been considered a fixed linear model approach with only 7 parameters for each State will provide a technically easier and simpler interpretation of an analysis of variance. For this *theoretical* case if y is a linear estimate subject to interviewer and party effects, the appropriate residual that would permit a valid and meaningful analysis would be $\hat{y} - y_0$, where $\hat{y}$ is the corresponding unbiased estimate free from all effects, and y_0 is the *true* value. In the context of the foregoing experimental design let

$$y_{apk} = y_0 + \alpha_a + \gamma_{ap} + (\hat{y}_{apk} - y_0) \tag{4.4}$$

where α_a is the effect of agency a, γ_{ap} is the effect of the pth party of agency a, and $\hat{y}_{apk} - y_0$ is the usual residual error for all a, p and k. The usual assumptions

Table 4
Analysis of variance of average area owned per household; landholdings enquiry, 1954–1955

State	Sum of squares			Mean square			F	
	agency d.f. = 1	party d.f. = 2	error d.f. = 8	agency	party	error	agency	party
(1)	(2)	(3)	(4)	(5)	(6)	(7)	(8)	(9)
1. Uttar Pradesh	0.252	0.176	0.150	0.252	0.088	0.019	13.26**	4.63*
2. Bihar	0.004	0.246	0.919	0.004	0.123	0.115	28.75(r)	1.07
3. Orissa	0.001	0.381	1.864	0.001	0.190	0.233	233.00(r)	1.23(r)
4. West Bengal	0.583	0.249	0.998	0.583	0.124	0.125	4.66	1.01(r)
5. Assam	1.480	0.050	17.776	1.480	0.025	2.222	1.50(r)	88.88*(r)
6. Andhra	0.064	0.381	2.740	0.064	0.190	0.342	5.34(r)	1.80(r)
7. Madras	0.059	0.080	1.230	0.059	0.040	0.154	2.61(r)	3.85(r)
8. Mysore	0.254	0.0016	7.281	0.254	0.0008	0.910	3.58(r)	1137.5**(r)
9. Travancore & Cochin	0.441	0.631	2.395	0.441	0.316	0.299	1.47	1.06
10. Bombay	0.175	0.919	3.778	0.175	0.460	0.472	2.70(r)	1.03(r)
11. Saurashtra	15.072	5.941	71.547	15.072	2.970	8.943	1.69	3.01(r)
12. Madhya Pradesh	0.037	1.673	4.935	0.037	0.836	0.617	16.68(r)	1.35
13. Madhya Bharat	0.134	3.266	9.278	0.134	1.633	1.160	8.66(r)	1.41
14. Hyderabad	4.673	2.422	12.922	4.673	1.211	1.615	2.89	1.33(r)
15. Vindhya Pradesh	6.120	1.703	16.394	6.120	0.852	2.049	2.99	2.40(r)
16. Rajasthan	1.649	4.819	45.394	1.649	2.410	5.674	3.44(r)	2.35(r)
17. Punjab	1.397	3.945	9.973	1.397	1.972	1.247	1.12	1.58
18. PEPSU	0.771	4.990	25.181	0.771	2.495	3.148	4.08(r)	1.26(r)
19. Jammu & Kashmir	0.0004	0.083	1.334	0.0004	0.042	0.167	417.50*(r)	3.98(r)

$F_{0.05}(1, 8) = 5.32$, $F_{0.01}(1, 8) = 11.26$, $F_{0.05}(8, 1) = 238.9$, $F_{0.01}(8, 1) = 5982$, $F_{0.05}(2, 8) = 4.46$, $F_{0.01}(2, 8) = 8.65$, $F_{0.05}(8, 2) = 19.37$, $F_{0.01}(8, 2) = 99.37$.
(r) indicates error/party or error/agency in F-ratio.
* Significant at 5 per cent level. ** Significant at 1 per cent level.

$$\alpha_1 + \alpha_2 = \gamma_{11} + \gamma_{12} = \gamma_{21} + \gamma_{22} = 0, \tag{4.5}$$

are made to enable the solution of the 7 least squares equations for y_0, the α's and the γ's. Then for each source of variation, with the same notation for sum of squares, it can be shown that

$$E(\text{SSA}) = V_y + \tfrac{8}{3}(\alpha_1 - \alpha_2)^2,$$
$$E(\text{SSP}/2) = V_y + \gamma_{11}^2 + \gamma_{12}^2 + 2(\gamma_{21}^2 + \gamma_{22}^2), \tag{4.6}$$
$$E(\text{SSE}/8) = V_y,$$

where $V(\hat{y}_{apk}) = V_y$ for all a, p and k.

Thus with linear estimates, (4.6) provides the basis for assessing differences between effects even without assumptions involving the F-distribution.

5. Summary and comments

The reason and purpose for a unified approach may be questioned. The reason is that it establishes results, with an economy of effort, true under all survey conditions for all sampling designs and all types of estimators specified. The purpose is to use these results to explain the technique of interpenetrating samples which, unlike the techniques of classical theory, explicitly recognizes ascertainment and other errors.

Consequently, it has been possible to demonstrate in Section 2.2 that under conditions of error there are no estimators, linear in the observed variates as defined in Section 2.1, with the least mean square error. Furthermore, a plausible conjecture that there are no ratio estimators with least mean square error, under the same conditions of error, has been made tenable. All this is important for the validation of estimators normally used with data obtained through the technique. The functional forms of such estimators are largely determined by the nature of the factual information often needed to address a variety of social and economic questions. And because there are no best estimators for any given estimand their use is at least justified.

The technique introduces simplicity in variance estimation which, under conditions of error, also makes possible almost unbiased estimation of total error variance, which is poorly estimated at greater expense with a single sample of equivalent size, despite the relative richness in the number of observations. Seemingly there are fewer degrees of freedom for estimating variance efficiently with the technique; but, if stratification is recognized, particularly with the sampling designs used in large-scale multi-purpose surveys, the number of degrees of freedom is usually adequate.[4]

The analysis of variance of data resulting from any given experimental design is a method in its own right. But combined with the technique, in the way demonstrated by Mahalanobis and Lahiri, it becomes a powerful tool even when ratio estimates are under analysis. It has yet to be demonstrated that other methods can accomplish equally well with a minimum of nonrestrictive assumptions. Fisher (1956) was fully aware of the importance of combining the technique of experimental design with that of sample surveys, both of which are founded on the fundamental principles of randomization, replication and control. There are literally thousands of experimental and sampling designs that may be candidates for investigational purposes, certainly without prohibitive cost. They await choice and application, almost without exception, to all survey problems where the differential effects stemming from causal factors other than those of random sampling need to be assessed or controlled.

Finally, it should be mentioned that besides descriptive uses, the data from any sample survey can be used for analytical purposes, in the sense defined by Deming (1950, 1953) and by Yates (1949, 1981), to investigate causal factors

---

[4] It should be mentioned that the writer became fully aware of this important fact only in 1961 from an answer to a question on this subject put to Sir Ronald Fisher.

behind an existing universe or to compare systems of such factors behind two or more strata or universes. With the technique of interpenetrating samples, when a field survey is conducted over a period of time, as in the example given in Section 3.2, comparisons between periods of time, to determine if causes have changed, are facilitated.

Acknowledgements

The author is grateful to the publishers of Sankhyā for permission to reproduce Table (E.1.11) from *Sankhyā Ser. B* **24** (1962) p. 181, appearing as Table 2, and to the International Statistical Institute for Table (12) from its *Bulletin* **38**(2) (1961) p. 427, appearing as Table 4.

I thank Mrs J. C. Koop for typing the chapter.

References

Deming, W. E. (1944). On errors in surveys. *Amer. Soc. Review* **9**, 359–369.
Deming, W. E. (1950). *Some Theory of Sampling*. Wiley, New York.
Deming, W. E. (1953). On the distinction between enumerative and analytic surveys. *J. Amer. Statist. Assoc.* **48**, 244–255.
Deming, W. E. (1956). On simplification of sampling design through replication with equal probabilities and without stages. *J. Amer. Statist. Assoc.* **51**, 24–53.
Fisher, R. A. (1956). *Statistical Methods and Scientific Inference*. Third edition published in 1971, Oliver and Boyd, Edinburgh.
Fisher, R. A. (1961). Discussion following the Mahalanobis–Lahiri paper. *Bull. Inter. Statist. Inst.* **38**(1), 110.
Flores, A. M. (1957). The theory of duplicated samples and its use in Mexico. *Bull. Inter. Statist. Inst.* **36**(3), 120–126.
Godambe, V. P. (1955). A unified theory of sampling from finite populations. *J. Roy. Statist. Soc. Ser. B* **17**, 269–278.
Hartley, H. O. and Ross, A. (1954). Unbiased ratio estimators. *Nature* **174**, 270–271.
Horvitz, D. G. and Thompson, D. J. (1952). A generalization of sampling without replacement from a finite universe. *J. Amer. Statist. Assoc.* **47**, 663–665.
Hubback, J. A. (1927). Sampling for rice yields in Bihar and Orissa. Bulletin No. 166. Agricultural Research Institute, Pusa. (Also in *Sankhyā* **7**, 281–294, 1946.)
Hansen, M. H. et al. (1951). Response errors in surveys. *J. Amer. Statist. Assoc.* **46**, 147–190.
Hansen, M. H., Hurvitz, W. N. and Bershad, M. (1961). Measurement errors in censuses and surveys. *Bull. Inter. Statist. Inst.* **38**(2), 359–374.
Jones, H. L. (1956). Investigating the properties of a sample mean by employing random subsample means. *J. Amer. Statist. Assoc.* **51**, 54–83.
Koop, J. C. (1960). On theoretical questions underlying the technique of replicated or interpenetrating samples. In: *Proceedings of the Social Statistics Section, American Statistical Association*, 196–205.
Koop, J. C. (1974). Notes for a unified theory of estimation for sample surveys taking into account response errors. *Metrika* **21**, 19–39.
Koop, J. C. (1976). Systematic sampling of two-dimensional surfaces and related problems. Research Triangle Institute, Research Triangle Park, NC.
Koop, J. C. (1979). On statistical inference in sample surveys and the underlying role of randomization. *Ann. Inst. Statist. Math.* **31**, 253–269.

Kpedekpo, G. M. L. and Chacko, V. J. (1972). The use of interpenetrating sample designs in Ghana with special reference to demographic studies. *Sankhyā Ser. B* **34**, 41–66.

Lahiri, D. B. (1954). Technical paper on some aspects of the development of the sample design: National Sample Survey Report No. 5. *Sankhyā* **14**, 264–316.

Lahiri, D. B. (1957a). Observations on the use of interpenetrating samples in India. *Bull. Inter. Statist. Inst.* **36**(3), 144–152.

Lahiri, D. B. (1957b). Recent developments in the use of techniques for assessment of errors in nation-wide surveys in India. *Bull. Inter. Statist. Inst.* **36**(2), 71–93.

Lahiri, D. B. (1964). Multi-subject sample-survey system some thoughts based on Indian experience. In: C. R. Rao, ed., *Contributions to Statistics. Presented to Professor P. C. Mahalanobis*. Statistical Publishing Society, Calcutta, 175–220.

Lieberman, M. D. (1958). The Phillipine statistical program development and the survey of households. *J. Amer. Statist. Assoc.* **53**, 76–88.

Madow, W. G. and Madow, L. (1944). On the theory of systematic sampling. *Ann. Math. Statist.* **15**, 1–24.

Mahalanobis, P. C. (1940a). A sample survey of acreage under jute in Bengal. *Sankhyā* **4**, 511–530.

Mahalanobis, P. C. (1940b). Discussion on planning of experiments. *Sankhyā* **4**, 530–531.

Mahalanobis, P. C. (1944). On large-scale sample surveys. *Phil. Trans. Roy. Soc. London Ser. B* **231**, 329–451.

Mahalanobis, P. C. (1946a). Recent experiments in statistical sampling in the Indian Statistical Institute. *J. Roy. Statist. Soc.* **109**, 325–378.

Mahalanobis, P. C. (1946b). Sample surveys of crop yields in India. *Sankhyā* **7**, 269–280.

Mahalanobis, P. C. (1950). Cost and securacy of results in sampling and complete enumeration. *Bull. Inter. Statist. Inst.* **32**(2), 210–213.

Mahalanobis, P. C. (1960). A method of fractile graphical analysis. *Econometrica* **28**, 325–351.

Mahalanobis, P. C. and Lahiri, D. B. (1961). Analysis of errors in censuses and surveys with special reference to experience in India. *Bull. Inter. Statist. Inst.* **38**(2), 401–433.

McCarthy, P. C. (1966). Replication: An approach to the analysis of data from complex surveys. PHS Publication No. 1000, Ser. E, No. 14, U.S. Govt. Printing Office, Washington, D.C.

McCarthy, P. C. (1969). Pseudo-replication: Half-samples. *Inter. Statist. Review* **37**, 239–264.

Mehok, W. J. (1963). An introduction to replicated sub-sampling. *Social Compass* **10**, 525–535.

Murthy, M. N. (1963). Assessment and control of non-sampling errors in censuses and surveys. *Sankhyā Ser. B* **25**, 263–282.

Murthy, M. N. (1964). On Mahalanobis' contributions to the development of sample survey theory and methods. In: C. R. Rao, ed., *Contributions to Statistics Presented to Professor Mahalanobis*: Statistical Publishing Society, Calcutta, 283–316.

Murthy, M. N. and Nanjamma, N. S. (1959). Almost unbiased ratio-estimates based on interpenetrating sub-sample estimates. *Sankhyā* **21**, 381–392.

Murthy, M. N. and Roy, A. S. (1975). Development of the Indian National Sample Survey during its first twenty-five rounds. *Sankhyā Ser. C* **37**, 1–42.

Narain, R. D. (1951). On sampling without replacement with varying probabilities. *J. Ind. Soc. Agric. Statist.* **3**, 169–174.

National Sample Survey (1961). Number 36: *Report on Land Holdings*. Cabinet Secretariat, Government of India. Eka Press, Calcutta.

National Sample Survey (1962). Number 33: Tables with notes on wages, employment, income and indebtedness of agricultural households in rural areas. *Sankhyā Ser. B* **24**, 43–200.

Pearson, K. (1902). On the mathematical theory of errors of judgement. *Phil. Trans. Roy. Soc. Lond. Ser. A* **198**, 235–299.

Plackett, R. L. and Burman, J. P. (1946). The design of multifactorial experiments. *Biometrika* **33**, 305–325.

Prabhu-Ajgaonkar, S. G. (1969). Determination of phonemic and graphemic frequencies by sampling techniques. Deccan College Postgraduate and Research Institute, Poona.

Savur, S. R. (1937). The use of the median in test of significance. *Proc. Ind. Academy Sc. A* **5**, 564–576.

Shaul, J. R. H. (1952). Sampling surveys in Central Africa. *J. Amer. Statist. Assoc.* **47**, 239–254.
Shaul, J. R. H. and Myburgh, C. A. L. (1948). A sample survey of the African population in Southern Rhodesia. *Population Studies* **2**, 339–353.
Shaul, J. R. H. and Myburgh, C. A. L. (1949). Provisional results of the sample survey of the African population of Southern Rhodesia. *Population Studies* **3**, 274–285.
Shaul, J. R. H. and Myburgh, C. A. L. (1951). Vital statistics of the African population of Southern Rhodesia in 1948. *Population Studies* **4**, 432–468.
Som, R. K. (1965). Use of interpenetrating samples in demographic studies. *Sankhyā Ser. B* **27**, 329–342.
Som, R. K. and Mukherjee, H. (1962). Analysis of variance of demographic variables. *Sankhyā Ser. B* **24**, 13–22.
Statistics Canada (1973). Advance Bulletin, 1971 Census of Canada: Agriculture Data from the Post-Sensus Agriculture Sample Survey, Ministry of Industry, Trade and Commerce, Ottawa.
Student (1908). The probable error of a mean. *Biometrika* **6**, 1–25.
Sukhatme, P. V. and Seth, G. R. (1952). Non-sampling errors in surveys. *J. Ind. Soc. Agric. Statist.* **4**, 5–41.
Tschuprow, A. A. (1928). The mathematical theory of statistical methods employed in the study of correlation in the case of three variables. *Trans. Camb. Phil. Soc.* **23**(12), 337–382.
United Nations (1949). *Recommendations for the Preparation of Sample Survey Reports.* Ser. C, No. 1. United Nations, New York.
Yates, F. (1949). *Sampling Methods for Censuses and Surveys.* Fourth edition, published in 1981. Charles Griffin, London.

Works not cited in text

Bailar, B. A. and Dalenius, T. (1969). Estimating the resposne variance components of the U.S. Bureau of the Census' survey model. *Sankhyā Ser. B* **31**, 341–360.
Bose, R. C. (1949). Least squares aspects of analysis of variance. N.C. Institute of Statistics Mimeo. Series No. 9.
Brewer, K. W. R. et al. (1977). Use of experimental design and population modelling in survey sampling. *Bull. Inter. Statist. Inst.* **47**(3), 173–190.
Cochran, W. G. (1969). Errors of measurement in statistics. *Technometrics* **10**, 637–666.
Cochran, W. G. (1977). *Sampling Techniques*, Third Edition, Wiley, New York.
Durbin, J. (1956). Non-response and call backs in surveys. *Bull. Inter. Statist. Inst.* **34**(2), 72–86.
Felligi, I. P. (1964). Response variance and its estimation. *J. Amer. Statist. Assoc.* **50**, 1016–1041.
Fellegi, I. P. and Holt, D. (1976). A systematic approach to automatic edit and imputation. *J. Amer. Statist. Assoc.* **71**, 17–35.
*Ghosh, B. (1949). Interpenetrating (networks of) samples. *Bull. Cal. Statist. Assoc.* **2**, 108–119.
*Ghosh, B. (1957). Enumerational errors in surveys. *Bull. Cal. Statist. Assoc.* **7**, 50–59.
*Kitagawa, T. (1963). Estimation after preliminary test of significance. *University of California Publications in Statistics* **3**(4), 147–186.
Koch, G. G. (1973). An alternative approach to multivariate response error models for sample survey data with applications to estimators involving subclass means. *J. Amer. Statist. Assoc.* **68**, 906–913.
Koop, J. C. (1967). Replicated (or interpenetrating) samples of unequal sizes. *Ann. Math. Stat.* **38**, 1142–1147.
Koop, J. C. (1972). On the derivation of expected value and variance of ratios without the use of infinite series expansions. *Metrika* **19**, 156–170.
*Mokashi, V. K. (1950). A note on interpenetrating samples. *J. Ind. Soc. Agric. Statist.* **2**, 189–195.
*Panse, V. G. and Sukhatme, P. V. (1948). Crop surveys in India I. *J. Ind. Soc. Agric. Statist.* **1**, 34–58.
*Panse, V. G. and Sukhatme, P. V. (1951). Crop surveys in India II. *J. Ind. Soc. Agric. Statist.* **3**, 97–168.
Platek, R., Singh, M. P. and Tremblay, V. (1977). Adjustment for nonresponse in surveys: A paper presented at the Second Symposium on Survey Sampling in Chapel Hill, NC, April 1977, and

subsequently published in *Survey Sampling and Measurement*, edited by N. K. Namboodiri, in 1978, Academic Press, New York.

Quenouille, M. H. (1949). Problems in plane sampling. *Ann. Math. Statist.* **20**, 355–375.

Rao, J. N. K. (1973). On double sampling for stratification and analytical surveys. *Biometrika* **60**, 125–133.

Roy, J. (1978). Sample surveys and computing. In: *Proceedings of the 65th Indian Science Congress, Ahmedabad*, 31–44.

Roy, J. (1981). On the National Sample Survey of India. *Sankhyā Ser. C* **38**, 1–2.

Sedransk, J. (1965). Analytical surveys with cluster sampling. *J. Roy Statist. Soc. Ser. B* **27**, 264–278.

Sen, A. R. (1953). On the estimate of the variance in sampling with varying probabilities. *J. Ind. Soc. Agric. Statist.* **5**, 119–127.

Singh, R. and Bansal, M. L. (1975). On the efficiency of interpenetrating sub-samples in simple random sampling. *Sankhyā Ser. C* **37**, 190–198.

Singh, R. and Bansal, M. L. (1978). A note on the efficiency of interpenetrating sub-samples in simple random sampling. *Sankhyā Ser. C* **40**, 174–176.

Sukhatme, P. V. (1952). Measurement of observational errors in surveys. *Inter. Statist. Inst. Review* **20**, 121–134.

Tin, M. (1965). Comparison of some ratio estimators. *J. Amer. Statist. Assoc.* **60**, 294–307.

Zarkovich, S. S. (1966). *Quality of Statistical Data*. Food and Agriculture Organization of the United Nations, Rome.

On the Use of Models in Sampling from Finite Populations

Ib Thomsen* and Dinke Tesfu

1. Introduction

During the last years, a series of articles have been published in which the finite population is assumed to have been generated as a random sample from an infinite population, called a superpopulation in what follows. This idea is not new as superpopulation models for several years have been used for comparing estimators and designs. However, the idea of using the superpopulation model in order to predict the not observed units in the population is new. If, for instance, the superpopulation can be specified as a Normal distribution, and a simple random sample is selected from the finite population, the best predictor for the values not in the sample is the sample mean. (In this simple case the prediction approach and the conventional approach produces the same estimator, but this is not always the case.)

At first glance, a practitioner may consider this approach primarily of theoretical interest, but in this paper we shall attempt to describe how we have been inspired by these new developments when discussing and solving practical problems.

The two main contributions of the predictive approach to practitioners are in our opinion:

(i) Using the techniques of prediction to sampling from finite population, makes communication with colleagues in other areas of applied statistics simpler.

(ii) It gives the sampler a possibility to use many of the techniques that are developed within other areas of statistics.

In Section 2 we shall outline some differences between the conventional approach and the prediction approach. In Sections 3 and 4 two practical applications of the prediction approach are given. In Section 5 we present a number of examples of practical problems, which in our opinion can be solved in a more efficient and elegant way by applying a model instead of applying the conventional approach.

---

* This work was partly done while the author was social statistics adviser to the Central Bureau of Statistics and UNICEF, Ethiopia.

In Sections 6 and 7 we turn our attention to non-sampling errors, which are at least as serious problems as sampling errors in practical survey work. Even though many problems in this area do not fall within prediction, we find it useful to apply models to cope with non-sampling errors.

The application of models in this area has a long history, but we feel that, like in the case of sampling theory, models have been used primarily to compare different designs and estimators rather than to develop optimal procedures.

Very few of the ideas discussed in this paper have a long history in sampling. Furthermore, we can not claim that the methods presented here are well established among practitioners, as several of them need further investigation. Our main objective is to give examples of areas within which we feel that the use of models offers a good framework for solving practical problems.

2. On the prediction approach

In several recent studies the finite population problems are formulated as prediction problems. Examples are Ericson (1969), who uses a Bayesian approach, and Kalbfleisch and Sprott (1969), whose approach is fiducial.

A more 'classical' superpopulation approach is also possible. In Royall (1970, 1971) some interesting results are obtained by using a linear model and the Gauss–Markov Theorem. As the applications presented in Sections 3 and 4 are inspired primarily by these two contributions, we shall give a short outline of their content.

Notation and terminology

Assume that the finite population consists of N units, with lable set $U = \{1, 2, \ldots, N\}$. With each unit is associated two numbers (y_i, x_i), where x_i is known for all units in the population. The aim is to estimate

$$Y = \sum_{i=1}^{N} y_i.$$

In order to estimate Y, a sample of size n, is selected from the population. A sample is defined as a subset of lables $s = \{i_1, i_2, \ldots, i_n\}$. A sampling design assigns a probability $p(s)$ to each $s \in \mathscr{S}$, where $\mathscr{S}$ denotes the set of all possible samples and

$$\sum_{s \in \mathscr{S}} p(s) = 1.$$

We shall study the following class of estimators:

$$\hat{Y} = \sum_{i \in s} a_i(s) y_i, \qquad (2.1)$$

where $\Sigma_{i\in s}$ denotes the summation over all lables in the sample, and $a_i(s)$ are constants depending on s and x_i, but not on y_i. Within the conventional approach, one will usually look for estimators which are *design-unbiased*, i.e. estimators for which

$$E(\hat{Y} - Y) = \sum_{s \in \mathscr{S}} p(s)(\hat{Y} - Y) = 0,$$

or in some cases try to find estimators with a small design-mean-square error, i.e. try to minimize:

$$E(\hat{Y} - Y)^2 = \sum_{s \in \mathscr{S}} p(s)(\hat{Y} - Y)^2.$$

One difficulty with this approach is that no minimum variance, unbiased estimator exist within this frame work, see Godambe (1955).

Within the prediction approach, the estimators are evaluated in a different way.

First a probabilistic model is introduced. It is often in consistence with experience to assume that $y_1, \ldots, y_N$ are considered to be realized values of independent random variables, $Y_1, \ldots, Y_N$, such that

$$Y_i = \beta x_i + U_i, \tag{2.2}$$

where β is an unknown constant and U_i are independent, random variables and

$$E(U_i) = 0, \qquad E(U_i^2) = \sigma_i^2.$$

In what follows we shall assume that the sample design is non-informative. An informal definition of a non-informative design is that for each possible $s \in \mathscr{S}$, the relationship between x_i and Y_i can be expressed by (2.2). In general it can be very difficult to determine whether a design is non-information or not, but in connection with the applications in Sections 3 and 4 the designs are non-informative. In Section 5.3, where we consider estimation of parameters in the superpopulation, we shall return to this problem.

For a given sample s, the estimator $\hat{Y}$ is said to be *model-unbiased* if

$$E_\xi \left(\sum_{i \in s} a_i(s) Y_i - \sum_{i=1}^N Y_i \right) = 0$$

where E_ξ is the expectation taken with respect to model (2.2), for a given sample. To demonstrate the difference between design and model-unbiasedness, let us look at one design, and two estimators. Under simple random sampling the simple expansion estimator

$$\frac{N}{n} \sum_{i \in s} y_i,$$

is a design-unbiased estimator of Y. However, it is not in general model-unbiased with respect to model (2.2). It is easily seen that

$$E_\xi\left(\frac{N}{n}\sum_{i\in s} Y_i - \sum_{i=1}^N Y_i\right) = N\beta(\bar{x}_s - \bar{x}),\tag{2.3}$$

where $\bar{x}_s$ and $\bar{x}$ denote the sample mean and the population mean, respectively. If, after the sample is selected, it is observed that there is a disproportionate number of units with large x-values, (2.3) suggests that the simple expansion estimator overestimates Y.

The second estimator we shall consider here, is the ratio estimator

$$\left(\sum_{i\in s} y_i / \sum_{i\in s} x_i\right)\sum_{i=1}^N x_i.\tag{2.4}$$

Under simple random sampling this is not design-unbiased. It is, however, model-unbiased under model (2.2).

The estimator (2.4) can be written as the simple expansion estimator, multiplied with a factor

$$\sum_{i=1}^N x_i \bigg/ \left(\frac{N}{n}\sum_{i\in s} x_i\right).$$

This factor 'adjusts' the expansion estimator in cases where there is an excess of units with large, or small x-values. In spite of this (2.4) is biased according to the conventional definition. The design-bias simply describe what will happen on the average over all possible samples, while the model-bias depends on the actual selected sample. We therefore feel that the model-bias is useful for inference after the sample is selected.

Another appealing consequence of the predictive approach is that it is often possible to find 'best' estimators.

Let us consider the class of linear estimators defined by (2.1). Then it is shown in Royall (1970) that for a given sample, the linear, model-unbiased estimator which with respect to model (2.2) minimizes the model-mean-square error,

$$E_\xi(\hat{Y} - Y)^2,$$

can be written as

$$\hat{Y}^* = \sum_{i\in s} y_i + \sum_{i\notin s} \hat{\beta} x_i,$$

where $\hat{\beta}$ is the weighted, least-square estimator of β. $\hat{Y}^*$ looks appealing as it tells us to use the observed y-values and predict the not observed with a predictor, which is natural under model (2.2).

When evaluating a given estimator within the prediction approach, we shall in the two following sections use a very simple approach, following Smith (1976).

After the sample is selected, the population total can be written as

$$Y = \sum_{i \in s} y_i + \sum_{i \notin s} y_i, \qquad (2.5)$$

where $\sum_{i \notin s} y_i$ is the sum of the not observed y-values.

Any estimator $\hat{Y}$ can similarily be written as

$$\sum_{i \in s} y_i + \left(\hat{Y} - \sum_{i \in s} y_i \right). \qquad (2.6)$$

It follows that when evaluating $\hat{Y}$ as an estimator of Y, one can evaluate the second term of (2.6) as a predictor of the not observed sum $\sum_{i \notin s} y_i$. By writing an estimator in its predictive form, (2.6) and evaluate the second term as a predictor, one can often identify whether the predictor is reasonable or whether it has a peculiar form. The following example is from Smith (1976).

Assume that the sample is selected with probability proportional to x_i. If Π_i is the inclusion probability, the widely applied Horvitz–Thompson estimator is

$$\sum_{i \in s} y_i / \Pi_i,$$

the predictive form of which is

where
$$\sum_{i \in s} y_i + \hat{\beta} \sum_{i \notin s} x_i,$$

$$\hat{\beta} = \left\{ \sum_{i \notin s} x_i \right\}^{-1} \sum_{i \in s} y_i \left(\frac{\sum_{i=1}^{N} x_i}{n x_i} - 1 \right).$$

Under model (2.2) this hardly seems a reasonable predictor.

We have emphasised how the prediction approach can be used to evaluate given estimators, and produce optimal ones. In addition, it has lead to a critique of the sampling variance as a measure of accuracy (Royall, 1970). This critique is of substantial practical interest, but in what follows we shall only use the prediction approach to evaluate given estimators of Y, and to find alternatives.

Another aspect which we shall only mention concerns the need for randomization.

Within the prediction approach inference only depends on the model, and not on the sample design. And the question is why random sampling is necessary.

In some practical situations a random sample is obviously suboptimal. For simple linear regression models through the origin it is more efficient to select the units with the largest x-values, and this design is used in practice in a few cases. In large, nation wide, multi-sectoral surveys, however, we feel that the designs recommended today are efficient and economical.

However, whenever the design is non-informative we do not feel that inferences have to be based on the randomization distribution. In experimental designs randomization is used when collecting the data but not in the analysis of them.

3. Application of the prediction approach to two-stage sampling

3.1. Background and sample design

In this section we shall demonstrate how the prediction approach can be used when estimating total from a two stage sample. The total we shall use as example is the population of Ethiopia.

The sample design of the Demographic Survey of Ethiopia is described in detail in Central Statistical Office (1982). Here it is sufficient to give a short outline:

The overall sample design is a two stage, stratified design. Each region in Ethiopia is treated as a 'Domain of Study', and is stratified, using the administrative division within each region. The smallest administrative unit in the rural areas is a Farmer's Association, FA, which was chosen as primary sampling unit. The size of the FAs vary from about 100 to 500 households, but most of them cover 150–300 households.

Before selecting the sample of FAs to be used in the survey, information concerning the number of households within each FA was collected administratively for all FAs in the country.

This information was not expected to be 100 percent correct, but was collected in order to get a measure of size of each FA. When selecting the sample of FAs, each FA was given a selection probability, which was proportional to the measure of size. Such designs are common, and are known to be more efficient than simple random samples of FAs.

In the second stage a simple random sample of 100 households was selected from each of the selected FAs.

At present the survey program only covers rural Ethiopia, which is divided into about 100 strata, from which about 500 FAs are selected.

In what follows we shall compare three estimators by writing them on their predictive form, and finally make an empirical comparison using information collected independently from the survey.

3.2. Choice of estimator

Within each stratum we shall use the following notation:

Mos_i: measure of size of the ith FA.
m_i : correct number of household in the ith FA.
y_{ij} : number of persons in the jth household in the ith FA.
m : number of household selected within each of the selected FAs. (In our case $m = 100$.)
N : number of FAs in the population.
n : number of FAs selected.

The purpose is to estimate

$$Y = \sum_{i=1}^{N} \sum_{j=1}^{m_i} y_{ij}.$$

Mos$_i$ is known for all FAs in the population. m_i is observed for all FAs selected in the first stage, and y_{ij} are observed for all households in the sample.

We shall adopt the following simple model:

Assume that $m_1, \ldots, m_N$, and y_{ij} ($i = 1, \ldots, N_{ij} = 1, \ldots, m_i$) can be considered to be realized values of random variables, M_i and Y_{ij}, such that

$$M_i = \beta \operatorname{Mos}_i + U_i \quad \text{and} \quad Y_{ij} = \mu + e_{ij}, \tag{3.1}$$

where β and μ are unknown constants, U_i and e_{ij} are independent random variables, and

$$E(U_i) = E(e_{ij}) = 0,$$

$$E(U_k, U_l) = \begin{cases} \sigma^2 \operatorname{Mos}_k & \text{if } k = l, \\ 0 & \text{otherwise}, \end{cases}$$

$$E(e_{ij}, e_{kl}) = \begin{cases} V^2 & \text{if } i = k \text{ and } j = 1, \\ \zeta^2 & \text{if } i = k, j \neq 1, \\ 0 & \text{if } i \neq k. \end{cases}$$

It should be noticed that model (3.1) is limited, especially the assumption that the number of persons in each household has the same expection and variance for all households within a stratum. In Scott and Smith (1969) a more general situation is considered using a Bayesian Approach.

We shall compare three estimators:

$$\hat{Y}_1 = \sum_{i \in s} m_i \bar{y}_i + \frac{\sum_{i \in s} m_i}{\sum_{i \in s} \operatorname{Mos}_i} \left(\frac{1}{n} \sum_{i \in s} \bar{y}_i \right) \sum_{i \notin s} \operatorname{Mos}_i,$$

$$\hat{Y}_2 = \frac{\sum_{i \in s} m_i \bar{y}_i}{\sum_{i \in s} \operatorname{Mos}_i} \sum_{i=1}^{N} \operatorname{Mos}_i,$$

$$\hat{Y}_3 = \frac{1}{n} \sum_{i \in s} \frac{m_i \bar{y}_i}{\operatorname{Mos}_i} \sum_{i=1}^{N} \operatorname{Mos}_i,$$

where $\Sigma_{i \in s}$ means the sum over all FAs in the sample, and $\bar{y}_i$ denotes the average household size in the ith FA, $\hat{Y}_3$ is the commonly used design-unbiased estimator of Y, while $\hat{Y}_1$ and $\hat{Y}_2$ are ratio-type estimators.

After the sample is selected, the total can be written:

$$Y = \sum_{i \in s} \sum_{j \in s_i} y_{ij} + \sum_{i \in s} \sum_{j \notin s_i} y_{ij} + \sum_{i \notin s} \sum_{j=1}^{m_i} y_{ij}, \qquad (3.2)$$

where $\sum_{j \in s_i} y_{ij}$ means the sum over all sampled households in the ith FA.

The first term of (3.2) is known, while the two other terms must be estimated. Writing the estimators on their predictive form, we find

$$\hat{Y}_1 = \sum_{i \in s} \sum_{j \in s_i} y_{ij} + \sum_{i \in s} (m_i - m)\bar{y}_i + \frac{\sum_{i \in s} m_i}{\sum_{i \in s} \text{Mos}_i} \left(\frac{1}{n} \sum_{i \in s} \bar{y}_i\right) \sum_{i \notin s} \text{Mos}_i.$$

The first term is that part of Y which is known, the second term estimates the number of persons in the not selected households within the selected FAs, and the third term estimates the total population in the FAs not in the sample.

Under model (3.1) this seems a reasonable estimator as $(\sum_{i \in s} m_i / \sum_{i \in s} \text{Mos}_i)$ is an estimator of β and $(1/n)\sum_{i \in s} \bar{y}_i$ is a natural estimator of μ, making the third term a natural predictor. In the second term, one could argue that $\sum_{i \in s}(m_i - m)(\sum_{i \in s}\bar{y}_i/n)$ is a better predictor, but the numerical difference would be neglectable. Similarily, we find

$$\hat{Y}_2 = \sum_{i \in s} \sum_{j \in s_i} y_{ij} + \sum_{i \in s} (m_i - m)\bar{y}_i + \frac{\sum_{i \in s} m_i \bar{y}_i}{\sum_{i \in s} \text{Mos}_i} \sum_{i \notin s} \text{Mos}_i.$$

Also this estimator can be recognized as a natural estimator under (3.1). The difference from $\hat{Y}_1$ is that instead of estimating β and μ separately, $(\sum_{i \in s} m_i \bar{y}_i)/\sum_{i \in s} \text{Mos}_i$ is an estimator of the product of the two parameters.

Finally, for $\hat{Y}_3$ we find that

$$\hat{Y}_3 = \sum_{i \in s} \sum_{j \in s_i} y_{ij} + \sum_{i \in s} (m_i - m)\bar{y}_i + \hat{\gamma} \sum_{i \notin s} \text{Mos}_i,$$

where

$$\hat{\gamma} = \left\{\sum_{i \in s} \text{Mos}_i\right\}^{-1} \left\{\sum_{i \in s} m_i \bar{y}_i \left(\frac{\sum_{i \in s} \text{Mos}_i}{n \,\text{Mos}_i} - 1\right)\right\} + \sum_{i \in s} \frac{m_i \bar{y}_i}{n \,\text{Mos}_i}.$$

It is hard to recognize $(\hat{\gamma} \sum_{i \notin s} \text{Mos}_i)$ as a natural predictor of the total population in the not-sampled FAs, under model (3.1).

It can be shown that

$$E(\hat{Y}_1 - Y)^2 \leq E(\hat{Y}_2 - Y)^2$$

and we shall therefore not study $\hat{Y}_2$ any further.

During the calculation of $\hat{Y}_1$, it was observed that for some FAs in the sample the difference between Mos_i and m_i was extremely large. It is known that in such cases a robust estimator of β should be applied instead of the weighted least-squares estimator. As no programme for robust regression was available, and because the number of outliers was limited, we chose an estimator, $\hat{Y}_r$, which is similar to $\hat{Y}_1$, but with the difference that all FAs for which

$$\left| \frac{\text{Mos}_i - m_i}{\text{Mos}_i} \right| > 0.33$$

were excluded when estimating β.

2.2. Empirical comparision of $\hat{Y}_1$ and $\hat{Y}_3$ and $\hat{Y}_r$

To supplement the theoretical investigation of the estimators given above, we shall compare Y_1, Y_3 and Y_r with population counts done in connection with the preparatory work for the coming population census. The comparison is limited to 22 strata, as counts were only available for these strata.

Table 1
$\hat{Y}_1$, $\hat{Y}_3$, $\hat{Y}_r$, and the result from the population count for the 22 strata

Stratum no.	Population count reduced with 3% '000 persons	Y_1 '000 persons	Y_3 '000 persons	Y_r '000 persons	Number of FAs selected
1	170	184	180	172	6
2	175	199	204	189	4
3	651	659	662	659	10
4	355	385	395	338	10
5	350	411	426	380	10
6	253	226	240	243	9
7	1041	1073	988	1056	13
8	1027	998	1031	1043	12
9	696	734	751	701	8
10	1078	1086	1092	1086	7
11	235	240	250	240	6
12	231	208	214	198	6
13	415	395	407	413	9
14	463	541	567	517	10
15	375	436	420	379	10
16	250	228	233	228	3
17	556	448	451	448	4
18	279	325	379	260	4
19	505	494	514	451	7
20	592	425	715	618	8
21	176	178	176	152	6
22	279	430	518	289	5
Total	10152	10303	10813	10060	167

In Table 1 the results from the population count have been reduced with 3 percent. This is done because the count was done about one year after the survey was taken, and the population growth for the whole country is about 3 percent. This limits the usefulness of the comparison substantially as the growth could vary from one stratum to another. However, we think the following conclusions can be made:

(i) For the 22 strata studied, $\hat{Y}_3$, which is the design unbiased estimator usually applied, give a somewhat higher estimate for the population than $\hat{Y}_1$ and $\hat{Y}_r$. (This has also been observed in other strata, which are not included in Table 1, as the results from the population count are not yet ready.) $\hat{Y}_r$ and $\hat{Y}_1$ seem to give the most reasonable results as compared with the results from the population count.

(ii) If we concentrate attention to strata where there is some difference between $\hat{Y}_r$ and $\hat{Y}_3$, which is the case in stratum nos. 2, 5, 7, 9, 14, 15, 18, 20 and 22, $\hat{Y}_r$ seems to give the most reasonable result as compared with the result from the population count.

When estimating the total population of Ethiopia large differences were observed between $\hat{Y}_r$ and $\hat{Y}_3$ in some regions. In such cases an $\hat{Y}_r$-type estimator was chosen. Unfortunately no results from the population count are ready for these regions, and we can not make empirical comparisons like for the 22 strata in Table 1.

4. Estimation in election surveys

4.1. Background

In many countries surveys are regularily undertaken in order to estimate the proportion of persons intending to vote for a specific party in the coming election. To simplify, let us assume that there are only two parties one can vote for either party A or party B. The problem is to estimate

P_A: The proportion of persons which would vote for party A.

One simple estimator would be

$\hat{p}_A$: The proportion of persons in the sample who report that they intend to vote for party A in case of an election.

Under many sample designs used in practice, $\hat{p}_A$ is a design-unbiased estimator of P_A.

The estimation problem in election survey arises from the fact that the proportion of people which voted for party A is known from previous elections, and the question is how to use this information in order to find an estimator, which is more efficient than $\hat{p}_A$.

Several estimators are suggested and used in practice. In this section we shall discuss and compare these estimators under a simple model, and finally suggest an estimator, which is reasonable in this simple situation.

4.2. Definitions and notations

To each person in the population is associated two binary variables.

$$x_i = \begin{cases} 1 & \text{if person } i \text{ voted for party A in the last election}, \\ 0 & \text{otherwise}, \end{cases}$$

$$y_i = \begin{cases} 1 & \text{if person } i \text{ says he/she will vote for party A in the coming election}, \\ 0 & \text{otherwise}. \end{cases}$$

We assume that $y_1, \ldots, y_N$ are realized values of independent random variables $Y_1, \ldots, Y_N$, and that:

$$P(Y_i = 1 | x_i = 1) = P_{11},$$
$$P(Y_i = 1 | x_i = 0) = P_{01}.$$
$$i = 1, 2, \ldots, N. \tag{4.1}$$

As N is assumed known, the problem is to select a sample of size n in order to estimate

$$Y = \sum_{i=1}^{N} y_i.$$

We shall study one sampling plan, and three estimators.
The first estimator is the usual inflated sample mean

$$\hat{Y} = \frac{N}{n} \sum_{i \in s} y_i.$$

For many of the commonly used sample plans, including simple random sampling, this is design-unbiased. The other estimator is the ratio-estimator

$$\hat{Y}_R = \left(\sum_{i \in s} y_i \bigg/ \sum_{i \in s} x_i \right) \sum_{i=1}^{N} x_i.$$

Under simple random sampling, this estimator is slightly biased.

The third estimator we shall consider is a post-stratified mean, where the sample is divided into two post-strata according to the value of x_i. This estimator can be written as

$$\hat{Y}_P = \frac{\Sigma_{i \in s} y_i x_i}{\Sigma_{i \in s} x_i} \sum_{i=1}^{N} x_i + \frac{\Sigma_{i \in s} y_i (1 - x_i)}{\Sigma_{i \in s} (1 - x_i)} \sum_{i=1}^{N} (1 - x_i).$$

Provided that the probability of empty post-strata is approximated with zero, $\hat{Y}_P$ is design-unbiased under simple random sampling.

4.3. Comparing the three estimators

In order to compare the three estimators, we shall write them in their predictive form

$$\sum_{i \in s} y_i + \sum_{i \notin s} \hat{Y}_i,$$

and evaluate $\Sigma_{i \notin s} \hat{Y}_i$ as predictor for the sum of the not observed y-values under model (4.1).

The predictive form of $\hat{Y}$ is

$$\sum_{i \in s} y_i + \sum_{i \notin s} \left(\frac{1}{n} \sum_{i \in s} y_i \right). \tag{4.2}$$

When evaluating the second term in (4.2) as a predictor of $\Sigma_{i \notin s} y_i$, we consider

$$E_\xi \left\{ \sum_{i \notin s} \left(\frac{1}{n} \sum_{i \in s} Y_i \right) \right\},$$

where E_ξ means the expectation with respect to model (4.1), given the sample. From (4.1) follows that

$$E_\xi(Y_i) = P_{11} x_i + P_{01}(1 - x_i)$$

and therefore,

$$E_\xi \left\{ \frac{N - n}{n} \sum_{i \in s} Y_i - \sum_{i \notin s} Y_i \right\} = (N - n) \{(\bar{x}_s - \bar{x}_{\bar{s}}) P_{11} + P_{01}(\bar{x}_{\bar{s}} - \bar{x}_s)\},$$

where

$$\bar{x}_s = \frac{1}{n} \sum_{i \in s} x_i \quad \text{and} \quad \bar{x}_{\bar{s}} = \frac{1}{N - n} \sum_{i \notin s} x_i.$$

It follows that the second term of (4.2) is not in general a model-unbiased predictor of the sum of the not observed y-values. In cases where $\bar{x}_s$ differs substantially from $\bar{x}_{\bar{s}}$, the bias can be serious.

Treating $\hat{Y}_R$ in the same way, we find that

$$\hat{Y}_R = \sum_{i \in s} y_i + \frac{\Sigma_{i \in s} y_i}{\Sigma_{i \in s} x_i} \sum_{i \notin s} x_i, \tag{4.3}$$

and

$$E_\xi \left\{ \frac{\Sigma_{i \in s} Y_i}{\Sigma_{i \in s} x_i} \sum_{i \notin s} x_i - \sum_{i \notin s} Y_i \right\} = P_{01}(N-n)\left(\frac{\bar{x}_{\bar{s}}}{\bar{x}_s} - 1\right).$$

Like in the case of $\hat{Y}$, the second term of $\hat{Y}_R$ is not in general a model-unbiased predictor for the sum of the not observed y-values.

Finally, for $\hat{Y}_P$ we find that the predictor of the sum of the not observed y-values to be

$$\frac{\Sigma_{i \in s} Y_i x_i}{\Sigma_{i \in s} x_i} \sum_{i \notin s} x_i + \frac{\Sigma_{i \in s} Y_i(1-x_i)}{\Sigma_{i \in s}(1-x_i)} \sum_{i \notin s}(1-x_i),$$

which is a model-unbiased predictor of the sum of the not observed y-values.

In Thomsen (1981), $\hat{Y}_P$ is shown to have the smallest model mean-square error among all linear, model unbiased estimators. For election surveys, there are, however, some problems connected with the use of $\hat{Y}_P$.

Two important problems are:

(i) Many election surveys use rather small samples. In countries with many parties, this means that many post-strata have to be introduced. In such cases the accuracy of $\hat{Y}_P$ can be very low.

(ii) In practice it is often observed that people tend to 'forget' what party they voted for in the last election. This can seriously affect the accuracy of $\hat{Y}_R$ and $\hat{Y}_P$.

The estimator $\hat{Y}_P$ is, to our knowledge, not widely used, but in Norway and Denmark it is being used by two opinion research institutes. Based on their experiences, it should be possible to give a more complete evaluation of $\hat{Y}_P$ in the future.

5. Other applications of models

In this section we shall give a short outline of the potentials of the use of models within two special areas, namely repeated surveys and small area statistics. Finally, we shall discuss some of the problems encountered in estimating parameters in the superpopulation model.

5.1. Repeated surveys

In some cases surveys are planned as a series of identical surveys to be done with regular intervals in order to follow the development of a variable over time. Labour force surveys are good examples of such surveys.

In connection with repeated surveys it is common to use rotating designs, which means that part of the sample is used in more than one survey. As a simple example involving overlaps of $\frac{1}{2}$ in successive samples, with two successive measurements on each new selection:

$$ab-bc-cd-de \cdots$$

The b half-sample appears in the first and second round; the c half-sample in the second and third, and so on. More complicated rotation plans are possible, and actually used.

The motives for using rotating samples are several. Only rotating samples and panels (meaning that the overlap is 100%) can give information about the gross change behind a net change. Generally one can say that rotating samples and panels have a larger potential for making sophisticated analysis of the data than independently selected samples (Markus, 1979).

Also when the aim is to follow the variation of one variable over time, there are clear gains from using rotating samples.

Let $\bar{x}(t)$ denote the sample mean observed at time t. Then

$$\text{Var}(\bar{x}(t+1) - \bar{x}(t)) = \text{Var}(\bar{x}(t+1)) + \text{Var}(\bar{x}(t)) - 2\,\text{cov}(\bar{x}(t+1), \bar{x}(t)).$$

Within the conventional approach the covariance between the two means is zero if the two samples are selected independently, but will often be positive if the two samples are overlapping. Within the conventional approach it is also shown that one can construct estimators for the population means and their difference, which are more efficient than the sample means and their difference (Patterson, 1950).

Within a model approach it seems more reasonable to assume that the variation of the population mean will show some regularity over time, which often can be described by a time series model. The following simple example illustrates the approach (Scott and Smith, 1974). Let θ_i, $t = 1, 2 \ldots$ denote the realized values of a stationary first-order auto-regressive process of the following type:

$$\theta_t = \lambda \theta_{t-1} + \varepsilon_t, \quad 0 \leq \lambda \leq 1, \tag{5.1}$$

where ε_t are uncorrelated, random variables with $E(\varepsilon_t) = 0$ and $E(\varepsilon_t^2) = \sigma^2$.

Let y_t denote an estimator of θ_t base on a sample at time t, such that:

$$y_t = \theta_t + e_t, \tag{5.2}$$

where e_t are independent variables, with $E(e_t) = 0$, and $E(e_t^2) = s_t^2$. In other words, the sampling variance is s_t^2, and y_t are uncorrelated, meaning that the samples are selected independently from one survey to another.

In Scott and Smith (1974) it is shown that the best linear unbiased estimator of θ_t, given $y_t, y_{t-1}, \ldots$ is an exponentially weighted moving average of the estimates $y_t, y_{t-1}, \ldots$,

$$\hat{\theta}_t = (1 - \pi) \sum_{j=0}^{\infty} \beta^j y_{t-j},$$

where π and β are known functions of the parameters in (5.1) and (5.2).

When trying to use time series models to repeated surveys in practice, a number of problems have been encountered, some of these are:

(i) In practice models are often more complicated than (5.1). Usually data show a clear seasonal variation, which leads to complications when trying to find optimal estimators.

(ii) For a given set of $\{y_t\}$ one has to identify the model for $\{\theta_t\}$. What are the consequences of choosing the wrong model for (θ_t)?

(iii) In the example we assumed that the parameters of the model for $\{\theta_t\}$ are known. What are the effects of using estimated values?

The advantage of this approach is that the problems encountered when studying surveys are treated within the framework of time series analysis, where they, in our opinion, rightly belong.

A time series approach is applied in connection with the Norwegian labour force surveys. See Dagsvik (1978). A Box–Jenkins procedure was applied to employment data for the period 1955–1970, during which time labour statistics were based on total counts. In order to estimate the auto-correlation function of the sampling error process, data from the surveys during the period 1967–1970 were used. Finally the filter coefficients and the mean-square error of the filter were estimated in the usual manner.

The empirical results in Dagsvik (1978) show that considerable improvements of the usually applied composite estimates can be obtained from relatively weak assumption about the θ_t-process.

In addition, the robustness properties of the filter is examined, and it is found that even quite rough estimates of the parameters produces filter estimates with higher precision than composite estimates.

One of the most important reasons why the filter approach seems so efficient in connection with the Norwegian labour force surveys is probably that relative small samples are used (about 10 000 persons every quarter of the year). Because of this the composite estimates have very large sampling errors.

5.2. Small area statistics

One of the serious limitations of most nation-wide surveys is that estimators for small areas are very unreliable. During the last years the need for statistics within small areas has increased, and in order to meet this demand, attempts have recently been made in order to develop methods for small area estimation. The methods which have been tried out combines auxiliary information, including models, with more directly observed data.

It is difficult to describe all methods in use, as they vary with respect to available auxiliary information and with respect to the models applied. We shall therefore give a short review of the methods involved when estimating the total for categorized variables within a small geographical area (Royall, 1979).

Assume the population is divided into A areas, and C classes. This gives $A \times C$ cells, with N_{ij} ($i = 1, 2, \ldots, C$; $j = 1, 2, \ldots, A$) units in each cell. N_{ij} is assumed known. Let y_{ijk} denote the value of y associated with unit k in cell (i, j). Denote the sample units of cell (i, j) by $s(i, j)$, and the non-sample units by $\bar{s}(i, j)$. Our aim is to estimate the total in area a,

$$Y_a = \sum_{i=1}^{C} \sum_{s(i,a)} y_{iak} + \sum_{i=1}^{C} \sum_{\bar{s}(i,a)} y_{iak}. \tag{5.3}$$

The first term of (5.3) is observed, while the second term must be predicted. Let us consider the following simple model: y_{ijk} is assumed to be the realized values of a random variable Y_{ijk}, and let

$$E(Y_{ijk}) = \varepsilon_{ij},$$

$$\text{cov}(Y_{ijk}, Y_{1,m,n}) = \begin{cases} \sigma_{ij}^2, & i = 1, j = m \text{ and } k = n, \\ \tau_{ij}^2, & i = 1, j = m \text{ and } k \neq n, \\ 0, & i \neq 1, j \neq m. \end{cases} \tag{5.4}$$

Under model (5.4) it seems reasonable to predict the non-sample y_{iak} by the cell mean, and the following estimator is produced:

$$\hat{Y}_a^I = \sum_{i=1}^{C} n_{ia} \bar{y}_{ia} + \sum_{i=1}^{C} (N_{ia} - n_{ia}) \bar{y}_{ia} = \sum_{i=1}^{C} N_{ia} \bar{y}_{ia}, \tag{5.5}$$

where n_{ia} is the sample size and $\bar{y}_{ia}$ is the sample mean within cell (i, a). The estimator (5.5) is denoted the post-stratified estimator, and is widely used. For small areas this estimator can be very unreliable, and often one introduces a more restrictive model in order to use sample units from other cells, and thereby get a more efficient estimator.

Let us modify (5.4) such that

$$E(Y_{ijk}) = \varepsilon_i, \tag{5.6}$$

and the $\text{cov}(Y_{ijk}, Y_{1,m,n})$ as in (5.4).

This model implies that for a given class all cells have a common expected value, independent of area. It is natural to use this, by estimating the not observed y-values within a class by a weighted sum of cell means within the same class. The corresponding estimator for the total in area a is,

$$\hat{Y}_a^{II} = \sum_{i=1}^{C} n_{ia} \bar{y}_{i,a} + \sum_{i=1}^{C} (N_{ia} - n_{ia}) \hat{\mu}_i \tag{5.7}$$

where

$$\hat{\mu}_i = \sum_{j=1}^{A} \frac{n_{ij}}{n_{\cdot j}} \bar{y}_{i,j}.$$

The estimator (5.7) is the synthetic estimator, which is used by several institutions when estimating small area estimates (Schaible and Brock, 1977; Laake, 1977).

In many situations $\hat{Y}_a^I$ and $\hat{Y}_a^{II}$ can be improved by combining them in the following way:

Under model (5.6) we have two unbiased predictors for the non-observed *y*-values. These are the mean of the observed *y*-values within the cell, and the predictor used in (5.7), $\hat{\mu}_i$. By taking a weighted linear combination of these two predictors, and adding up over the classes within an area, one arrives at what is often called the composite estimator. The efficiency of the different estimators depends on σ_{ij} and τ_{ij}, and shall not be discussed here. Empirical studies seem to indicate that the composite estimator is the most efficient in many applications.

A serious practical problem is due to the fact that the model has to be identified, and the parameters have to be estimated from the sample. Due to the small sample size the estimates are likely to be very unreliable, and as a consequence the final result can be of very low quality.

In connection with the Norwegian labour force surveys, a composite estimator is used in order to estimate the rate of employment within small geographical areas (fylke). In order to determine the weights to be applied, a model was fitted to data from the population census, and the optimal weights were found. In addition the mean-square errors of three different estimators were calculated. The results for one small area are presented in Figure 1. It is seen that large reductions of the mean-square error can be gained by using a composite estimator instead of the directly observed area mean or the synthetic estimator. It was also found that these results are robust against the choice of weights in the composite estimate. Similar results were found for most other small areas.

5.3. Estimating parameters in the superpopulation

Till now the superpoulation model has been used in order to find optimal estimators for finite population quantities. In many cases, however, one wants to estimate a parameter in the superpopulation, and the question is to what extent the sample design influences such analysis.

A simple example is the estimation of price and income elasticities based on a household income and expenditure survey. In many countries the selection probabilities varies from one region to another, and the question is whether one should apply a weighted estimation procedure, where the weights are inversely proportional to the probability of selection.

It is hard to give a general answer to this question. It is clear that if the model in the sample is identical to the model in the whole population meaning that the design is uninformative, there are no problems. The model alone dictates the analysis. Let, for instance,

$$C_i = \beta I_i + \alpha + U_i, \tag{5.8}$$

where

C_i = is the consumption of some commodity for household *i*,
I_i = is the income of household *i*, and
U_i = is a random variable with $E(U_i) = 0$, $E(U_i^2) = \sigma^2$, U_i independent of I_i.

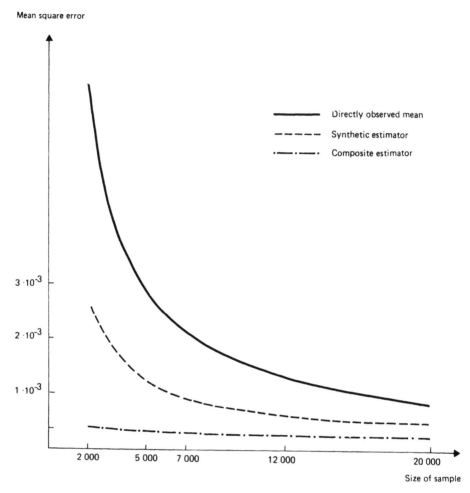

Fig. 1. Mean-square error of the directly observed mean, the synthetic estimator, and the composite estimator of the rate of employment in finnmark.

Under (5.8) the following situation may occur:

If income is the stratifying variable the distribution of U_i is not affected as U_i is assumed independent of I_i. In this case the selection probabilities should be disregarded. If U_i depends on I_i, however, the model in the sample will not be the same as (5.8). The same is the case if the stratifying variable is C_i.

The problem, as we see it is that for a given model and for a given design it is in many cases difficult to say whether the model in the sample is the same as the model for the whole population.

In cases with unequal selection probabilities, simulation studies presented in Holt, Smith and Winter (1980) seem to indicate that weighted regression is superior to unweighted in many practical situations.

Another problem, which has received some attention in the literature is the effect of the clustering of the sample (Kish and Frankel, 1974). We feel that this problem can be handled more simple, by introducing a correlation in the error term between units in the same cluster. One can then test whether this correlation is different from zero, and make the necessary corrections.

This problem is not a result of the design of the sample but follows from the model. Data from a Census would have to be analyzed the same way.

This field has been given considerable attention during the last years, and it is to be expected that several theoretical and empirical studies will appear in the journals in the coming years.

6. Models for response errors

The models introduced until now in this paper almost completely neglect response errors. This is a definitive shortcoming since this type of errors may have serious consequences in practical situations.

In many developing countries measurement errors are particularly serious: Most vital events are underreported, making it necessary to adjust observed fertility and mortality rates before publication. Age reporting is another area with serious response errors. Observed age-distributions also have to be adjusted before publication.

In agricultural surveys in Ethiopia a serious problem arises from the fact that each region in the country uses its own measurement unit for area under crop. In addition the unit is often descriptive rather than quantitative. One unit usually used in hilly areas is a 'terrace'. The actual size of a 'terrace' naturally depends on the slope of the hill side, making it inadequate for statistical purpose. In order to overcome this problem, objective measures are being used extensively in Ethiopia. This, on the other hand, increases cost per unit considerably as compared to subjective measures reported by the farmer. Because of the high unit cost, the sample size is usually small, leading to large sampling errors.

On this background it is surprising to see that only a few response error models are suggested, and even fewer are actually applied in survey work. There is a vast literature covering different techniques for adjustment of demographic data. These techniques are, however, seldom developed within explicitly formulated models. The reliability of adjusted results is therefore completely unknown in most cases.

When response errors are studied, essentially only one type of model is applied, namely the one formulated by Hansen, Hurwitz and Bershad (1961). One recent application of this model was done in connection with the World Fertility Survey (O'Muircheartaigh and Marckwart, 1980).

What we think is needed is a number of 'taylor made' models, designed for specific types of surveys. Such models should combine auxiliary information and observed data, including reinterview data when available. In what follows we outline two simple models and their potentials when studying response errors.

6.6. A model for analyzing reinterview surveys

When studying response errors, one often performs a reinterview survey, in which each unit or a sample of units is interviewed a second time. The design of the reinterview depends on the purpose of the study. Usually one of the two following kinds of designs is applied:

(i) The reinterview is as similar to the first interview as is practically possible.

(ii) The reinterview is made with an a priori more accurate (and usually more expensive) measurement instrument.

Design (i) is often chosen when the aim is to estimate the accuracy of a given method, while design (ii) usually is chosen to compare an expensive measurement instrument with a less expensive. In what follows we shall look at a model for design (i).

Let x_i denote the 'true' value of a dichotomous variable for unit i; x_i is 1 or 0. Let Y_i denote the reported value of x_i; Y_i is 1 or 0.

We now assume that for each measurement, the following simple model is realistic:

$$P(Y_i = 1 \mid x_i = 1) = p_{11} \quad \text{and} \quad P(Y_i = 1 \mid x_i = 0) = p_{01}. \tag{6.1}$$

The aim of the main survey and the reinterview survey is to estimate the population mean $\bar{x}$, p_{11} and p_{01}.

The data from the two surveys can be presented in a so-called consitency table, Z_{11}, Z_{12}, Z_{21} and Z_{22}, where

Z_{11} is the proportion of persons reporting '1' in both surveys,

Z_{12} is the proportion of persons reporting '1' in the main survey and '0' in the reinterview survey,

Z_{21} is the proportion of persons reporting '0' in the main survey and '1' in the reinterview survey,

Z_{22} is the proportion of persons reporting '0' in both surveys.

Assuming that the response error model is the same in the two surveys, and assuming independence between the two measurements, we have that

$$E(Z_{11}) = \bar{x} p_{11}^2 + (1 - \bar{x}) p_{01}^2,$$

$$E(Z_{21}) = E(Z_{12}) = \bar{x} p_{11}(1 - p_{11}) + (1 - \bar{x}) p_{01}(1 - p_{01}), \tag{6.2}$$

$$E(Z_{22}) = \bar{x}(1 - p_{11})^2 + (1 - \bar{x})(1 - p_{01})^2,$$

where $\bar{x}$ is the sample mean, and expectations are with respect to repeated measurements given the sample.

If we define

$$Z = Z_{11} + \frac{Z_{21} + Z_{12}}{2},$$

equations (6.2) are equivalent with

$$E(Z) = p_{11}\bar{x} + (1 - \bar{x})p_{01}, \tag{6.4}$$

and

$$E(Z_{11}) = p_{11}^2\bar{x} + (1 - \bar{x})p_{01}^2. \tag{6.5}$$

Equations (6.4) and (6.5) include three unknown parameters, and $\bar{x}$, p_{11} and p_{01} are therefore not identifiable.

One parameter of some practical interest is the response variance, which is defined as the variance of the sample mean, $\bar{y}$, with respect to repeated measurements for a given, fixed sample. (This definition differs slightly from the definition given in Hansen, Hurwitz and Bershad, 1961.) Under the model the response variance of $\bar{y}$ is

$$\mathrm{Var}(\bar{z}) = \frac{1}{n}(p_{11}(1 - p_{11})\bar{x} + p_{01}(1 - p_{01})(1 - \bar{x})).$$

From (6.2) follows that

$$(Z_{12} + Z_{21})/2n$$

is an unbiased estimator of the response variance.

The main problem is, however, the identification problem in (6.4) and (6.5). If additional information is available this can be used. In age reporting, for instance, information from model life tables can be used. If information from more than two surveys is available, model (6.1) can be modified slightly, and general results from reliability models for categorical data applied (Dempster, Laid and Rubin, 1977).

Even with no additional information available, equations (6.4) and (6.5) together with the inequalities $0 \leq p_{01}, p_{11} \leq 1$, define an interval within which $\bar{x}$ must take its value.

From (6.4) it follows that

$$p_{11} = \frac{E(Z)}{\bar{x}} + \frac{p_{01}(1 - \bar{x})}{\bar{x}}. \tag{6.6}$$

As $0 \leq p_{11} \leq 1$, we have that

$$p_{01} \geq \frac{\bar{x}}{1 - \bar{x}}\{(E(Z)/\bar{x}) - 1\}, \tag{6.7}$$

and

$$p_{01} \leq E(Z)/(1 - \bar{x}). \tag{6.8}$$

From (6.6) and (6.5) it follows that

$$p_{01} = E(Z) \pm \sqrt{\frac{\bar{x}}{1-\bar{x}} \{E(Z_{11}) - [E(Z)^2]\}}. \qquad (6.9)$$

If

$$p_{01} = E(Z) + \sqrt{\frac{\bar{x}}{1-\bar{x}} \{E(Z_{11}) - [E(Z)]^2\}}$$

is eligible, it follows from (6.9) and (6.8) that

$$\bar{x} \geq 1 - \frac{(E(Z))^2}{E(Z_{11})}. \qquad (6.10)$$

As $p_{01} \leq 0$, it also follows that

$$\frac{\bar{x}}{1-\bar{x}} \leq \frac{1 - (E(Z))^2}{E(Z_{11}) - (E(Z))^2}.$$

If

$$p_{01} = E(Z) - \sqrt{\frac{\bar{x}}{1-\bar{x}} (E(Z_{11}) - (E(Z))^2)}$$

is eligible, we find in a similar way that

$$\frac{\bar{x}}{1-\bar{x}} \geq \frac{E(Z_{11}) - (E(Z))^2}{(1 - E(Z))^2} \quad \text{and} \quad \frac{\bar{x}}{1-\bar{x}} \leq \frac{E(Z))^2}{E(Z_{11}) - (E(Z))^2}.$$

It is seen that the interval endpoints are all estimable from the data, thereby giving an indication of the size of the response bias of the observed proportion $\bar{y}$.

6.3. Outline of a response error model for agricultural surveys

In connection with agricultural surveys the choice between objective and subjective measurements is crucial. Objective measurements reduce the response errors. However, because they are more costly than subjective measurements, the size of the sample is usually reduced when using objective measurements.

In order to find a balance between the sample errors and response errors, we shall outline a simple response error model:

Let S_i denote a subjective measurement of area under crop for farmer i, and o_i denote an objective measurement of area under crop for farmer i.

We assume that o_i is measured without measurement error, and that

$$S_i = \beta o_i + e_i, \qquad (6.11)$$

where $E(e_i) = 0$, $\text{Var}(e_i) = \sigma^2$, and $\text{cov}(e_i, e_j) = 0$. Expectation, variance and covariance are with respect to repeated measurements.

The following approach is now natural under model (6.11):

Select a sample, s, of size n, for which subjective measurements are taken. In a subsample, s_1, of size n_1, objective measurements are taken in addition to the subjective ones. For simplicity we assume both samples to be simple random samples. Then it is natural to consider the following kind of estimates for total area under crop:

$$\hat{T} = \frac{N}{n} \left[\sum_{i \in s_1} o_i + \sum_{i \in s \setminus s_1} \hat{o}_i \right],$$

where N is the number of farmers in the population. The first term is simply the observed objective measurements in subsample s_1, while $\hat{o}_i$ is a prediction of the not observed objective measurements based on the subjective measurements in sample $s \setminus s_1$. Methods for predicting o_i based on S_i are discussed within calibration theory. See for instance Brown (1979).

The total mean-square error $\hat{T}$ can now be found as a function of the parameters in model (6.11), $(1/N) \Sigma_{i=1}^{N} (o_i - v)^2$, and n_1. Finally this expression can be minimized with respect to n_1. The total mean-square error should be taken with respect to model (6.11) as well as with respect to repeated samples.

7. A probabilistic model for nonresponse

7.1. Background

During the last 10 years, response rates in sample surveys have decreased in most industrialized countries. The reasons for this are only partly known, but the following two factors are important in most countries: Most people seem to spend less time at home today than previously, and are therefore hard to find. Secondly, the general public is becoming more sceptical towards the usefulness of the surveys, which leads to an increase in refusal rates.

In order to reduce refusal rates, most survey organizations spend considerable resources on informing the general public about how the results from the surveys are applied by governmental and other organizations, and what is done to improve confidentiality, safeguards and computer security within the survey organization. To formulate a model in order to measure the efficiency of such activities is very difficult.

The most common and successful means of reducing the nonresponse rate is to make several attempts to obtain responses from the nonrespondants. In particular, the proportion of not-at-homes is reduced significantly. Callbacks are also a very heavy burden on the survey budget, and it is therefore important to find reasonably good rules for allocating resources between the initial sample and the efforts to reduce nonresponse by callbacks.

In this section we present a probabilistic model for nonresponse. This model gives rise to an estimable variance, and an estimable nonrespose bias, after each callback. It is therefore, possible to study the relationship between the mean-square error and the number of callbacks, and the results can be used to determine how many callbacks it is reasonable to make. The model can also be used to find estimation methods, which reduce the nonresponse bias. This application will, however, not be illustrated here. See Thomsen and Siring (1983).

7.2. The model

When an interviewer makes an attempt to interview a selected household, there are three possible outcomes:

(i) The interviewer gets a response.

(ii) The interviewer gets no response and decides to call back.

(iii) The interviewer gets no response and decides to categorize the household as non-response (refusal).

In practice it is well known that it is difficult for the interviewer to distinguish between permanent and temporary refusals, but in this connection the important thing is that such a categorization is done after each visit.

Let p denote the probability that outcome (i) occurs in the first call, and let f denote the probability that outcome (iii) occurs in the first visit. We now assume that f is constant in the successive visits, but that outcome (i) occurs with probability Δp in the second and following calls. We also expect Δ to be larger than one because the interviewers use ingenuity. They find out from neighbours or parents when the people now absent will be available. They make appointments, etc. The result of this is, expectedly, that the probability of getting a response should increase after the first call.

In Figure 2 below the model is shown in a diagram.

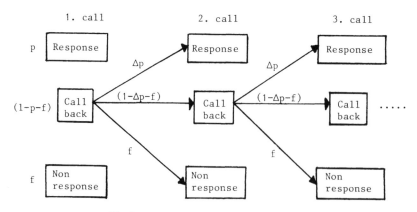

Fig. 2. A probabilistic model for non-response.

Let C denote the outcome that the interviewer gets a response from a selected household in the Cth visit, then

$$P(C = 1) = p,$$

$$P(C = 2) = (1 - p - f)\Delta p,$$

$$P(C = 3) = (1 - p - f)(1 - \Delta p - f)\Delta p,$$

$$\vdots$$

$$P(C = c) = \begin{cases} p & \text{if } c = 1, \\ (1 - p - f)(1 - \Delta p - f)^{c-2}\Delta p & \text{if } c \geq 2. \end{cases}$$

The parameters p, Δ and f can be estimated from the sample, assuming that information concerning the number of calls for each selected household is available.

The model can be generalized by allowing p to vary between the households or persons. One possibility is to assume that p is generated by a Beta-distribution, while another is to assume that p is constant within certain subclasses in the population, but varies between them.

In what follows we shall assume that the parameters are constant within certain subclasses, called post-strate, but vary between them.

7.3. An application of the model

We shall now demonstrate the use of the model in connection with the Norwegian fertility survey. In fertility surveys an important aim is to estimate the mean number of live births in the population. When estimating this parameter, nonresponses can result in a serious bias, because response rates is a function of household size, and therefore, number of children in the household. It is therefore, of particular importance to determine the effect of callbacks on the bias and the mean-square error.

When applying the model we define 7 post-strata. Post-stratum i consists of women with i live birth, $i = 0, 1, 2, \ldots, 6$. Post-stratum 6 includes women with 6 or more live births.

We assume that p and Δ are constant within each post-stratum, but vary between them. In addition, we assume that f is constant in the whole sample. (This last assumption is known to be reasonable.)

To estimate the mean number of live births in the population, we must estimate the number of women in each post-stratum.

Assume that N women have been selected in the sample, and let N_i denote the number of women selected in post-stratum i. Then the expected number of responses in the jth call in post-stratum i is $N_i P(C_i = j)$, where

$$p(C_i = c_i) = \begin{cases} p_i & \text{if } c_i = 1, \\ (1 - p_i - f)(1 - \Delta_i p_i - f)^{c_i - 2}\Delta_i p_i & \text{if } c_i \geq 2. \end{cases}$$

Let Z_{ij} denote the observed number of responses in post-stratum i in the jth call.

Several estimation methods are now available, but in order to make the calculations simple, we shall use a simple estimation method. The method is based on data from the three first calls, and the estimators of N_i, p_i and Δ_i are found as the solution to the following equations:

$$N_i p_i = Z_{i1},$$

$$N_i(1 - p_i - f)\Delta_i p_i = Z_{i2}, \qquad i = 0, 1, 2, \ldots, 6), \qquad (7.1)$$

$$N_i(1 - p_i - f)(1 - \Delta_i p_i - f)\Delta_i p_i = Z_{i3},$$

$\Sigma N_i = 5047$, as the total sample size was 5047.

The solution is presented in Table 2. Table 3 shows the number of responses in each of the first three calls.

Table 2
Number of responses in each cell

Post-stratum	Call		
	1	2	3
Total...	1 483	1 345	610
0...	311	387	199
1...	258	248	134
2...	487	410	158
3...	261	199	88
4...	107	79	30
5...	37	15	9
6...	12	7	3

Table 3
Estimates of N_i, Δ_i, and p_i

Post-stratum	$\hat{N}_i$	$\hat{\Delta}_i$	$\hat{p}_i$	
0...	1.380	1.840	0.226	
1...	1.049	1.470	0.245	
2...	1.433	1.489	0.347	
3...	771	1.356	0.339	$f = 0.099$
4...	287	1.398	0.373	
5...	96	0.783	0.383	
6...	30	1.717	0.400	

We now need an estimate of the bias and the variance of the respondent mean. For this purpose, let X_i denote the number of live births for the ith selected woman, and let

$$y_i = \begin{cases} 1 & \text{if the } i\text{th selected women is a respondent}, \\ 0 & \text{otherwise}. \end{cases}$$

The respondent mean can now be written as

$$\overline{X}_s = \frac{\sum_{i=1}^{n} X_i y_i}{\sum_{i=1}^{n} y_i},$$

where n is the sample size.

We now assume that the sample is post-stratified as above, and let

P (a woman in the sample belongs to post-stratum j | the woman is a respondent) $= q_j$; $j = 0, 1, 2, \ldots, 6$.

We have that after the first call

$$q_j = \frac{N_j p_j}{\sum_{j=0}^{6} N_j p_j}, \quad j = 0, 1, 2, \ldots, 6.$$

After $k (\geq 2)$ calls, we have

$$q_j = \frac{N_j p_j + \sum_{i=2}^{k} N_j (1 - p_j - f)(1 - \Delta_j p_j - f)^{i-2} \Delta_j p_j}{\sum_{j=0}^{6} N_j p_j + \sum_{j=0}^{6} \sum_{i=2}^{k} N_j (1 - p_j - f)(1 - \Delta_j p_j - f)^{i-2} \Delta_j p_j},$$

$$j = 0, 1, \ldots, 6. \quad (7.2)$$

We now have that

$$E(\overline{X}_s) \approx \sum_{j=1}^{5} j q_j + 6.5 q_6, \quad (7.3)$$

and

$$\text{Var}(\overline{X}_s) \approx \left[\sum_{j=0}^{5} q_j (j - E(\overline{X}_s))^2 + q_6 (6.5 - E(\overline{X}_s))^2 \right] / E\left[\sum_{i=1}^{n} y_i \right], \quad (7.4)$$

where he average number of live births per woman in post-stratum 6 is assumed equal to 6.5.

Substituting N_j, p_j, Δ_j, and f with $\hat{N}_j$, $\hat{p}_j$, $\hat{\Delta}_j$, and $\hat{f}$ found above we find estimates for q_j, $\hat{q}_j$, which inserted in (7.3) and (7.4) give us estimates of $E(\overline{X}_s)$ and $\text{Var}(\overline{X}_s)$.

From official registers the mean in the total selected sample $\overline{X}$, is known, such that

$$\text{Bias}(\overline{X}_s) = E(\overline{X}_s - \overline{X})$$

can be found.

Table 4
Estimated bias and mean-square errr by number of calls

	Number of cells									
	1	2	3	4	5	6	7	8	9	10
Bias	0.339	0.207	0.166	0.144	0.133	0.126	0.123	0.122	0.121	0.120
Mean square error	0.1162	0.0435	0.0281	0.0212	0.0182	0.0166	0.0156	0.0153	0.0151	0.0149

In Table 4 the estimated bias and mean-square error is given as a function of the number of calls.

It is seen that the bias and mean-square error are both substantially reduced after 3 to 4 calls, and that the effects of making further callback are rather small.

References

Brown, G. H. (1979). An optimization criterion for linear inverse estimation. *Technometrics* **21**, 575–579.
Central Statistical Office of Ethiopia (1982). *The rural household survey programme*. Statist. Bulletin 34. Central Statistical Office, Addis Abeba.
Dagsvik, J. (1978). Filter estimation in repeated sample surveys. Part one: Numerical results. Unpublished manuscript. Central Statistical Office of Norway.
Ericson, W. A. (1969). Subjective Bayesian models in sampling finite populations, I. *J. Roy. Statist. Soc. B* **31**, 195–234.
Godambe, V. P. (1955). A unified theory of sampling from finite populations. *J. Roy. Statist. Soc. B* **17**, 269–278.
Hansen, M. H., Hurwitz, W. N. and Bershad, M. A. (1961). Measurement errors in censuses and surveys. *Bull. Int. Statist. Inst.* **38**(2), 359–374.
Holt, D., Smith, T. M. F. and Winter, P. D. (1980). Regression analysis of data from complex surveys. *J. Roy. Statist. Soc. A* **143**, 474–487.
Kalbfleisch, J. D. and Sprott, D. A. (1969). Applications of likelihood and fiducial probability to sampling finite populations. In: *New Developments in Survey Sampling*. Wiley-Inter Science, New York.
Kish, L. and Frankel, M. R. (1974). Inference from complex samples (with discussion). *J. Roy. Statist. Soc. B* **36**, 1–37.
Laake, P. (1977). An evaluation of synthetic estimates of employment. *Scand. J. Statist.* **5**, 57–60.
Markus, G. (1979). *Analyzing panel data*. Sage University Paper Series on Quantitative Applications of the Social Sciences. 07–001. Sage Publications, Beverly Hills and London.
O'Muircheartaigh, C. A. and Marckwart, A. M. (1980). An assessment of the reliability of WFS data. In: *Proceedings from World Fertility Conference 1980. Vol. 3*, 313–360.
Patterson, H. D. (1950). Sampling on successive occasions with partial replacement of units. *J. Roy. Statist. Soc. B* **12**, 241–255.
Royall, R. M. (1970). On finite population sampling theory under certain linear regression models. *Biometrika* **57**, 377–387.
Royall, R. M. (1971). Linear regression models in finite population sampling theory. In V. P. Godambe and D. A. Sprott, eds. Holt, Rinehart and Winston, Toronto.
Royall, R. M. (1979). Prediction models in small area estimation. In: *Synthetic Estimates for small areas*. Research Monograph Series 24. National Institute on Drug Abuse.

Schaible, W. L. and Bock, D. B. (1977). An empirical comparison of the simple inflation, synthetic, and composite estimators for small area statistics. *Proceedings of the Social Statistics Section of the Amer. Statist. Ass.*

Scott, A. J. and Smith, T. M. F. (1969). Estimation in multistage surveys. *J. Amer. Statist. Ass.* **64**, 830–840.

Scott, A. J. and Smith, T. M. F. (1974). Analysis of repeated surveys using time series methods. *J. Amer. Statist. Ass.* **69**, 674–678.

Smith, T. M. F. (1976). The foundations of survey sampling: A review (with discussion). *J. Roy. Statist. Soc. A* **139**, Part 2, 183–204.

Thomsen, I. (1981). The use of markov chain models in sampling from finite populations. *Scand. J. Statist.* **8**, 1–9.

Thomsen, I. and Siring, E. (1983). On the causes and effects of nonresponse Norwegian experiences. In: *Incomplete data in sample surveys. Vol. 3, Session 1.* Academic Press, New York.

The Prediction Approach to Sampling Theory

Richard M. Royall

1. Introduction

A finite population is a collection of units, such as people, businesses, file cards, etc. Associated with the ith unit is a number y_i. If N is the total number of units in the population, the general problem is to choose some of the units as a sample, to observe their y-values, and to estimate the value of a function $h(y_1, y_2, \ldots, y_N)$ of all the y's. The prediction approach to this problem views the numbers, $y_1, y_2, \ldots, y_N$ as realizations of random variables $Y_1, Y_2, \ldots, Y_N$. After the sample has been observed, estimating h entails predicting a function of the unobserved Y's. Relationships among the variables are expressed in a model for their joint probability distribution, and the predictions are made with reference to this model. We begin with an example.

Figure 1 shows the number of beds (x) and the number of patients discharged (y) during one month from each of 32 hospitals. These 32 are a sample from a listed population of $N = 393$ short-stay hospitals. The list shows the number of beds in each of the 393 hospitals, but we know the y-values for those in the sample only. To estimate the total number of patients discharged, $T = \Sigma_1^N y_i$, we begin by writing T as the sum of the known sample y's plus the sum of the rest, $T = \Sigma_s y_i + \Sigma_r y_i$, where s is the set of 32 hospitals in the sample and r is the set of 361 non-sample hospitals.

If the number of patients discharged, y_i, is a realization of a random variable Y_i, $i = 1, 2, \ldots, N$, then estimating T is equivalent to predicting the value $\Sigma_r y_i$ of the unobserved random variable $\Sigma_r Y_i$. Formal theory for this prediction is derived from a model for the joint probability distribution of the Y's. For example, we might consider the simple proportional regression model,

$$E(Y_i) = \beta x_i, \quad i = 1, 2, \ldots, N,$$

$$\operatorname{Cov}(Y_i, Y_j) = \begin{cases} \sigma^2 x_i, & i = j, \\ 0, & \text{else}. \end{cases} \quad (1)$$

Here the expected number of patients discharged is proportional to the number

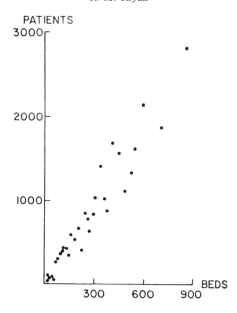

Fig. 1. Number of patients discharged and number of beds in 32 short-stay U.S. hospitals—June 1968.

of beds. This model seems to be compatible with the observations in Figure 1, but it is certainly not unique in this respect. Thus it might be reasonable to adopt (1) as a *working model*, using it to generate and evaluate estimators, confidence intervals, etc., while fully aware that other models might be more appropriate. Robustness to failure of the working model must be an important consideration.

Under the model (1) the statistic $(\Sigma_s Y_i/\Sigma_s s_i) \Sigma_r x_i$ is the linear unbiased predictor of $\Sigma_r Y_i$ having the smallest possible error variance. The corresponding estimator of T, $\hat{T} = \Sigma_s Y_i + (\Sigma_s Y_i/\Sigma_s x_i) \Sigma_r x_i = (\Sigma_s Y_i/\Sigma_s x_i) \Sigma_1^N x_i$ (the popular ratio estimator) is thus the BLU (best linear unbiased) estimator under (1). We can estimate the error variance, $\text{Var}(\hat{T} - T)$, under model (1) and use this to set approximate confidence intervals for T based on the asymptotic normality of the standardized error under the model. The theory behind these results will be sketched in the following sections, where we also examine some questions of robustness. For now we simply emphasize the equivalence, after sampling, between estimating T and predicting $\Sigma_r y_i$, and the central role of probability models in providing a framework for this prediction.

Within this framework we can use other approaches than the one just described. With some additions to the model (1) we can use Bayesian prediction techniques (Ericson, 1969), calculating the distribution of T, given the sample y's. Or we can try fiducial (Kalbfleisch and Sprott, 1969) or likelihood prediction (Royall, 1976a). What all of these approaches have in common is the recognition that (i) after the sample units have been chosen and their y-values observed, estimating T is equivalent to predicting the value $\Sigma_r y_i$ of the unobserved random variable $\Sigma_r Y_i$, and (ii) the model for the joint probability distribution of $Y_1, Y_2, \ldots, Y_N$ provides the basis for this prediction.

By contrast, the approach which has dominated finite population sampling theory for the last half-century is not based on prediction models but on the probability distributions created when the choice of a sample is left to chance. This approach, like the prediction approach based on linear regression models, produces a theory in which inferences are made in terms of bias, variance, and approximate normality. In the next section we will examine the linear prediction theory. Then we will compare it briefly with the conventional (probability sampling) theory.

2. The linear prediction approach

The ratio estimator has been a valuable guide in the development of the prediction approach, and it will be used as a key example here. This estimator is simple enough that analysis is easy and results are transparent, yet many of the results illustrate more general truths. Optimality of the ratio estimator under model (1) is established in the following Theorem 1, which gives the best linear unbiased (BLU) estimator and its error variance for the general linear model.

The prediction approach treats the unknown N-vector y as a realization of a random vector Y. Also associated with each unit is some auxiliary information in the form of a known p-dimensional row vector x_i. These N p-vectors form the rows of an $N \times p$ matrix X. For a given sample s we can rearrange the units, listing those in s first, and partition according to sample and non-sample units:

$$Y = \begin{pmatrix} Y_s \\ Y_r \end{pmatrix}, \quad X = \begin{pmatrix} X_s \\ X_r \end{pmatrix}.$$

2.1. Inference under linear regression models

The first theorem, borrowed from standard prediction theory (Whittle 1963, Chapter 4), gives the BLU estimator for any linear combination, $l'y = l'_s y_s + l'_r y_r$, of the y's under a general linear regression model relating Y to X:

$$E(Y) = X\beta, \quad \text{Var}(Y) = V\sigma^2. \tag{2}$$

Here the matrix V is given, but the p-vector β and the scalar σ^2 are unknown. Under this model a linear estimator $h'_s Y_s$ is unbiased if $E(h'_s Y_s - l'Y) = 0$, and the error variance is

$$\text{Var}(h'_s Y_s - l'Y) = \sigma^2[(h_s - l_s)' V_{ss}(h_s - l_s) - 2(h_s - l_s)' V_{sr} l_r + l'_r V_{rr} l_r]. \tag{3}$$

Note that (3) is the error variance for a particular sample s. It is often easy to find a sample which is optimal in the sense of minimizing this quantity.

The weighted least-squares estimator of β, $\hat{\beta} = (X'_s V_{ss}^{-1} X_s)^{-1} X'_s V_{ss}^{-1} Y_s$ will play an important role.

THEOREM 1. *Among linear unbiased estimators, the statistic*

$$\hat{h}'_s Y_s = l'_s Y_s + l'_r [X_r \hat{\beta} + V_{rs} V_{ss}^{-1} (Y_s - X_s \hat{\beta})] \qquad (4)$$

has the smallest error variance

$$\mathrm{Var}(\hat{h}'_s Y_s - l' Y) = \sigma^2 [l'_r (X_r - V_{rs} V_{ss}^{-1} X_s)(X'_s V_{ss}^{-1} X_s)^{-1}$$
$$\times (X_r - V_{rs} V_{ss}^{-1} X_s)' l_r + l'_r (V_{rr} - V_{rs} V_{ss}^{-1} V_{sr}) l_r]. \qquad (5)$$

According to this theorem the BLU estimator is calculated by adding to the known term $l'_s y_s$ the value of the BLU predictor of $l'_r Y_r$, $l'_r [X_r \hat{\beta} + V_{rs} V_{ss}^{-1} (Y_s - X_s \hat{\beta})]$. If the sample and non-sample units are uncorrelated (V_{rs} is the zero matrix) the predictor is simply $l'_r X_r \hat{\beta}$, the BLU estimator of $E(l'_r Y_r)$. In this case the error variance (5) also simplifies considerably.

An unbiased estimator, v_L, of the error variance can be obtained by replacing σ^2 in (5) by the usual formula

$$\hat{\sigma}^2 = r' V_{ss}^{-1} r / (n - p) \qquad (6)$$

where r is the vector of residuals, $r = y_s - X_s \hat{\beta}$.

EXAMPLES. We apply Theorem 1 with $l' = (1, 1, \ldots, 1)$ to produce estimators of the population total under some simple models with uncorrelated Y's. In the simplest case the Y's have common mean $E(Y_i) = \beta$ and variance $\mathrm{Var}(Y_i) = \sigma^2$. Then $\hat{\beta} = \bar{y}_s$ and the BLU estimator of T is $N\bar{y}_s$, the expansion estimator, with error variance $(N/f)(1 - f)\sigma^2$, where f denotes the sampling fraction n/N. Note that the error variance is the same for all samples of the same size. An unbiased estimator is

$$v_L = (N/f)(1 - f) \sum_s (y_i - \bar{y}_s)^2 / (n - 1).$$

Under the proportional regression model (1), $\hat{\beta}$ is $\hat{\beta}_R = \bar{y}_s / \bar{x}_s$, the ratio of sample means. The BLU estimator of T is then

$$\hat{T}_R = \sum_s y_i + \hat{\beta}_R \sum_s x_i = N \bar{x} \bar{y}_s / \bar{x}_s,$$

the ratio estimator. The error variance (5) is

$$\mathrm{Var}(\hat{T}_R - T) = (N/f)(1 - f)(\bar{x} \bar{x}_r / \bar{x}_s) \sigma^2. \qquad (7)$$

The sample minimizing the error variance is obviously the one for which $\bar{x}_s$ is maximized—the optimal sample consists of the n units whose x values are largest. At the other extreme, the units whose x values are smallest make up the worst possible sample. An unbiased estimator of the error variance is

$$v_{RL} = (N/f)(1 - f)(\overline{xx}_r/\bar{x}_s) \sum_s (y_i - \hat{\beta} x_i)^2 / [x_i(n - 1)]. \tag{8}$$

Another simple model which is often used has $E(Y_i) = \beta_0 + \beta_1 x_i$ and $\mathrm{Var}(Y_i) = \sigma^2$. Under this model the BLU estimator (4) of T is the 'linear regression estimator' $\hat{T}_L = N[\bar{y}_s + b(\bar{x} - \bar{x}_s)]$, where $b = \Sigma_s(x_i - \bar{x}_s)y_i/\Sigma_s(x_i - \bar{x}_s)^2$. The error variance is $(N/f)(1 - f)\{1 + [(\bar{x}_s - \bar{x})^2/(1 - f)g(s)]\}\sigma^2$, where $g(s) = \Sigma_s (x_i - \bar{x}_s)^2/n$. Samples in which the error variance is small are those in which $(\bar{x} - \bar{x})^2/g(s)$ is small, so that the error variance is minimized when $\bar{x}_s = \bar{x}$. An unbiased estimator of error variance is obtained when σ^2 is replaced by $\hat{\sigma}^2 = \Sigma_s[y_i - \bar{y}_s - b(x_i - \bar{x}_s)]^2/(n - 2)$.

For any linear estimator $h'_s y_s$, whether BLU or not, the error $h'_s y_s - l' y$, is a linear function of the elements of y. Thus under a model for which the estimator is unbiased, the standardized error $(h'_s Y_s - l' Y)/[\mathrm{Var}(h'_s Y_s - l' Y)]^{1/2}$ has an asymptotic standard normal distribution when appropriate conditions are met. (Sufficient conditions can be derived from central limit theorems in Serfling (1980) and Dvoretsky (1972).) If v is a consistent estimator of the error variance ($v/\mathrm{Var} \to 1$ in probability) then large-sample confidence intervals for $l' y$ can be based on the approximate standard normal distribution of $(h'_s Y_s - l' Y)/v^{1/2}$.

The ratio estimator illustrates the kind of conditions required. The error, $\hat{T}_R - T$, can be written $N(1 - f)[(\bar{x}_r/\bar{x}_s)\bar{Y}_s - \bar{Y}_r]$, and division by the square root of the variance, (7), yields

$$[Nf(1 - f)]^{1/2}[(\bar{x}_r/\bar{x}_s)\bar{Y}_s - \bar{Y}_r]/[(\overline{xx}_r/\bar{x}_s)\sigma^2]^{1/2}. \tag{9}$$

Suppose new units are added to the population and some of these new units are added to the sample, so that both n and $N - n$ grow. Under model (1) the standardized error (9) will converge in law to a standard normal random variable provided that the Y's are independent and the population's growth is stable in that both $\bar{x}_s$ and $\bar{x}_r$ converge to non-zero constants, while the Y's in s and the Y's in r satisfy Lindeberg conditions.

At first glance the preceding results might appear to provide a reasonable complete practical solution to the estimation problem. They give, for a very general linear model, the BLU estimator, a formula for the error variance showing what features of the sample are desirable, an unbiased estimator of the error variance, and large-sample confidence intervals.

One obvious limitation is that the BLU estimator (4) is available only when the covariance matix is known to within a constant scalar σ^2. In applications where the Y's are correlated the covariance matrix usually contains unknown parameters which appear in (4). A practical procedure might use (4) anyway, choosing from some convenient approximate values for these parameters (Royall, 1976b). The resulting estimator remains unbiased, but is no longer optimal, and its error variance is not given by (5). Thus although Theorem 1 may suggest useful estimators in such problems, it does no more.

But even with simple working models like (1) leading to explicit BLU estimators, the above results must be carefully studied before they can be judged useful. This is because of uncertainties about the models. We shall see that for some working models the samples minimizing the error variance leave the estimator extremely vulnerable to bias when the models are inaccurate, so that in choosing the sample we must often sacrifice efficiency for robustness. We shall also see that the variance estimator v_L can be badly biased by model inaccuracies, so that one of several bias-robust alternatives should be considered.

2.2. Model failure and robust inference

Simple probability models can represent real world processes only approximately. The process generating the number of patients discharged from a hospital might follow the working model (1) approximately. But a large sample of hospitals would reveal that neither the expected value of Y nor its variance is strictly proportional to the size x. Our inferences must be valid under (1), but they must remain valid under plausible departures from (1) as well.

2.2.1. Departures from the model covariance matrix

An important general result illustrated by the ratio estimator concerns the non-robustness of the variance estimator obtained by the standard procedure of replacing σ^2 in (5) by its estimator $\hat{\sigma}^2$ given in (6). This statistic v_L can be badly biased, as an estimator of the error variance, if the actual covariance matrix is not proportional to the one, V, use in the working model.

For the ratio estimator the working model is (1), so that $V = \text{diag}\{x_i\}$. Suppose the true covariance matrix is proportional to $\text{diag}\{x_i^2\}$. That is, we are using (1) as a working model, but it is inaccurate—the true model has V replaced by $\text{diag}\{x_i^2\}$. This does not affect the expected value of the error—$\hat{T}_R$ remains unbiased. But now the actual error variance is not expression (7), but

$$\text{Var}(\hat{T}_R - T) = \left\{\sum_s x_i^2[(N-n)\bar{x}_r/n\bar{x}_s]^2 + \sum_r x_i^2\right\}\sigma^2,$$

while for the estimator v_{RL} we have

$$E(v_{RL}) = [N\bar{x}(N-n)\bar{x}_r/n\bar{x}_s]\left[n\bar{x}_s - \sum_s x_i^2/n\bar{x}_s\right]\sigma^2/(n-1).$$

If n is a large but n/N is small then the relative bias in v_{RL}, $[E(v_{RL}) - \text{Var}(\hat{T}_R - T)]/\text{Var}(\hat{T}_R - T)$, is approximately $-\Sigma_s(x_i - \bar{x}_s)^2/\Sigma_s x_i^2$. Thus if the average and the dispersion of the sample x-values remain stable as n grows, v_{RL} has a negative relative bias which does not vanish. This implies that for large n and N the approximate confidence interval $\hat{T}_R \pm z(v_{RL})^{1/2}$ will have actual coverage probability smaller than its nominal value. By contrast, if the true variance matrix is proportional to the identity (Var(Y_i) constant) then the relative bias is approximately $(\bar{x}_s/n)\Sigma_s(1/x_i) - 1$, which is positive, so that the actual coverage probability will exceed the nominal value. Only when the matrix V in the working model is correct is the approximate confidence interval based on v_{RL} valid.

Fortunately there are robust alternatives to v_{RL}. They are asymptotically unbiased and consistent estimators of Var$(\hat{T}_R - T)$ under very mild restrictions on the true covariance matrix W. Their existence depends on two facts:

(i) If $W = \text{diag}\{w_i\}$ then the squared residual $r_i^2 = (Y_i - \hat{\beta}_R x_i)^2$ is an approximately unbiased estimator of w_i. Specifically,

$$E(r_i^2) = w_i - 2w_i(x_i/n\bar{x}_s) + (x_i/n\bar{x}_s)^2 \sum_s w_i, \qquad (10)$$

and this equals $w_i + O(1/n)$ under mild conitions on the sequences of x's and w's of sample units.

(ii) The error variance, Var$(\hat{T}_R - T) = [(N-n)/n]^2(\bar{x}_r/\bar{x}_s)^2 \Sigma_s w_i + \Sigma_r w_i$, is dominated, for large n and small n/N, by the first term, for which an approximately unbiased estimator can be obtained using (10). Thus

$$v = [(N-n)/n]^2(\bar{x}_r/\bar{x}_s)^2 \sum_s r_i^2 \qquad (11)$$

is asymptotically unbiased and consistent in that Var$(\hat{T}_R - T)/E(v) \to 1$ and $v/E(v) \xrightarrow{P} 1$ as $n \to \infty$ and $(n/N) \to 0$.

The kernel (11) can be adjusted so that it is strictly unbiased under the working model without altering its large sample properties. One approach (Chew, 1970) uses the system of equations defined by (10), replacing $E(r_i^2)$ by r_i^2 and solving for the w_i's to obtain strictly unbiased estimators, d_i^2, to replace the r_i^2 in (11). Then to this adjusted version of (11) is added an estimator of $\Sigma_r \text{Var}(Y_i)$, $(\Sigma_r x_i)(\Sigma_s d_i^2/\Sigma_s x_i)$, which is unbiased under the working model. The result is an estimator of Var$(\hat{T}_R - T)$,

$$v_V = (N/f)(1-f)(\bar{x}\bar{x}_r/\bar{x}_s^2) \sum_s d_i^2/n,$$

which, like v_{RL}, is strictly unbiased under the working model, but which unlike v_{RL}, is asymptotically unbiased and consistent much more generally. A related approach, using the fact that under the working model $E(r_i^2) = \sigma^2 x_i[1 - (x_i/n\bar{x}_s)]$, defines $r_i^{*2} = r_i^2/[1 - (x_i/n\bar{x}_s)]$, so that $E(r_i^{*2}) = \text{Var}(Y_i)$ under (1). Then when r_i^{*2} replaces d_i^2 in v_V the result is a statistic v_D which is also strictly unbiased under (1) and asymptotically equivalent to v_V quite generally.

A complete different approach, jackknife variance estimation, produces a statistic v_J which, although not strictly unbiased under (1), is asymptotically equivalent to v_D and v_V (Royall and Eberhardt, 1975).

These results extend easily to general linear models. If $\hat{Y}_i$ is an unbiased predictor of Y_i with $\mathrm{Var}(\hat{Y}_i)$ of order n^{-1}, then for a linear estimator $h'_s Y_s$ of $l'Y$, the statistic $\Sigma_s(h_i - l_i)^2(Y_i - \hat{Y}_i)^2$, which generalizes (11), is an approximately unbiased estimator of the dominant term in the error variance, $\mathrm{Var}(h'_s Y_s - l'Y)$. As with the ratio estimator, this kernel can be adjusted to produce an estimator of the error variance which is strictly unbiased under a specific working model and which is asymptotically unbiased and consistent much more generally (Royall and Cumberland, 1978). Empirical studies for simple estimators (ratio, linear regression, Horvitz–Thompson) in real populations have been published by Royall and Cumberland (1981a, b) and Cumberland and Royall (1981).

2.2.2. Departure from model regression function

Unbiasedness of an estimator $\hat{T}$ under the working model (2) is unaffected by changes in the variance structure. The same is not true of changes in the condition that $E(Y) = X\beta$. Inaccuracy in this part of the model can introduce a severe bias. For example, if instead of $E(Y_i) = \beta x_i$ in model (1), the actual regression function is $E(Y_i) = \beta_0 + \beta_1 x_i$, then the ratio estimator has a bias

$$E(\hat{T}_R - T) = N\beta_0(\bar{x} - \bar{x}_s)/\bar{x}.$$

This bias is a function of $\bar{x}_s$, the average size of sample units. It vanishes when the sample is *balanced* on x, i.e. when $\bar{x}_s = \bar{x}$, in which case $\hat{T}_R$ becomes simply $N\bar{y}_s$, the expansion estimator. Protection against the bias caused by a non-zero intercept β_0 is assured when a balanced sample is chosen. Another way to protect against this bias is to replace $\hat{T}_R$ with the estimator which is BLU under the more general model having $E(Y_i) = \beta_0 + \beta_1 x_i$. It is remarkable that in balanced samples this other estimator also becomes simply $N\bar{y}_s$.

More generally, let $\hat{T}^*$ denote the BLU estimator of T given by (4) with $l' = (1, 1, \ldots, 1)$. Suppose the covariance matrix V is diagonal, and for any matrix B let $\mathcal{M}(B)$ denote the linear manifold generated by the columns of B. Define a *balanced sample* to be one in which $l'_s X_s/n = l'X/N$—for each of the regressors, the sample mean equals the population mean.

THEOREM 2. (Royall and Herson, 1973a; Tallis, 1978). $\hat{T}^* = N\bar{y}_s$ if and only if (i) the sample is balanced and (ii) $V_s l_s \in \mathcal{M}(X_s)$.

Theorem 2 implies that some estimators which are appropriate under very simple linear models such as the expansion and ratio estimators are actually the BLU estimators under much more complicated regression models when the sample is well balanced on the regressors. This shows one way to obtain simple estimators which are robust in the sense of being approximately unbiased under rather general models.

Simple random sampling (SRS) can be a valuable technique for choosing samples which are approximately balanced. Suppose we are using the expansion estimator $N\bar{y}_s$ under the simple working model with constant mean and variance when in fact $E(Y_i) = \beta + \gamma z_i$, where z_i is a regressor that we have overlooked. Our estimator has a bias, $N\gamma(\bar{z}_s - \bar{z})$, that vanishes only in samples which are balanced on z. Under SRS the probability of choosing a sample showing a given degree of imbalance, $|\bar{z}_s - \bar{z}| > \varepsilon$, is small when n is large. Thus given a sample which was selected using SRS, we have reason for confidence that it is not badly balanced on an important overlooked regressor, so that our estimator is protected against the bias which can be caused by such a regressor.

The limitations of this argument appear when it is applied where there is a known auxiliary variable. For example, in an application of the ratio estimator the argument still applies to an overlooked regressor z. But since the x's are all known, we can *check* the sample for balance on x, x^2, etc., and when balance is not achieved, $\hat{T}_R$ is biased if the regression function has a non-zero intercept, a quadratic term, etc. High probability of good balance on x is insufficient. Our inferences must be made from the unique sample actually observed, and if by bad luck SRS delivers one that is badly balanced on x, there is little consolation (and no robustness) to be derived from the fact that this outcome was improbable. What is needed is demonstrated balance on the low moments of variables like x which are known to be important, and reasonable assurance of approximate balance on other (perhaps unidentified) variables like z. This might be achieved by some form of restricted randomization, such as Wallenius' (1980) 'basket method'.

Another shortcoming of SRS is that although we can make the probability of getting a sample with $|\bar{x}_s - \bar{x}| > \varepsilon$ small by taking n large, the larger n becomes the smaller ε must be if the bias caused by imbalance is not to upset the coverage properties of confidence intervals. For example, if $E(Y_i) = \beta_0 + \beta_1 x_i$ and $|\beta_0|$ is large then the ratio estimator has a substantial bias in badly balanced samples. If balance improves with increasing sample size, as it tends to do when simple random sampling is used, the absolute bias decreases. However, the importance of the bias must be judged in relation to the standard error, and the ratio of bias to standard error is $\beta_0 n^{1/2}(\bar{x} - \bar{x}_s)/[(1-f)\bar{x}_r\bar{x}_s\bar{x}\sigma^2]^{1/2}$, which vanishes as n grows only if $n^{1/2}(\bar{x}_s - \bar{x}) \to 0$. Thus to ensure that the bias caused by the non-zero intercept β_0 becomes negligible, relative to the standard error, balance must improve ($\bar{x}_s - \bar{x}$ must approach zero) *faster* than the $n^{-1/2}$ rate associated with the simple random sampling. Later in this section we show that when balance improves at just the $n^{-1/2}$ random sampling rate the bias distorts the coverage probabilities of confidence intervals even in large samples (Cumberland and Royall, 1985).

Stratifying on the regressors is another important technique for robust estimation (Royall and Herson, 1973b; Holt and Smith, 1979). Suppose the 393 hospitals are divided according to size (x) into K strata. Let $\bar{x}_k$ be the average size of the N_k hospitals in stratum k and let $\bar{y}_{s_k}$ and $\bar{x}_{s_k}$ be the sample averages. The separate ratio estimator

$$\hat{T}_{SR} = \sum_{1}^{K} (\bar{y}_{s_k}/\bar{x}_{s_k}) N_k \bar{x}_k,$$

is the BLU estimator under model (1) with different parameters (β_k, σ_k^2) in each stratum. When the sample from each stratum is balanced $(\bar{x}_{s_k} = \bar{x}_k)$, $\hat{T}_{SR}$ becomes simply $\Sigma_1^K N_k \bar{y}_{s_k}$. Then the estimator is unbiased under any model specifying a linear regression relation within each stratum, and, as Theorem 2 shows, it is BLU when each stratum has a different regression line (with variance proportional to a linear function of x). Even when balance is not achieved within each stratum, stratification itself limits the degree of imbalance possible, so that $\hat{T}_{SR}$ is protected from extreme bias under piecewise linear models. This is important because such models can approximate very general smooth regression functions.

Examination of the ratio estimator's bias under models different from (1) showed the importance of balance for protecting from bias. Similar analysis show, for other estimators and models, which sample characteristics are most important with respect to bias. For example, like the ratio estimator, the mean-of-ratios estimator $\hat{T}_{HT} = N\bar{x} \Sigma_s (Y_i/x_i)/n$ is unbiased under (1) but has a bias when $E(Y_i) = \beta_0 + \beta_1 x_i$:

$$E(\hat{T}_{HT} - T) = N\beta_0 \left[\sum_s (\bar{x}/nx_i) - 1 \right].$$

The bias vanishes when $\Sigma_s (1/nx_i) = 1/\bar{x}$, a condition which is analogous balance. Again the bias vanishes with increasing sample size when a random sampling plan is used. (In this case the inclusion probability of unit i must be proportional to its size x_i.) And again random sampling does not force the bias to zero fast enough to eliminate coverage problems in large sample confidence intervals (Cumberland and Royall, 1981).

We now consider what effect departure from the working model's regression structure has on variance estimators. Suppose $\hat{T}$ is a linear estimator that is unbiased under the working model (2). If in fact

$$E(Y) = X\beta + Z\gamma, \quad \text{Var}(Y) = V\sigma^2, \qquad (12)$$

then $\hat{T}$ is no longer unbiased, in general, but its error variance is unchanged. Let the expectation and variance operators under the *working* model be denoted with asterisks, E^* and Var*, and let the operators without asterisks refer to the true model. Then $E^*(\hat{T} - T) = 0$, $E(\hat{T} - T) \neq 0$, and $\text{Var}(\hat{T} - T) = \text{Var}^*(\hat{T} - T)$.

Now many variance estimators have the form

$$v = \sum_s k_i r_i^2$$

where the k_i are positive weights and the r_i are linear functions of Y_s with $E^*(r_i) = 0$. If v is unbiased under (2), $E^*(v) = \text{Var}^*(\hat{T} - T)$, then it has a nonnegative bias under (12), since $E(v) = \Sigma_s k_i \text{Var}(r_i) + \Sigma_s k_i [E(r_i)]^2$, and the first term equals $\text{Var}(\hat{T} - T)$.

Since $\hat{T}$ itself is biased under (12) it might be more appropriate to compare $E(v)$ to the mean squared error, $E(\hat{T} - T)^2$, than to the error variance. In that case either quantity, $E(v)$ or $E(\hat{T} - T)^2$, can be larger. But useful inequalities can be found for some important cases. A simple example occurs when $\hat{T}$ is the BLU estimator under (2) and $V \in \mathcal{M}(X)$. Then the bias vanishes when the sample is balanced on the columns of X and Z, so that $E(v) \geqslant E(\hat{T} - T)^2$, and the effect of Z is to give v a conservative bias.

The above results can be combined to describe the large-sample behavior of the Studentized error, $(\hat{T} - T)/v^{1/2}$, when $\hat{T}$ is a linear estimator which is unbiased under a specific working model whose regression and variance structure are imperfect. For example consider the estimator $\hat{T}$ given by (4) with $l' = (1, 1, \ldots, 1)$. This is the BLU estimator of T under the working model (2). Let the model's covariance matrix V be diagonal, and suppose v is a bias robust estimator, such as v_V, of the error variance. Suppose the distribution of Y is actually described, not by (2) but by

$$E(Y) = X\beta + Z\gamma, \quad \text{Var}(Y) = W,$$

where W is diagonal and the quantities $E\{[Y_i - E(Y_i)]/W_i^{1/2}\}^4$ are uniformly bounded. Now suppose that new units are added to the population and some of these new units are added to the sample, so that $n \to \infty$ and $n/N \to 0$. Under conditions which ensure that, as this growth occurs, certain key characteristics of the sample and the population remain stable (e.g. $\bar{v}_s$ and $\bar{v}_r$ converge to positive constants and $X_s' V_{ss}^{-1} W_{ss} V_{ss}^{-1} X_s/n$ converges to a positive definite matrix) it can be proved that $(\hat{T} - T)/v^{1/2}$ has an asymptotic normal distribution $N(\theta/(1 + \Delta)^{1/2}, 1/(1 + \Delta))$. Here Δ is a positive constant and $\theta = \lim E(\hat{T} - T)/\{\text{Var}(\hat{T} - T)\}^{1/2}$ (Royall and Cumberland, 1978). This shows that the limiting coverage probability of the large-sample confidence interval $\hat{T} \pm zv^{1/2}$ depends strongly on the limit, θ, of the standardized bias. Unless the bias vanishes as fast as the standard error, $|\theta| = \infty$ and the limiting coverage probability is zero. When the bias is asymptotically negligible, relative to the standard error ($\theta = 0$), the mean of the limiting normal distribution is zero. In that case, since the variance is less than unity, the associated confidence interval is conservative in that the actual coverage probability exceeds the nominal confidence coefficient.

For example, in case the working model is (1) and $\hat{T}$ is the ratio estimator, if the true model has the Y's independent with

$$E(Y_i) = \beta_0 + \beta_1 x_i, \quad \text{Var}(Y_i) = w_i,$$

then $\Delta = \beta_0^2 = \lim \Sigma_s [(x_i - \bar{x}_s)/\bar{x}_s]^2/[(n-1)\bar{w}_s]$; $\theta = (\beta_0/\bar{x}) \lim [n^{1/2}(\bar{x} - \bar{x}_s)/\bar{w}_s^{1/2}]$. The limiting coverage probability of the interval $\hat{T}_R \pm zv^{1/2}$ is zero if the sequence of samples is such that $\bar{x}_s$ is bounded away from $\bar{x}$ (as would be the case if s were chosen to maximize $\bar{x}_s$). If balance improves at the random sampling rate, say $\bar{x}_s = \bar{x} - k[\Sigma_s(x_i - \bar{x}_s)^2/n(n-1)]^{1/2}$, then the limiting distribution is $N(k[\Delta/(1 + \Delta)]^{1/2}, 1/(1 + \Delta))$. In this case, the limiting coverage probability of $\hat{T}_R \pm zv^{1/2}$ is a function of $f(k)$ which is maximized at $k = 0$ and which decreases with increasing $|k|$. When balance is achieved $k = 0$ and the coverage probability exceeds the nominal confidence coefficient, $\Phi(z) - \Phi(-z)$. But as $|k|$ increases the coverage probability falls below the nominal value. When the variability about the regression line is small, the limiting coverage probability is approximated by the step function

$$f_0(k) = \begin{cases} 1, & |k| < z, \\ 0, & |k| > z. \end{cases}$$

Royall and Cumberland (1983) have illustrated these results in an empirical study.

3. The approach based on probability sampling distributions

A probability sampling plan is a scheme for choosing the sample so that every subset of units s has known probability $p(s)$ of selection. Under the prediction approach, probability sampling appears as an important tool, useful, for example, for protecting against unconscious bias and for providing some assurance of approximate balance on overlooked regressors. But even when a probability sampling plan is used to select the sample, the prediction approach bases its inferences from that sample on a model for the distribution of Y, not on the probabilities used in deciding which elements of Y to observe.

Another approach is much more popular. In that approach the basic probabilities used in inference are derived from the sampling plan, not from a prediction model. Definitions of bias, variance, etc. are stated in terms of the selection probabilities. Thus the bias in the ratio estimator $\hat{T}_R(s) = N\bar{x}\bar{y}_s/\bar{x}_s$ is defined as

$$E_p(\hat{T}_R - T) = \sum p(s)[\hat{T}_R(s) - T],$$

where the summation is over all possible samples s. The variance is similarly defined:

$$\mathrm{Var}_p(\hat{T}_R) = \sum_s p(s)[\hat{T}_R(s) - E_p(\hat{T}_R)]^2.$$

In this approach the essential random element is the set s. Sometimes the vector y is represented as a realization of a random vector Y, but the probability distribution of Y, which in this context is usually called a 'superpopulation model', is allowed no essential role in inference. Thus the probability sampling approach can be used in the presence of a prediction model, just as the prediction approach can be used with a probability sample. The essential distinction concerns which probability distribution is used in inference, not what procedure is used for choosing the sample on which that inference is based.

We referred to the prediction model as a 'working model' to emphasize that it is tentative and approximate. By contrast our control and knowledge of the probability sampling distribution is complete, at least in principle. Thus it seems prudent to base inferences on the latter, if possible, keeping them independent of fallible prediction models.

However, another argument suggests the opposite conclusion—that prediction models are *necessary* in inference. After the sample is selected, observations on the units in s must be used to make inferences, or predictions, about the unobserved y values. The strength of the probability sampling approach, its freedom from assumptions about the y's, now becomes a weakness. The probability sampling plan does not describe or establish any relationship between the y's of sample units and the unobserved y's, except that the non-sample units also had a chance to be the ones selected. Inference from sample to non-sample units requires a stronger relationship than this. Prediction models specify probabilistic structures which link the two sets of units and provide a basis for inference, as detailed in Theorem 1 for example. This line of reasoning leads to the conclusion that the probability sampling distribution is inadequate as a basis for inference. Additional structure, e.g. a prediction model, is required.

Which conclusion is correct? Should inferences be based on probability sampling distributions or on prediction models? ... or on some combination of the two (Hartley and Sielken, 1975)? Two general inference principles address these questions, and their answer is unequivocal—inference should be based on the prediction model only. The Likelihood Principle implies that the probability sampling distribution is irrelevant for inference, since the likelihood function does not depend on the sampling plan (Basu, 1969; Godambe, 1966). The Conditionality Principle implies the same conclusion, since the sample s is an ancillary statistic and the conditionsal distribution, given s, is just the prediction model.

While the probability sampling approach defines bias and variance so as to associate with a sample s the average characteristics of the set of samples from which s was drawn, the prediction approach forces us to condition inferences on the sample actually observed. For example, with the ratio estimator and SRS, probability sampling theory associates one standard error and one bias which are averages over all possible samples and are the same regardless of which sample was selected. On the other hand, the prediction approach recognizes that under a very general model a sample with large $\bar{x}_s$ provides a more stable ratio estimator than a sample with small $\bar{x}_s$. It further recognizes that when the assumption of approximate proportionality of y to x, which led to adoption of the ratio estimator in the first place, fails, a sample balanced on x and x^2 provides less bias than a badly balanced sample under quite general conditions. The constant bias and standard error from probability sampling theory are undeniably correct, mathematically. This fact appears to be not only irrelevant but positively misleading with reference to the goal of robust inference from the particular sample actually selected.

The question of whether inferences should be based on distributions created by deliberate randomization or on probability models arises also in interpreting data from designed experiments, clinical trials, observational studies, etc. The Randomization Principle, which asserts that the random sampling distribution provides the only valid inferences, has its proponents in those areas (Kempthorne, 1955). But there the principle has never achieved the general acceptance which has enabled it to dominate finite population sampling theory (Cornfield, 1971; Basu, 1980; Royall, 1976c). There is no apparent reason why finite population inference should be different and isolated from the rest of statistics (Smith, 1976). The prediction approach bridges the gap.

References

Basu, D. (1969). Role of the sufficiency and likelihood principles in sample survey theory. *Sankhyā A* **31**, 441–454.

Basu, D. (1980). Randomization analysis of experimental data: the Fisher randomization test. *J. Amer. Statist. Assoc.* **75**, 575–595.

Chew, V. (1970). Covariance matrix estimation in linear models. *J. Amer. Statist. Assoc.* **65**, 173–181.

Cornfield, J. (1971). The University Group Diabetes Program: A further statistical analysis of the mortality findings. *J. Amer. Med. Assoc.* **217**, 1676–1687.

Cumberland, W. G. and Royall, R. M. (1981). Prediction models in unequal probability sampling. *J. Roy. Statist. Soc. Ser. B* **43**, 353–367.

Cumberland, W. G. and Royall, R. M. (1985). Does simple random sampling provide adequate balance? Submitted to *J. Roy. Statist. Soc. Ser. B*.

Dvoretsky, A. (1972). Asymptotic normality for sums of dependent random variables. In: L. M. LeCam, J. Neyman and E. Scott, eds., *Proceedings of the Sixth Berkeley Symposium on Mathematical Statistics and Probability, Volume II: Probability Theory*. University of California Press, Berkeley, 513–535.

Ericson, W. A. (1969). Subjective Bayesian models in sampling finite populations. *J. Roy. Statist. Soc. Ser. B* **31**, 195–233.

Godambe, V. P. (1966). A new approach to sampling from finite populations I: Sufficiency and linear estimation. *J. Roy. Statist. Soc. Ser. B* **28**, 310–319.

Hartley, H. O. and Sielken, R. L. Jr. (1975). A 'superpopulation' viewpoint for finite population sampling. *Biometrics* **31**, 411–422.

Holt, D. and Smith, T. M. F. (1979). Post stratification. *J. Roy. Statist. Soc. Ser. A* **142**, 33–46.

Kalbfleisch, J. D. and Sprott, D. A. (1969). Application of likelihood and fiducial probability to sampling finite populations. In: N. L. Johnson and H. Smith, Jr., eds., *New Developments in Survey Sampling*. Wiley, New York, 358–389.

Kempthorne, O. (1955). The randomization theory of experimental inference. *J. Amer. Statist. Assoc.* **50**, 946–967.

Royall, R. M. (1976a). Likelihood functions in finite population sampling theory. *Biometrika* **63**, 605–614.

Royall, R. M. (1976b). The linear least-squares prediction approach to two-stage sampling. *J. Amer. Statist. Assoc.* **71**, 657–664.

Royall, R. M. (1976c). Current advances in sampling theory: implications for human observational studies. *Amer. J. Epid.* **104**, 463–474.

Royall, R. M. and Cumberland, W. G. (1978). Variance estimation in finite population sampling. *J. Amer. Statist. Assoc.* **73**, 351–358.

Royall, R. M. and Cumberland, W. G. (1981a). An empirical study of the ratio estimator and estimators of its variance. *J. Amer. Statist. Assoc.* **76**, 66–77.

Royall, R. M. and Cumberland, W. G. (1981b). The finite population linear regression estimator and estimators of its variance—An empirical study. *J. Amer. Statist. Assoc.* **76**, 924–930.

Royall, R. M. and Cumberland, W. G. (1983). Conditional coverage properties of finite population confidence intervals. *J. Amer. Statist. Assoc.* **80**, 355–359.

Royall, R. M. and Eberhardt, K. R. (1975). Variance estimates for the ratio estimator. *Sankhyā C* **37**, 43–52.

Royall, R. M. and Herson, J. (1973a). Robust estimation in finite populations I. *J. Amer. Statist. Assoc.* **68**, 880–889.

Royall, R. M. and Herson, J. (1973b). Robust estimation in finite populations II: Stratification on a size variable. *J. Amer. Statist. Assoc.* **68**, 890–893.

Serfling, R. J. (1980). *Approximation Theorems of Mathematical Statistics*. Wiley, New York.

Smith, T. M. F. (1976). The foundations of survey sampling: a review. *J. Roy. Statist. Soc. Ser. A* **139**, 183–204.

Tallis, G. M. (1978). Note on robust estimation in finite populations. *Sankhyā C* **40**, 136–138.

Wallenius, K. T. (1980). Statistical methods in sole source contract negotiation. *J. Undergrad. Math. and Appl.* **10**, 35–47.

Whittle, P. (1963). *Prediction and Regulation by Linear Least-Squares Methods*. The English Universities Press, London.

Sample Survey Analysis: Analysis of Variance and Contingency Tables

Daniel H. Freeman, Jr.

1. Introduction

Data obtained from sample surveys provide health, medical and social scientists unprecedented opportunities to study large human populations. The opportunity derives from the ability to make statistical inferences from samples of economically reasonable size to much larger populations of interest, when the sample has a known probability structure. It is this simple fact which has led to the implementation of probability samples both for small areas such as municipalities and for national descriptive surveys such as the National Health Survey. Survey theoreticians have focused on developing both selection plans and estimators with the highest possible precision at the lowest possible cost. This focus has led to progressively more complex designs which ultimately yield very good estimators even for small or rare sub-populations.

When data from these surveys are put to analytical purposes the care with which the surveys are constructed often becomes a hindrance. This is because the survey designs systematically violate the assumptions of simple random sampling which are central to most conventional statistical methodologies. As a result, data analysts are placed in the following, somewhat paradoxical position. If a purposive or convenience sample had been utilized then normal scientific thinking leads to assuming the data are based on simple random sampling and analysis can proceed. Of course, since the sample has no underlying probability structure the inferences cannot be strictly valid from a statistical point of view. On the other hand, if a 'true' probability sampling scheme is utilized then inevitably at some stage a clustering plan is introduced and independence of the observations is destroyed. Hence, standard methods seem inappropriate. This suggests that by trying to attain a more rigorous foundation in probability the scientist surrenders analytical flexibility.

This work was supported in part under contract to the State of Connecticut Department of Health and the National Heart Lung and Blood Institute. All conclusions are solely the responsibility of the author.

A number of approaches to this paradox are found in the literature. The most prominent is a renewed defense of purposive sampling. This is developed in great detail by Royall (1968, 1971), Royall and Cumberland (1981a, b), Cumberland and Royall (1981), Holt, Smith and Winter (1980), and Little (1983). It is clear that when robust models of the population can be specified then estimators with superior precision can be utilized for appropriately designed samples. The approach does not go on to justify ignoring the underlying sample design when a statistical analysis is undertaken. There remains considerable controversy over the appropriateness of purposive selection designs. As noted by J. Neyman (1971): '... to arrange the sampling scheme the validity of which depends on these (or any other) unverified hypotheses, seems dangerous (p. 277)'.

A parallel approach is to examine what occurs for standard estimators when the design is ignored. As has been known for many years this can lead to substantial underestimates of sampling variances (Kish, 1965). In an attempt to adjust for this a variety of summary measures or 'design effect' corrections have been proposed (Kish, 1965; Pfefferman and Nathan, 1981). But as noted by Scott and Rao (1981) no simple adjustment seems to work for the full range of designs in current use.

Another approach is to develop parallel weighted least squares and ordinary least squares estimators and test for significant differences. Dumouchel and Duncan (1983) argue that systematic departures from the regression model can be detected using standard methods of residual analysis. When these departures are adjusted for, the simpler model can be utilized. However this approach ignores a real source of difficulty, namely dependencies induced by the clustering of the ultimate sampling units.

An alternative approach is to attempt incorporating the sampling design into the analysis directly. This is particularly relevant when several sub-populations are to be compared and contrasted. The method (KFF) proposed by Koch, Freeman and Freeman (1975) has been extensively explored in a variety of analysis of variance and contingency table applications (Freeman et al., 1976; Freeman and Brock, 1978; NCHS et al., 1982). These investigations show the methodology can be reasily implemented for virtually any survey designs as long as appropriate variance estimators are available. In addition, for certain selection designs the correlation structure created by the clustering is reflected in the test statistics used for comparing sub-populations.

In this chapter we will review the KFF methodology. The use of a matrix formulation will simplify the presentation but for a variety of applications the estimators and test statistics can be computed using standard weighted least squares software. An example taken from the Connecticut Blood Pressure Surveys is used to illustrate the method.

2. The KFF methodology for analysis of variance models

Suppose a sampled population consists of N elements or ultimate sampling units (usu's) indexed by $i = 1, 2, \ldots, N$. The survey analyst wishes to compare J domains or sub-populations. As noted by Cochran (1963, pp. 37–38) this means testing whether the domains are drawn from the same infinite population. Hence the finite population correction factors should be ignored. The sample is to be of total size n but the individual domain sizes are unknown. Define an indicator variable to characterize the selection of the ith unit:

$$U(i) = \begin{cases} 1 & \text{if } i\text{th unit is selected}, \\ 0 & \text{otherwise}. \end{cases} \qquad (1)$$

Note $\sum_{i=1}^{N} U(i) = n$.

Attention is restricted to probability samples so assume

$$E(U(i)) = p_i > 0 \qquad (2)$$

is either known or conceptually knowable appropriate resources. Complex sampling plans can be characterized in terms of the pair-wise selection probabilities:

$$E(U(i)U(i')) = p_{ii'} \geq 0, \quad i, i' = 1, \ldots, N. \qquad (3)$$

Now suppose two variables are asociated with each unit: $G(i, j)$, $Y(i)$ where,

$$G(i, j) = \begin{cases} 1 & \text{if } i\text{th-unit is in domain } j, \\ 0 & \text{otherwise}; \end{cases} \qquad (4)$$

$Y(i) =$ a value for ith unit.

The population mean and total are given by

$$\overline{Y} = Y/N = \sum_{i=1}^{N} Y(i)/N. \qquad (5)$$

The domain values to be compared are:

$$\overline{Y}_j = Y_j/N_j = \sum_{j=1}^{N} G(i, j)Y(i) \Big/ \sum_{i=1}^{I} G(i, j),$$

$$Y_j = \sum_{i=1}^{N} G(i, j)Y(i), \quad N_j = \sum_{i=1}^{I} G(i, j), \qquad j = 1, \ldots, J. \qquad (6)$$

The Horvitz and Thompson (1952) estimators of the domain values are

$$y_j = \sum_{i=1}^{N} U(i)G(i,j)Y(i)/p_i, \qquad n_j = \sum_{i=1}^{N} U(i)G(i,j)/p_i. \qquad (7)$$

The ratio estimators of $\bar{Y}_j$ are then given by

$$\bar{y}_j = y_j/n_j, \quad j = 1, \ldots, J. \qquad (8)$$

The comparison across domains requires that we estimate the variances and covariances associated with y_j and n_j. It is easy to show (Koch, 1973) the general form of the sampling variance is given by:

$$\text{var}(y) = \sum_{i=1}^{N} \sum_{i'=1}^{N} Y_i Y_{i'} (p_{ii'} - p_i p_{i'})/p_i p_{i'}. \qquad (9)$$

If the variances can be estimated then as noted by KFF (1975), the covariances both across domains and between numerators and denominators may be obtained directly:

$$2\,\text{cov}(y_j, n_j) = [v(y_j + n_j) - v(y_j) - v(y_j)],$$

$$2\,\text{cov}(y_j, y_{j'}) = [v(y_j + y_{j'}) - v(y_j) - v(y_{j'})], \qquad (10)$$

$$2\,\text{cov}(n_j, n_{j'}) = [v(n_j + n_{j'}) - v(n_j) - v(n_{j'})].$$

Thus it is in principle easy to construct the full set of variances and covariances for the $\bar{y}_j$. Unfortunately, this requires not only the p_j but also the full set of pair-wise selection probabilities, $p_{ii'}$. For a variety of multi-stage surveys with clustering of the usu's this has proved an intractable problem which is exacerbated by various post-sampling adjustments such as for non-response and under coverage.

A number of indirect approaches are available to circumvent this difficulty. These are based on some form of random sub-sampling or replication (Deming, 1960). Kish and Frankel (1974) and Bean (1975) have compared the approaches and found no important differences. One of the most common indirect methods is known as 'balanced repeated replication' or BRR (McCarthy, 1966, 1969a, b). We present a summary here only because it illustrates the basic methodology.

Assume the sample can be partioned into K half samples each of which is a mirror image of the original selection design except for size. Specifically we require that each stratum and clustering pattern is represented in each half sample and that the half samples be constructed according to a pattern of orthogonal replicates. The sample statistics, y_j and n_j are computed in each half sample:

$$y_{jk} = \sum_{i=1}^{N} U(i)G(i,j)Y(i)H(i,k)/p_i,$$

$$n_{jk} = \sum_{i=1}^{N} U(i)G(i,j)H(i,k)/p_i, \qquad (11)$$

where

$$H(i, k) = 2 \text{ (or some other appropriate weight)}$$
$$\text{if the } i\text{th observation is in the } k\text{th half sample},$$
$$= 0 \text{ otherwise}.$$

The estimated variances are then of the form

$$v_B^2(y_j) = \sum_{k=1}^{K} (y_{jk} - y_j)^2/K, \quad v_B^2(n_j) = \sum_{k=1}^{K} (n_{jk} - n_j)^2/K. \tag{12}$$

The major gain in computational efficiency comes when the covariances are also of interest. Let

$$y' = (y_1, n_1, y_2, n_2, \ldots, y_J, n_J), \quad y_k' = (y_{1k}, n_{1k}, y_{2k}, n_{2k}, \ldots, y_{Jk'}, n_{Jk}).$$

The complete variance covariance matrix is estimated by (Koch and Lemeshow, 1972):

$$V_B(y) = \sum_{k=1}^{K} (y_k - y)(y_k - y)'/K. \tag{13}$$

A Taylor series estimator for $\bar{y}_j$ follows immediately by noting

$$\bar{y}_j = y_j/n_j = \exp\{(1 - 1)\log[(y_j, n_j)']\}. \tag{14}$$

Letting $\bar{j}' = (\bar{y}_1, \ldots, \bar{y}_J)$, following Grizzle, Starmer and Koch (1969), we obtain

$$V_T(\bar{y}) = D_{\bar{y}} A D_y^{-1} V_B(y) D_y^{-1} A' D_{\bar{y}}, \tag{15}$$

where D_y is a diagonal matrix with y on the diagonal, $A = [(1 - 1) \otimes I_J]$, and $\otimes$ denotes a left direct product.

An obvious alternative is to apply BRR to $\bar{y}$ directly. Freeman et al. (1975a), Kish and Frankel (1974), and Bean (1975) all conclude two approaches yield equivalent results. The important point is that some method of estimating the complete variance–covariance matrix be available.

The analyst can now address two separate issues:
(1) How large is the design or variance inflation due to the selection design?
(2) Is there significant inter-domain correlation?

These are addressed by recalling the correlation coefficient has the form

$$\rho(x, y) = \text{cov}(x, y)/\sqrt{\text{var}(x)\,\text{var}(y)}.$$

Hence the estimated variance–covariance matrix can be written

$$V(\bar{y}) = S(\bar{y}) R(\bar{y}) S(\bar{y}) \tag{16}$$

where $S(\bar{y})$ is a diagonal matrix with the standard errors of $\bar{y}_j$ on the diagonal, and $R(\bar{y})$ is the estimated correlation matrix associated with $\bar{y}$:

$$R(\bar{y}) = \begin{bmatrix} 1 & r_{12} & r_{13} & \cdots & r_{1J} \\ r_{12} & 1 & r_{23} & \cdots & r_{2J} \\ r_{13} & r_{23} & 1 & \cdots & r_{3J} \\ \vdots & \vdots & \vdots & & \vdots \\ r_{1J} & r_{2J} & r_{3J} & \cdots & 1 \end{bmatrix}$$

If $S(\bar{y})$ is close to what would be obtained under simple random sampling the design effect is negligible. If $R(\bar{y})$ is close to the identity, the correlation effect can be ignored. At this time appropriate test statistics for these hypotheses are not known. However, a descriptive inspection of R and S should prove useful. This will be illustrated in the next section.

The final step is to compare the means across the domains. The KFF method consists of fitting an analysis of variance type model to the vector of means. Let X be a $J \times M$ 'design matrix' which reflects hypothesized differences among the domains. That this model characterizes the variation corresponds to

$$H_0: \quad E(\bar{y}) = XB \tag{17}$$

where $E(\cdot)$ denotes expectation and B is an unknown vector of parameters. A 'best asymptotically normal' set of estimators is given by the usual (Neyman, 1949; Bhapkar, 1966; Grizzle et al., 1969) weighted least square computation:

$$b = (X'V^{-1}X)^{-1}X'V^{-1}\bar{y}.$$

If $R = I$ this simplifies to

$$b = (X^1 S^{-2} X)^{-1} X S^{-2} \bar{y} \tag{19}$$

where S^{-2} is the inverse of the diagonal matrix with the estimated variances of $\bar{y}_j$ on the diagonal. The Neyman chi-square statistic for the hypothesis (17) is

$$Q_e = (\bar{y} - Xb)' V^{-1} (\bar{y} - Xb) \tag{20}$$

which will follow chi-square distribution with $J - M$ degrees of freedom when the n_j are reasonably large (≥ 25) and H_0 is true.

The final analytical step is to test hypothesis about the components of B:

$$H_c: \quad CB = 0 \tag{21}$$

where C is a $c \times M$ contrast matrix of rank c. The form of the test statistics, which follow a chi-square distribution (df = c) when (21) is true, is given by

$$Q_c = (Cb)' [C(X'V^{-1}X)^{-1}C']^{-1} CB. \tag{22}$$

3. An example: Connecticut Blood Pressure Survey

The preceeding illustrates that the key elements for the analysis of survey data are a vector of domain estimates and the corresponding variance–covariance matrix estimate. For variety of surveys the former and a vector of standard errors, is readily available. The corresponding correlation matrix, R, is less commonly published. As shown in the formulation for either replicated surveys or surveys which lead themselves to 'pseudo-replication' (McCarthy, 1966, 1969a, b) the correlation matrix is an immediate byproduct of the variance estimation process. In this section the process of testing hypothesis and estimating model parameters is illustrated with data from the first and second Connecticut Blood Pressure Surveys (CBPS-I and CBPS-II).

Details of CBPS-I and CBPS-II have been previously published (Freeman et al., 1983, 1985). An important point is the survey were designed specifically to permit intra-survey comparisons so as to allow inferences about the civilian adult population of Connecticut in the periods 1978 and 1982. Both surveys were designed so as to have 20 orthogonal half samples in accordance with the methodology described by McCarthy. In both surveys primary sampling units (PSU) were selected with probability proportional to size. In the first survey this was accomplished by selecting one PSU from each of 32 strata. After sampling, the 8 PSU's from certainty strata were split according to replicates within the PSU's. The remaining 24 strata were paired so as to form 12 additional 'super-strata'. Together these formed the required 20 pseudo-replicates each of which reflected the overall sample design in all aspects except size.

The second survey was similarly constructed. A total of 34 strata were used of which 6 contained only one PSU. Paired selections of PSU's were then made from the remaining 28 strata. Again replication within PSU's was used throughout, so the certainty strata could be appropriately split. This again permitted forming 20 pseudo-replicates each of which mirrored the overall survey design except for size. It should be noted that in CBPS-II interviewer assignments were randomized across the replicates so the between interviewer effects can also be controlled.

An example of the final tabulations of the survey data is shown in Table 1. Here the prevalence of controlled blood pressure per 100 hypertensives is shown for adult males by obesity, income, and period. Since obesity and income have 2 and 3 levels respectively there are up to 6 domains in each time period. No obese, upper income males with controlled blood pressure were detected in the first survey so only a total of 11 domains could have directly estimated prevalences. The corresponding standard errors were estimated using BRR according to (13).

Table 1
Percent of hypertensive adult males with controlled blood pressure by obesity and income: Connecticut 1978, 1982

Obesity	Income $1000's	Domains	Period			
			1978		1982	
			Estimate	SE	Estimate	SE
No	<15	1	9.69	3.66	13.66	5.41
	15–29	2	8.57	2.72	11.81	4.12
	30+	3	11.76	3.55	19.83	3.26
Yes	<15	4	4.77	2.13	15.14	5.14
	15–29	5	7.94	2.34	8.13	3.34
	30+	6	–	–	14.23	3.41

– No observations.
Obesity: Quetelet exceeds 75th percentile for males 18–29.
Hypertension: One or more of: SBP $\geq$ 140, DBP $\geq$ 90, or report of current treatment.
Controlled blood pressure: SBP < 140, DBP < 90, and report of current treatment.

Here $Y(i)$ is a dummy variable indicating the presence or absence of controlled blood pressure. Only males with hypertension were considered in the denominator. The $G(i, j)$ were used to group individuals into the $J = 6$ domains. It should be noted that the standard errors were estimated directly rather than using the two-step Taylor Series approach (15). The standard errors for each survey form the diagonal elements of S in equation (16).

The correlation matrix, R, is shown in Table 2. This was generated as a by product of the BRR procedure so there was no additional cost associated with obtaining it. Since the surveys were conducted on independent samples it was assumed there would be '0' between survey correlation. The between domain correlations are shown in the upper triangle of Table 2 for CBPS-I and in the

Table 2
Correlation matrices for Table 1: 1978 above and 1982 below the diagonal

Domains	Domains					
	1	2	3	4	5	6
1	1.	–0.255	0.210	0.002	–0.172	–
2	–0.305	1.	0.068	–0.114	–0.098	–
3	0.137	–0.009	1.	–0.110	0.169	–
4	–0.370	0.358	–0.223	1.	–0.326	–
5	0.101	–0.593	0.210	–0.457	1.	–
6	–0.002	–0.193	0.184	0.090	0.083	1.

– No observations.

lower triangle for CBPS-II. For CBPS-I the mean correlation is -0.063 and CBPS-II the mean correlation is -0.066. This suggests first that the surveys are quite similar and second using the full covariance matrix (16) may produce a more powerful analysis. The latter effect has been noted by Freeman et al. (1976) and Freeman and Brock (1978).

The analysis proceeds by fitting a simple analysis of variance model to the rates in Table 1. Let $\bar{y}_{ijk}$ denote the estimated prevalences where $i = 1, 2$ indexes obesity, $j = 1, 2, 3$ indexes income, and $k = 1, 2$ indexes period. The model is then of the form

$$E(\bar{y}_{ijk}) = m + O_i + I_j + P_k \tag{23}$$

where m is a baseline for 1982, 30+ income, and obese:

$$O_2 = 0, \quad I_3 = 0 \quad \text{and} \quad P_2 = 0.$$

The corresponding test statistics are shown in Table 3, where the estimated R is on the left and $R = I$ is assumed on the right.

Table 3
Analysis of variation and parameter estimates for percent of controlled blood pressure among hypertensives

Source		R estimated			$R = I$	
		Analysis of variation				
	df	Q	p-value		Q	p-value
Model	4	18.74	0.00		16.77	0.00
Period	1	6.41	0.01		4.39	0.04
Obesity	1	2.49	0.11		2.06	0.15
Income	2	4.68	0.10		3.12	0.21
Error	6	4.61	0.59		3.16	0.79
Total	10	23.40*	0.01		19.93*	0.03
		Estimates				
		B	SE		B	SE
Intercept		14.27	2.47		14.96	2.45
1978 vs 1982		-3.89	1.54		-4.54	2.17
Normal vs Obese		3.17	2.01		2.90	2.02
<15 vs 30+		-3.57	2.75		-4.50	2.90
15–29 vs 30+		-4.89	2.26		-4.43	2.64

* Model + Error ≠ Total because of rounding error.

The first concern is whether the model is appropriate. The Q_e statistics (20) are shown on the 'Error' line. There are 6 degrees of freedom since only 11 domains have estimates and 5 parameters are estimated. Regardless of which R is used the model 'fits the data' quite well. The next hypothesis concerns whether there a is significant variation across the domains. This corresponds to the hypothesis (20) where

$$C = \begin{bmatrix} 0 & 1 & 0 & 0 & 0 \\ 0 & 0 & 1 & 0 & 0 \\ 0 & 0 & 0 & 1 & 0 \\ 0 & 0 & 0 & 0 & 1 \end{bmatrix} \quad H_0: O_1 = I_1 = I_2 = P_1 = 0.$$

This yields Q_c (22) which are highly significant, as shown in the line labelled 'Model'. This is the broken into the components: Period ($P_1 = 0$), Obesity ($O_1 = 0$), Income ($I_1 = I_2 = 0$). Only the first of these is significant.

The estimated model parameters (18) and (19) are shown in the lower panel of Table 3. The only significant effect corresponds to the comparison of 1978 to 1982 which indicates control rates were nearly 4 percentage points lower in 1978 than in 1982. Using the model is is also possible to estimate the prevalence among obese upper, income, adult male hypertensives in 1978:

$$\hat{Y}_{231} = 14.27 - 3.89 = 10.29 \quad (R \text{ estimated}),$$

$$\hat{Y}_{231} = 14.96 - 4.54 = 10.42 \quad (R = I).$$

4. Summary

This chapter has reviewed the KFF approach to survey analysis. The procedure is appropriate for both analysis variance of cell means and contingency table analysis. It allows the survey analysis to take into account the complexity of the selection design by which the data are gathered. This is particularly appropriate for large surveys which include clustering at one or more stages. The procedure may be used for any survey which permits estimation of standard errors. Moreover, for a variety of estimation techniques the inter-domain correlations can be generated at no additional cost. When these are available more powerful analyses may be undertaken. Fortunately, for many surveys, such as the example in Section 3, this additional step may be unnecessary. When only the standard errors are available, the analysis may be undertaken using standard weighted least squares soft-ware. The important caveat is that while the usual F-tests are inappropriate, the sums of squares can be interpreted as Neyman chi-square test statistics.

A final note concerns the computations. The example used fairly simple procedures. In this case all computations including the estimation of the variance–covariance matrix using BRR were performed with PROC MATRIX (SAS, 1982). The standard error estimation required approximately 5 times as much CPU time as the computation of the estimates. The correlation matrix is a 'free' by product.

Acknowledgement

The author wishes to thank V. A. Richards and T. O'Connor for their research assistance, and the staff of the Connecticut Bload Pressure Surveys.

References

Bean, J. A. (1975). Distribution and properties of variance estimators for complex multi-stage probability samples: an empirical distribution. *Vital and Health Services Statistics* 2:65, National Center for Health Statistics, Rockville, MA.
Bhapkar, V. P. (1966). A note on the equivalence of two test criteria for hypotheses in categorical data. *J. Am. Statist. Assoc.* **61**, 28–235.
Cochran, W. G. (1963). *Sampling Techniques, Second Edition*. Wiley, New York.
Cumberland, W. G. and Royall, R. M. (1981). Prediction models and unequal probability samping. *J. Am. Statist. Assoc.* **76**, 341–353.
Deming, W. E. (1960). *Sample Design in Business Research*. John Wiley, New York.
DuMouchel, W. H. and Duncan, G. J. (1983). Using sample survey weights in multiple regression analyses of stratified samples. *J. Am. Statist. Assoc.* **78**, 535–543.
Freeman, D. H. and Brock, D. B. (1978). The role of covariance matrix estimation in the analysis of complex sample survey data. In: K. Namboodiri, ed., *Survey Sampling and Measurement*. Academic Press, New York, 121–140.
Freeman, D. H., D'Atri, D. A., Hellenbrand, K., Osfelt, A. M., Papke, E., Pioreen, K., Richards, V. A., Sardinas, A. (1983). The prevalence distribution of hypertension: Connecticut Adults 1978–79, *J. Chronic Disease* **36**, 171–183.
Freeman, D. H., Freeman, J. L., Brock, D. B., Koch, G. G. (1976a). Strategies in the multivariate analysis of data from complex Surveys II: An application to the United States National Health Interview Survey. *Int. Statist. Rev.* **44**, 317–330.
Freeman, D. H., Freeman, J. L., Koch, G. G., Brock, D. B. (1976b). An analysis of physician visit data from a complex sample survey. *Am. J. Public Health* **66**, 979–983.
Freeman, D. H., Ostfeld, A. M., Hellenbrand, K., Richards, V. A., Tracy, R. (1985). Changes in the prevalence distribution of hypertension: Connecticut Adults 1978–79 to 1982. *J. Chronic Disease* **38**, 157–164.
Grizzle, J. E., Starmer, C. F., and Koch, G. G. (1969). Analysis of categorical data by linear models. *Biometrics* **25**, 489–504.
Holt, D., Smith, T. M. F., and Winter, P. D. (1980). Regression analysis of data from complex surveys. *J. Royal Statist. Soc. Ser. A* **143**, 474–487.
Kish, L. (1965). *Survey Sampling*. Wiley, New York.
Kish, L. and Frankel, M. A. (1974). Inferences from complex samples. *J. Royal Statist. Soc. Ser. B* **36**, 1–37.
Koch, G. G. (1973). An alternative approach to multivariate response error models for sample survey data with application involving subclass means. *J. Am. Statist. Assoc.* **68**, 906–913.
Koch, G. G., Freeman, D. H., and Freeman, J. L. (1975). Strategies in the multivariate analysis of data from complex surveys. *Int. Statist. Rev.* **43**, 59–78.

Koch, G. G. and Lemeshow, S. (1972). An application of multivariate analysis to complex sample survey data. *J. Am. Statist. Assoc.* **67**, 780–782.

Little, R. J. A. (1983). Estimating a finite population mean from unequal probability samples. *J. Am. Statist. Assoc.* **78**, 596–604.

McCarthy, P. J. (1966). Replication: An approach to the analysis of data from complex surveys. *Vital and Health Statistics* 2 : 14, National Center for Health Statistics, Rockville, MA.

McCarthy, P. J. (1969a). Pseudo-replication: Further evaluation and application of the half-sample technique. *Vital and Health Statistics* 2 : 31, National Center for Health Statistics, Rockville, MA.

McCarthy, P. J. (1969b). Pseudo-replication: Half samples. *Rev. Int. Statist. Inst.* **37**, 239–264.

National Center for Health Statistics, Landis, J., Lepkoweski, Eklund, S. and Stehouwer, D. (1982). A statistical methodology for analyzing data from complex surveys: The First National Health and Nutrition Examination Survey. *Vital and Health Statistics Series* 2 : 92. National Center for Health Statistics, Hyattsville, MA.

Neyman, J. (1949). Contributions to the theory of the χ^2 test. In: J. Neyman, ed.), *Proc. 1st. Berkeley Symposium on Mathematical Statistics and Probability*. pp. 230–273, U. Col. Press, Berkeley, CA.

Neyman, J. (1971). Comments on: Linear regression models in finite population sampling theory. In: V. P. Godambe and D. A. Sprott, eds., *Foundations of Statistical Inference*. Holt, Rinehart and Winston, Toronto, 276–279.

Pfeffermann, D. and Nathan, G. (1981). Regression analysis of data from a cluster sample. *J. Am. Statist. Assoc.* **76**, 681–689.

Royall, R. M. (1968). An old approach to finite population sampling theory. *J. Am. Statist. Assoc.* **63**, 1269–1279.

Royall, R. M. (1971). Linear regression models in finite population sampling theory. In: V. P. Godambe and D. A. Sprott, eds., *Foundations of Statistical Inference*. Holt, Rinehart and Winston, Toronto, 259–274.

Royall, R. M. and Cumberland, W. G. (1981a). An empirical study of the ratio estimator and estimators of its variance. *J. Am. Statist. Assoc.* **76**, 66–76.

Royall, R. M. and Cumberland, W. G. (1981b). A finite population linear regression estimator of its variance. *J. Am. Statist. Assoc.* **76**, 924–930.

Scott, A. J. and Rao, J. N. K. (1981). Chi-squared tests for contingency tables with proportions estimated from survey data. In: D. Krewski, R. Platek and J. N. K. Rao, eds., *Current Topics in Survey Sampling*. Academic Press, New York, 274–265.

Variance Estimation in Sample Surveys

J. N. K. Rao

Introduction

A substantial part of sample survey theory is devoted to the derivation of mean square errors (or variances) of estimators of a population total and their estimators. The mean square error (MSE) enables us to compare the efficiencies of alternative sample designs and to determine the optimal allocation of the sample size, the optimal points of stratification, etc. under a specified design. On the other hand, an estimate of MSE or variance provides us with a measure of uncertainty in the estimate and a large-sample confidence interval for the parameter. It is a common practice to report the estimates in a tabular form along with the corresponding estimates of coefficient of variation (c.v.) or design effect (deff), where c.v. = $\sqrt{\text{MSE}}$/(estimate) and deff is defined as the ratio of MSE under the specified design to the variance with a simple random sample of the same size (Kish, 1965, p. 162).

This article provides a review of some recent work pertaining to mean square error (or variance) estimation. In particular, the following topics are covered: (1) A unified approach to deriving MSE of linear estimators of a population total and their nonnegative unbiased estimators. (2) Variance estimation methods for nonlinear statistics like ratio, regression and correlation coefficients. The methods investigated include linearization (Taylor expansion), jackknife, balanced repeated replication and bootstrap techniques. (3) Modelling mean square errors.

The reader is referred to Wolter's (1985) book for an introduction to variance estimation in sample surveys and for an account of some recent developments in this area.

We assume throughout the paper that nonsampling errors (measurement errors, nonresponse, coverage errors, etc.) are absent. Moreover, the conventional probability set-up is used throughout the article.

1. Unified approach for linear statistics

To simplify the discussion, we consider the variance estimation for a single stratum, but it is straight-forward to combine the variance estimates from two or more strata. Subsection 1.1 deals with the unistage designs while the extension to general multistage designs is provided in Subsection 1.2.

1.1. Unistage designs

Suppose that the finite population (stratum) consists of a known number, N, of units numbered $1, 2, \ldots, N$ with corresponding values $y_1, \ldots, y_N$ of a character of interest, y. The population total $Y = \sum y_i$ is to be estimated by selecting a sample s of units (not necessarily distinct) with associated probability of selection $p(s)$ and observing the corresponding y-values $\{y_i, i \in s\}$. The number of distinct units in s is denoted by $v(s)$. A general linear (in the y_i) estimator, $\hat{Y}$, of Y is given by (Godambe, 1955)

$$\hat{Y} = \sum_{i \in \tilde{s}} d_i(\tilde{s}) y_i, \tag{1.1}$$

where $\tilde{s}$ takes the value s with probability $p(s)$ for $s \in S$, the set of all samples s with $p(s) > 0$, and the known weights $d_i(s)$ can depend both on s and i ($i \in s$). In the conventional set-up, the values $y_1, \ldots, y_N$ are regarded as unknown constants and the probabilities $p(s)$ provide the stochastic input. Hence, the mean square error (MSE) of $\hat{Y}$ is defined as

$$\mathrm{MSE}(\hat{Y}) = \sum_s p(s)(\hat{Y}_s - Y)^2 = E(\hat{Y} - Y)^2,$$

where $\hat{Y}_s$ is the value of $\hat{Y}$ for the sample s. An estimator of $\mathrm{MSE}(\hat{Y})$ is denoted by $\mathrm{mse}(\hat{Y})$, taking the values $\mathrm{mse}(\hat{Y}_s)$ with probabilities $p(s)$. The estimator $\mathrm{mse}(\hat{Y})$ is unbiased if

$$E[\mathrm{mse}(\hat{Y})] = \sum_s p(s)[\mathrm{mse}(\hat{Y}_s)] = \mathrm{MSE}(\hat{Y}).$$

Similarly, the estimator $\hat{Y}$ is unbiased if

$$E(\hat{Y}) = \sum_s p(s)\hat{Y}_s = Y \quad \text{or} \quad \sum_{\{s: i \in s\}} p(s) d_i(s) = 1, \quad i = 1, \ldots, N.$$

Rao (1979) obtained the following results on $\mathrm{MSE}(\hat{Y})$ and its unbiased estimators.

THEOREM 1. *Suppose that* $\mathrm{MSE}(\hat{Y})$ *becomes zero when* $y_i = cw_i$ *for some known nonzero constants* w_i, *where* c ($\neq 0$) *is arbitrary. Then, letting* $z_i = y_i/w_i$,

(a) $MSE(\hat{Y})$ reduces to

$$MSE(\hat{Y}) = -\sum_{\substack{i<j \\ i,j=1}}^{N} d_{ij} w_i w_j (z_i - z_j)^2, \quad (1.2)$$

where

$$d_{ij} = \sum_s p(s)(d_i(s) - 1)(d_j(s) - 1); \quad (1.3)$$

(b) *a nonnegative unbiased quadratic (in the y_i) estimator of $MSE(\hat{Y})$ is necessarily of the form*

$$mse(\hat{Y}) = -\sum_{\substack{i<j \\ i,j \in s}} d_{ij}(\tilde{s}) w_i w_j (z_i - z_j)^2, \quad (1.4)$$

where the known weights $d_{ij}(s)$ can depend both on s and $(i, j) \in s$, and satisfy the unbiasedness conditions

$$\sum_{\{s:i,j \in s\}} p(s) d_{ij}(s) = d_{ij} \quad (i < j = 1, \ldots, N). \quad (1.5)$$

Note that the expressions (1.2) and (1.4) for $MSE(\hat{Y})$ and $mes(\hat{Y})$ both involve a negative sign. The negative sign arise naturally in the derivation of Theorem 1, but it should be noted that d_{ij}, given by (1.3), is usually negative. For instance, under simple random sampling without replacement ($v(s) = n$ for all s) with the usual unbiased estimator $\hat{Y} = Nn^{-1} \sum_{i \in s} y_i$, we have

$$d_i(s) = Nn^{-1} \quad \text{for } i \in s \quad \text{and} \quad d_{ij} = d = -(N-n)[n(N-1)]^{-1} < 0.$$

Several useful results follow from Theorem 1.

(1) The formula (1.2) clearly shows that the evaluation of $MSE(\hat{Y})$ depends only on the d_{ij} which in turn depends only on the sample design, $p(s)$. Thus, our attention is focussed on the properties of the sample design and unnecessary manipulations involving the y-values avoided. Similarly, (1.4) shows that $mse(\hat{Y})$ depends only on the $d_{ij}(s)$.

(2) The estimators (1.4) of $MSE(\hat{Y})$ are the only quadratic unbiased estimators which can be nonnegative; those outside the class (1.4) can take negative values. However, it follows from (1.5) that infinitely many choices of $d_{ij}(s)$ exist except in the special case of $v(s) = 2$ for all s, provided the joint inclusion probabilities $\pi_{ij} = \sum_{\{s:i,j \in s\}} p(s) > 0$ for all $i < j$. In the special case of sample size $v(s) = 2$, the only possible nonnegative unbiased quadratic estimator (NNUQE) of $MSE(\hat{Y})$ is given by

$$mse(\hat{Y}) = -\frac{d_{ij}}{\pi_{ij}} w_i w_j (z_i - z_j)^2, \quad (1.6)$$

provided s is defined as a subset $\{i, j\}$ of the population. In the general case of $v(s) > 2$, other considerations such as nonnegativity, computational simplicity and stability might be necessary to arrive at a suitable choice of $\text{mse}(\hat{Y})$.

(3) If $d_{ij} = d$ for all $i < j$, then (1.2) reduces to

$$\text{MSE}(\hat{Y}) = -d \sum_{i=1}^{N} \tilde{w}_i \left(\frac{y_i}{\tilde{w}_i} - Y\right)^2, \tag{1.7}$$

where $\tilde{w}_i = w_i/W$ and $W = \sum w_i$. Further, the choice

$$d_{ij}(s) = \theta a_i(s) a_j(s), \tag{1.8}$$

where θ is determined by (1.5) and $\sum_{i \in s} a_i(s) w_i = 1$, reduces $\text{mse}(\hat{Y})$ to

$$\text{mse}(\hat{Y}) = -\theta \sum_{i \in s} a_i(\tilde{s}) w_i (z_i - \bar{z}_a)^2, \tag{1.9}$$

where $\bar{z}_a = \sum_{i \in \tilde{s}} a_i(\tilde{s}) w_i z_i$. Two well-known sample designs satisfy (1.7) and (1.9): (a) Simple random sampling (SRS) without replacement ($v(s) = n$) and $d_i(s) = N/n = a_i(s)$, $w_i = 1$ leads to $d = -(N-n)/[n(N-1)]$, $\theta = -(N-n)/(n-1)$. (b) Unequal probability sampling with probabilities p_i and with replacement and $d_i(s) = t_i(s)/(np_i) = a_i(s)$, $w_i = p_i$ leads to $d = -1/n$, $\theta = -1/(n-1)$, where $t_i(\tilde{s})$ is the number of times ith unit selected in a sample of size n. The random group method of Rao, Hartley and Cochran (1962) also satisfies (1.7) and (1.9), as shown by Rao (1979).

(4) The well-known Horvitz–Thompson (H–T) unbiased estimator, $\hat{Y}_{HT}$, of Y is a special case of (1.1) with $d_i(s) = d_i = 1/\pi_i$, where $\pi_i = \sum_{\{s:i \in s\}} p(s)$ is the inclusion probability for unit i. In the case of fixed $v(s) = n$, we can choose $w_i = \pi_i$ and d_{ij} reduces to $d_{ij} = \pi_{ij}/(\pi_i \pi_j) - 1$ and the formule (1.2) reduces to the well-known Sen–Yates–Grundy form for the variance of the H–T estimator. Rao (1979) proposed two general choices for $d_{ij}(s)$:

$$d_{ij}(s) = d_{ij}/\pi_{ij} \tag{1.10}$$

valid for any $\hat{Y}$, and

$$d_{ij}(s) = d_i(s) d_j(s) - f_{ij}(s)/p(s) \tag{1.11}$$

valid only for unbiased estimators $\hat{Y}$, where $f_{ij}(s)$ is any choice such that $\sum_{\{s:i,j \in s\}} f(s) = 1$. The choice (1.10) or (1.11) with $f_{ij}(s) = p(s)/\pi_{ij}$ for the H–T estimator with $v(s) = n$ leads to the well-known Sen–Yates–Grundy variance estimator $v(\hat{Y}_{HT})$:

$$v(\hat{Y}_{HT}) = \sum \sum_{i<j \in s} \frac{\pi_i \pi_j - \pi_{ij}}{\pi_{ij}} \left(\frac{y_i}{\pi_i} - \frac{y_j}{\pi_j}\right)^2, \tag{1.12}$$

where $\hat{Y}_{HT} = \Sigma_{i \in s} y_i/\pi_i$. Several useful sample designs for which $v(\hat{Y}_{HT})$ is nonnegative have been identified in the literature (see Brewer and Hanif's (1983) monograph on sampling with unequal probabilities). The choice (1.11) with $f_{ij}(s) = 1/M_2$, where $M_2 = (N-2)^C(n-2)$, leads to a computationally simpler variance estimator not involving π_{ij}, but designs for which it is nonnegative have not been identified. The choice (1.11) also leads to variance estimators for two other well-known methods: (c) Murthy's unbiased estimator $\hat{Y}$ in unequal probability sampling without replacement with $d_i(s) = p(s|i)/p(s)$ and the choice (1.11) with $f_{ij}(s) = p(s|i, j)$ lead to Murthy's nonnegative unbiased variance estimator. Here $p(s|i)$ and $p(s|i, j)$ are the conditional probabilities of getting s given that ith unit was selected in the first draw and ith and jth units were selected in the first two draws, respectively. (d) The classical ratio estimator $\hat{Y}$ in probability proportional to aggregate size (PPAS) sampling, i.e. $p(s) \propto \Sigma_{i \in s} p_i$, with $d_i(s) = d(s) = (\Sigma_{i \in s} p_i)^{-1}$ and the choice (1.11) with $f_{ij}(s) = 1/M_2$ lead to a computationally simple unbiased variance estimator (Rao and Vijayan, 1977). Here $p_i = x_i/X$ and x_i are the values of a bench mark variable x with known population total X.

(5) If the computation of d_{ij} requires summing over all s containing both i and j, as in the case of the classical ratio estimator under SRS, then no simplification is achieved in using (1.2) in comparison to the definition $MSE(\hat{Y}) = E(\hat{Y} - Y)^2$. Considerable simplification, however, can be achieved by treating the estimator $\hat{Y}$ as a nonlinear function of unbiased estimators permitting simple variance calculations and then using a large-sample approximation to MSE. For instance, the MSE of ratio estimator under SRS is approximately equal to the variance of the simple unbiased estimator $\hat{Y} = (N/n) \Sigma_{i \in s} y_i$ with y_i replaced by $y_i - Rx_i$, where $R = Y/X$.

1.2. Multistage designs

Suppose that the population consists of N primary sampling units (psu) or clusters, numbered $1, 2, \ldots, N$ with corresponding totals $Y_1, Y_2, \ldots, Y_N$ ($\Sigma Y_j = Y$). In multistage sampling, unbiased variance estimation is greatly simplified if the clusters are sampled with replacement with probabilities p_j and subsampling is done independently each time a cluster is selected. An unbiased estimator of Y is then given by

$$\hat{Y}_{pps} = \sum_{i=1}^{n} \hat{Y}'_i/(np'_i) \qquad (1.13)$$

with an unbiased variance estimator

$$v(\hat{Y}_{pps}) = \frac{1}{n(n-1)} \sum_{i=1}^{n} \left(\frac{\hat{Y}'_i}{p'_i} - \hat{Y}_{pps}\right)^2, \qquad (1.14)$$

where $\hat{Y}'_i$ is an unbiased estimator of the total, Y'_i, for the cluster chosen at the *i*th sample draw ($i = 1, \ldots, n$), and p'_i is the corresponding p_j-value. The result (1.14) is valid for any subsampling method as long as it permits unbiased estimators of the sample cluster totals Y'_i. The variance of $\hat{Y}_{\text{pps}}$, however, is not simple since it depends also on the within-cluster variabilities, unlike (1.14). Despite the simplicity of with-replacement sampling, it is a common practice to sample the clusters without replacement, primarily to avoid the possibility of selecting the same cluster more than once. Moreover, the variance of the estimated total can be reduced by sampling without replacement at the first-stage.

Suppose $\tilde{s}$ denotes a sample of clusters with associated probabilities of selection $p(s)$. Given $\tilde{s}$, let $\hat{Y}_i$ denote an unbiased linear estimator of Y_i based on subsampling of cluster i ($i \in \tilde{s}$). Then, a general linear unbiased estimator $\hat{Y}_m$ of Y is given by

$$\hat{Y}_m = \sum_{i \in \tilde{s}} d_i(\tilde{s}) \hat{Y}_i, \tag{1.15}$$

where $\tilde{s}$ takes the value s with probability $p(s) > 0$ and the weights $d_i(s)$ satisfy the unbiasedness conditions $\sum_{\{s : i \in s\}} d_i(s) p(s) = 1$, $i = 1, \ldots, N$. The variance of $\hat{Y}_m$ is given by

$$V(\hat{Y}_m) = V(\hat{Y}) + E\left\{\sum_{i \in \tilde{s}} d_i(\tilde{s}) V_i(\tilde{s})\right\}, \tag{1.16}$$

where $\hat{Y}$ is given by (1.1) and y_i replaced by the cluster total Y_i, and $V_i(\tilde{s})$ is the conditional variance of $\hat{Y}_i$ for a given $\tilde{s}$ ($i \in \tilde{s}$). The uni-stage sampling result (1.2) gives $V(\tilde{Y})$ whereas $V_i(\tilde{s})$ can be evaluated by repeated application of (1.16) to second and subsequent stages of sampling, again using the result (1.2).

Turning to unbiased variance estimation, Rao (1975) obtained a general formula:

$$v^{(1)}(\hat{Y}_m) = f(\underset{\sim}{Y}) + \sum_{i \in \tilde{s}} [d_i^2(\tilde{s}) - b_i(\tilde{s})] \hat{V}_i(\tilde{s}), \tag{1.17}$$

where $f(\underset{\sim}{Y})$ is an unbiased quadratic estimator of $V(\hat{Y})$ under unistage cluster sampling,

$$b_i(s) = -\sum_{\substack{j \neq i \\ i,j \in s}} d_{ij}(s) w_j / w_i \tag{1.18}$$

and $\hat{V}_i(s)$ is an unbiased estimator of $V_i(s)$ for a given s. The unistage sampling result (1.4) gives $f(\underset{\sim}{Y})$ whereas $\hat{V}_i(s)$ is evaluated by repeated application of (1.17) to second and subsequent stages of sampling, again using (1.4).

In the special case $V_i(s) = V_i$, i.e. $V_i(s)$ is independent of s, another unbiased variance estimator is given by Raj (1966):

$$v^{(2)}(\hat{Y}_m) = f(\underset{\sim}{Y}) + \sum_{i \in \tilde{s}} d_i(\tilde{s}) \hat{V}_i(\tilde{s}). \tag{1.19}$$

Raj's formula (1.19) covers many commonly used multistage designs, satisfying $V_i(s) = V_i$, and it suggests the following simple computational rule: Get a copy of the unbiased estimator of variance in unistage cluster sampling by substituting for the cluster total Y_i its estimator $\hat{Y}_i$. Also obtain a copy of the estimator $\hat{Y}$ in unistage cluster sampling by substituting $\hat{V}_i(\tilde{s})$ for Y_i, $i \in s$. The sum of the two copies is an unbiased estimator of the variance in multistage sampling. An advantage of $v^{(2)}(\hat{Y}_m)$ over $v^{(1)}(\hat{Y}_m)$ is that it does not require the evaluation of $b_i(s)$ given by (1.18). If $V_i(s)$ is of the form $W_i(s) - F_i$, then a 'hybrid' of the variance estimators (1.17) and (1.19) is given by

$$v^{(3)}(\hat{Y}_m) = f(\hat{\underline{Y}}) + \sum_{i \in \tilde{s}} [d_i^2(\tilde{s}) - b_i(\tilde{s})] \hat{W}_i(\tilde{s}) - \sum_{i \in \tilde{s}} d_i(s) \hat{F}_i(\tilde{s}), \qquad (1.20)$$

where $\hat{W}_i(s)$ and $\hat{F}_i(s)$ are unbiased estimators of $W_i(s)$ and F_i, for a given s, respectively.

Bellhouse (1980) has developed an efficient computer program to calculate the above variance estimates, utilizing the general variance formulae for unistage designs in Subsection 1.1. The desired variance stimates are obtained with only one pass through the data file.

The total variance (1.16) can be expressed as a sum of variances due to different stages of sampling. A knowledge of these variance components is also needed in the efficient design of multistage sample surveys. We now provide general formulae for unbiased estimators of the components of $V(\hat{Y}_m)$. The first-stage component, $A_1 = V(\hat{Y})$, is unbiasedly estimated by

$$\hat{A}_1 = v(\hat{Y}_m) - \sum_{i \in \tilde{s}} d_i^2(\tilde{s}) \hat{V}_i(\tilde{s}), \qquad (1.21)$$

where $v(\hat{Y}_m)$ is the unbiased estimator of $V(\hat{Y}_m)$ given by (1.17) or, in the special cases, by (1.19) or (1.20). The second-stage component, A_2, is obtained from the last term of (1.16) by decomposing $V_i(\tilde{s})$ as

$$V_i(\tilde{s}) = V_2 \left(\sum_{j \in \tilde{s}_i} d_j(\tilde{s}_i) Y_{ij} \right) + E_2 \left(\sum_{j \in \tilde{s}_i} d_j^2(\tilde{s}_i) V_j(\tilde{s}_i) \right) \qquad (1.22)$$

with an obvious extension of the previous notation ($\hat{Y}_i = \sum_{j \in \tilde{s}_i} d_j(\tilde{s}_i) \hat{Y}_{ij}$, etc.) leading to

$$\hat{A}_2 = \sum_{i \in \tilde{s}} d_i^2(\tilde{s}) \left[\hat{V}_i(\tilde{s}) - \sum_{j \in \tilde{s}_i} d_j^2(\tilde{s}_i) \hat{V}_j(s_i) \right]. \qquad (1.23)$$

Here

$$\hat{V}_i(\tilde{s}) = f_i(\hat{\underline{Y}}_i) + \sum_{j \in \tilde{s}_i} [d_j^2(\tilde{s}_i) - b_j(\tilde{s}_i)] \hat{V}_j(\tilde{s}_i), \qquad (1.24)$$

where $\hat{V}_j(\tilde{s}_i)$ is an unbiased estimator of $V_j(\tilde{s}_i)$ based on sampling at the third and subsequent stages, and $f_i(\underset{\sim}{Y}_i)$ is the unbiased estimator of the first component of $V_i(\tilde{s})$ based on uni-stage cluster sampling within the ith psu. The above procedure is repeated until all the components of $V(\hat{Y}_m)$ are estimated.

EXAMPLE. Consider a three-stage design with unequal probability sampling and without replacement at the first two stages and SRS without replacement at the last stage (fixed sample sizes at each stage). The H–T estimator of Y is given by

$$\hat{Y}_{HT,m} = \sum_{i \in \tilde{s}} \frac{1}{\pi_i} \sum_{j \in \tilde{s}_i} \frac{1}{\pi_{j|i}} \hat{Y}_{ij},$$

where π_i and $\pi_{j|i}$ are the inclusion probabilities for cluster i and jth second stage unit within cluster i, respectively and $\hat{Y}_{ij} = M_{ij} \bar{y}_{ij}$. Here M_{ij} is the number of second-stage units within cluster i and $\bar{y}_{ij}$ is the sample mean based on m_{ij} third-stage units. Hence, in the general notation $d_i(s) = 1/\pi_i$, $d_j(s_i) = 1/\pi_{j|i}$, and, using the Sen–Yates–Grundy variance estimator (1.12), we get

$$f(\underset{\sim}{Y}) = \sum_{\substack{i < i' \\ i, i' \in \tilde{s}}} \left(\frac{\pi_i \pi_{i'}}{\pi_{ii'}} - 1 \right) \left(\frac{Y_i}{\pi_i} - \frac{Y_{i'}}{\pi_{i'}} \right)^2$$

and

$$f_i(\underset{\sim}{Y}_i) = \sum_{\substack{j < j' \\ j, j' \in \tilde{s}_i}} \left(\frac{\pi_{j|i} \pi_{j'|i}}{\pi_{jj'|i}} - 1 \right) \left(\frac{Y_{ij}}{\pi_{j|i}} - \frac{Y_{ij'}}{\pi_{j'|i}} \right)^2,$$

where $\pi_{jj'|i}$ are the joint inclusion probabilities within cluster i. Finally,

$$\hat{V}_j(s_i) = M_{ij}^2 \left(\frac{1}{m_{ij}} - \frac{1}{M_{ij}} \right) s_{ij}^2(y),$$

where $s_{ij}^2(y)$ is the sample mean square based on the m_{ij} sampled third-stage units. Note that

$$\hat{A}_3 = \sum_{i \in \tilde{s}} \pi_i^{-2} \sum_{j \in \tilde{s}_j} \pi_{j|i}^{-2} \hat{V}_j(s_i).$$

1.3. Simplified variance estimators (multistage designs)

Durbin (1967) and Rao and Lanke (1984) have developed selection procedures and simplified unbiased variance estimators in multi-stage sampling with $n = 2$ psu's per stratum. These variance estimators avoid the calculation of within-cluster estimated variance, $\hat{V}_i(s)$, with a high probability. The procedure for selecting $n = 2$ psu's in a stratum consisting of N psu's is as follows: (a) Divide the N clusters into R groups such that each group contains at least three clusters. A procedure for partitioning the psu's into R groups such that the group sizes

$P_t = \Sigma_i p_{ti}$ are approximately equal is given by Rao and Lanke (1984), where p_{ti} is the selection probability associated with the ith cluster in the tth group ($\Sigma\Sigma p_{ti} = 1$, $t = 1, \ldots, R$). In the case of a method with inclusion probability π'_{ti} proportional to p_i in step (b) below, we also require that $\max_i p_{ti} \leq \tfrac{1}{2} P_t$ for each t. (b) Select two groups with probabilities P_t and with replacement. If different groups G_t and G_r are selected, draw one cluster from G_t with probability p_{ti}/P_t. If the same group, G_t, is selected twice, draw $n = 2$ clusters from G_t by any desired method of sampling without replacement with probabilities p_{ti}/P_t; in particular, any method with $\pi'_{ti} = 2p_{ti}/P_t$ can be used in conjunction with the H–T estimator. The selected clusters are subsampled in the usual manner, leading to conditionally unbiased estimators $\hat{Y}_{ti}$ of cluster totals Y_{ti}.

The estimator of $Y = \Sigma\Sigma Y_{ti}$ is given by

$$\hat{Y}_m(G) = \begin{cases} \tfrac{1}{2}\left(\dfrac{\hat{Y}_{ti}}{p_{ti}} + \dfrac{\hat{Y}_{rj}}{p_{rj}}\right) & \text{if } G_t \text{ and } G_r \text{ selected } (t \neq r), \\ \hat{Y}_m(t)/P_t & \text{if } G_t \text{ selected twice}, \end{cases} \quad (1.25)$$

where $\hat{Y}_m(t)$ is of the form (1.15) and conditionally unbiased for $Y_t = \Sigma_i Y_{ti}$, given that G_t is selected twice in step (b). If the H–T estimator is used for $\hat{Y}_m(t)$ and $\pi'_{ti} = 2p_{ti}/P_t$, then $\hat{Y}_m(G)$ reduces to $\tfrac{1}{2}(\hat{Y}_{ti}/p_{ti} + \hat{Y}_{rj}/p_{rj})$ and the over-all inclusion probability $\pi_{ti} = 2p_{ti}$, i.e. $\hat{Y}_m(G) = \hat{Y}_{HT,m}$ in this case. An unbiased variance estimator is obtained as

$$v[\hat{Y}_m(G)] = \begin{cases} \tfrac{1}{4}\left(\dfrac{\hat{Y}_{ti}}{p_{ti}} - \dfrac{\hat{Y}_{rj}}{p_{rj}}\right)^2 & \text{if } G_t \text{ and } G_r \text{ selected } (t \neq r), \\ v_2(\hat{Y}_m(t))/P_t^2 & \text{if } G_t \text{ selected twice}, \end{cases} \quad (1.26)$$

where $v_2(\hat{Y}_m(t))$ is of the form (1.17), (1.19) or (1.20) and unbiased for the conditional variance of $\hat{Y}_m(t)$ given that G_t is selected twice. It is clear from (1.26) that the variance estimator $v[\hat{Y}_m(G)]$ depends only on the estimated psu totals $\hat{Y}_{ti}$ whenever two different groups are selected in step (b). The procedure of Rao and Lanke (1984), for partitioning the psu's into R groups, maximises the probability of selecting two different groups in step (b). This probability increases with R, suggesting the use of as many groups as possible. However, empirical results for unistage designs indicate that the coefficient of variation of the grouping variance estimator increases with R, although the increase is not large especially for the Murthy method (Rao and Lanke, 1984).

Frequently, the simplified variance estimator (1.14), appropriate for sampling clusters with replacement with probabilities p_i, is used as an approximation to the unbiased variance estimator of $\hat{Y}_{HT,m}$ under any method with $\pi_i = np_i$. This procedure leads to over-estimation, and the bias may not be negligible unless the first-stage sampling fraction is small. In fact, the bias is $n/(n-1)$ times the reduction in variance obtained by sampling without replacement, instead of sampling with replacement, in uni-stage cluster sampling (Durbin, 1953).

2. Nonlinear statistics

Variance estimation methods in the case of nonlinear statistics, like ratios and regression coefficients, have received considerable attention in recent years. The methods that have been proposed include the well-known linearization (Taylor expansion) method, and sample reuse techniques like the jackknife, balanced repeated replication (BRR) and the bootstrap. The linearization method provides a ready extension of results in Section 1 to nonlinear statistics. It is applicable to general sample designs, but involves a separate variance formula for each nonlinear statistic. On the other hand, the sample reuse methods employ a single variance formula for all nonlinear statistics. However, the jackknife and BRR methods are applicable only to those designs in which the clusters within strata are sampled *with* replacement or the first-stage sampling fraction is negligible. The bootstrap seems to be more generally applicable, but it is computationally cumbersome and its properties have not been fully investigated.

2.1. Linearization method

We denote the linear estimator $\hat{Y}$ of a population total Y as a function $\hat{Y}(y_t)$. Similarly, we denote the variance estimator, $v(\hat{Y})$, of a linear statistic $\hat{Y}$ as a function $v(y_t)$, where y_t is the value of an ultimate unit, t, in the sample. The linearization method provides a variance estimator for a non-linear statistic $\hat{\theta}$ as $v(z_t)$ for a suitably defined synthetic variable z_t which depends on the form of $\hat{\theta}$.

Ratio-type statistics. A ratio of two population totals, $R = Y/X$, is estimated by $\hat{R} = \hat{Y}/\hat{X} = \hat{Y}(y_t)/\hat{Y}(x_t)$. The linearization variance estimator of $\hat{R}$ is given by

$$v(\hat{R}) = v(y_t - \hat{R}x_t)/\hat{X}^2 = v(z_t), \tag{2.1}$$

where $z_t = (y_t - \hat{R}x_t)/\hat{X}$. Similarly, the ratio estimator of a population total Y is given by $\hat{Y}_r = X(\hat{Y}/\hat{X})$ when X is known. The linearization method leads to two variance estimators given by

$$v(\hat{Y}_r) = \left(\frac{X}{\hat{X}}\right)^2 v(y_t - \hat{R}x_t) \tag{2.2}$$

and

$$v^*(\hat{Y}_r) = v(y_t - \hat{R}x_t), \tag{2.3}$$

respectively. These variance estimators are asymptotically equivalent, but theoretical and empirical investigations indicate that $v(\hat{Y}_r)$ may be preferable to $v^*(\hat{Y}_r)$ (see Rao and Rao, 1971; and Chapter 15 by R. M. Royall). It may be noted that a class of asymptotically equivalent variance estimators can be obtained as $(X/\hat{X})^g v^*(\hat{Y})$, where the parameter $g \geq 0$ generates the class (Wu, 1982, 1985). In particular, the choice $g = 1$ leads to a ratio-type variance estimator.

By taking $x_t = 1$ for all t, $\hat{R}$ reduces to the estimator of the population mean $\bar{Y} = Y/M$ when the population size M is unknown, Similarly, replacing x_t and y_t by $_i a_t$ and $_i y_t$ respectively, where $_i a_t = 1$ and $_i y_t = y_t$ if the ultimate unit t belongs to domain (subpopulation) i, and $_i a_t = _i y_t = 0$ otherwise, $\hat{R}$ reduces to $_i \hat{Y}/_i \hat{M}$, the estimator of ith domain mean $_i \bar{Y} = _i Y/_i M$. If the domain size $_i M$ is known, the ratio estimator $\hat{Y}_r$ reduces to the estimator of the domain total $_i Y$.

The estimator of the finite population, simple regression coefficient, $B = \Sigma (x_t - \bar{X})(y_t - \bar{Y})/\Sigma (x_t - \bar{X})^2$, can be expressed as

$$\hat{B} = \hat{Y}(z_{1t})/\hat{Y}(z_{2t})$$

where

$$z_{1t} = (y_t - \hat{\bar{Y}})(x_t - \hat{\bar{X}}), \quad z_{2t} = (x_t - \hat{\bar{X}})^2,$$

$$\hat{\bar{X}} = \hat{Y}(x_t)/\hat{Y}(a_t), \quad \hat{\bar{Y}} = \hat{Y}(y_t)/\hat{Y}(a_t)$$

and $a_t = 1$ for all t. Hence,

$$v(\hat{B}) = v(z_{1t} - \hat{B} z_{2t})/[\hat{Y}(z_{2t})]^2 \tag{2.4}$$

using (2.1).

The estimator of a finite population correlation coefficient (simple or multiple) can also be expressed as a ratio of estimated totals. Let

$$z_{1t} = y_t - \hat{\bar{Y}} - \sum_{i=1}^{p} \hat{B}_i(x_{ti} - \hat{\bar{X}}_i) \quad \text{and} \quad z_{2t} = (y_t - \hat{\bar{Y}})^2,$$

where $(x_1, \ldots, x_p)$ are the regressor variables with corresponding regression coefficient estimates

$$\hat{B} = (\hat{B}_1, \ldots, \hat{B}_p)', \quad \hat{\bar{Y}} = \hat{Y}(y_t)/\hat{Y}(a_t) \quad \text{and} \quad \hat{\bar{X}}_i = \hat{Y}(x_{ti})/\hat{Y}(a_t).$$

(The formula for $\hat{B}$ is given below.) Then the estimator of multiple correlation coefficient, denoted by Δ, may be expressed as $1 - \hat{\Delta}^2 = \hat{Y}(z_{1t})/\hat{Y}(z_{2t})$ with estimator of variance

$$v(\hat{\Delta}^2) = v(z_{1t} - (1 - \hat{\Delta}^2)z_{2t})/[\hat{Y}(z_{2t})]^2 \tag{2.5}$$

(Fuller, 1980). The linearization method also provides the estimated covariance matrix of $\hat{B}$. We can express $\hat{B}$ as $\hat{B} = \hat{Z}_2^{-1} \hat{Z}_1$, where $\hat{Z}_2$ is a $p \times p$ nonsingular matrix with elements $\hat{Z}_{2ij} = \hat{Y}(z_{2tij})$ and $\hat{Z}_1$ is a p-vector with elements $\hat{Z}_{1i} = \hat{Y}(z_{1ti})$, where $z_{2tij} = (x_{ti} - \hat{\bar{X}}_i)(x_{tj} - \hat{\bar{X}}_j)$ and $z_{1ti} = (y_t - \hat{\bar{Y}})(x_{ti} - \hat{\bar{X}}_i)$.

The estimated covariance matrix of $\hat{B}$ is given by

$$\hat{D}(\hat{B}) = \hat{Z}_2^{-1} \hat{A} \hat{Z}_2^{-1}, \tag{2.6}$$

where $\hat{A}$ is a $p \times p$ matrix with elements

$$\hat{a}_{ij} = \tfrac{1}{2}[v(u_{ti} + u_{tj}) - v(u_{ti}) - v(u_{tj})] \quad \text{and} \quad u_{ti} = z_{1ti} - \sum_{l=1}^{p} \hat{B}_l z_{2til}$$

(Fuller, 1975; Folsom, 1974). Using the estimated covariance matrix (2.6), one could test a linear hypothesis H on the finite population regression vector B. For instance, if H is given by $C(B - B_0)$, where C is a known $d \times p$ matrix $(d < p)$ of rank d and B_0 is a specified vector, then the Wald statistic

$$(\hat{B} - B_0)' C' (C D(\hat{B}) C')^{-1} C (\hat{B} - B_0) \tag{2.7}$$

is asymptotically distributed as a χ^2 variable with d degrees of freedom, under H.

Nonlinear statistics of the form $\hat{\theta} = \Sigma l_i \hat{R}_i$ are often encountered in practice, where l_i are specified constants and

$$\hat{R}_i = \hat{Y}_i / \hat{X}_i = \hat{Y}(y_{ti}) / \hat{Y}(x_{ti}), \quad i = 1, \ldots, p.$$

For instance, if $p = 2$, $l_1 = 1$, $l_2 = -1$, then $\hat{\theta}$ reduces to $\hat{R}_1 - \hat{R}_2$, the difference between two estimated ratios. An important special case of $R_1 - R_2$ is the difference between two domain means. The linearization method leads to

$$v\left(\sum l_i \hat{R}_i\right) \doteq v\left(\sum_i l_i z_{ti}\right) \quad \text{with} \quad z_{ti} = (y_{ti} - \hat{R}_i x_{ti}) / \hat{X}_i. \tag{2.8}$$

General functions of estimated totals. Suppose the nonlinear statistic $\hat{\theta}$ can be expressed as an explicit function, $g(\hat{Y})$, of a vector of estimated totals, $\hat{Y} = (\hat{Y}_1, \ldots, \hat{Y}_q)'$. We assume that the first derivatives $g_i(t) = \partial g(t)/\partial t_i$, $i = 1, \ldots, q$, exist. Then the linearization method gives the variance estimator

$$v(\hat{\theta}) = v(z_t) \quad \text{with} \quad z_t = \sum_{i=1}^{q} y_{ti} g_i(\hat{Y}) \tag{2.9}$$

(Woodruff, 1971). One drawback of the linearization method is that the evaluation of partial derivatives may be difficult for certain statistics. However, useful approximations to the desired partial derivatives can be obtained using numerical methods (Woodruff and Causey, 1976). Besides, the special case of ratio-type statistics covers many commonly used nonlinear statistics and we have already given explicit formulae for $v(\hat{\theta})$, not involving the partial derivatives g_i.

Sometimes, the estimator $\hat{\theta}$ cannot be expressed explicitly as a function of estimated totals, but defined implicitly as the solution of a set of nonlinear equations. As an example, consider a loglinear model on the population cell proportions, π_t, in a multiway table ($\Sigma\, \pi_t = 1$) given by

$$\mu = a(\theta)\mathbf{1} + X\theta \tag{2.10}$$

where μ is a T-vector of log probabilities $\mu_t = \ln \pi_t$, X is a known $T \times r$ matrix of full rank $r (\leq T - 1)$, θ is an r-vector of parameters, $\mathbf{1}$ is a T-vector of 1's, $X'\mathbf{1} = \mathbf{0}$, $a(\theta)$ is a normalizing factor to ensure that $\pi'\mathbf{1} = 1$, and $\pi = (\pi_1, \ldots, \pi_T)'$. The survey estimator of π, denoted by $\hat{p}$, typically is a ratio estimator with estimated covariance matrix $\hat{\Sigma}$, say. Due to difficulties in deriving appropriate likelihoods with survey data, the estimates $\hat{\theta}$ of θ and $\hat{\pi}$ of $\pi = \pi(\theta)$ are obtained by solving the likelihood equations based on multinomial sampling: $X'\hat{\pi} = X'\hat{p}$. The solution $\hat{\theta}$ or $\hat{\pi}$ of these nonlinear equations is obtained by iterative calculations. Using a standard asymptotic expansion for $\hat{\theta} - \theta$ in terms of $\hat{p} - \pi$, it can be shown that the estimated asymptotic covariance matrix of $\hat{\theta}$ is given by

$$\hat{D}(\hat{\theta}) = (X'\hat{P}X)^{-1}(X'\hat{\Sigma}X)(X'\hat{P}X)^{-1}, \tag{2.11}$$

where $\hat{P} = \text{diag}(\hat{\pi}) - \hat{\pi}\hat{\pi}'$. Similarly, the estimated asymptotic covariance matrix of smoothed estimates $\hat{\pi}$ is obtained as

$$\hat{D}(\hat{\pi}) = \hat{P}X\hat{D}(\hat{\theta})X'\hat{P} \tag{2.12}$$

(Imrey et al., 1982; Rao and Scott, 1984). The smoothed estimates $\hat{\pi}$ can be considerably more efficient than the survey estimates $\hat{p}$ if the model (2.10) fits the data well. Rao and Scott (1984) provide appropriate test statistics for goodness-of-fit of the model (2.10).

Binder (1983) derived a general formula for $\hat{D}(\hat{\theta})$ when $\hat{\theta}$ is defined as the solution of a set of r nonlinear equations given by

$$w(\hat{\theta}) = u(\hat{\theta}) - v(\hat{\theta}) = \mathbf{0}. \tag{2.13}$$

Here $\hat{u} = u(\hat{\theta})$ in an r-vector with elements $u_i(\hat{\theta}) = \hat{Y}(u_{ti})$, $u_{ti} = u_i(y_t, \hat{\theta})$, $y_t = (y_{t1}, \ldots, y_{tp})'$ is the vector of observations attached to the tth sample unit, and $v(\hat{\theta})$ is an r-vector with elements $v_i(\hat{\theta}) = v_i$. Denoting the estimated covariance matrix of $u(\hat{\theta})$ by $\hat{D}(\hat{u})$, the estimated asymptotic covariance matrix of $\hat{\theta}$ is obtained as

$$\hat{D}(\hat{\theta}) = \left(\frac{\partial w(\hat{\theta})}{\partial \hat{\theta}}\right)^{-1} \hat{D}(\hat{u}) \left(\frac{\partial w(\hat{\theta})}{\partial \hat{\theta}}\right)^{-1}, \tag{2.14}$$

where $(\partial w(\hat{\theta})/\partial \hat{\theta})$ is the $r \times r$ matrix with elements $\partial w_i(\hat{\theta})/\partial \theta_j$. Note that the (i, j)th element of $\hat{D}(\hat{u})$ is given by $\frac{1}{2}[v(u_{ti} + u_{tj}) - v(u_{ti}) - v(u_{tj})]$. Binder's formulation covers the generalized linear models of Nelder and Wedderburn (1972) for which

$$u_{ti} = x_{ti}[\mu(x_t'\hat{\theta}) - y_t], \quad x_t = (x_{t1}, \ldots, x_{tr})', \quad x_{t1} = 1, \quad v_i = 0 \quad (2.15)$$

and y_t is the value of a response variable and $(x_{t2}, \ldots, x_{tr})$ are the values of $r - 1$ explanatory variables attached to the tth sample unit. The link function $\mu(\cdot)$ specifies the model. For instance, the choice $\mu_t = \mu(x_t'\theta) = x_t'\theta$ gives the linear regression model while $\ln[\mu_t/(1 - \mu_t)] = x_t'\theta$ leads to a logistic regression model. The partial derivatives $\partial w_i(\hat{\theta})/\partial\theta_j$ reduce to $\hat{Y}(u_{tij}^*)$, where

$$u_{tij}^* = x_{ti}x_{tj}\mu'(x_t'\theta) \quad (2.16)$$

and $\mu'(a) = \partial\mu(a)/\partial a$. In the linear regression case $\mu'(a) = 1$ while $\mu'(a) = \mu(a)(1 - \mu(a))$ in the logistic regression case.

Post-stratification. Post-stratification is commonly used in large-scale sample surveys to increase the efficiency of estimators utilizing known strata sizes. For instance, in the Canadian Labor Force Survey the survey estimates are adjusted for post-stratification using the projected census age–sex distribution at the provincial level. The post-stratified estimator of the total Y is given by

$$\hat{Y}_{(p)} = \sum_{h=1}^{L} M_h \frac{\hat{Y}(_h y_t)}{\hat{Y}(_h a_t)} = \sum_{h=1}^{L} M_h \frac{\hat{Y}_h}{\hat{M}_h} = \sum_{h=1}^{L} M_h \hat{\bar{Y}}_h, \quad (2.17)$$

where M_h is the total number of elements in the hth stratum assumed known, and $_h y_t, _h a_t$ are as defined before for a domain. Noting that $\hat{Y}_{(p)}$ is a special case of the statistic $\Sigma l_h \hat{R}_h$ with $l_h = M_h$, we obtain the variance estimator from (2.8) as

$$v(\hat{Y}_{(p)}) = v\left(\sum_h \frac{M_h}{\hat{M}_h}(_h y_t - \hat{\bar{Y}}_{hh}a_t)\right). \quad (2.18)$$

Another variance estimator, asymptotically equivalent to $v(\hat{Y}_{(p)})$, is given by

$$v^*(\hat{Y}_{(p)}) = v\left(y_t - \sum_h \hat{\bar{Y}}_{hh}a_t\right), \quad (2.19)$$

noting that $M_h/\hat{M}_h$ converges in probability to 1 and $\Sigma_h \,_h y_t = y_t$. The results for $\hat{Y}_r$, however, suggest that $v(\hat{Y}_{(p)})$ is likely to perform better than $v^*(\hat{Y}_{(p)})$. Moreover, in the special case of simple random sampling without replacement $v(\hat{Y}_{(p)})$ agrees with the unbiased estimator of conditional variance of $\hat{Y}_{(p)}$ for given strata sample sizes $(n_1, \ldots, n_L)$, provided the finite population corrections are negligible; the conditional variance is more relevant than the unconditional variance for inferential purposes. However, such conditional inferences are not easy to implement with complex survey designs.

The corresponding variance formulae for nonlinear functions of post-stratified estimators $\hat{Y}_{1(p)}, \ldots, \hat{Y}_{q(p)}$, say, are readily obtained if we replace y_{ti} by $\Sigma_h (M_h/\hat{M}_h)(_h y_{ti} - \hat{\bar{Y}}_{hih} a_t)$ or by $y_{ti} - \Sigma_h \hat{\bar{Y}}_{hih} a_t$ and $\hat{Y}_i$ by $\hat{Y}_{i(p)}$ in the previous variance formulae for the case of no post-stratification, where $_h y_{ti} = y_{ti}$ if the tth ultimate unit belongs to stratum h and $_h y_{ti} = 0$ otherwise, etc. It may be noted that $\hat{M}_{(p)} = \hat{Y}_{(p)}(a_t)$ reduces to the known population size, M.

2.2. The jackknife method

Suppose that n clusters are selected with replacement with probabilities p_i and subsampling is done independently each time a cluster is selected. Then the estimator (1.13) of Y may be written as a sample mean, $\hat{Y}_{pps} = \Sigma r_i/n = \bar{r}$ of n i.i.d. random variables $r_i = \hat{Y}_i/p_i'$ with $E(r_i) = Y$. The corresponding estimator of θ is given by $\hat{\theta} = g(\bar{r})$, where $\bar{r}$ is the vector of such means for q characters. If the sample of clusters is divided at random into k groups each of size l ($kl = n$), then each group provides an independent estimator $\hat{\theta}_j$ of θ and the variance of the mean $\hat{\theta}. = \Sigma \hat{\theta}_j/k$ can be simply estimated by

$$v(\hat{\theta}.) = \Sigma (\hat{\theta}_j - \hat{\theta}.)^2/[k(k-1)]. \tag{2.20}$$

The variance estimator (2.20) is also applicable to samples consisting of k independent, interpenetrating subsamples, each selected by the same sample design, however complex it may be.

The estimator $\hat{\theta}.$, however, is not a consistent estimator of θ unless $l \to \infty$ as $n \to \infty$ whereas $\hat{\theta}$ is consistent if $n \to \infty$. Moreover, the coefficient of variation of $v(\hat{\theta}.)$ could be quite large if k is small. The jackknife variance estimator circumvents these difficulties. Let $\hat{\theta}^i$ denote the estimator of θ computed from the sample after omitting the ith sampled cluster, then the jackknife variance estimator of $\hat{\theta}$ is given by

$$v_j(\hat{\theta}) = \frac{n-1}{n} \sum_{i=1}^{n} \left(\hat{\theta}^i - \frac{1}{n} \Sigma \hat{\theta}^i \right)^2. \tag{2.21}$$

If $\hat{\theta} = g(\bar{Y})$, then the variance estimator (2.21) is consistent as $n \to \infty$, but the jackknife method fails to provide a consistent variance estimator when θ is the finite population median or a quantile. It may be noted that in the linear case, $\hat{\theta} = \hat{Y}_{pps} = \bar{r}$, the jackknife variance estimator (2.21) reduces to the customary unbiased variance estimator, (1.14).

Many large-scale sample surveys employ large numbers of strata, L, with relatively few clusters, n_h, sampled within each stratum, h. In fact, it is a common practice to select $n_h = 2$ clusters within each stratum to permit maximum degree of stratification of clusters consistent with the provision of a valid variance estimator. Suppose that n_h clusters are selected with replacement with probabilities p_{hi} in stratum h and subsampling is done independently each time a cluster is selected. Then an unbiased estimator of $Y = \Sigma Y_h$ is given by $\hat{Y}_{pps} =$

$\Sigma \hat{Y}_{\mathrm{pps},h} = \Sigma \bar{r}_h$ with an unbiased variance estimator

$$v(\hat{Y}_{\mathrm{pps}}) = \sum_h v(\hat{Y}_{\mathrm{pps},h}) = \sum_h \sum_i (r_{hi} - \bar{r}_h)^2/[n_h(n_h - 1)]. \tag{2.22}$$

Here Y_h is the stratum h population total, $\bar{r}_h = \Sigma_i r_{hi}/n_h$, $r_{hi} = \hat{Y}'_{hi}/p'_{hi}$ and $\hat{Y}'_{hi}$ is an unbiased estimator of the total for the cluster chosen at the ith sample draw ($i = 1, \ldots, n_h$) in the hth stratum ($h = 1, \ldots, L$) and p'_{hi} is the corresponding p-value. The corresponding estimator of $\hat{\theta}$ is given by $\hat{\theta} = g(\Sigma \bar{r}_h)$, where $\bar{r}_h$ is the vector of means for q characters in stratum h.

Let $\hat{\theta}^{hi}$ denote the estimator of θ computed from the sample $\{r_{hi}\}$ after omitting r_{hi}, the value for ith sample cluster in hth stratum. Then, a jackknife variance estimator of $\hat{\theta}$ is given by

$$v_J(\hat{\theta}) = \sum_h \frac{n_h - 1}{n_h} \sum_i (\hat{\theta}^{hi} - \hat{\theta}^h)^2, \tag{2.23}$$

where $\hat{\theta}^h = \Sigma_i \hat{\theta}^{hi}/n_h$. Several variations of $v_J(\hat{\theta})$ can be obtained; for instance, $\hat{\theta}^h$ in (2.22) may be replaced by $\hat{\theta}$ (see Krewsky and Rao, 1981). In the linear case, $\hat{\theta} = \hat{Y}_{\mathrm{pps}} = \Sigma \bar{r}_h$, all these jackknife variance estimators reduce to the customary variance estimator, (2.22).

2.3. Balanced repeated replication (BRR)

McCarthy (1969) proposed the BRR method for the important special case of $n_h = 2$ for all h. This method of variance estimation is based on a set of R 'balanced' half-samples (BHS) formed by deleting one cluster from the sample in each stratum. This set may be defined by an $R \times L$ design matrix (δ_h^r), $1 \leq r \leq R$, $1 \leq h \leq L$, where $\delta_h^r = +1$ or -1 according as whether the first or second sample cluster in hth stratum is in rth half-sample, and $\Sigma_r \delta_h^r \delta_{h'}^r = 0$ for all $h \neq h'$. A minimal set of R ($L + 1 \leq R \leq L + 4$) BHS may be constructed from Hadamard matrices (Plackett and Burman, 1946).

Let $\hat{\theta}^{(j)}$ and $\hat{\theta}_c^{(j)}$ respectively denote the estimators of θ from the jth half-sample and its complement. Then two BRR variance estimators are given by

$$v_{\mathrm{BRR-H}} = \sum_{j=1}^R (\hat{\theta}^{(j)} - \hat{\theta})^2/R \tag{2.24}$$

and

$$v_{\mathrm{BRR-D}} = \sum_{j=1}^R (\hat{\theta}^{(j)} - \hat{\theta}_c^{(j)})^2/(4R). \tag{2.25}$$

Note that $v_{\mathrm{BRR-D}}$ requires the computation of twice as many half-sample estimates as $v_{\mathrm{BRR-H}}$. In the linear case, both the variance estimators reduce to the customary variance estimator, (2.22).

The BRR method has been extended to $n_h > 2$, provided $n_h = q$ for all h and q is a prime or power of a prime (Gurney and Jewett, 1975), but no such method is at present available for arbitrary n_h.

2.4. Bootstrap method

The bootstrap method for the i.i.d. case has been extensively studied (Efron, 1982). Rao and Wu (1984) provided an extension to the previous stratified multistage design, involving a large number, L, of strata with relatively few clusters, n_h, selected within strata ($\Sigma n_h = n$). The estimator of θ is $\hat{\theta} = g(\Sigma \bar{r}_h)$, as before. The method is as follows:

(i) Draw a simple random sample $\{r_{hi}^*\}_{i=1}^{m_h}$ of size m_h with replacement from $\{r_{hi}\}_{i=1}^{n_h}$, independently for each h. Calculate

$$\tilde{r}_{hi} = \bar{r}_h + [m_h/(n_h - 1)]^{1/2}(r_{hi}^* - \bar{r}_h), \quad \tilde{r}_h = m_h^{-1} \sum_{i=1}^{m_h} \tilde{r}_{hi}, \qquad (2.26)$$

and

$$\tilde{\theta} = g(\Sigma \tilde{r}_h).$$

(ii) Independently replicate step (i) a large number, B, of times and calculate the corresponding estimates $\tilde{\theta}^1, \ldots, \tilde{\theta}^B$.

(iii) The bootstrap variance estimator of $\hat{\theta}$ is given by

$$v_{\text{BOOT}} = \frac{1}{B} \sum_{b=1}^{B} (\tilde{\theta}^b - \hat{\theta})^2. \qquad (2.27)$$

One can also replace $\hat{\theta}$ in the formula (2.27) by $\Sigma \tilde{\theta}^b/B$.

In the linear case, $\hat{\theta} = \Sigma \bar{r}_h$, v_{BOOT} reduces to the customary variance estimator, (2.22), for any choice m_h. The bootstrap method is computationally more intensive than the jackknife or the BRR, but more generally applicable than the latter. For instance, Rao and Wu (1984) have extended the bootstrap to general unistage unequal probability sampling without replacement and two-stage simple random sampling without replacement. It might also lead to better confidence intervals than those based on the normal approximation, as indicated by some empirical results for the ratio estimator under simple random sampling (Hinkley and Wei, 1984). The normal $(\hat{\theta} - \theta)/\sqrt{v(\hat{\theta})}$ is approximately $N(0, 1)$, where $v(\hat{\theta})$ denotes a variance estimator.

2.5. Comparison of the methods

The work on the comparison of linearization, jackknife and BRR methods has been largely empirical until recently. Utilizing the data from the U.S. Current Population Survey (CPS) and sample designs with $n_h = 2$ sample clusters from each of $L = 6$, 12 and 30 strata, Kish and Frankel (1974) evaluated the empirical coverage probability of the $(1 - \alpha)$-level confidence intervals, $\hat{\theta} \pm t_{\alpha/2}\sqrt{v(\hat{\theta})}$, for these three methods, where $t_{\alpha/2}$ is the upper $\alpha/2$-point of a t-variable with L degrees of freedom. The BRR performed consistently better than the jackknife which in turn was better than the linearization method, but the observed differences were small for relatively simple statistics $\hat{\theta}$ such as ratios. Moreover, the performance of the methods with regard to stability of variance estimator was found to be in the reverse order. Bean (1975), Campbell and Meyer (1978), Lemeshow and Levy (1978), Shah, Holt and Folsom (1977) and others have made further empirical studies.

Krewski and Rao (1981) povided an asymptotic justification for the three methods by showing that the variance estimators, $v(\hat{\theta})$, are asymptotically consistent and that $(\hat{\theta} - \theta)/\sqrt{v(\hat{\theta})}$ is asymptotically $N(0, 1)$ as $L \to \infty$, within the context of a sequence of finite populations $\{\Pi_L\}$ with L strata in Π_L. They have also obtained limited exact analytical results on the bias and stability of the variance estimators in the case of a ratio, assuming a linear regression model.

Since the first order asymptotic result of Krewski and Rao (1981) does not enable us to distinguish among the three methods, Rao and Wu (1985) made a second order analysis of the variance estimators. Their main results include the following: (a) The jackknife variance estimators are asymptotically equal to higher order terms; more precisely, $n[v_J^{(1)}(\hat{\theta}) - v_J^{(2)}(\hat{\theta})] = O_p(n^{-2})$ denotes terms of order n^{-2} for any two jackknife variance estimators, say $v_J^{(1)}(\hat{\theta})$ and $v_J^{(2)}(\hat{\theta})$ where $O_p(n^{-2})$ in probability. (b) If $n_h = 2$ for all h, $v_L(\hat{\theta})$ and any $v_J(\hat{\theta})$ are also asymptotically equal to higher order terms, indicating that the choice between the linearization and the jackknife methods in this important special case should depend more on other considerations like computational ease. (c) The BRR variance estimators ($n_h = 2$ for all h) are not asymptotically equal to $v_L(\hat{\theta})$ to higher order terms, but $v_{\text{BRR-D}}$ is 'closer' to $v_L(\hat{\theta})$ than $v_{\text{BRR-H}}$.

The properties of bootstrap method are currently under investigation, but Rao and Wu (1984) have shown that $n[v_{\text{BOOT}} - v_L(\hat{\theta})] = O_p(n^{-1})$ for any choice m_h. They have also given some guidelines on the choice of m_h, but further work is needed.

3. Modelling mean square errors

Large-scale repetitive surveys, like the monthly CPS, produce large numbers of estimates at regular intervals of time. For such surveys, computation of standard error for each estimate, even after simplification of variance estimation procedures, is not feasible due to time and cost constraints. Hence it is necessary to develop models for standard errors, using a small representative subset of the estimates and the associated standard errors. Such a model can be used to produce smoothed standard errors for estimates fo which direct computations have not been made. It also enables us to present standard errors in a concise form (e.g. graphs) in published reports.

In the CPS, the following model has been used for standard errors of domain totals $_i\hat{Y}$:

$$_i\hat{C}^2 = a + b \, _i\hat{Y}^{-1} . \tag{3.1}$$

Here $_i\hat{C}$ is the estimated coefficient of variation of $_i\hat{Y}$, and the constants a and

b are obtained by fitting the model to a representative subset of estimates $_i\hat{Y}$. Usually, the survey variables are separated into a number of homogeneous groups and a separate model is fitted for each group. The U.S. National Center for Health Statistics has also used this model. Cohen (1979) studied the reliability of model (3.1), using data from the U.S. National Medical Care Expenditure Survey.

An alternative model, used at the Australian Bureau of Statistics, is of the form

$$\ln {_i\hat{C}} = \tilde{a} + \tilde{b} \ln {_i\hat{Y}}. \tag{3.2}$$

Ghangurde (1981) made an evaluation of the model (3.2) and related models, utilizing data from the monthly Canadian Labour Force Survey.

The smoothed standard errors for proportions, ratios or differences can be approximately obtained from the model (3.1) (or (3.2)). For instance, if $_i\hat{p}(1) = {_i\hat{M}(1)}/{_i\hat{M}}$ denotes the proportion possessing a certain characteristic in a domain i, then the smoothed CV of $_i\hat{p}(1)$ from (3.1) is approximately given by

$$\mathrm{CV}(_i\hat{p}(1)) \doteq \left[\frac{b}{_i\hat{M}} \frac{(1 - {_i\hat{p}})}{_i\hat{p}} \right]^{1/2} \tag{3.3}$$

(Cohen, 1979). Graphs of (3.3) can be prepared for selected values of $_i\hat{M}$ and the CV for any desired $_i\hat{M}$ can be obtained from the graphs by interpolation.

Another approach to modelling standard errors is based on the design effect, d, i.e., the ratio of MSE under the specified design to the variance under a simple random sample of the same size (see Kalton, 1977). The design effect (deff) for an estimated total $\hat{Y}$ is expressed as

$$d = 1 + (\bar{m} - 1)\rho, \tag{3.4}$$

where $\bar{m}$ is the average number of sampled ultimate units in a cluster and ρ stands for 'rate of homogeneity' within a cluster. The parameter ρ is more 'portable' than d in terms of dependence on cluster sample size, m. Hence, it is possible to use the survey estimate of ρ from (3.4) to impute a ρ-value, $_i\rho$, for a desired domain and then compute the deff for that domain. For instance, Kish et al. (1976) found empirically that $_i\rho$ is roughly equal to 1.2ρ. The deff for a domain total $_i\hat{Y}$ is given by

$$_id = 1 + (_i\bar{m} - 1){_i\rho}, \tag{3.5}$$

where $_i\bar{m}$ is the average number of sampled ultimate units belonging to domain i in a cluster. The imputed standard error of a proportion $_i\hat{p} = {_i\hat{M}}/\hat{M}$ can be computed knowing only the estimate, $_i\hat{d}$ of $_id$ and $_i\hat{p}$:

$$\mathrm{s.e.}(_i\hat{p}) = {_i\hat{d}} \left\{ \frac{_i\hat{p}(1 - {_i\hat{p}})}{_in} \right\}^{1/2}, \tag{3.6}$$

where $_i n$ is the domain sample size. On the other hand, the imputed standard error of an estimated domain total $_i\hat{Y}$ requires the knowledge of variance of $_i\hat{Y}$ under simple random sampling in addition to $_i\hat{d}$.

Brewer et al (1977) noted that it is more realistic to regard ρ as a function of $\bar{m}$, and suggested the model $\rho \doteq \bar{m}^{2\gamma-2}$. Using the survey estimate of ρ one obtains an estimate of the portable parameter γ from this formula which in turn gives $_i\rho \doteq {}_i\bar{m}^{2\gamma-2}$.

The concept of design effect has been found useful in studying the impact of sample design on standard chisquared tests with categorical survey data. Suitable corrections to chisquared tests in terms of design effects can also be obtained. The reader is referred to Chapter 10 (by G. Nathan) for details.

Acknowledgements

This work was supported by a research grant from the Natural Sciences and Engineering Research Council of Canada.

References

Bean, J. A. (1975). Distribution and properties of variance estimators for complex multistage probability samples. Vital and Health Statistics. Ser. 2, No. 65, U.S. Government Printing Office, Washington, DC.

Bellhouse, D. R. (1980). Computation of variance-covariance estimates for general multistage sampling designs. *COMPSTAT Proceedings*, 57–63.

Binder, D. A. (1983). On the variances of asymptotically normal estimators from complex surveys. *International Statistical Review* **51**, 279–292.

Brewer, K. R. W. and Hanif, Muhammad (1983). *Sampling with Unequal Probabilities*, Springer-Verlag, New York.

Brewer, K. R. W., Foreman, E. K., Mellor, R. W. and Trewin, D. J. (1977). Use of experimental designs and population modelling in survey sampling. *Bulletin of the International Statistical Institute* **47**(3), 173–190.

Campbell, C. and Meyer, M. (1978). Some properties of T confidence intervals for survey data. *Proceedings of the Section on Survey Research Methods*. American Statistical Association, Washington, DC, 437–442.

Cohen, S. B. (1977). An assessment of curve smoothing strategies which yield variance estimates from complex survey data. *Proceedings of the Section on Survey Research Methods*. American Statistical Association, Washington, DC, 101–104.

Durbin, J. (1953). Some results in sampling theory when the units are selected with unequal probabilities. *Journal of the Royal Statistical Society Series B* **15**, 262–269.

Durbin, J. (1967). Design of multistage surveys for the estimation of sampling errors. *Applied Statistics* **16**, 152–164.

Efron, B. (1982). *The Jackknife, the Bootstrap and Other Resampling Plans*, SIAM, Philadelphia.

Folsom, R. E., Jr. (1974). National assessment approach to sampling error estimation. Sampling Error Monograph, National Assessment of Education Progress (First Draft).

Fuller, W. A. (1975). Regression analysis for sample survey. *Sankhyā Series C* **37**, 117–132.

Fuller, W. A. (1980). Personal Communication.

Godambe, V. P. (1955). A unified theory of sampling from finite populations. *Journal of the Royal Statistical Society Series B* **17**, 269–278.

Ghangurde, P. D. (1981). Models for estimation of sampling errors. *Survey Methodology* **7**, 177–191.
Gurney, M. and Jewett, R. S. (1975). Constructing orthogonal replications for variance estimation. *Journal of the American Statistical Association* **71**, 819–821.
Hinkley, D. and Wei, B. (1984). Improvements of jackknife confidence limit methods. *Biometrika* **71**, 331–339.
Imrey, P. B., Koch, G. G. and Stokes, M. E. (1982). Categorical data analysis: some reflections on the log linear model and logistic regression, Part II. Data analysis. *International Statistical Review* **50**, 35–64.
Kalton, G. (1977). Practical methods for estimating survey sampling errors. *Bulletin of the International Statistical Institute* **47**(3), 495–514.
Kish, L. (1975). *Survey Sampling.* Wiley, New York.
Kish, L. and Frankel, M. R. (1974). Inference from complex samples (with discussion). *Journal of the Royal Statistical Society Series B* **36**, 1–37.
Kish, L., Groves, R. M. and Krotki, K. P. (1976). Sampling errors for fertility surveys. Occasional Papers No. 17, World Fertility Survey, London.
Krewski, D. and Rao, J. N. K. (1981). Inference from stratified samples: Properties of the linearization, jackknife and balanced repeated replication methods. *Annals of Statistics* **9**, 1010–1019.
Lemeshow, S. and Levy, P. (1978). Estimating the variance of ratio estimates in complex sample surveys with two primary primary units per stratum—a comparison of balanced replication and jackknife techniques. *Journal of Statistical Computation and Simulation* **8**, 191–205.
McCarthy, P. J. (1969). Pseudo-replication: half samples. *Review of the International Statistical Institute* **37**, 239–264.
Nelder, J. and Wedderburn, R. (1972). Generalized linear models. *Journal of the Royal Statistical Society Series A* **135**, 370–384.
Plackett, R. L. and Burman, J. P. (1946). The design of optimum multifactorial experiments. *Biometrika* **33**, 305–325.
Raj, D. (1966). Some remarks on a simple procedure of sampling without replacement. *Journal of the American Statistical Association* **61**, 391–396.
Rao, J. N. K. (1975). Unbiased variance estimation for multistage designs. *Sankhyā Series C* **37**, 133–139.
Rao, J. N. K. (1979). On deriving mean square errors and their non-negative unbiased estimators in finite population sampling. *Journal of the Indian Statistical Association* **17**, 125–136.
Rao, J. N. K., Hartley, H. O. and Cochran, W. G. (1962). On a simple procedure of unequal probability sampling without replacement. *Journal of the Royal Statistical Society Series B* **24**, 482–491.
Rao, J. N. K. and Lanke, J. (1984). Simplified unbiased variance estimation for multistage designs. *Biometrika* **71**, 387–395.
Rao, J. N. K. and Scott, A. J. (1984). On chi-squared tests for multiway contingency tables with cell proportions estimated from survey data. *Annals of Statistics* **12**, 46–60.
Rao, J. N. K. and Wu, C. F. J. (1984). Bootstrap inference for sample surveys. *Proceedings of the Section on Survey Research Methods*, American Statistical Association, Washington, D.C. 106–112.
Rao, J. N. K. and Vijayan, K. (1977). On estimating the variance in sampling with probability proportional to aggregate size. *Journal of the American Statistical Association* **72**, 579–584.
Rao, J. N. K. and Wu, C. F. J. (1985). Inference from stratified samples: second order analysis of three methods for nonlinear statistics. *Journal of the American Statistical Association* **80**(3), 620–630.
Shah, B. V., Holt, M. M. and Folsom, R. E. (1977). Inference about regression models from sample survey data. *Bulletin of the International Statistical Institute* **47**(3), 43–57.
Wolter, K. M. (1985). *Introduction to Variance Estimation.* Springer-Verlag, New York.
Woodruff, R. S. (1971). Simple methods for approximating the variance of a complex estimate. *Journal of the American Statistical Association* **66**, 411–414.
Woodruff, R. S. and Causey, B. D. (1976). Computerized methods for approximating the variance of a complex estimate. *Journal of the American Statistical Association* **71**, 315–321.
Wu, C. F. J. (1982). Estimation of variance of the ratio estimator. *Biometrika* **69**, 183–189.
Wu, C. F. J. (1985). Variance estimation for the combined ratio and combined regression estimators. *Journal of the Royal Statistical Society Series B* **47**, 147–154.

Ratio and Regression Estimators

Poduri S. R. S. Rao

1. Introduction

For a finite population of size N, let (x_i, y_i), $i = 1, 2, \ldots, N$, denote two characteristics of interest, like the yield of a crop and the acreage, incomes from employment and from all other sources, expenditures on education and on food and clothing, an so on.

The population totals are $X = (\Sigma_1^N x_i)$ and $Y = (\Sigma_1^N y_i)$, and the means are $\bar{X} = (X/N)$ and $\bar{Y} = (Y/N)$. The population variances are

$$S_x^2 = \sum_1^N (x_i - \bar{X})^2/(N-1) \quad \text{and} \quad S_y^2 = \sum_1^N (y_i - \bar{Y})^2/(N-1).$$

The population covariance is

$$S_{xy} = \sum_1^N (x_i - \bar{X})(y_i - \bar{Y})/(N-1)$$

and the correlation coefficient is $\rho = S_{xy}/S_x S_y$. Let $C_{xx} = S_x^2/\bar{X}^2$ and $C_{yy} = S_y^2/\bar{Y}^2$ denote the squares of the coefficients of variation of x and y, and let $C_{xy} = S_{xy}/\bar{X}\bar{Y}$.

Consider a simple random sample of size n drawn without replacement from the N units. The sample totals are $x = (\Sigma_1^n x_i)$ and $y = (\Sigma_1^n y_i)$, and the means are $\bar{x} = (x/n)$ and $\bar{y} = (y/n)$. The sample variances are

$$s_x^2 = \sum_1^n (x_i - \bar{x})^2/(n-1) \quad \text{and} \quad s_y^2 = \sum_1^n (y_i - \bar{y})^2/(n-1).$$

Let $b = s_{xy}/s_x^2$ denote the sample regression coefficient.

An estimator for $R = (\bar{Y}/\bar{X})$ is $\hat{R} = (\bar{y}/\bar{x})$, and the ratio estimator for $\bar{Y}$ is

$$\hat{\bar{Y}}_R = \hat{R}\bar{X}. \tag{1.1}$$

The regression estimator for $\bar{Y}$ is

$$\hat{\bar{Y}}_L = \bar{y} + b(\bar{X} - \bar{x}). \tag{1.2}$$

The literature on sample surveys is abundant with illustrations in which either of the above procedures have been used to estimate $\bar{Y}$ or the total $Y = N\bar{Y}$. For instance, Cochran (1978) points out that as long ago as 1802, Laplace's procedure for estimating the population of France resulted in a ratio type of estimator. In the modern statistical literature, these procedures were considered almost fifty years ago. For instance, Cochran (1939) describes the regression method for applications and refers to the earlier work by Yates and by Watson (1937).

As is well known, for large n, MSE $(\hat{\bar{Y}}_R) < V(\bar{y})$ if ρ is positive and large, but MSE$(\hat{\bar{Y}}_R) \geq V(\hat{\bar{Y}}_L)$. However, as has been mentioned in the literature, the regression method may become computationally cumbersome, especially in the case of multistage sampling designs. We quote from Yates (1960): 'The ratio method is simpler computationally, but the regression method is in certain circumstances more accurate... When the supplementary variate x represents size of unit, the regression line generally passes through the origin, ...'. In the latter case, the ratio method, of course, results in the optimum estimator. The superiority of $\hat{\bar{Y}}_R$ over $\bar{y}$ and its simplicity relative to $\hat{\bar{Y}}_L$ have been two of the reasons for giving considerable amount of attention to the ratio method of estimation.

Cochran (1977) presents comprehensive summaries and discussions on the research related to the ratio and regression methods of estimation. J. N. K. Rao (1986) presents a brief discussion on some of the ratio methods of estimation, for the cases of single and multistage designs with equal and unequal probabilities of selection. In this article, several aspects of the ratio and regression methods of estimation are presented for the case of simple random sampling without replacement.

2. Bias and mean square error of the ratio estimator

For large n, approximations to the bias and the variance or mean square error of $\hat{\bar{Y}}_R$ are

$$B(\hat{\bar{Y}}_R) \doteq \frac{(1-f)}{n} \bar{Y} C_x (C_x - \rho C_y). \tag{2.1}$$

and

$$V(\hat{\bar{Y}}_R) \doteq \frac{(1-f)}{n} S_d^2, \tag{2.2}$$

where $C_x = \sqrt{C_{xx}}$ and

$$S_d^2 = \sum_1^N (y_i - Rx_i)^2/(N-1) = S_y^2 + R^2 S_x^2 - 2RS_{xy}.$$

An estimator $v_1(\hat{\bar{Y}}_R)$ for the variance is obtained by replacing S_d^2 in (2.2) by

$$s_d^2 = \sum_1^n (y_i - \hat{R}x_i)^2/(n-1) = s_y^2 + \hat{R}^2 s_x^2 - 2\hat{R}s_{xy}.$$

The bias in (2.1) would be small if C_x is small. Empirical investigation by Kish, Namboodiri and Pillai (1962) showed that $|B(\hat{\bar{Y}}_R)|/\sqrt{V(\hat{\bar{Y}}_R)}$ is small unless the sample size is very small. Hartley and Ross (1954) showed that the exact bias of $\hat{\bar{Y}}_R$ does not exceed $C_{\bar{x}} = \sqrt{(1-f)C_{xx}/n}$.

3. Bias reduction

Several interesting procedures have been developed to reduce the bias in $\hat{R}$ or $\hat{\bar{Y}}_R$ or to completely remove the bias from them. We describe them below along with a brief description of the investigations regarding the merits of these procedures.

3.1. The jackknife procedure

Quenouille (1956) proposed a method for reducing the bias in an estimator. Since it may be used for many purposes like bias reduction, variance estimation, testing hypotheses, and setting confidence limits, Tukey (1958) named it the *jackknife*.

In this procedure, the sample is divided randomly into g groups of size $m = n/g$ each. Let $(\bar{x}_j, \bar{y}_j)$, $j = 1, 2, \ldots, g$, denote the means of these groups. Let

$$\bar{x}'_j = (n\bar{x} - m\bar{x}_j)/(n-m) \quad \text{and} \quad \bar{y}'_j = (n\bar{y} - m\bar{y}_j)/(n-m)$$

denote the means of the $n - m$ observations, obtained by deleting the observations of the jth group. Let

$$\hat{R}'_j = (\bar{y}'_j/\bar{x}'_j) \quad \text{and} \quad \hat{R}' = \left(\sum_1^g \hat{R}'_j\right)\!\bigg/g.$$

Now, for estimating R, consider

$$\hat{R}_Q = g\hat{R} - (g-1)\hat{R}'. \tag{3.1}$$

For large n, the bias of this estimator will contain only terms of order $1/n^2$ and smaller.

If N is not large, as shown by Jones (1963), see also Rao and Rao (1971),

$$\hat{R}_Q = w\hat{R} - (w-1)\hat{R}' \tag{3.2}$$

where $w = g(N - n + m)/N$, will have bias of $O(n^{-2})$. In either case, the estimator for $\overline{Y}$ is $\hat{\overline{Y}}_Q = \overline{X}\hat{R}_Q$.

Since $\hat{\overline{Y}}_R$ can be written as

$$\hat{\overline{Y}}_R = \hat{R}\overline{X} = \bar{y} + \hat{R}(\overline{X} - \bar{x}), \tag{3.3}$$

and since it is based due to the estimation of R, Rao (1979) considered the alternative estimator

$$\hat{\overline{Y}}_Q^* = \bar{y} + \hat{R}_Q(\overline{X} - \bar{x}). \tag{3.4}$$

Quenouille (1956) demonstrated that the bias in an estimator can be further reduced by reapplying the jackknife procedure and considering the second and higher order estimators. For $g = 2$, Schucany, Gray and Owen (1971) suggested alternative estimators through the reapplication and Rao (1974) extended them to more than two groups.

3.2. Alternative estimators with reduced bias

De Pascual (1961), Beale (1962), and Tin (1965) reduce the bias in $\hat{\overline{Y}}_R$ to $O(n^{-2})$. The estimators suggested by them are respectively given by

$$\hat{\overline{Y}}_D = \hat{\overline{Y}}_R + \frac{\bar{y} - \bar{r}\bar{x}}{n - 1}, \tag{3.5}$$

$$\hat{\overline{Y}}_B = \frac{1 + (1 - f)C_{xy}/n}{1 + (1 - f)C_{xx}/n} \hat{\overline{Y}}_R, \tag{3.6}$$

and

$$\hat{\overline{Y}}_T = \left[1 - \frac{(1 - f)}{n}(C_{xx} - C_{xy})\right]\hat{\overline{Y}}_R. \tag{3.7}$$

where $r_i = y_i/x_i$ and $\bar{r} = (\Sigma_1^n r_i)/n$.

Murthy and Nanjamma (1959) consider k independent interpenetrating subsamples each of the same size $m = n/k$. The means $(\bar{x}_i, \bar{y}_i)$ of these samples are unbiased for $(\overline{X}, \overline{Y})$. Let $\overline{X} = (\Sigma_1^k \bar{x}_i)/k$ and $\hat{\overline{Y}} = (\Sigma_1^k \bar{y}_i)/k$. They show that, for large k,

$$\hat{\overline{Y}}_{MN} = \frac{k}{k-1} \frac{\hat{\overline{Y}}}{\overline{X}} \overline{X} - \frac{1}{k-1}\left(\frac{1}{k}\sum_1^k \frac{\bar{y}_i}{\bar{x}_i}\right)\overline{X} \tag{3.8}$$

is approximately unbiased for $\overline{Y}$.

3.3. Unbiased estimators

Lahiri (1951) shows that $(\bar{y}/\bar{x})$ is unbiased for R if the sample (x_i, y_i), $i = 1, 2, \ldots, n$, is drawn with probability proportional to $(\Sigma_i^n x_i)$. He suggests a procedure for drawing the sample. In the alternative procedure suggested by Midzuno (1951), the first unit of the sample is drawn with probability proportional to x_i, and the rest of the $n - 1$ units are drawn with equal probability and without replacement from the remaining $N - 1$ units. The estimator for $\bar{Y}$ is $\bar{Y}_{PS} = (\bar{y}/\bar{x})\bar{X}$. Des Raj (1954, 1968) and J. N. K. Rao and Vijayan (1977) present variance estimators for $\hat{\bar{Y}}_{PS}$ obtained by this method.

With simple random sampling, the unbiased estimator considered by Hartley and Ross (1954) and Goodman and Hartley (1958) is

$$\hat{\bar{Y}}_{HR} = \bar{r}\bar{X} + \frac{(N-1)n}{N(n-1)}(\bar{y} - \bar{r}\bar{x}). \tag{3.9}$$

Utilizing the jackknife procedure, Mickey (1959) and J.N.K. Rao (1967) propose the unbiased estimator

$$\hat{\bar{Y}}_M = w\bar{y} - (w\bar{x} - \bar{X})\hat{R}'. \tag{3.10}$$

With the interpenetrating subsamples, an unbiased estimator of the Hartley–Ross type is

$$\hat{\bar{Y}}_{MN} = \frac{\bar{X}}{k}\sum_1^k \frac{\bar{y}_i}{\bar{x}_i} + \frac{k}{k-1}\left(\bar{y} - \frac{\bar{X}}{k}\sum_1^k \frac{\bar{y}_i}{\bar{x}_i}\right). \tag{3.11}$$

For a discussion on the estimators in (3.8) and (3.11), see T. J. Rao (1966).

4. Relative merits of the estimators

While the estimators in Section 3 are intended to have smaller biases than $\hat{\bar{Y}}_R$, their MSE's may be larger. Extensive studies have been conducted to examine this problem.

J. N. K. Rao and Beagle (1967) and J. N. K. Rao (1968, 1969) compared several of these estimators through Monte Carlo studies and with data from natural populations.

Some of the investigations have been conducted through the model

$$y_i = \alpha + \beta x_i + \varepsilon_i, \tag{4.1}$$

$i = 1, 2, \ldots, n$, with the assumptions that $E(\varepsilon_i|x_i) = 0$, $V(\varepsilon_i|x_i) = \delta x_i^t$, and $E(\varepsilon_i \varepsilon_j | x_i, x_j) = 0$. In practice, the value of t is often found to lie between zero and 2. It is further assumed that x has a gamma distribution $f(x) = e^{-x} x^{h-1}/(h-1)!$.

The following studies used the above model. Durbin (1959) compared $\hat{\bar{Y}}_R$ with $\hat{\bar{Y}}_Q$ when $g = 2$ and $t = 0$. J. N. K. Rao and Webster (1966) compared $\hat{\bar{Y}}_R$ with $\hat{\bar{Y}}_Q$, for $g = n$. J. N. K. Rao (1965) showed that when x has a normal distribution $g = n$ is optimum for $\hat{\bar{Y}}_Q$, and $\hat{\bar{Y}}_M$ was studied by him in J. N. K. Rao (1967). Rao (1969) compared the MSE's of $\hat{\bar{Y}}_R$, $\hat{\bar{Y}}_D$, $\hat{\bar{Y}}_{HR}$ and $\hat{\bar{Y}}_Q$ with $g = 2$, when $0 \leqslant t \leqslant 2$. Rao and Rao (1971) examined the performances of $\hat{\bar{Y}}_R$, $\hat{\bar{Y}}_Q$, $\hat{\bar{Y}}_M$ and $\hat{\bar{Y}}_T$, when $0 \leqslant t \leqslant 2$. Hutchinson (1971) considered the above model and the lognormal distribution for x for comparing $\hat{\bar{Y}}_R$ with $\hat{\bar{Y}}_Q$.

Conclusions from the above investigations can be summarized as follows.

(1) The classical estimator $\hat{\bar{Y}}_R$ should be preferred to the remaining estimators when α is close to zero and t is around one.

(2) The bias of $\hat{\bar{Y}}_T$ is small and its MSE is smaller than that of the remaining estimators, except that of $\hat{\bar{Y}}_R$ for $t = 1$.

(3) Beale's estimator $\hat{\bar{Y}}_B$ does not differ much from $\hat{\bar{Y}}_T$, unless n is very small.

(4) If reducing the bias, and not necessarily the MSE, is of importance, $\hat{\bar{Y}}_Q$ and $\hat{\bar{Y}}_M$ may be preferred to the remaining estimators.

(5) The unbiased estimator $\hat{\bar{Y}}_{HR}$ may not perform well under the above model; a different model suitable for this estimator was suggested by Hartley and Ross (1954).

Rao's (1974) study through the above model indicates that while the second-order jackknife estimators have smaller biases than the first order estimators, their MSE's may be larger.

Rao (1968) considered the N units of the finite population to be a sample from a superpopulation and found that when $\alpha \neq 0$ and $0 \leqslant t \leqslant 2$, $\varepsilon[V(\overline{Y_{PS}})] \leqslant \varepsilon[\mathrm{MSE}(\hat{\bar{Y}}_R)]$, where the operator ε represents expectation over the model in (4.1).

5. Variance estimation and confidence limits

5.1. Estimators for the variance

As mentioned in Section 2, an estimator for $V(\hat{\bar{Y}}_R)$ in (2.2) is

$$v_1(\hat{\bar{Y}}_R) = \frac{(1-f)}{n} s_d^2. \tag{5.1}$$

Another estimator that has been considered is

$$v_2(\hat{\bar{Y}}_R) = \frac{(1-f)}{n} \left(\frac{\overline{X}}{\overline{x}}\right)^2 s_d^2. \tag{5.2}$$

The jackknife estimator with g groups is

$$v_Q(\hat{\bar{Y}}) = \frac{(1-f)(g-1)}{g} \bar{X}^2 \sum_1^g (\hat{R}'_j - \hat{R}')^2 . \tag{5.3}$$

J. N. K. Rao (1969) examined the biases and MSE's of these three estimators with data from eight populations. Rao and Rao (1971) examined them through the model in (4.1). Further investigations on the relative merits of these variance estimators were conducted by J. N. K. Rao and Kuzik (1974), Krewski and Chakrabarti (1981), and Rao (1981).

As regards the bias in estimating $V(\hat{\bar{Y}}_R)$, these studies indicate that (1) $v_1(\hat{\bar{Y}}_R)$ underestimates $V(\hat{\bar{Y}}_R)$, (2) $v_2(\hat{\bar{Y}}_R)$ overestimates when $t = 0$ and underestimates when $t = 1$ and 2, and (3) $v_Q(\hat{\bar{Y}}_R)$ overestimates.

With the model in (4.1), Wu (1982) showed that $v_3(\hat{\bar{Y}}_R) = (1-f)(\bar{X}/\bar{x})s_d^2/n$ has smaller bias than $v_1(\hat{\bar{Y}}_R)$ and $v_2(\hat{\bar{Y}}_R)$ for $0.6 \leq t \leq (6/7)$.

Unless the sample size is too small, the practice of estimating $V(\hat{R})$ by $(1-f)s_d^2/n\bar{x}^2$, especially when $\bar{X}$ is unknown, and estimating $V(\hat{\bar{Y}}_R)$ by $(1-f)s_d^2/n$ may be satisfactory. For small samples, the above studies have favored $v_2(\hat{\bar{Y}}_R)$ for estimating $V(\hat{\bar{Y}}_R)$.

5.2. Confidence limits

Confidence limits for $\bar{Y}$ may be obtained by treating $(\hat{\bar{Y}}_R - \bar{Y})/\sqrt{v(\hat{\bar{Y}}_R)}$, with the denominator obtained by any one of the above estimators, as the standard normal variable or as the t-statistic with $n-1$ degrees of freedom. Scott and Wu (1981) studied the asymptotic normality of the above standardized variable, with different estimators for the variance. For an empirical study in the case of multistage designs, see Kish and Frankel (1974).

In the alternative approach suggested by Fieller (1932), confidence limits for R are obtained by equating $(\bar{y} - R\bar{x})/\sqrt{v(\bar{y} - R\bar{x})}$, where $v(\bar{y} - R\bar{x}) = (1-f)(s_y^2 + R^2 s_x^2 - 2Rs_{xy})/n$, to the t-statistic with $n-1$ degrees of freedom and solving the resulting quadratic equation in R.

6. Regression through the origin and the ratio estimator

6.1. Infinite population

Consider the model

$$y_i = \beta x_i + \varepsilon_i, \tag{6.1}$$

$i = 1, 2, \ldots, n$, with the same assumptions for ε_i as for the model in (4.1). Denote $1/x_i^t$ by w_i.

For estimating β, it is well known that, among all linear unbiased estimators of the form $\Sigma_1^n l_i y_i$, the estimator

$$b = \frac{\Sigma_1^n w_i x_i y_i}{\Sigma_1^n w_i x_i^2} \qquad (6.2)$$

has minimum variance. The variance of this estimator is $V(b) = \delta/(\Sigma_1^n w_i x_i^2)$, and the estimator of variance is $v(b) = \hat{\delta}/(\Sigma_1^n w_i x_i^2)$, where

$$\hat{\delta} = \frac{\Sigma_1^n w_i (y_i - b x_i)^2}{n-1}. \qquad (6.3)$$

Thus, for large populations, the optimum estimator for $\bar{Y}$ is $\hat{\bar{Y}}_w = b\bar{X}$, which has variance $V(\hat{\bar{Y}}_w) = \bar{X}^2 V(b)$ and estimator of variance $v(\hat{\bar{Y}}_w) = \bar{X}^2 v(b)$.

When $t = 1$, $\hat{\bar{Y}}_w$ is the same as the ratio estimator $\hat{\bar{Y}}_R = \hat{R}\bar{X}$, and the estimator of variance is

$$v(\hat{\bar{Y}}_R) = \frac{\bar{X}^2}{n(n-1)\bar{x}} \sum_1^n \frac{(y_i - \hat{R} x_i)^2}{x_i}. \qquad (6.4)$$

6.2. Model-unbiasedness

Consider the finite population to be a sample from a superpopulation with mean $\varepsilon(y_i | x_i) = \beta x_i$ and variance $V(y_i | x_i) = \delta x_i^t$. Thus, the model in (6.1) is assumed to be valid for $i = 1, 2, \ldots, N$. For estimating $\bar{Y}$, consider a linear estimator $\hat{\bar{Y}} = \Sigma_1^n l_i y_i$. Now $\varepsilon(\hat{\bar{Y}}) = \beta \Sigma_1^n l_i x_i$ and its variance is $\delta \Sigma_1^n l_i^2 x_i^t$. Furthermore, $\varepsilon(\bar{Y}) = \beta \bar{X}$. The 'model-unbiasedness' condition is $\varepsilon(\hat{\bar{Y}}) = \varepsilon(\bar{Y})$, that is, $\Sigma l_i x_i \equiv \bar{X}$. Minimizing the variance of $\hat{\bar{Y}}$ with this condition, as shown by Cochran (1977), the estimator for $\bar{Y}$ is

$$\hat{\bar{Y}} = \frac{\Sigma_1^n w_i x_i y_i}{\Sigma_1^n w_i x_i^2} \bar{X}. \qquad (6.5)$$

This 'model dependent' approach is independent of the actual procedure of drawing the sample (x_i, y_i), $i = 1, 2, \ldots, n$, from the finite population. From the model in (6.1), the MSE of the above estimator is

$$V(\hat{\bar{Y}} - \bar{Y}) = \varepsilon(\hat{\bar{Y}} - \bar{Y})^2 = \left[\frac{\bar{X}^2}{\Sigma_1^n x_i^{2-t}} + \frac{\Sigma_1^n x_i^t}{N^2} - 2\frac{N\bar{X}\bar{x}}{N\Sigma_1^n x_i^{2-t}} \right] \delta$$

$$= \frac{(N\bar{X} - n\bar{x})^2 + (\Sigma_1^N x_i^t)(\Sigma_1^n x_i^{2-t}) - n^2 \bar{x}^2}{N^2 \Sigma_1^n x_i^{2-t}} \delta. \qquad (6.6)$$

The estimator of variance $V(\hat{\bar{Y}})$ is obtained by substituting $\hat{\delta}$ from (6.3) in (6.6). For large N, $\hat{\bar{Y}}$ becomes the same as $\hat{\bar{Y}}_w$.

When $t = 1$, $\hat{\bar{Y}}$ in (6.5) is the same as $\hat{\bar{Y}}_R = \hat{R}\bar{X}$. Its MSE from (6.6) is

$$\text{MSE}(\hat{\bar{Y}}) = \frac{(1-f)}{n} \frac{\bar{X}\bar{X}'}{\bar{x}} \hat{\delta}, \tag{6.7}$$

where $\bar{X}' = (N\bar{X} - n\bar{x})/(N - n)$. The estimator of variance is

$$v(\hat{\bar{Y}}) = \frac{(1-f)}{n} \frac{\bar{X}\bar{X}'}{\bar{x}} \hat{\delta}, \tag{6.8}$$

where $\hat{\delta} = \Sigma_1^n [(y_i - \hat{R}x_i)^2/x_i]/(n-1)$.

Royall and Eberhardt (1975) consider a modification to $v(\bar{Y})$ and Royall and Cumberland (1978) conduct an empirical study to evaluate its merits. Rao (1981) compared $v(\hat{\bar{Y}})$ with $v_1(\hat{\bar{Y}}_R)$ and $v_2(\hat{\bar{Y}}_R)$ through the model in (4.1).

For the total, Royall (1970) considers

$$\hat{Y} = \sum_1^n y_i + \hat{\beta}\left(X - \sum_1^n x_i\right), \tag{6.9}$$

where $X = \Sigma_1^n x_i$ and $\hat{\beta}$ is an estimator of β. For the above model with $\varepsilon(y_i|x_i) = \beta x_i$ and $V(y_i|x_i) = \delta x_i^t$, he shows that minimizing the mean square error of $\bar{Y}$ with the model-unbiasedness condition results in the estimator

$$\hat{Y} = \sum_1^n y_i + b\left(X - \sum_1^n x_i\right) \tag{6.10}$$

where b is given in (6.5). The corresponding estimator for $\bar{Y}$ is

$$\hat{\bar{Y}} = b\bar{X} + \frac{n}{N}(\bar{y} - b\bar{x}). \tag{6.11}$$

For large n, this estimator is the same as $b\bar{X}$. Brewer (1963) has earlier considered the above type of model dependent estimators.

7. Stratification and the ratio estimators

When the population is divided into L strata, let the additional subscript $h = 1, 2, \ldots, L$ denote the strata. The population means are $\bar{X} = \Sigma_1^L W_h \bar{X}_h$ and $\bar{Y} = \Sigma_1^L W_h \bar{Y}_h$, where $W_h = N_h/N$ is the proportion of the units in the hth stratum. A simple random sample of size n_h, $n = \Sigma n_h$, is drawn from the hth stratum. Two types of estimators for $\bar{Y}$ are considered.

7.1. The separate ratio estimator

The ratio estimator for the mean $\bar{Y}_h$ of the hth stratum is $\hat{\bar{Y}}_{Rh} = \hat{R}_h \bar{X}_h = (\bar{y}_h/\bar{x}_h)\bar{X}_h$. Its bias $B(\hat{\bar{Y}}_{Rh})$, variance $V(\hat{\bar{Y}}_{Rh})$ and estimator of variance can be obtained from the results given in Section 2 for a single stratum. The separate estimator for the population mean $\hat{\bar{Y}}_{RS} = \Sigma_1^L W_h \hat{\bar{Y}}_{Rh}$. Its bias is given by $\Sigma_1^L W_h B(\hat{\bar{Y}}_{Rh})$. The bias will be small if the sample sizes n_h in the strata are large, and the coefficients of variation of x in the strata are small. The variance of $\hat{\bar{Y}}_{RS}$ is $\Sigma_1^L W_h^2 V(\hat{\bar{Y}}_{Rh})$ and the estimator of variance is $\Sigma_1^L W_h^2 v(\hat{\bar{Y}}_{Rh})$.

7.2. The combined ratio estimator

This type of estimator for $\bar{Y}$ considered by Hansen, Hurwitz and Gurney (1946) is

$$\hat{\bar{Y}}_{RC} = \frac{\Sigma W_h \bar{y}_h}{\Sigma W_h \bar{x}_h} \bar{X} = \frac{\bar{y}_{st}}{\bar{x}_{st}} \bar{X} = \hat{R}\bar{X}. \qquad (7.1)$$

The bias of this estimator will be small if the sample size n is large and the coefficient of variation of $\Sigma_1^L W_h \bar{x}_h$ is small. Let

$$S_{dh}'^2 = = S_{yh}^2 + R^2 S_{xh}^2 - 2R\rho_h S_{yh} S_{xh} \qquad (7.2)$$

and

$$s_{dh}'^2 = s_{yh}^2 + \hat{R}^2 s_{xh}^2 - 2\hat{R} r_h s_{yh} s_{xh},$$

where ρ_h is the correlation coefficient in the hth stratum and r_h is its sample estimator.

An approximation to the variance of $\hat{\bar{Y}}_{RC}$ is

$$V(\hat{\bar{Y}}_{RC}) \doteq \sum_1^L W_h^2 (1 - f_h) \frac{S_{dh}'^2}{n_h} = \sum_1^L a_h S_{dh}'^2. \qquad (7.3)$$

where $a_h = W_h^2(1 - f_h)/n_h$. For the estimator of variance, $S_{dh}'^2$ in (7.3) is replaced by $s_{dh}'^2$. A comparison of $\hat{\bar{Y}}_{RS}$ and $\hat{\bar{Y}}_{RC}$ was made by J. N. K. Rao and Ramachandran (1974).

The combined ratio estimator is widely used in multistage designs with stratified sampling. For estimating the variance of the combined estimator, the Taylor's series expansion, jackknife, balanced repeated replication and bootstrapping can be utilized. Relative performances of these procedures are examined by McCarthy (1969), Lemeshow and Levy (1978), Krewski and J. N. K. Rao (1981), J. N. K. Rao (1985), and J. N. K. Rao and Wu (1985); see Chapter 17 by J. N. K. Rao in this Volume.

8. Regression estimator

8.1. Infinite population

For the classical regression model in (4.1) for an infinite population it is usually assumed that

$$E(\varepsilon_i|x_i) = 0 \quad \text{and} \quad V(\varepsilon_i|x_i) = \sigma_y^2(1 - \rho^2).$$

For this model, $\mu_y = \alpha + \beta\mu_x$, where $\mu_y = E(y)$ and $\mu_x = E(x)$. The least squares estimator for β is $b = s_{xy}/s_x^2$. When μ_x is known, an estimator for μ_y is

$$\hat{\mu}_y = \bar{y} + b(\mu_x - \bar{x}). \tag{8.1}$$

This estimator is unbiased, and its variance for given x_i is

$$V(\hat{\mu}_y|x_1, x_2, \ldots, x_n) = \left[\frac{1}{n} + \frac{(\bar{x} - \mu_x)^2}{\Sigma(x_i - \bar{x})^2}\right]\sigma_e^2. \tag{8.2}$$

The estimator of this variance is obtained by replacing σ_e^2 by its unbiased estimator

$$\hat{\sigma}_e^2 = \frac{\Sigma_1^n[(y_i - \bar{y}) - b(x_i - \bar{x})]^2}{n - 2}. \tag{8.3}$$

When x_i has a normal distribution with mean μ_x and variance σ_x^2, the expectation of the variance in (8.2) over x_i is

$$E[V(\hat{\mu}_y|x_1, x_2, \ldots, x_n)] = \left[\frac{1}{n} + \frac{1}{n(n-3)}\right]\sigma_e^2 = \frac{(n-2)}{n(n-3)}\sigma_e^2. \tag{8.4}$$

8.2. Finite population

For the mean $\bar{Y}$ of a finite population, we can consider the analogous estimator $\hat{\bar{Y}} = \bar{y} + B(\bar{X} - \bar{x})$, where B is a constant. This estimator is unbiased and its variance is minimized when $B = S_{xy}/S_x^2 = \rho(S_y/S_x)$.

With the sample regression coefficient $b = s_{xy}/s_x^2$, the estimator for $\bar{Y}$ is

$$\hat{\bar{Y}}_L = \bar{y} + b(\bar{X} - \bar{x}). \tag{8.5}$$

For large samples, the bias of this estimator is small, and its variance is

$$V(\hat{\bar{Y}}_L) \doteq \frac{(1-f)}{n}S_e^2 = \frac{(1-f)}{n}S_y^2(1 - \rho^2). \tag{8.6}$$

An unbiased estimator for this variance is $v(\hat{\bar{Y}}_L) = (1-f)s_e^2/n$, where s_e^2 is obtained from (8.3) with the sample from the finite population. J. N. K. Rao (1968) compared the ratio and regression estimators for small samples.

The variance in (8.6) is smaller than $V(\hat{\bar{Y}}_R)$ in (2.2). These two variances are equal only when the population regression coefficient $\beta = S_{xy}/S_x^2$ becomes equal to $R = \bar{Y}/\bar{X}$. Notice that when N is large $V(\hat{\bar{Y}}_L)$ does not differ much from the average variance in (8.4).

Consider the finite population to be a sample from an infinite population with the model in (4.1), and consider the estimator $\Sigma\, l_i y_i$ for $\bar{Y}$. Minimizing $V(\Sigma\, l_i y_i)$ with the model-unbiasedness condition $\varepsilon(\Sigma\, l_i y_i) = \varepsilon(\bar{Y})$, we find that $\hat{\bar{Y}}_L$ is the optimum estimator for $\bar{Y}$.

8.3. Stratification

The separate estimator for $\bar{Y}$ is

$$\hat{\bar{Y}}_{LS} = \sum W_h \hat{\bar{Y}}_{Lh} = \sum W_h[\bar{y}_h + b_h(\bar{X}_h - \bar{x}_h)] \tag{8.7}$$

where $b_h = s_{xyh}/s_{xh}^2$. For large n_h, the bias of this estimator becomes small, and its approximate variance is

$$V(\hat{\bar{Y}}_{lS}) = \sum a_h S_{eh}^2 \tag{8.8}$$

where $a_h = W_h^2(1-f_h)/n_h$ and $S_{eh}^2 = (1-\rho_h^2)S_{yh}^2$. The estimator of variance is obtained by replacing S_{eh}^2 by

$$s_{eh}^2 = \frac{\sum_1^{n_h}[y_{hi} - \bar{y}_h) - b_h(x_{hi} - \bar{x}_h)]^2}{n_h - 2} = \frac{n_h - 1}{n_h - 2} s_{yh}^2(1 - r_h^2). \tag{8.9}$$

Jones (1974) considers the jackknife method to reduce the bias in $\hat{\bar{Y}}_{LS}$. With a common slope, we can consider

$$\hat{\bar{Y}} = \sum_1^L W_h(\bar{y}_h + B(\bar{X}_h - \bar{x}_h)] = \bar{y}_{st} + B(\bar{X} - \bar{x}_{st}).$$

The variance of this unbiased estimator is minimized when

$$B = \left(\sum_1^L a_h S_{xyh}\right) \bigg/ \left(\sum_1^L a_h S_{xh}^2\right).$$

With the sample estimator $b_c = (\Sigma_1^L a_h s_{xyh})/(\Sigma_1^L a_h s_{xh}^2)$, the estimator for $\bar{Y}$ is

$$\hat{\bar{Y}}_{LC} = \bar{y}_{st} + b_c(\bar{X} - \bar{x}_{st}). \tag{8.10}$$

The bias of this estimator becomes small for large n. An approximation to the variance is

$$V(\hat{\bar{Y}}_{LC}) \doteq \sum a_h(S_{yh}^2 - B^2 S_{xh}^2). \qquad (8.11)$$

An estimator for the variance is $v(\hat{\bar{Y}}_{LC}) = \sum a_h s_{eh}^{\prime 2}$, where $s_{eh}^{\prime 2}$ is obtained from (8.9) by replacing b_h by b_c. If n is large, this variance estimator becomes equal to $\sum a_h(s_{yh}^2 - b_c^2 s_{xh}^2)$.

9. Multivariate ratio, product and regression estimators

9.1. Ratio estimator

When p auxiliary variables $x_1, x_2, \ldots, x_p$ are available, let $(\bar{Y}; \bar{X}_1, \ldots, \bar{X}_p)$ denote the population means. Let $(\bar{y}; \bar{x}_1, \ldots, \bar{x}_p)$ denote the means of a simple random sample of size n. Olkin (1958) considered the estimator

$$\hat{\bar{Y}}_{MR} = \left(W_1 \frac{\bar{X}_1}{\bar{x}_1} + \cdots + W_p \frac{\bar{X}_p}{\bar{x}_p} \right) \bar{y}, \qquad (9.1)$$

where the weights $(W_1, \ldots, W_p)$ have to be determined.

Let

$$S_{00} = S_y^2 = \sum_1^N (y_k - \bar{Y})^2/(N-1), \quad S_{0i} = \sum_1^N (y_k - \bar{Y})(x_{ik} - \bar{X}_i)/(N-1),$$

and

$$S_{ij} = \sum_1^N (x_{ik} - \bar{X}_i)(x_{jk} - \bar{X}_j)/(N-1).$$

Let

$$a_{ij} = (S_{00} - R_i S_{00} - R_j S_{0j} + R_i R_j S_{ij}), \quad i, j = 1, 2, \ldots, p.$$

For large samples,

$$V(\hat{\bar{Y}}_{MR}) \doteq \frac{(1-f)}{n} \sum_1^p \sum_1^p W_i W_j a_{ij} = \frac{(1-f)}{n} W'AW \qquad (9.2)$$

where $W' = (W_1, W_2, \ldots, W_p)$ and A is the $p \times p$ matrix with elements a_{ij}.

Minimizing the above variance with $\sum_1^p W_i = 1$, the optimum value of W is $(1/e'A^{-1}e)A^{-1}e$, where e is a $(p \times 1)$ vector of unities. The minimum variance is equal to $(1-f)/(ne'A^{-1}e)$.

Let

$$s_{00} = s_y^2 = \sum_1^n (y_k - \bar{y})^2/(n-1), \quad s_{0i} = \sum_1^n (y_k - \bar{y})(x_{ik} - \bar{x}_i)/(n-1)$$

and

$$s_{ij} = \sum_1^n (x_{ik} - \bar{x}_i)(x_{jk} - \bar{x}_j)/(n-1),$$

and let $\hat{R}_i = (\bar{y}/\bar{x}_i)$. An estimator for the variance in (9.2) is obtained by substituting these quantities for S_{00}, S_{0i}, S_{ij} and R_i respectively.

An alternative estimator

$$\hat{\bar{Y}}_{MR} = \bar{y} \left(\sum_1^p W_i \bar{X}_i \right) \Big/ \left(\sum_1^p W_i \bar{x}_i \right)$$

was considered by Shukla (1966).

9.2. Product estimator

With a single auxiliary variable x, Murthy (1964) considered the product estimator

$$\hat{\bar{Y}}_p = \frac{\bar{y}\bar{x}}{\bar{X}}. \tag{9.3}$$

For large samples, an approximation to the variance of this estimator is

$$V(\hat{\bar{Y}}_p) \doteq \frac{(1-f)}{n} (S_y^2 + R^2 S_x^2 + 2R\rho S_x S_y). \tag{9.4}$$

This variance is smaller than $V(\bar{y})$ if $\rho < -(C_x/2C_y)$.

A comparison of $\hat{\bar{Y}}_p$ with $\hat{\bar{Y}}_R$ was made by Chaubey, Dwivedi and Singh (1984).

With p auxiliary variables, a multivariate product estimator is

$$\hat{\bar{Y}}_{MP} = \left(W_1 \frac{\bar{x}_1}{\bar{X}_1} + \cdots + W_p \frac{\bar{x}_p}{\bar{X}_p} \right) \bar{y}. \tag{9.5}$$

Let

$$b_{ij} = S_{00} + R_i S_{0i} + R_j S_{0j} + R_i R_j S_{ij}, \quad i, j = 1, 2, \ldots, p.$$

For large samples,

$$V(\hat{\bar{Y}}_{MP}) \doteq \frac{(1-f)}{n} W'BW \tag{9.6}$$

where B is the $p \times p$ matrix with elements b_{ij}. The value of W that minimizes the variance with $\Sigma_1^p W_i = 1$ is $W = (1/e'B^{-1}e)B^{-1}e$, and the minimum variance is $(1-f)/(ne'B^{-1}e)$.

An estimator for the variance in (9.6) is obtained through the procedure described at the end of the last section.

9.3. Ratio-product estimator

To include both positively and negatively correlated auxiliary variables, Rao and Mudholkar (1967) considered

$$\hat{\bar{Y}}_{RP} = W_1 \frac{\bar{y}_1}{\bar{x}_1} \bar{X}_1 + \cdots W_p \frac{\bar{y}_p}{\bar{x}_p} \bar{X}_p + W_{p+1} \frac{\bar{y}_{p+1}\bar{x}_{p+1}}{\bar{X}_{p+1}}$$

$$+ \cdots + W_{p+q} \frac{\bar{y}_{p+q}\bar{x}_{p+q}}{\bar{X}_{p+q}}; \quad (9.7)$$

see also Srivastava (1965).

Let a_{ij}, $i, j = 1, 2, \ldots, p$, be as defined in Section 9.1, and let b_{ij} be as defined in Section 9.2 for $i, j = p+1, p+2, \ldots, p+q$. Let

$$c_{ij} = S_{00} - R_i S_{0i} + R_j S_{0j} - R_i R_j S_{ij},$$

$i = 1, 2, \ldots, p$ and $j = p+1, p+2, \ldots, p+q$. Let A denote the $p \times p$ matrix with elements a_{ij} as before, let B denote the $q \times q$ matrix with elements b_{ij}, and let C denote the $p \times q$ matrix with elements c_{ij}. Further, let

$$D = \begin{pmatrix} A & C \\ C' & B \end{pmatrix}. \quad (9.8)$$

For large n,

$$V(\hat{\bar{Y}}_{RP}) \doteq \frac{(1-f)}{n} W'DW \quad (9.9)$$

where W is now a $(p+q) \times 1$ vector and D is a $(p+q) \times (p+q)$ matrix.

As before, the optimum value of W is $(1/e'De^{-1})D^{-1}e$, and the minimum variance is $(1-f)/(ne'D^{-1}e)$. The variance in (9.9) can be estimated through the procedures described in the last two sections.

Singh (1965, 1967) considers the estimator in (9.5) and also alternative estimators by combining ratio and product type of estimators.

9.4. Regression estimator

Let y denote the $n \times 1$ vector of the deviations $(y_i - \bar{y})$, let x_1 denote the $n \times 1$ vector of the deviations $(x_{1i} - \bar{x}_1)$, and so on. Let x denote the $n \times p$ matrix with $(x_1, \ldots, x_p)$ as its column vectors. Let $b = (x'x)^{-1}x'y$ denote the $p \times 1$ vector of the sample regression coefficients. The regression estimator for $\bar{Y}$ is

$$\hat{\bar{Y}}_{ML} = \bar{y} + b_1(\bar{X}_1 - \bar{x}_1) + \cdots + b_p(\bar{X}_p - \bar{x}_p). \quad (9.10)$$

Let y_N and x_N denote the corresponding $N \times 1$ vector and $N \times p$ matrix for the N population units. Note that the first column of x_N consists of the N values of $(x_{1i} - \overline{X_1})$, and so on. Let

$$S_e^2 = \frac{y_N'[I_N - x_N(x_N'x_N)^{-1}x_N']y_N}{N-1} . \tag{9.11}$$

For large samples,

$$V(\hat{\overline{Y}}_{ML}) \doteq \frac{(1-f)}{n} S_e^2 . \tag{9.12}$$

This variance would be smaller than the variances of the three multivariate estimators in the previous sections.

Let

$$s_e^2 = \frac{y'y - b'x'xb}{n-k-1} = \frac{y'[I - x(x'x)^{-1}x']y}{n-k-1} .$$

The estimator of variance $v(\hat{\overline{Y}}_{ML})$ is obtained by replacing S_e^2 in (9.12) by s_e^2.

10. Two phase sampling for ratio and regression estimators

When the mean $\overline{X}$ for the ratio or regression estimator is not known, it is estimated by $\overline{x}_1$ obtained from an initial simple random sample of size n_1. A subsample of size n from the first sample gives the means $(\overline{x}, \overline{y})$ and also the regression coefficient. Cochran (1977, p. 343) discusses the procedure for finding the optimum values of n_1 and n_2 for the ratio and regression estimators. Rao (1981) examined the performance of several double sampling ratio estimators. Khan and Tripathi (1967) considered the double sampling procedure for the multiple regression estimator.

Bose (1943) considered the second sample to be drawn independently of the first. Tikkiwal (1960) considered the regression estimator with independent samples. Rao (1972, 1975a, 1975b) suggested improved ratio and regression estimators for the case of independent samples.

11. Further developments

Williams (1961, 1962) and J. N. K. Rao (1964) consider unbiased ratio and regression estimators for the case of *multistage sampling*.

J. N. K. Rao and Pereira (1968) examine the *double ratio estimator* for the relative value of two ratios. Yates (1960) presents this procedure and mentions that Keyfitz described it earlier.

Rao (1986, 1987) considered Hansen and Hurwitz's (1946) procedure of subsampling the *nonrespondents* and suggested different ratio and regression estimators for $\overline{Y}$. He has also suggested ratio type of estimators when the population consists of *hard-core* nonrespondents.

Chakrabarti (1979) considers estimators of the form $C\overline{y} + (1 - C)\hat{\overline{Y}}_R$ for $\overline{Y}$, and determines the optimum value of C for different values of the correlation coefficient.

Swain (1964) and Singh (1967) examine the effect of *systematic sampling* on the ratio method of estimation.

Acknowledgements

The author would like to thank Professor C. R. Rao for his interest in this topic and Professor J. N. K. Rao for his valuable comments on the first draft of this chapter.

References

Beale, E. M. L. (1962). Some uses of computers in operations research. *Industrielle Organization* **31**, 51–52.
Bose, C. (1943). Note on the sampling error in the method of double sampling. *Sankhyā* **6**, 330.
Brewer, K. W. R. (1963). Ratio estimation in finite populations. Some results deductible from the assumption of an underlying stochastic process. *Australian Journal of Statistics* **5**, 93–105.
Chakrabarty, R. P. (1979). Some ratio type estimators. *Journal of the Indian Society of Agriculture Statistics* **31**, 49–62.
Chaubey, Y. P., Dwivedi, T. D. and Singh, M. P. (1984). An efficiency comparison of product and ratio estimator. *Communications in Statistics Ser. A* **13**, 699–709. Errata (1985) **14**, 1249–1250.
Cochran, W. G. (1939). The use of the analysis of variance in enumeration by sampling. *Journal of the American Statistical Association* **34**, 492–510.
Cochran, W. G. (1977). *Sampling Techniques*. John Wiley & Sons, New York.
Cochran, W. G. (1978). Laplace's ratio estimator. In: H. A. David, ed., *Contributions to Survey Sampling and Applied Statistics*. Academic Press, New York, 3–10.
De Pascual, N. (1961). Unbiased ratio estimators in stratified sampling. *Journal of the American Statistical Association* **56**, 70–87.
Des Raj (1954). Ratio estimation in sampling equal and unequal probabilities. *Journal of the Indian Society of Agricultural Statistics*, 6, 127–138.
Des Raj (1968). *Sampling Theory*. McGraw-Hill, New York.
Durbin, J. (1959). A note on the application of Quenouille's method of bias reduction to the estimation of ratios. *Biometrika* **46**, 477–480.
Fieller, E. C. (1932). The distribution of the index in a normal bivariate population, *Biometrika* **24**, 428–440.
Goodman, L. A. and Hartley, H. O. (1958). The precision of unbiased ratio-type estimators. *Journal of the American Statistical Association* **53**, 491–508.
Hansen, H. H. and Hurwitz, W. N. (1946). The problem of nonresposne in sample surveys. *Journal of the American Statistical Association* **41**, 517–529.
Hansen, M. H., Hurwitz, W. N. and Gurney, M. (1946). Problems and methods of the sample surveys of business. *Journal of the American Statistical Association* **41**, 173–189.
Hartley, H. O. and Ross, A. (1954). Unbiased ratio estimates. *Nature* **174**, 270–271.

Hutchinson, M. C. (1971). A Monte Carlo comparison of some ratio estimators. *Biometrika* **58**, 313–321.

Jones, H. L. (1963). The jack-knife method. *Proceedings of the IBM Scientific Computing Symposium in Statistics*.

Jones, H. L. (1974). Jackknife estimation of functions of stratum means. *Biometrika* **61**, 343–348.

Khan, S. and Tripathi, T. P. (1967). The use of multiauxiliary information in double-sampling. *Journal of the Indian Statistical Association* **5**, 42–48.

Kish, L., Namboodiri, N. K. and Pillai, R. K. (1962). The ratio bias in surveys. *Journal of the American Statistical Association* **57**, 863–876.

Kish, L. and Frankel, M. R. (1974). Inference from complex samples. *Journal of the Royal Statistical Society Ser. B* **36**, 1–37.

Krewski, D. and Rao, J. N. K. (1981). Inference from stratified samples: properties of the linearization, jackknife and balanced repeated replication methods. *Annals of Statistics* **9**, 1010–1019.

Krewski, D. and Chakrabarti, R. P. (1981). On the stability of the jackknife variance estimator in ratio estimation. *Journal of Statistical Planning and Inference* **5**, 71–79.

Lahiri, D. B. (1951). A method of sample selection providing unbiased ratio estimates. *Bulletin of the International Statistical Institute* **33**(2), 133–140.

Lemeshow, S. and Levy, P. (1978). Estimating the variance of the ratio estimates in complex sample surveys with two primary units per stratum – A comparison of balanced replication and jackknife techniques. *Journal of Statistical Computing and Simulation* **8**, 191–205.

McCarthy, P. J. (1969). Pseudoreplication: Half-samples. *Review of the International Statistical Institute* **37**, 239–264.

Mickey, M. R. (1959). Some finite population unbiased ratio and regression estimators. *Journal of the American Statistical Association* **54**, 594–612.

Midzuno, H. (1951). On the sampling system with probability proportionate to sum of sizes. *Annals of the Institute of Statistical Mathematics* **2**, 99–108.

Murthy, M. N. and Nanjamma, N. S. (1959). Almost unbiased ratio estimates based on interpenetrating sub-sample estimates. *Sankhyā* **21**, 381–392.

Murthy, M. N. (1964). Product method of estimation. *Sankhyā Ser. A* **26**, 69–74.

Olkin, I. (1958). Multivariate ratio method of estimation for finite populations, *Biometrika* **45**, 154–165.

Quenouille, M. H. (1965). Notes on bias in estimation. *Biometrika* **43**, 353–360.

Rao, J. N. K. (1964). Unbiased ratio and regression estimators in multi-stage sampling. *Journal of the Indian Society of Agriculture Statistics* **14**, 175–188.

Rao, J. N. K. (1965). A note on the estimation of Quenouille's method. *Biometrika* **52**, 647–649.

Rao, J. N. K. and Webster, J. T. (1966). On two methods of bias reduction in the estimation of ratios. *Biometrika* **53**, 571–577.

Rao, J. N. K. (1967). The precision of Mickey's unbiased ratio estimator. *Biometrika* **54**, 321–324.

Rao, J. N. K. and Beagle, L. D. (1967). A Monte Carlo study of some ratio estimators. *Sankhyā Ser. B* **29**, 47–56.

Rao, J. N. K. (1968). Some small sample results in ratio and regression estimation. *Journal of the Indian Statistical Association* **6**, 160–168.

Rao, J. N. K. and Pereira, N. P. (1968). On double ratio estimators. *Sankhyā Ser. A* **30**, 83–90.

Rao, J. N. K. (1969). Ratio and regression estimators. In: N. L. Johnson and H. Smith, Jr., eds., *New Development in Survey Sampling*. Wiley, New York, 213–234.

Rao, J. N. K. and Kuzik, R. A. (1974). Sampling errors in ratio estimation. *Sankhyā Ser. C* **36**, 43–58.

Rao, J. N. K. and Ramachandran, V. (1974). Comparison of the separate and combined ratio estimates. *Sankhyā Ser. C* **36**, 151–156.

Rao, J. N. K. and Vijayan, K. (1977). On estimating the variance in sampling with probability proportional to aggregate size. *Journal of the American Statistical Association* **72**, 579–584.

Rao, J. N. K. (1985). Inference from stratified samples: Second order analysis of three methods for nonlinear statistics. *Journal of the American Statistical Association* **80**, 620–630.

Rao, J. N. K. (1986). Ratio estimators. In: S. Kotz and N. L. Johnson, eds., *Encyclopedia of Statistical Sciences*. J. Wiley, New York.

Rao, J. N. K. (1988). Variance estimation in sample surveys. *Handbook of Statistics*, Vol. 6. North-Holland, Amsterdam, 427–447.

Rao, P. S. R. S. and Mudholkar, G. S. (1967). Generalized multivariate estimator for the mean of finite populations. *Journal of the American Statistical Association* **62**, 1009–1012.

Rao, P. S. R. S. (1968). On three procedures of sampling from finite populations. *Biometrika* **55**, 438–440.

Rao, P. S. R. S. (1969). Comparison of four ratio-type estimates. *Journal of the American Statistical Association* **64**, 574–580.

Rao, P. S. R. S. and Rao, J. N. K. (1971). On the jack-knife estimator in finite populations. *Bulletin of the International Statistical Institute*.

Rao, P. S. R. S. and Rao, J. N. K. (1971). Small sample results for ratio estimators, *Biometrika* **58**, 625–630.

Rao, P. S. R. S. (1972). On two phase regression estimator. *Sankhyā Ser. A* **33**, 473–476.

Rao, P. S. R. S. (1974). Jackknifing the ratio estimator. *Sankhyā Ser. C* **36**, 84–97.

Rao, P. S. R. S. (1975). Hartley–Ross type estimator with two phase sampling. *Sankhyā Ser C* **37**, 140–146.

Rao, P. S. R. S. (1975). On the two-phase ratio estimator in finite populations. *Journal of the American Statistical Association* **70**, 839–845.

Rao, P. S. R. S. (1979). On applying the jack-knife procedure to the ratio estimator. *Sankhyā Ser. C* **41**, 115–126.

Rao, P. S. R. S. (1981). Efficiencies of nine two-phase ratio estimators for the mean. *Journal of the American Statistical Association* **76**, 434–442.

Rao, P. S. R. S. (1981). Estimation of the mean square error of the ratio estimator. In: D. Krewski, R. Platek and J. N. K. Rao, eds., *Current Topics in Survey Sampling*. Academic Press, New York, 305–315.

Rao, P. S. R. S. (1983). Randomization approach. In: W. G. Madow, I. Olkin and D. B. Rubin, eds., *Incomplete Data in Sample Surveys*, Theory and Bibliographies, Vol. 2. Academic Press, New York, 97–105.

Rao, P. S. R. S. (1986). Ratio estimation with subsampling the nonrespondents. *Methodology symposium on missing data in surveys*. Statistics Canada and Carleton University. Survey Methodology.

Rao, P. S. R. S. (1987). Ratio and regression estimators with subsampling the nonrespondents. *International statistical Institute Meetings*, 46th Session, 360–361.

Rao, T. J. (1966). On certain unbiased ratio estimators. *Annals of Institute of Statistical Mathematics* **18**, 117–121.

Robson, D. S. (1957). Application of multivariate polykeys to the theory of unbiased ratio-type estimators. *Journal of the American Statistical Association* **52**, 511–522.

Royall, R. M. (1970). On the finite population sampling theory under certain linear regression models. *Biometrika* **57**, 377–387.

Royall, R. M. and Eberhardt, K. R. (1975). Variance estimates for the ratio estimator. *Sankhyā Ser. C* **37**, 43–52.

Royall, R. M. and Cumberland, W. G. (1981). *Journal of the American Statistical Association* **76**, 66–77.

Schucany, W. R., Gray, H. L. and Owen, D. B. (1971). On bias reduction in estimation. *Journal of the American Statistical Association* **66**, 524–533.

Scott, A. J. and Wu, C. F. (1981). On the asymptotic distribution of ratio and regression estimators. *Journal of the American Statistical Association* **76**, 98–102.

Shukla, G. K. (1966). An alternative multivariate ratio estimate for finite population. *Calcutta Statistical Association Bulletin*, 127–133.

Singh, M. P. (1965). On the Estimation of ratio and product of population parameters. *Sankhyā Ser. B* **27**, 321–328.

Singh, M. P. (1966). Efficient use of systematic sampling in ratio and product estimation. *Metrika* **10**, 199–205.

Singh, M. P. (1967). Multivariate product method of estimation for finite populations. *Journal of the Indian Society of Agricultural Statistics* **109**, 1–10.

Srivastava, S. K. (1965). An estimate of the mean of a finite population using several auxiliary variables, *Journal of the Indian Statistical Association* **52**, 511–522.

Sukhatme, B. V. (1962). Some ratio-type estimators in two-phase sampling. *Journal of the American Statistical Association* **57**, 628–632.

Swain, A. K. P. C. (1964). The use of systematic sampling in ratio-estimate. *Journal of the Indian Statistical Association* **2**, 160–164.

Tikkiwal, B. D. (1960). Classical regression and double sampling estimation. *Journal of the Royal Statistical Society Ser.* **22**, 131–138.

Tin, M. (1965). Comparison of some ratio estimators. *Journal of the American Statistical Association* **60**, 294–307.

Tukey, J. H. (1958). Bias and confidence in not-quite large samples (Abstract). *Annals of Mathematical Statistics* **29**, 614.

Watson, D. J. (1937). The estimation of leaf areas. *Journal of Agricultural Science* **27**, 474.

Williams, W. H. (1961). Generating unbiased ratio and regression estimators. *Biometrics* **17**, 267–274.

Williams, W. H. (1962). On two methods of unbiased estimation with auxiliary variates. *Journal of the American Statistical Association* **57**, 184–186.

Yates, F. (1960). *Sampling Methods for Census and Surveys*. Charles Griffin and Company, London, 3rd ed.

Role and Use of Composite Sampling and Capture-Recapture Sampling in Ecological Studies

M. T. Boswell, K. P. Burnham and G. P. Patil[1]

Introduction

The physical mixing of samples with other samples or with the population has turned out to be a basis of some important sampling procedures.

Sampling with replacement may be interpreted as returning a sample to the original population and thoroughly mixing it before the next sample is selected. This type of sampling has been quite common in practice.

A relatively recent sampling procedure, called composite sampling, involves physically mixing of samples before measuring, counting, or otherwise analyzing the composite sample. Pertinent statistical analysis is able to extract most of the information from the composite sample that can otherwise be extracted from the measurements on the individual original samples before they are physically mixed. The savings in the cost of measuremental analyses can be substantial.

Another sampling procedure, called capture-recapture sampling, involves physical mixing of a sample back into the original population. While composite sampling, and sampling with replacement, are used to estimate the population density/abundance, capture-recapture sampling is used to estimate population size and survival at of individuals.

Both composite sampling and capture-recapture sampling techniques have been refined and adapted in response to the varying needs involving different kinds of parameters of the populations of interest. The purpose of this paper is to provide a perspective of these sampling procedures. The material for composite sampling is taken from Boswell and Patil (1987).

---

[1] Prepared in part as a Visiting Professor of Biostatistics, Department of Biostatistics, Harvard School of Public Health and Dana-Farber Cancer Institute, Harvard Medical School, Harvard University, Boston, MA.

1. Composite sampling

1.1. Introduction

A composite sample is formed by taking a number of individual samples and physically mixing them. Composite samples are used for different purposes. Usually the goal is to obtain the desired information in the original samples but at reduced cost or effort.

There are two general types of applications: estimating the mean of a stochastic process and identifying the individuals with a certain trait. Composite sampling can also be used to estimate the fraction of a population that possesses a given trait while maintaining the privacy or confidentiality of the individuals.

Composite samples have been formed from water (Becker, 1977; Roskopf, 1968; Schaeffer and Janardan, 1978; Schaeffer et al., 1980) and from bulk materials such as soils and fertilizers (Rohde, 1979; Brown and Robson, 1975; Hueck, 1976). In these cases, a single measurement on the composite sample is used in place of the average of individual measurements of the original samples.

In most of the applications, it is hoped that the mean of the process can be estimated by a single measurement taken on the composite sample, thereby saving the cost of taking many measurements. Considerable savings in the overall cost may be realized. What is lost is the ability to estimate the variability in the measurement process. However, by processing several composite samples, the variance may be estimated, still at considerable savings.

Instead of analyzing the entire composite, subsamples or aliquots are drawn and processed separately. This allows the measurement errors to be taken into account. Further, several composite samples may be formed from subsamples or 'increments' drawn from the original samples. This allows the variance of the original observation to be estimated.

1.2. Composite samples of known proportions

1.2.1. Composite samples for the detection of a certain trait

Laboratory procedures (such as blood tests for a disease), to see if a certain trait is present or not, can use compositing of samples to reduce costs (see, for example, Feller, 1968, p. 239; Garner, Stapanian and Williams, 1987). If the problem is to identify every individual with a rare trait, compositing several individual samples and testing the individual samples only when the composite sample tests are positive, has the potential of greatly reducing the number of tests required. Feller gives references to several generalizations involving two stage sampling.

The expected number of tests required for a procedure that composites n samples and tests all n samples when the composite sample exhibits the trait, can be calculated explicitly in terms of the unknown incidence of the trait, p. The expected number of tests is

$$E[N] = 1 \cdot (1-p)^n + (n+1)[1-(1-m)^n] = (n+1) - n(1-p)^n,$$

and the relative factor (RCF) equals $1 + 1/n - (1-p)^n$ tests per individual.

For p less than about 0.29, the graph of RCF versus n first decreases to a global minimum less than 1, then increases to a local maximum, and then decreases to an asymptote greater than the minimum. For larger values of p the graph steadily decreases but never falls below 1; thus, composite sampling is not advantageous for large p. If the RCF = 0.5, then compositing results in a saving of 50% of the tests required if compositing is not done.

1.2.2. Estimation of incidence of a trait with composite sampling for confidentiality

When testing for a trait is likely to cause embarassment to an individual, confidentially can be assured by compositing samples from several individuals and only testing the composite. If the composite has the trait, then the individual(s) who have the trait are still unknown. Let p be the unknown incidence of the trait. A composite sample of n individual samples will exhibit the trait with probability $1 - (1 - p)^n$. Test a (large) number, m, of composite samples, and let x be the total and p_m be the fraction of these that exhibit the trait. Then

$$\hat{p} = 1 - (1 - p_m)^{1/n}$$

is an estimator of p with a positive bias (see Boswell and Patil, 1987).

1.2.3. Composite sampling to test for compliance

As in Section 1.2.1, it is desired to identify cases having a certain trait. The measurement is a continuous variable which must satisfy a certain criterion. All cases that do not satisfy this criterion are to be identified. If the compliance rate is high, considerable savings can be realized by composite sampling.

Let c be a criterion value that should not be exceeded. For example, the measure may be the level of a toxicant in fish or other food products which are routinely tested. If the level exceeds c, then the product can not be used.

For a composite of n samples, if any one sample exceeds c, then the composite sample of n individuals will exceed c/n. Of course, it is possible that none of the samples will exceed c and the composite will still exceed c/n. If the composite exceeds c/n, then it is necessary to test each individual sample. On the other hand, if the composite does not exceed c/n, then *none* of the individual samples need to be tested.

Let $F(x)$ be the distribution function of the individual samples; then the n-fold convolution of F, $F^{(n)}(x) = F(x) * F(x) * \cdots * F(x)$, is the distribution function of the sum. The probability that a composite sample of n individual samples exceeds c/n is $1 - F^{(n)}(c)$. The cost per individual is

$$\frac{F^{(n)}(c) + (n + 1)[1 - F^{(n)}(c)]}{n} = 1 + 1/n - F^{(n)}(c).$$

As n increases, $F^{(n)}(c)$ decreases to a limiting value of zero. The optimal composite size depends on the distribution of the sum.

1.2.4. Compositing to reduce variance in the presence of inexact analytical procedures

In the above discussions, the only variance of the estimator was that associated with the heterogeneity of the population and sampling. If the outcome of the test applied to a sample has an error with mean zero and variance σ_t^2, and if the population has a variance of σ^2, then the variance of $\overline{X}$, the average of n samples, is $(\sigma^2 + \sigma_t^2)/n$. The variance of a composite sample of n individual samples is $\sigma^2/n + \sigma_t^2$. Thus, the variance of the composite sample is larger than the variance of the average of n samples; however, the composite involves only 1 measurement. Further, if the variance σ_t^2 of the test is small in comparison to σ^2, the variance due to the heterogeneity of the population, then these two methods of estimating the mean may have variances that are not very different. If k composite samples of size n are formed, then the estimates based on the average of these k measurements will have variance $\sigma^2/(kn) + \sigma_t^2/k$ which can be significantly smaller than $(\sigma^2 + \sigma_t^2)/n$, and the cost of estimation may still be reduced.

1.2.5. Compositing to reduce the cost of estimating the mean

The usual method of estimating the mean of a stochastic process is to take n observations and calculate the sample mean. Nearly the same information can be obtained by mixing the n observations and taking a single measure from the composite sample. Some laboratory procedures are time consuming or costly. In such cases, being able to make a single measurement instead of n measurements, is a great benefit.

Let $X_1, \ldots, X_n$ be measurements on n independent samples, and let Z be the measurement of the composite of the n samples. Then

$$Z = \sum_{i=1}^{n} w_i X_i, \quad E[Z] = \sum_{i=1}^{n} w_i \mu = \mu, \quad \text{and} \quad V(Z) = \sum_{i=1}^{n} w_i^2 \sigma^2,$$

where w_i is the fraction of the composite sample coming from the ith sample. If each sample consists of the same amount of material, then $w_i = 1/n$ and $Z = \overline{X}$, and the estimator of μ based on the composite sample is exactly the same as that based on the n samples.

The procedure described above assumes the fractions $w_1, w_2, \ldots, w_n$ to be fixed and known. Examples of this kind of compositing include:

(1) Stratified random samples are of this form, showing that a measurement of the composite sample formed by combining the individual samples gives the same result.

(2) Compositing of soil samples to obtain the overall or average fertility of the soil. Other bulk samples include sampling of fertilizers, coffee beans and concrete mixes.

(3) Compositing of samples taken from waste sites to obtain the overall concentration of hazardous material.

(4) Compositing of filtrate (assuming a known amount of water is filtered) to estimate the abundance of various plankton species.

1.3. Estimation of the mean using composite samples of random proportions

Example 4 above could be a situation where a net is towed through a water body for a constant time period at a constant speed. The amount of water filtered depends on many random conditions, such as wind speed, wave height and water currents. The fraction of each sample in the composite is then a random quantity. The measurement on the composite sample is $Z = \sum_{i=1}^{n} W_i X_i$, where W_i is a random variable representing the fraction of the composite sample from the ith sample. This same formula holds if the composite sample is formed by combining only a part of a subsample ('increment') from each sample. Using this procedure, several composite samples can be formed from the same collection of samples. Figure 1 illustrates the formation of composite samples taken from 'increments' of the original samples. Subsamples taken from each composite sample and tests or measurements made on these subsamples are also illustrated (heterogeneity within all of these samples can be accounted for).

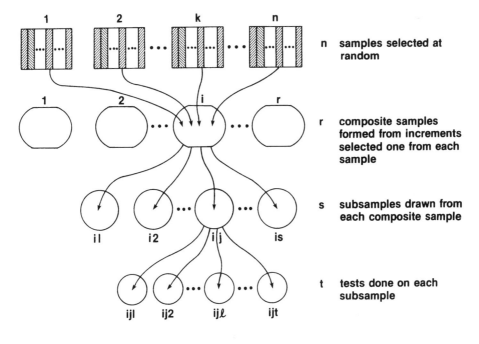

Fig. 1.

Rohde (1976) develops the theory for a single composite sample formed from n samples. The proportions of each sample used to make the composite sample are either fixed (known) or random. Elder (1977) (see also Elder, Thompson, and Myers, 1980) generalizes to r composite samples, each formed from n increments, and s subsamples taken from each composite sample. Further, t analyses or tests are done on each subsample. This generalization takes into account the variability in dividing the original samples into increments before compositing, variability due to non-perfect mixing of the composite samples and the selection of subsamples, and, finally, the variability of the test procedure itself.

The properties of the various composite-sample procedures are studied in the sections below. Formulas for the mean and variance of the measurement or average of measurements from composite samples are obtained. If the means of the observations vary, as they do in stratified sampling, then stratified means are estimated. On the other hand, if the means are constant, then these procedures give unbiased estimators of the common means.

It would be desirable to estimate the variance of the observations. This information is lost when composite samples are made. A possible solution is discussed in Section 1.3.2.

The means and variances of the composite-sample estimators depend on the distributions of the weights and the observations. Two statistical distributions for the weights are studied. Rohde argues for the Dirichlet distribution, and Elder (1977) formulates a model for the weights which gives the multivariate hypergeometric distribution which is the model originally used by Brown and Fisher (1972). Elder points out that both this distribution and the Dirichlet distribution converge to singular multivariate normal distribution under certain suitable conditions. He suggests this to be a reasonable approximation in many cases because of the physical averaging that occurs in blending a composite sample.

1.3.1. A single composite sample

Let $Z = \sum_{i=1}^{n} W_i X_i$ be the measurement on the composite where W_i is the random proportion of the ith sample in the composite and X_i is the measurement on the ith individual sample. We assume the W_i's and the X_i's are independent.

In the matrix notation $Z = W'X$. Let the means be $E[W] = \mu_w$ and $E[X] = \mu$; let the variance–covariance matrices be $C_w = \text{Cov}(W)$ and $C_x = \text{Cov}(X)$. Then $E[Z] = \mu_w' \mu_x$. Further, the variance of Z is

$$V(Z) = \mu_w' C_x \mu_w + \mu_x' C_w \mu_x + \text{tr}(C_w C_x).$$

1.3.2. Several composite samples

If composite samples are formed by taking subsamples, also called increments or aliquots, from the n original samples, the joint distribution of the fractions of the original measurements represented in the composite sample must have the properties given above. If $Z = W'X$ and $Y = U'X$ are two composite samples formed from aliquots of the original samples, then

$$\text{Cov}(Y, Z) = \mu_x' C_{u,w} \mu_x + \mu_u' C_x \mu_w + \text{tr}(C_{u,w} C_x)$$

where $C_{u,w}$ is the covariance $\text{Cov}(U, W) = (\text{Cov}(U_i, W_j))$.

In many cases, it is reasonable to assume the random weights are exchangeable random variables. Then $E[W_i] = 1/n$ and $\text{Cov}(W_i, W_j) = \sigma_w^2 \rho$ is a constant. This implies that

$$C_w = \sigma_w^2 \begin{bmatrix} 1 & \rho & \cdots & \rho \\ \rho & 1 & \ddots & \vdots \\ \vdots & \ddots & \ddots & \rho \\ \rho & \cdots & \rho & 1 \end{bmatrix}.$$ (∗)

As Rohde (1976) pointed out, the symmetric Dirichlet distribution has these properties. Now suppose that r measurements are all that can be made. We then construct r composite samples and compare the results with r measurements on original samples. As before, let $X_1, X_2, \ldots, X_n$ be the values associated with n samples. By subsampling r times from each sample we form composite samples by combining one subsample from each sample. Then $Z_i = W_i' X$ is the ith composite sample for $i = 1, 2, \ldots, r$. We compare $\bar{Z} = (1/r) \Sigma_{i=1}^r Z_i$ with the average $\bar{X} = (1/r) \Sigma_{i=1}^r X_i$ r of the original samples. Since

$$\bar{Z} = (1/r)(W_1 + \cdots + W_r)' X = V'X, \text{ say,}$$

$\bar{Z}$ represents a composite sample measurement and the formulas presented above apply. Both $\bar{X}$ and $\bar{Z}$ are unbiased estimators of μ. The variance $V(\bar{X}) = \sigma_x^2/r$. Let C_x, C_w and C_z be the variance–covariance matrices of X, W and Z, respectively. The variance of $\bar{Z}$ is

$$V(\bar{Z}) = \mu_w' C_x \mu_w + [\mu_x' C_w \mu_x + \text{tr}(C_w C_x)]/r.$$

Now assume $X_1, X_2, \ldots, X_n$ to be independent identically distributed with mean μ and variance σ^2. It follows from (∗), above, that

$$V(\bar{Z}) = \frac{\sigma_x^2}{n} + \frac{n \sigma_w^2 \sigma_x^2}{r} = \frac{\sigma_x^2}{n} \left[1 + \frac{n^2 \sigma_w^2}{r} \right].$$

Assume W to have a Dirichlet distribution with parameters **1**; then

$$V(\bar{Z}) = \frac{\sigma_x^2}{n} \left[1 + \frac{n-1}{r(n+1)} \right].$$

Also $V(\bar{X}) = \sigma_x^2/r$. The ratio of the variances is

$$V(\bar{Z})/V(\bar{X}) = \frac{r}{n} \left[1 + \frac{n-1}{r(n+1)} \right].$$

If n is large compared to r, then the variance of $\bar{Z}$ is much smaller than the variance of $\bar{X}$.

It seems that choosing n large in comparison to r would result in the best compositing situation. However, when a composite sample is made, often the physical mixing is imperfect, resulting in more variability. The optimal choice may also depend on the cost of sampling and the cost of analysis of samples.

Rohde (1976) points out that the sample variance of the composite samples is a biased estimator. In fact, $E[S_z^2] = n\sigma_x^2\sigma_w^2$. Rohde suggests two approaches. If an independent estimate of σ_w^2 is available, then S_z^2 can be used to estimate σ_x^2. The other approach is to assume some model, such as the Dirichlet distribution, which reduces the number of parameters that need to be estimated. For further details, see Rohde (1976) and Elder (1977).

1.3.3. Subsampling of the composite samples

The comparisons given above are based on the models discussed by Rohde (1976). Elder, Thompson and Myers (1980) generalize Rohde's work to allow for within aliquot variability. That is, when the composite samples are formed by subsampling from the original samples, different concentrations are possible. Thus, different composite samples can be formed from different random variables.

Further, Elder et al. also generalize to include the testing error. Suppose r composite samples are formed as before. From each composite sample, select s subsamples and run t analyses on each subsample. The estimator of the mean of the original population is the overall average. Let the result of the kth test from the jth subsample from the ith composite sample be represented by Z_{ijk}. Then $Z_{ijk} = \sum_{l=1}^{n} W_{ijl} X_{ijl} + \varepsilon_{ijk}$, where X_{ijl} depends on the composite index i. This assumption incorporates the variability from dividing a sample into r aliquots; this is called within increment variability. The variability of the testing procedure results in ε_{ijk}. The $r \cdot s \cdot t$ tests are assumed to have independent identically distributed errors with $E[\varepsilon] = 0$ and $\text{Cov}(\varepsilon) = \sigma_t^2 I$. The technique used to find the mean and variance of the estimator $\bar{Z}$ is similar to that described above. The final result is that $\bar{Z}$ is an unbiased estimator of μ_x with variance

$$V(\bar{Z}) = \frac{\sigma_x^2}{rn} + \frac{n\sigma_w^2\sigma_x^2}{rs} + \frac{\sigma_t^2}{rst}.$$

If the increments used to form the composite are either made up of discrete (equal sized) pieces or are divided into equal sized pieces then a hypergeometric distribution can be used to model the number of pieces from the various increments that make up the composite samples that finally end up in a subsample (Brown and Fisher, 1972; see also Elder, 1977). Assume each increment is composed of g equal-sized pieces and that each subsample is formed from a random selection of G pieces. Taking the limit as g and G go to infinity, while keeping g/G constant results in the distributions of the weights approaching a singular multivariate normal distribution, see Elder (1977).

The best values of n, r and t depend on many factors. Elder (1977) discusses this, pointing out that closed form solutions do not exist. Generally, if the cost of sampling is relatively small, then n can be much larger than r. If the within increment variability is small, then r is small. If the test procedure has small errors, then t is small. Often r and t can be taken as one or two.

1.4. Examples

1.4.1. Power plant effect on environment

Consider a situation presented by Rohde (1979) where a small scale study was made to see if compositing would be useful in studying the effect of power plant construction on the environment. For the Crane Power Plant near Baltimore City, data was collected on the density of plankton before entrainment. A transect near the intake to the plant was selected starting in deep water and ending in shallow water near shore. Samples were taken at six locations along the transect. The first two, in deep water, were sampled near the top and near the bottom. The density of plankton seemed to be smaller near the shore and near the bottom. All locations were sampled with equal effort by pumping water for 10 minutes at a rate of 37.8 gal/minute into a collecting net. The filtrate was washed into a 250 ml. bottle from which 1 ml. subsamples were selected. Actual measurements of the samples and of composite samples are given in Table 1. It is clear that about the same results were obtained from the composite samples as from the average of the original samples. The average of the eight observations was 203.2 organisms per liter while the density in the composite sample was 209.3 organisms per liter.

Table 1

Location	Density organisms/liter		Density of average/of composite	
1 top	269.3	260.8/253.3	211.8/219.0	203.2/209.3
1 bottom	252.3			
2 top	195.0	167.2/170.3		
2 bottom	139.3			
3	282.3	262.8/286.0		
4	243.3			
5	126.0	122.0/131.3		
6	118.0			

1.4.2. Heavy metal pollution

A study of heavy metal pollution of aquatic environments was reported by Hueck (1976). In this study, mussels from relatively unpolluted water were transplanted to various locations under study. After a period of time, tissue from the mussels was analyzed for heavy metal content. Since there was a great local variation within a region, pooling of results from adjacent localities seemed reasonable. Composite samples were formed by homogenizing the tissue of several specimens. The amount of heavy metal in the sample was used as an indicator of the pollution of the region. A map of coastal region of France, Belgium, Netherlands, Germany, Denmark and Great Britain giving the results for mercury is given in Hueck (1976). The measurements varied from 60 to 1650 ppb. Other maps show the regional change in mercury, copper and zinc.

2. Capture–recapture sampling

2.1. Introduction

Capture–recapture sampling is used in ecology and wildlife studies to estimate animal population size (N) and associated demographic parameters such as survival rate. The basic procedure is to mark and release a sample of individuals into the target population and then, assuming a thorough mixing of this sample with the population, recapture marked and unmarked individuals in one or more subsequent samples. Capture rates and marked-to-unmarked ratios are the basis for parameter estimation and inference.

In the simplest study, an initial capture effort results in the release of n_1 marked animals. A short interval of time later (relative to the population dynamics of survival, recruitment and dispersal), a second sample is captured, of size n_2. The second sample will consist of both previously marked animals (m_2) and unmarked individuals (u_2), thus $n_2 = m_2 + u_2$. Let p_2 be the average capture probability for the second sample. Assuming that marking has not affected recapture, then $E(n_2) = Np_2$, $E[m_2/n_1 | n_1] = p_2$, and the classical Petersen estimator of N is $\hat{N} = n_1 n_2 / m_2$. The Petersen estimator provides the heuristic basis of most estimators of wildlife population size, and from this humble beginning (Peterson, 1896) a very large capture–recapture literature has developed (see, e.g., Seber, 1982).

We use capture–recapture here in a generic sense to refer to any ecological methodology that uses the release and recapture of marked indiciduals in order to estimate a wildlife population size or a survival rate. Statistical models for these studies are said to be either closed or open. Often, the sampling occurs during a short enough time interval that the wildlife population can be assumed to be closed: no unknown additions or removals. Then the only parameter of interest is the population size, N. In contrast, the open population models emphasize estimating survival rates. For example, bird banding studies involve the yearly release of thousands of banded waterfowl. Bands are returned by hunters and these release-recovery data are used to estimate waterfowl annual survival rates, S. The statistical models for the band 'recaptures' are conditional on the known releases (see, e.g., Brownie et al., 1985). However, if we can also model the process of capturing unmarked individuals, which are then marked and released, we can also estimate yearly population sizes.

Capture–recapture does involve sampling populations. However, this methodology does not correspond well to the usual finite population sampling paradigm of statistics. Definition of the sampled population is a major problem. There is no frame, and estimation of N is often a primary study objective. For more critical than unknown N is that the sampling probabilities (p) cannot be determined in advance, nor can they be exactly determined after sampling. Rather, these probabilities must be estimated, and factors that can affect them must be considered in the models. Instead of classical statistical sampling theory, the basis for analysis of capture–recapture data is sampling models (primarily, binomial, multinomial, hypergeometric and multiple hypergeometric) incorporating some, or all, of the three types of parameters N, p and S. The central issue in the analysis of capture–recapture data is what model to use (see e.g., Cormack, 1979, p. 248).

2.2. Closed models

The literature on closed models for capture–recapture has been recently summarized by White et al. (1982). They elaborate a series of increasingly general models and provide a computer program to fit these models to data. In wildlife and ecology we can have $k = 5$ to 10, and sometimes more, capture occasions on successive days before the assumption of closure becomes untenable.

In the case of 2 occasions the basic statistical model for N is the hypergeometric distribution:

$$\Pr\{m_2 | n_1, n_2, N\} = \binom{n_1}{m_2}\binom{N - n_1}{n_2 - m_2} / \binom{N}{n_2};$$

here we condition on the initial releases n_1 and total captures at time 2, n_2. For a complete sampling model of the data we can start with

$$\Pr\{n_1, n_2, m_2 | N, p_1, p_2\} = \left[\binom{N}{n_1}(p_1)^{n_1}(1 - p_1)^{N - n_1}\right]\left[\binom{n_1}{m_2}(p_2)^{m_2}(1 - p_2)^{n_1 - m_2}\right]$$

$$\times \binom{N - n_1}{n_2 - m_2}(p_2)^{n_2 - m_2}(1 - p_2)^{(N - n_1) - (n_2 - m_2)}$$

which factors into two binomials and the hypergeometric given above.

Chapman (1948, 1951) has thoroughly investigated the hypergeometric model. In particular his recommended point estimator is $\hat{N} = (n_1 + 1)(n_2 + 1)/(m_2 + 1) - 1$. Confidence limits should be computed as N_L and N_U satisfying

$$\sum_{i=0}^{m} \binom{n_1}{i}\binom{N_U - n_1}{n_2 - i}/\binom{N_U}{n_2} = \alpha_U, \qquad \sum_{i=m}^{\min\{n_1, n_2\}} \binom{n_1}{i}\binom{N_L - n_1}{n_2 - i}/\binom{N_L}{n_2} = \alpha_L$$

(Chapman, 1948).

We illustrate this estimator with data used by Buckland (1984). The size of the workers component of an ant colony was estimated by initially capturing, marking and releasing 500 ants. A second sample had $n_2 = 189$ and $m_2 = 17$ recaptures: $\hat{N} = (501)(190)/(18) - 1 = 5287$. Using $\alpha_U = \alpha_L = 0.025$, the solution of the non-linear equations for N_U and N_L gives 3603 to 9330 as the 95% CI on N.

When there are multiple recapture occasions the model can allow for possible variations in capture probabilities due to factors such as time, animal behavior (response to capture) and individual differences in capture probabilities (heterogeneity). White et al. (1982) elaborate a sequence of models which incorporate these features. For example, if there are k capture occasions, the Petersen model generalizes to Darroch's model (Darroch, 1958) with parameters $N, p_1, \ldots, p_k$, where p_i = the capture probability on occasion i. The complete data are the capture histories of each animal captured, provided unique marks are used. A minimal sufficient statistic is $n_1, \ldots, n_k$ and M where n_i = the number of animals captured on occasion i and M = the total number of different individuals captures over all days. The maximum likelihood estimator is not closed-form. An approximation to the MLE is the solution of

$$M = N\left[1 - \prod_{i=1}^{k}\left(1 - \frac{n_i}{N}\right)\right],$$

which derives from $E(M) = N[1 - \Pi(1 - p_i)]$ and $E(n_i) = Np_i$. Complete theory on this model is in Darroch (1958).

Another common situation is removal sampling, especially electrofishing studies (White et al., 1982, Chapter 4). In removal sampling the data are $u_1, \ldots, u_k$, where u_i = number of unmarked individuals captured on occasion 1. Under the assumption that removal probability is constant, the data is multinomial:

$$\Pr\{u_1, \ldots, u_k | N, p\} = \binom{N}{u_1, \ldots, u_k}\left(\prod_{i=1}^{k}(1-p)^{i-1}p\right)^{u_i}(1-p)^{k(N - \Sigma_{i=1}^{k} u_i)}$$

The MLE is not closed form if $k > 2$. For $k = 2$, $\hat{N} = u_1/1 - (u_2/u_1))$. This removal model was first rigorously studied by Zippin (1956).

More complex closed models exist, in particular ones that allow for heterogeneity of individual capture probabilities (see, e.g., Burnham and Overton, 1978), and both behavioral response to first capture and heterogeneity (White et al., 1982). Program CAPTURE (White et al., 1982) incorporates all these models and considers model selection, and thus provides for near-comprehensive analysis of closed population capture data.

Log-linear methodology also provides a basis for comprehensive analysis of closed population capture data (Cormack, 1979). The two approaches overlap to a large extent, but not totally. There are log-linear models that have no counterparts in White et al. (1982); also some models of White et al. have no log-linear counterparts. However, our key point is that comprehensive model systems do exist for the analysis of capture–recapture data from closed wildlife populations.

2.3. Open models

It has been mostly since the early 1960's that a thorough foundation has appeared in the literature for the estimation of parameters based on capture–recapture sampling of open populations. One branch of the literature includes bird-banding and fish-tagging studies that yield data from a single, terminal, harvest-related recovery, as reviewed by Brownie et al. (1985). Another branch deals with multiple (live) recaptures of marked animals; we will refer to this as Jolly–Seber sampling (and Jolly–Seber models) after two of the leading contributors. Seber (1982, 1986) reviewed this literature. The links between these two major developments were outlined by Brownie et al. (1985). Until recently, the two approaches were developed separately, usually with different notations and contexts. Recently, however, there has been growing recognition that most open capture–recapture models are special cases of a more general theory. Now the entire subject of open populations, capture–recapture can be unified under one umbrella of theory (see e.g., Burnham et al., 1987).

2.3.1. Bird-banding

In a typical banding study a sample of birds is captured, banded and released just before the annual hunting season for a number of consecutive years. The data are symbolized as

N_i = the number of birds released (banded) in year i,
R_{ij} = the number of band recoveries from birds shot in hunting season j from birds released in year i, $i = 1, \ldots, k$; $j = i, \ldots, k$.

A convenient tabular representation is

Releases	Recoveries				
N_1	R_{11}	R_{12}	R_{13}	$\cdots$	R_{1k}
N_2		R_{22}	R_{23}	$\cdots$	R_{2k}
N_3			R_{33}	$\cdots$	R_{3k}
$\vdots$				$\ddots$	$\vdots$
N_k					R_{kk}

For example, from Brownie et al. (1985, p. 21):

Year	Releases	Recoveries		
1964	1603	124	44	37
1965	1595		62	76
1966	1197			82

These are data from a banding study on adult male wood duck.

The model structure under the assumption that recovery rate and survival rate parameters are time-specific only is

$$E(R_{ij}) = \begin{cases} N_i f_i, & j = i;\ i = 1, \ldots, k, \\ N_i S_i, \cdots, S_{j-1} f_j, & j = i+1, \ldots, k. \end{cases}$$

We assume that the k cohorts are each multinomial random varuables, i.e., $R_{ii}, \ldots, R_{ik}, N_i - R_i$ are mult$(N_i;\ \pi_{ii}, \ldots, \pi_{ik}, 1 - \lambda_i)$ where

$$\pi_{ij} = E(R_{ij})/N_i, \quad \lambda_i = 1 - \sum_{j=i}^{k} \pi_{ij}, \quad \text{and} \quad R_i = \sum_{j=i}^{k} R_{ij}.$$

The statistical theory of this, and other banding models, is well developed in, e.g., Brownie et al. (1985). The minimal sufficient statistic can be taken as $R_1, \ldots, R_k, C_1, \ldots, C_{k-1}$ where the $C_i = \Sigma_{j=1}^{i} R_{ij}$, are the column sums, $i = 1, \ldots, k$. Given these row and column sums, we define block totals $T_1 = R_1$, $T_2 = T_1 - C_1 + R_2, \ldots, T_{i+1} = T_i - C_i + R_{i+1}$, and $T_k = C_k$; T_i is all recoveries, at and after recovery year i, from birds released in, or before, year i.

The ML estimators and their sampling variances are

$$\hat{f}_i = \frac{R_i}{N_i} \frac{C_i}{T_i}, \qquad i = 1, \ldots, k,$$

$$\hat{S}_i = \frac{R_i}{N_i}\left[1 - \frac{C_i}{T_i}\right] \bigg/ \frac{R_{i+1}}{N_{i+1}}, \qquad i = 1, \ldots, k-1,$$

$$\widehat{\text{var}}(\hat{f}_i) = (\hat{f}_i)^2 \left[\frac{1}{R_i} - \frac{1}{N_i} + \frac{1}{C_i} - \frac{1}{T_i}\right], \qquad i = 1, \ldots, k,$$

$$\widehat{\text{var}}(\hat{f}_i) = (\hat{S}_i)^2 \left[\frac{1}{R_i} - \frac{1}{N_i} + \frac{1}{R_{i+1}} - \frac{1}{N_{i+1}} + \frac{1}{T_{i+1} - R_{i+1}} - \frac{1}{T_i}\right],$$

$$i = 1, \ldots, k-1.$$

For the wood duck example

i	R_i	C_i	T_i	$\hat{f}_i$	$\text{se}(\hat{f}_i)$	$\hat{S}_i$	$\text{se}(\hat{S}_i)$
1	208	127	208	0.079	0.0067	0.58	0.079
2	138	106	219	0.042	0.0045	0.63	0.094
3	82	195	195	0.071	0.0076	–	–

Numerous other models are well developed for analysis of band-recovery data. For birds banded as adults, one can consider the case of all annual survival rates being constant ($S_i \equiv S$) and the further restriction of constant recovery rates

($S_i \equiv S$, $f_i \equiv f$). Generalizations also exist to allow for possible response of birds to being captured, banded and released; primarily this could cause a different recovery or survival rate in the first year after release. Age-specific models exist for the case of two or more age classes being banded at the same time. Consideration of model selection and tests of assumptions are summarized in Brownie et al. (1985). There is a well-developed literature for the statistical analysis of band-recovery data.

2.3.2. The Jolly–Seber model

The extension of capture–recapture to fully stochastic, open models was not entirely successful until 1965 (Jolly, 1965; Seber, 1965). The Jolly–Seber model assumes time specific parameters; it is analogous to the time-specific banding model considered in Section 2.3.1. The complete specification of the model entails a substantial amount of notation; we present here only the summary statistics needed for parameter estimation. Moreover, we present the model in a manner that leads to a unified theory of open models, but using here the notation that has arisen in Jolly–Seber modelling (see, e.g., Jolly, 1965; Pollock, 1975). That notation is very different from the banding notation.

Let releases of marked animals be $R_1, \ldots, R_{k-1}$ at occasions $i = 1, \ldots, k-1$. The released R_i animals constitute a cohort from which animals are lost by either being recaptured, dying, or permanently leaving the study area. Let m_{ij}, $j = i+1, \ldots, k$ be the number of animals recaptured *for the first time* from the R_i releases at time i. Upon recapture at time j, an animal may be re-released as part of the cohort releases R_j.

As typically done, capturing and releasing occur simultaneously for k occasions. The parameters of interest are population sizes N_i and capture probabilities p_i, $i = 1, \ldots, k$, and ϕ_i the probability of surviving from occasion i to $i+1$, for $i = 1, \ldots, k-1$. The ϕ_i and p_i apply to all animals in the population at risk of capture on occasion i. The estimable parameters are $\phi_1, \ldots, \phi_{k-2}, p_2, \ldots, p_{k-1}$, $(\phi_{k-1}p_k)$, $N_2, \ldots, N_{k-1}$, $(N_1 p_1)$ and $(N_k p_k)$. The estimation of $N_2, \ldots, K_{k-1}$ requires assuming that marked and unmarked animals have equal capture probabilities, this often fails.

In addition to the releases, R_i, and (first) recaptures, m_{ij}, we define $u_1, \ldots, u_k$, the number of unmarked animals captured at times $i = 1, \ldots, k$. These summary data are conveniently represented as

i	u_i	R_i	m_{ij}
1	u_1	R_1	$m_{12}\ m_{13} \cdots m_{1k}$
2	u_2	R_2	$m_{23} \cdots m_{2k}$
$\vdots$	$\vdots$	$\vdots$	$\ddots\ \vdots$
$k-1$	u_{k-1}	R_{k-1}	$m_{k-1,k}$

The minimal sufficient statistic is the row and column sums, $r_1, \ldots, r_{k-1}$, $m_2, \ldots, m_{k-1}$ and the initial captures $u_1, \ldots, u_k$. Here $r_i = \sum_{j=i+1}^{k} m_{ij}$, $m_j = \sum_{i=1}^{j-1} m_{ij}$, $j = 2, \ldots, k$. We also define a block total, z_j, $j = 2, \ldots, k-1$, $z_2 = r_1 - m_2$, and in general, $z_{i+1} = r_i + z_i - m_{i+1}$, $i = 2, \ldots, k-2$. The total number of animals captured on occasion j is $n_j = m_j + u_j$ ($m_1 = 0$); the difference between captures and releases, $n_j - R_j$, is losses on capture.

The row and column sums here are exactly like the row and column sums for banding data. Also, the block total $m_j + z_j = T_j$ is exactly like the block total defined for banding data.

Given the releases, R_i, the $m_{i,i+1}, \ldots, m_{ik}$, $R_i - r_i$ are multinomially distributed with cell probabilities

$$\pi_{ij} = \begin{cases} (\phi_i p_{i+1}), & j = i+1, \\ (\phi_i q_{i+1}) \cdots (\phi_{j-2} q_{j-1})(\phi_{j-1} p_j), & j > i+1. \end{cases}$$

If one defines $\phi_i p_{i+1} = f_i$ and $\phi_i q_{i+1} = S_i$, and makes allowance for the different indexing here, then this capture–recapture model is identical to the band recovery model.

The MLE's are

$$\hat{\phi}_i = \frac{r_i}{R_i}\left[\frac{m_{i+1}}{T_{i+1}} + \frac{z_{i+1}}{T_{i+1}} \bigg/ \frac{R_{i+1}}{r_{i+1}}\right], \quad i = 1, \ldots, k-2,$$

$$\hat{p}_i = \frac{m_i}{m_i + z_i R_i/r_i}, \quad i = 2, \ldots, k-1, \quad \widehat{\phi_{k-1} p_k} = \frac{r_{k-1}}{R_{k-1}},$$

$$\hat{N}_i = n_i/\hat{p}_i, \quad i = 2, \ldots, k-1, \quad \widehat{N_1 p_1} = u_1 \equiv n_1 \quad \text{and} \quad \widehat{N_k p_k} = n_k.$$

Brownie and Robson (1983) give the Jolly–Seber estimators in this formulation (see also Burnham et al., 1987). Usually, however, the Jolly–Seber model is conceptualized in terms of N_i and M_i, M_i being the number of marked animals still alive at risk of capture just before capture occasion i. The Petersen estimator of N_i is $\hat{N}_i = n_i M_i/m_i$. A second application of the Petersen estimator uses $E[z_i/(M_i - m_i)] = E[r_i/R_i]$. The recaptures z_i all come from the $M_i - m_i$ animals marked in the population just after occasion i, and the capture probability of any of these $M_i - m_i$ animals is the same as that for the new releases of R_i animals. Hence, $\hat{M}_i = m_i + z_i R_i/r_i$, and $\hat{p}_i = m_i/\hat{M}_i$.

Just as with the banding models, there are many other Jolly–Seber (i.e., open capture–recapture) models. They are still proliferating and have not quite yet been systematized. Pollock (1975) presents a series of Jolly–Seber models which allow behavioral response to capture in both survival rates and capture rates. More recently, Pollock (1981) has produced Jolly–Seber models that incorporate age-effects. Brownie et al. (1986) consider models with survival and/or capture parameters constant over occasions. There is also the extension by Cormack (1981) of log-linear models to open model capture–recapture. Seber (1986) reviews recent developments in open-population capture–recapture models.

2.4. Summary and discussion

A large literature on capture–recapture in ecology and wildlife has developed in the past 100 years; most of it is from about 1950 and the true proliferation of the subject started in the mid 1960's. The literature for closed models is simpler, and more complete, partly because the only biological parameter of interest is one population size, N.

With open populations, the parameters of interest are survival, and population size; these are possibly different at each sampling time. When recapture is by harvest, as in bird-banding, only survival rate is estimable, then the number of models is tractable and a good comprehensive reference exists: Brownie et al. (1985). Capture–recapture on open populations has a very large and diverse literature. There has not yet been an accepted unification of this literature (let alone all open population capture literature). A cutting edge in capture–recapture research now is to produce such a unification, and statistical theory that is easier to use when developing new models.

Most of the models now in use for capture data require computer programs for their use. This will become increasingly true. A second cutting edge of research is to generate good computer programs for micro computers that allow the users to specify model structures. Flexibility and ease of use are the keys to good software implementation of currently available sophisticated capture–recapture models. There has been progress in this direction; see, for example, Arnason and Baniuk (1980), Clobert and Lebreton (1986), White (1983), Burnham et al. (1987), Cormack (1985), Conroy and Williams (1984).

We have a final set of comments on the extensions of capture–recapture models and methods to problems other than wildlife; there are numerous such applications and extensions. In ecology, capture–recapture models can be used to estimate species diversities (Burnham and Overton, 1979), and abundance of sessile objects such as nests (Magnusson et al., 1978). Both open and closed models have been applied to paleobiology problems of estimating species abundance and extinction rates (Nichols and Pollock, 1983).

Applications of capture–recapture modeling and analysis methods are frequently encountered in human populations in regards to estimating the size of 'hidden' subpopulations. Application areas include epidemiology and demography where the data base tends to be incomplete dual, or multiple record systems (El-Kharozaty et al., 1977). There has been estimation of the size of criminal populations from arrest records (Greene and Stollmack, 1981). The Census Bureau uses capture–recapture models to estimate the undercount problem (Cowan and Malec, 1986). Applications of capture–recapture methods are also found in numismatology (Chao, 1984), quality control (Jewell, 1985), vocabulary of authors (Efron and Thisted, 1976) and remote sensing (Maxim et al., 1981).

Capture–recapture methods motivated and initiated by seemingly simple wildlife problems have expanded to become a major class of statistical methods with application far beyond just the estimation of wildlife numbers.

Acknowledgements

Preparation of this paper has been partially supported by the NOAA research grants to the Penn State Center for Statistical Ecology and Environmental Statistics under the auspices of the Northeast Fisheries Center. The NCI grant CA 23415 to Dana–Farber Cancer Institute provided partial support to G. P. Patil. Our thanks are due to R. C. Hennemuth and C. Taillie for interesting discussions on various themes of the paper.

References

Arnason, A. N. and Baniuk, L. (1980). A computer system for mark–recapture analysis of open populations. *Wildlife Management* **44**, 325–332.

Becker, H. B. (1977). Composite sampling in aquatic environments. CM 1975/L:5, Plankton Committee, International Council for the Exploration of the Sea.

Boswell, M. T. and Patil, G. P. (1987). A perspective of composite sampling. *Communications in Statistics*. (To appear.)

Brown, G. H. and Fisher, N. I. (1972). Subsampling a mixture of sampled materials. *Technometrics* **14**, 663–668.

Brown, G. H. and Robson, D. S. (1975). The estimation of mixing proportions by double sample using bulk measurements. *Technometrics* **17**, 119–126.

Brownie, C. and Robson, D. S. (1983). Estimation of time-specific survival rates from tag-resighting samples: A generalization of the Jolly–Seber model. *Biometrics* **39**, 437–453.

Brownie, C. and Anderson, D. R., Burnham, K. P. and Robson, D. S. (1985). *Statistical Inference from Band Recovery Data—A Handbook*. 2nd edition. U.S. Fish and Wildlife Service Resource Publication 156.

Brownie, C., Hines, J. E. and Nichols, J. D. (1986). Constant parameter capture–recapture models. *Biometrics* **42**, 561–574.

Buckland, S. T. (1984). Monte Carlo confidence intervals. *Biometrics* **40**(3), 811–817.

Burnham, K. P. and Overton, W. S. (1978). Estimation of the size of a closed population when capture probabilities vary among animals. *Biometrika* **65**(3), 625–633.

Burnham, K. P. and Overton, W. S. (1979). Robust estimation of population size when capture probabilities vary among animals. *Ecology* **60**(5), 927–936.

Burnham, K. P., Anderson, D. R., White, G. C., Brownie, C. and Pollock, K. H. (1987). Design and Analysis Methods for Fish Survival Experiments Based in Release–Recapture. American Fisheries Society Monograph 5, American Fisheries Society, Bethesda, MA.

Chao, A. (1984). Nonparametric estimation of the number of classes in a population. *Scand. J. Statist.* **11**, 265–270.

Chapman, D. G. (1948). A mathematical study of confidence limits of salmon populations calculated from sample tag ratios. International Pacific Salmon Fisheries Commission, Bulletin II, 69–85.

Chapman, D. G. (1951). Some properties of the hypergeometric distribution with application to zoological sample censuses. University of California, Publ. Stat. **1**(7), 131–160.

Clobert, J. and Lebreton, J. D. (1986). User's Manual for Program SURGE, Version 2.0. C.E.P.E./L.N.R.S., BP 5051. Centre L. Emberger, Montpellier, France, 24 pp.

Conroy, M. J. and Williams, B. K. (1984). A general methodology for maximum likelihood inference from band-recovery data. *Biometrics* **40**, 739–748.

Cormack, R. M. (1979). Models for capture–recapture. In: R. M. Cormack, G. P. Patil and D. S. Robson, eds., *Sampling Biological Populations*. Statistical Ecology, Vol. 5, International Co-operative Publishing House, Fairland, MA, 217–255.

Cormack, R. M. (1981). Loglinear models for capture–recapture experiments on open populations. In: R. W. Hiorns and D. Cooke, eds., *The Mathematical Theory of the Biological Populations* II. Academic Press, London, 217–235.

Cormack, R. M. (1985). Examples of the use of GLIM to analyse capture–recapture studies. In: B. J. T. Morgan and P. M. North, eds., *Statistics in Ornithology. Lecture Notes in Statistics* **29**, 243–273. Springer, New York.

Cowan, C. D. and Malec, D. (1986). Capture–recapture models when both sources have clustered observations. *J. Am. Stat. Assoc.* **81**, 347–353.

Darroch, J. N. (1958). The multiple–recapture census: I. Estimation of a closed-population. *Biometrika* **45**(3/4), 343–359.

Efron, B. and Thisted, R. (1976). Estimating the number of unseen species: how many words did Shakespeare know? *Biometrika* **64**(3), 435–447.

Elder, R. S. (1977). Properties of composite sampling procedures. Ph.D. Dissertation, Virginia Polytechnic Institute and State University, Blacksburg, VA.

Elder, R. S., Thompson, W. O. and Myers, R. H. (1980). Properties of Composite sampling procedures. *Technometrics* **22**(2), 179–186.

El-Khorozoty, M. N., Imrey, P. B., Koch, G. C. and Wells, H. B. (1977). Estimating the total number of events with data from multiple record systems: A review of methodological strategies. *International Statistical Review* **45**, 129–157.

Feller, W. (1968). *An Introduction to Probability Theory and Its Applications, Vol. I*. Third Edition, Wiley, New York.

Garner, F. C., Stapanian, M. A. and Williams, L. R. (1986). Composite sampling for environmental monitoring. To appear in *Proc. Amer. Chem. Soc.* **139**, NACS Meeting, Principles of Environmental Sampling, April 1987.

Greene, M. A. and Stollmack, S. (1981). Estimating the number of criminals. In: J. A. Fox, ed., *Models in Quantitative Criminology*. Academic Press, New York.

Hueck, H. J. (1976). Active surveillance and use of bioindicators. In: *Principles and Methods for Determining Ecological Criteria on Hydrobiocenoses*. Pergamon Press, New York, 275–286.

Janardan, K. G. and Schaeffer, D. J. (1977). Sampling frequency and comparison of grab and composite sampling programs for effluents. Invited Paper for Satelite A, International Statistical Ecology Program, Berkely, CA.

Jewell, W. S. (1985). Bayesian estimation of undetected errors. In: J. M. Bernardo, M. H. DeGroot, D. V. Lindley and A. F. M. Smith, eds., *Bayesian Statistics 2*. Elsevier, New York, 663–671.

Jolly, G. M. (1965). Explicit estimates from capture–recapture data with both death and immigration-stochastic models. *Biometrika* **52**, 225–247.

Magnusson, W. E., Caughley, G. J. and Grigg, G. C. (1978). A double-survey estimate of population size from incomplete counts. *J. Wildlife Management* **42**(1), 174–176.

Maxim, L. D., Harrington, L. and Kennedy, M. (1981). A capture–recapture approach for estimation of detection probabilities in aerial surveys. *Photogrammatic Engineering and Remote Sensing* **47**(6), 779–788.

Nichols, J. D. and Pollock, K. H. (1983). Estimating taxonomic diversity, extinction rates, and speciation rates from fossil data using capture–recapture models. *Paleobiology* **9**(2), 150–163.

Petersen, K. H. (1896). The yearly immigration of young plaice into Limfjord from the German Sec. *Rep. Danish Biol. Sta.* **6**, 1–48.

Pollock, K. H. (1975). A K-sample tag-recapture model allowing for unequal survival and catchability. *Biometrika* **62**, 577–583.

Pollock, K. H. (1981). Capture–recapture models allowing for age-dependent survival and capture rates. *Biometrics* **37**, 521–529.

Rohde, Charles A. (1976). Composite sampling. *Biometrics* **32**, 273–282.

Rohde, Charles A. (1976). Composite sampling. *Biometrics* **32**, 273–282.
Rohde, Charles A. (1979). Batch, bulk and composite sampling. In: R. M. Cormack, G. P. Patil and D. S. Robson, eds., *Sampling Biological Populations*. Statistical Ecology, Volume 5, International Co-operative Publishing House, Fairland, MD, 365–377.
Roskopf, R. (1968). A composite-grab of water pollution control sampling. *J. Water Pollution Control Fed.* **40**, 492–498.
Seber, G. A. F. (1965). A note on the multiple-recapture census. *Biometrika* **52**, 249–259.
Seber, G. A. F. (1982). *The Estimation of Animal Abundance and Related Parameters*. 2nd edition, Macmillan, New York.
Seber, G. A. F. (1986). A review of estimating animal abundance. *Biometrics* **42**(2), 267–292.
Schaeffer, D. J. and Janardan, K. G. (1978). Theoretical comparison of grab and composite sampling programs. *Biom. J.* **20**, 215–227.
Schaeffer, D. J., Kerster, H. W. and Janardan, K. G. (1980). Grab versus composite sampling: A primer for managers and engineers. *Environmental Management* **4**, 157–163.
White, G. C. (1983). Numerical estimation of survival rates from band-recovery and biotelemetry data. *J. Wildlife Management* **47**, 716–728.
White, G. C., Anderson, D. R., Burnham, K. P. and Otis, D. L. (1982). Capture–recapture and removal methods for sampling closed populations. Los Alamos National Laboratory, LA-8787-NERP, Los Alamos, NM.
Zippin, C. (1956). An evaluation of the removal method of estimating animal populations. *Biometrics* **12**, 163–169.

Data-based Sampling and Model-based Estimation for Environmental Resources

G. P. Patil, G. J. Babu, R. C. Hennemuth, W. L. Myers, M. B. Rajarshi and C. Taillie

1. Introduction

Orthodox doctrine of sampling in applied statistics begins with the assumption of a well-defined population of interest, the study of which proceeds through acquisition of data having known probability of procurement defined by the sampling design. The key to this classical protocol for empirical extension of knowledge lies in the well-defined nature of the population and in the assigned probability of sampling its members.

Populations in nature, however, are defined by natural processes and one cannot quite construct the probability of encountering their members under arbitrarily assumed axiomatic behaviors. When populations are thus incompletely specified or incompletely accessible, it becomes necessary to use available data, and data that become available, for progressively more precise formulation and parameterization of population models. The population models, in turn, provide expectations concerning encounters that lead to estimators for parameters.

The purpose of the present paper is to provide insights into the approaches by which the interactions between data and design evolve into the enhanced knowledge of environmental attributes and ecological processes, while avoiding the cyclic problem of new information becoming non-information.

2. Data-based definitions of populations

Environmental investigations often begin at a stage where the kinds of entities that could comprise populations of interest and the extent of their occurrence are incompletely known. We begin instead with a universe of concern which may contain populations of interest. The universe of concern frequently has a strong spatial context, and can often be spatially delimited as a particular region of the biosphere. The potential populations of interest tend to be physical or biological phenomena or objects. For example, possible occurrence of stress conditions in organisms may be of concern as an indication of environmental degradation—even in the absence of prior knowledge regarding the kinds and numbers of organisms present.

Pursuit of such environmental concerns typically begins with a search for existing data that might provide indications of variation within the universe of concern. There is, of course, the realization that encountered data probably include components of variance that are not of present interest. Nevertheless, possibilities that the variance structures reveal differences of interest usually make such data sets deserving of careful examination. Interest lies in statistical techniques that allow the investigator to determine which aspects of encountered variation are relevant (information) and which are extraneous (noise).

In a sense similar to the use of the term 'generic' in referring to drugs, we might think of generic data in the present context as being data likely to have utility for several types of problems in environmental analysis. For example, modern technologies have made remotely sensed data one of the most readily available generic sources of environmental information. The generic character of remotely sensed data arises from the fact that spectrally specific reflectance of environmental surfaces is largely determined by the composition and condition of those surfaces. The data are 'encountered' in the sense that there is seldom an opportunity to have the sensors custom configured in terms of spectral bands, resolution, and time of data collection. Remote sensing thus provides an illustration of spatially specific data that may be encountered by the environmental scientist concerned with determining possible occurrence and extent of phenomena that might indicate need for managerial or regulatory action. One strategy for statistical analysis puts the scientist precisely in the position of asking what population the sample represents.

2.1. A null hypothesis addressed in a meta-analytical mode

A simple null hypothesis can be posed as a point of beginning for analysis. It is that there are no differences in the environmental surfaces exposed to the sensor over the region of concern. Quantitative evidence bearing on the acceptance or rejection of this null hypothesis resides in spectral measurements made by the sensor. Since sensors usually make simultaneous measurements in several spectral bands, the data have a multivariate character. The decision to accept or reject the hypothesis is largely subjective, however, due to the lack of ability to construct a probability-based test criterion. Nevertheless, the rather wide array of quasi-statistical techniques encompassed by the term 'cluster analysis' provides an assessment of possibilities for arranging the spectral observations in groups that are 'well-separated' according to some index of discrimination. Since the grouping is done with the intent of creating groups or 'clusters' that are well-separated by subjective factors of choice, there is no probability basis for declaring that the degree of grouping so observed is meaningful. The spatial specificity of the spectral measurements coupled with the tendency of environmental phenomena to exhibit spatial contagion provides a means of partially avoiding the specter of circular reasoning. Before pursuing this avenue of inquiry, however, it would be well to note the need for subsampling the remotely sensed data in connection with cluster analysis.

The resolution level of modern remote sensors provides a capability for generating millions of observations over areas of sizes routinely studied for purposes of environmental analysis. A moderate level of resolution, for instance, involves one observation for each 30-meter square on the environmental surface. The more sophisticated algorithms for cluster analysis typically require all possible comparisons between subgroups of observations, and several even require all possible comparisons over the entire set of observations. Even though remote sensors typically provide complete areal coverage in terms of observations, it would clearly be a monumental task for the largest computers to run such a data set through one of these algorithms in its entirety. Therefore, it becomes necessary to take a sample of the remotely sensed observations in order to make cluster analysis practical from a computational standpoint.

Another possibility for reducing computational burden lies in data compression relative to variates. It is commonly observed that measurements made in adjacent spectral bands are highly correlated. Principal component analysis serves to compress information onto fewer variates by segregating correlated aspects of variation. This often has a beneficial effect on the cluster analysis by removing implicit weighting arising from collinearity among variables. A further benefit is often realized from the action of principal component analysis as a noise filter.

For reasons indicated earlier, cluster analysis is more descriptive than inferential. Ecologists are well aware, however, that environmental phenomena tend to be contagiously distributed in a spatial sense. If the clustering could be made to encompass the entire set of spectral observations, then one might hope to distinguish environmentally induced aspects of clustering from artifacts of random variation according to spatial pattern. When the clusters are depicted on an image display device in correct spatial perspective, one would expect to find environmentally induced clusters occurring in a patchwork arrangement whereas artifacts of random variation should also be more or less random in their spatial distribution. We are reminded at this point that spatial patterns will be poorly expressed, if at all, when cluster assignments have been made for only a sparse sample of spectral observations.

In order to extend the assignment of cluster membership beyond the original sample for which clustering was performed, the analyst can treat the clusters as class prototypes and invoke statistical classification methods. A suite of such classification methods is available, differing primarily with respect to assumptions regarding distributional differences among classes. The less restrictive the assumptions, the more computationally complex (and consequently expensive) the algorithm. After extending the clustering through classification, the analyst is able to view an image-like representation of spatial distribution in an index of spectral non-uniformity over the environmental surface. If the display indicates a perceivable presence of spatial pattern over the environmental surface, the analyst would reject the null hypothesis of environmental uniformity and would proceed to inquire regarding the causal basis for the perceived differences. It would be possible to conduct probability-based tests of nonrandomness in the spatial distribution of clusters by sampling the map through randomly located quadrats in parallel with the methods used by field ecologists. Perceived as discontinuities or intermittancies, spatial contagion is usually of such a degree as to be obvious. What we need is statistics to elaborate the not so obvious.

2.2. Sampling to identify populations

The foregoing scenario has presumably served to partition or stratify the universe of concern, with the partitions being relatively homogeneous in terms of ecological setting. The presumption of ecological homogeneity is, of course, based on similarity of spectral expression in the domain of the sensor. This presumption in turn becomes a null hypothesis of no differences within cluster unit to be tested objectively. Likewise, knowledge is lacking as to the specific ecological character of any given cluster unit. At this stage, then, the clusters become potential populations of interest, and the task becomes one of collecting data to determine the ecological identity of each cluster and decide whether it warrants further study.

In the terminology of sampling frames, each cluster becomes a stratum. With respect to sampling scenarios, the situation is essentially one of double sampling for (sub)stratification. The determination to be made on a sampled unit is a decision whether or not that unit is of further interest. If the proportion of sampled units in a stratum is less than some predetermined value, then the stratum as a whole will be deemed not to constitute a population of interest. In order to localize the effort involved in obtaining sample data for a stratum, it may be appropriate to treat spatially disjoint patches as primary sampling units in a multistage design.

It is possible that at this point the situation assumes the character of a normal sampling design problem in preparation for collection of new data to answer the questions posed. Alternatively, there may be a further appeal to existence of higher resolution remotely sensed data available from archives of agencies. Very possibly it will be in order to obtain photographic images from which objects can be identified visually by a human interpreter as opposed to computer-processed multispectral data (multivariate spectral measurements made by nonphotographic sensors).

Having located the populations of environmental interest in this manner, the investigator proceeds to focus on each of these populations individually and elaborate a data acquisition scheme for completion of the analysis. These data acquisition schemes may be, in turn, multiphase and/or multistage—possibly involving several different kinds of determinations made from encountered data. An important point is that encountered data should be incorporated in such a way that it does not constrain the capacity for ultimate inference.

3. Sampling design as design of encounters—The case of living marine resources

Study of marine ecology has been strongly motivated by the development and conservation of fisheries and the longstanding value to society of fish as food. The fishermen themselves sample the resource, using a variety of gear, and were the first to observe the inherent variability of the success of such encounters. They designed their encounters to maximize the catch per unit of fishing time and effort. The fisheries biologist's first knowledge of the distribution and abundance of the resources came from observations of the fisherman's practices and results.

The desire to better define the fluctuation in resource abundance and the reasons for it inevitably led to the conduct of sampling by the researchers themselves, which would overcome the recognized biases of the fishermen's catch. The rapid post-World-War-II expansion of fisheries elevated the concern for conservation of the resources and management of fisheries, and led to increased resource 'survey' activity. The development of biometrics in agriculture was adopted as the basis for designing and analysis of the surveys. The same approach was also taken for studies of most other natural living resources such as forest and wildlife, but at least the degree, if not the fundamental nature, of difficulties of encountering and observing marine resources is different from that of observing terrestrial resources (Darr and Hennemuth, 1985).

The research surveys of marine resources cover all life stages—adults, eggs, larvae, and juveniles. Different types of sampling gear are used for each of the life stages and for different species, depending on the habitat occupied and the behavioral characteristics. For adults, the gear is mostly an adaptation, to some degree, of that used in commercial fisheries. For other life stages, the gear has usually been developed to meet variously defined standards of representativeness and consistency constrained by physical and logistical factors. Thus, the design of the sampling gear is the initial, and important, stage of designing encounters with the organisms, the characteristics of which we want to observe.

The second stage of design has been to determine how the gear is to be employed in time and space. The first and second stages described here involve some of the same factors, but the second stage is more related to the consideration of precision, whereas accuracy is taken to have been solved in the first stage—e.g., the size or species bias. The second stage is then concerned with selection of a statistical design of sampling and the associated data analysis and inference. Traditional developments in current statistical texts on design and analysis of surveys and experiments are usually applied.

These surveys have proven very useful in providing a fishery-independent measure of the relative magnitude and biological attributes of the stocks. The surveys have met the need for comprehensive qualitative descriptions; and the inclusion of species for which there is not a directed fishery and, hence, poor fishery statistics, has been particularly valuable. Use of the survey data for quantitative analysis has been more difficult; the validity of statistical inferences has frequently enough been questioned, and with some justification.

The areas sampled are large in most cases. The sampling fractions considered only in two dimensions (bottom or surface area) are often of the order of 10^{-6}. The relatively high variability of point estimates (coefficients of variation of 100% are not unusual), which seems characteristic of most survey data is, perhaps, tolerable providing that it does reflect 'true' error variance. In other words, our observations are a representative sample for the hypothesized model. We are then imprecise, but can draw inferences that are at least vaguely right. What is certainly less tolerable is to be inaccurate, no matter how precise. This latter phenomenon is evident at times, and we can suitably qualify our conclusions, but even infrequent large and unexplainable deviations cast doubt on the soundness of our advice based on such data and the data analyses.

The process of 'designing' the surveys must, therefore, deal with the assumption of traditional, industrial statistics of design and analysis. We may need to have protocols for assuring consistency in our sampling gear and the way it is applied, so that bias due to this factor can be minimized. This will not, however, suffice for designing our surveys. A purpose of this section is to try to define what might suffice. This will be done in the context of encountered data in marine resource surveys. Improved ecological knowledge is really necessary for 'solving' the problem. In the mean time, there is a need for methods that deal with unpredictable and seemingly sporadic encounters.

Ancillary observations of the physical environment are made while sampling the resources, and direct surveys are also taken to provide information on environmental factors which might affect their distribution and availability. Very little attention is paid to the statistical distribution of these natural physical variables—e.g., temperature, water density, currents. Much more concern has been directed to determining concentrations and distribution of pollutants. Many of the same difficulties of encountered data apply to these aspects.

3.1. The issues

In marine ecology, the primary issue is how to develop an appropriate basis for drawing valid conclusions from the observations we have obtained. This applies to experiments, to surveys, and to monitoring, nearly equally. Of principal concern is accuracy or bias; in the context of this article, precision is involved to the extent that eliminating bias enables one to estimate error variance correctly.

In marine ecology, we are faced with a very dynamic physical and biological system over which the scientist has no means of control. All observations are taken under a different set of state conditions. For the most part, the targets are moving in an opaque fluid. The system may be said to be variable only in the sense that weather and people's reaction to it are said to be variable, i.e., there are physical and biological laws operating, but the multiplicity of causal factors and interactions make the system appear stochastic.

The unexplained variance—really differences between what we observe and what we might expect cannot be incorporated in a stochastic variable in the sense of 'noise'. This would be as great a mistake as assuming that an observation came from a set of replicable events. With attribute sampling, a binomial error term (or trinomial) is assumed. Often, however, there remains an 'extraneous' portion of variation not accounted for by the model (Moore, 1987). The extraneous variation has been found to be upwards of 70% in samples of length classes.

Thus, the factors which affect catchability (or sightability, etc.) must be incorporated in the statistical analysis in a manner which distinguishes between accuracy and precision.

Most living organisms are found in clusters; i.e., they have a spatial dimension which results from some fundamental behavior mechanisms reacting to environmental factors. Statistical distributions were developed to express the behavior of variables in relation to an artificial set of postulates, and they cannot be expected to describe spatial attributes of organisms.

Our observations of an organism's density are a consequence of the encounter of the sampling gear with the aggregations of the organism. We may capture all or part of the aggregation depending on the relative size of the gear to the aggregation, or all or parts of several aggregations. The aggregation, and aggregations of aggregations, change their characteristics with changing activities—feeding, spawning, etc.—and changing magnitudes of the population. Environmental physical factors will also affect them.

If our 'sampler' is large relative to the aggregations, there may be a tendency to approach a (0, 1) variable (i.e., we are likely to encounter all of the aggregation or none at all). At another extreme, say with fewer, very large aggregations, the same sampler, now relatively smaller, would tend to generate a more continuous variable by including parts of the larger aggregations. The possibilities are infinite. The tendency is generally to fix the sampler (and method of application), thus we cannot detect such changes.

Much of the survey activity leads to a time series, from which inferences about the change on some process are drawn. Seasonal and annual series are common. It is often impossible to 'randomize' observations on the time series and the alternative is to fix a time of observation relative to some attribute of the environment or activity of the organism.

The natural—and now increasingly man-induced—events that effect real changes in populations and also the accuracy of observations may indeed be repetitious, but operate on various time and space scales in varying degrees. Thus, the seasons always come and go, and within less than epochal time scale, it will be warmer in summer than winter. However, each month, year, and decade will likely be different enough so that the biological response will be significantly altered.

The statistical issue is how to analyze the data when fixed samples are taken from a variable event. The question of what is a random independent sample in these circumstances seems moot. How can valid statistical inferences be drawn?

Thus, the sampling and analysis must begin from the perspective that the effect we are trying to estimate is not observed randomly over replicate conditions. Instead, there is a nested set of repetition of events, each varying with its own pattern that is, by and large, unpredictable. That does not mean that the conditions or state of the system is not foreordained. We are observing the results of both the biological and physical processes, and their interactions—i.e., ecology. Under these circumstances, the traditional concepts of error, mean square, bias, and randomness, must be modified to deal with the reality. This comes down to correct identification of expectations, and using statistical models and analysis which fit the observed conditions rather than the other way around.

3.2. Some examples

Most marine organisms have a cyclical pattern of reproduction. The period may be several cycles per year for phytoplankton; a more continuous activity through the year for tropical fish; or shorter-term, distinct, spawning, egg, and larval stages within a year for polar and temperate fish.

A variable of great interest is the number of survivors of an annual spawning which 'recruit'—i.e., become available—to the fishery. The annual recruitment determines to a great extent the possible fishery yields over the near surface. There are two aspects of particular interest (Hennemuth, 1979):

First, the short-term advice to fishery managers is based on estimates of the fishable biomass which depend on the recruitment estimates of mortality, and gain in fish weight. In the longer run, determining the causal factors for success of recruitment, particularly the effect of spawning stock, has been a traditional concern. Lacking knowledge of cause of yearly differences in recruitment, the time series may provide knowledge of the expected year-to-year changes and longer-run consequences in a probabilistic context (Hennemuth and Avtges, 1982).

One gets only one observation per year, however, and the number of observations in the time-series is limited—generally well below that which is dear to a statistician's heart. Further, the multiple factors which cause each year's results are difficult to observe, and replicability of them is an unrealizable concept.

One attempt to provide a valid statistical analysis is described in Section 6.

Seasonal trawsl surveys (inventories) of fish have been conducted off the northeast coast of the U.S. for many years. These are stratified random trawl hauls. The operational limitations of one vessel over 72 000 n.m.^{2} reduces spatial and temporal synopticity relative to the more remote sensing which introduces this article. The spatial density distribution of the fish is unknown in advance; in fact, it is unknown in general, except that it is 'contagious'. Doubtless, we are faced with nested distributions, in the 'burst' context described by Mandelbrot (1977). The aggregates are the result of the biological behavior—pairs at spawning, aggregation of pairs in physically optimal locales, aggregations of these aggregations over the species-preferred regions. Furthermore, the spatial density distribution is a variable itself, depending on environmental conditions that change in time and space.

Catchability of the trawl for each species, the age (life stage) of the fish, and activity of the fish—e.g., feeding, spawning, etc.—different. This is generally a seasonally changing phenomenon.

The season has been defined by conducting the survey within a fixed time, e.g., October and November in the 'autumn'. One knows, however, that the 'seasonal' factors which control the biological behavioral responses of the fish are different from year to year, and predictable only within some time span which would encompass the whole range of possible seasonality.

Thus, there is some 'probability' distribution (a time series) of the time of the start of a season, and another probability distribution of the rate of change of events and their duration. One possible approach to designing our analysis, if not encounters, would be to incorporate these distributions.

An interesting consequence of seasonal variability is illustrated by the Georges Bank haddock. Schuck (1949) had used the observed seasonal (intra-annual) decrease in commercial catch per unit effort to estimate the actual weight of a year class removed by fishing. Subsequent analysis of survey catches indicated, however, that the catchability of haddock changed seasonally, and that this change was different for different age groups (Hennemuth et al., 1985). Thus, a significant part of the observed intra-annual decrease in catch per unit effort was due to catchability and not the removal of fish from the population.

Further analysis has indicated that intra-annual, seasonal variation in catchability is not consistent from year to year. This would preclude combining seasonal surveys unless additional information about causal factors can be obtained.

It is important to consider that the probability density functions which represent instantaneous spatial distributions contain variables (as parameters or functions) to accommodate changes through time, in environment or population states. This may be in the obvious-but-impossible category.

4. Sampling design as design of encounters—The bias in fisheries harvest data

4.1. Introduction

Non-response is a rather common feature in surveys of human subjects. One of the usual assumptions for addressing the problem of non-response, is that the characteristic of the population members on which information is being sought, does not differ between the groups of respondents and non-respondents. This allows us to regard the sample of those who respond as a random sample (of reduced size) from the entire population. Another problem which sample survey scientists often face is the possibility of getting a biased or inaccurate response. The combined effect of these two factors is that the mean of the numerical characteristic under investigation cannot be estimated by standard methods.

Indeed, we need to model characteristics that make a population member respond and/or give a biased response (if selected in the sample). What we observe is not a random sample. The event of response itself has a chance mechanism that may be related to the quantity being investigated. Statistical scientists need to modify both the interpretation of the sampling design and the estimation strategies. Following is an account of such an attempt in a survey of fishermen who report their commercial catch and data related to fishing efforts.

The Maryland Department of Natural Resources and the Virginia Marine Resources Commission conduct sample surveys to estimate the total catch of commercial as well as recreational fishery. These estimates find a number of important applications in resource management: providing indices of species abundance; documenting economic value; modelling fish population dynamics; etc. See Bonzek, Myers, Parolari and Patil (1986).

Agencies in charge of the surveys send mail questionnaires to a randomly selected sample of fishermen on a periodic basis. The samples are drawn by following a stratified sampling design, strata being based on license type and counties of residence of fishermen, so that no fisherman receives the questionnaire in two consecutive months.

A rather crude picture, that emerges from a study of the surveys over the years is that a number of fishermen do not respond at all (the non-response rate varies from 20% to 40%), and that among those who report, some report zero catch. Under-reporting is also suspected.

The commercial fishery survey offers two rather significant departures from the standard survey approaches. These amount to regarding the sample of fishermen who respond as a random sample (of a reduced size) from the target population. The fishing pattern of non-respondents may be dramatically different from that of the respondents.

Under-reporting by fishermen who report may have to do with an imagined connection with the federal and state government agencies responsible for regulation and revenue. Even when the surveying agency assures confidentiality of reported catches with the associated information on fishing effort, an application of randomized response survey may be in order.

Warner (1965), in a pioneering paper, introduced randomized response surveys to increase the response rate in surveys which involve questions that are sensitive either because of a social-religious stigma attached to them or because of possible legal implications of the response. Warner's technique can be illustrated as follows. Suppose we are interested in estimating the proportion of persons who submitted an incorrect tax return last year. The randomized response questionnaire has two questions:

(1) Did you submit an incorrect tax return last year?
(2) Did you submit a correct tax return last year?

A respondent selects (1) with probability p and (2) with probability $q = 1 - p$, p being known to the survey management. It is expected that the respondent, once he selects a question randomly, responds truthfully, in view of the assurance of confidentiality (since the interviewer/management does not know the question the respondent is answering). The observed proportion of 'yes' answers in the sample is linearly related to the unknown proportion π of those who submitted incorrect returns last year. This allows a numerical estimate of π.

Randomized response technique known as *unrelated question design* is due to Horwitz, Shah and Simmons (1967). Again, the respondent selects either of the two questions based on the outcome of a random experiment. However, only one of the questions is related to the sensitive character and the other is not. The true proportion of 'yes' answers to the unrelated question would be either known or can be estimated from an independent survey. Unlike Warner's technique, the respondent has a feeling of assurance that he answered in a way, an absolutely unrelated question, resulting in more protection than Warner's technique.

Since Warner's introduction of the randomized response technique, there has been extensive work on statistical and logical aspects of randomized response techniques. (See Emrich (1983) and Fox and Tracy (1986). If the answer to the sensitive question is a numerical score, see Pollock and Beck (1976), Duffy and Waterton (1984), and others.) These techniques include extension of unrelated question design to the case of numerical response. In these methods, the respondent is asked to add or multiply the true score by a random number and report the final number.

Success of randomized response technique depends upon several factors including how the respondent views the randomization device, and assesses the degree of protection offered to him. Since the randomized response technique estimates have a larger standard error, we need larger sample size. Also, a respondent may not view the survey as a serious attempt to obtain relevant information because it incorporates something so very extraneous!

It appears that in these surveys, community interest and mistrust are at issue, and not a procedural guarantee of *personal* confidentiality. A statistical technique which assures the respondent of confidentiality or offers anonymity may be successful in sociological surveys on illegal abortion, drunken driving, child abuse, etc. But, if the respondent feels that the data gathered through such surveys are likely to be used against himself or his community, then the success of randomized response cannot be taken for granted.

5. Sample-based modeling of populations—An approach with weighted distributions

5.1. Introduction

The concept of weighted distributions can be traced to the study of the effects of methods of ascertainment upon estimation of frequencies by Fisher in 1934. It was formulated by Rao (1965) as a problem of specification of the model for observed data. Since then, weighted distributions have served as a very useful tool in the selection of an appropriate model for the observational process, especially when samples are drawn without a (proper) frame. See Patil and Rao (1977), Patil (1981), Rao (1985), Patil, Rao and Zelen (1987) and the cited references for interesting discussions on weighted distributions and their applications.

Let $w(x, \alpha) \geq 0$ be a weight function, and let $f(x; \theta)$ be a probability density function (pdf), such that

$$\int w(x, \alpha) f(x; \theta) \, dx < \infty .$$

The distribution with pdf

$$f^w(x; \alpha, \theta) = \frac{w(x, \alpha)f(x; \theta)}{E[w(X, \alpha)]} \tag{5.1.1}$$

is known as the weighted version of the distribution with pdf $f(x; \theta)$. Weighted distributions corresponding to discrete distributions can be similarly defined. The weight function $w(x, \alpha)$ is considered to be a factor explaining the probability by which the unit with value x enters the records/observational process. For example, the weight function $w(x, \alpha) = x^\alpha$ with $\alpha = \frac{1}{2}$ was found to give a good fit to data on the number of albino children in a family ascertained through affected children (Rao, 1965, 1985). The weight function $w(x, \alpha) = x$ (i.e. $\alpha = 1$) gives a distribution which is known as the *size-biased* distribution of X. This weight function has been found applicable in many diverse areas of applications (Patil and Rao, 1977, 1978; Zelen and Feinleib, 1969).

In the above formulation, it is *understood* that the weight function does not depend upon the parameters of the random variable X. In other words, $w(x, \alpha)$ completely describes the properties of the observational process. Recently, we have discovered an application of weighted distributions where it is fruitful to extend the concept of the weight function by allowing it to depend upon θ, a parameter of the distribution of X. Thus, the probability that a value x enters its data base depends upon the relative position of x vis-à-vis θ. As a result, we are able to extend the scope of weighted distributions where we model the observational process in relation to the parameter-related characteristics of the distribution. Models of overdispersion discussed by Diaconis and Efron (1985) and Efron (1986) provide an interesting application as briefly discussed below.

5.2. Double exponential family of distributions

Diaconis and Efron (1985) introduce a double exponential family of distributions in order to model overdispersion. Let

$$f_{\mu, n}(x) = \exp\{n(\eta x - B(\mu) + C(x))\} = f(x) \tag{5.2.1}$$

be the pdf of a one-parameter exponential family of distributions with mean μ. Diaconis and Efron define the double exponential family corresponding to this distribution by

$$g_{\mu, \theta, n}(x) = c(\mu, \theta, n)\theta^{1/2}[f_{\mu, n}(x)]^\theta [f_{x, n}(x)]^{1-\theta} = g(x) \text{ (say)}. \tag{5.2.2}$$

The parameter n in (5.2.1) and (5.2.2) is to be regarded as a sample size parameter. Rewrite (5.2.2) as

$$g(x) = \frac{w(x, \mu, \theta)f(x)}{E[w(X, \mu, \theta)]} \tag{5.2.3}$$

where
$$w(x, \mu, \theta) = [f_{x, n}(x)/f_{\mu, n}(x)]^{1-\theta}. \tag{5.2.4}$$

The distribution $g(x)$ is a weighted distribution with a weight function depending on the parameter μ of the original distribution.

We now discuss the implications of the weight function (5.2.4) in terms of the selection bias.

Notice that for the family $f_{\mu, n}(x)$, x is the unique maximum likelihood estimator of the mean μ. Thus, if $\theta < 1$, $w(x, \mu, \theta) > 1$ for all x. This offers an interesting interpretation of the density (5.2.2). Having realized x, imagine that one carries out a likelihood ratio test for the hypothesis that the true mean is μ and incorporates the observation with a probability proportional to weight function (5.2.4). That is, the observations which reject the true parameter value μ, are selected in the sample with a high probability; further, the smaller the p-values of these observations, the higher is their chance of being recorded in the sample.

More directly, this can be seen via $I(x, \mu)$, the Kullback–Leibler distance function between x and μ. For, it is known that

$$\frac{f_{x, n}(x)}{f_{\mu, n}(x)} = \exp\{nI(x, \mu)\} \tag{5.2.5}$$

(see Efron, 1986, p. 712); so that the weight function (5.2.4) reduces to

$$w(x, \mu, \theta) = \exp\{n(1 - \theta)I(x, \mu)\}. \tag{5.2.6}$$

Thus, the distribution (5.2.2) admits the values x in the sample space with probability proportional to their distance from the true parameter. The larger the distance, the larger is the chance of these observations being recorded. This explanation is entirely consistent with the usual understanding of the weighted distributions, in that the probability of recording an x is not the same for all x in the sample space. See also Royal and Cumberland (1981a, b) for some discussion on randomization and non-randomization.

In this connection, it is interesting to note that in his analysis of the data on disease *toxoplasmosis* in 34 cities of El Salvador, Efron (1986) argues that 'genuine random sampling was infeasible and the subjects may have been obtained in clumps'. In the data sets that he discusses, the frequency of very high counts as well as the frequency of very low counts is larger than expected when compared to the binomial and Poisson distributions. While carrying out the regression analysis of these counts on a covariate, such as the rainfall in his case, we need to attend to the selection bias. Efron's approach does this by bringing into the model a weight factor which accounts for possible non-random sampling.

6. Combining recruitment data and kernel approach

6.1. Background

For several years the Northeast Fisheries Center (NEFC) has been assembling recruitment series for a large number of oceanic fish stocks. Recruitment is defined by the number of fish of 'catchable' size entering a fish stock. Estimation of recruitment distributions is important for the assessment and prediction of long term frequencies of good and poor year classes. In this connection, several parametric distributional models have been fitted to each of the available recruitment data sets (Hennemuth, Palmer, and Brown, 1980; Patil and Taillie, 1981). The small sample sizes prevented reliable assessment of goodness-of-fit. It also proved difficult to effectively discriminate between competing models, e.g., between the gamma and the lognormal distribution.

In view of the above, it was suggested that the recruitment data for the various stocks be combined into a single large data set and analyzed with the two-fold purpose:

(i) to better assess the fitting performance of the different methods and models, and

(ii) to arrive at a fairly precise estimate for a 'universal' recruitment distribution.

6.2. Combining the data sets

Recruitment series for 18 stocks were selected for analysis. The data and histograms for the individual stocks appear in Figure 6.2.1. Sample sizes range from 10 for North Sea mackerel to 43 for Georges Bank haddock. On the whole, the data exhibit strong positive skewness with the occasional occurrence of large positive values corresponding to the appearance of a strong year class.

When combining data, the various data sets must have some common features (or there would be no reason to combine) as well as some differences (or the matter would be trivial). The trick is to model the common features and to suitably adjust the data for the differences before combining. The large combined data set is then used to draw reliable inferences concerning the common features.

In our case, it is hypothesized that the pth recruitment data set can be described as a random sample from a scale-parameter family of distributions with cumulative distribution function (cdf)

$$F(x, \theta_p) \equiv F(x/\theta_p). \tag{6.2.1}$$

Here the scale parameter θ_p is allowed to vary from stock to stock. The functional form of the cdf F is assumed to be the same for all stocks and therefore represents a 'universal' recruitment distribution. The pth data set is adjusted by dividing through by a suitable scale statistic. The arithmetic mean (divided by 5) was used in the present analysis.

6.3. Estimating the universal recruitment distribution

Having combined the descaled recruitment values, the next step is estimation of the common cdf F. Here a nonparametric approach has been adopted. In passing, it may be noted that the problem would be trivial if we were prepared to assume a parametric form for F. In fact, if $F(\cdot) = G(\cdot, \varphi)$ where G is a *known* distribution and φ is a vector of unknown parameters, then from (6.2.1) the pth data set is a random sample from $G(x/\theta_p, \varphi)$. From this, the joint likelihood can be written down and parameters estimated in the usual way.

The nonparametric method employed is a variation of the kernel technique. The distribution to be estimated is approximated by a mixture of lognormal distributions. There is one lognormal (known as a *kernel*) for each available observation. Each kernel is 'centered' so that its geometric mean is located at the corresponding observation. The various kernels are taken to have the same logarithmic standard deviation, which is known as the *bandwidth*.

The central theme in application of kernel methodology is the determination of suitable bandwidths. Overly small bandwidths yield estimated pdf's whose graphs have a rough, jagged appearance. Excessively large bandwidths smooth the probability mass over a wide interval, losing most of the local features of the data. There is extensive literature on kernel methods (Wertz and Schneider, 1979; B. L. S. Prakasa Rao, 1983; Devroye and Gyorfi, 1985; Silverman, 1986) and we will not here dwell upon the technical aspects except to point out that bandwidth determination was done through cross-validation.

The histogram of the combined data set is shown in Figure 6.2.2. Superimposed are the fitted lognormal distribution, the fitted gamma distribution and the kernel estimate. Parameter estimation for the gamma and lognormal was done by the method of maximum likelihood. Inadequacy of the lognormal fit is readily apparent. The kernel fit, while generally acceptable, does exhibit a leftward bias for small year classes. We have been able to remove most of this bias by either of two techniques: (i) variable bandwidths and (ii) regression toward the mean of the kernel centers. The gamma distribution also shows a leftward bias, and has a right hand tail that is much too short to give an adequate fit.

6.4. Estimating individual recruitment distributions—Interpretation of the kernel estimator

The simplest estimate of the recruitment distribution for a particular stock is formed by rescaling the universal recruitment distribution. This raises the question of whether the limited data that is available on a particular stock can be used to improve the estimate. Here the James–Stein (1961) paradigm may offer some guidance. Envision the separate (descaled) recruitment distributions as forming a cloud of points in the space of all probability distributions. The universal curve estimates the center of this cloud. Use the available data to obtain, perhaps by the kernel method, a low quality estimate $\hat{F}_p$ for a particular distribution. For the final estimate, use a convex linear combination of the imprecise estimate $\hat{F}_p$ and the precise but inaccurate universal estimate.

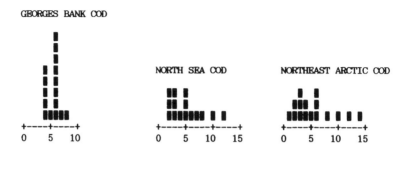
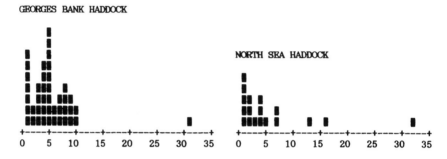
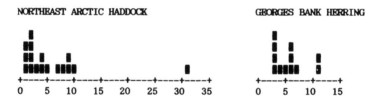
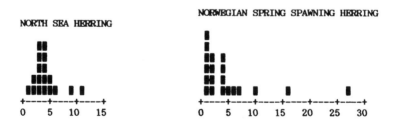

Fig. 6.2.1. Recruitment data.

Data-based sampling and model based estimation for environmental resources 505

GEORGES BANK MACKEREL

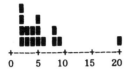

NORTH SEA MACKEREL

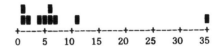

NORTH SEA SAITHE

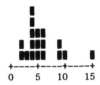

NORTH SEA WHITING

GEORGES BANK
SILVER HAKE

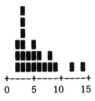

PERUVIAN ANCHOVY

SOUTH AFRICAN PILCHARD

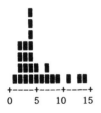

SOUTH AFRICAN
ANCHOVY

SOUTH AFRICAN
ROUND HERRING

Fig. 6.2.1. (continued).

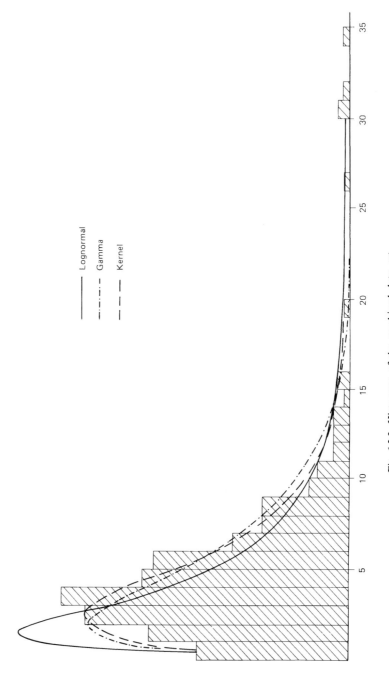

Fig. 6.2.2. Histogram of the combined data set.

It may be of interest to close this section with an interpretation of the kernel estimator. The recruitment process is governed by many factors, both environmental and biological. Currently there is little understanding of what these factors are, how they operate quantitatively and how they interact. The kernel method attempts to account for the annual variability in recruitment without developing a detailed explanatory model. Consider the multidimensional space of all relevant factors and let this space be partitioned into N subsets, one for each available recruitment value; the subsets occur with the same long term relative frequency of $1/N$. Conditional upon a particular partitioning set, there is still residual environmental variability within that set and a corresponding variability in recruitment. It is this variability that is represented by the lognormal kernels. Each kernel is centered at the corresponding observation, in effect treating each observation as typical for its partition set.

7. Encountered ecotoxicological data and chronic effects thresholds

In problems related to fisheries, we often find ourselves interested in, say, two possible parameters (L, M). To conduct experiments on a population of various species and to estimate M may be very expensive and time consuming, whereas, L may be obtained with comparatively little effort and cost. If there is a functional relation between M and L, then M can be extrapolated for a given species from that of the estimated value of L. In order to establish such a functional relationship, we need data on (L, M). Due to the time and cost involved, 'observations' are 'collected' from experiments previously conducted in various laboratories. The data obtained do not constitute a random sample, and, therefore, standard statistical techniques, such as regression, cannot be applied. In this section, we discuss a statistical approach to deal with such problems and illustrate it with an example. For details, see Linder et al. (1986) and Suter et al. (1987).

Ecological effects of chemicals on commercial fish are commonly assessed by estimating a 'safe' exposure level, below which no observed effect on the growth, reproductivity and mortality of the fish occurs. These effects usually vary over species and toxicants. For a particular species-toxicant pair, the effects of chemical levels are significant at various life cycle stages of the species. The largest concentration level, below which no significant effect is observed, is estimated. This is a very expensive and time consuming process. It is not possible to conduct a test for every possible toxicant and species of interest. The proposed approach consists of looking at the following two kinds of effects.

(1) *CHRONIC* Long term low-level effect; Maximum Acceptable Toxic Concentration (MATC), the largest concentration level below which no significant effect is observed.
(2) *ACUTE* Short term high-level effect. Achieve 50% mortality in 96 hours (LC50).

If these paired results are available corresponding to several species-toxicant pairs, a functional relationship, the so-called acute-chronic extrapolation, can be estimated. This relationship can then be used for extrapolation from LC50 to MATC.

This procedure involves curve fitting for the data points (X_i, Y_i), the LC50–MATC pair for a particular toxicant species combination. The main features of the 'data set' are:

(i) Each point (or pair) represents reported results from two bio-assay experiments.

(ii) Different points result from different studies reported in literature. Hence, the (X_i, Y_i) do not constitute a random sample from the population of all possible LC50–MATC pairs.

(iii) Since (X_i, Y_i) are *estimates* of threshold concentrations, they are themselves random quantities. There is considerable uncertainty about their 'true' values.

Even though the pairs (X_i, Y_i) are independent, the classical least squares method cannot be applied for curve fitting as the data is an encountered one, and with possible errors. It is not a random sample. This can be viewed as independent observations on the population $\Omega = \{(s, t): s$ denotes species and t denotes toxicant$\}$. For each ω in Ω, let $(L(\omega), M(\omega))$ denote the actual values of LC50 and MATC. What we observe, for some of these species, is $X = L +$ error and $Y = M +$ error. This leads us to the errors-in-variables model.

7.1. *The errors-in-variables model*

In the errors-in-variables (EIV) model, (X_i, Y_i) are assumed to have been recorded with error $(\delta_i, \varepsilon_i)$. They represent unknown mathematical quantities (U_i, V_i). Linearity is assumed between U_i and V_i, resulting in the model:

$$X_i = U_i + \delta_i, \qquad Y_i = V_i + \varepsilon_i, \qquad V_i = \alpha + \beta U_i \qquad (7.1.1)$$

and

$$(\delta_i, \varepsilon_i) \sim \text{i.i.d.} \left\{ \mathbf{0}, \begin{pmatrix} \sigma_\delta^2 & 0 \\ 0 & \sigma_\varepsilon^2 \end{pmatrix} \right\}. \qquad (7.1.2)$$

The $\{\delta_i, \varepsilon_j : i, j\}$ are independent identically distributed random variables with zero means and with σ_δ^2 and σ_ε^2 as variances.

Two structures are possible.

(i) Structural EIV model; and (ii) Functional EIV model

In (i) U_i are assumed to be independent identically distributed from an unknown distribution with mean μ and variance σ^2, and are independent of $\{(\delta_k, \varepsilon_j)\}$. In this case, (X_i, Y_i) is a sequence of independent identically distributed vectors with mean $(\mu, \alpha + \beta\mu)$ and dispersion matrix

$$\begin{pmatrix} \sigma^2 + \sigma_\delta^2 & \beta\sigma^2 \\ \beta\sigma^2 & \beta^2\sigma^2 + \sigma_\varepsilon^2 \end{pmatrix}.$$

The classical asymptotic theory is applicable for the structural model. But the more interesting model in the present context is model (ii), where the $U_1, \ldots, U_n$ are treated as unknown nuisance parameters. These two EIV models have been studied extensively (Kendall and Stuart, 1979; Gleser, 1983, 1985). For purposes of identifiability, both models assume $\lambda = \sigma_\varepsilon^2/\sigma_\delta^2$ to be known.

The maximum liklihood estimators of the slope β and the intercept α (under normality assumptions) for both the structural and the functional models are:

$$\hat{\beta} = h + \text{sign}(SXY)\sqrt{(h^2 + \lambda)}, \qquad \hat{\alpha} = \overline{Y} - \hat{\beta}\overline{X}, \qquad (7.1.3)$$

where $SXY = \sum_{i=1}^{n}(X_i - \overline{X})(Y_i - \overline{Y})$ and $h = (SYY - \lambda SXX)/2SXY$,

$$\hat{\sigma}_\delta^2 = (1/(\lambda + \hat{\beta}^2))\sum_{i=1}^{n}(Y_i - \hat{\alpha} - \hat{\beta}X_i)^2$$

where bars denote averages and where SXX, and SYY are the usual corrected sums of squares.

7.2. Simulation results and data analysis

The following table gives the simulation results to compare OLS and Structural EIV models:

TRUE VALUES $\alpha = 0$, $\beta = 0.5$, $\lambda = \sigma_\varepsilon = \sigma_\delta = 1$; 20 observations, 30 simulations;

Average of OLS $\hat{\beta}$	Average of EIV $\hat{\beta}$	Distribution
0.479	0.501	$U_i \sim N(0, 25)$
0.489	0.500	$U_i \sim U(10, 10)$

For the data described above, a more realistic model may be a modified version of the functional EIV model with $(\delta_i, \varepsilon_i)$ independent but not identically distributed. Further, it may not be unrealistic to assume that $(\delta_i, \varepsilon_i)$ have the same dispersion matrix. Without loss of generality, assume that $E(\delta_i) = E(\varepsilon_i) = 0$. (If they are not zero, merge these values in U_i and V_i.) It can be shown, under these assumptions that, if $\mu_n = (1/n)\Sigma U_i$ converges to μ and if $s_n^2 = (1/n)\Sigma(U_i - \mu_n)^2$ converges to $\sigma^2 > 0$, then $\hat{\beta}$ and $\hat{\alpha}$ are asymptotically consistent. Using the Lindeberg's theorem and the so-called δ-method we can show that $\sqrt{n}(\hat{\beta} - \beta)$ is asymptotically normal.

For the applicability of the EIV method, the results (absolute magnitudes) have to be evaluated in terms of their biological meaning. This has been done extensively (Suter et al., 1986) by comparing extrapolated MATC's with measured MATC's for toxicant-species combinations, where the results are available. In addition, the method was compared to the traditional methods which use fathead minnow as a basis for extrapolation. It has been generally found to out perform the old methods.

Often the sampling distributions of the estimates of α and β are skewed. So the normal approximation is not reliable for moderate size samples, as it ignores the skewness factor. This is where the resampling method like bootstrap helps. It turns out that the bootstrap is not straightforward in the case of EIV model. The next section explains the bootstrap method for the EIV model.

7.3. The bootstrap method

Babu (1984) and Babu and Singh (1984) discuss the asymptotic properties of bootstrap. In order to use the bootstrap method, we need to 'imitate' the random structures involved. This amounts to choosing proper fitted values $\hat{U}_i$ and $\hat{V}_i$ so that on the average $|X_i - \hat{U}_i|$ is close to $|Y_i - \hat{V}_i|\sqrt{\lambda}$. It turns out that the maximum likelihood estimates do not satisfy this restriction under normal assumptions. The adjusted residuals are given by

$$d_i = -C_i \, \text{sign}(\hat{\beta})/\sqrt{\lambda + \hat{\beta}^2} \quad \text{and} \quad e_i = -\sqrt{\lambda} \, d_i \, \text{sign}(\hat{\beta}),$$

where $C_i = Y_i - \hat{\alpha} - \hat{\beta} X_i$ with $\hat{U}_i = X_i + C_i/(\hat{\beta} + \sqrt{\lambda} \, \text{sign} \, \hat{\beta})$ and $\hat{V}_i = \hat{\alpha} + \hat{\beta} \hat{U}_i$.

If we sample with replacement from (d_i, e_i) in order to use the bootstrap method, the structure of the resulting residuals does not match that of $(\delta_1, \varepsilon_1)$. Further modification of the bootstrap sampling procedure is required. Need for such modifications is also pointed out in the case of autoregressive processes (see Babu and Bose, 1986).

The asymptotic theory plays an important role in suggesting the necessary adjustments and refinements for the bootstrap method. Thus let $r_1^*, \ldots, r_{2n}^*$ be a sample of size $2n$ from $d_1, \ldots, d_n$ with replacement. The sample $\{(r_i^*, -\sqrt{\lambda} r_{i+n}^*)\}$ behaves roughly like a sample with replacement from the unknown quantities $(\delta_1, \varepsilon_1), \ldots, (\delta_n, \varepsilon_n)$. Then,

$$X_i^* = \hat{U}_i + r_i^* \quad \text{and} \quad Y_i^* = \hat{V}_i - \sqrt{\lambda} r_{i+n}^*$$

have essentially the same structure as those of (7.1.1) and (7.1.2). Since β^* is computed from (X_i^*, Y_i^*) as $\hat{\beta}$ is from (X_i, Y_i), as $n \to \infty$,

$$\sup_x |P^*(\sqrt{n}(\beta^* - \hat{\beta}) < xs^*) - P(\sqrt{n}(\hat{\beta} - \beta) < xs)| \to 0$$

for almost all sample sequences, where s^* and s are the estimated standard deviations of $\beta^* - \hat{\beta}$ and $\hat{\beta} - \beta$ respectively. Even though no closed form expression for the distribution of $\hat{\beta}$ is known, the bootstrap method gives a good approximation for the distribution of $\hat{\beta}$, even in the non-normal case. Using Edgeworth expansions, we can show that the convergence here is faster than the convergence of the normal approximation.

8. Synthesis

As in the foregoing examples, it is common that classical theory of sampling cannot be directly applied in situations calling for quantification of environmental resources. Definitions of populations, and uncertainties relative to the stochastics of encounters, constitute fundamental problems with respect to the application of classical methodology in sample design and estimation. It becomes necessary to use existing data bases in conjunction with descriptive models of behaviors and processes to arrive at a workable, it not always exact, approach to further quantification.

Utilization of existing data may be considered as the ultimate lack of ability to design encounters. The contents of the data base have already been collected, so there is no design opportunity whatever. The data can only be subjected to (perhaps secondary) analysis. The focus turns to exploiting any ordinations common to both the data base and the field environment. Ordinations, such as space and time, carry universal relevance because of ecological linkages causing extensive covariation of environmental variables. The more the variables in a data base share a pattern of covariation in space and time, the more compelling the conjecture that the environmental spectrum extends to other variables. In other words, such covariance provides evidence of environmental regionalization. Such regions provide a natural basis for stratification, and the dimensions of ordination may provide a basis for a future sampling frame.

The pervasiveness of covariation in the spatial context among environmental resources serves to underscore the need for advances in multivariate techniques of regionalization. One can envision algorithmic examination of spatially specific environmental data bases for automated delineation of uniform regions to serve as strata. Likewise, cases where environmental variables recorded in a data base exhibit lack of coherence in space and time, indicate need for intensive data collection to determine causes of inconsistency. The latter are likely to be ecologically complex environmental settings and/or instances of alteration. In either case, such areas are of central concern for environmental monitoring.

If one cannot know the probability of observing a potential sampling unit, the next best thing is to have a means of modeling the probability. Knowledge of behavioral paterns and processes, whether or not formalized, provides a basis for modeling the probabilities of encounters.

Finally, the commonalities among populations may be exploited for purposes of combined estimation and extrapolation. Some approaches of this kind are among the illustrative examples.

Acknowledgements

Preparation of this paper has been partially supported by the NOAA research grants to the Penn State Center for Statistical Ecology and Environmental Statistics under the auspices of the Northeast Fisheries Center, the Chesapeake Bay

Stock Assessment Committee, and the Oceans Assessments Division. The NCI grant CA 23415 to Dana-Farber Cancer Institute provided partial support to G. P. Patil. Our thanks are due to Marllyn Boswell, Ernst Linder, and Joel O'Connor for interesting discussions on the contents of this paper from time to time.

References

Babu, G. J. (1984). Bootstrapping statistics with linear combinations of chi-squares as weak limit. *Sankhyā A* **46**, 85–93.

Babu, G. J. and Bose, A. (1986). Bootstrap confidence intervals with applications to autoregressive models. Department of Statistics, The Pennsylvania State University. (Manuscript.)

Babu, G. J. and Singh, K. (1984). Asymptotic representations related to jackknifing and bootstrapping L-statistics. *Sankhyā A* **46**, 195–206.

Bell, J. and Atterbury, T. (eds.) (1983). *Renewable Resource Iventories for Monitoring Changes and Trends. Proc. Ihntl. Conf., Aug. 15–19, 1983, Corvallis, Oregon.* Soc. of American Foresters, 737 pp.

Bonzek, C. R., Myers, W. L., Parolari, B. W. and Patil, G. P. (1986). Sources of bias in surveys for marine fisheries. In: *Oceans 86 Proceedings: Vol. 3: Monitoring Strategies Symposium.* Washington, DC, 908–913.

Darr, David R. and Hennemuth, Richard C. (1985). Forecasting for resources in forestry and marine fisheries in the year 2000. In: Proc. Eighth Sympos. on Statistics, Law, and the Environment. *Amer. Statist.* **39**, 384–392.

Devroye, L. and Gyorfi, L. (1985). *Nonparametric Density Estimation: The L_1 View.* Wiley, New York.

Diaconis, P. and Efron, B. (1985). Testing the independence of a two-way table: New interpretations of the chi-square statistic. *Ann. Statist.* **13**, 845–913.

Duffy, J. C. and Waterton, J. J. (1984). Randomized response model for estimating the distribution function of a quantitative character. *Int. Statist. Rev.* **52**, 165–171.

Efron, B. (1986). Double exponential families and their use in generalized regression models. *J. Amer. Statist. Assoc.* **81**, 709–721.

Emrich, L. (1983). Randomized response techniques. In: W. G. Madow, H. Nisselson, and I. Olkin, eds., *Incomplete Data in Sample Surveys, Vol. 2.* Academic Press, New York, 43–80.

Fox, James A. and Tracy, Paul E. (1986). *Randomized Response: A Method for Sensitive Surveys.* Sage Publications, Beverly Hills, CA.

Gleser, L. J. (1983). Functional, structural, and ultrastructural errors-in-variables models. *Proc. Bus. Econ. Statist. Sect.* Washington, DC, 57–66.

Gleser, L. J. (1985). A note on G. R. Dolby's unreplicated ultrastructural model. *Biometrika* **72**, 117–124.

Hennemuth, R. C. (1979). Man as predator. In: G. P. Patil and M. Rosenzweig, eds., *Statistical Ecology Series Volume 12: Contemporary Quantitative Ecology and Related Ecometrics.* International Co-operative Publishing House, Fairland, MD, 507–532.

Hennemuth, R. C. and Avtges, S. M. (1982). Effects of variability of recruitment of management advice. *Int. Counc. Explor. Sea,* C.M. 1982/H:22, mimeo, 11 pp.

Hennemuth, R. C., Palmer, J. E. and Brown, B. E. (1980). A statistical description of recruitment in eighteen selected fish stocks. *J. Northwest Atlantic Fisherby Science* **1**, 101–111.

Hennemuth, R. C., Patil, G. P., and Taillie, C. (1985). Can we design our encounters? CM 1985/D:9, Int. Counc. Expl. Sea (Note: unpublished meeting document available from authors).

Horwitz, D. G., Shah, B. V. and Simmons, W. R. (1967). The unrelated question randomized response model. *Proc. Amer. Statist. Assoc., Social Statistics Section,* pp. 65–72.

James, W. and Stein, Charles (1961). Estimation with quadratic loss. In: *Proc. 4th Berkeley Symp. on Math. Statist. and Prob.* **1**, 361–379. University of California Press, Berkeley.

Kendall, M. G. and Stuart, A. (1979). *The Advanced Theory of Statistics, Vol. 2,* Chapter 29, 4th Edition. MacMillan, New York.

Linder, E., Patil, G. P., Suter, G. W. and Taillie, C. (1986). Effects of toxic pollutants on aquatic resources using statistical models and techniques to extrapolate acute and chronic effects benchmarks. In: *Oceans 86 Proceedings: Vol. 3: Monitoring Strategies Symposium*, Washington, DC, 960–963.

Mandel, J. (1984). Fitting straight lines when both variables are subject to error. *J. Qual. Technol.* **16**, 1–14.

Mandelbrot, Benoit B. (1977). *Fractals—Form, Chance, and Dimension*. Freeman, San Francisco, 365 pp.

McCullagh, P. and Nelder, J. (1983). *Generalized Linear Models*. Chapman and Hall, London.

Moore, D. F. (1987). Modelling the extraneous variation in the presence of extra binomial variation. *Appl. Statist.* **36**, 8–14.

Patil, G. P. (1981). Studies in statistical ecology involving weighted distributions. In: J. K. Ghosh and J. Roy, (eds.), *Statistics Applications and New Directions: Proceedings of ISI Golden Jubilee International Conference*. Statistical Publishing Society, Calcutta, India, 478–503.

Patil, G. P., Babu, G. J., Boswell, M. T., Chatterjee, K., Linder, E. and Taillie, C. (1986). Statistical issues in combining ecological and environmental studies with examples in marine fisheries research and management. Invited paper at ASA/EPA Conference on Statistical Issues in Combining Environmental Studies, Washington, DC.

Patil, G. P. and Rao, C. R. (1977). Weighted distributions: a survey of their applications. In: P. R. Krishnaiah, ed., *Applications of Statistics*, North-Holland, Amsterdam, 383–405.

Patil, G. P. and Rao, C. R. (1978). Weighted distributions and size biased sampling with applications to wildlife populations and human families. *Biometrics* **34**, 179–189.

Patil, G. P., Rao, C. R. and Zelen, M. (1987). Weighted distributions. In: S. Kotz and N. L. Johnson, eds., *Encyclopedia of Statistics Sciences*, Wiley, New York.

Patil, G. P. and Taillie, C. (1981). Statistical analysis of recruitment data for eighteen marine fish stocks. Invited paper presented at the Annual Meetings of the American Statistical Association, Detroid, MI.

Pollock, K. H. and Beck, Y. (1976). A comparison of three randomized response models for quantitative data. *J. Amer. Statist. Assoc.* **71**, 884–886.

Prakasa Rao, B. L. S. (1983). *Nonparametric Functional Estimation*. Academic Press, New York.

Rao, C. R. (1965). On discrete distributions arising out of methods of ascertainment. In: G. P. Patil, ed., *Classical and Contagious Discrete Distributions*, Pergamon Press and Statistical Publishing Society, Calcutta, 320–332.

Rao, C. R. (1985). Weighted distributions arising out of methods of ascertainment: What population does a sample represent? In: A. C. Atkinson and S. E. Fienberg, eds., *Celebration of Statistics, The 151 Centenary Volume*, Chapter 24, 543–569.

Royal, R. M. and Cumberland, W. G. (1981a). Rejoinder. *J. Amer. Statist. Assoc.* **76**(373), 87–88.

Royal, R. M. and Cumberland, W. G. (1981b). The finite-population linear regression estimator and estimators of its variance—An empirical study. *J. Amer. Statist. Assoc.* **76**(376), 924–930.

Schreuder, H. T. and Thomas, C. E. (1985). Efficient sampling techniques for timber sale surveys and inventory updates. *Forest Sci.* **31**(4), 857–866.

Schuck, Howard A. (1949). Relationship of catch to changes in population size of New England haddock. *Biometrics* **5**, 213–231.

Silverman, B. W. (1986). *Density Estimation for Statistics and Data Analysis*. Chapman and Hall, New York.

Suter, G. W., II, Rosen, A. E. and Linder, E. (1986). Analysis of extrapolation error. In: L. W. Barnthouse and G. Suter, eds., *Users Manual for Ecological Risk Assessment*, ORNL-6251, Oak Ridge National Laboratory, Oak Ridge, TN.

Suter, G. W., II, Rosen, A. E., and Linder, E. (1987). Endpoints for responses of fish to chronic toxic exposures. *Environmental Toxicology and Chemistry*. (To appear.)

Warner, S. L. (1965). Randomized response: A survey technique for eliminating evasive answer bias. *J. Amer. Statist. Assoc.* **60**, 63–69.

Wertz, W. and Schneider, B. (1979). Statistical density estimation: A bibliography. *International Statistical Review* **47**, 155–175.

Zelen, M. and Feinleib, M. (1969). On the theory of screening for chronic diseases. *Biometrika* **56**, 601–614.

On Transect Sampling to Assess Wildlife Populations and Marine Resources

F. L. Ramsey, C. E. Gates, G. P. Patil[1] and C. Taillie

1. Introduction

Frequently a statistician must examine subjects that appear in the sample as results of selection with quite different probabilities. Non-response in sample surveys is an obvious instance. In attempting to estimate numbers of heroin addicts, it is also clear that one will be much more likely to observe (in hospitals, clinics, courtrooms, etc.) those that have more extreme addictions. The problem of estimating total numbers present in such situations is the objective of Transect Surveys, arising from the literature on wildlife sampling.

Crudely put, observers go afield to seek wildlife and return to tell the statistician how many animals they have found. It is then the statistician's task to determine how many animals they did NOT find. This task is not impossible, so long as the statistician insists that observers record, for each sample animal, some measurement to which its probability of detection is directly related.

An example: Imagine that you are addressing an audience and that you ask all those to stand who come from a family where all children have the same sex. Eleven persons rise. Using those eleven, can you guess how many persons are in your audience?

Interview those eleven to determine what the number of children is in the family of each. Suppose that two are 'only' children, six come from two-child families, two have families with three children and one comes from a family with five children. These numbers form column (2) in Table 1.

Notice that every member of your audience who comes from a one-child family must have been counted, so your audience has exactly two persons from such families. If any given child has equal chances of being female or male, then half of all two-child families would have both children of the same sex. It is therefore reasonable to assume that the six you found represent about half the two-child-

---

[1] Prepared in part as a Visiting Professor of Biostatistics, Department of Biostatistics, Harvard School of Public Health and Dana-Farber Cancer Institute, Harvard Medical School, Harvard University, Boston, MA.

family members in your audience. More generally, the chances of your having 'detected' in this way a person in your audience who comes from a k-child family is 2^{-k+1}. If n_k are detected, then you may estimate the total number of audience members with k-child families to be $n_k 2^{k-1}$. Column (3) of the table contains the chances of detection and column (4) has the projected numbers of persons in the audience from families with various numbers of children. Summing column (4) leads to an estimated audience size of 38. Indeed, this is the familiar Horvitz–Thompson estimator (see Cochran, 1977, p. 259) from probability sampling.

Table 1

(1) Family size	(2) Number detected	(3) Chance detection	(4) Estimated number	(5) % of families	(6) % of persons	(7) Expected % detected
1	2	1	2	24	9.6	9.6
2	6	1/2	12	38	30.4	15.2
3	2	1/4	8	18	21.6	5.4
4	–	1/8	–	10	16.0	2.0
5	1	1/16	16	5	10.0	0.6
6	–	1/32	–	4	9.6	0.3
7	–	1/64	–	1	2.8	0.0

What can be done if the probabilities in column (3) are not so accessible? One possibility is to let them be unknown parameters and estimate them from the data. Unfortunately, the maximum likelihood estimates of the detection probabilities are all 'one'; i.e., what you saw is all there is. This solution is not particularly interesting. To see how we proceed, suppose that a demographic study of the area determined the various proportions of families having different numbers of children to be column (5). Column (6) shows the proportions of *people* who then come from those families. Approximately 9.6% of your audience are from single-child families, and all are detected. About 30.4% of your audience come from two-child families and you should detect half of them, making another 15.2% of your audience detected. Continuing, the law of total probability determines that—overall—the chance of an arbitrarily selected person in the audience being detected is the sum of column (7) (= column (3) times column (6)), or 33.1%. A quicker estimate of audience size is obtained by dividing your eleven observed by this overall detection probability. Thus, you estimate the audience size to be 33.

This 'simplification' actually involves not only knowing the conditional chances of detection given a family size but also the chances of audience members having come from families of various sizes. In wildlife surveys, it is assumed that one of these columns, (6), is known: the distribution of animals over a region is assumed to be uniform by dint of animal behavior or clever placement of observers. And whereas it is not assumed that the conditional probabilities of detection—column (3)—are known exactly, strong assumptions about the form of the detection probabilities are made. With suitable assumptions, the overall detection probability becomes a single unknown parameter which can be estimated with relative ease.

This review paper will give a short overview of the theory of transect sampling with emphasis on the concept of an effective area in Section 2, a summary of current methods in Section 3 and a discussion of dealing with certain departures from underlying assumptions in Section 4. Sections 5 and 6 utilize the concept of an effective distance and illustrate the techniques using data on terrestrial and marine animals, respectively.

2. Transect surveys

The following treatment is fully developed in several articles. (See Seber, 1982; Burnham and Anderson, 1976; Gates, 1979.) Field procedures differ for surveying different kinds of animals in different kinds of terrain. Where terrain is open and detectability conditions are uniform, Line Transects (LT) are appropriate. An observer transverses a straight-line path of length L through the target region, going at a fixed speed, continuously recording perpendicular distances from the transect of detected animals. Where terrain is rugged and/or detectability conditions variable, Variable Circular Plots (VCP) provide a similarly simple technique for covering a large area. An observer transverses a straight-line path as above, but stops at regular intervals to survey a station for fixed periods of time, usually 8–10 minutes (Reynolds et al., 1980). The observer in this case records the radial distance from station for any detected animal. What is common to these cases is that each records distances as a substitute for the time-consuming task of delineating regions of fixed area to be surveyed intensively. The design anticipates that animals will be missed, and the analysis compensates.

Because animals are assumed to be distributed uniformly over area, we transform detection distances, d, to detection *areas* y, where $y = 2Ld$ for LTs and $y = \pi d^2$ for VCPs. Detectability is generally assumed to decrease with increasing distance, so y may be viewed as the area of the region where detectability is at least as good as it is at the point where a particular animal was detected. This general description of a detection area is applicable to any survey we might invent where distances form the 'family size' measurement to which detectability is related. The transformation to detection areas allows one to analyze all such surveys within the same theoretical framework (Burnham et al., 1980).

Conditional chances of detection, given an animal is located at area y, are incorporated in a detectability function $g(y) = \Pr\{\text{detection}|y\}$ for $y \geq 0$. The procedure requires that you know the value of $g(y)$ for at least one value of y. Ordinarily, this is at $y = 0$ where $g(0) = 1$. Indeed, the choice of survey method is dictated to a large extent by considerations guaranteeing the validity of this assumption. Define

$$\alpha = \int_0^\infty g(y)\,dy.$$

This parameter of the detectability function, called the effective area surveyed, plays the role of the overall detection probability in the example above. It is preferred to the unconditional probability because it is horizon-free: the probability of detecting a randomly located animal is $\theta = \alpha/A$, where A is the total area of the target region and most authors take the infinite-boundary view that θ is zero for any specific animal.

Burnham and Anderson (1976) concentrate on the parameter $1/\alpha = f(0)$, where $f(y) = g(y)/\alpha$ is the probability density function of a single detection area (obtainable by Bayes'π theorem). In our view, there are several reasons for preferring to use effective area.

(1) Effective area *has natural units*. The total number of detections, n, has $E[n] = D\alpha$, where D is the population density. This and the above definition make it clear that α is a dimensioned parameter and that its dimension is area.

(2) Effective area *suggests pooling strategies*. When detectability and animal density vary widely, sample sizes in cells where both are constant may be insufficient to estimate anything (Burnham et al., 1980, suggest a minimum of 30 detections). However, data from cells with detectability patterns that a priori should allow equal area coverage can be pooled to estimate the common area covered, regardless of whether these data come from cells with different animal densities.

(3) Many environmental factors that affect detectability do so only by changing the amount of area effectively surveyed. Ramsey et al. (1987) express $\log(\alpha)$ as a linear model in factors such as time-of-day, observer indicator, slope and weather conditions arriving at ways to adjust effective area within the framework of general linear models.

(4) Effective area is a valuable tool for *design*. With experienced observers and known terrain structure, it is possible to use past data to obtain very reasonable estimates of how much area can be surveyed for important target species with a given amount of survey effort. Knowing this allows one to design a survey that will have specified minimum probability of detecting at least a certain fraction of a population with any stated density. By extension, effective area forms an important basis for designing a monitoring plan to follow up a survey.

3. Estimation procedures

Methods exist for estimating effective area within parametric frameworks and within a non-parametric framework. We review the latter first.

(A) *Non-parametric estimators*. The two common methods of estimation that follow make no explicit assumptions about the actual shape of the detectability curve, although the first is quasi-parametric and the second assumes a detectability curve that is non-decreasing in y.

3.1. The Fourier series estimator

Perhaps the most widely-used non-parametric estimator in current use is the Fourier Series estimator developed by Crain et al. (1978) and available in the programs TRANSECT (Burnham et al., 1980) and LINETRAN (Gates, 1980). See also Patil et al. (1979b). The method approximates the probability density function, $f(y)$, of a detection area with a Fourier series of low order near $y = 0$. The estimator is

$$\hat{\alpha}_{FS} = 1 \Big/ \Big[(1/w^*) + \sum_{k=1}^{m^*} b_k \Big],$$

where

$$b_k = [2/nw^*] \sum_{j=1}^{N^*} \cos(k\pi y_j/w^*), \quad k = 1, 2, \ldots,$$

are the estimated coefficients in a Fourier expansion of $f(y)$. Burnham et al. (1980) suggest that the horizon w^* be chosen to eliminate $1-3\%$ of the largest detection. N^* is the number of retained detections (not exceeding w^*). The recommended number of terms is determined from the data: m^* is given by

$$m^* = \min\{m: |b_{m+1}| \leq \sqrt{2}/[w^* \sqrt{(n+1)}]\}.$$

Burnham et al. (1980) emphasize that one should resist the temptation to set the horizon w^* at some large value beyond which no detections occur. Because the horizon is the half-period of the fundamental harmonic in the Fourier series representation of the density function, a given curve would require larger frequencies with a large horizon, and hence more terms in a Fourier expansion. It is also important, surprisingly, that w^* be determined from the data by setting it equal, for example, to the 98th percentile in the detection area distribution. With this choice, the estimator is a true scale-parameter estimate, in the sense that conversion of the data from acres, say, to hectares would simply change the effective area from acres to hectares.

3.2. The cumulative distribution estimator

An estimator based upon the empirical cumulative distribution function, $F_n(y)$, of detection areas was described by Wildman and Ramsey (1985). This estimator was designed for use with VCP surveys of birds, where time on station allows observers to have high detectability in a reasonably large region surrounding each station. The form of the estimator is

$$\hat{\alpha}_{CD} = n y_{(k)}/k,$$

where n is the total number of detections, $y_{(k)}$ is the kth smallest detection area, and where k is determined from the data in such a way that one has reasonable confidence that detection is near certain in the region with area $y_{(k)}$ surrounding each station. Wildman and Ramsey (1985) describe a protocol for determining k from the data.

The estimator $\hat{\alpha}_{CD}$ is so-named because the procedure mimics how one might draw in a line on a plot of the empirical distribution function to approximate the slope of the function at zero. (Slope = $f(0) = 1/\alpha$).

3.3. Other non-parametric estimators

A great many other suggestions have been made for non-parametric estimation. For example, one based upon a Hermite Polynomial model, developed by Buckland (1985), may be less sensitive than the FS estimator due to the indefinite nature of a cutoff parameter. A current review of other techniques is available in Seber (1986). A strong advantage the FS and CD estimators share over many others is their lack of reliance on data grouped into histogram form.

(B) *Parametric estimators.* Various authors (Gates, et al., 1968; Ramsey et al., 1979; Emlen and DeJong, 1981; Hayes and Buckland, 1983) attempt to model the actual mechanism of detection. As such, the models naturally resemble hazard models, because detection is the result of many factors that can operate against detection. While the approach has some appeal, the models have had limited acceptance in practice so far.

Proposed models for detectability are mathematically convenient functions that possess many of the characteristics one ordinarily expects of a detectability function. For example, most LT models have $g(0) = 1$, have $g(y)$ non-increasing in y, have $g(y)$ continuous and integrable. As these traits are shared by the reverse probability distribution functions, Ramsey (1979) offers the following prescription.

Consider any non-negative random variable X with finite expectation μ. Integration by parts establishes that the function $h(t) = \Pr\{X \geq \mu t\}$ satisfies the conditions (i) $h(0) = 1$; (ii) $h(t)$ is non-decreasing; and (iii) $h(t)$ integrates to 1. Such a function is a detectability curve kernel, and a detectability curve is constructed by setting $g(y) = h(y/\alpha)$, thus making α a scale parameter in the distribution of detection areas.

3.4. The exponential power MLE

When X has a two-parameter Weibull distribution, the result is the *exponential power* family of detectability curves, by far the most popular of the parametric models (Pollock, 1978; Ramsey, 1979; Quinn and Gallucci, 1980) given by

$$g(y) = \exp\{-[\Gamma(1 + 1/\gamma)(y/\alpha)]^\gamma\}$$

where $\gamma > 0$ is a shape parameter. The family includes the exponential model for $\gamma = 1$ (Gates et al., 1968) and the half-normal model for $\gamma = 2$ (Eberhardt, 1967). To estimate effective area, maximize the likelihood function $\mathscr{L}(\alpha, \gamma) = \Pi\ [g(y_i)/\alpha]$. The estimate for effective area is

$$\hat{\alpha}_{EP} = \hat{\gamma}^{1/\hat{\gamma}}\, \Gamma(1 + 1/\hat{\gamma})\{T_{\hat{\gamma}}\}^{1/\hat{\gamma}} \tag{3.4.1}$$

where

$$T_\gamma = (1/n) \sum_{j=1}^{n} (y_j)^\gamma .$$

Wildman (1983) emphasized the importance of estimating γ from the distribution of a maximal invariant statistic under scale change. Without this precaution, $\hat{\alpha}_{EP}$ does not possess the scale-parameter behavior. The maximal invariant $Z = \{y_1/y_n, \ldots, y_{n-1}/y_n\}$ leads to solving this likelihood equation, iteratively

$$V_\gamma/T_\gamma + 1/n = (1/\gamma)[\log(T_\gamma) + \Omega(1 + 1/\gamma) - \Omega(n/\gamma)],$$

where $\Omega(\cdot)$ is the digamma function and

$$V_\gamma = (1/n) \sum_{j=1}^{n} (y_j)^\gamma \log(y_j) .$$

3.5. Comparison of estimators

No single estimation procedure dominates the others. Generally speaking, if one knows that the detectability function belongs to some parametric class, it is usually best to use the parametric estimator. With detectability functions that have near-perfect detectability over a considerable region around the observer—which is the aim of VCP surveys—use of the CD estimator is recommended. With detectability functions that have short shoulders and long tails, the FS estimator performs well. These latter shapes appear most commonly in LT surveys.

Readers interested in standard errors for these estimators of effective area are referred to the papers cited above, plus Buckland (1984) and Gates (1981).

4. Variable detectability conditions

A wide range of factors affect detectability. For convenience, it seems reasonable to form some broad categories of such factors, according to the prescriptions for dealing with them. Three categories come to mind: (D) those factors that must be dealt with by the design of the sampling procedure itself; (K) those factors that affect the shape or kernel of the detectability function; and (S) factors that affect detectability through scale.

Category (D): Many factors influence detectability in ways that are not discernible from the distribution of detection areas. These must be anticipated and dealt with in the survey's design. Here are some examples.

(1) *Swamping* can be a serious problem in transect surveys, particularly with VCP's (Verner, 1985; Scott and Ramsey, 1981). It occurs when too many different species or too many individuals must be recorded. There is confusion over which animals are different. Increasing density therefore decreases the chances of

adequate coverage. One method for dealing with swamping is to divide the responsibilities for recording different species among different observers. For single, super-abundant species, cameras are also useful.

(2) Rapid *animal movement* relative to the observer, is also problematic (Scott and Ramsey, 1981). Fast-flying birds such as swifts can be over-counted because they move into detectable positions from initially distant locations. For rapidly-moving species, the time interval for detection of such species should be reduced.

(3) Some *secretive animals* simply defy detection by a neutral observer. Oral cues, as supplied with a cassette tape player for example, may elicit responses in some species. Other species may have to be detected by 'signs' they leave in the territories they occupy.

(4) *Heaping* or clumping of distance measurements occurs when certain values are over-represented and adjacent values are under-represented. The terminology is from psychology, but Burnham et al. (1980) were the first to note its applicability in LT applications. A strict interpretation of the underlying LT assumptions would seem to be violated whenever heaping occurs, but that does not appear to be true if distances are estimated unbiasedly. Gates et al. (1984) studied the phenomenon in detail in several very large (>1000 sightings for the same species) and exhibited dramatic examples. They showed that the FS series estimator was not biased by clumping, while another non-parametric estimator, the polynomial estimator, was biased substantially upward by the clumps at 0. A surprising note is that the effect of clumping can be ameliorated for influenced estimators by grouping the data into intervals and splitting the observations at the heap points, 5, 10, 20 m, etc.

Category (K): This category is by far the most difficult to cope with (Burnham et al., 1980), because the factors usually operate to make detectability less than certain at zero distance from the observer. The shape of the detectability function typically has a 'donut'—a region of low detectability very near the observer. To cope with such problems at all, one must inject a considerable amount of prior knowledge about the nature of the detectability function and how the factors operate. Here are two examples.

(1) *Observer avoidance.* The presence of an observer causes animals nearby to either move off or to hide. Suppose that, without this factor, the detectability function would be the exponential $g(y) = \exp[-(y/\alpha)]$, for $y > 0$. Suppose further that an animal at area y from the observer will make itself undetectable with probability $\psi \exp[-(y/\xi)]$, where $\xi \ll \alpha$ and $0 \leq \psi \leq 1$. Then the correct probability of detection at y is

$$g^*(y) = \{1 - \psi \exp[-(y/\xi)]\} \exp[-(y/\alpha)],$$

and the true effective area surveyed is $\alpha^* = \alpha\{1 - \psi\xi/(\alpha + \xi)\}$. The probability density function of a detection area is then $f(y) = g^*(y)/\alpha^*$. Notice in this case that $f(0) = (1 - \psi)/\alpha^*$, which is not the inverse of the correct effective area.

(2) *Measurement errors* in detection areas may occur when observers estimate

distances. This is unavoidable, for example, if detections are made on vocal cues alone. Experiments by Scott et al. (1981) and Verner (1985) show that a plausible model for such errors is to assume that one actually records $X = Y/W$, where Y is the true area and where W is a random corruption. Wildman (1983) argued that the probability density function for recorded areas is

$$f_X(x) = \int_0^\infty w f_y(xw) \, dF_W(w)$$

and in particular, $f_x(0) = f_y(0)E\{W\}/\alpha$. Again, unless $E\{W\} = 1$ the density at zero is not the inverse of the correct effective area surveyed (which is unaffected by measurement errors).

Category (S): Many factors have little effect on the shape of the detectability function; rather they simply make the area covered smaller or larger. Such factors include observer differences, weather-related variables, slope, habitat density, observer speed, time-of-day, season, time-on-station. Ramsey et al. (1987) showed that analysis in such situations is relatively straightforward. Efficient parametric estimators are developed for regression parameters in a detectability model of the form

$$\log(\alpha) = \beta_0 + \beta_1 X_1 + \cdots + \beta_p X_p.$$

Furthermore, it was demonstrated that unbiased estimates of the parameters β_1, $\beta_2, \ldots, \beta_p$ result from an ordinary least-squares fit of a model expressing the expectation of the logarithm of a detection area as the right-hand-side of the above expression.

Table 2
Bob-white flushings. (x = right angle distances in m, f = frequency, L = length of transect = 790 km, and $n = 1262$. Source: Gates et al. 1984).

x	f	x	f	x	f
0.25	141	12	12	24	10
1	96	13	66	28	24
2	103	14	23	29	2
3	115	15	19	30	2
4	85	16	8	31	1
5	82	17	21	37	1
6	95	18	8	39	2
7	44	19	5	44	4
8	86	20	34	46	11
9	37	21	2	51	2
10	45	22	5	69	21
11	25	23	17	94	8

5. Bob-white flushing

Table 2 gives the raw data for an extensive set of Bob-white sightings (Gates et al. 1984). Originally, the observers recorded the 'sighting angle' and the 'radial distance'; the right angle distances were subsequently computed to the nearest meter. The original radial distances showed evidence of extensive 'heaping', which is partially disguised when right angle distances are computed.

Burnham et al. (1980) showed tha tfor both parametric and non-parametric methods of estimation, density is estimated as

$$\hat{D} = n\hat{f}(0)/2L = n/(2L\hat{\omega}).$$

Here we deviate from the notation of the preceding sections, letting ω and x stand for 'effective distance' and 'sighting distance' rather than 'effective area' and 'sighting area'. Equation (3.4.1) with $\gamma = 1$ and 2, still gives maximum likelihood estimators for the effective distance ω, for the exponential and half-normal models respectively. From the full data set, $n = 1262$, $\Sigma x = 11\,773.25$ and $\Sigma X^2 = 326\,958.813$. Thus, using (3.4.1) (for width),

$$\hat{\omega}_{exp} = 11\,773.25/1262 = 9.329 \text{ m}$$

and

$$\hat{\omega}_{hnorm} = [(\pi/2)\{326\,958.813/1262\}]^{1/2} = 20.173 \text{ m}.$$

These estimates give respective estimates of density equal to 85.62 Bob-white per km^2 and 39.59 Bob-white per km^2.

Following Burnham et al. (1980), we may wish to truncate approximately 2% of the data. This means truncating Table 1 at 51 m. The resulting estimate of density with the exponential model is 87.06 Bob-white per km^2. Or, we may also wish to treat the data as being grouped and truncated, in which case iterative maximum likelihood techniques give 98.68 Bob-white per km^2 for the exponential model. The Fourier series estimator, also using 2% truncation, gives 83.6 Bob-white per km^2. The observed density of detections in the first distance class is 178.5 Bob-white per km^2. Because this is significantly higher than the densities in all other classes this becomes the cumulative distribution estimate of density.

Such a wide range of estimates comes about because the exponential and, particularly, the half-normal models do not fit this data well.

6. Deep-sea red crab

The National Marine Fisheries Service applied the line transect technique to a quantitative underwater photographic survey of the deep-sea red crab off the North Eastern Coast of the United States. The primary purpose of the survey was to obtain quantitative estimates of the number and biomass of red crabs in that region. Secondary purposes were to assess the size composition of this species, and to obtain additional information on its distribution, life history, and ecology. The purpose of this section is to review the statistical aspects of line transect theory as it pertains to this survey. For more details, see Patil, Taillie, and Wigley (1979a, b).

Photographs of the sea bottom were obtained to determine the density of red crabs. Stations were pre-selected according to a stratified random design. The camera was mounted on an underwater sled and was programmed to obtain a photograph every 10 sec; thus the maximum number of photographs during one tow was approximately 400. A total of 18 000 photographs was obtained at 33 stations. Of this total, 8 262 photographs representing the best quality for counting crabs were selected for quantitative analyses.

The counts of red crabs in this study first increase and then decrease with the right angle distance. This unusual feature can be explained, however, by decreasing visibility coupled with the expanding field of view of the camera with distance from the sled. This composite effect can be represented by a weighted visibility function, which we use to develop methods for estimating population density.

6.1. Weighted visibility function and effective area

What appears to be a rectangle in a photograph is in fact a trapezoid on the ocean floor as displayed in Figure 1. Let a and H be the base and the height of the trapezoid. The length of the trapezoid at distance x is of the form $a + bx$, where b is a dimensionless constant. For our setup, the values of a, b, and H are 2.868 m, 0.961, and 5.49 m respectively.

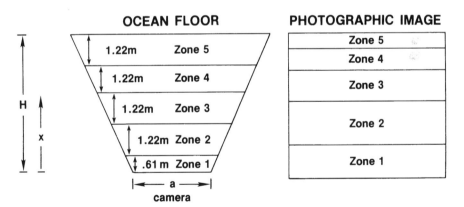

Fig. 1. Schematic diagrams of the acutal field of view on the ocean floor and its photographic image.

Initially, when a crab was detected in the photograph, its right angle distance x was measured by superimposing a grid and counting the number of squares to the crab. This procedure was tedious and time-consuming, however. So, instead, the field of view was divided into five zones as indicated in Figure 1, and the number of crabs was counted for each zone. Originally all the five zones were of equal width. But sediment clouds obscured the visibility near the sled, and, therefore, only the upper half of the first zone has been analyzed.

If $g(x)$ is the visibility function, and $n(x)$ is the number of crabs sighted in the trapezoid strip $(x, x + \Delta x)$,

$$E[n(x)] = sD[a + b(x + \Delta x/2)]\Delta x g(x), \qquad (1)$$

where s is the number of frames (photographs) and D is the density of crabs per square meter. This implies that the probability density function $f(x)$ of recorded x is proportional to the weighted visibility function, $(a + bx)g(x)$, giving

$$f(x) = \frac{(a + bx)g(x)}{w}, \quad 0 \leq x \leq H, \qquad (2)$$

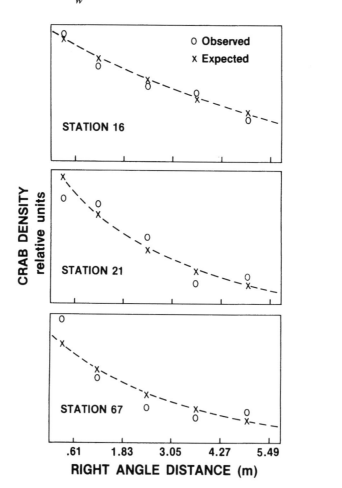

Fig. 2. Observed and expected crab density versus right angle distance at three stations.

where

$$w = \int_0^H (a + bx)g(x)\,dx. \qquad (3)$$

Further, from equation (1) we can estimate density D by

$$\frac{n(x)}{s(a + bx)g(x)\Delta x}.$$

The minimum variance pooled estimate is then given by

$$\hat{D} = \frac{n}{s\int_0^H (a + bx)g(x)\,dx} = \frac{n}{sw}, \qquad (4)$$

where n is the total number of crabs recorded on s frames. Because of the nature of equation (4), w is called the effective area per frame.

In order to utilize (4), we need to know w, which appears as a parameter in the probability density function of the recorded right angle distances as given in equation (2). The parameter w can be estimated in principle from these distances. We discuss two possible methods in the next two sections.

6.2. Exponential estimators

The exponential sighting function has received considerable attention in the line transect literature. In the present study we use

$$g(x) = e^{-\lambda x}, \quad 0 \leqslant x \leqslant H. \qquad (5)$$

The graph of the corresponding weighted visibility function $(a + bx)e^{-\lambda x}$ increases for all small x and decreases for large x as needed in our present study. While the appropriateness of the exponential sighting functon for our data is discussed later, we provide the necessary results for estimating the density here. The effective area given by

$$w = \frac{a}{\lambda}(1 - e^{-\lambda x}) + \frac{b}{\lambda^2}(1 - e^{-\lambda H} - \lambda H e^{-\lambda H}). \qquad (6)$$

The maximum likelihood estimate of λ is the solution of

$$\bar{x} = -\frac{1}{w}\left(\frac{dw}{d\lambda}\right) = \frac{2}{\lambda} - \frac{1}{w}\left[\frac{a}{\lambda^2}(1 - e^{-\lambda H}) + \frac{(a + bH)H}{\lambda}e^{-\lambda H}\right] \qquad (7)$$

where $\bar{x}$ is the average x. Using the method of statistical differentials (see Johnson and Kotz, 1969), the asymptotic variance of $\hat{D}$ in (4) is given by

$$\text{Var}(\hat{D}) = \frac{D}{sW} \frac{\text{Var}(n)}{E[n]} + \frac{\left(\frac{dw}{d\lambda}\right)^2}{\frac{d^2w}{d\lambda^2} w - \left(\frac{dw}{d\lambda}\right)^2}. \tag{8}$$

The variance-to-mean ratio is known to be related to spatial pattern. It is one for Poisson pattern and exceeds one for aggregated pattern. Although there is indication of some aggregation in our data, in the analysis section we have set the variance-to-mean ratio to be one, thus giving an underestimate for the asymptotic variance of $\hat{D}$. We feel, however, that the bias is not serious.

6.3. Cox estimators

As Eberhardt (1978) has discussed, the Cox estimators provide estimates of the density when no parametric form is assumed for the sighting function. Putting $x = 0$ in equation (2) gives

$$f(0) = \frac{a}{w} \tag{9}$$

provided $g(0)$ can be assumed to be unity. The Cox method obtains the estimate of w by estimating $f(0)$ as follows. Consider the first two zones with widths Δ and 2Δ respectively. Let n_1 and n_2 be the observed counts of the crabs in these zones. Then $f(\Delta/2)$ and $f(2\Delta)$ are estimated by $n_1/n\Delta$ and $n_2/2n\Delta$. The linear extrapolation back to the origin gives the estimate

$$\hat{f}(0) = \frac{8n_1 - n_2}{6n\Delta}. \tag{10}$$

Combining (4), (9) and (10) gives the Cox estimator

$$\hat{D} = \frac{1}{sa\Delta}\left(\frac{8n_1 - n_2}{6}\right). \tag{11}$$

If the spatial pattern is random, n_1 and n_2 are independent and follow Poisson distributions. It follows that $\text{Var}(\hat{D})$ can be estimated by

$$\hat{\text{Var}}(\hat{D}) = \frac{1}{(sa\Delta)^2}\left(\frac{64n_1 + n_2}{36}\right). \tag{12}$$

6.4. Data analysis

We illustrate the above methodology using zone counts from three stations. Computations and estimates are summarized in Table 3. Note that $\bar{x}$ is computed from grouped data in the form of zone counts. The effect of grouping on the estimates is not sizable, as discussed in Patil, Taillie and Wigley (1979a).

Table 3
Density estimation using zone counts and exponential sighting function (estimates are ± one standard deviation)

Station	Photo zone	Zone mid-point (meters)	Number of crabs		Estimates
			Observed	Expected	
16	1	0.305	11	10.7	Exponential estimates
Depth: 530–530 m	2	1.220	22	22.5	$\hat{\lambda} = 0.207$ m^{-1}; $\hat{w} = 16.47$ m^2
	3	2.440	22	22.6	$\hat{D} = 149 \pm 30$ crabs/ha
394 frames	4	3.660	23	21.5	Cox estimates
	5	4.880	19	19.8	$\hat{w} = 15.42$ m^2
					$\hat{D} = 160 \pm 65$ crabs/ha
		$\bar{x} = 2.69$	97	$X_3^2 = 0.17$	
21	1	0.305	17	20.5	Exponential estimates
Depth: 393–412 m	2	1.220	40	35.9	$\hat{\lambda} = 0.414$ m^{-1}; $\hat{w} = 9.92$ m^2
	3	2.440	34	28.0	$\hat{D} = 299 \pm 48$ crabs/ha
404 frames	4	3.660	12	20.7	Cox estimates
	5	4.880	17	14.8	$\hat{w} = 13.12$ m^2
					$\hat{D} = 226 \pm 79$ crabs/ha
		$\bar{x} = 2.20$	120	$X_3^2 = 6.3$	
67	1	0.305	18	14.0	Exponential estimates
Depth: 412–960 m	2	1.220	26	26.5	$\hat{\lambda} = 0.323$ m^{-1}; $\hat{w} = 12.24$ m^2
	3	2.440	18	23.1	$\hat{D} = 189 \pm 36$ crabs/ha
423 frames	4	3.660	17	19.1	Cox estimates
	5	4.880	19	15.3	$\hat{w} = 8.72$ m^2
					$\hat{D} = 266 \pm 77$ crabs/ha
		$\bar{x} = 2.41$	98	$X_3^2 = 3.5$	

We have assumed the exponential visibility function for a parametric approach. This assumption appears reasonable, at least according to the usual chi-square test. The Cox estimates are based on non-parametric procedures that do not assume a form for the visibility function. Comparing the exponential estimates with Cox estimates, the Cox estimates have larger standard errors as expected of non-parametric procedures. The two density estimates are similar for Station 16, but for Station 21, the Cox estimate is smaller that the exponential estimate, whereas the reverse is true for Station 67. For this reason, we will examine more closely the validity of an exponential visibility function.

6.5. Appropriateness of exponential visibility function

For each zone, the observed and expected number of crabs were divided by the zonal area. These densities are plotted in Figure 2 against the zone midpoints. The observed and expected points closely match for Station 16. For Station 21, the observed plot appears to be more nearly half-normal than exponential. The use of exponential sighting function in such a case tends to over-estimate the population density. This may explain the smaller value of the non-parametric Cox estimate for Station 21. At Station 67, the observed points fall off more rapidly and are more convex that the expected points. In this case, the exponential estimate is likely to be an underestimate. This may explain the larger value of the non-parametric Cox estimate for Station 67.

In view of the above observations, a richer and more flexible family of visibility functions which includes a wide variety of shapes is desirable. A promising possibility lies in the exponential-power-family discussed in Section 3.4. The family includes the exponential ($\gamma = 1$), half normal shapes ($\gamma > 1$) and shapes more convex than exponential ($0 < \gamma < 1$). See Patil, Taillie and Wigley (1979b) for results of fitting this family to the red crab data.

7. Summary

Variation and the resulting uncertainty in wildlife population studies is huge. Considering the problem, 'How many did they NOT find?', one should, however, be prepared for large measures of uncertainty whenever one attempts to implement these variable-area surveys. Nevertheless, they have been used with considerable success to estimate sizes and trends of populations as diverse as kangaroos in the Australian outback, fruit-bats in Samoa and Yap, nesting shorebirds in the arctic tundra, fish inhabiting coral reefs around Hawaii, and various others.

Acknowledgements

Preparation of this paper has been partially supported by the NOAA research grants to the Penn State Center for Statistical Ecology and Environmental Statistics under the auspices of the Northeast Fisheries Center. The NCI grant CA 23415 to Dana-Farber Cancer Institute provided partial support to G. P. Patil. Our thanks are due to M. T. Boswell, K. P. Burnham, R. C. Hennemuth, and R. L. Wigley for interesting discussions on the contents of this paper from time to time.

References

Buckland, S. T. (1985). Perpendicular distance models for line transect sampling. *Biometrics* **41**, 177–195.

Burnham, K. P. and Anderson, D. R. (1976). Mathematical models for non-parametric inferences from line transect data. *Biometrics* **32**, 325–337.

Burnham, K. P., Anderson, D. R. and Laake, J. L. (1980). Estimation of density for line transect sampling of biological populations. *Wildlife Monographs* **72**.

Cochran, W. G. (1977). *Sampling Techniques*. 3rd ed., Wiley, New York.

Crain, B. R., Burnham, K. P., Anderson, D. R. and Laake, J. L. (1978). *A Fourier series estimator of population density for line transect sampling*. Utah State University Press, Logan.

Eberhardt, L. L. (1978). Transect methods for population studies. *J. Wildlife Management* **42**, 1–31.

Emlen J. T. and DeJong, M. J. (1981). The application of song detection threshold distance to census operations. In: C. J. Ralph and J. M. Scott, eds., *Estimating the Numbers of Terrestrial Birds*. Stud. Avian Biology **6**, 399–404.

Gates, C. E. (1979). Line transect and related issues. In: R. M. Cormack, G. P. Patil and D. S. Robson, eds., *Sampling Biological Populations, Statistical Ecology Series, Volume 5*. International Co-operative Publishing House, Fairland, MD, pp. 71–154.

Gates, C. E. (1980). LINETRAN, a general computer program for analyzing line transect data. *J. Wildlife Management* **44**, 658–661.

Gates, C. E. (1981). Optimizing sampling frequency and numbers of transects and stations. In: C. J. Ralph and J. M. Scott, eds., *Estimating the Numbers of Terrestrial Birds*. Stud. Avian Biology **6**, 399–404.

Gates, C. E., Evans, W., Gober, D. R., Guthery, F. S. and Grant, W. E. (1984). Line transect estimation of animal densities from large data sets. In: S. L. Beason and S. R. Robertson, eds., *Game Harvest Management*. Caesar Kleberg Wildlife Research Institute, Kingsville, TX, 37–50.

Gates, C. E., Marshall, H. and Olson, D. P. (1968). Line transect method of estimating grouse population densities. *Biometrics* **24**, 135–145.

Hayes, R. J. and Buckland, S. T. (1983). Radial-distance models for the line-transect method. *Biometrics* **39**, 29–42.

Johnson, N. L. and Kotz, S. (1969). *Discrete Distributions*. Wiley, New York, 29.

Patil, G. P., Taillie, C. and Wigley, R. L. (1979a). Density estimation of the deep-sea red crab using the line-transect method. In: F. D. Diemer, F. J. Vernberg and D. Z. Mirkes, eds., *Advanced Concepts in Ocean Measurements for Marine Biology*. University of South Carolina Press, 493–504.

Patil, G. P., Taillie, C. and Wigley, R. L. (1979b). Transect sampling methods and their applications to the deep-sea red crab. In: J. Cairns, G. P. Patil and W. E. Waters, eds., *Environmental Biomonitoring, Assessment, Prediction and Management*. International Co-operative Publishing House, Fairland, Maryland, 51–75.

Pollock, K. H. (1978). A family of density estimators for line transect sampling. *Biometrics* **34**, 475–478.

Quin, T. J. and Gallucci, V. F. (1980). Parametric models for line transect estimators of abundance. *Ecology* **61**, 293–302.

Ramsey, F. L. (1979). Parametric models for line-transect surveys. *Biometrika* **66**, 505–512.

Ramsey, F. L. and Scott, J. M. (1978). Estimating population density for variable circular plot surveys. In: R. M. Cormack, G. P. Patil and D. S. Robson, eds., International Co-operative Publishing House, Fairland, MD, 155–181.

Ramsey, R. L., Wildman, V. J. and Engbring, J. (1987). Covariate adjustments to effective area in wildlife surveys. *Biometrics* **43**, 1.

Reynolds, R. T., Scott, J. M. and Nussbaum, R. A. (1980). A variable circular-plot method for estimating bird numbers. *Condor* **82**, 309–313.

Scott, J. M. and Ramsey, F. L. (1981). Length of count period as a possible source of bias in estimating bird densities. In: C. J. Ralph and J. M. Scott, eds., *Estimating the Numbers of Terrestrial Birds*. Stud. Avian Biology **6**, 409–413.

Scott, J. M., Ramsey, F. L. and Kepler, C. B. (1981). Distance estimation as a variable in estimating bird numbers from vocalizations. In: C. J. Ralph and J. M. Scott, eds., *Estimating the Numbers of Terrestrial Birds*. Stud. Avian Biology **6**, 334–340.

Seber, G. A. F. (1982). *The estimation of Animal Abundance and Related Parameters*. 2nd ed., Griffin, London.

Seber, G. A. F. (1986). A review of estimating animal abundance. *Biometrics* **42**, 267–292.
Verner, J. (1985). Assessment in counting techniques. In: R. F. Johnstone, ed., *Current Ornithology, Vol. II*. Plenum, New York.
Wildman, V. J. (1983). *A New Estimator of Effective Area Surveyed in Wildlife Studies*. Ph.D. Dissertation, Department of Statistics, Oregon State University, Corvallis, OR 97331.
Wildman, V. and Ramsey, F. L. (1985). Estimating effective area surveyed with the cumulative distribution function. *TR* #106, Dept. of Statistics, Oregon State University, Corvallis, OR 97331.

A Review of Current Survey Sampling Methods in Marketing Research (Telephone, Mall Intercept and Panel Surveys)

Raja Velu and G. M. Naidu

1. Introduction

Survey sampling methods play an important role in marketing research. The discipline of marketing itself draws its techniques from various social and physical sciences and any advances made in sampling methods in these areas almost always find an application in marketing research. Recognizing the importance of the topic, the first special issue (August 1977) of the *Journal of Marketing Research* was devoted to survey research. The articles in that special issue addressed three aspects of survey research, namely, sampling design, questionnaire preparation and data collection. Our objective here is to briefly review and update the aspect of sampling design with special focus on telephone, mall intercept, and panel surveys. A bibliography follows and, while not exhaustive, the listing of books and other references should provide a starting point for an iterative search.

Marketing researchers have been aware that subjective sampling procedures must be avoided in favor of probability methods of selection in order to make valid inferences about the target segments. Because of the inherent diversity of the marketing discipline, there has been a growing demand for all types of data necessitating more complicated marketing surveys. Also, during the past two decades, the household (the nucleus of most consumer surveys) has undergone dramatic changes in terms of its composition and size. More women have joined the workforce, have become economically independent, and are making buying decisions. As the environment was changing, techniques for sample surveys were also changing. The high cost of personal household interviews has led to the development and use of more efficient sample designs and less expensive data collection methods such as the use of telephone and mall intercept interviews. At the same time the public has become increasingly concerned about invasion of privacy and the maintenance of confidentiality of the information obtained.

Some developments (see Frankel and Frankel, 1977) of interest to marketing researchers include (i) techniques related to the manipulation of sampling frames,

(ii) techniques related to respondent selection, (iii) methods for minimizing the total survey error, and (iv) improving the quality of nonprobability sampling. We shall focus in Section 2 on sampling frames and procedures for telephone household surveys, a topic receiving a great deal of attention over the past decade. Judgment or nonprobability sampling procedures, still viable in marketing research, are convenient to carry out and less expensive compared to other methods. In Section 3 we shall briefly comment on mall intercept surveys, which are used increasingly by several marketing research firms. The consumer panel studies are reviewed in Section 4.

2. Telephone surveys

The telephone has become the most important tool for the collection (approximately 60%) of marketing survey data in the USA. While it has been used in the past mainly for short follow-up interviews, to clarify the information provided in personal or mail interviews, marketing researchers have now resorted to using the telephone more than ever due to the increasing cost of surveys. A distinct advantage of the method is accessibility to the respondent. Some major disadvantages are the limited time a respondent may want to spend with a physically absent interviewer and the inability of the respondent to actually 'see' the product in question as in surveys where the interviewer can display the product and obtain observational data.

In the United States telephone numbers have three parts: a three-digit area code, a three-digit central office code or prefix, followed by a four-digit suffix. A telephone exchange contains one or more prefixes and is a geographically defined unit, regulated by state public service commissions. The prefixes do not usually cross exchange boundaries within the same area code. The list of all area code-central office code combinations currently in service can be obtained from the telephone companies and there are more than 36 000 such combinations. Numbers to exclude from such a list are those of (i) the telephone company central offices (such as 555 used for directory assistance) and (ii) other central offices used solely by government or businesses. We shall refer to groups of consecutive numbers starting with 0, 00 or 000 within the suffix as 'banks of numbers'. For the operational convenience of the exchange, only certain banks of numbers are assigned to users.

2.1. Sampling frames for telephone households

The sampling unit for most marketing investigations has been primarily a household and it is implicitly assumed when telephone sampling is used that a single telephone serves a single household. This is not necessarily the case in practice. Some households have more than one telephone and more than one number. With the call forward option, business calls are sometimes automatically transferred to home phones. On the other hand, more than one household may have access to only a single telephone. Despite these problems, it is estimated that more than 90% of the households in the USA can be contacted via telephone.

Households with telephones are different from households without telephones, as shown by several demographic and economic variables. People without home phones are more likely to be male, young and black and more likely to be employed in low income occupations. These demographic and economic differences seem to be manifested in attitudinal differences as well (see Wolfle, 1979). However, because of the small number of telephone nonsubscribers, the bias introduced by surveying only subscribers is minimal, particularly if the inferences to be made are about the characteristics of the total population that are not related to the characteristics of nonsubscribers.

There are basically two kinds of sampling frames used for telephone surveys. The rest are minor variants of these two frames. One is the list of numbers and names in published telephone directories. The other is the set of all possible four-digit suffixes within the existing central office codes. The latter is used in Random Digit Dialing (RDD) methods. There are some advantages and disadvantages in both frames. It is difficult to obtain telephone directories for some areas and duplication within and between directories is not readily detected. The most important drawback of this frame, however, is that it excludes working telephone numbers that are not listed in the directory. Also, telephone directories are outdated, on the average, by at least seven to eight months. The percentage of unlisted numbers varies by regions with roughly 30% of numbers in large metropolitan areas of the USA unlisted. Households with unlisted telephone numbers tend to differ from households with listed telephone numbers on key demographic characteristics (see Moberg, 1982). Brunner and Brunner (1971) found significant differences between the two groups on certain product ownership, usage and purchase patterns.

The disadvantage of the second frame is the large number of nonworking telephone numbers that may be sampled with unrestricted random sampling. In the USA only about 1 in 5 dialings will connect with a useable residential household. The effort to identify these numbers adds considerably to the cost of a survey. Waksberg (1978) reports that this spade work is done by marketing rsearch firms and the more 'useful' sampling frames are developed by these firms at considerable expense and are not available to the general public. Most researchers cannot afford to duplicate such a costly task. It is important to narrow the frame used for RDD. The designs to be discussed in what follows are expected to reduce the proportion of unused numbers sharply.

To emphasize the inherent differences between the two frames and their variants, it is useful to mention the problems in determining the status of a given number. Dialing a working number can result in (i) a completed call, (ii) unanswered rings, (iii) a busy signal, or (iv) wrong or no connection because of misdialing or technical problems. Unless the call results in a contact, it is impossible to determine whether the number belongs to a household or a nonhousehold. In RDD sampling, a nonworking number is not always easily determined. Dialing such a number can result in (i) a recorded message stating that the call cannot be completed as dialed, (ii) no connection, (iii) unanswered rings, or (iv) connection with a number other than the number that was dialed. The last possibility introduces biases in RDD sampling, because the telephone system equipment is not normally designed to receive a nonworking number. Note that the households reached in this manner have a greater probability of inclusion in the sample.

2.2. Telephone sample designs

Telephone sample designs can be broadly divided into directory-based and Random Digit Dialing (RDD) methods. In addition to the directory with the usual alphabetic listing of subscribers, one can obtain a street address telephone directory arranged by geographic location. This directory is more expensive and is available only for large metropolitan areas. We shall briefly discuss directory assisted designs followed by RDD procedures and finally discuss the concept of dual frame designs.

(i) Directory based methods

(A) *Direct selection from directory.* This is the most basic of all the directory assisted methods. A sample of directory lines is selected using either systematic or simple random sampling. One could also use cluster sampling for easy execution. The cluster consists of a randomly selected line and the next k lines. To avoid actually counting lines, directory column inches can be used. This method yields an equal probability sample of all listed numbers with a minimal percentage of wasted dialings due to nonworking numbers. A disadvantage of the cluster sampling method is that names listed together in a directory might belong to the same community, religion, etc., and if they are homogeneous with respect to the variables being estimated, the design is inefficient as compared to simple random sampling of lines. The major disadvantage of the Directory method is that it does not give any chance for unlisted working telephone numbers to appear in the sample. The bias may be significant in certain surveys and the following procedures are proposed to correct partially for the bias.

(B) *Addition of a constant to a listed number.* A number is randomly selected from the directory and an integer, either fixed or randomized (between 0 and 9), is added to the directory number. This gives a chance for inclusion of possibly unlisted numbers in the sample. Some variants of the above procedure involve randomization of the last r (2, 3, or 4) digits of a directory number. Two drawbacks of these procedures are (i) when r increases, the number of wasted dial rings will increase, and (ii) all telephone numbers do not have an equal chance of inclusion, since the probability of selection of a number would be proportional to the number of directory listed numbers in the same rth bank. If the numbers are not in the directory, they automatically eliminate the possibility that numbers which follow them will be in the sample. A method suggested by Sudman (1973) to correct for (ii) is described below.

(C) *Sudman's method.* A random sample of listed numbers is selected and the last (usually $r = 3$) digits are ignored. This results in banks of numbers selected with probability proportionate to the number of listed numbers in the bank. Calls are made using random digit dialing within the bank until a predetermined number of households with *listed* numbers have been reached. The predetermined number is fixed so that the resulting sample is self-weighting. If we let N = total number of household telephones, N_L = number of telephones among N that are listed, n = sample size, m = number of selected banks of working numbers, and N_{L_i} = number of listed telephones in the ith bank, then

$$\text{probability of inclusion of a number in the sample} = \left(N_{L_i} \frac{m}{N_L}\right)\left(\frac{N_L n}{NmN_{L_i}}\right) = \frac{n}{N}. \quad (1)$$

REMARK. This probability is exact (and the sample is self-weighting) only if (a) the proportion of listed households numbers in the ith bank is equal to the overall proportion (N_L/N) of listed household numbers and the predetermined number of sampled listed households in a bank is n/m, or if (b) the predetermined number of sampled listed households in a bank is fixed as $= nN_i N_L/NmN_{L_i}$ where N_i = number of household telephones in ith bank.

The first bracketed term indicates the probability of inclusion of bank i in the sample and the second term, that of selecting a number within the bank. The procedure is unbiased and self-weighting. As Waksberg (1978) points out, this method also has several problems. Ascertaining whether a number dialed is listed or not can be difficult. For example, in a national survey, the procedure requires the use of a large number of telephone directories. Finally, because the numbers are clustered, a large proportion of them may occur in relatively empty banks, resulting in unequal numbers of households per cluster.

(ii) *Random digit dialing methods*

These methods are used to obtain equal probability samples of all telephone numbers both listed and unlisted. As mentioned before, an unrestricted application of the procedure will lead to the inefficient use of survey resources. Therefore, it is important to narrow the sampling frame by eliminating nonworking numbers. If information on nonworking numbers is available (e.g., which banks are not assigned), random digits within these banks could be excluded from the sample. Some telephone companies will provide information about working banks. However, this information is usually not available, forcing researchers to use directories to determine working banks. Typically those banks with less than three listed phone numbers are eliminated. Currently these represent approximately 56% of all banks in active exchanges in the USA. The incidence of telephone households in the sample can be increased by eliminating the business telephones listed in the yellow pages of the telephone directory. It is evident that all these efforts require a considerable investment of time and, unless the frame is used repeatedly, the cost may be prohibitive for a small survey.

(A) *Waksberg–Mitofsky design*. The (RDD) selection procedure proposed by Waksberg (1978) is as follows. Obtain from the telephone companies all area code-central office code combinations currently in service. Append all possible two digits and treat the resulting eight-digit numbers as primary sampling units (PSU). Randomly select a PSU and the next two digits. If the ten-digit number is for a residential address, the PSU is retained in the sample and if not, it is rejected. If retained, additional pairs of random numbers to identify the two last digits are selected within the same PSU and dialed until a set number of residential telephones are reached. This process is repeated until a predetermined number of PSUs are chosen. This design produces an equal probability sample of working telephone numbers. The procedure of selecting PSUs is similar to Lahiri's (1951) selection procedure for probability proportionate to size (pps), although the latter requires a prior estimate of cluster size. This procedure which selects PSUs with probability proportional to working numbers differs from Sudman's method which selects PSUs proportional to listed working numbers. The stopping rule for the Walksberg–Mitofsky design also refers to working numbers and is not restricted to listed numbers. It is important to note that this procedure uses a cluster size of 100, a practical advantage over a cluster of 1000.

Since this sample design is the one that is most often used in practice, we shall briefly outline its design parameters. The following notation, in addition to what has been already defined, will be used: Let M = number of PSUs, $k + 1$ = cluster size in sampled PSU (number of residential telephones selected), $n = m(k + 1)$, ρ = measure of homogeneity (intraclass correlation) with prefixes, N_i = number of residential numbers in ith PSU, both listed and unlisted, $\pi = \Sigma N_i/100M$, proportion of residential numbers in the population, $100\pi M = N$ = total number of household telephones, t = proportion of PSUs with no residential numbers, C_u = cost of an unproductive call to a non-residential number, and C_p = cost of a productive call.

A crucial problem in this procedure is the large value of t. Because all possible choices of two-digit numbers are appended to area code-central office code combinations to arrive at the PSU, it is possible that a large number of PSU's may not contain any residential numbers. It is important to obtain an estimate of the proportion of PSUs with no residential numbers. In the absence of such information, optimal procedures suggest $k = 0$, which implies no clustering and selects n different PSUs. This holds for cluster sampling with PSUs selected with (i) equal probabilities, and (ii) probabilities proportional to size (pps) and within cluster rate set to produce a self-weighting sample. Assume $N_i \geqslant k + 1$. In both instances it can be shown that the expected value of the total calls is n/π which is independent of k. But for cluster sampling with pps, and a known proportion t of PSUs having no residential numbers (for which $N_i = 0$), if it is assumed that for the remaining $(1 - t)M$ areas, $N_i \geqslant k + 1$, we can show the expected total calls is $(m/\pi)(1 + (1 - t)k)$ (see Waksberg, 1978, p. 43) and the optimum cluster size is

$$k + 1 = \left[\frac{t}{\pi(C_p/C_u) + (1 - \pi - t)} \left(\frac{1 - \rho}{\rho} \right) \right]^{1/2}. \qquad (2)$$

The value of t can be expected to be smaller for urban than rural areas. An estimate based on a national U.S. study is given by Groves (1978) as $t = 0.65$. A reasonable value for π at a national level can be arrived at as follows: Currently there are an estimated 87 million households in the United States of which roughly 93% (81 million) have telephones. Hence, among the 360 million possible numbers (based on the current estimate of 36 000 exchanges), at least 81 million telephone numbers are assigned to households. Thus a probable value for $\pi = 81/360 = 0.22$. Once the working banks are identified, it is estimated that on the average there are 43 listed numbers per active bank. Hence, the cost of identifying a productive telephone call could be taken to be twice the cost of an unproductive call. The value for the ratio $C_p/C_u = 2$ appears to be reasonable. Most marked research data refer to attitudinal variables and Groves (1978) reports an average value for $\rho = 0.08$. With this value for ρ, the optimal size given in (2) is $k + 1 = 4$. These numbers are used mainly to illustrate the calculations and the researcher should make his/her own judgment on these values. For many geographical areas, it is possible that $1 - \pi - t \simeq 0$, thus yielding the simplified formula

$$k + 1 \simeq \left[\frac{t}{\pi C_p/C_u} \left(\frac{1-\rho}{\rho} \right) \right]^{1/2} \qquad (3)$$

which largely depends on the value of t. Information on t can be obtained from local telephone companies or can be estimated from the directories.

(B) *Stratified element sample.* An alternative design is discussed in Groves (1978). The procedure initially groups together all central office codes in the same exchange and then groups together exchanges in the same area code. Size categories of the exchanges are then formed based on the number of central office codes in an exchange with the number of central office codes acting as a proxy to population density. Within each size category, exchanges are ordered geographically within an area code and similarly area codes are then ordered geographically. Given this ordering of the frame, a systematic sample of central office codes is drawn. A four-digit random number is generated and appended to a selected central office code, yielding a 10-digit sample telephone number. Groves (1978) observes that only about one-fifth of the numbers were confirmed as working household numbers, while in Waksberg's design a roughly threefold increase in identifying working household number is possible. The main attraction for using this design would be when ρ, the measure of homogeneity, is large. This design can be treated as a simple random sample when the stratification introduced based on the exchange size is rather weak.

(C) *Dual frame sample design.* The two stage cluster design, proposed by Waksberg (1978), is better than directory-based designs in terms of coverage rates and over stratified element sampling in terms of cost. But the design requires a new selection from the same PSU for each nonworking number encountered and thus adds to the cost of screening numbers to identify residences. It is difficult to distinguish nonworking numbers from unanswered residential numbers. Another problem is the low response rates for telephone surveys attempted without prior contact. It is found that persons with listed numbers are more likely to cooperate than those with unlisted numbers. Groves and Lepkowski (1986) consider dual frame designs as proposed by Hartley (1962) to be useful when the target segment forms a majority of elements in one incomplete list frame (directory listings) but a minority in another complete frame (RDD generated numbers). The poststratified estimator suggested by Casady, Snowden and Sirken (1981), which mixes the estimates from each of the two frames, is investigated by Groves and Lepkowski (1986) and Lepkowski and Groves (1986). If we let p denote the proportion of the unlisted telephone population and θ denote a mixing parameter, the estimator of the mean is

$$\bar{y} = p\bar{y}_{\text{UL, RDD}} + (1-p)[\theta\bar{y}_{\text{L, RDD}} + (1-\theta)\bar{y}_{\text{L, DL}}] \qquad (4)$$

where $\bar{y}_{\text{UL, RDD}}$ is the estimate for the unlisted population chosen by RDD, $\bar{y}_{\text{L, RDD}}$ is the estimate of the listed population chosen by RDD and $\bar{y}_{\text{L, DL}}$ is the estimate of the listed population chosen from the directory frame cases. The cost advantage of the dual frame derives from the list frame in identifying the working numbers. Several survey research firms (Metromail Corporation, Survey Sampling, Inc.) maintain a computerized data bank of all published directories and in one test for the state of Michigan, Groves and Lepkowski (1986) report 88% of numbers on the list were found to be working numbers as compared to 59% for the selection of samples within the PSU in RDD design. From the form of the poststratified estimator, it can be seen that the crucial parameters are p and θ which depend on the geographical region and the type of marketing research investigation. It is estimated that roughly 64% of contacted RDD sample households are in directory listings, but the proportion of RDD numbers not contacted but found in the listing is around 66%. At the national level it is not known what proportion of these noncontacted numbers are working residential numbers. This may be influenced by large metropolitan areas where a low rate of list frame coverage is known to exist. Thus, the dual frame design can result in increased coverage (than list frame) and also increased precision (than the cluster RDD) by following simple random/stratified element designs on the list frame, thereby avoiding homogeneity due to clustering. To evaluate the dual design more thoroughly, the marketing investigator must know several cost elements and the relative nonresponse bias. The nonresponse bias is typically measured by the difference between two group means, where only one group receives an advance letter. Based on a simulation study for the U.S. National Crime Survey, Groves and Lepkowski (1986) suggest optimal allocations between 35% and 80% to the list frame.

2.3. Respondent selection in telephone surveys

There are a number of other issues to be addressed in telephone surveys. Some households (roughly 2.4% in the USA) have more than one telephone number, making it necessary to obtain this information during the interview so that appropriate estimation weights could be constructed. In any telephone survey, ambiguities exist about no answers, uncertain rings, busy signals, etc. Any stopping rule for classifying these is bound to introduce some bias in sample selection. A more serious problem from a marketing researcher's point of view is that the person answering the telephone is not necessarily the same person who makes the purchase decisions. As shown in the literature on consumer behavior, buying decisions result from an interaction of all family members. To retain the characteristics of a probability sample, the person to be interviewed should be selected at random. We discuss a few approaches to the problem in the following.

A selection procedure suggested by Kish (1967) in the context of area probability samples requires all eligible respondents within a household to be listed by sex and by age within sex categories. The interviewer then selects one respondent by using a random number table (see Kish, 1967, Section 11). This procedure is difficult to use in telephone surveys where most refusals to participate occur at the beginning of the interview. The procedure is time consuming and could present problems establishing rapport. For example, asking for the number of adult males in residence could be perceived as insensitive to single women living alone. Because rapport with the respondent is so vital to telephone surveys, Troldahl and Carter (1964) adapted the Kish format but based the selection on only two easy-to-answer questions: (i) How many adults live in your household, counting yourself, and (ii) how many of them are men? Using four selection matrices rotated randomly over the sample, a respondent is selected. This procedure does not significantly reduce refusals when compared to the Kish strategy (see Frey, 1983, p. 80). Bryant (1975) suggested dropping one of the four matrices every second time it appeared in the rotation. This would result in the selection of more male respondents and the procedure takes into account increases in one-person households and households headed by women. It must be noted that these alternative strategies assign unequal probabilities of selection to some eligible respondents such as middle-aged adults. Another variation used by Groves and Kahn (1979) is to modify (ii) 'how many of them are women?' A recent investigation by Czaja, Blair and Sebestik (1982) reveals no major differences in cooperation rates and demographic characteristics across the three methods.

Two procedures reported recently seem to be effective in terms of operational use and eliciting higher response rates. Basically, these two avoid asking household composition questions before beginning the interview. The first procedure is suggested by Hagan and Collier (1983). The designated respondent is predetermined to be one of four possibilities: oldest man, youngest man, oldest woman or youngest woman. After the initial introduction, interviewers simply ask for the designated respondent (randomly chosen and printed on the interview form a priori) and when a respondent of that designation does not live in the household, the opposite sex is interviewed. In single-person households, the age designation is irrelevant. Based on a national study, the authors suggest that this procedure is an improvement in terms of lower refusal rate. The second procedure given by O'Rourke and Blair (1983) selects the adult who had the 'most recent birthday'. This is a probability selection method and ascertaining the birthday is considerably easy. Comparing this with Kish's procedure, based on a survey, the authors found the major difference in refusal rate occurred at the preselection stage. Once the respondents agreed to participate, it did not matter which procedure was used to continue the interview.

2.4. Randomized response techniques in telephone sampling

The randomized response technique originally introduced by Warner (1965) to obtain estimates of behaviour that is usually underreported, is found to be useful for personal interviews. A randomizing device is used to choose a statement and the respondent is asked to provide a response to the one selected. The interviewer is not shown the outcome of the device nor is informed of which statement is answered. The most difficult aspect of a telephone application of the randomized response technique for sensitive questions is the provision of a randomization device. As Stern and Steinhorst (1984) observe, there are two main problems (i) the device is not readily available to many respondents, and (ii) the complexity of instructions necessary to provide a satisfactory distribution may inhibit respondent cooperation. Also, suggestions from a 'faceless' voice to flip a coin may be regarded as foolish by some respondents. However, an advantage of using a respondent-supplied randomizer is that it eliminates the respondent's suspicions that the interviewer has 'fixed' the randomizer. A potential disadvantage is that it does not provide a known probability distribution.

There are several randomizers suggested in the literature including credit card number, street address, occurrence of events, etc. (see Orwin and Boruch, 1982). The one that is tested on a limited basis is the last digit of randomly selected telephone numbers. This provides a known distribution for both the selection of sensitive and nonsensitive questions and the generation of surrogate answers (see Stern and Steinhorst, 1984). Although this method is considered to be successful on the issue of response privacy, the nonresponse is still high. This method also requires both the interviewer and respondent to have access to the same telephone directory. Each geographical area served by a different telephone exchange and telephone directory would be sampled as a separate stratum. At a national level, this may create some operational problems. Other randomizers such as the last digit of street address are supposed to overcome this problem, but in the absence of a known distribution of the last digits, they are not statistically attractive to use.

2.5. Locating a special population using RDD

In many instances, the researcher may be interested in locating a subclass of the total population. Blair and Czaja (1982) show how Waksberg's two-stage cluster design can be modified, if it is known that this special population clusters geographically. This modification takes advantage of the fact that the telephone central office codes are assigned to well-defined geographic locations. It works as follows: select a simple random sample from all possible telephone numbers. These numbers are then called and only those working residential numbers of a household with the appropriate special characteristic are retained. The first eight digits of each retained number are then defined as a PSU. Using each retained telephone number as a random start in the PSU it created, numbers are then sequentially generated and screened. This procedure is continued until a certain cluster size is identified. The formula given in (2) for optimum cluster size for a general population holds also for a special population, with all the symbols defined for special population; e.g., N_i = residential numbers for special population in ith PSU, etc. (see Sudman and Kalton, 1986, p. 416).

As Waksberg (1983) notes, this procedure has some serious statistical implications which in many situations may reduce the efficiency. Observe that $N_i \geq k + 1$ in Waksberg's procedure and in practice, when the average value of N_i is 65 per cluster, the choice of k could go up to 40. But in the case of special populations, PSUs could exist in which it is not possible to reach the predetermined cluster size even if the 100 numbers are used. The special population households associated with clusters that are smaller than the specified cluster size have a lower probability of selection than the rest of the special population. Hence, to produce an unbiased estimate for the total population, we must adjust for unequal probabilities which increase the sampling variance (see Kish, 1967, p. 430).

2.6. Ring policy in telephone surveys

Each telephone call is composed of two-second rings followed by five seconds of silence. Survey research firms on the average allow six rings per call, thus the amount of time taken to reach a potential respondent is on the averge 37 seconds. Smead and Wilcox (1980) questioned how long the phone should be allowed to ring based on a telephone survey using the members of a major university consumer panel. Ten rings and three callbacks were used. The average answer time for the 219 respondents was 8.7 seconds with a standard deviation of 6.3 seconds. The answer times followed a gamma distribution and suggested that only four rings (or 23 seconds) were necessary to reach 97%.

2.7. Telephone sampling: Recent developments and other uses

The use of telephone interviewing is widespread because of its major cost advantages. However, there are still many situations that require face-to-face interviewing, particularly those that deal with special subgroups of the general population. This involves screening, and the rarer the group the more costly the screening. However, in general telephone screening costs are lower than face-to-face screening. Sudman (1978), based on a realistic cost model, has shown that telephone screening will be an optimum procedure unless (i) ρ is small, (ii) the density of interviews is low, and (iii) locating and screening costs are small relative to interviewing costs. From the discussion in Section 1, it follows that (iii) could be an important consideration in using RDD. However, directory-based telephone screening might be cost effective.

Many survey research firms have data bases constructed from the telephone directories supplemented with auto registration data. These are useful for mail samples. Information collected from other sources such as census records are sorted by area code and telephone exchange which provides a faster way to reach a target population such as low income families, Hispanic groups, etc. The yellow page listings are used for business samples, because the directory category headings are broad and easy to use by marketing researchers.

Computer-assisted telephone interviewing (CATI) was used first by market research agencies in the private sector. The concept was proposed by the American Telephone and Telegraph Company to measure customer evaluation of telephone services. CATI is now very popular in other types of organizations as well. Interview responses are quickly processed and by accumulating counts of key respondent characteristics while interviewing, quota targets, i.e., desired sample sizes in strata in RDD sampling, can be tracked. By adding visual monitoring to telephones from supervisory terminals, CATI provides efficient control in the interview process (see Nichols II and Groves, 1986).

3. Shopping center sampling and interviewing

Interviewing shoppers in shopping malls started in the early 60's when the development of totally enclosed shopping centers provided researchers access to a large number of shoppers from a wide geographic area. Prior to the mall intercept, surveys were mostly conducted in supermarkets, discount stores, train stations and places where large concentrations of people could be found. More than 170 malls have permanent market research facilities, some of which are equipped with interviewing stations, video tape equipment and food preparation facilities for conducting taste tests. A large number of malls permit intercepts on a temporary basis but many prohibit interviewing because they see it as an inconvenience to their shoppers.

The two major advantages of a mall intercept interview are cost and control and it has many of the advantages associated with personal interviewing. Also, it is the only way to conduct most taste tests and ad tests requiring movie projectors or video tape equipment. However, there are a number of disadvantages. The important one is that shoppers are frequently in a hurry and may not respond carefully. It may be difficult to maintain a controlled interviewing environment in the presence of the respondent's children, relatives, etc. In spite of these problems, mall intercept interviews are increasingly used in market research. It is estimated that, of those who had participated in any form of a survey, 18% were contacted through mall intercept interviews as compared to 12% through personal interviews (see Gates and Solomon, 1982). Because of the administrative efficiency, it has some potential for growth.

3.1. Sampling issues

Samples for most shopping center interviews are selected haphazardly and do not reflect the general population. The effect and sources of biases are not properly understood and are not taken into account. If the investigation is at the early stages of product development, it may not be necessary to follow rigorous sampling procedures. But if the objective is to generalize to the population, it is important to follow rigorous sampling schemes. Shopping center sampling can be compared to sampling mobile populations. The major interest in studies related to mobile populations has been in estimating the size of the population but little attention has been paid to sampling time and location. Sudman (1980) provides some procedures that take these aspects into account.

The key assumption in the mall samples is that all households have a nonzero (but not equal) probability of being found in a shopping center. This assumption may not be realistic and the bias introduced for some special groups such as lower income or older households may be substantial. Secondly, because the probability of selection is a function of the frequency of visits, that frequency must be estimable. This may strain respondent memory and may introduce some biases.

Sudman's procedure works as follows: First select the shopping centers using the same basic random sampling procedures used in the selection of locations in a multistage area probability sample with probability proportional to a size measure such as total annual dollar volume. The optimum number of shopping centers and the number of respondents can be determined using the formulae for area cluster samples,

$$n_{opt} = \left[\frac{C_1}{C_2}\left(\frac{1-\rho}{\rho}\right)\right]^{1/2} \tag{5}$$

where C_1 is the set up cost at a shopping center and C_2, the cost per interview and with a total budget $C = C_1 m + C_2 mn$, where m is the sampled number of shopping centers, n is the number of interviews per shopping center and ρ is the intraclass correlation coefficient between shoppers within shopping centers. Because C_1 is generally much larger than C_2, large samples are selected from each center; but the heavy clustering increases the sampling variance.

The respondents can be selected either when they arrive at the center or as they move around within it. For the latter, we require information on how much time they have spent in the center because persons spending more time shopping have a higher probability of selection. To select an unbiased sample of entrances, it is important to know the fraction of customers the entrances attract from previous counts. This size measure can be used to sample entrances with probability proportional to size and is much more efficient than sampling them with equal probability. While the less used entrances will be sampled fewer times than the more heavily used entrances, the sampling rate would be higher at the less used entrances if a self-weighting sample is desired. Establishing rules for within shopping center sampling is more difficult than entrance sampling. Identical traffic patterns in all parts of the center cannot be assumed since the location of discount stores is more likely to attract customers different from those who shop at fashion centers.

It is important to use careful time sampling procedures, to avoid biases against certain types of customers, for example, working women who mostly shop in the evenings and weekends. Selecting an eligible time period with equal probability is not an efficient design. The solution is identical-sampling of time periods with probabilities proportionate to the number of customers expected in the time period. Sudman (1980) suggests forming time-location clusters, based on past data and selecting these clusters with probability proportional to past size.

The above procedures are far more sophisticated than those procedures used in the past. There are still problems in their implementation and generalizability. We suggest using the dual frame concept. For each shopping center, we may obtain trade area maps showing geographic areas from which stores draw their trade, because shopping centers generally attract those households nearest to it. These maps are sometimes drawn from shopper surveys (see Blair, 1983) intended for a different use by the retail merchants. With such a map, we may have a sampling frame from which we can draw an independent sample by telephone that can be combined with the mall sample. For a related discussion, see Bush and Hair, Jr. (1985).

4. Consumer panels

The panel has become an important tool for monitoring market factors ever since Jenkins (1938) and Lazarsfeld and Fiske (1938) used them to study brand preferences and reader reactions to a magazine (*Women's Home Companion*). Since then, the use of panels to study the purchase behavior of nondurable consumer goods has gained importance in North America and some Western European countries. See Hardin and Johnson (1971) for various applications of panels in marketing research. Marketing Research Corporation of America (MRCA) followed with a panel of 7500 households in 1941 and focused on the consumer purchase behavior of grocery, health and personal care, and textile products. Today the use of panels in marketing studies is much more widespread and there are hundreds of consumer and industrial panels mostly located in North America and Western Europe. Nevertheless, some of the initial sampling problems related to panels still remain. This section will briefly review some of these problems from a sample design perspective. The problems related to panel sample design are not usually covered in discussions of Sample Survey Methods. Sudman and Ferber (1979) identified three critical areas likely to induce bias in panel sample design. These are: (i) bias created by initial refusals, (ii) bias created by subsequent mortality, and (iii) bias created through conditioning. A brief discussion of these areas follows. It must be recognized that there are other critical areas, such as aging of the panel and possible changes in the population that are not represented in the sample, which are not discussed here.

A consumer panel measures purchases of a product at any given point and over a period of time. This has been used to measure market trends, seasonal effects and the effects of marketing strategies. Panel data from the *Chicago Tribune*, National Panel Diary (NPD), National Family Opinion (NFO), Marketing Research Corporation of America (MRCA), Intercontinental Marketing Services (IMS), etc., focus on different product lines and industries. The majority specialize on consumer products, mostly nondurables distributed through grocery stores, while industrial panels such as those from IMS focus on hospital equipment, supplies, and doctors' prescriptions. Alternatively, store audits are used to estimate market size and trends (A. C. Nielsen) and with the advent of electronic scanners of Universal Product Codes (UPC), purchase data have become much more reliable and offer extensive detail on product/brand purchases as well as profiles of sample buyers. Information Resources, Inc., with headquarters in Chicago provide Infoscan and Behaviorscan services to business clients. Each panel member receives a member identification card that is presented to the store clerk at the time of checkout. All purchases are electronically recorded, eliminating the need for written diaries. This method has distinct advantages, as its popularity is growing both in North America and abroad (Information Resources has operations in Australia, Canada, France, Great Britain, Japan and West Germany). These sources also study consumer brand preferences and brandswitchings over a period of time. Panel data have been extensively used in the formulation and evaluation of pricing strategies (see Montgomery, 1971). Segmentation by usage, package size, effectiveness of 'marketing mix' variables have been studied by, among others, Blattberg and Sen (1976). Models are developed to predict market penetration based on repeat buying rates (see Eskin, 1973). With the information provided by panels on both purchasing and media exposure, efforts were made to estimate the effectiveness of advertising particularly for new products (see Nakanashi, 1973). Carefoot (1982) and Information Resources, Inc., have used scanners to evaluate the effectiveness of advertising. MRCA's panel data have been utilized to sense changing food habits leading to the modification of existing products and the development/introduction of new products to better serve the consumer.

4.1. Bias created by initial refusals

Refusals, non-cooperation and nonresponse are to be expected in any survey. The level of cooperation attained is dependent on recruiting methods used and the nature of tasks required by the panel members. Often higher rates of cooperation are achieved if the expected effort from the respondent is lower. Panels recruited by face to face contact tend to have higher rates of cooperation than those recruited by telephone or mail. Oversamples are drawn initially to balance demographic variables such as geography, household size, income, education of the head of the household, etc. Even if the panel fits all these demographics, there is no assurance that the panel results are bias free if willingness to cooperate on a panel and purchase of a product are related to a variable such as lifestyle. Panel cooperation seems to be closely associated with family size; for example, households with two or more members tend to cooperate more readily than single person households. From the studies of U.S. Department of Agriculture (1953) and additional investigations ('Panel bias reviewed', 1976), the following patterns emerge:

• Single person households have a higher tendency to be noncooperators or 'not-at-homes'. They have less interest in food purchases and maintain records on an irregular basis.

• The older the housewife (after 55 years), the lower the chances of joining the panel. This may be related to education and the ability to keep records.

• Homeowners are more likely to cooperate than tenants. This again may be related to household size.

• Working wives are less likely to join the consumer panel than nonworking wives.

• Panel cooperators tend to be more 'price conscious' than noncooperators.

• The income distribution of panel members and that of the U.S. population tend to be very similar except at the lower end where a smaller percentage of lower income households are represented in the panel.

Except for household size, the differences between cooperators and noncooperators tend to be negligible with respect to demographic profiles. However, the differences could be significant with respect to socio-psychographic characteristics such as organization, recordkeeping, and price consciousness. With new developments (Infoscan and Behaviorscan) the need to keep records by the panel members is eliminated, reducing potential errors in reporting, recall, and recordkeeping. Atwood consumer panels in Great Britain and Germany show no significant differences between panel members and the general population with respect to readership of magazines and newspapers and selected psychological and buying variables (Sudman and Ferber, 1979).

In summary, the evidence from the U.S. and European studies indicates that some biases in consumer panels such as household size, age of the housewife, and level of education of the head of the household are possible. In panels requiring less effort, the refusal rate is lower resulting in lower sample bias. Panels that require more effort and those recruited by mail or telephone often tend to have a higher percentage of non-cooperators resulting in higher bias.

The ratio method of estimation has often been used to obtain better estimates of the population. Under-representation of smaller households or a specific geographic region is overcome by the application of suitable post-stratified weights in deriving the population estimates.

4.2. Bias due to attrition/mortality/formation of new households

A panel should be representative of a target population. While the population itself may not change drastically from year to year, some changes do occur over time. Dissolution of old households, formation of new households, household moves, etc., are examples of changing population characteristics. Potential problems are (i) panel member dropouts, (ii) household moves, (iii) household dissolutions, and (iv) new household formations. We will discuss each of them briefly.

(i) Dropouts: Panel dropouts or attrition is often estimated to be 5% to 10% from one period to the next in the USA. Charlton and Ehrenberg (1976) reported that 88% of their limited sample completed the 25-week panel. Farley, Howard, and Lehman (1976) reported a 43% dropout rate from waves of interviewing spanning 18 months. Personal situations such as illness in the family, birth of a child, enlistment in the army, etc., are often reasons for dropout. Two methods have been used to overcome this problem. An oversample could be made in anticipation of an expected dropout rate. However, in practice, it may not be possible to maintain large oversamples (European panel operators tend to follow this procedure). Besides, this would lead to sampling bias. The second method is to replace the dropout household with a new household of similar characteristics by a method of imputation in the field. The problem of noncooperation of a newly selected household is similar to that of initial recruiting. A prepared list of substitute households is searched until a replacement is found. Even if replacements are representative with respect to selected socio-economic and demographic variables, they could differ on behavioral variables such as purchase quantity, degree of brand loyalty, private brand proneness, etc. Winer (1983) suggested that replacements be made with due consideration to selected behavior variables.

Sobol (1959) and Bucklin and Carman (1967) demonstrated that attrition introduces potential bias in panel based market research. Hausman and Wise (1979) have designed a model of attrition and proposed a maximum likelihood method of estimation of parameters. They estimated the parameters in the presence of attrition as well as bias due to attrition. Winer (1980, 1983) and Olsen (1980) developed procedures for estimation of attrition bias in the absence of replacement of dropouts.

Maintaining a representative panel is not easy. Most panel operators recognize the importance of suitable compensation and effective communication with panel households as essential factors in keeping morale high and turnover rate at a minimum.

(ii) Household moves: When a household moves, it is a generally accepted principle to follow it. The only exception is if the panel is confined to a specific geographic area and the move takes the household out of that target area. Following the panel wherever they go ensures continuous representativeness of the panel including the patterns of mobility inherent in the population.

(iii) Household dissolutions: In the event that all members of a panel household die, the household is often replaced with a similar household. If one of the spouses dies and the other joins a nursing home, the household is dropped from the panel.

(iv) New household formations: The panels are continuously monitored as to the size of the household. If a new household is formed through marriage, the new household is recruited with probabilities inversely proportional to the number of persons who will constitute the new household. Thus, in the case of new households resulting from a marriage, half the split-offs are recruited. This way the panel recruits younger households to maintain their representativeness in the population.

4.3. Bias created through conditioning

The term 'conditioning' refers to stimuli in a broad sense and includes all contacts between panel operators and panel households such as initial recruiting calls, instructions/training, diary keeping, compensation, and newsletter or other forms of communication whether personal or mail. Sudman and Ferber (1979) classified the effects of the stimuli into three categories: immediate, short-term and long-term. These effects could be in terms of purchase behavior affecting brand choice, store choice, quantities purchased, number of shopping trips per unit time, expenditures on a product per unit of time, etc. For example, keeping a 'time use' diary might cause a person to use a different pattern of time utilization than the 'usual'. Besides changes in behavior, it might also change attitudes and beliefs affecting future behavior.

Studies focusing on the immediate effect of the acceptance of an invitation to join the panel on a household have used 'recall' techniques to assess the differences in purchase behavior before and after joining the panel. The results, however, were inconclusive. The effect of short-term conditioning seems to be evident based on empirical studies. A 1973–1979 study conducted by the Survey Research Laboratory at the University of Illinois on medical diaries found that first month reportings were 14% higher than the subsequent records of the following two months. Similarly, Sudman (1962) found that a panel diary method used to collect data on ten product purchases over an eight-week period reported that first week purchases were 20% higher than the eight-week average and second week expenditures were 8% below the average. The experiences of U.S. Bureau of Census (1972–1973) also support the evidence of the existence of a short-term conditioning effect on the behavior of panel households. As a result, many practitioners ignore the first period as 'trial' data or omit it in the trend analysis.

Substantial evidence exists that a special stimulus can result in major changes in reported purchase behavior. A sticker reminder in a diary and a postcard reminder to record all soft drink purchases resulted in an increase of more than 30% in reported purchases. A similar study on reporting purchases of citrus products (special form included for reporting) showed that the experimental group had a significantly higher incidence of purchase records of citrus products during the first month than the control group. However, the initial conditioning effect seemed to have disappeared in later months.

Some researchers have speculated that keeping diary records could sensitize households over time and cause them to be better shoppers. One panel study indicated that an average household made 2.7 trips per week for grocery shopping during the first three months of data collection period and 2.6 trips per week in the next three months. The differences are not statistically significant, and any conditioning effect is negligible. Ehrenberg (1960) using a British consumer panel and Cordell and Rahmel (1962) using A. C. Nielsen panel for television viewing habits concluded that there may be a slight short-term effect of panel conditioning but it disappears over the long term.

Long-term effects on households serving as panel members is of major concern as they could develop fatigue or become uninterested in keeping diaries. Interestingly, there is no evidence to support such a hypothesis. Ehrenberg (1960) described several studies and pointed out that over a ten-year period the Atwood consumer panel compared 'old' and 'new' panel members and found no significant differences. The general conclusion was that the length of panel membership did not systematically affect the reported results. Any 'conditioning' that may exist in the early period of panel membership is likely to wear off or stabilize over a reasonably short time.

Some form of compensation is very common for most continuing panels and is often in the form of money, gifts or other forms of motivation (participation in lotteries, etc.). The amount or value of compensation seem to vary widely depending on the type of respondent. For most consumer nondurables, the compensation has been in the range of $10 to $60 a year. For physician panels, the compensation was several hundred dollars. Both European and Japanese panels seem to receive better compensation than those in North America. Ferber and Sudman (1974) and Sudman and Ferber (1971) reported that the households receiving compensation provided better quality data than those who did not. Their conclusion was that compensation in sufficient amounts is necessary to ensure initial and continuing cooperation as well as quality of reporting. There is no evidence that the form of compensation has any major impact on cooperation (Ferber and Sudman, 1974).

4.4. Recent developments in consumer panels

A recent study by Grootaert (1986) on the estimation of household expenditures in Hong Kong using the panel diary method suggested the use of multiple diaries—each member of the household maintains a separate diary of daily expenditures. This method resulted in more accurate reporting of expenditures particularly on 'personal' products such as clothing, shoes and services. The reporting arrangements depend on family structure, size and decision making process within a household. As such, the results are not usually generalizable to other countries.

With high-tech electronic methods of data collection using scanners, the need to maintain written diaries is diminishing. As increasingly more retail stores are equipped with UPC scanners, data collection using panel method has become increasingly important for various marketing experiments. This has led to what is called 'single source' research where many promotional experiments can be tested out by following the panel members from their TV sets to checkout counters.

It is also easy to measure accurately the effect of promotional campaigns via this high-tech research. Information Resources, Inc. (IRI) monitors 3 000 households in 8 small town markets. The micro computers record when the television is on and which station it is tuned to. IRI sends out special test commercials via cable channels. The single source research has its drawbacks. The size of the panels is still relatively small due to the high cost nature of data collection and hence it is doubtful how generalizable the results would be to the entire market. Secondly, how do we know viewers are actually watching the test commercials. The change in the buying behavior is also questionable when the panel members are probably conscious of being in the panel. Brand loyalties are somewhat difficult to change by a short-term advertising. But this research may be useful for new products (see Kessler, 1986).

Acknowledgements

We wish to thank B. N. Chinnappa, Director, Business Survey Methods Division, Statistics Canada, whose comments were very helpful in revising the chapter.

References

Blair, E. (1983). Sampling issues in trade area maps drawn from shopper surveys. *Journal of Marketing* **47** (Winter), 98–106.

Blair, J. and Czaja, R. (1982). Locating a special population using random digit dialing. *Public Opinion Quarterly* **46**, 585–590.

Blattberg, R. C. and Sen, S. K. (1976). Market segments and stochastic brand choice models. *Journal of Marketing Research* **13**, 34–45.

Brunner, J. A. and Brunner, G. A. (1971). Are voluntarily unlisted telephone subscribers really different? *Journal of Marketing Research* **8**, 121–124.

Bryant, B. E. (1975). Respondent selection in a time of changing household composition. *Journal of Marketing Research* **12**, 129–135.

Bucklin, L. B. and Carman, J. (1967). The design of consumer research panels: Conception and administration of the Berkeley Food Panel. Institute of Business & Economic Research, University of California, Berkeley, CA.

Bush, A. J. and Hair, Jr., J. F. (1985). An assessment of the mall intercept as a data collection method. *Journal of Marketing Research* **22**, 158–167.

Carefoot, J. (1982). Copy testing with scanners. *Journal of Advertising Research* **1**(22), 25–27.

Casady, R. J., Snowden, C. B. and Sirken, M. G. (1981). A study of dual frame estimators for the national health interview survey. *Proceedings of the Survey Research Section, American Statistical Association*, 444–447.

Charlton, P. and Ehrenberg, A. (1976). An experiment in brand choice. *Journal of Marketing Research* **13** (May), 152–160.

Cordell, W. and Rahmel, H. (1962). Are Nielsen ratings affected by noncooperation, conditioning, or response error? *Journal of Adversing Research* **2** (September), 45–49.

Czaja, R., Blair, J. and Sebestik, J. P. (1982). Respondent selection in a telephone survey: A comparison of three techniques. *Journal of Marketing Research* **19**, 381–385.

Ehrenberg, A. (1960). A study of some potential biases in the operation of a consumer panel. *Applied Statistics* **9** (March), 20–27.

Eskin, G. (1973). Dynamic forecasts of new product demand using a depth of repeat model. *Journal of Marketing Research* **10** (May), 115–129.

Farley, J., Howard, J. and Lehman, D. (1976). A working system model for car buyer behavior. *Management Science* **23** (November), 235–247.

Ferber, R. and Sudman, S. (1974). Effects of compensation in consumer expenditure studies. *Annals of Economic and Social Measurement* 3 (April), 319–331.

Frankel, M. R. and Frankel, L. R. (1977). Some recent developments in sample survey design. *Journal of Marketing Research* 14, 280–293.

Frey, J. H. (1983). *Survey Research by Telephone*. Sage Publications, Beverly Hills, CA.

Gates, R. and Solomon, P. (1982). Research using the mall intercept: State of the art. *Journal of Advertising Research* 22(4), 43–49.

Grootaert, C. (1986). The use of multiple diaries in a household expenditure survey in Hong Kong. *Journal of the American Statistical Association* 396(81), 938–944.

Groves, R. M. (1978). An empirical comparison of two telephone sample designs. *Journal of Marketing Research* 15, 622–631.

Groves, R. M. and Kahn, R. L. (1979). *Surveys by Telephone*. Academic Press, New York.

Groves, R. M. and Lepkowski, J. M. (1986). An experimental implementation of a dual frame telephone sample design. *Proceedings of the Survey Research Section, American Statistical Association*.

Hagan, D. E. and Collier, C. M. (1983). Must respondent selection procedures for telephone surveys be invasive? *Public Opinion Quarterly* 47, 547–556.

Hardin, D. and Johnson, R. (1971). Patterns of use of consumer purchase panels. *Journal of Marketing Research* 8 (August), 364–367.

Hartley, H. O. (1962). Multiple frame surveys. *Proceedings of the Social Statistics Section, American Statistical Association*, 203–206.

Hausman, J. and Wise, D. (1979). Attrition bias in experimental and panel data: The Gary income maintenance experiment. *Econometrica* 2(47), 455–473.

Jenkins, J. G. (1938). Dependability of psychological brand barometers: I. The problem of reliability. *Journal of Applied Psychology* 22, 1–7.

Kessler, F. (1986). High-tech stocks in ad research. *Fortune*, July 7, 58–60.

Kish, L. (1967). Survey sampling. Wiley, New York.

Lahiri, D. B. (1951). A method of sample selection providing unbiased ratio estimates. *Bulletin International Statistical Institute* 33, 133–140.

Lazarsfeld, P. and Fiske, M. (1938). The 'panel' as a new tool for measuring opinion. *Public Opinion Quarterly* 2 (October), 596–612.

Lepkowski, J. M. and Groves, R. M. (1986). A mean squares error model for dual frame, mixed mode survey design. *Journal of the American Statistical Association* 81, 930–937.

Moberg, P. E. (1982). Biases in unlisted phone numbers. *Journal of Advertising Research* 22(4), 51–55.

Montgomery, D. (1971). Consumer characteristics associated with dealing: An empirical example. *Journal of Marketing Research* 8 (February), 118–120.

Nakanishi, M. (1973). Advertising and promotion effects on consumer response to new products. *Journal of Marketing Research* 10 (August), 242–249.

Nichols II, W. L. and Groves, R. M. (1986). The status of computer-assisted telephone interviewing: Part I – Introduction and impact on cost and timelines of survey data. *Journal of Official Statistics* 2, 93–115.

Olsen, R. (1980). A least squares correction for selectivity bias. *Econometrica* (November), 1815–1820.

O'Rourke, D. and Blair, J. (1983). Improving random respondent selection in telephone surveys. *Journal of Marketing Research* 20, 428–432.

Orwin, R. G. and Boruch, R. F. (1982). RRT meets RDD: Statistical strategies for assuring response privacy in telephone surveys. *Public Opinion Quarterly* 46, 560–571.

'Panel bias reviewed; results inconclusive' (1976). The Sampler from *Response Analysis* 7 (fall), 2.

Smead, R. J. and Wilcox, J. (1980). Ring policy in telephone surveys. *Public Opinion Quarterly* 44, 115–116.

Sobol, M. (1959). Panel mortality and panel bias. *Journal of the American Statistical Association* 54 (March), 52–68.

Stern, Jr., D. E. and Steinhorst, R. K. (1984). Telephone interview and mail questionnaire applications of the randomized response model. *Journal of the American Statistical Association* 79, 555–564.

Sudman, S. (1962). On the accuracy of recording of consumer panels. Unpublished PhD. dissertation, University of Chicago.

Sudman, S. (1973). The uses of telephone directories for survey sampling. *Journal of Marketing Research* **10**, 204–207.

Sudman, S. (1978). Optimum cluster designs within a primary unit using combined telephone screening and face-to-face interviewing. *Journal of the American Statistical Association* **73**, 300–304.

Sudman, S. (1980). Improving the quality of shopping center sampling. *Journal of Marketing Research* **17**, 423–431.

Sudman, S. and Ferber, R. (1971). Experiments in obtaining consumer expenditures by diary methods. *Journal of the American Statistical Association* **66** (December), 725–735.

Sudman, S. and Ferber, R. (1974). A comparison of alternative procedures for collecting consumer expenditure data for frequently purchased products. *Journal of Marketing Research* **11** (May), 128–135.

Sudman, S. and Ferber, R. (1979). *Consumer Panels*. American Marketing Association, Chicago.

Sudman, S. and Kalton, G. (1986). New developments in the sampling of special populations. *Annual Review of Sociology* **12**, 401–429.

Troldahl, V. C. and Carter, Jr., R. E. (1964). Random selection of respondents within households in phone surveys. *Journal of Marketing Research* **1**, 71–76.

U.S. Department of Agriculture (1953). Establishing a national consumer panel from a probability sample. Marketing Research Report No. 40, U.S. Government Printing Office, Washington, D.C.

Waksberg, J. (1978). Sampling methods for random digit dialing. *Journal of the American Statistical Association* **73**, 40–46.

Waksberg, J. (1983). A note on 'locating a special population using random digit dialing'. *Public Opinion Quarterly* **47**, 576–578.

Warner, S. L. (1965). Randomized response: A survey technique for eliminating evasive answer bias. *Journal of the American Statistical Association* **60**, 63–69.

Winer, R. (1980). Estimation of a longitudinal model to decompose the effects of an advertising stimulus on family consumption behavior. *Management Science* **26** (May), 471–482.

Winer, R. (1983). Attrition bias in econometric models estimated with panel data. *Journal of Marketing Research* **20**, 177–186.

Wolfle, L. M. (1979). Characteristics of persons with and without home telephones. *Journal of Marketing Research* **16**, 421–425.

Observational Errors in Behavioural Traits of Man and their Implications for Genetics

P. V. Sukhatme

1. Introduction

The importance of controlling and estimating observational errors has long been recognised by survey practitioners. Several models for the study of such errors have been presented in the statistical literature. Among these, the model which is found suitable for observations over time is made up of four uncorrelated components as follows:

$$Y_{ijk} = X_i + \alpha_j + \delta_{ij} + e_{ijk} \tag{1}$$

where

y_{ijk} is the value reported by jth enumerator on the ith unit for the kth occasion, $i = 1, 2, \ldots, h$; $j = 1, 2, \ldots, m$; $k = 0, 1, 2, \ldots, n_{ij}$;
X_i ($i = 1, 2, \ldots, h$) denotes the true value of the characteristic on the ith unit in a simple random sample of h units drawn from N units;
α_j is the bias of the jth enumerator in repeated observations on all units;
δ_{ij} is the interaction of the jth enumerator with the ith unit;
e_{ijk} is the error when the jth enumerator reports on the ith unit on the kth occasion.

This model is especially instructive in the study of behavioural traits of hierarchically organised structures such as societies and individuals. Thus, societies are made of men and women and individuals are made of organs which have a functional unity of their own, organs in turn are composed of cells which have a certain measure of autonomy. This structure helps to understand as to why a society should disintegrate before any of its constituent parts and why a human being should die before its individual organs like heart, kindney or eye loose their capacity to work in another human being. Indeed transplants would not be feasible if this was not so.

The model is equally powerful in gaining an insight into how a society can be undermined from within through corruptive influences. Thus, a man can be tempted to under-report his income whenever he has a chance to benefit from the local programs announced by the Government to assist the poor. Equally, he can be depended upon to over-report the gains from such benefits whenever such programmes are appraised at the end of a specified period. It is the failure to involve the ultimate community as a whole rather than selected individuals that explains loss of cohesion of the society. In this paper we shall describe a couple of applications of this model for the study of behavioural traits in man and its genetic significance.

2. Current theory and its limitations: The sample mean and its variance

Since we cannot make more than one observation at any given point in time and the notion of time is implied in the study of behavioural traits in man, we are forced to combine the third and the fourth components into a simple component and write (1) as

$$y_{ij} = x_i + \alpha_j + e_{ij} \tag{2}$$

with

$n_{ij} = 1$ or 0;
$n_{i\cdot} = \Sigma' n_{ij}$ the number of observations on the ith unit;
$n_{\cdot j} = \Sigma_i^h n_{ij}$, the number of observations made by the jth enumerator;
$\bar{y}_{\cdot j}$ = the mean of all the $n_{\cdot j}$ observations made by the jth enumerator;
$\bar{y}_{\cdot\cdot}$ = the mean of all the n observations made on the h units in the sample.

The expected value of the sample mean will depend upon how the enumerators are selected for making observations on the selected sample and the way the sample is distributed amongst them. Again, the units in the sample may be allotted to the enumerators either randomly, or the enumerators may be assigned the units falling within the respective geographical areas from which they are drawn. We shall assume:

(i) the m enumerators are a simple random sample out of the population of M enumerators;

(ii) the h units in the sample are randomly allotted to the different enumerators.

(iii) $n_{\cdot j} = n/m = \bar{n}$, say, that is, each enumerator makes an equal number of observations;

(iv) $n_{i\cdot} = n/h = p$, say, that is, equal number of observations is made on each unit in the sample.

We can then write, under model (2), the sample mean as

$$\bar{y}_{\cdot j} = \frac{1}{\bar{n}} \sum_i^h n_{ij} x_i + \alpha_j + \frac{1}{\bar{n}} \sum_i^h n_{ij} e_{ij}, \tag{3}$$

$$\bar{y}_{\cdot\cdot} = \frac{1}{h} \sum_i^h x_i + \frac{1}{m} \sum_j^m \alpha_j + \frac{1}{n} \sum_j^m \sum_i^h n_{ij} e_{ij}, \tag{4}$$

and we have

$$E(\bar{y}_{.j}) = \frac{1}{N} \sum_{i=1}^{N} x_i + \frac{1}{m} \sum_{j=1}^{m} \alpha_j = \mu + \bar{\alpha}, \qquad (5)$$

$$E(\bar{y}_{..}) = \mu + \bar{\alpha}, \qquad (6)$$

where N is the total number of units in the population, μ is the population mean of the true values and $\bar{\alpha}$ is the population mean of enumerators' biases. As indicated by (6) $\bar{y}_{..}$ is not an unbiased estimator of μ, unless α's vary in such a way that $\bar{\alpha}$ equals zero.

Now, to find the variance of $\bar{y}_{.j}$ we have

$$V(\bar{y}_{.j}) = V\left(\frac{1}{n}\sum_{i}^{h} n_{ij} x_i\right) + V(\alpha_j) + V\left(\frac{1}{n}\sum_{i}^{h} n_{ij} e_{ij}\right)$$

$$= \left(\frac{1}{n} - \frac{1}{N}\right) S_x^2 + \left(1 - \frac{1}{M}\right) S_\alpha^2 + \frac{S_e^2}{n} \qquad (7)$$

where

$$S_x^2 = \frac{1}{(N-1)} \sum_{i=1}^{N} (x_i - \mu)^2, \qquad (8)$$

$$S_\alpha^2 = \frac{1}{(M-1)} \sum_{j=1}^{M} (x_j - \bar{\alpha})^2. \qquad (9)$$

For large values of N and M,

$$V(\bar{y}_{.j}) \doteq \frac{1}{n}(S_x^2 + S_e^2) + S_\alpha^2. \qquad (10)$$

The variance of $\bar{y}$ is similarly given by

$$V(\bar{y}_{..}) = \left(\frac{1}{h} - \frac{1}{N}\right) S_x^2 + \left(\frac{1}{m} - \frac{1}{M}\right) S_\alpha^2 + \frac{S_e^2}{n} \qquad (11)$$

and, if N and M are large,

$$V(\bar{y}_{..}) = \frac{S_x^2}{h} + \frac{S_\alpha^2}{m} + \frac{S_e^2}{n}. \qquad (12)$$

Particularly, when $p = 1$, which is usually the case in practice, $m = hp = h$ and $V(\bar{y}_{..})$ of (12) becomes

$$V(\bar{y}) = \frac{S_x^2 + S_e^2}{h} + \frac{S_\alpha^2}{m}. \qquad (13)$$

Since the variance of a single observation drawn from an infinite population when M is large is

$$S_y^2 = S_x^2 + S_\alpha^2 + S_e^2, \qquad (14)$$

the expression for the variance given by (13) becomes

$$V(\bar{y}..) = \frac{S_x^2 + S_\alpha^2 + S_e^2}{h} + S_\alpha^2 \left(\frac{1}{m} - \frac{1}{h}\right) = \frac{S_y^2}{h} + S_\alpha^2 \left(\frac{1}{m} - \frac{1}{h}\right). \qquad (15)$$

Yet another alternative expression for $V(\bar{y}..)$, due to Hansen et al. (1951) is worth noting here. They give the variance in terms of the correlation $\bar{r}$ between responses obtained by the same enumerator. By definition

$$\bar{r} S_y^2 \left(1 - \frac{1}{N}\right) = E[\{y_{ij} - E(y_{ij})\}\{y_{i'j} - E(y_{i'j})\}]$$

$$= E[\text{Cov}(y_{ij}, y_{i'j})|j] + \text{Cov}[E(y_{ij}|j), E(y_{i'j}|j)]$$

$$= E[\text{Cov}(x_i, x_{i'})] + \text{Cov}(\mu + \alpha_j, \mu + \alpha_j)$$

$$= -\frac{S_x^2}{N} + \left(1 + \frac{1}{M}\right) S_\alpha^2 \qquad (16)$$

so that

$$\bar{r} S_y^2 \cong S_\alpha^2 \qquad (17)$$

and M and N are sufficiently large.
   Hence from (15) we have

$$V(\bar{y}..) = \frac{S_y^2}{h} \left\{1 + \bar{r}\left(\frac{h}{m} - 1\right)\right\} = \frac{S_y^2}{h} [1 + \bar{r}(\bar{n} - 1)]. \qquad (18)$$

The fundamental formulas, (15) and (18) for the variance of $\bar{y}..$ clearly show that the sampling variance of the estimator is not entirely due to errors arising from chance variation in the selection of the sample of h units, but is inflated by the variability in the biases of the enumerators. This emphasizes that it is not sufficient in a survey to ensure that α_j cancel each other over all the enumerators. In fact if $\bar{\alpha} = 0$, the bias in the estimator of μ is eliminated as is seen from (6), however, the enumerators' contribution to the variance as given by (15) does not vanish. From (18) it is clear that $V(\bar{y}..) = S_y^2/h$, if $\bar{n} = 1$ or $\bar{r} = 0$ or if α_j is a constant for each j. Also if α_j is constant for each j, then $\bar{r} = 0$, as seen from (17). Normally, however, $\bar{r}$ will be greater than zero, since every enumerator will have his own tendency towards bias of either under-estimating or over-estimating the characteristics under study. Even a small tendency towards bias will contribute appreciably to the variance of the estimator, since the sample size is usually many times larger than the number of enumerators.

3. Estimation of the different components

Let S_e^2 denote the mean square error between the means of m enumerators, defined by

$$s_e^2 = \frac{1}{(m-1)} \sum_{j=1}^{m} (\bar{y}_{.j} - \bar{y}_{..})^2, \tag{19}$$

we have

$$(m-1)E(s_e^2) = E\left[\sum_{j}^{m} \bar{y}_{.j}^2 - m\bar{y}_{..}^2\right] = \sum_{j}^{m} V(\bar{y}_{.j}) - mV(\bar{y}_{..}). \tag{20}$$

Substituting from (7) and (11) in (20) and writing $n = hp$, we see

$$E(s_e^2) = \frac{m(m-p)}{hp(m-1)} S_x^2 + S_\alpha^2 + \frac{m}{hp} S_e^2. \tag{21}$$

Similarly, denoting by s_{e0}^2 the mean square between observations within enumerators defined by

$$s_{e0}^2 = \frac{1}{n-m} \sum_{j}^{m} \sum_{i}^{\bar{n}} (y_{ij} - \bar{y}_{.j})^2, \tag{22}$$

we get

$$(n-m)E(s_{e0}^2) = \sum_{j}^{m} \sum_{i}^{\bar{n}} V(y_{ij}) - \bar{n} \sum_{j}^{m} V(\bar{y}_{.j}). \tag{23}$$

Now letting $\bar{n} = 1$ in (7) we have

$$V(y_{ij}) = S_x^2 \left(1 - \frac{1}{N}\right) + S_\alpha^2 \left(1 - \frac{1}{M}\right) + S_e^2. \tag{24}$$

Hence using (7) and (24) in (23) we get

$$E(s_{e0}^2) = S_x^2 + S_e^2 \tag{25}$$

As it is not possible to obtain separate estimates of S_x^2, S_e^2 and S_α^2 from (21) and (25), we shall consider yet another mean square, namely, that between unit means, S_u^2, given by

$$s_u^2 = \frac{1}{h-1} \sum_{i=1}^{h} (\bar{y}_{i\cdot} - \bar{y}_{\cdot\cdot})^2. \tag{26}$$

Now using (26) and (11) we see that

$$E(s_u^2) = S_x^2 + \frac{S_e^2}{p} + \frac{h(m-p)}{(h-1)mp} S_\alpha^2. \tag{27}$$

The set of three equations (21), (25) and (27) provide estimators of S_x^2, S_α^2 and S_e^2. In particular, we obtain

$$\text{Est } S_\alpha^2 = \frac{p(m-1)(h-1)}{pmh - ph - pm + m} \left\{ s_e^2 + \frac{m}{h(m-1)} s_u^2 - \frac{m^2}{hp(m-1)} s_{e0}^2 \right\}. \tag{28}$$

In practice however, $p = 1$, and (21), (25) and (27) simplify as follows:

$$E(s_e^2) = \frac{m}{h} (S_x^2 + S_e^2) + S_\alpha^2, \tag{29}$$

$$E(s_{e0}^2) = S_x^2 + S_e^2, \tag{30}$$

$$E(s_u^2) = S_x^2 + S_e^2 + \frac{h}{(h-1)} \frac{(m-1)}{m} S_\alpha^2. \tag{31}$$

Notice that $S_x^2 + S_e^2$ occurs together in the above equations and the three equations are no longer linearly independent. In fact when $p = 1$,

$$(h-1)s_u^2 \equiv (m-1) \frac{h}{m} s_e^2 + (h-m)s_{e0}^2. \tag{32}$$

Now using s_{e0}^2 as an unbiased estimator of $S_x^2 + S_e^2$ we see from (29) that an unbiased estimator of S_α^2 is

$$\text{Est } S_\alpha^2 = s_e^2 - \frac{m}{h} s_{e0}^2. \tag{33}$$

If N and M are large and $p = 1$, an unbiased estimator of $V(\bar{y}_{\cdot\cdot})$ is obtained as

$$\text{Est } V(\bar{y}..) = \frac{s_e^2}{m} = \frac{s_u^2}{h} + \frac{h-m}{m-1}\frac{1}{h}(s_u^2 - s_{e0}^2). \tag{34}$$

Thus it is clear from (34) that s_u^2/h does not give an unbiased estimate of the variance of the estimated mean but it is inflated by $[h-m)/(m-1)h] (s_u^2 - s_{e0}^2)$, which vanishes when the differential biases are absent.

4. Application to longitudinal studies

The formulae developed in the previous section, particularly (15) and (18) have great significance for interpreting the results of longitudinal studies. This is particularly so in studies of behavioural traits in man, where observations are made by the observer on himself and therefore constitute an occasion for experience. Available evidence shows that when this is the case, experience gets integrated into the motion of the system with the process simulating the autoregressive pattern. Space does not allow us to develop the topic except for illustrating through the following example.

Example 1. This example is taken from the longitudinal observations of energy intake taken on 5 young army recruits engaged in group activities at depot Centre A in England during the 2nd, 5th and 8th weeks of their training. The intake of the subjects was measured using the 'weigh as you eat' method from a common table and expenditure was calculated by timing activities and from energy expended during each activity as estimated from the amount of oxygen consumed

Table 1.1
Analysis of daily energy intake and daily energy expenditure in Kcal/kg of body weight (period: Monday to Friday)

	Intake, kcal/kg				Expenditure, kcal/kg			
	df	MS	F-value	Estimates of true value	df	MS	F-value	Estimates of true variance
Between subjects	4	50			4	300		–
Between periods within subjects	10	666	3.4[a]	91	10	179	2.4[a]	21
Between days within periods	60	188		188	60	72		72
σ_w^2		279				93		
σ_w		17				10		
Kcal/kg		62				69		
% CV		27				14		
Var. of daily mean		129				35		
% SE		18				9		

[a] Significant at 5% level.

using integrated motor pneumotacograph (IMP) devised for the purpose (Edholm, Adam, Healy, Wolff, Goldsmith and Best, 1970). Table 1.1 shows the analysis of variance for daily energy intake and daily energy expenditure in Kcal/kg body weight.

(1) Estimate the different components of true variance.

(2) Calculate the CV of daily intake and daily expenditure and of daily mean intake and daily mean expenditure for 5 day periods.

(3) Test for significance the MS for between periods relative to differences between days within periods.

(4) Work out the relationship of the intra-class correlation coefficient $\bar{r}$ and the serial correlation coefficient ρ_1. Use the results to estimate from Table 1.1 the value of $\bar{r}$ and hence of ρ_1.

(5) What interpretation would you put on the significance in Step 3 above in terms of $\bar{r}$?

(6) Draw a graph showing how the variance of the daily mean intake changes with the length of the period.

In the usual hierarchical model (1) of analysing the variance, the expectations of the three mean squares shown in Table 1.1 respectively are given by

$$pd\,\sigma_b^2 + d\,\sigma_p^2 + \sigma_d^2, \quad d\,\sigma_p^2 + \sigma_d^2 \quad \text{and} \quad \sigma_d^2,$$

where σ_b^2 stands for the true variance between subjects, σ_p^2 stands for the true variance between periods within subjects, and σ_d^2 denotes the true variance between days within periods. d denotes the length of the period in days and p denotes the number of periods during which each subject was observed for his intake and expenditure. Numbers preceding σ_b^2 and σ_p^2 represent the number of days observed for each subject and during given period. Using the above expressions for the expected mean squares estimates of the three components of variance can be obtained. The variance of a single observation for any subject is given by

$$\sigma_w^2 = \sigma_p^2 + \sigma_d^2 \tag{35}$$

and that of the mean intake and expenditure per kg of body weight over d days by

$$\sigma_p^2 + \frac{\sigma_d^2}{d}. \tag{36}$$

Table 1.1 shows estimates of the three components of variance as also of σ_w^2 and of the variance of mean intake and expenditure per kg of body weight together with the respective coefficients of variation.

It will be seen that the mean square between periods is significantly larger than the mean square between days within periods, giving a positive value for the estimate of σ_p^2. The implication is that even when the daily values are averaged over a period of five days, the differences from period to period for the same individual persist. The coefficient of variation of daily mean intake is seen to be 18 per cent. If day-to-day variations were random, resulting from errors of measurement, the coefficient of variation of mean daily values over time in the same individual would be smaller. It would appear that the variation over time in the same individual is not random.

Unlike intake, the variation in the case of energy expenditure is small. The coefficient of variation for daily expenditure is 14 percent and that of the mean daily expenditure 9 percent. The smaller magnitude of the coefficient of variation in the case of energy expenditure is to be expected for the army recruits were engaged in fixed tasks from day to day, but they had choice of intake on all days. The analysis suggests that the body regulates its energy balance on a range of intakes by varying the efficiency of utilisation of its intake.

As the series reported by Edholm et al. (1970) is limited to 3 non-continuous weeks, it does not permit a direct study of auto-regressive pattern to verify the inference that energy requirement is self regulated over a range of intakes. However, Sukhatme (1978) and Sukhatme and Margen (1982) have used indirect methods to show that the series can be adequately represented by auto-regressive series of order one, comprising two components—one a short term component arising from the current value of the process at the previous time point and the other a long term component in the form of errors of measurement given by

$$w_t = \rho_1 w_{t-1} + e_t \tag{37}$$

where w_t is the balance on the ith day, ρ_1 is the serial correlation of order one between w_t and w_{t-1}, and e_t is a random variable distributed around zero with variance σ_e^2. The meaning of this process called the stochastic stationary Markov process is that if it were possible to repeat the circumstances which gave rise to the observed value of balance on any day, t, then the balance will be distributed around zero within limits called homeostasis, which are independent of t, given by

$$\pm \frac{2\sigma_e}{\sqrt{1-\rho^2}}. \tag{38}$$

In particular, Sukhatme and Margen (1982) computed the variance of the mean balance when the daily balance is averaged over 2, 3 or more successive days and showed that the daily balance is distributed in a stochastic stationary manner of the Markovian type with serial correlation of the first order equal to 0.3. The results of this exercise for Depot D are given in Table 1.2. It will be seen that the variance of the mean balance does not vary inversely as the length of the period, but that it decreases slowly, thus confirming that successive values are serially correlated. There is a suggestion that the variance of the mean is stabilised as the number of observations over which the mean is taken is increased.

Table 1.2
Variance of an individual mean energy intake, expenditure and balance based on d successive days as proportion of unit variance for $d = 1$. Depot Centre D

Period in days	Observed			Calculated	
	Intake	Expenditure	Balance	$\bar{r} = 0.00$	$\bar{r} = 0.30$
1	1.00	1.00	1.00	1.00	1.00
3	0.43	0.63	0.55	0.33	0.49
4	0.27	0.37	0.31	0.25	0.39
5	0.28	0.29	0.29	0.20	0.32
6	0.21	0.32	0.25	0.16	0.27

Sukhatme has interpreted the results to mean that specialized environment in which an individual is brought up interacts with the genetic entities in man to keep the variance constant. Edholm et al.'s data on energy balance on army recruits must therefore be interpreted to mean that although intake may not be equal to expenditure even when averaged over a week, man is in balance every day in a probabilistic sense, with varying intervals between peaks and troughs and varying amplitudes in daily balance. This can also be expressed by saying that intake on successive days will continue to vary in the same univariate distribution. To put it differently, the fact that an individual on any given day is observed to have a specified intake cannot be taken to mean that he will continue to have the same intake on successive days, but that the successive values will be correlated, each varying within stationary limits given by (38). Finding that energy balance is regulated is consistent with the evidence that humans possess a physiological regulatory mechanism for controlling appetite and energy expenditure. However, the process with which the control is exercised is weak and that ancillary cofactors also play a part in the control system. It is even possible that the control is not triggered into action until the accumulated energy balance acquires a set point value.

The homeostatic model shown in equation (37) is an autoregressive (AR) model of order 1. It is given here more to bring out the regulatory homeostatic character of intake and energy balance than as an exact description of the phenomenon.

Edholm et al.'s data bring out another feature, viz. the genetic significance of the process of day-to-day variation in intake and the cause of why the variation persists even after averaging over several days. Used to analysing data in terms of the hierarchical additive model, one is apt to forget that the two components of phenotype, viz. nature (g) and nurture (e) need not necssarily be independent, but interact, giving for the variance of the phenotypic value the expression

$$V(g) + V(e) + 2 \operatorname{Cov}(e, g). \tag{39}$$

As we shall shortly show, we have a data situation over time which precisely contributes to the covariance component in this expression.

We have already seen that the variance of the mean daily intake shown in equation (36) does not decrease inversely as the number of observations. It is only the environmental component of the variance which decreases inversely with the length of the period, the other part representing the true variance between periods remaining intact.

The longitudinal observations carried out in several periods on the same individual thus present a data situation similar to that where observations reported by the same observer are correlated with each other. We can, therefore, interpret data in the same individual by the method of intraclass correlation and state that for given total observed variance, higher the association $\bar{r}$ between the different values within the same period, smaller would be σ_d^2 and greater would be σ_p^2. In other words, the ratio of σ_p^2 to the total observed variation σ_w^2 can be said to represent the fraction of the phenotypic genetics, this ratio is called the 'repeatability' of the character and sets an upper limit to the heritable portion of an individual's variance. Since the observations refer to the same individual, the heritable portion must be interpreted as arising from interaction between genetic entities in individual and the environment in which he is brought up.

Substituting then

$$\sigma_p^2 = \bar{r}\sigma_w^2 \tag{40}$$

in equation (36) and noting further that $\sigma_d^2 = (1 - \bar{r})\sigma_w^2$ since $\sigma_d^2 + \sigma_p^2 = \sigma_w^2$, we may rewrite for the variance of the mean intake per kg of body weight the expression

$$\sigma_w^2 \left\{ \bar{r} + \frac{1 - \bar{r}}{d} \right\}. \tag{41}$$

Table 1.1 shows the estimated values for σ_p^2 and σ_w^2. It will be seen that σ_p^2 is small relative to the total variance but is far from negligible and accounts for some 20 to 25 percent of the total variance.

It is now easy to establish the genetic significance of the autocorrelated process when successive observations are represented by autoregressive series of order one; the intra-class correlation coefficient $\bar{r}$ can be written in terms of the serial correlation coefficient ρ_1 as

$$\bar{r} = \frac{\sum_{m-1}^{d-1}(d - m)\rho_1^m}{\sum_{m-1}^{d-1}(d - m)}, \tag{42}$$

which, on simplification, gives (Sukhatme and Narain, 1982a, b, 1983)

$$\bar{r} = \frac{2\rho_1}{(d - 1)(1 - \rho_1)} \left[\frac{1 - \rho_1^d}{d(1 - \rho_1)} \right]. \tag{43}$$

and can be approximated as

$$\bar{r} = \frac{2\rho_1}{(d-1)(1-\rho_1)}. \tag{44}$$

Substituting in equation (41), we get for the variance of the daily mean based on d values, the expression

$$\frac{\sigma_w^2}{d} \cdot \frac{1+\rho_1}{1-\rho_1}. \tag{45}$$

The enhanced variability that we observe in mean daily intake in the same individual can thus be explained in two ways: (1) in terms of auto-correlation of order one, with its implication for regulation and (2) in terms of intra-class correlation $\bar{r}$ with its implication for genetic significance as described above. It would appear that the capacity to adapt within the threshold limits has a heritable basis and implies adaptive regulation.

As Table 1.3 shows, the heritable portion of the variance arising from interaction with environment is admittedly small. Although small, the resulting value of the auto-correlation is consistent with the hypothesis of auto-regressive model of order one explaining the cause of the persistent nature of intra-individual variation. In the nature of things $\bar{r}$ and with it ρ_1 are expected to be small when the main concern is to study the influence of day-to-day change on the performance of the next day within the framework of ontogenic growth. It is for this reason that the evidence presented in Table 1.1 assumes significance. If intra-individual variation persists even after data are averaged over 5 day period, it is then because the first term in the expression for the variance of the mean in equation (36) is not only related to the parameter of heritability as shown in (40), but also with the auto-regressive process in the manner shown in expression (44). Figure 1 shows that the two move in line with each other after the length of the period exceeds 5. When ρ_1 is zero, we get the usual situation where errors are random.

Table 1.3
Estimation of intra-class correlation $\bar{r}$ and of serial correlation of order one in daily intake kcal/kg and expenditure kcal/kg

	Intake kcal/kg	Expenditure kcal/kg
$\bar{r}$	0.33	0.23
ρ_1	0.40	0.32

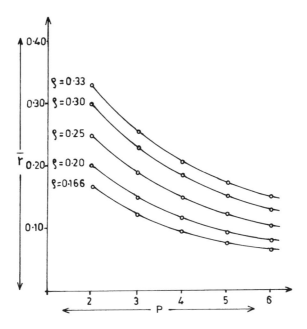

Fig. 1. Relationship between autocorrelation and intraclass correlation.

It is thus the presence of ρ_1 that anables us to separate healthy well-nourished. It is again the presence of ρ_1 that gives the threshold character to the phenomenon of nutritional status implying that man can have his intake anywhere in the range represented by the process and yet can engage himself in defined tasks.

The fact that man is able to bring for himself a change in the rate of energy flow, speeding it sometimes and slowing it down on other occasions to maintain homeostasis would suggest that energy balance is under the physiological control of the body. No one can change the genetic blue-print with which one is born but it appears that he can influence its development under the impact of sustained perturbations in environment. The result is consistent with the experimental evidence that not all DNA are transcribed into messenger RNA and that indeed upto 40 percent remain untranscribed without interrupting the functions of the genetic code of directing the assembly of aminoacids in prescribed sequence.

Example 2. Space does not permit us to give another detailed example. Instead, we shall give a brief description of a longitudinal study of growth increment in children reported by Roche and Himes (1980). The children in this study were all born at the University of Arizona Hospital and were measured for weight using carefully standardised methods and detecto beam balance six times in the first year and at six months intervals thereafter. The number of children covered was 62; all were normal infants. Original data are not reported by the authors; instead charts showing percentiles are presented to describe the distribution of growth increments for each period from birth to 36 months.

A longitudinal analysis of growth patterns showed that two-thirds of the babies had at least one six month period which showed extreme departure from the median growth for that period, falling either below the 3rd percentile or above the 97th percentile. More than half the babies had 2 or more such extreme six month increments. Usually, they were compensatory periods of extreme growth, meaning thereby that if a child grew fast in one period, his growth was slow in another. The initial extreme period was as likely to be rapid as slow. None of the children showed a consistently high or low growth increment, though about a third moved around median within $\pm$ the standard deviation for the respective period. None of the infants experienced severe insult of malnutrition. The observed growth rate patterns accordingly represent the dynamics of growth rates under almost ideal conditions.

So large were the differences in growth increments between one period of six months and another in the same child that some babies were found to be growing at twice the average annual rate in one period and hardly growing during other 6 month period. Thus, some girl infants recorded a growth increment as large as 5.3 kg from birth to 6 months followed by a growth increment as small as 1.8 kg during 6 to 12 months. Again, some others showed a growth increment close to median value of 4 kg from 0 to 6 months and continued growing at about that rate or slightly lower during 6 to 12 months. However, their growth increment during 12–18 months was only 0.5 kg.

The charts also show that children differed markedly in their growth increments during each 6 month period much in the same way as army recruits differed in their energy balance during same week. Thus some grew twice as fast as others during 0 to 6 months and some grew thrice as fast as others from 6 to 12 months. Again a child who remained on a specified percentile in one period did not remain on the same percentile in the following periods. This means that the observed variation in growth increments of children in any period, like that of energy balance in any week, cannot be fully explained by variation in their intake. Seasonal differences could account for part of the observed differences but they were found to be small. The conclusion is that the relationship of body weight with intake changed over time in the same child and also changed from child to child.

Age to age correlation reported in the study confirm that children enjoying health changed paths of development over a fairly wide though limited range as a matter of course, as they grow. Since none of the children surveyed experienced an episode of malnutrition or was short of food, the pattern observed in growth increments over time must be taken as the pattern characteristic of healthy children under ideal conditions. This pattern is the pattern of uneven growth. There is no evidence in it that children with body weight below 90 percent of the ideal weight are all malnourished as assumed under Gomez classification in current nutrition literature. It is the steady growth rate near the extreme that is more likely to be cause for concern in ensuring homeostasis. It follows that the current practice of classifying children with body weight between 75 to 90 percent of the ideal weight (e.g. 12 kg for children 1 year old) as mildly malnourished, grossly overstates the dimensions of the problem. We have several other data reported to us in personal correspondence by Dr K. Sheth in Milwaukee which confirms the conclusion reached above.

To compare current weight with the ideal weight, as sought to be done under Gomez classification, also cannot tell whether the current weight is the result of inadequate intake unless 90 percent of ideal weight is interpreted as the outside limit of normal variation in healthy well-nourished children which we saw if not the case. The comparison is particularly faulty under conditions in developing countries where living conditions are very different from those in USA. Thus water over large area is unsafe for drinking, personal hygiene and sanitation almost totally lacking, flies and mosquitos are almost constant companions and toilett facilities are unavailable. If some of us do not relish the perspect of living in villages, it is because of these unsatisfactory living conditions. And yet it is not difficult to bring about a social change in their conditions through education around appropriate ocial action such as a fence around a school, a fence around a well, biogas plants fed with dung, garbage and agricultural waste. To emphasise feeding programs in this situation, as ICDS does, under the false pretext that children with weight, between 75 to 90 percent of the ideal weight, numbering over one-third, are malnourished for lack of adequate diet is to increase the dependendence of people on charity and the processed food.

5. Genetic implications

The significance of the changes in body weight increment and in energy balance reported in the previous examples for changes in development path have genetic implications which need especial emphasis. First and foremost it needs to be stressed that this behaviour is not prefigured in the genetic program but that it is built in the course of development through interaction. Thus, an individual may respond by reducing energy expenditure on basal functions (BMR), thereby reducing energy requirement through reduction of dissipated heat. This is indeed what experiments in metabolic unit has shown. Or again an individual may respond by choosing metabolic pathways that may increase short term efficiency of energy utilisation. Yet again in times of abundance, an individual may consciously decide to enjoy more food, dissipating more energy in the process and even put on fat. It is the stability of the genetic component of intra-individual variation arising from interaction between man and environment which enables man engaged in fixed tasks to maintain energy balance on a range of intakes and a child to deviate over a wide range in growth rate from the course of median growth. The process has a built in feedback mechanism which ensures that in case of disruption, the effectors controlling appetite and expenditure receive signals to absorb the disruptive force into the motion of child's system.

These implications have their origin principally in the variable relationship between phenotypic value for body weight and intake from individual to individual i.e. genotype to genotype and over time in the same individual. The situation is illustrated in Figure 2. As a consequence, a fixed supplement of calories, say 300 cannot be expected to produce the same change in body weight in all individuals even when they are similar. Again, it is common experience that at similar intakes and similar levels of activity, some adults gain in weight, others do not. On biological grounds also, the differences between phenotypes are not expected to remain the same in all environments. It is now nearly 50 years since Waddington concluded that careful selection of the environmental conditions at critical periods in development can give rise to a wide array of phenotypes, so different that they appear as if they are gene mutations. There is also confirmatory evidence in humans which shows that large amounts of DNA do not get transcribed and translated in the development of an individual. The variability is found not only in the sequence of amino acids but also in the nucleotide sequences of the introns.

It must be concluded that when a trait is a quantitative trait, as incremental growth is, with several genes contributing to its expression and is also under the influence of a variety of environmental factors such as variety of food served, drinks and company, plasticity is to be expected. It is this plasticity arising from the polygenic character of the genotype that has endowed man with a wide latitude of homeostatic range for his intake and enabled him to indulge in food under conditions of affluence without being overweight and obese. Equally, the same plasticity has served well people in the third world in that it has enabled them to regulate energy balance without detriment to their health and reduction of work output over a fairly large range. It follows that we must allow for this variation in any inference regarding the genetic determinism (or heredity) of energy requirement of man. If we do not do this, it would be tantamount to assuming

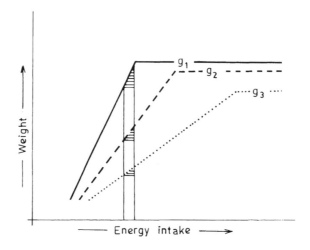

Fig. 2. Relationship of weight with intake.

that the components g and e in the phenotypic value p are independent and their variances are additive as shown below

$$V(p) = V(g) + V(e) \tag{46}$$

when infact we should have added a covariance term to the right hand side as shown in (39). It is this omission that has led Widdowson (1947), Rand and Scrimshaw (1984), Beaton and others to conclude that $V(e)$ is negligible relative to $V(g)$, thereby grossly overstating the physiological deterministic basis of individual's intake and energy balance in influencing body weight under current nutrition theory. At the other extreme it is also the omission of the same term that led Pavlov to overestimate the influence of experience and environment in shaping the behaviour of man. These views underscore the revisions that our conception of human nature and the method of moulding it have undergone during the present century.

There is no known way of estimating the covariance between the two components g and e when individuals are observed at a single point or period of time. It is necessary to study them over time in order to estimate the true variance σ_p^2 and obtain coefficient of heritability therefrom. The importance of σ_p^2 lies in its deterministic nature of representing genetic component of variation of within variability. In plants an individual genotype can be replicated and its development studied over a range of environmental conditions. This approach is not possible in man but given a determined and conscious effort on his part, as in Yoga, it can be exploited to regulate and control one's eating and expending behaviour. A significant value for σ_p^2 thus not only establishes the non-random nature of daily variation, but also implies adaptive regulation of evolutionary character in an individual; not adaptation. In other words, we can expect an individual to evolve over time in the course of generation using non-genetic channels of communication arising from interaction between environment and genotype.

The principal reason of stabilisation of variance is that while both intake and expenditure vary from day to day, the latter varies less than the former, being more physiological than behavioural in nature. As a consequence there is a variable time lag and energy balance is found to cumulate and vary from week to week in the same individual. This means that the body can tolerate energy balance of considerably larger magnitude than is apparent from limits of observed variation in body weight. The phenomenon is suggestive of the presence of an homeostatic mechanism which works for the good of the whole body and yet enables man to control body weight in the process. Errors of measurement, including non-sampling errors, often present a difficulty in establishing the validity of the auto-regulatory process showing that requirement evolves in a self-regulated pattern but this is inevitable in the absence of more sophisticate equipment such as human calorimeter to take observations. Space precludes further discussion.

Nothing can better illustrate these implications than the use of Intelligence Quotient (IQ) for personality testing. It is well known that IQ testing is extensively used by Educational Institutions, by Industry and by Army for selection of personnel. And yet, there is no agreement among scientists on the utility and significance of this practice. Indeed the disagreement is so large that no scientific controversy of our time can be said to have raged longer than IQ controversy. We have read the public criticism of this practice which appeared in the London Times a few years back. There are two schools of thought. One claims that IQ is determined principally by heredity. Cyril Burt of London and his followers are the principal exponents of this view. The other school is led by Kamin of Princeton Unviversity. It is the environment, the upbringing and the culture in Kamin's view which principally determines man's IQ. Briefly, the issue is 'nature versus nurture'.

The finding as reported by the London Times was that Sir Cyril Burt had falsified his experimental data. The names of students who are said to have carried out research from his department for Ph.D. were found to be fictitious. Sir Cyril Burt is now dead, but his work is being carried forward by Eysenck and other followers. Prof. Kamin has entered the field relatively recently. He has a sounder case but has little objective evidence to offer on the many issues that have figured in this debate. Therefore his arguments also fail to carry conviction. Burt and Eysenck mainly base their case on 3 beliefs: '(i) anything that can be measured and numerically summarised in the form of index is to them the foundation of science; (ii) when a large number of such measurements are made and presented in the form of statistical distribution, they call it a natural law; and (iii) the indiscriminate use of statistical methods, particularly of multivariate factor analysis aided by computer, are held by this school to give their activity the status of atural science'. As Medawar (1977) points out, no scientist in modern times can accept these beliefs as the basis of a natural science. It is therefore no surprise that Kamin should have serious objections to accepting this viewpoint expressed by Burt and Eysenck. In this situation there is no way out other than initiating extensive research to resolve the issue. What lends urgency to the research in this area is the tendency to spring into political postures on the part of both schools. Thus the Tories hold that differences in abilities arise from differences in man's genetic make up whereas socialists, while acknowledging that individual's differ in their genotype, hold that man is what his environment and culture make him. As another example, Burt's school asserts that whites are superior to blacks because they have higher IQ. Galton went so far as to advocate the use of the science of Eugenics in improving the quality of the human race.

The failure of both schools lies in their inability to formulate a precise hypothesis. This is shown by the manner in which each school tries to express their contribution of nature and nurture to differences of IQ without taking note of covariance. It is an illusion to think that answer to the problem can be stated in such a simple fashion. My own hypothesis based on my work on the genetics of behavioural traits described in the preceding sections is that it is the interaction between the genotype and environment that will matter in resolving the nature–nurture issue. One cannot help feeling that to draw such far reaching conclusions without understanding the limitations of IQ is clearly not only unscientific but dangerous. Even the concept of education as a process has not been clearly set forth by either school. Unless this is done, we will not know how our education should be reformed to mould national character. Even the policy for reservation of jobs for scheduled castes in the developing countries as in India depends on the outcome of this research. There is a real need to set up long term experiments to study this problem. We hope to be able to report on the progress from time to time.

References

Beaton, G. H. (1983). Atwater memorial lecture. *Nutrition Review* **41**(11), 325.
Endholm, O. G., Adam, J. H., Healy, M. J. R., Wolff, R. S., Goldsmith, R. and Best, T. W. (1970). Food intake and energy expenditure of army recruits. *Br. J. Nutr.* **24**, 1091.
Eyesenck, H. J. (1962). *A Model for Intelligence*. Springer, Berlin.
Eysenck, H. J. (1981). *The IQ Controversy*. Wiley-Interscience, New York.
FAO/WHO (1973). Energy and protein requirements. Nutrition Meetings Report, Series No. 522, Geneva.
Harsen, M. H., Hurwitz, W. N., Marks, E. S. and Mauldin, W. P. (1951). Response errors in surveys', *J. Amer. Stat. Assoc.* **46**, 147.
Kamin, L. (1981). *The IQ Controversy*. Wiley-Interscience, New York.
Medawar, P. B. (1984). *Pluto's Republic*. Oxford University Press, New York.
Roche, A. F. and Himes, J. H. (1980). Incremental growth charts, *Am. J. Clin. Nutr.* **33**, 2041.
Rand, W. M. and Scrimshaw, N. S. (1984). Protein and energy requirements—Insights from long-term studies. *Bull. of NFI* **5**(4).
Sukhatme, P. V. (1961). The World's hunger and future needs in food supplies. *J. Roy. Stat. Soc., Series A.* **124**, 463.
Sukhatme, P. V. (1982). Poverty and malnutrition. *Newer Concepts in Nutrition and their Implications for Policy.* MACS, Pune.
Sukhatme, P. V. and Margen, S. (1982). Autoregulatory homeostatic nature of energy balance. *Am. J. Clin. Nutr.* **35**, 355.
Sukhatme, P. V. and Prem Narain (1982). The genetic significance of intra individual variation in energy requirement. *Research Work of William G. Cochran*, Memorial Volume, edited by P. S. R. S. Rao and J. Sedransk. Wiley, New York.
Sukhatme, P. V. and Seth, G. R. (1952). Non-sampling errors in Surveys. *J. Ind. Soc. Agri. Stat.* **5**.
Sukhatme, P. V., Sukhatme, B. V., Sukhatme, Sukhatme, S. and Asok, C. (1984). *Sampling Theory of Surveys with Applications*. 3rd edition, published by Indian Society of Agricultural Statistics, New Delhi and Iowa State University Press, Ames.
Sukhatme, P. V. (1985). The nature of energy requirement and ts implications for measurement of undernutrition. *Bulletin of International Stat Institute L1 Book 4*.
Widdowson, L. M. (1947). A study of individual children's diets. *Special Report series, Medical Research Council* **257**, London, HMSO.

Designs in Survey Sampling Avoiding Contiguous Units

A. S. Hedayat[1], C. R. Rao[2] and J. Stufken[1]

1. Introduction

Various ideas to obtain improved estimates of population characteristics in the presence of some additional knowledge on the sampling units are available in the literature. In this chapter we discuss one of such ideas, recently introduced in a paper by Hedayat, Rao and Stufken (1988). It applies if there exists some ordering of the units under which contiguous units are anticipated to provide similar data. In such a situation, it is intuitively appealing that more information on the population can be obtained if the sample avoids pairs of contiguous units. The required ordering of the units will in most such situations be induced by a natural entity, such as time, location, and so on. It may be an ordering in one or more dimensions. As an example, small neighboring plots at a dump site for chemical waste tend to give similar measurements on the present amount of some chemical compound. With the plots as the sampling units, location induces a two-dimensional ordering of the units.

Section 2 provides a more detailed description of and introduction to the preceding idea. A class of sampling designs appropriate for this set-up is introduced, and a discussion on alternative sampling designs is included. Section 3 contains results on the existence and construction of the desired sampling designs, while Section 4 provides a brief discussion on the implementation of such designs.

2. Sampling designs excluding contiguous units

Consider a population of N identifiable units, labeled as $1, 2, \ldots, N$. For technical convenience and expositive simplicity we will assume a one-dimensional ordering of the units, in which units 1 and N are also considered as contiguous units.

---

[1] Research sponsored by Grant No. AFOSR 85-0320.
[2] Research sponsored by Grant No. AFOSR F49620-85-C-0008.

In other words, we assume a cyclic ordering of the units. Let y_i denote the characteristic of interest associated with unit i, and let $\mu = (y_1 + \cdots + y_N)/N$ denote the corresponding population mean.

A sample of size $n < N$ is formally defined as a collection of pairs

$$\{(i, y_i) : i \in s\},$$

where s is a subset of $\{1, 2, \ldots, N\}$ with cardinality n. For notational convenience we will often just use s to identify a sample. A sampling design of fixed size n is a pair,

$$d = (S_d, P_d),$$

where S_d consists of subsets of $\{1, 2, \ldots, N\}$, all with cardinality n, and P_d provides a probability distribution over S_d.

A sampling strategy for estimating a population characteristic, in our case μ, consists of a sampling design and an estimator for the characteristic. Throughout this chapter we will restrict our attention to sampling strategies with the following two properties:

(1) The sampling design is of fixed size n.
(2) The estimator is the celebrated Horvitz–Thompson (1952) estimator.

The condition in (1) is very reasonable in most practical situations. The condition in (2) is more restrictive. Although the Horvitz–Thompson estimator is a popular estimator with various desirable statistical properties, it is not inconceivable to utilize information about the sampling units in the selection of an estimator as well as a sampling design. For a discussion on this we refer to Rao (1975, 1987), where it is suggested that if in any observed sample contiguous (or close to each other in some sense) units occur, they may be collapsed into a single unit with the corresponding response as the average observed response over these units. An estimate of the unknown parameter is then made on the basis of such a reduced sample.

Throughout this chapter we will denote the first and second order inclusion probabilities by π_i and π_{ij}, respectively. In the preceding description of a sampling strategy, all the relevant features of a sampling design are contained in these inclusion probabilities. Under the assumed ordering of the units, obtained through some natural entity or otherwise, and the assumption that neighboring units will yield approximately the same data, it seems advisable to use a sampling design for which the π_{ij}'s are non-decreasing in the distance between the units i and j in the ordering. However, this concept has limitations, due to the question of existence of sampling designs with the required π_{ij}'s. An extreme example that nicely illustrates the limitations, is a systematic sampling design. If we demand that $\pi_{ii'} = 0$ for units i and i' that are a small distance apart in our ordering, and also demand that $\pi_{ij} = 1$, for some units i and j further apart, we are forced to accept that $\pi_{i'j} = 0$, although i' and j may be far apart.

A simple random sampling design, in which all π_{ij}'s are equal, can be considered as another extreme example that fits this concept. These two examples are in some sense two opposite extremes. As an alternative to these extremes, one could consider fixed size sampling designs for which $\pi_{ij} = 0$ if there are less than α units between i and j, for some positive integer α, while all other π_{ij}'s are equal to an appropriate constant. As an example, we exhibit such a sampling design for $\alpha = 2$, $N = 17$ and $n = 3$ in Table 1. The columns are the samples, all of which have a probability of selection equal to 1/34. Thus, for example $\pi_{1(16)} = \pi_{1(17)} = \pi_{12} = \pi_{13} = 0$, while $\pi_{1j} = 1/34$ for all units $j \in \{4, \ldots, 15\}$.

Table 1
Samples for a design with $N = 17$, $n = 3$, $\alpha = 2$

1	1	1	1	1	1	2	2	2	2	2	2	3	3	3	3	3	3
4	5	6	7	8	10	5	6	7	8	9	11	6	7	8	9	10	12
9	12	15	11	14	13	10	13	16	12	15	14	11	14	17	13	16	15

4	4	4	4	4	5	5	5	5	6	6	6	7	7	8	9
7	8	10	11	13	8	9	11	14	9	10	12	10	13	11	12
12	15	14	17	16	13	16	15	17	14	17	16	15	17	16	17

Although various results and techniques presented in this chapter have a simple generalization to $\alpha \geq 2$, we will confine ourselves here to $\alpha = 1$. Hedayat, Rao and Stufken (1988) named these designs Balanced Sampling designs Excluding Contiguous units (BSEC). Thus, if it exists, such a design has the properties that

$$\pi_i = n/N, \quad i \in \{1, \ldots, N\},$$

and, with $i \neq j$,

$$\pi_{ij} = \begin{cases} 0 & \text{if } i - j \equiv \pm 1 \pmod{N}, \\ n(n-1)/N(N-3) & \text{otherwise}. \end{cases}$$

Table 2 gives an example for $N = 15$ and $n = 4$. All samples in that table have probability of selection 1/15.

Table 2
Samples for a BSEC with $N = 15$, $n = 4$

1	1	1	1	2	2	2	2	3	3	3	4	4	5	6
3	4	5	7	4	5	6	8	5	7	9	6	10	7	8
6	8	11	9	7	9	12	10	8	13	11	9	12	10	11
10	14	13	12	11	15	14	13	12	15	14	13	15	14	15

In comparing this sampling strategy with alternatives, it is not surprising that its performance depends on the population to be studied. With the notations

$$\mu = \sum_{i=1}^{N} y_i/N, \qquad \sigma^2 = \sum_{i=1}^{N} (y_i - \mu)^2/N,$$

and

$$\rho_1 \sigma^2 = \sum_{i=1}^{N} (y_i - \mu)(y_{i+1} - \mu)/N,$$

where the indices of the y's are read modulo N, the variance of the Horvitz–Thompson estimators for μ equals

$$V_{\text{BSEC}} = \frac{\sigma^2}{n}(1 - (1 + 2\rho_1)(n-1)(N-3)^{-1}).$$

In comparing this with the strategy that utilizes a Simple Random Sampling (SRS) design, for which

$$V_{\text{SRS}} = \frac{\sigma^2}{n} \cdot \frac{N-n}{N-1},$$

it is easily seen that excluding the contiguous units as suggested provides a superior strategy if and only if $\rho_1 > -1/(N-1)$. This inequality will usually be satisfied if contiguous units provide similar observations, but will also be satisfied under various other conditions. It may, for example, be shown that for any given set of y's, an ordering with $\rho_1 \geq -1/(N-1)$ exists.

A theoretical comparison of V_{BSEC} with variances under alternative sampling designs, such as systematic sampling or stratified sampling with one unit per stratum, is less illuminating. Numerical examples can be used to gain more insight into the performance of BSEC's in relation to these alternatives. A simple example is given in Hedayat, Rao and Stufken (1988). If the assumption of similar observations from contiguous units is met, we can anticipate that a BSEC is a reasonable sampling design. However, it will not always be better than the possible alternatives. Intuitively, it seems desirable to exclude more pairs of units by using a sampling design with $\alpha \geq 2$ if N is very large compared to n. More research is needed to clarify some of these points.

Since $\pi_{ij} = 0$ for some units i and j, it follows that V_{BSEC} can not be estimated unbiasedly. Estimable approximations of V_{BSEC} may be obtained by using the knowledge that contiguous units have similar y-values. Estimates of these approximations may then be used as estimates for V_{BSEC}. Two such estimates are given by

$$\hat{V}_1 = \{(N-3n)/(2n^2(n-1)N)\} \sum\sum_{\substack{i_1, i_2 \in s \\ i_1 \neq i_2}} (y_{i_1} - y_{i_2})^2,$$

and

$$\hat{V}_2 = \{(N - 2n) \sum_{i \in s} y_i^2 - \{(N - 3n)/(n - 1)\} \sum_{\substack{i_1, i_2 \in s \\ i_1 \neq i_2}} y_{i_1} y_{i_2}$$

$$- \{n(N - 3)/(n - 1)\} \left\{ \sum_{\substack{i \\ i, i+2 \in s}} y_i y_{i+2} \right\} / (n^2 N).$$

Clearly, $\hat{V}_1$ is a nonnegative variance estimate, provided that $N \geq 3n$. This latter condition is not restrictive at all, since it is a necessary condition for the existence of a BSEC as will be seen shortly.

3. Results on the existence and construction

In this section we discuss the existence and construction of balanced sampling designs excluding contiguous units. The proofs of the indicated results in this section can easily be translated to the language of design theory by using the correspondence between sampling designs and block designs, as first pointed out by Chakrabarti (1963).

The existence of the desired sampling designs is obvious for $n = 2$ and $N \geq 4$. We simply use the same selection probability for each of the $N(N - 3)/2$ samples of size 2 without contiguous units. For $n \geq 3$ the existence is less clear. The following result gives a necessary condition.

THEOREM 1. *If $n \geq 3$, a balanced sampling excluding contiguous units exists only if $N = 3n$.*

Although the condition in Theorem 1 is necessary, it is not sufficient. It can be shown that the required design with $N = 3n$ exists only for $n = 2, 3$ and 4. In fact, for $n = 4$ this sampling plan is unique, and is given in Table 3. Each sample in that table has selection probability 1/18.

Table 3
Samples for a BSEC with $N = 12$, $n = 4$

1	1	1	1	1	1	2	2	2	2	2	2	3	3	3	3	4	4
3	3	4	4	5	5	4	4	5	5	6	6	5	5	6	7	6	7
6	7	6	8	7	8	7	8	7	9	8	9	8	9	8	10	10	9
10	9	9	11	11	10	11	10	10	12	12	11	12	11	11	12	12	12

As a first attempt to construct the desired sampling designs we could select all of the $\{N/(N - n)\} \{(N - n)!/(n!(N - 2n)!)\}$ distinct samples without contiguous units, and assign equal selection probabilities to them. This, however, will only give the desired design for $n = 2$. In fact, using different non-zero selection probabilities may not be successful either, as the uniqueness of the sampling design in Table 3 illustrates. In that design only 18 of the 105 possible samples receive a positive selection probability.

In general, we are faced with the problem of determining a support for the sampling design and corresponding selection probabilities for the samples in this support. This is not an easy task. The construction that is used to provide the following existence result, is often useful to solve this problem.

THEOREM 2. *Let $A_1, A_2, \ldots, A_r$ be subsets of $\{1, \ldots, N\}$ with cardinality n. For $A_i = \{a_{i1}, \ldots, a_{in}\}$, compute the $rn(n-1)$ differences*

$$\pm (a_{ij} - a_{ik}), \quad j \neq k, \quad i = 1, \ldots, r,$$

modulo N. If the residues 0, 1 and $N - 1$ do not appear among these differences, while all others appear equally often, then a balanced sampling design excluding contiguous units based on N and samples of size n exists.

To illustrate the construction, let $N = 23$ and $n = 5$. We use $r = 1$, and take $A_1 = \{1, 3, 7, 16, 19\}$. The 20 differences 2, 3, ..., 21 all appear once. Now take the samples in the design as $A_1 + j = \{1 + j, 3 + j, 7 + j, 16 + j, 19 + j\}$, $j = 0, 1, \ldots, 22$, where the addition is modulo 23. These samples are given in Table 4. Each has a selection probability of 1/23.

Table 4
Samples for a BSEC with $N = 23$, $n = 5$

1	2	3	4	5	1	2	3	1	2	3	4	5	6	7	8	9	1	2	3	4	1	2
3	4	5	6	7	6	7	8	4	5	6	7	8	9	10	11	12	10	11	12	13	5	6
7	8	9	10	11	8	9	10	9	10	11	12	13	14	15	16	17	13	14	15	16	14	15
16	17	18	19	20	12	13	14	11	12	13	14	15	16	17	18	19	18	19	20	21	17	18
19	20	21	22	23	21	22	23	15	16	17	18	19	20	21	22	23	20	21	22	23	22	23

The selection probabilities may not always be uniform over the support of the sampling design. Indeed, some of the constructed samples $A_i + j$ may coincide. In general, if a sample can be written as $A_i + j$ for f different combinations of i and j, its selection probability will be $f/(rN)$.

Finding appropriate sets $A_1, A_2, \ldots, A_r$ is not always easy. Techniques to find them should be developed. This is certainly worthwhile due to the generality of Theorem 2, as highlighted in the following result.

THEOREM 3. *If a balanced sampling design excluding contiguous units based on N units and sample size n exists, then such a design can be obtained through the method of Theorem 2.*

Not only can this be used to obtain nonexistence results for certain values of N and n, but it shows that in the search for the desired designs we may restrict our attention to the construction method of Theorem 2. Of course, if simpler techniques for certain values of N and n can be developed, this should be exploited. For $N \equiv 0 \pmod{3}$, $N \geqslant 3n$ and $n = 3$ or 4 the following techniques may be used. Denote by $S_N(\alpha_0, \alpha_1, \alpha_2)$, where $\alpha_0 + \alpha_1 + \alpha_2 = n$, all samples of size n based on N units with the properties that no contiguous units are included and that α_i units with a label equal to $i \pmod 3$ are included, $i = 0, 1, 2$. For example, $S_{12}(2, 2, 0)$ consists of the 6 samples in Table 5.

The following two results describe the use of such collections of samples in the construction of BSEC's.

Table 5
Samples in $S_{12}(2, 2, 0)$

1	1	1	3	4	4
3	3	4	7	6	7
6	7	6	10	10	9
10	9	9	12	12	12

THEOREM 4. *If $N \equiv 0 \pmod{3}$, $N \geqslant 9$, then the $\binom{N/3}{2}(N-6)$ samples in $S_N(2, 1, 0) \cup S_N(1, 0, 2)$, all with the same selection probability, form a balanced sampling design excluding contiguous units.*

THEOREM 5. *If $N \equiv 0 \pmod{3}$, $N \geqslant 12$, then the $3\binom{N/3}{2}\binom{(N-6)/3}{2}$ samples in $S_N(2, 2, 0) \cup S_N(0, 2, 2) \cup S_N(2, 0, 2)$, all with the same selection probability, form a balanced sampling design excluding contiguous units.*

We will see in Section 4 that the implementation of these sampling designs is quite simple.

Other useful methods of construction are recursive methods. One such method allows us to construct a BSEC based on $N + 3$ units in samples of size n from a BSEC based on N units in samples of size n. Starting with the latter plan, we replace unit N by $N + 1$. The differences within the samples computed modulo $N + 3$ do then satisfy the requirements in Theorem 2. Thus, the method corresponding to that theorem can be used to construct the desired plan.

For $n = 3$, BSEC's were constructed in Hedayat, Rao and Stufken (1988) for $N = 9, 10$ and 11; for $n = 4$ constructions were given for $N = 12, 13$ and 14. From Theorem 1 and the preceding recursive construction, we can thus reach the following conclusion.

THEOREM 6. *For $n = 3$ or 4, a balanced sampling design excluding contiguous units exists if and only if $N \geqslant 3n$.*

More research on the existence and construction of the desired sampling designs is needed. One example of a recursive construction that would be of interest is one in which the sample size can be increased. How large this increase is will be fairly irrelevant, since it is easy to obtain BSCE's with a smaller sample size from one with a larger sample size. This is expressed in the following result.

THEOREM 7. *The existence of a balanced sampling design excluding contiguous units for N units and fixed sample size n, implies the existence of such a design for N units and fixed sample size $n' \leqslant n$.*

The latter design is obtained by replacing each sample of size n by all $\binom{n}{n'}$ subsets of size n', where the selection probability of the original sample is equally divided over all these subsets. Formally, if $d = (S_d, P_d)$ is the original design, the new design $d' = (S_{d'}, P_{d'})$ is defined by

$$S_{d'} = \{s' : s' \text{ has cardinality } n', \text{ and there exists an } s \in S_d \text{ such that } s' \subset s\},$$

and

$$P_{d'}(s') = \sum_{s' \subset s \in S_d} P_d(s)/\binom{n}{n'}, \quad \text{for } s' \in S_{d'}.$$

The implementation of d' is simple, provided that the implementation of d is simple.

4. Implementation

In this section we look into the implementation of the designs obtained through the methods of Section 3. For most of these, this is quite simple. Before looking into the various methods, one may in a first attempt try to use a rejective sampling scheme. Randomly draw n units from the population, and let these n units consist of the sample unless contiguous units are included. In the latter case, start all over again. This procedure will however not lead to the desired sampling design, unless $n = 2$. Otherwise it will, for example, have the undesirable property that π_{13} exceeds π_{14}.

The construction in Theorem 2 allows for an extremely simple implementation after the initial samples $A_1, \ldots, A_r$ have been determined. It required the selection of one or two random numbers, for $r = 1$ or $r \geqslant 2$, respectively. Start by randomly selecting a number from $\{1, \ldots, r\}$, say r_0. If $r = 1$, this step can obviously be omitted. Next randomly select a number from $\{1, \ldots, N\}$, say i_0. Now declare $A_{r_0} + i_0 = \{a_{r_0 1} + i_0, \ldots, a_{r_0 n} + i_0\}$ as the selected sample, the addition being modulo N. Thus, for the design in Table 4, where $r = 1$, if our random selection from $\{1, \ldots, 23\}$ results in 8, say, the sample will be $\{1 + 8, 3 + 8, 7 + 8, 16 + 8, 19 + 8\} \equiv \{1, 4, 9, 11, 15\}$.

The designs as constructed in Theorems 4 and 5 also allow a simple implementation. There we randomly select four and five digits, respectively. We illustrate it here for $N = 78$ and $n = 3$. Start by selecting a digit from $\{0, 1, 2\}$. Next randomly select two distinct digits from $\{1, 2, \ldots, N/3\}$, and select a third one from the same set, distinct from the previous two. Suppose the entire selection process results in 2, and 5, 19, and 16. The first selected digit determines the collection from which the sample will be selected; in this case from $S_{78}(1, 0, 2)$.

The next two digits determine the elements in the modulo class with two representatives; in this case $3(5) - 1 = 14$, and $3(19) - 1 = 56$. The last digit determined the final unit in the sample, here $3(16) = 48$. Thus, the sample is $\{14, 48, 56\}$. It should be pointed out that the representation for each modulo class should be fixed before selecting the last three digits. We used for this collection $3t - 1$ for those that are $2 \pmod 3$ and $3t$ for those that are $0 \pmod 3$. We could have used $3t + 2$ and $3t + 3$ instead, but should not use $3t + 2$ and $3t$. In the latter case a sample that contains, for example, units 5 and 3 will never be selected.

The implementation of designs constructed through the recursive method, preceding Theorem 6, is more cumbersome. If not too many steps are needed in this construction, it is however, still fairly simple. As an example, consider $N = 29$, $n = 5$. Such a design can be obtained from the design in Table 4 upon applying the suggested technique twice. For the implementation of this design we could proceed as follows. First make a random selection from $\{1, ..., 23\}$, then from $\{1, ..., 26\}$, and finally from $\{1, ..., 29\}$. Say the three selections result in 4, 10 and 10, respectively. The first number tells us to start with $A_1 + 4 = \{5, 7, 11, 20, 23\}$. The addition at this stage is modulo 23. If 23 is contained in this set, as it is in our example, replace it by 24. This yields $\{5, 7, 11, 20, 24\}$. Now we add the second random number, 10, to each of these. This addition is modulo 26 and results in $\{4, 8, 15, 17, 21\}$. If 26 had been in this set, it would have been replaced by 27. Now add the third digit, again 10, where the addition is modulo 29. This results in $\{2, 14, 18, 25, 27\}$, which is the sample to be used.

Finally, the implementation corresponding to Theorem 7 is quite simple for d', provided that it is simple for d. To illustrate this, we will use the design in Table 4 to obtain a sample for $N = 23$ and $n = 3$. To implement d, all we have to do is to select a digit from $\{1, ..., 23\}$. Say this results in 5. This yields the set $A_1 + 5 = \{1, 6, 8, 12, 21\}$. Now select randomly three out of these five elements. This may result in $\{1, 6, 12\}$. This then is the sample to be used.

References

[1] Chakrabarti, M. C. (1963). On the use of incidence matrices in sampling from finite populations. *J. Indian Statist. Assoc.* **1**, 78–85.
[2] Hedayat, A. S., Rao, C. R. and Stufken, J. (1988). Sampling plans excluding contiguous units. *J. Statist. Plann. Infer.*, to appear.
[3] Horvitz, D. G. and Thompson, D. J. (1952). A generalization of sampling without replacement from a finite universe. *J. Amer. Statist. Assoc.* **47**, 663–685.
[4] Rao, C. R. (1975). Some problems of sample surveys. Proceedings of the conference on directions for mathematical statistics, University of Alberta, Edmonton, Canada. Special Supplement to *Adv. Appl. Prob.* **7**, 50–61.
[5] Rao, C. R. (1987). Strategies of data analysis. *Proc. 46th Session of the International Statistical Institute, Tokyo.*

Subject Index

Accuracy needed, 21
Admissibility, 51
Analysis of variance, 360
Analysis, 247
Ascertainment errors, 333
Assumptions of simple random sampling, 415
Asymptotically normal, 294
Asymptotic efficiency, 97, 101
Asymptotic formulae, 127
Asymptotic normality, 97, 100, 104, 108
Asymptotic results, 175
Asymptotic theory, 84
Asymptotics in systematic sampling, 135
Auto-regressive process, 561
Autocorrelated populations, 134
Average of ratios estimators, 233

Balanced design, 55
Balanced repeated replication, 418, 422, 427
Balanced sample, 406, 577
Bayesian inference, 213
Bayesian methodology, 268, 283
Bayesian sufficient, 81
Best asymptotically normal estimators, 420
Beta-binomial distribution, 278, 285
Beta distribution, 271, 278, 285
Bias reduction, 451
Binomial distribution, 272
Bird-banding, 481
Block designs, 577
BLU estimator, 401
Bootstrap method, 443, 510
Brownian bridge, 294, 300

Capture–Mark–Release–Recapture (CMRR), 292
Capture–recapture, 469
Capture–recapture closed models, 479
Capture–recapture open models, 481

Capture–recapture sampling, 478
Categorical data analysis, 251
Central limit theorem in sampling, 5
Classification methods, 491
Cluster analysis, 490
Cluster sampling, 32, 268, 277
Combining data, 502
Complete class, 58
Complex sample design, 247
Composite estimator, 191
Composite samples of known proportions, 470
Composite samples of random proportions, 473
Composite sampling to test for compliance, 471
Composite sampling, 469
Compositing to reduce the cost of estimating the mean, 472
Compositing to reduce variance, 472
Computer programs, 181
Conditional exchangeability, 229
Conditionality, 411
Connecticut blood pressure survey, 421
Consumer panels, 546
Contiguous units, 575
Control, 333
Coupon collector's problem, 316, 327
Coupon extension, 327
Coupon waiting time, 318

Data to be collected, 19
Degree, 293
Design-based, 247
Design-based variance estimation, 140
Design effect, 416, 445
Detectability curve, 520
Detection probability, 518
Directory based, 536
Dirichlet distribution, 248
Dirichlet-multinomial distribution, 284
Dirichlet process, 286

Distinct sample, 338
Domains, 417, 422
Double sampling, 291
Dual frame, 539

Effective area, 517
Effective distance, 517
Effective sample size, 338
Element sampling, 24
Elementary estimates, 197
Empirical Bayes, 206
Empirical cumulative distribution function, 270, 281
End corrections estimators, 129
Erdös–Rényi–Hájek condition, 97, 100, 108
Errors-in-variables, 508
Estimand, 339
Estimation of incidence of a trait with composite sampling for confidentiality, 471
Estimation of population size, 304
Estimation theory, 340
Estimator, 195
Exchangeability, 217, 218
Exponential power estimator, 520

Finite population, 47
Fourier series estimator, 519
Frame, 338
Function EIV, 508

Generalized linear models, 439
Godambe, 213
Growth pattern in children, 567

Hájek's measure of disparity, 98, 105
Hansen–Hurwitz estimator, 313
Heavy metal pollution, 477
Heritability, 567
Hierarchic structure, 535
History of science, 1
Hoeffding inequality, 297
Homeostasis, 563
Horvitz–Thompson estimator, 315, 418, 430

Implementation, 582
Inclusion probability, 314
Indicator variable, 417, 422
Inference, 49, 247
Interaction between genotype and environment, 565
Interactive covariance, 343
Interactive variance, 343
Interpenetrating sampling, 291
Intra-class correlation, 565

Intra-individual variation, 569
IQ, 571

Jackknife, 451
Jackknife method, 427
Jackknifed estimator, 295
Jolly–Seber model, 483

Kalman filter, 200
Kernel estimator, 503
KFF method, 416

Labels, 47
Likelihood, 79, 400, 411
Likelihood function, 214
Linear Bayes estimators, 216
Linearization method, 427
Linear models, 251
Linear regression model, 401
Linear trend, 128
Linear unbiased, 195
Loglinear model, 438

Mail questionnaires, 498
Maximum likelihood estimator, 306
Measurement errors, 43
Methods of ascertainment, 499
Minimal sufficiency, 58
Minimaxity, 67
Mixing of samples, 469
Model-based, 247
Model-based variance estimation, 138
Model-based versus design-based inference, 12
Model failure, 400
Model-unbiasedness, 456
Modelling standard errors, 445
Moment inequality, 297
Monetary unit sampling, 111, 119
Multi-stage sampling, 39
Multinomial distribution, 283, 286
Multistage designs, 431
Multistage, 465
Multivariate, 461
Multivariate analysis, 251

Nature vs nurture, 572
Negman chi-square, 420
Non-response, 41, 491
Non-sampling errors, 41, 571
Nonlinear statistics, 436
Nonrespondents, 465
Nonresponse, 391
Normal distribution, 269, 281
Nuisance parameters, 111

Subject index

Optimal randomization in experimental design, 137
Optimal strategies, 47
Optimality, 47
Optimum stratified design, 221
Optimum two-stage design, 226
Order statistics, 267, 273, 287
Overdispersion, 500

Paradigms, 1
Parameter, 293
Periodic variation, 132
Permutation model, 69
Poisson law, 322
Population characteristics to estimate, 20
Population mean, 292
Population of interest, 19
Population parameters, 20
Population variance, 293
Post-stratification, 379, 440
Power plant effect on environment, 477
PPS systematic sampling, 127
Prediction approach, 399, 412
Prediction, 83, 369
Prediction model, 401
Principal component analysis, 491
Probability inequality, 297
Probability sampling, 17
Processing errors, 44
Proportional sample size allocation, 271
Purposive sample, 415–416
Purposive selection versus randomization, 6

Random digit dialing, 537
Randomization, 49, 111, 220, 333
Randomization in sampling, 4
Randomization principle, 411
Randomized response survey, 498
Random permutation models, 133
Random replacement sampling, 298
Random sampling, 333
Ratio and regression estimation, 227
Ratio estimator, 233, 303
Recruitment, 502
Regression, 251
Regression estimator, 231
Rejective sampling, 315
Remote sensing, 490
Replicated or interpenetrating samples, 333
Replication, 333
Representative samples, 2
Respondent selection, 541
Response covariance, 344
Response error and bias, 233

Response errors, 387
Response variance, 343
Reverse martingale property, 298
Robust inference, 404, 411
Robustness, 49
Role of randomization, 219
Rotation group bias, 198
Rotation sampling, 187
Rules of association, 22

Samford–Durbin sampling plan, 315
Sample design, 24
Sample information on the bias, 237
Sample survey, 47, 247
Sampling a time series, 138
Sampling design, 338, 561
Sampling designs for regression models, 117
Sampling fraction, 291
Sampling frame, 21, 291
Sampling frames, 492
Sampling procedure, 338
Sampling scheme, 24
Sampling strategy, 47
Sampling system, 24
Sampling unit, 291, 337
Sampling variance, 343
Sampling with unequal probabilities of selection, 30
Sen–Yates–Grundy variance estimator, 430
Sequential sampling tagging, 308
Shopping center sampling, 544
Simple random sampling, 97, 578
Simplified variance estimators, 434
Size-biased distribution, 500
Spatial contagion, 491
Spatial sampling, 141
SRSWOR, 291
SRSWR, 291
SRSWR distinct unit, 296
Stages in sampling, 339
Statespace model, 202
Stationary variance, 571
Statistical independence, 339
Statistical sampling, 16
Statistics, 58
Statistics needed, 18
Stochastic processes, 294
Stopping time/variable, 308
Stratification, 339, 457
Stratified random sampling, 268, 281
Stratified sampling, 27, 220
Structural EIV, 508
Subject-matter problem, 19
Subsampling of the composite samples, 476

Subsampling with varying probabilities, 292, 324
Successive sampling, 291
Sufficiency, 58
Super-population, 75, 215, 454
Superpopulation models, 175, 369
Survey populations, 49
Survey sampling, 16
Survey sampling design, 44
Symmetric Dirichlet distribution, 475
Symmetric sampling plan, 297
Symmetric stopping rule, 105
Systematic sampling, 36, 147

Taylor series estimator, 419
Telephone surveys, 534
Tied-down Brownian sheet, 304
Time series analysis, 200, 382
Total error variance, 344
Transect sampling, 517
Two phase, 464
Two stage cluster sampling, 268, 277
Two stage sampling, 223, 374

Unequal probability sampling, 9, 111
Uni-cluster designs, 50
Uniform admissibility, 51
Unistage designs, 428
Units in random order, 133
Universe, 337
U-statistic, 294
Urn model, 310

Value of tainted observations, 36
Variance components, 433
Variance covariance matrix, 419
Variance estimation, 158, 427, 454
Variance structures, 490
Variations in systematic sampling, 165
Varying probability sampling, 291, 313

Weighted distributions, 499
Weighted least squares, 416, 420
Wood duck example, 482
Working model, 400

Handbook of Statistics
Contents of Previous Volumes

Volume 1. Analysis of Variance
Edited by P. R. Krishnaiah
1980 xviii + 1002 pp.

1. Estimation of Variance Components by C. R. Rao and J. Kleffe
2. Multivariate Analysis of Variance of Repeated Measurements by N. H. Timm
3. Growth Curve Analysis by S. Geisser
4. Bayesian Inference in MANOVA by S. J. Press
5. Graphical Methods for Internal Comparisons in ANOVA and MANOVA by R. Gnanadesikan
6. Monotonicity and Unbiasedness Properties of ANOVA and MANOVA Tests by S. Das Gupta
7. Robustness of ANOVA and MANOVA Test Procedures by P. K. Ito
8. Analysis of Variance and Problems under Time Series Models by D. R. Brillinger
9. Tests of Univariate and Multivariate Normality by K. V. Mardia
10. Transformations to Normality by G. Kaskey, B. Kolman, P. R. Krishnaiah and L. Steinberg
11. ANOVA and MANOVA: Models for Categorical Data by V. P. Bhapkar
12. Inference and the Structural Model for ANOVA and MANOVA by D. A. S. Fraser
13. Inference Based on Conditionally Specified ANOVA Models Incorporating Preliminary Testing by T. A. Bancroft and C.-P. Han
14. Quadratic Forms in Normal Variables by C. G. Khatri
15. Generalized Inverse of Matrices and Applications to Linear Models by S. K. Mitra
16. Likelihood Ratio Tests for Mean Vectors and Covariance Matrices by P. R. Krishnaiah and J. C. Lee
17. Assessing Dimensionality in Multivariate Regression by A. J. Izenman

18. Parameter Estimation in Nonlinear Regression Models by H. Bunke
19. Early History of Multiple Comparison Tests by H. L. Harter
20. Representations of Simultaneous Pairwise Comparisons by A. R. Sampson
21. Simultaneous Test Procedures for Mean Vectors and Covariance Matrices by P. R. Krishnaiah, G. S. Mudholkar and P. Subbaiah
22. Nonparametric Simultaneous Inference for Some MANOVA Models by P. K. Sen
23. Comparison of Some Computer Programs for Univariate and Multivariate Analysis of Variance by R. D. Bock and D. Brandt
24. Computations of Some Multivariate Distributions by P. R. Krishnaiah
25. Inference on the Structure of Interaction in Two-Way Classification Model by P. R. Krishnaiah and M. Yochmowitz

Volume 2. Classification, Pattern Recognition and Reduction of Dimensionality
Edited by P. R. Krishnaiah and L. N. Kanal
1982 xxii + 903 pp.

1. Discriminant Analysis for Time Series by R. H. Shumway
2. Optimum Rules for Classification into Two Multivariate Normal Populations with the Same Covariance Matrix by S. Das Gupta
3. Large Sample Approximations and Asymptotic Expansions of Classification Statistics by M. Siotani
4. Bayesian Discrimination by S. Geisser
5. Classification of Growth Curves by J. C. Lee
6. Nonparametric Classification by J. D. Broffitt
7. Logistic Discrimination by J. A. Anderson
8. Nearest Neighbor Methods in Discrimination by L. Devroye and T. J. Wagner
9. The Classification and Mixture Maximum Likelihood Approaches to Cluster Analysis by G. J. McLachlan
10. Graphical Techniques for Multivariate Data and for Clustering by J. M. Chambers and B. Kleiner
11. Cluster Analysis Software by R. K. Blashfield, M. S. Aldenderfer and L. C. Morey
12. Single-link Clustering Algorithms by F. J. Rohlf
13. Theory of Multidimensional Scaling by J. de Leeuw and W. Heiser
14. Multidimensional Scaling and its Application by M. Wish and J. D. Carroll
15. Intrinsic Dimensionality Extraction by K. Fukunaga
16. Structural Methods in Image Analysis and Recognition by L. N. Kanal, B. A. Lambird and D. Lavine

17. Image Models by N. Ahuja and A. Rosenfeld
18. Image Texture Survey by R. M. Haralick
19. Applications of Stochastic Languages by K. S. Fu
20. A Unifying Viewpoint on Pattern Recognition by J. C. Simon, E. Backer and J. Sallentin
21. Logical Functions in the Problems of Empirical Prediction by G. S. Lbov
22. Inference and Data Tables and Missing Values by N. G. Zagoruiko and V. N. Yolkina
23. Recognition of Electrocardiographic Patterns by J. H. van Bemmel
24. Waveform Parsing Systems by G. C. Stockman
25. Continuous Speech Recognition: Statistical Methods by F. Jelinek, R. L. Mercer and L. R. Bahl
26. Applications of Pattern Recognition in Radar by A. A. Grometstein and W. H. Schoendorf
27. White Blood Cell Recognition by E. S. Gelsema and G. H. Landweerd
28. Pattern Recognition Techniques for Remote Sensing Applications by P. H. Swain
29. Optical Character Recognition—Theory and Practice by G. Nagy
30. Computer and Statistical Considerations for Oil Spill Identification by Y. T. Chien and T. J. Killeen
31. Pattern Recognition in Chemistry by B. R. Kowalski and S. Wold
32. Covariance Matrix Representation and Object–Predicate Symmetry by T. Kaminuma, S. Tomita and S. Watanabe
33. Multivariate Morphometrics by R. A. Reyment
34. Multivariate Analysis with Latent Variables by P. M. Bentler and D. G. Weeks
35. Use of Distance Measures, Information Measures and Error Bounds in Feature Evaluation by M. Ben-Bassat
36. Topics in Measurement Selection by J. M. Van Campenhout
37. Selection of Variables Under Univariate Regression Models by P. R. Krishnaiah
38. On the Selection of Variables Under Regression Models Using Krishnaiah's Finite Intersection Tests by J. L. Schmidhammer
39. Dimensionality and Sample Size Considerations in Pattern Recognition Practice by A. K. Jain and B. Chandrasekaran
40. Selecting Variables in Discriminant Analysis for Improving upon Classical Procedures by W. Schaafsma
41. Selection of Variables in Discriminant Analysis by P. R. Krishnaiah

Volume 3. Time Series in the Frequency Domain
Edited by D. R. Brillinger and P. R. Krishnaiah
1983 xiv + 485 pp.

1. Wiener Filtering (with emphasis on frequency-domain approaches) by R. J. Bhansali and D. Karavellas
2. The Finite Fourier Transform of a Stationary Process by D. R. Brillinger
3. Seasonal and Calendar Adjustment by W. S. Cleveland
4. Optimal Inference in the Frequency Domain by R. B. Davies
5. Applications of Spectral Analysis in Econometrics by C. W. J. Granger and R. Engle
6. Signal Estimation by E. J. Hannan
7. Complex Demodulation: Some Theory and Applications by T. Hasan
8. Estimating the Gain of A Linear Filter from Noisy Data by M. J. Hinich
9. A Spectral Analysis Primer by L. H. Koopmans
10. Robust-Resistant Spectral Analysis by R. D. Martin
11. Autoregressive Spectral Estimation by E. Parzen
12. Threshold Autoregression and Some Frequency-Domain Characteristics by J. Pemberton and H. Tong
13. The Frequency-Domain Approach to the Analysis of Closed-Loop Systems by M. B. Priestley
14. The Bispectral Analysis of Nonlinear Stationary Time Series with Reference to Bilinear Time-Series Models by T. Subba Rao
15. Frequency-Domain Analysis of Multidimensional Time-Series Data by E. A. Robinson
16. Review of Various Approaches to Power Spectrum Estimation by P. M. Robinson
17. Cumulants and Cumulant Spectral Spectra by M. Rosenblatt
18. Replicated Time-Series Regression: An Approach to Signal Estimation and Detection by R. H. Shumway
19. Computer Programming of Spectrum Estimation by T. Thrall
20. Likelihood Ratio Tests on Covariance Matrices and Mean Vectors of Complex Multivariate Normal Populations and their Applications in Time Series by P. R. Krishnaiah, J. C. Lee and T. C. Chang

Volume 4. Nonparametric Methods
Edited by P. R. Krishnaiah and P. K. Sen
1984 xx + 968 pp.

1. Randomization Procedures by C. B. Bell and P. K. Sen
2. Univariate and Multivariate Multisample Location and Scale Tests by V. P. Bhapkar
3. Hypothesis of Symmetry by M. Hušková
4. Measures of Dependence by K. Joag-Dev
5. Tests of Randomness against Trend or Serial Correlations by G. K. Bhattacharyya
6. Combination of Independent Tests by J. L. Folks
7. Combinatorics by L. Takács
8. Rank Statistics and Limit Theorems by M. Ghosh
9. Asymptotic Comparison of Tests – A Review by K. Singh
10. Nonparametric Methods in Two-Way Layouts by D. Quade
11. Rank Tests in Linear Models by J. N. Adichie
12. On the Use of Rank Tests and Estimates in the Linear Model by J. C. Aubuchon and T. P. Hettmansperger
13. Nonparametric Preliminary Test Inference by A. K. Md. E. Saleh and P. K. Sen
14. Paired Comparisons: Some Basic Procedures and Examples by R. A. Bradley
15. Restricted Alternatives by S. K. Chatterjee
16. Adaptive Methods by M. Hušková
17. Order Statistics by J. Galambos
18. Induced Order Statistics: Theory and Applications by P. K. Bhattacharya
19. Empirical Distribution Function by E. Csáki
20. Invariance Principles for Empirical Processes by M. Csörgő
21. M-, L- and R-estimators by J. Jurečková
22. Nonparametric Sequantial Estimation by P. K. Sen
23. Stochastic Approximation by V. Dupač
24. Density Estimation by P. Révész
25. Censored Data by A. P. Basu
26. Tests for Exponentiality by K. A. Doksum and B. S. Yandell
27. Nonparametric Concepts and Methods in Reliability by M. Hollander and F. Proschan
28. Sequential Nonparametric Tests by U. Müller-Funk
29. Nonparametric Procedures for some Miscellaneous Problems by P. K. Sen
30. Minimum Distance Procedures by R. Beran
31. Nonparametric Methods in Directional Data Analysis by S. R. Jammalamadaka
32. Application of Nonparametric Statistics to Cancer Data by H. S. Wieand

33. Nonparametric Frequentist Proposals for Monitoring Comparative Survival Studies by M. Gail
34. Meteorological Applications of Permutation Techniques based on Distance Functions by P. W. Mielke, Jr.
35. Categorical Data Problems Using Information Theoretic Approach by S. Kullback and J. C. Keegel
36. Tables for Order Statistics by P. R. Krishnaiah and P. K. Sen
37. Selected Tables for Nonparametric Statistics by P. K. Sen and P. R. Krishnaiah

Volume 5. Time Series in the Time Domain
Edited by E. J. Hannan, P. R. Krishnaiah and M. M. Rao
1985 xiv + 490 pp.

1. Nonstationary Autoregressive Time Series by W. A. Fuller
2. Non-Linear Time Series Models and Dynamical Systems by T. Ozaki
3. Autoregressive Moving Average Models, Intervention Problems and Outlier Detection in Time Series by G. C. Tiao
4. Robustness in Time Series and Estimating ARMA Models by R. D. Martin and V. J. Yohai
5. Time Series Analysis with Unequally Spaced Data by R. H. Jones
6. Various Model Selection Techniques in Time Series Analysis by R. Shibata
7. Estimation of Parameters in Dynamical Systems by L. Ljung
8. Recursive Identification, Estimation and Control by P. Young
9. General Structure and Parametrization of ARMA and State-Space Systems and its Relation to Statistical Problems by M. Deistler
10. Harmonizable, Cramér, and Karhunen Classes of Processes by M. M. Rao
11. On Non-Stationary Time Series by C. S. K. Bhagavan
12. Harmonizable Filtering and Sampling of Time Series by D. K. Chang
13. Sampling Designs for Time Series by S. Cambanis
14. Measuring Attenuation by M. A. Cameron and P. J. Thomson
15. Speech Recognition Using LPC Distance Measures by P. J. Thomson and P. de Souza
16. Varying Coefficient Regression by D. F. Nicholls and A. R. Pagan
17. Small Samples and Large Equation Systems by H. Theil and D. G. Fiebig